ORGANIZING A STATISTICAL PROBLEM: THE FOUR-STEP PROCESS

STATE: What is the practical question in the context of the real-world setting?

FORMULATE: What specific statistical operations does this problem call for?

SOLVE: Make the graphs and carry out the calculations needed for this problem.

CONCLUDE: Give your practical conclusion in the setting of the real-world problem.

CONFIDENCE INTERVALS: THE FOUR-STEP PROCESS

STATE: What is the practical question that requires estimating a parameter?

FORMULATE: Identify the parameter and choose a level of confidence.

SOLVE: Carry out the work in two phases:

(a) **Check the conditions** for the interval you plan to use.

(b) Calculate the **confidence interval.**

CONCLUDE: Return to the practical question to describe your results in this setting.

TESTS OF SIGNIFICANCE: THE FOUR-STEP PROCESS

STATE: What is the practical question that requires a statistical test?

FORMULATE: Identify the parameter and state null and alternative hypotheses.

SOLVE: Carry out the test in three phases:

(a) **Check the conditions** for the test you plan to use.

(b) Calculate the **test statistic.**

(c) Find the **P-value.**

CONCLUDE: Return to the practical question to describe your results in this setting.

The Practice of Statistics in the Life Sciences

Brigitte Baldi
University of California, Irvine

David S. Moore
Purdue University

W. H. Freeman and Company
New York

Senior Publisher: **Craig Bleyer**
Publisher: **Ruth Baruth**
Development Editor: **Shona Burke**
Senior Media Editor: **Roland Cheyney**
Associate Editor: **Brendan Cady**
Assistant Editor: **Brian Tedesco**
Editorial Assistant: **Katrina Wilhelm**
Marketing Coordinator: **Dave Quinn**
Photo Editor: **Bianca Moscatelli**
Cover and Text Designer: **Vicki Tomaselli**
Senior Project Editor: **Mary Louise Byrd**
Illustration Coordinator: **Bill Page**
Production Manager: **Susan Wein**
Composition: **ICC Macmillan Inc.**
Printing and Binding: **RR Donnelley and Sons**

Library of Congress Control Number: 2007938574

ISBN-13: 978-1-4292-1876-4

ISBN-10: 1-4292-1876-2

Printed in the United States of America

Third printing

W. H. Freeman and Company
41 Madison Avenue
New York, NY 10010
Houndmills, Basingstoke RG21 6XS, England
www.whfreeman.com

Brief Contents

*Starred material is optional.

Contents ──

─────────────

*Starred material is optional.

PART IV
Optional Companion Chapters (available on the *PSLS* CD and Web site: www.whfreeman.com/psls)

To the Instructor: About This Book

The Practice of Statistics in the Life Sciences (*PSLS*) is an introduction to statistics for college and university students interested in the quantitative analysis of life science problems. Statistics has penetrated the life sciences pervasively with a specific set of application challenges, such as observational studies with confounding variables or limited sample sizes. Consequently, students can clearly benefit from teaching of statistics that is explicitly applied to their major. All examples and exercises in *PSLS* are drawn from diverse areas of biology, such as physiology, brain and behavior, health and medicine, nutrition, ecology, and microbiology. Instructors can choose to either cover a wide range of topics or select examples and exercises related to a particular field.

PSLS focuses on the applications of statistics rather than the mathematical foundation. The book is adapted from David Moore's bestselling introductory statistics textbook, *The Basic Practice of Statistics* (*BPS*). *BPS* was the pioneer in presenting a modern approach to statistics in a genuinely elementary text. Like *BPS*, *PSLS* emphasizes balanced content, working with real data, and statistical ideas. It does not require any specific mathematical skills beyond being able to read and use simple equations and can be used in conjunction with almost any level of technology for calculating and graphing.

In the following we describe in further detail for instructors the nature and features of *PSLS*.

Guiding principles

PSLS is based on three principles: balanced content, experience with data, and the importance of ideas.

Balanced content. The actual practice of statistics includes the exploratory analysis of data, the design of data production, and probability-based inference all integrated into a coherent approach. The teaching of statistics should reflect these to form a coherent science of data. There are also good pedagogical reasons for beginning with data analysis (Chapters 1 to 6), then moving to data production (Chapters 7 and 8), and then to probability (Chapters 9 to 13) and inference (Chapters 14 to 27). In studying data analysis, students learn useful skills immediately and get over some of their fear of statistics. Data analysis is a necessary preliminary to inference in practice, because inference requires clean data. Designed data production is the surest foundation for inference, and the deliberate use of chance in random sampling and randomized comparative experiments motivates the study of probability in a course that emphasizes data-oriented statistics. *PSLS* gives a full presentation of basic probability and inference (19 of the 27 chapters) but places it in the context of statistics as a whole.

Experience with data. The study of statistics is supposed to help students work with data in their varied academic disciplines and in their unpredictable later employment. This is particularly important for students in the life sciences, because they are asked to collect and analyze data in their laboratory courses and elective undergraduate research. *PSLS* prepares students by providing real data from many areas of the life sciences. Data are more than mere numbers—they are numbers with a context that should play a role in making sense of the numbers and in stating conclusions. Examples and exercises in *PSLS* give enough background to allow students to consider the meaning of their calculations. Statistics, more than mathematics, depends on judgment for effective use. *PSLS* begins to develop students' judgment about statistical studies.

The importance of ideas. A first course in statistics introduces many skills, from making a histogram and calculating a correlation to choosing and carrying out a significance test. In practice (even if not always in the course), calculations and graphs are automated. Moreover, anyone who makes serious use of statistics will need some specific procedures not taught in her college stat course. *PSLS* therefore tries to make clear the larger patterns and big ideas of statistics in the context of learning specific skills and working with specific data. Many of the big ideas are summarized in graphical outlines, and a handy "Caution" icon in the margin calls attention to common confusions or pitfalls in basic statistics. Several discussions placed throughout the book cover in greater detail some important concepts, such as the identification and treatment of outliers and the scientific approach to planning and interpreting studies. Formulas without guiding principles do students little good once the final exam is past, so it is worth the time to slow down a bit and explain the ideas.

These three principles are widely accepted by statisticians concerned about teaching. In fact, statisticians have reached a broad consensus that first courses should reflect how statistics is actually used. Figure 1 is an outline of the consensus as summarized by the Joint Curriculum Committee of the American Statistical Association and the Mathematical Association of America.[1]* More recently, the College Report of the Guidelines for Assessment and Instruction in Statistics Education (GAISE) Project has emphasized exactly the same themes.[2] Fostering active learning is the business of the teacher, and an emphasis on working with data helps. *PSLS* is guided by the content emphasis of the modern consensus. In the language of the GAISE recommendations, these are: develop statistical thinking, use real data, stress conceptual understanding.

Practice of statistics through exercises

Students learn to work with data by working with data. *PSLS* provides about 50 examples and exercises per chapter, using real data drawn from various areas of

---

*All notes are collected in the Notes and Data Sources section at the end of the book.

1. **Emphasize the elements of statistical thinking:**

    (a) the need for data;
    (b) the importance of data production;
    (c) the omnipresence of variability;
    (d) the measuring and modeling of variability.

2. **Incorporate more data and concepts, fewer recipes and derivations. Wherever possible, automate computations and graphics.** An introductory course should:

    (a) rely heavily on *real* (not merely realistic) data;
    (b) emphasize *statistical* concepts, for example, causation versus association, experimental versus observational, and longitudinal versus cross-sectional studies;
    (c) rely on computers rather than computational recipes;
    (d) treat formal derivations as secondary in importance.

3. **Foster active learning,** through the following alternatives to lecturing:

    (a) group problem solving and discussion;
    (b) laboratory exercises;
    (c) demonstrations based on class-generated data;
    (d) written and oral presentations;
    (e) projects, either group or individual.

FIGURE 1 Outline of the consensus for statistical teaching as summarized by the Joint Curriculum Committee of the American Statistical Association and the Mathematical Association of America.

biology. The wealth of exercises allows instructors to emphasize some statistical topics or biological themes to tailor the content to their specific learning objectives.

Within chapters, exercises progress from straightforward applications to comprehensive review problems. To facilitate the learning process, content is broken into digestible bites of material followed by a few "Apply Your Knowledge" exercises for a quick check of basic mastery. After a summary of the chapter's key concepts, a set of "Check Your Skills" multiple-choice items with answers in the back of the book lets students assess their grasp of basic ideas and skills. (Alternatively, these problems can be used in an "i-clicker" Classroom Response System for class review.) End-of-chapter exercises then integrate all aspects of the chapter, while exercises in the three review chapters enlarge the statistical context beyond that of the immediate lesson. (Many instructors will find that the review chapters appear at the right points for pre-exam review.)

Some of the examples and exercises are presented in the context of a "four-step process" (State, Formulate, Solve, Conclude) intended to teach students how to work on realistic statistical problems. See Figure 2 for an overview. The process emphasizes a major theme in *PSLS*: Statistical problems originate in a real-world setting ("State") and require conclusions in the language of that setting ("Conclude"). Translating the problem into the formal language of statistics

APPLY YOUR KNOWLEDGE ———

Check Your Skills ———

ORGANIZING A STATISTICAL PROBLEM: THE FOUR-STEP PROCESS

STATE: What is the practical question, in the context of the real-world setting?

FORMULATE: What specific statistical operations does this problem call for?

SOLVE: Make the graphs and carry out the calculations needed for this problem.

CONCLUDE: Give your practical conclusion in the setting of the real-world problem.

CONFIDENCE INTERVALS: THE FOUR-STEP PROCESS

STATE: What is the practical question that requires estimating a parameter?

FORMULATE: Identify the parameter and choose a level of confidence.

SOLVE: Carry out the work in two phases:

1. **Check the conditions** for the interval you plan to use.
2. Calculate the **confidence interval.**

CONCLUDE: Return to the practical question to describe your results in this setting.

TESTS OF SIGNIFICANCE: THE FOUR-STEP PROCESS

STATE: What is the practical question that requires a statistical test?

FORMULATE: Identify the parameter and state null and alternative hypotheses.

SOLVE: Carry out the test in three phases:

1. **Check the conditions** for the test you plan to use.
2. Calculate the **test statistic.**
3. Find the **P-value.**

CONCLUDE: Return to the practical question to describe your results in this setting.

FIGURE 2 Overview of the "four-step process" used in *PSLS*.

("Formulate") is a key to success. The graphs and computations needed ("Solve") are essential but are not the whole story. A marginal icon helps students see the four-step process as a thread throughout the text. The four-step process appears whenever it fits the statistical content. Its repetitive use should foster the ability to address statistical problems independently.

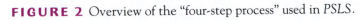

Practice of statistics with technology

Automating calculations increases students' ability to complete problems, reduces their frustration, and helps them concentrate on ideas and problem recognition rather than mechanics. *All students should have at least a "two-variable statistics" calculator* with functions for correlation and the least-squares regression line, as well as for the mean and standard deviation. Because students have calculators, the text doesn't discuss out-of-date "computing formulas" for the sample standard deviation or the least-squares regression line.

Using technology

Many instructors will take advantage of more elaborate technology, as ASA/MAA and GAISE recommend. And many students will find themselves using (for example) Excel on the job. *PSLS* does not assume or require use of software, except in the last few chapters, where the work is otherwise too tedious. However, there are regular "Using Technology" sections throughout the text to train students to read and use output from almost any source. The output always concerns one of the main teaching examples, so that students can compare text and output. These sections display and comment on output from different technologies, representing graphing calculators (the Texas Instruments TI-83 or TI-84), spreadsheets (Microsoft Excel), and statistical software (CrunchIt!, Minitab, and SPSS). CrunchIt! statistical software is available free online with each new copy of *PSLS*. Developed by Webster West of Texas A&M University, CrunchIt! is particularly easy to use and offers capabilities well beyond those needed for a first course. We encourage teachers who have avoided software in the past for reasons of availability, cost, or complexity to consider CrunchIt!

A different use of technology appears in the interactive applets created to our specifications and available online and on the text CD. These are designed primarily to help in learning statistics rather than in doing statistics. An icon calls attention to comments and exercises based on the applets. We suggest using selected applets for classroom demonstrations even if you do not ask students to work with them. The *Correlation and Regression, Confidence Interval,* and *P-value* applets, for example, convey core ideas more clearly than any amount of chalk and talk.

Why did you do that?

There is no single best way to organize the presentation of statistics to beginners. That said, our choices reflect thinking about both content and pedagogy. Here are comments on several "frequently asked questions" about the order and selection of material in *PSLS*.

Why does the distinction between population and sample not appear in Part I? This is a sign that there is more to statistics than inference. In fact, statistical inference is appropriate only in rather special circumstances. The chapters in Part I present tools and tactics for describing data—any data. Many data sets in these chapters do not lend themselves to inference, because they represent an entire population. John Tukey of Bell Labs and Princeton, the philosopher of modern data analysis, insisted that the population-sample distinction be avoided when it is not relevant, and we agree with him.

Why not begin with data production? It is certainly reasonable to do so—the natural flow of a planned study is from design to data analysis to inference. But in their future employment, most students will use statistics mainly in settings other than planned research studies. We place the design of data production (Chapters 7 and 8) after data analysis to emphasize that data-analytic techniques apply to any data. One of the primary purposes of statistical designs for producing data is to make inference possible, so the discussion in Chapters 7 and 8 opens Part II and motivates the study of probability.

Why not delay correlation and regression until late in the course, as is **traditional?** *PSLS* begins by offering experience working with data and gives a conceptual structure for this nonmathematical but essential part of statistics. Students profit from more experience with data and from seeing the conceptual structure worked out in relations among variables as well as in describing single-variable data. Correlation and least-squares regression are very important descriptive tools and are often used in settings where there is no population-sample distinction, such as studies based on state records or average species data. Perhaps most important, the *PSLS* approach asks students to think about what kind of relationship lies behind the data (confounding, lurking variables, association doesn't imply causation, and so on), without overwhelming them with the demands of formal inference methods. Inference in the correlation and regression setting is a bit complex, demands software, and often comes right at the end of the course. Delaying all mention of correlation and regression to that point often impedes the mastering of the basic uses and properties of these methods. We consider Chapters 3 and 4 (correlation and regression) essential and Chapter 23 (regression inference) optional when time constraints limit the amount of material that can be taught. For similar reasons, two-way tables are introduced first in the context of exploratory data analysis before moving on to inference with the chi-square test in Part III.

What about probability? Chapters 9, 11, and 13 present in a simple format the ideas of probability and sampling distributions that are needed to understand inference. These chapters go from the idea of probability as long-term regularity through concrete ways of assigning probabilities to the idea of the sampling distribution of a statistic. The law of large numbers and the central limit theorem appear in the context of discussing the sampling distribution of a sample mean. What is left for the *optional* Chapters 10 and 12 is mostly "general probability rules" (including conditional probability) and the binomial and Poisson distributions.

We suggest that you omit these optional chapters unless they represent important concepts for your particular audience. Experienced teachers recognize that students find probability difficult, and research has shown that this is true even of professionals. If a course is intended for med or premed students, for instance, the concept of conditional probability is quite relevant, because it is a key part of diagnosis that both doctors and patients have difficulty interpreting.[3] However, attempting to present a substantial introduction to probability in a data-oriented statistics course for students who are not mathematically trained is a very difficult challenge. Instructors should keep in mind that formal probability does not help students master the ideas of inference as much as we teachers often imagine, and it depletes reserves of mental energy that might better be applied to essential statistical ideas.

Why use the z procedures for a population mean to introduce the reasoning of inference? This is a pedagogical issue, not a question of statistics in practice. Some time in the golden future we will start with resampling methods as permutation tests make the reasoning of tests clearer than any traditional approach. For now the main choices are z for a mean and z for a proportion.

The z procedures for means are pedagogically more accessible to students. We can say up front that we are going to explore the reasoning of inference in an overly simple setting. Remember, exactly Normal population and true simple random sample are as unrealistic as known σ, especially in the life sciences. All the issues of practice—robustness against lack of Normality and application when the data aren't an SRS as well as the need to estimate σ—are put off until, with the reasoning in hand, we discuss the practically useful t procedures. This separation of initial reasoning from messier practice works well.

In contrast, starting with inference for p introduces many side issues: no exact Normal sampling distribution, but a Normal approximation to a discrete distribution; use of $\hat{p}$ in both the numerator and denominator of the test statistic to estimate both the parameter p and $\hat{p}$'s own standard deviation; loss of the direct link between test and confidence interval. In addition, we now know that the traditional z confidence interval for p is often grossly inaccurate. See the following section for an explanation.

Why does the presentation of inference for proportions go beyond the traditional methods? Recent computational and theoretical work has demonstrated convincingly that the standard confidence intervals for proportions can be trusted only for very large sample sizes. It is hard to abandon old friends, but the graphs in Section 2 of the paper by Brown, Cai, and DasGupta in the May 2001 issue of *Statistical Science* are both distressing and persuasive.[4] The standard intervals often have a true confidence level much less than what was requested, and requiring larger samples encounters a maze of "lucky" and "unlucky" sample sizes until very large samples are reached. Fortunately, there is a simple cure: Just add two successes and two failures to your data. (Therefore, no additional software tool is required for this procedure.) We present these "plus four intervals" in Chapters 19 and 20, along with guidelines for use.

Why didn't you cover Topic X? Introductory texts ought not to be encyclopedic. Including each reader's favorite special topic results in a text that is formidable in size and intimidating to students. The topics covered in *PSLS* were chosen because they are the most commonly used in the life sciences and they are suitable vehicles for learning broader statistical ideas. Two chapters available on CD and online cover more advanced inference procedures. Students who have completed the core of *PSLS* will have little difficulty moving on to more elaborate methods.

We are grateful to the many colleagues from two-year and four-year colleges and universities who commented on successive drafts of the manuscript. Special thanks are due to Patricia Humphrey (Georgia Southern University), who read the manuscript line by line and checked its accuracy, and to Clifford Anderson-Bergman (University of California, Irvine) and Peter Sprangers (Ohio State University), who worked on the back-of-book short answers. Others who offered comments are:

Ahmed A. Arif
  Texas Tech University
Bassam H. Atieh
  Murray State University
Emilia Bagiella
  Columbia University
Priya Banerjee
  SUNY Brockport
Toby Bennett
  St. Gregory's University
Robin M. Bush
  University of California Irvine
Erica A. Corbett
  Southeastern Oklahoma State University
Dorin Dumitrascu
  The University of Arizona
David Elashoff
  University of California, Los Angeles
Jane Grosset
  Community College of Philadelphia
Jeremiah N. Jarrett
  Central Connecticut State University
Aubrey Dale Magoun
  University of Louisiana at Monroe
Holly E. Menzel
  Marietta College

Bhramar Mukherjee
  University of Michigan
Leonard C. Onyiah
  St. Cloud State University
Gordon E. Robertson
  University of Ottawa
James M. Sinacore
  Loyola University Chicago
Carl R. Spitznagel
  John Carroll University
Sukanya Subramanian
  Collin College
Andrew Tierman
  Saginaw Valley State University
Elizabeth J. Walters
  Loyola College in Maryland
Paul Weiss
  Emory University
John W. Wilson
  University of Pittsburgh
Ken Yasukawa
  Beloit College
George L. Zimmermann
  Richard Stockton College of New Jersey

We are also particularly grateful to Craig Bleyer, Ruth Baruth, Mary Louise Byrd, Pam Bruton, Amy Fass, Vicki Tomaselli, Bianca Moscatelli, Suzanne Slope, and the large editorial and design team who have helped make this textbook a reality.

Brigitte Baldi and David S. Moore

Media and Supplements

For Students

Online Study Center (OSC): www.whfreeman.com/osc/psls. (Access code required. Available for purchase online.) In addition to all the offerings available on the companion Web site (see below), the OSC offers:

- StatTutor Tutorials
- Stats@Work Simulations
- Study Guide
- Statistical Software Manuals

Study Guide by Brigitte Baldi, University of California, Irvine. This printed guide helps students sharpen their analytical and problem-solving skills as they review material from the text and prepare for quizzes and exams. Written in an easily accessible style, the Study Guide provides step-by-step solutions to selected text exercises along with summaries of the key concepts needed to solve the problems. Each chapter begins with an overview of each section's basic concepts, followed by guided solutions to selected problems in that section. These solutions provide hints for setting up and thinking about each exercise and explain both how the answer was reached and why it was done that way. ISBN: 1-4292-0165-7

Companion Web Site: www.whfreeman.com/psls. Seamlessly integrates topics from the text. On this open-access Web site, students can find:

- **Interactive statistical applets** that allow students to manipulate data and see the corresponding results graphically.
- **Data sets** in ASCII, Excel, JMP, Minitab, TI, SPSS, and S-PLUS formats.
- **Interactive exercises and self-quizzes** to help students prepare for tests.
- **Key tables and formulas** summary sheet.
- **Optional Companion Chapters 26 and 27** covering two-way analysis of variance and follow-up tests, and nonparametric tests.
- **CrunchIt!** statistical software is available via an access-code-protected Web site. Access codes are available in every new text or can be purchased online.
- **EESEE** case studies are available via an access-code-protected Web site. Access codes are available in every new text or can be purchased online.

Interactive Student CD-ROM: Included with every new copy of *PSLS*, the CD contains access to the applets, companion chapters, and data sets.

Special Software Packages: Student versions of JMP, Minitab, S-PLUS, and SPSS are available on a CD-ROM packaged with the textbook. This software is not sold separately and must be packaged with a text or a manual. Contact your W. H. Freeman representative for information or visit www.whfreeman.com.

SMARTHINKING Online Tutoring: W. H. Freeman and Company is partnering with SMARTHINKING to provide students with free access-code-protected online tutoring and homework help from specially trained, professional educators. Twelve-month subscriptions are available to be packaged with *PSLS*. (Access code is included with every new text.)

For Instructors

The Instructor's Web site: www.whfreeman.com/psls. Requires user registration as an instructor and features all of the student Web materials, plus:

- Instructor version of **EESEE** (Electronic Encyclopedia of Statistical Examples and Exercises), with solutions to the exercises in the student version.
- The **Instructor's Guide,** including full solutions to all exercises in PDF format.
- **PowerPoint slides** containing all textbook figures and tables.

Instructor's Guide with Solutions by Brigitte Baldi, University of California, Irvine. This printed guide includes full solutions to all exercises and provides additional examples and data sets for class use, Internet resources, and sample examinations. It also contains brief discussions of the *PSLS* approach for each chapter. ISBN: 1-4292-0162-2.

Test Bank The test bank contains hundreds of multiple-choice questions to generate quizzes and tests. Available in print as well as electronically on CD-ROM (for Windows and Mac), where questions can be downloaded, edited, and resequenced to suit each instructor's needs.

Printed Version, ISBN: 1-4292-0163-0.

Computerized (CD) Version, ISBN: 1-4292-0164-9.

Enhanced Instructor's Resource CD-ROM: Allows instructors to **search** and **export** (by key term or chapter) all the material from the student CD, plus:

- All text images and tables.
- Statistical applets and data sets.
- Instructor's Guide with full solutions.
- PowerPoint files and lecture slides.
- Test bank files.

ISBN: 1-4292-1862-2

Course Management Systems: W. H. Freeman and Company provides courses for Blackboard, WebCT (Campus Edition and Vista), and Angel course management systems. These are completely integrated solutions that you can easily customize and adapt to meet your teaching goals and course objectives. Upon request, we also provide courses for users of Desire2Learn and Moodle. Visit www.bfwpub.com/lmc for more information.

i-clicker

i-clicker Radio Frequency Classroom Response System: Offered by W. H. Freeman and Company, in partnership with i-clicker, and created by Educations for Educations, i-clicker's system is the hassle-free way to make class time more interactive. Visit www.iclicker.com for more information.

To the Student: Statistical Thinking

Statistics is about data. Data are numbers, but they are not "just numbers." **Data are numbers with a context.** The number 10.5, for example, carries no information by itself. But if we hear that a friend's new baby weighed 10.5 pounds at birth, we congratulate her on the healthy size of the child. The context engages our background knowledge and allows us to make judgments. We know that a baby weighing 10.5 pounds is quite large and that a human baby is unlikely to weigh 10.5 ounces or 10.5 kilograms. The context makes the number informative.

Statistics is the science of data. To gain insight from data, we make graphs and do calculations. But graphs and calculations are guided by ways of thinking that amount to educated common sense. Let's begin our study of statistics with an informal look at some principles of statistical thinking.

DATA BEAT ANECDOTES

An anecdote is a striking story that sticks in our minds exactly because it is striking. Anecdotes humanize an issue, but they can be misleading.

Does living near power lines cause leukemia in children? The National Cancer Institute spent 5 years and $5 million gathering data on this question. The researchers compared 638 children who had leukemia with 620 who did not. They went into the homes and measured the magnetic fields in the children's bedrooms, in other rooms, and at the front door. They recorded facts about power lines near the family home and also near the mother's residence when she was pregnant. Result: no connection between leukemia and exposure to magnetic fields of the kind produced by power lines. The editorial that accompanied the study report in the *New England Journal of Medicine* thundered, "It is time to stop wasting our research resources" on the question.[1]

Now compare the effectiveness of a television news report of a 5-year, $5 million investigation against a televised interview with an articulate mother whose child has leukemia and who happens to live near a power line. In the public mind, the anecdote wins every time. A statistically literate person knows better. **Data are more reliable than anecdotes, because they systematically describe an overall picture rather than focus on a few incidents.**

ALWAYS LOOK AT THE DATA

Yogi Berra said it: "You can observe a lot by just watching." That's a motto for learning from data. **A few carefully chosen graphs are often more instructive than great piles of numbers.**

Let's look at some data. Figure 1 displays a histogram of the body lengths of 56 perch (*Perca fluviatilis*) caught in a Finnish lake.[2] Each bar in the graph

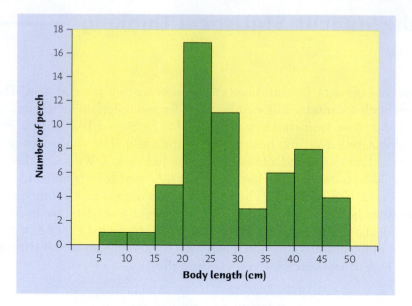

FIGURE 1 Body lengths of 56 perch from a Finnish lake. Always look at the data: Two separate clusters are clearly visible.

represents how many perch had a body length between two values on the horizontal axis. For example, the tallest bar indicates that 17 perch had a body length between 20 and 25 centimeters (cm).

We see great variability in perch body lengths, from about 5 to 50 cm, but we also notice two clear clusters: one group of smaller fish and one group of larger fish. This suggest that the data include two different groups of perch, possibly males and females or perch of different ages. In fact, observations in the wild indicate that perch can become much larger if they survive to a very old age (up to 22 years).

Any attempt to summarize the data in Figure 1 without a close inspection first would miss the two distinct clusters. Failure to examine any data carefully can lead to misleading or absurd results. As humorist Des McHale put it, "the average human has one breast and one testicle."

BEWARE THE LURKING VARIABLE

The Kalamazoo (Michigan) Symphony once advertised a "Mozart for Minors" program with this statement: "Question: Which students scored 51 points higher in verbal skills and 39 points higher in math? Answer: Students who had experience in music."[3] *Who would dispute that early experience with music builds brainpower?* The skeptical statistician, that's who. Children who take music lessons and attend concerts tend to have prosperous and well-educated parents. These same children are also likely to attend good schools, get good health care, and be encouraged to study hard. No wonder they score well on tests.

We call family background a *lurking variable* when we talk about the relationship between music and test scores. It is lurking behind the scenes, unmentioned in the symphony's publicity. Yet family background, more than anything else we can measure, influences children's academic performance. Perhaps the Kalamazoo Youth Soccer League should advertise that students who play soccer score higher on tests. After all, children who play soccer, like those who have experience in music, tend to have educated and prosperous parents. **Almost all relationships between two variables are influenced by other variables lurking in the background.**

WHERE THE DATA COME FROM IS IMPORTANT

Stockbyte/PictureQuest

The advice columnist Ann Landers once asked her readers, "If you had it to do over again, would you have children?" A few weeks later, her column was headlined "70% OF PARENTS SAY KIDS NOT WORTH IT." Indeed, 70% of the nearly 10,000 parents who wrote in said they would not have children if they could make the choice again. *Do you believe that 70% of all parents regret having children?*

You shouldn't. The people who took the trouble to write Ann Landers are not representative of all parents. Their letters showed that many of them were angry at their children. All we know from these data is that there are some unhappy parents out there. A statistically designed poll, unlike Ann Landers's appeal, targets specific people chosen in a way that gives all parents the same chance to be asked. Such a poll showed that 91% of parents *would* have children again. Where data come from matters a lot. If you are careless about how you get your data, you may announce 70% "No" when the truth is about 90% "Yes."

Here's another question: *Should episiotomy be a routine part of childbirth?* Episiotomy is the surgical cut of the skin and muscles between the vagina and the rectum sometimes performed during childbirth. Until recently, it was one of the most common surgical procedures in women in the United States, performed routinely to speed up childbirth and in the hope that it would help prevent tearing of the mother's tissue and possible later incontinence. Episiotomy rates vary widely by hospital and by physician, based principally on personal beliefs about the procedure's benefits.

However, recent clinical trials and epidemiological studies have shown no benefit of episiotomy unless the baby's health requires accelerated delivery or a large natural tear seems likely. In fact, these studies indicate that episiotomy is associated with longer healing times and increased rates of complications, including infection, extensive tearing, pain, and incontinence.[4]

To get convincing evidence on the benefits and risks of episiotomy we need unbiased data. Proper clinical trials and epidemiological studies rely on randomness of patient or treatment selection to avoid bias. The careful studies of the risks and benefits of routine episiotomy could be trusted because they had sound data collection designs. As a result, rates of episiotomy in the United States have dropped substantially.

The most important information about any statistical study is how the data were produced. Only statistically designed opinion polls and surveys can be trusted. Only experiments can give convincing evidence that an alleged cause really does account for an observed effect.

VARIATION IS EVERYWHERE

Many students in biology lab courses are surprised to obtain somewhat different results when they repeat an experiment. Yet variability is ubiquitous.

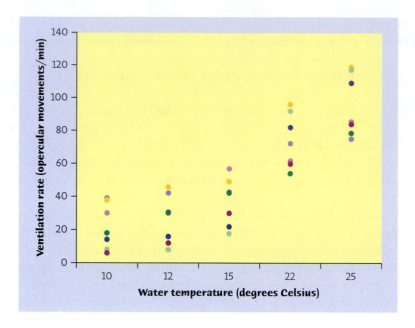

FIGURE 2 Variation is everywhere: the ventilation rates of 7 goldfish placed in tanks with varying water temperature.

Larry Larimer/Brand X/Corbis

Figure 2 plots the ventilation rate of 7 goldfish placed in tanks of varying temperature. Each goldfish is represented by one color.[5] While there is a clear overall pattern in this figure, there is also a lot of variability, with ventilation rates ranging overall from 6 to 119 opercular movements per minute. Some of that variability can be attributed to lack of precision in measurements, some to fish physiology, and some to slight differences in circumstances at the time of each measurement (such as movements around the fish tanks that might stress the fish). Before the experiment, the fish were kept in tanks set to 22 degrees Celsius. At that temperature, the fish's ventilation rates range from about 60 to 100 opercular movements per minute. Yet when the same 7 fish are observed in warmer water, their ventilation rate increases overall. And when the fish are observed at lower temperatures, the ventilation rates decrease quite dramatically. These data show that ventilation rate in goldfish varies both from fish to fish and as a function of water temperature.

Students are not the only ones with a tendency to underestimate variability in real data. Here is Arthur Nielsen, head of the country's largest market research firm, describing his experience:

> *Too many business people assign equal validity to all numbers printed on paper. They accept numbers as representing Truth and find it difficult to work with the concept of probability. They do not see a number as a kind of shorthand for a range that describes our actual knowledge of the underlying condition.*[6]

Variation is everywhere. Individuals vary; repeated measurements on the same individual vary; almost everything varies over time. One reason we need to know some statistics is that statistics helps us deal with variation.

Most women who reach middle age have regular mammograms to detect breast cancer. *Do mammograms reduce the risk of dying of breast cancer?* Doctors rely on clinical trials that compare different ways of screening for breast cancer. The conclusion from 13 such experiments is that mammograms reduce the risk of death in women aged 50 to 64 years by 26%.[7]

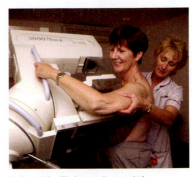

By Ian Miles-Flashpoint Pictures/Alamy

On the average, then, women who have regular mammograms between the ages of 50 and 64 are less likely to die of breast cancer. But because variation is everywhere, the results are different for different women. Some women who have yearly mammograms die of breast cancer, and some who never have mammograms live a long, breast-cancer-free life. Statistical conclusions are "on average" statements only. Well, then, can we be certain that mammograms reduce risk on average? No. We can be very confident, but we can't be certain.

Because variation is everywhere, conclusions are uncertain. Statistics gives us a language for talking about uncertainty that is used and understood by statistically literate people everywhere. In the case of mammograms, doctors use that language to tell us that "mammography reduces the risk of dying of breast cancer by 26 percent (95 percent confidence interval, 17 to 34 percent)." That 26% is, in Arthur Nielsen's words, a "shorthand for a range that describes our actual knowledge of the underlying condition." The range is 17% to 34%, and we are 95 percent confident that the truth lies in that range. We will soon learn to understand this language. We can't escape variation and uncertainty. Learning statistics enables us to live more comfortably with these realities.

Statistical Thinking and You

What lies ahead in this book The purpose of *The Practice of Statistics in the Life Sciences* (*PSLS*) is to give you a working knowledge of the ideas and tools of practical statistics. We will divide practical statistics into three main areas:

1. **Data analysis** concerns methods and strategies for exploring, organizing, and describing data using graphs and numerical summaries. Only organized data can illuminate reality. Only thoughtful exploration of data can defeat the lurking variable. Part I of *PSLS* (Chapters 1 to 6) discusses data analysis.

2. **Data production** provides methods for producing data that can give clear answers to specific questions. Where the data come from really is important. Basic concepts about how to select samples and design experiments are the most influential ideas in statistics. These concepts are the subject of Chapters 7 and 8.

3. **Statistical inference** moves beyond the data in hand to draw conclusions about some wider universe, taking into account that variation is everywhere and that conclusions are uncertain. To describe variation and uncertainty, inference uses the language of probability, introduced in Chapters 9, 11, and 13. Because we are concerned with practice rather than theory, we need only

a limited knowledge of probability. Chapters 10 and 12 offer more probability for those who want it. Chapters 14 and 15 discuss the reasoning of statistical inference. These chapters are the key to the rest of the book. Chapters 17 to 21 present inference as used in practice in the most common settings. Chapters 22 to 24, and the Optional Companion Chapters 26 and 27 on the text CD or online, concern more advanced or specialized kinds of inference.

STEP

Because data are numbers with a context, doing statistics means more than manipulating numbers. You must **state** a problem in its real-world context, **formulate** the problem by recognizing what specific statistical work is needed, **solve** the problem by making the necessary graphs and calculations, and **conclude** by explaining what your findings say about the real-world setting. We'll make regular use of this four-step process to encourage good habits that go beyond graphs and calculations to ask, "What do the data tell me?"

Statistics does involve lots of calculating and graphing. The text presents the techniques you need, but you should use a calculator or software to automate calculations and graphs as much as possible. Because the big ideas of statistics don't depend on any particular level of access to computing, *PSLS* does not require software. Even if you make little use of technology, you should look at the "Using Technology" sections throughout the book. You will see at once that you can read and use the output from almost any technology used for statistical calculations. The ideas really are more important than the details of how to do the calculations.

Using technology

You will need a calculator with some built-in statistical functions. Specifically, your calculator should find means and standard deviations and calculate correlations and regression lines. Look for a calculator that claims to do "two-variables statistics" or mentions "regression." You will get the most, though, out of a graphing calculator or statistical software.

Because graphing and calculating are automated in statistical practice, the most important assets you can gain from the study of statistics are an understanding of the big ideas and the beginnings of good judgment in working with data. *PSLS* tries to explain the most important ideas of statistics, not just teach methods. Some examples of big ideas that you will encounter (one from each of the three areas of statistics) are "always plot your data," "randomized comparative experiments," and "statistical significance." Some particularly important ideas (such as how to treat outliers, and the scientific approach) are given more extensive treatment, with real life examples in discussion topics placed throughout the book.

You learn statistics by doing statistical problems. As you read, you will see several levels of exercises, arranged to help you learn. Short "Apply Your Knowledge" problem sets appear after each major idea. These are straightforward exercises that help you solidify the main points as you read. Be sure you can do these exercises before going on. The end-of-chapter exercises begin with multiple-choice "Check Your Skills" exercises (with all answers in the back of the book). Use them to check your grasp of the basics. The regular "Chapter Exercises" help you combine all the ideas of a chapter. Finally, the three part review chapters look back over major blocks of learning, with many review exercises. At each step you are given

APPLY YOUR KNOWLEDGE ⎯⎯⎯

Check Your Skills ⎯⎯⎯

less advance knowledge of exactly what statistical ideas and skills the problems will require, so each type of exercise requires more understanding.

The part review chapters (and the individual optional chapters on CD or on-line) include point-by-point lists of specific things you should be able to do. Go through that list, and be sure you can say "I can do that" to each item. Then try some of the review exercises. The book ends with a review titled "Statistical Thinking Revisited," which you should read and think about no matter where in the book your course ends.

The key to learning is persistence. The main ideas of statistics, like the main ideas of any important subject, took a long time to discover and take some time to master. The gain will be worth the pain.

Plantography/Alamy

PART

1

PART I: Exploring Data

The first step in understanding data is to hear what the data say, to "let the statistics speak for themselves." Numbers speak clearly only when we help them speak by organizing, displaying, and summarizing. That's *data analysis*. The six chapters in Part I present the ideas and tools of statistical data analysis. They equip you with skills that are immediately useful whenever you deal with numbers.

These chapters reflect the strong emphasis on exploring data that characterizes modern statistics. Although careful exploration of data is essential if we are to trust the results of inference, data analysis isn't just preparation for inference. To think about inference, we carefully distinguish between the data we actually have and the larger universe we want conclusions about.

The National Center for Health Statistics, for example, has data about each member of the 40,000 households contacted by its National Health Interview Survey. The center wants to draw conclusions about the health status of household members for all 113 million U.S. households. That's a complex problem. From the viewpoint of data analysis, things are simpler. We want to explore and understand only the data in hand. The distinctions that inference requires don't concern us in Chapters 1 to 6. What does concern us is a systematic strategy for examining data and the tools that we use to carry out that strategy.

Part of that strategy is to first look at one thing at a time and then at relationships. In Chapters 1 and 2 you will study **variables and their distributions.** Chapters 3, 4, and 5 concern **relationships among variables.** Chapter 6 reviews this part of the text, along with more comprehensive exercises.

Boris Karpinski/Alamy

Picturing Distributions with Graphs

Statistics is the science of data. The volume of data available to us is overwhelming. The Human Genome Project uncovered, after 13 years of international coopera- tion, the complete sequence of the 3 billion DNA bases of the human genome. In the United States, the Census Bureau constantly gathers information about the nation. One example is the yearly survey of 40,000 households that is produced for the National Center for Health Statistics. This National Health Interview Survey records facts about each household, including details about the family structure and even whether the household members eat together. It also records facts about each person in the household—age, sex, weight, income, medical or behavioral in- capacitation, access to health insurance, detailed health history, and much more. The first step in dealing with such a flood of data is to organize our thinking about data.

Individuals and variables

Any set of data contains information about some group of *individuals*. The infor- mation is organized in *variables*. The techniques of descriptive statistics covered in Part I of this book apply equally to data sets obtained from all individuals in a given **population** or from only those individuals in a smaller **sample.** The distinc- tion between population and sample is an important one for inference, and we will address it in more detail in Chapter 7.

population
sample

INDIVIDUALS AND VARIABLES

Individuals are the objects described by a set of data. Individuals may be people, but they may also be animals or things.

A **variable** is any characteristic of an individual. A variable can take different values for different individuals.

Gesundheit!

Researchers who study treatments against the common cold must find ways to assess their effectiveness. Effectiveness may be either preventing a cold or reducing its symptoms. Researchers can record simply whether or not each person exposed to the common-cold virus developed cold symptoms. They can also count the number of cold symptoms detectable, count the number of tissues used in a day, or record the duration of symptoms. Smart studies often look at a problem from more than just one angle.

A botanist's plant data base, for example, includes data about various aspects of the plants examined. The plants are the individuals described by the data set. For each individual, the data contain the values of variables such as stem length, number of flower petals, and flower color. An example of what such a data base might look like is shown in Figure 1.1. In practice, any set of data is accompanied by background information that helps us understand the data. When you plan a statistical study or explore data from someone else's work, ask yourself the following questions:

1. **Who?** What **individuals** do the data describe? **How many** individuals appear in the data?

2. **What?** How many **variables** do the data contain? What are the **exact definitions** of these variables? In what **units of measurement** is each variable recorded? Lengths, for example, might be recorded in inches, in yards, or in meters.

3. **Why?** **What purpose** do the data have? Do we hope to answer some specific questions? Do we want to draw conclusions about individuals other than the ones we actually have data for? Are the variables suitable for the intended purpose?

Some variables, like a plant's flower color, simply place individuals into categories. Others, like stem length and number of flower petals, take numerical values with which we can do arithmetic. It makes sense to give an average stem length for plants of a given environment, but it does not make sense to give an "average" flower color. We can, however, count the numbers of yellow, pink, and white flowers and do arithmetic with these counts.

Microsoft Excel

	A	B	C	D
1	Specimen ID	Stem length (cm)	Number of flower petals	Flower color
2	1	3.1	28	yellow
3	2	2.6	5	white
4	3	8.2	8	red
5	4	5.9	12	pink
6	5	14.7	5	yellow

Book 1 — Sheet1 / Sheet2 / Sheet3

FIGURE 1.1 Example of a botanical data base displayed in a spreadsheet. Each column includes data about a different variable. Each row represents data for one plant specimen.

> **CATEGORICAL AND QUANTITATIVE VARIABLES**
>
> A **categorical variable** places an individual into one of several groups or categories.
>
> A **quantitative variable** takes numerical values for which arithmetic operations such as adding and averaging make sense. The values of a quantitative variable are usually recorded in a **unit of measurement** such as seconds or kilograms.

Further distinction is sometimes made beyond simply categorical or quantitative. Some quantitative variables, like stem length, are **continuous** variables that can take any real numerical value over an interval. **Discrete** variables, on the other hand, are quantitative variables that can take only a limited, finite number of values, like the number of petals in a flower. Categorical variables can also be broken down into nominal and ordinal. **Nominal** variables are purely qualitative and unordered, like flower color, whereas **ordinal** data can be ranked, such as with star ratings or the Likert scales commonly used in psychology (for example, "rate from 0 to 5, with 0 for really dislike and 5 for really like"). Although ordinal data can be ranked, they are not true quantitative variables, because the intervals between consecutive ranks are often not identical.

continuous
discrete

nominal
ordinal

─── **EXAMPLE 1.1**   Health of the nation ───────

The National Center for Health Statistics (NCHS) is the nation's principal health statistics agency. Every year, it collects extensive information on each member of 40,000 households for its National Health Interview Survey. The persons interviewed are the *individuals* for which we have information in the form of many *variables*. A few of the 2006 survey variables were:[1]

FMX	Family Serial Number (a code protecting identities)
AWEIGHTP	Weight without shoes (pounds)
AHEIGHT	Total height (inches)
CIGDAMO	Number of days smoked in past 30 days
CIGQTYR	Tried to quit smoking in past year
SLEEP	Hours of sleep
ADENLONG	Time since last saw a dentist (months)
CANEV	Ever told by a doctor you had cancer
AHCAFYR1	Can't afford prescription medicine in past 12 months

In addition to the serial number, eight variables are displayed. Three are categorical variables (tried to quit smoking, ever told you had cancer, and can't afford prescription medicine). *Whether the answers are recorded as "no/yes" or as "0/1" makes no difference because these are only labels (and therefore never true numerical values).* The other five variables are quantitative. Their values do have units. These variables are weight in pounds, height in inches, number of days smoked in past 30 days, hours of sleep, and time in months since last saw a dentist.

CAUTION

The *purpose* of the National Health Interview Survey is to collect data that represent the entire nation in order to guide government actions and policies to improve the health of the American people. To do this, the households contacted are chosen at random from all households in the country. We will see in Chapter 7 why choosing at random is a good idea.

spreadsheet   Collected data are often organized in a table in which each row is an individual and each column is a variable, as in Figure 1.1. **Spreadsheet** programs that have rows and columns readily available are commonly used to enter and store data and to do simple calculations.

APPLY YOUR KNOWLEDGE

1.1   A medical study. Data from a medical study contain values of many variables for each of the people who were the pie subjects of the study. Which of the following variables are categorical and which are quantitative?

(a) Gender (female or male)

(b) Age (years)

(c) Race (Asian, black, white, or other)

(d) Smoker (yes or no)

(e) Systolic blood pressure (millimeters of mercury)

(f) Level of calcium in the blood (micrograms per milliliter)

1.2   Cereal content. Here is a small part of an EESEE data set, "Nutrition and Breakfast Cereals," that describes the nutritional content per serving of 77 brands of breakfast cereals:[2]

Brand name	Manufacturer	Cold/ hot	Calories (Cal)	Sugars (grams)	Fibers (grams)
⋮					
All Bran	K	C	70	5	9
All Bran with Extra Fiber	K	C	50	0	14
Almond Delight	R	C	110	8	1
Apple Cinnamon Cheerios	G	C	110	10	2
Apple Jacks	K	C	110	14	1
⋮					

(a) What are the individuals in this data set?

(b) For each individual, what variables are given? Which of these variables are categorical and which are quantitative?

Categorical variables: pie charts and bar graphs

exploratory data analysis   Statistical tools and ideas help us examine data in order to describe their main features. This examination is called **exploratory data analysis.** Like an explorer crossing unknown lands, we want first to simply describe what we see. Here are two principles that help us organize our exploration of a set of data.

> **EXPLORING DATA**
>
> 1. Begin by examining each variable by itself. Then move on to study the relationships among the variables.
> 2. Begin with a graph or graphs. Then add numerical summaries of specific aspects of the data.

We will also follow these principles in organizing our learning. Chapters 1 and 2 present methods for describing a single variable. We study relationships among several variables in Chapters 3 to 5. In each case, we begin with graphical displays, then add numerical summaries for more complete description.

The proper choice of graph depends on the nature of the variable. To examine a single variable, we usually want to display its *distribution*.

> **DISTRIBUTION OF A VARIABLE**
>
> The **distribution** of a variable tells us what values it takes and how often it takes these values.
>
> The values of a categorical variable are labels for the categories. The **distribution of a categorical variable** lists the categories and gives either the count or the percent of individuals that fall in each category.

Convenient graphs

The movie *An Inconvenient Truth* won the Oscar for Best Documentary in 2006, and it is the third-highest grossing documentary in U.S. history. The movie features scientific evidence of global warming, supported by a detailed presentation of about 30 graphs and data tables.

┌ **EXAMPLE 1.2** Leading causes of death ─────────────

What are the leading causes of death in America? The National Center for Health Statistics reports such information on a yearly basis. Here is a breakdown of the top 10 causes of death for the year 2001 in the United States:[3]

Top 10 causes of death	Count of deaths	Percent of top 10 causes
Heart disease	700,142	37
Cancer	553,768	29
Cerebrovascular disease	163,538	9
Chronic respiratory disease	123,013	6
Accidental death	101,537	5
Diabetes	71,372	4
Flu and pneumonia	62,034	3
Alzheimer's disease	53,852	3
Kidney disorder	39,480	2
Septicemia	32,238	2
Total	1,900,974	100

roundoff error

It's a good idea to check data for consistency. The counts should add to 1,900,974, the total number of deaths from the top 10 causes in the year 2001. They do. The percents should add to 100% or very nearly 100%, as each percent is rounded to the nearest integer. **Roundoff errors** don't point to mistakes in our work, just to the effect of rounding off results.

pie chart

Columns of numbers take time to read. The **pie chart** in Figure 1.2 shows the distribution of the top 10 causes of death more vividly. For example, the "cancer" slice makes up 29% of the pie because 29% of all those who died from a top 10 cause in the United States in 2001 died from cancer. Pie charts are awkward to make by hand, but software will do the job for you. *A pie chart must include all the categories that make up a whole. Use a pie chart only when you want to emphasize each category's relation to the whole.* Here the pie chart displays all 10 categories making up the top 10 causes of death. The graph shows more clearly than the table of raw numbers the predominance of heart disease and cancer as leading causes of death.

bar graph

We could also make a **bar graph** that represents each cause of death by the height of a bar. Pie charts must include all the categories that make up a whole, but bar graphs are more flexible: They can compare any set of quantities that are measured in the same units. Bar graphs are particularly clear in pointing to the order and the relative importance of the different categories. Figure 1.3 shows two possible bar graphs of the data in Example 1.2. The bar graph shown in (a) is sorted alphabetically: Accidental death is the first bar in the graph because this category starts with "A." A more interesting and more intuitive way of representing the same data is shown in (b). This bar graph is sorted by magnitude: The tallest bar appears first, followed by the second-tallest bar, etc. It makes very clear that heart disease and cancer are by far the most important causes of death in the modern-day United States.

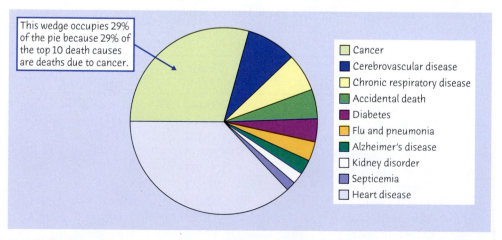

This wedge occupies 29% of the pie because 29% of the top 10 death causes are deaths due to cancer.

- Cancer
- Cerebrovascular disease
- Chronic respiratory disease
- Accidental death
- Diabetes
- Flu and pneumonia
- Alzheimer's disease
- Kidney disorder
- Septicemia
- Heart disease

FIGURE 1.2 You can use either a pie chart or a bar graph to display the distribution of a categorical variable. Here is a pie chart of the top 10 causes of death in the United States for 2001.

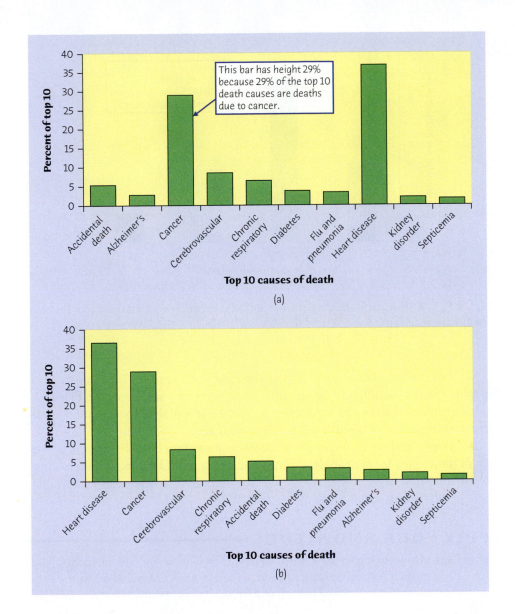

FIGURE 1.3 The data from Figure 1.2 on the top 10 causes of death in the United States for 2001 are displayed in a bar graph format. (a) The bars in the graph are sorted alphabetically. (b) The bars are now sorted according to their height (by order of importance).

EXAMPLE 1.3 Who uses marijuana?

The 2002 National Household Survey on Drug Use and Health reported the percent of current users of marijuana and hashish in each of four age groups:[4]

Age group (years)	Percent current users
12–17	8.2
18–25	17.3
26–34	7.7
35+	3.1

It's clear that young adults between 18 and 25 years of age show the most interest in consuming marijuana and hashish.

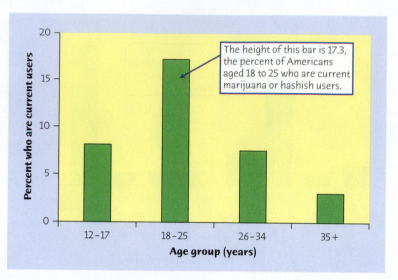

FIGURE 1.4 Bar graph showing the percent of current users of marijuana or hashish in each of four age groups in 2002.

We can't make a pie chart to display the data in Example 1.3. Each percent in the table refers to a different age group, not to parts of a single whole. Figure 1.4 is a bar graph of the data in Example 1.3. We see at a glance that, among adults, current marijuana and hashish use declines with age.

Bar graphs and pie charts help an audience grasp data quickly. They are, however, of limited use for data analysis because it is easy to understand data on a single categorical variable without a graph. We will move on to quantitative variables, where graphs are essential tools.

APPLY YOUR KNOWLEDGE

1.3   Cigarette smoking in the United States. The Centers for Disease Control and Prevention report the percent of adult males (18 years old and over) who declared that they smoked in each of their surveys between 1999 and 2002:[5]

Year	Percent smoking
1999	25.2
2000	25.2
2001	24.7
2002	24.8

(a) Make a bar graph of the data.

(b) Would it be correct to make a pie chart of these data?

1.4   Never on Sunday? Births are not, as you might think, evenly distributed across the days of the week. Here are the average numbers of babies born on each day of the week in 2002:[6]

Day	Births
Sunday	7,526
Monday	11,453
Tuesday	12,823
Wednesday	12,083
Thursday	12,366
Friday	12,285
Saturday	8,573

Present these data in a well-labeled bar graph. Would it also be correct to make a pie chart? Suggest some possible reasons why there are fewer births on weekends.

Quantitative variables: histograms

Quantitative variables often take many values. The distribution tells us what values the variable takes and how often it takes these values. A graph of the distribution is clearer if nearby values are grouped together. The most common graph of the distribution of one quantitative variable is a **histogram.**

histogram

—— **EXAMPLE 1.4**   Making a histogram: sharks ——

The great white shark, *Carcharodon carcharias*, is a large ocean predator at the top of the food chain. Here are the lengths in feet of 44 great whites:[7]

18.7	12.3	18.6	16.4	15.7	18.3	14.6	15.8	14.9	17.6	12.1
16.4	16.7	17.8	16.2	12.6	17.8	13.8	12.2	15.2	14.7	12.4
13.2	15.8	14.3	16.6	9.4	18.2	13.2	13.6	15.3	16.1	13.5
19.1	16.2	22.8	16.8	13.6	13.2	15.7	19.7	18.7	13.2	16.8

Boris Karpinski/Alamy

The *individuals* in this data set are individual sharks and the *variable* is body length in feet. To make a histogram of the distribution of this variable, proceed as follows:

Step 1. Choose the classes. Divide the range of the data into classes of equal width. The data range from 9.4 to 22.8 feet, so we decide to use these classes:

$$9.0 < \text{individuals with body length} \leq 11.0$$

$$11.0 < \text{individuals with body length} \leq 13.0$$

$$\vdots$$

$$21.0 < \text{individuals with body length} \leq 23.0$$

In this example we chose to exclude the lower bound and include the upper bound in making the histogram classes. Choosing, instead, to include the lower bound and exclude the upper bound would also have been a valid option. What matters is to be sure to specify the classes precisely so that each individual falls into exactly one class. You can specify the nature of the class boundaries in the legend accompanying your histogram.

Based on the chosen class definition, the smallest shark, measuring 9.4 feet, falls into the first class; a shark measuring exactly 11.0 feet would still fall in that first class, but a shark measuring 11.1 feet would fall into the second class.

Step 2. Count the individuals in each class. Here are the counts:

Class	Count
9.1 to 11.0	1
11.1 to 13.0	5
13.1 to 15.0	12
15.1 to 17.0	15
17.1 to 19.0	8
19.1 to 21.0	2
21.1 to 23.0	1

Check that the counts add to 44, the number of individuals in the data (the 44 great white sharks).

Step 3. Draw the histogram. Mark the scale for the variable whose distribution you are displaying on the horizontal axis. That's the shark body length in feet. The scale runs from 9 to 23 feet because that is the span of the classes we chose. The vertical axis contains the scale of counts. Each bar represents a class. The base of the bar covers the class, and the bar height is the class count. The horizontal axis represents a continuum of values (here body lengths), and therefore histograms do not leave any horizontal space between the bars unless a class is empty (in which case the bar has height zero). Figure 1.5(a) is our histogram.

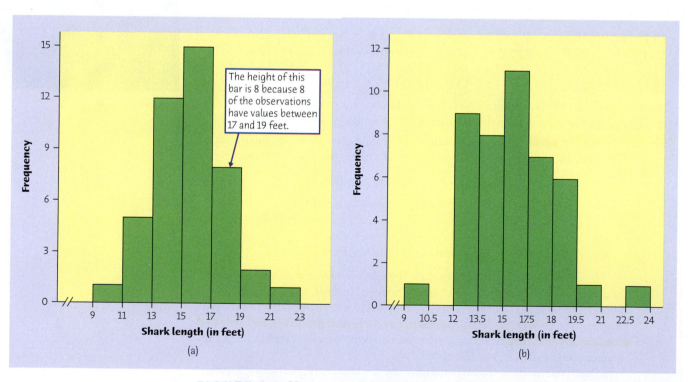

FIGURE 1.5 Histograms of the body length in feet for 44 great white sharks. (a) This histogram has 7 classes. (b) This histogram has 10 classes and shows more detail.

Although histograms resemble bar graphs, their details and uses are different. A histogram displays the distribution of a quantitative variable. The horizontal axis of a histogram is marked in the units of measurement for the variable. A bar graph compares the size of different items. The horizontal axis of a bar graph needs not have any measurement scale but simply identifies the items being compared. These may be the categories of a categorical variable, but they may also be separate, like the age groups in Example 1.3. Always make sure to leave some blank space between the bars in your bar graphs to separate the items being compared. In contrast, always draw histograms with no extra space between consecutive classes, to indicate that all values of the variable are covered.

Our eyes respond to the *area* of the bars in a histogram.[8] Because the classes are all the same width, area is determined by height and all classes are fairly represented. There is no one right choice of the classes in a histogram. Too few classes will give a "skyscraper" graph, with all values in a few classes with tall bars. Too many will produce a "pancake" graph, with most classes having one or no observations. Neither choice will give a good picture of the shape of the distribution. You must use your judgment in choosing classes to display the shape. Statistics software will choose the classes for you (but *you should be aware that different software programs use different conventions for displaying the class boundaries*). The software's choice is usually a good one, but you can change it if you want. The histogram function in the *One-Variable Statistical Calculator* applet on the text CD and Web site allows you to change the number of classes by dragging with the mouse, so that it is easy to see how the choice of classes affects the histogram.

APPLY YOUR KNOWLEDGE

1.5 **Healing of skin wounds.** Biologists studying the healing of skin wounds measured the rate at which new cells closed a razor cut made in the skin of an anesthetized newt. Here are the sorted data from 18 newts, measured in micrometers (millionths of a meter) per hour:[9]

Royalty-Free/CORBIS

| 11 | 12 | 14 | 18 | 22 | 22 | 23 | 23 | 26 |
| 27 | 28 | 29 | 30 | 33 | 34 | 35 | 35 | 40 |

Make a histogram of the healing rates using classes of width 5 micrometers/hour starting at 10 micrometers/hour. (Make this histogram by hand even if you have software, to be sure you understand the process. You may want to compare your histogram with your software's choice.)

1.6 **Choosing classes in a histogram.** The data set menu that accompanies the *One-Variable Statistical Calculator* applet includes a data set called "healing of skin wounds." Choose these data, then click on the "Histogram" tab to see a histogram.

(a) How many classes does the applet choose to use? (You can click on the graph outside the bars to get a count of classes.)

(b) Click on the graph and drag to the left. What is the smallest number of classes you can get? What are the lower and upper bounds of each class? (Click on the bar to find out.) Make a rough sketch of this histogram.

(c) Click and drag to the right. What is the greatest number of classes you can get? How many observations does the largest class have?

(d) You see that the choice of classes changes the appearance of a histogram. Drag back and forth until you get the histogram you think best displays the distribution. How many classes did you use?

Interpreting histograms

Making a statistical graph is not an end in itself. The purpose of the graph is to help us understand the data. After you make a graph, always ask, "What do I see?" Once you have displayed a distribution, you can see its important features as follows.

EXAMINING A HISTOGRAM

In any graph of data, look for the **overall pattern** and for striking **deviations** from that pattern.

You can describe the overall pattern of a histogram by its **shape, center,** and **spread.**

An important kind of deviation is an **outlier,** an individual value that falls outside the overall pattern.

We will learn how to describe center and spread numerically in Chapter 2. For now, we can describe the center of a distribution by its *midpoint,* the value with roughly half the observations taking smaller values and half taking larger values. We can describe the spread of a distribution by giving the *smallest and largest values.*

EXAMPLE 1.5   Describing a distribution: sharks ─────────

Look again at the histogram in Figure 1.5(a).

SHAPE: The distribution has a *single peak,* which represents sharks with a body length between 15 and 17 feet, and it is *symmetric.* In mathematics, the two sides of symmetric patterns are exact mirror images. Real data are almost never exactly symmetric. We are content to describe the histogram in Figure1.5(a) as symmetric.

CENTER: The counts in Example 1.4 show that 18 of the 44 sharks have a body length of 15 feet or less, but the count increases to 33 out of 44 for lengths of 17 feet or less. So the midpoint of the distribution is about 15 to 17 feet.

SPREAD: The spread is from 9.4 to 22.8 feet, but only 1 shark is shorter than 11 feet and only 1 is more than 21 feet long.

OUTLIERS: In Figure 1.5(a), the observations greater than 21 feet or less than 11 feet are part of the continuous range of body lengths and do not stand apart from the overall distribution. This histogram, with only 7 classes, hides some of the detail in the distribution. Look at Figure 1.5(b), a histogram of the same data with 10 classes. It reveals that the shortest and longest sharks, at 9.4 and 22.8 feet, do stand apart from the other observations.

Once you have spotted possible outliers, look for an explanation. Some outliers are due to mistakes, such as typing 19.4 as 9.4. Other outliers point to the special nature of some observations. The smallest shark might be a juvenile exhibiting some adult features, for instance. An outlier could also simply be an unusual but perfectly legitimate observation. For instance, the largest shark could be just that: an unusually large shark. Human height, for instance, has a somewhat homogeneous distribution within a gender and ethnicity, but some individuals, such as the famous basketball player Shaquille O'Neal, clearly stand out from the norm.

Comparing Figures 1.5(a) and (b) reminds us that *the choice of classes in a histogram can influence the appearance of a distribution*. Both histograms portray a symmetrical distribution with one peak, but only Figure 1.5(b) shows the mild outliers.

When you describe a distribution, concentrate on the main features. Look for major peaks, not for minor ups and downs in the bars of the histogram. Look for clear outliers, not just for the smallest and largest observations. Look for rough *symmetry* or clear *skewness*.

CAUTION

SYMMETRIC AND SKEWED DISTRIBUTIONS

A distribution is **symmetric** if the right and left sides of the histogram are approximately mirror images of each other.

A distribution is **skewed to the right** if the right side of the histogram (containing the half of the observations with larger values) extends much farther out than the left side. It is **skewed to the left** if the left side of the histogram extends much farther out than the right side.

The vital few

Skewed distributions can show us where to concentrate our efforts. Ten percent of the cars on the road account for half of all carbon dioxide emissions. A histogram of CO_2 emissions would show many cars with small or moderate values and a few with very high values. Cleaning up or replacing these cars would reduce pollution at a much lower cost than that of programs aimed at all cars. Statisticians who work at improving quality in industry make a principle of this: distinguish "the vital few" from "the trivial many."

Here are more examples of describing the overall pattern of a histogram.

EXAMPLE 1.6   Guinea pig survival times

Figure 1.6 displays the survival times in days (reported in the table below) of 72 guinea pigs after they were injected with infectious bacteria in a medical experiment.[10]

43	45	53	56	56	57	58	66	67	73	74	79
80	80	81	81	81	82	83	83	84	88	89	91
91	92	92	97	99	99	100	100	101	102	102	102
103	104	107	108	109	113	114	118	121	123	126	128
137	138	139	144	145	147	156	162	174	178	179	184
191	198	211	214	243	249	329	380	403	511	522	598

The distribution is *single-peaked* and obviously *skewed to the right*. Most guinea pigs have a short survival time, between 50 and 150 days, and they make up the peak of the distribution (a single peak can span several classes just like the top of a mountain can spread across country borders). However, some animals survive longer, so that the graph extends to the right of its peak much farther than it extends to the left. Survival times, whether of machines under stress or cancer patients after treatment, usually have distributions that are skewed to the right. The center (half above, half below) lies between 101 and 150 days (the third class). The survival times range from 43 to 598 days.

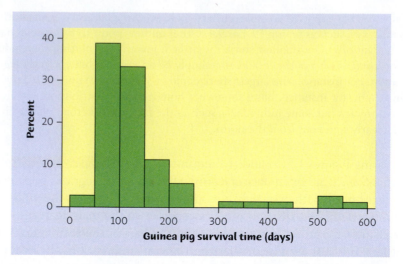

FIGURE 1.6 Histogram of guinea pig survival time (in days) after inoculation with a pathogen. Classes exclude the lower-bound value but include the upper-bound value.

However, almost all infected guinea pigs die within 250 days. A few guinea pigs are outliers and survive for much longer, maybe because they had some immunity to the bacteria used for the experiment.

Notice that the vertical scale in Figure 1.6 is not the *count* of guinea pigs but the *percent* of guinea pigs from the experiment falling in each histogram class. A histogram of percents rather than counts is convenient when we want to compare several distributions. It would, for instance, allow us to compare the survival times obtained in this experiment with the results of another experiment or with the survival time of a control group of guinea pigs not injected with the infectious bacteria. We will see in Chapter 8 that using a control group is very helpful in interpreting the findings of an experiment.

EXAMPLE 1.7   Fish sizes

Figure 1.7 displays data on the lengths in centimeters (cm) of 56 perch caught in a lake in Finland.[11] As is often the case, we can't call this irregular distribution either symmetric or skewed. The big feature of the overall pattern is the presence of two peaks, a **bimodal** distribution corresponding to two **clusters** of fish—small fish and large fish.

bimodal
clusters

Clusters suggest that several types of individuals are mixed in the data set. The first cluster has a clear peak at 20 to 25 cm, while the peak of the second cluster is at 40 to 45 cm. These two clusters might reflect a sexual dimorphism (one sex is typically larger than the other) or the presence of two varieties of perch in the lake.

Giving a unique center and spread for this distribution is not very useful because the data appear to mix two kinds of fish. It would be better to describe the two groups separately.

The overall shape of a distribution is important information about a variable. Some variables have distributions with predictable shapes. Many biological measurements on specimens from the same species and sex—lengths of bird bills, heights of young women—have symmetric distributions. On the other hand, survival times—patients following an organ transplant, lab animals following an

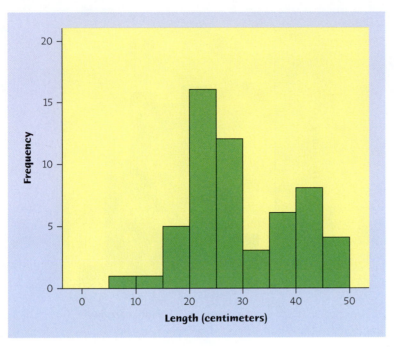

FIGURE 1.7 Histogram of body length in centimeters for 56 perch. Notice the two separate peaks around 25 and 40 cm.

experimental inoculation such as in the guinea pig example—have distributions that are typically strongly skewed to the right. Many distributions have irregular shapes that are neither symmetric nor skewed. Some data show other patterns, such as the clusters in Figure 1.7. *Do not try to artificially manipulate the classes of a histogram so that the data appear more symmetrical or more homogeneous.* Instead, accept that not all data have a distribution that follows a "neat" pattern, even if you could obtain a larger data set. Use your eyes, describe what you see, and then try to explain what you see.

APPLY YOUR KNOWLEDGE

1.7 **Healing of skin wounds.** In Exercise 1.5, you made a histogram of the healing rates of skin wounds in newts. The shape of the distribution is a bit irregular. Is it closer to symmetric or skewed? Would you say that it is unimodal or bimodal?

1.8 **Age onset for anorexia nervosa.** Anorexia nervosa is an eating disorder characterized, in part, by the fear of becoming fat and by voluntary starvation; it often leads to death from suicide or organ failure. A Canadian study surveyed 691 adolescent girls diagnosed with anorexia nervosa and their families.[12] Figure 1.8 is a histogram of the distribution of age at onset of anorexia for the 691 girls in the study (note that, if the onset occurred on the 12th birthday, for example, the onset is reported as 12.01 for the purpose of the histogram, to be coherent with customary age descriptions). Describe the shape of this distribution. Within which class does the midpoint of the distribution lie?

AP Photo/Lauren Greenfield/VII

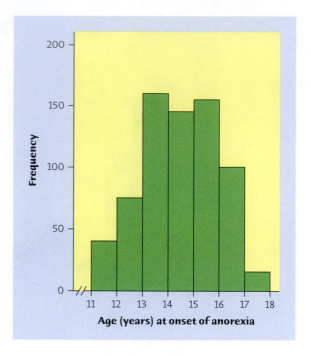

FIGURE 1.8 Histogram of age at onset (in years of age) of anorexia nervosa for 691 Canadian girls diagnosed with the disorder, for Exercise 1.8. The first class includes 11-year-old girls but excludes 12-year-olds.

Quantitative variables: stemplots and dotplots

Histograms are not the only graphical display of distributions. For small data sets, *stemplots* and *dotplots* are quicker to make and present more detailed information.

STEMPLOT

To make a **stemplot:**

1. Separate each observation into a **stem,** consisting of all but the final (rightmost) digit, and a **leaf,** the final digit. Stems may have as many digits as needed, but each leaf contains only a single digit.
2. Write the stems in a vertical column with the smallest at the top, and draw a vertical line at the right of this column.
3. Write each leaf in the row to the right of its stem, in increasing order out from the stem.

EXAMPLE 1.8 *Making a stemplot: skin wounds*

Exercise 1.5 presents the rate at which new cells closed a razor cut made in the skin of 18 anesthetized newts (in micrometers per hour). The data are all two digits long. To make a stemplot of these sorted data, take the tens part of the value as the stem and the final digit (ones) as the leaf. The lowest healing rate, 11 micrometers/hour, has stem 1 and leaf 1. There are three more values in the 10s (12, 14, 18). Place the final digit of these values next to the first leaf so that the first stem (1) has a total of four leaves (1, 2,

4, and 8) all on the same row in the stemplot. Next, there are 8 values in the 20s, and these all go on the second row, with stem 2. The first two values in the 20s range are 22 and 22. Both must be listed, as 2 and 2 on the leaves side. Here is the complete stemplot for the healing rates data:

```
1 | 1 2 4 8
2 | 2 2 3 3 6 7 8 9
3 | 0 3 4 5 5
4 | 0
```

You can also **split stems** to double the number of stems when all the leaves would otherwise fall on just a few stems. Each stem then appears twice. Leaves 0 to 4 go on the upper stem, and leaves 5 to 9 go on the lower stem. If you split the stems in the stemplot above, the new stemplot becomes

splitting stems

```
1 | 1 2 4
1 | 8
2 | 2 2 3 3
2 | 6 7 8 9
3 | 0 3 4
3 | 5 5
4 | 0
4 |
```

Data with many significant digits (for example, 21,422) need to be **rounded** (for example, to the nearest hundred, 21,400; then drop the last two zeros) before they can be used in a stemplot. Rounding and splitting stems are matters for judgment, like choosing the classes in a histogram.

rounding

In fact, a stemplot looks very much like a histogram turned on end: The stems serve the same function as the classes of a histogram. One distinction is that you can choose the classes in a histogram at will but the classes (the stems) of a stemplot are given to you by the actual numerical values. Compare the stemplots in the healing rates example above with the histograms of the same data that you built in Exercises 1.5 and 1.6. The *One-Variable Statistical Calculator* applet on the text CD and Web site allows you to decide whether to split stems, so that it is easy to see the effect.

Histograms are more flexible than stemplots because you can choose the beginning (anchor) and size of the classes. But the stemplot, unlike the histogram, preserves the actual value of each observation. *Stemplots do not work well for large data sets, where each stem must hold a large number of leaves.*

DOTPLOT

To make a **dotplot:**

1. Sort the data set and plot each observation according to its numerical value along a scaled horizontal axis.

2. Identical observations are either superimposed (flat dotplot) or stacked (asymmetrical dotplot).

EXAMPLE 1.9   Making a dotplot: skin wounds

Exercise 1.5 presents the rate (in micrometers/hour) at which new cells closed a razor cut made in the skin of 18 anesthetized newts. To make a dotplot of these sorted data, create a one-dimensional graph with healing rate on the horizontal axis (in micrometers/hour). This is the only axis on this graph. Figure 1.9 shows the dotplot of this data set. Data points with the same numerical value are shown as stacked dots. For example, two newts had a healing rate of 35 micrometers/hour.

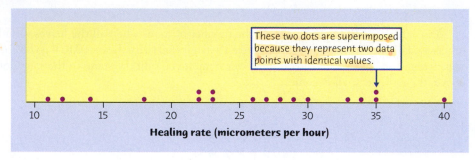

FIGURE 1.9 Dotplot of healing rate (in micrometers per hour) of skin wounds in 18 anesthetized newts.

Like a histogram or a stemplot, the dotplot shows the shape, center, and spread of the distribution. However, because the dotplot is one-dimensional, shape and center are indicated by the density of dots rather than the height of the histogram bar or length of the stemplot leaf. Notice how the dots in Figure 1.9 are fairly evenly spread out, with a slightly higher density in the center of the graph.

Like the stemplot, the dotplot has the advantage of retaining the information about individual observations. However, it is not well suited for very large data sets, which benefit from the summarizing offered by a histogram.

APPLY YOUR KNOWLEDGE

1.9   **Glucose levels.** People with diabetes must monitor and control their blood glucose level. The goal is to maintain "fasting plasma glucose" between about 90 and 130 milligrams per deciliter (mg/dl). Here are the fasting plasma glucose levels for 18 diabetics enrolled in a diabetes control class, five months after the end of the class:[13]

141	158	112	153	134	95	96	78	148
172	200	271	103	172	359	145	147	255

(a) Make a stemplot of these data and describe the main features of the distribution. (You will want to round and also split stems.) Are there outliers? How well is the group as a whole achieving the goal for controlling glucose levels?

(b) Because the stemplot preserves the actual values of the observations, it is easy to find the midpoint (9th and 10th of the 18 observations in order) and the spread. What are they?

(c) Make a dotplot of these data. Compare with the stemplot and explain the differences and similarities.

1.10 **Glucose levels.** The study described in the previous exercise also measured the fasting plasma glucose levels of 16 diabetics who were given individual, rather than group, instruction on diabetes control. Here are the data:

128 195 188 158 227 198 163 164
159 128 283 226 223 221 220 160

Make a **back-to-back stemplot** to compare the class and individual instruction groups. That is, use one set of stems with two sets of leaves, one to the right and one to the left of the stems. (Draw a line on either side of the stems to separate stems and leaves.) Order both sets of leaves from smallest at the stem to largest away from the stem. How do the distribution shapes and success in achieving the glucose control goal compare?

back-to-back stemplot

Florence Nightingale

Florence Nightingale (1820–1910) dedicated her life to promoting and improving the field of nursing. She recorded data and displayed the information in graphs that policymakers could understand. During the Crimean War, she showed that more soldiers died because of poor hospital conditions than from battle wounds. She also documented how casualties drastically dropped after sanitation was improved. Nightingale was the first woman elected to the Royal Statistical Society, honoring her outstanding contributions.

Time plots

Many variables are measured at intervals over time. We might, for example, measure the height of a growing child or precipitation at the end of each month. In these examples, our main interest is change over time. To display change over time, make a *time plot.*

> **TIME PLOT**
>
> A **time plot** of a variable plots each observation against the time at which it was measured. Always put time on the horizontal scale of your plot and the variable you are measuring on the vertical scale. Connecting the data points by lines helps emphasize any change over time.

┌─ **EXAMPLE 1.10** Water levels in the Everglades ─────────

Water levels in Everglades National Park are critical to the survival of this unique region. The photo shows a water-monitoring station in Shark River Slough, the main path for surface water moving through the "river of grass" that is the Everglades. Figure 1.10 is a time plot of water levels at this station from mid-August 2000 to mid-June 2003.[14]

When you examine a time plot, look once again for an overall pattern and for strong deviations from the pattern. Figure 1.10 shows strong **cycles,** regular up-and-down movements in water level. The cycles show the effects of Florida's wet season (roughly June to November) and dry season (roughly December to May). Water levels are highest in late fall. In April and May of 2001 and 2002, water levels were less than zero—the water table was below ground level and the surface was dry. If you look closely, you can see year-to-year variation. The dry season in 2003 ended early, with the first-ever April tropical storm. In consequence, the dry-season water level in 2003 never dipped below zero.

Courtesy U.S. Geological Survey

cycles

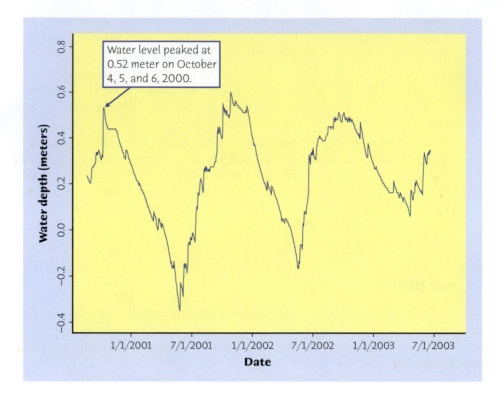

FIGURE 1.10 Time plot of water depth at a monitoring station in Everglades National Park over a period of almost three years. The yearly cycles reflect Florida's wet and dry seasons.

EXAMPLE 1.11 Mississippi River

Table 1.1 lists the volume of water discharged by the Mississippi River into the Gulf of Mexico for each year from 1954 to 2001.[15] The units are cubic kilometers of water—the Mississippi is a big river. Both graphs in Figure 1.11 describe these data. The histogram in Figure 1.11(a) shows the distribution of the volume discharged. The histogram is

---

TABLE 1.1 Yearly discharge of the Mississippi River (cubic kilometers of water)

Year	Discharge	Year	Discharge	Year	Discharge	Year	Discharge
1954	290	1966	410	1978	560	1990	680
1955	420	1967	460	1979	800	1991	700
1956	390	1968	510	1980	500	1992	510
1957	610	1969	560	1981	420	1993	900
1958	550	1970	540	1982	640	1994	640
1959	440	1971	480	1983	770	1995	590
1960	470	1972	600	1984	710	1996	670
1961	600	1973	880	1985	680	1997	680
1962	550	1974	710	1986	600	1998	690
1963	360	1975	670	1987	450	1999	580
1964	390	1976	420	1988	420	2000	390
1965	500	1977	430	1989	630	2001	580

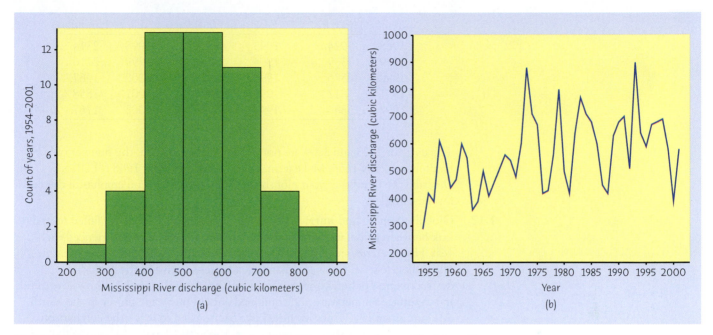

FIGURE 1.11 Histogram (a) and time plot (b) of the volume of water discharged by the Mississippi River over the 48 years from 1954 to 2001. Data from Table 1.1.

symmetric and **unimodal**, with center near 550 cubic kilometers. We might think that the data show just chance year-to-year fluctuation in river level about its long-term average.

unimodal

Figure 1.11(b) is a time plot of the same data. For example, the first point lies above 1954 on the "Year" scale at height 290, the volume of water discharged by the Mississippi in 1954. The time plot tells a more interesting story than the histogram. There is a great deal of year-to-year variation, but there is also a clear increasing **trend** over time. That is, there is a long-term rise in the volume of water discharged. A trend in a time plot is a long-term upward or downward movement over time. The trend in the Mississippi data reflects a climate change: Rainfall and river flows have been increasing over most of North America.

trend

Histograms and time plots give different kinds of information about a variable. The time plot in Figure 1.11(b) presents **time series data** that show the change in volume of water discharged by the Mississippi River into the Gulf of Mexico over time. A histogram displays the distribution of data, such as the volume of water, regardless of time.

time series

APPLY YOUR KNOWLEDGE ─────────────────────

1.11   **Vanishing landfills.** Garbage that is not recycled is buried in landfills. Here are time series data that emphasize the need for recycling: the number of landfills still operating in the United States in the years 1988 to 2002.[16]

Year	Landfills	Year	Landfills	Year	Landfills
1988	7924	1993	4482	1998	2314
1989	7379	1994	3558	1999	2216
1990	6326	1995	3197	2000	1967
1991	5812	1996	3091	2001	1858
1992	5386	1997	2514	2002	1767

The data for 2003 and 2004 are not available. But in 2005, the number of U.S. landfills was only 1654. Make a time plot of the data from 1988 till 2005. Describe the trend that your plot shows. Why does the trend emphasize the need for recycling?

1.12 **Global warming and hurricane strength.** Scientists have so far failed to find a link between global warming and the number of hurricanes experienced each year. However, new research indicates a possible link with hurricane strength. A climatologist at MIT computed an index representative of the potential destructiveness of hurricanes based on surface winds. Figure 1.12 shows the trends in sea surface temperature and combined annual hurricane power in the North Atlantic since 1950 (the two lines have been scaled to make the comparison easier, which is why there are no units on the vertical axis).[17] Describe the trend for each variable, then compare both trends. Explain why the data suggest an impact of global warming on hurricane strength.

CHAPTER 1 SUMMARY

A data set contains information on a number of **individuals.** Individuals may be people, animals, or things. For each individual, the data give values for one or

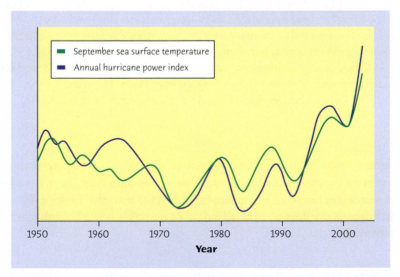

FIGURE 1.12 Time plot of sea surface temperature and hurricane strength in the North Atlantic since 1950, for Exercise 1.12. The two lines have been scaled to make the comparison easier, which explains the absence of units on the vertical axis.

more **variables.** A variable describes some characteristic of an individual, such as a person's height, sex, or age.

Some variables are **categorical** and others are **quantitative.** A categorical variable places each individual into a category, like male or female. A quantitative variable has numerical values that measure some characteristic of each individual, like height in centimeters or age in years.

Exploratory data analysis uses graphs and numerical summaries to describe the variables in a data set and the relations among them.

After you understand the background of your data (individuals, variables, units of measurement), the first thing to do is almost always **plot your data.**

The **distribution** of a variable describes what values the variable takes and how often it takes these values. **Pie charts** and **bar graphs** display the distribution of a categorical variable. Bar graphs can also compare any set of quantities measured in the same units. **Histograms, stemplots,** and **dotplots** graph the distribution of a quantitative variable.

When examining any graph, look for an **overall pattern** and for notable **deviations** from the pattern.

Shape, center, and spread describe the overall pattern of the distribution of a quantitative variable. Some distributions have simple shapes, such as **symmetric** or **skewed.** Not all distributions have a simple overall shape. Describing the shape of a distribution when there are few observations can be particularly challenging.

Outliers are observations that lie outside the overall pattern of a distribution. Always look for outliers and try to explain them.

When observations on a variable are taken over time, make a **time plot** that graphs time horizontally and the values of the variable vertically. A time plot can reveal **trends, cycles,** or other changes over time.

CHECK YOUR SKILLS

1.13 A study records the sex and weight (in kilograms) of 30 recently born bear cubs in Alaska. Which of the following statements is true?

(a) Sex and weight are both categorical variables.

(b) Sex and weight are both quantitative variables.

(c) Sex is a categorical variable and weight is a quantitative variable.

The Statistical Abstract of the United States, prepared by the Census Bureau, provides the number of single-organ transplants for the year 2003, by organ. The following two exercises are based on this table:

Heart	2,057	Kidney	15,129
Lung	1,085	Pancreas	502
Liver	5,671	Intestine	116

1.14   The data on single-organ transplants can be displayed in

(a) a pie chart but not a bar graph.

(b) a bar graph but not a pie chart.

(c) either a pie chart or a bar graph.

1.15   Kidney transplants represented what percent of single-organ transplants in 2003?

(a) Nearly 62%.

(b) Nearly 17%.

(c) This percent cannot be calculated from the information provided in the table.

Figure 1.7 (page 17) is a histogram of the lengths of 56 perch caught in a lake in Finland. The following two exercises are based on this histogram.

1.16   The number of perch with length covered by the rightmost bar in the histogram is

(a) 4.   (b) 7.   (c) 9.

1.17   The rightmost bar in the histogram covers lengths ranging from about

(a) 5 to 50 cm.   (b) 45.1 to 50 cm.   (c) 50 to 55 cm.

1.18   Here are the IQ test scores of 10 randomly chosen fifth-grade students:

    145   139   126   122   125   130   96   110   118   118

To make a stemplot of these scores, you would use as stems

(a) 0 and 1.

(b) 09, 10, 11, 12, 13, and 14.

(c) 96, 110, 118, 122, 125, 126, 130, 139, and 145.

1.19   The population of the United States is aging, though less rapidly than in other developed countries. Here is a stemplot of the percents of residents aged 65 and older in the 50 states, according to the 2000 census. The stems are whole percents and the leaves are tenths of a percent.

```
 5 | 7
 6 |
 7 |
 8 | 5
 9 | 6 7 9
10 | 6
11 | 0 2 2 3 3 6 7 7
12 | 0 0 1 1 1 3 4 4 5 7 8 9
13 | 0 0 0 1 2 2 3 3 3 4 5 5 6 8
14 | 0 3 4 5 7 9
15 | 3 6
16 |
17 | 6
```

There are two outliers: Alaska has the lowest percent of older residents, and Florida has the highest. What is the percent for Florida?

(a) 5.7%   (b) 17.6%   (c) 176%

1.20   Ignoring the outliers, the shape of the distribution in Exercise 1.19 is

(a) strongly skewed to the right.

(b) roughly symmetric.

(c) strongly skewed to the left.

1.21   The center of the distribution in Exercise 1.19 is close to

(a) 12.7%.   (b) 13.5%.   (c) 5.7% to 17.6%.

1.22   Multiple myeloma is a cancer of the bone marrow currently without effective cure. It affects primarily older individuals: It is almost never diagnosed in individuals under 40 years old, and its incidence rate (the number of diagnosed malignant cases per 100,000 individuals in the population) is highest among individuals 70 years of age and older.[18] The distribution of incidence rate of multiple myeloma by age at diagnosis is

(a) skewed to the left.

(b) roughly symmetric.

(c) skewed to the right.

CHAPTER 1 EXERCISES

1.23   **Endangered species.** Bald eagles are an endangered bird species suffering from loss of habitat and pesticide contamination of rivers. A field biologist studying the reproduction of bald eagles records the following variables. Which of these variables are categorical, and which are quantitative?

(a) Number of eggs laid

(b) Incubation period (in days)

(c) Parental care (mostly mother, mostly father, both parents)

(d) Nest size (in centimeters)

(e) Presence of pesticides in samples of local waters (yes/no)

1.24   **Eating habits.** You are preparing to study the eating habits of elementary school children. Describe two categorical variables and two quantitative variables that you might record for each child. Give the units of measurement for the quantitative variables.

Ron Sanford/CORBIS

1.25   **Mercury in lakes.** Mercury is a metal that is highly toxic to the nervous system. Below is a small part of an EESEE data set ("Mercury in Bass") from a study that assessed the water quality of 53 representative lakes in Florida in the early 1990s:[19]

Lake name	pH	Chlorophyll (mg/l)	Avg. mercury in fish (parts per million)	Number of fish sampled	Age of data
Alligator	6.1	0.7	1.23	5	year old
Annie	5.1	3.2	1.33	7	recent
Apopka	9.1	128.3	0.04	6	recent
Blue Cypress	6.9	3.5	0.44	12	recent
Brick	4.6	1.8	1.20	12	year old
Bryant	7.3	44.1	0.27	14	year old
⋮					

(a) What individuals does this data set describe?

(b) In addition to the lake's name, how many variables does the data set contain? Which of these variables are categorical and which are quantitative?

1.26   **Deaths among young people.** The number of deaths among persons aged 15 to 24 years in the United States in 2003 due to the leading causes of death for this age group were: accidents, 14,966; homicide, 5148; suicide, 3921; cancer, 1628; heart disease, 1083; congenital defects, 425.[20]

(a) Make two bar graphs of these data, one with bars ordered alphabetically (by death type) and the other with bars in order from tallest to shortest. Comparisons are easier if you order the bars by height (a bar graph ordered from tallest to shortest bar is sometimes called a **Pareto chart**).

Pareto chart

(b) What additional information do you need to make a pie chart?

1.27   **Garbage.** The formal name for garbage is "municipal solid waste." Exercise 1.11 (page 24) presented the number of landfills operating in the United States in the years 1988 to 2000. Here is a breakdown of the materials that made up American municipal solid waste in 2000:[21]

(a) The weights of the nine materials add to 231.9 million tons. What is the weight of the "Other" solid waste not included in the eight categories specified?

(b) What percent of the total municipal solid waste is made up of paper and paperboard?

(c) Make a bar graph sorted by weight. What does it show?

(d) Could you display these data in a pie chart? Explain why.

Material	Weight (million tons)
Food scraps	25.9
Glass	12.8
Metals	18.0
Paper, paperboard	86.7
Plastics	24.7
Rubber, leather, textiles	15.8
Wood	12.7
Yard trimmings	27.7
Other	
Total	231.9

1.28   **The overweight problem.** The 2002 National Health Interview Survey by the NCHS provides weight categorizations for adults aged 18 years old and older based on their body mass index:

Weight category	Percent of adults
Underweight	2.0
Healthy weight	39.4
Overweight	35.0
Obese	23.5

(a) Make a bar graph of these percents. What do you conclude about the weight problem in America?

(b) Could you make a pie chart for these data? If so, explain why. What would it emphasize?

1.29 The overweight problem. The same NCHS survey (see previous example) breaks down the sampled individuals by age group. Here are the percents of obese individuals in the 2002 survey for each age group:

Age group	Percent obese
18 to 44	21.4
45 to 64	27.8
65 to 74	· 27.3
75 and over	15.9

(a) Make a bar graph of these percents. What do you conclude about the weight problem in America?

(b) Could you make a pie chart for these data? If so, explain why. What would it emphasize?

1.30 Do adolescent girls eat fruit? We all know that fruit is good for us. Many of us don't eat enough. Figure 1.13 is a histogram of the number of servings of fruit per day claimed by 74 seventeen-year-old girls in a study in Pennsylvania.[22] Describe the shape, center, and spread of this distribution. What percent of these girls ate fewer than two servings per day?

1.31 IQ scores. Figure 1.14 is a histogram of the IQ ("intelligence quotient") scores of 60 fifth-grade students chosen at random from one school.[23] The histogram classes include the lower bound but not the upper bound, so that the first class represents IQ scores of 75 or more but less than 85.

(a) Describe the shape, center, and spread of this distribution. Notice that the distribution looks roughly "bell-shaped."

(b) What percent of the fifth-graders have an IQ below 105? What percent have an IQ between 95 and 135 (95 or more but less than 135)?

1.32 What graph? Of the following data sets, which would you display with a bar graph and which would you display with a histogram? Explain why.

(a) A record of gender of selected individuals (labeled as male = 0, female = 1)

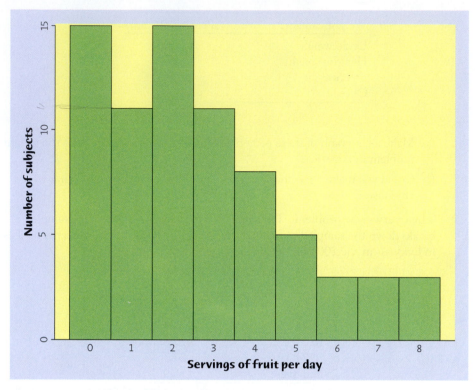

FIGURE 1.13 The distribution of fruit consumption in a sample of 74 seventeen-year-old girls, for Exercise 1.30.

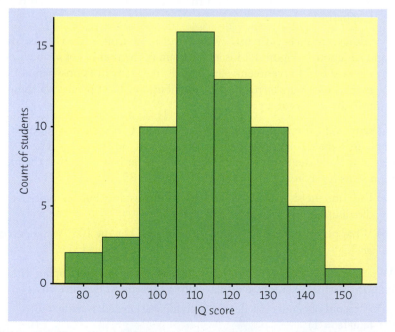

FIGURE 1.14 Histogram of IQ scores for 60 fifth-grade students, for Exercise 1.31.

(b) A record of age of selected parents (in years) at birth of first child

(c) A record of the height of selected individuals (in inches)

(d) A record of the blood type of selected individuals (A, B, AB, or O).

1.33   Acid rain. Changing the choice of classes can change the appearance of a histogram. Here is an example in which a small shift in the classes, with no change in the number of classes, has an important effect on the histogram. The data are the acidity levels (measured by pH) in 105 samples of rainwater. Distilled water has pH 7.00. As the water becomes more acid, the pH goes down. The pH of rainwater is important to environmentalists because of the problem of acid rain.[24]

4.33	4.38	4.48	4.48	4.50	4.55	4.59	4.59	4.61	4.61
4.75	4.76	4.78	4.82	4.82	4.83	4.86	4.93	4.94	4.94
4.94	4.96	4.97	5.00	5.01	5.02	5.05	5.06	5.08	5.09
5.10	5.12	5.13	5.15	5.15	5.15	5.16	5.16	5.16	5.18
5.19	5.23	5.24	5.29	5.32	5.33	5.35	5.37	5.37	5.39
5.41	5.43	5.44	5.46	5.46	5.47	5.50	5.51	5.53	5.55
5.55	5.56	5.61	5.62	5.64	5.65	5.65	5.66	5.67	5.67
5.68	5.69	5.70	5.75	5.75	5.75	5.76	5.76	5.79	5.80
5.81	5.81	5.81	5.81	5.85	5.85	5.90	5.90	6.00	6.03
6.03	6.04	6.04	6.05	6.06	6.07	6.09	6.13	6.21	6.34
6.43	6.61	6.62	6.65	6.81					

(a) Make a histogram of pH with 14 classes, using class boundaries 4.2, 4.4, . . . , 7.0. How many peaks does your histogram show? More than one peak suggests that the data contain groups that have different distributions.

(b) Make a second histogram, also with 14 classes, using class boundaries 4.14, 4.34, . . . , 6.94. The classes are those from (a) moved 0.06 to the left. How many peaks does the new histogram show?

1.34   Where are the doctors? Table 1.2 gives the number of active medical doctors per 100,000 people in each state.[25]

(a) Why is the number of doctors per 100,000 people a better measure of the availability of health care than a simple count of the number of doctors in a state?

(b) Make a histogram that displays the distribution of doctors per 100,000 people. Write a brief description of the distribution. Are there any outliers? If so, can you explain them?

1.35   Carbon dioxide emissions. Burning fuels in power plants or motor vehicles emit carbon dioxide (CO_2), which contributes to global warming. Table 1.3 displays CO_2 emissions per person from countries with populations of at least 20 million.[26]

(a) Why do you think we choose to measure emissions per person rather than total CO_2 emissions for each country?

(b) Make a stemplot to display the data of Table 1.3. Describe the shape, center, and spread of the distribution. Which countries are outliers?

TABLE 1.2 Medical doctors per 100,000 people, by state (2002)

State	Doctors	State	Doctors	State	Doctors
Alabama	202	Louisiana	258	Ohio	248
Alaska	194	Maine	250	Oklahoma	163
Arizona	196	Maryland	378	Oregon	242
Arkansas	194	Massachusetts	427	Pennsylvania	291
California	252	Michigan	230	Rhode Island	341
Colorado	236	Minnesota	263	South Carolina	219
Connecticut	360	Mississippi	171	South Dakota	201
Delaware	242	Missouri	233	Tennessee	250
Florida	237	Montana	215	Texas	204
Georgia	208	Nebraska	230	Utah	200
Hawaii	280	Nevada	174	Vermont	346
Idaho	161	New Hampshire	251	Virginia	253
Illinois	265	New Jersey	305	Washington	250
Indiana	207	New Mexico	222	West Virginia	221
Iowa	178	New York	385	Wisconsin	256
Kansas	210	North Carolina	241	Wyoming	176
Kentucky	219	North Dakota	228	District of Columbia	683

TABLE 1.3 Carbon dioxide emissions (metric tons per person)

Country	CO_2	Country	CO_2	Country	CO_2
Algeria	2.3	Iran	3.8	Poland	8.0
Argentina	3.9	Iraq	3.6	Romania	3.9
Australia	17.0	Italy	7.3	Russia	10.2
Bangladesh	0.2	Japan	9.1	Saudi Arabia	11.0
Brazil	1.8	Kenya	0.3	South Africa	8.1
Canada	16.0	Korea, North	9.7	Spain	6.8
China	2.5	Korea, South	8.8	Sudan	0.2
Colombia	1.4	Malaysia	4.6	Tanzania	0.1
Congo	0.0	Mexico	3.7	Thailand	2.5
Egypt	1.7	Morocco	1.0	Turkey	2.8
Ethiopia	0.0	Myanmar	0.2	Ukraine	7.6
France	6.1	Nepal	0.1	United Kingdom	9.0
Germany	10.0	Nigeria	0.3	United States	19.9
Ghana	0.2	Pakistan	0.7	Uzbekistan	4.8
India	0.9	Peru	0.8	Venezuela	5.1
Indonesia	1.2	Philippines	0.9	Vietnam	0.5

1.36 Rock sole in the Bering Sea. "Recruitment," the addition of new members to a fish population, is an important measure of the health of ocean ecosystems. Here are data on the recruitment of rock sole in the Bering Sea from 1973 to 2000:[27]

Sarkis Images/Alamy

Year	Recruitment (millions)	Year	Recruitment (millions)	Year	Recruitment (millions)	Year	Recruitment (millions)
1973	173	1980	1411	1987	4700	1994	505
1974	234	1981	1431	1988	1702	1995	304
1975	616	1982	1250	1989	1119	1996	425
1976	344	1983	2246	1990	2407	1997	214
1977	515	1984	1793	1991	1049	1998	385
1978	576	1985	1793	1992	505	1999	445
1979	727	1986	2809	1993	998	2000	676

Make a stemplot to display the distribution of yearly rock sole recruitment. (Round to the nearest hundred and split the stems.) Describe the shape, center, and spread of the distribution and any striking deviations that you see.

1.37 Coevolution. Different varieties of the tropical flower *Heliconia* are fertilized by different species of hummingbirds. Over time, the lengths of the flowers and the form of the hummingbirds' beaks have evolved to match each other. Here are data on the lengths in millimeters of two varieties of these flowers on the island of Dominica:[28]

			H. caribaea red				
41.90	42.01	41.93	43.09	41.47	41.69	39.78	40.57
39.63	42.18	40.66	37.87	39.16	37.40	38.20	38.07
38.10	37.97	38.79	38.23	38.87	37.78	38.01	

			H. caribaea yellow				
36.78	37.02	36.52	36.11	36.03	35.45	38.13	37.1
35.17	36.82	36.66	35.68	36.03	34.57	34.63	

(a) Make a back-to-back stemplot to compare the two samples. That is, use one set of stems with two sets of leaves, one to the right and one to the left of the stems. (Draw a line on either side of the stems to separate stems and leaves.) Order both sets of leaves from smallest at the stem to largest away from the stem.

(b) Report the approximate midpoints of both groups. What are the most important differences among the two varieties of flower?

1.38 Rock sole in the Bering sea. Make a time plot of the rock sole recruitment data in Exercise 1.36. What does the time plot show that your stemplot in Exercise 1.36 did not show? When you have time series data, a time plot is often needed to understand what is happening.

1.39   **Marijuana and traffic accidents.** Researchers in New Zealand interviewed 907 drivers at age 21. They had data on traffic accidents and they asked their subjects about marijuana use. Here are data on the numbers of accidents caused by these drivers at age 19, broken down by marijuana use at the same age:[29]

	Marijuana Use per Year			
	Never	1–10 times	11–50 times	51+ times
Drivers	452	229	70	156
Accidents caused	59	36	15	50

(a) Explain carefully why a useful graph must compare *rates* (accidents per driver) rather than *counts* of accidents in the four marijuana use classes.

(b) Make a graph that displays the accident rate for each class. What do you conclude? (You can't conclude that marijuana use *causes* accidents, because risk takers are more likely both to drive aggressively and to use marijuana.)

1.40   **Milk versus soft drink.** The 2004 *Statistical Abstract of the United States* reports the annual per capita consumption of milk and carbonated soft drinks (in gallons) between 1980 and 2000:[30]

Year	1980	1985	1990	1995	1998	1999	2000
Milk	27.6	26.7	25.7	23.9	23	22.9	23.3
Carb. soft drinks	35.1	35.7	46.2	47.4	47.9	49.7	49.3

Make two time plots in the same graph to compare the per capita annual consumption of milk and carbonated soft drinks. What are the most important conclusions you can draw from your graph?

1.41   **Watch those scales!** The Centers for Disease Control and Prevention report the percent of adults 18 years and older in the population who were current smokers in a given year. Here is a subset of this data set:

Year	1965	1974	1979	1983	1987	1990	1993	1997	2000	2002
Percent	41.9	37.0	33.3	31.9	28.6	25.3	24.8	24.6	23.1	22.4

Make a time plot of these data. The impression that a time plot gives depends on the scales you use on the two axes. If you stretch the vertical axis and compress the time axis, change appears to be more rapid. Compressing the vertical axis and stretching the time axis make change appear slower. Make two more time plots of these data, one that makes the percent of adult smokers appear to decrease very rapidly and one that shows only a gentle decrease. The moral of this exercise is: pay close attention to the scales when you look at a time plot.

1.42   **Alligator attacks.** Here are data on the number of unprovoked attacks by alligators on people in Florida over a 33-year period:[31]

Year	Attacks	Year	Attacks	Year	Attacks	Year	Attacks
1972	5	1981	5	1990	18	1999	15
1973	3	1982	6	1991	18	2000	23
1974	4	1983	6	1992	10	2001	17
1975	5	1984	5	1993	18	2002	14
1976	2	1985	3	1994	22	2003	6
1977	14	1986	13	1995	19	2004	11
1978	5	1987	9	1996	13		
1979	2	1988	9	1997	11		
1980	4	1989	13	1998	9		

Make two graphs of these data to illustrate why you should always make a time plot for data collected over time.

(a) Make a histogram of the counts of attacks. What is the overall shape of the distribution?

(b) Make a time plot. What overall pattern does your plot show? (The main reason for the time trend is the continuing increase in Florida's population.)

1.43 **Orange prices.** Figure 1.15 is a time plot of the average price of fresh oranges each month from March 1995 to March 2005.[32] The prices are "index numbers" given as percents of the average price during 1982 to 1984.

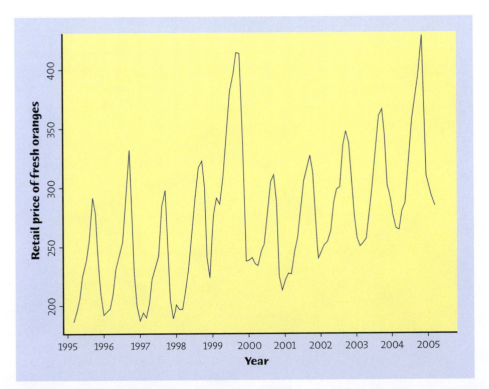

FIGURE 1.15 Time plot of the monthly retail price of fresh oranges from March 1995 to March 2005, for Exercise 1.43.

(a) The most notable pattern in this time plot is yearly cycles. At what season of the year are orange prices highest? Lowest? (To read the graph, note that the tick mark for each year is at the beginning of the year.) The cycles are explained by the time of the orange harvest in Florida.

(b) Is there a longer-term trend visible in addition to the cycles? If so, describe it.

1.44   To split or not to split.   The data sets in the *One-Variable Statistical Calculator* applet on the text CD and Web site include the "guinea pig survival times" data from Example 1.6. Use the applet to make stemplots with and without split stems. Which stemplot do you prefer? Explain your choice.

Craig Tuttle/CORBIS

Describing Distributions with Numbers

Examining nature closely reveals surprising variability. The needles in a given pine tree species are not all the same size, for instance. So it is the distribution of lengths that characterizes the species. Here are the lengths in centimeters of 15 needles taken at random from different parts of several Aleppo pine trees located in Southern California:[1]

7.2 7.6 8.5 8.5 8.7 9.0 9.0 9.3 9.4 9.4 10.2 10.9 11.3 12.1 12.8

Here is a stemplot of these data:

```
 7 | 2 6
 8 | 5 5 7
 9 | 0 0 3 4 4
10 | 2 9
11 | 3
12 | 1 8
```

The distribution is single peaked and slightly skewed to the right without outliers. Our goal in this chapter is to describe with numbers the center and spread of this and other distributions of *quantitative variables*. Categorical variables are simply described by the proportion p of the data made up by a given category, as we have already seen in Chapter 1.

Measuring center: the mean

The most common measure of center is the ordinary arithmetic average, or *mean*.

THE MEAN $\overline{x}$

To find the **mean** of a set of observations, add their values and divide by the number of observations. If the n observations are $x_1, x_2, \ldots, x_n$, their mean is

$$\overline{x} = \frac{x_1 + x_2 + \cdots + x_n}{n}$$

or, in more compact notation,

$$\overline{x} = \frac{1}{n} \sum x_i$$

The $\sum$ (capital Greek sigma) in the formula for the mean is short for "add them all up." The subscripts on the observations x_i are just a way of keeping the n observations distinct. They do not necessarily indicate order or any other special facts about the data. The bar over the x indicates the mean of all the x-values. Pronounce the mean $\overline{x}$ as "x-bar." This notation is very common. When writers who are discussing data use $\overline{x}$ or $\overline{y}$, they are talking about a mean.

EXAMPLE 2.1 Pine needle lengths

The mean needle length of our 15 Aleppo pine needles is

$$\begin{aligned}
\overline{x} &= \frac{x_1 + x_2 + \cdots + x_n}{n} \\
&= \frac{7.2 + 7.6 + \cdots + 12.8}{15} \\
&= \frac{143.9}{15} = 9.59 \text{ cm}
\end{aligned}$$

Craig Tuttle/CORBIS

In practice, you can key the data into your calculator or software and ask for $\overline{x}$. You don't have to actually add and divide. But you should know that this is what the calculator or software is doing.

Notice that only 5 of the 15 needle lengths are larger than the mean. This unbalanced allocation of data around the mean mirrors the slight right-skew in the distribution.

APPLY YOUR KNOWLEDGE

2.1 **Food poisoning.** Here are data on 18 people who fell ill from an incident of food poisoning.[2] The data give for each person the incubation period (the time in hours between eating the infected food and the first signs of illness).

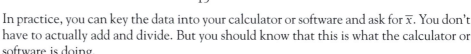

13 46 43 34 20 20 18 72 19 36 48 44 21 32 86 48 28 16

(a) Make a stemplot of these data. Describe the shape, center, and spread of the distribution.

(b) Find the mean incubation period. How many of the 18 individuals had an incubation period less than the mean?

Measuring center: the median

In Chapter 1, we used the midpoint of a distribution as an informal measure of center. The *median* is the formal version of the midpoint, with a specific rule for calculation.

THE MEDIAN M

The **median M** is the midpoint of a distribution, the number such that half the observations are smaller and the other half are larger. To find the median of a distribution:

1. Arrange all observations in order of size, from smallest to largest.
2. If the number of observations n is odd, the median M is the center observation in the ordered list. Find the location of the median by counting $(n + 1)/2$ observations up from the bottom of the list.
3. If the number of observations n is even, the median M is the mean of the two center observations in the ordered list. The location of the median is again $(n + 1)/2$ from the bottom of the list.

Note that the formula $(n + 1)/2$ does *not* give the median, just the location of the median in the ordered list. Medians require little arithmetic, so they are easy to find by hand for small sets of data. Arranging even a moderate number of observations in order is very tedious, however, so that finding the median by hand for larger sets of data is unpleasant. Even simple calculators have an $\bar{x}$ button, but you will need to use software or a graphing calculator to automate finding the median.

EXAMPLE 2.2   Finding the median: odd n

What is the median length for our 15 Aleppo pine needles? Here are the data again, arranged in order:

$$7.2 \quad 7.6 \quad 8.5 \quad 8.5 \quad 8.7 \quad 9.0 \quad 9.0 \quad \mathbf{9.3} \quad 9.4 \quad 9.4$$
$$10.2 \quad 10.9 \quad 11.3 \quad 12.1 \quad 12.8$$

The count of observations $n = 15$ is odd. The bold **9.3** is the center observation in the ordered list, with 7 observations to its left and 7 to its right. This is the median, $M = 9.3$ cm.

Because $n = 15$, our rule for the location of the median gives

$$\text{location of M} = \frac{n + 1}{2} = \frac{16}{2} = 8$$

That is, the median is the 8th observation in the ordered list. It is faster to use this rule than to locate the center by eye.

EXAMPLE 2.3   Finding the median: even n

We also have the lengths of 18 needles from trees of the noticeably different Torrey pine species.[3] What is the median length for these 18 pine needles? The ordered data are now

$$21.2 \quad 21.6 \quad 21.7 \quad 23.1 \quad 23.7 \quad 24.2 \quad 24.2 \quad 25.5 \quad \mathbf{26.6} \quad \mathbf{26.8}$$
$$28.9 \quad 29.0 \quad 29.7 \quad 29.7 \quad 30.2 \quad 32.5 \quad 33.7 \quad 33.7$$

There is no center observation, but there is a center pair. These are the bold **26.6** and **26.8,** with 8 observations before them in the ordered list and 8 after them. The median is midway between these two observations:

$$M = \frac{26.6 + 26.8}{2} = 26.7 \text{ cm}$$

With $n = 18$, the rule for locating the median in the list gives

$$\text{location of } M = \frac{n + 1}{2} = \frac{19}{2} = 9.5$$

The location 9.5 means "halfway between the 9th and 10th observations in the ordered list."

Comparing the mean and the median

In Examples 2.1 and 2.2 we studied the length of 15 Aleppo pine needles. We found that the mean and median were fairly similar, at 9.59 cm and 9.3 cm, respectively. The distribution of the 15 pine needles was only very slightly skewed and did not have any outliers. But what would happen to the relationship between the mean and the median of a data set with a marked skew or extreme outliers?

EXAMPLE 2.4   Guinea pig survival times

Figure 1.6 (page 16) shows the survival times of 72 guinea pigs after inoculation with infectious bacteria. The distribution is noticeably skewed to the right and has some potential high outliers. The mean survival time is 141.9 days, whereas the median survival time is only 102.5 days. The mean is pulled toward the right tail of this right-skewed distribution.

If the longest survival time were increased tenfold to 5980 days rather than 598 days, the mean would increase to more than 216 days, but the median would not change at all. The artificially created outlier would count as just one observation above the center, no matter how far above the center it would lie. In contrast, the mean uses the actual value of each observation and is, therefore, very sensitive to any extreme values.

Example 2.4 illustrates an important fact about the mean as a measure of center: it is sensitive to the influence of a few extreme observations such as in a skewed distribution or outliers. Because the mean cannot resist the influence of extreme observations, we say that it is not a **resistant measure** of center. In contrast, the median is influenced only by the total number of data points and the numerical value of the point or points located at the center of the distribution. The *Mean and Median* applet is an excellent way to compare the resistance of M and $\bar{x}$.

resistant measure

APPLET

> **COMPARING THE MEAN AND THE MEDIAN**
>
> The mean and median of a symmetric distribution are close together. If the distribution is exactly symmetric, the mean and median are exactly the same. In a skewed distribution, the mean is usually farther out in the long tail than is the median.[4]

Many biological variables have distributions that are skewed to the right. Survival times, such as in the guinea pig inoculation experiment, are typically right-skewed. Most individuals die within a certain period of time, but a few will survive much longer. When dealing with strongly skewed distributions, it is somewhat customary to report the median ("midpoint") rather than the mean ("arithmetic average"). However, a health organization or a government agency may need to include all survival times, and thus calculate the mean, to estimate the cost of medical care for a given disease and plan medical staffing appropriately. Relying only on the median would result in underestimating the medical and financial needs. The mean and median measure center in different ways, and both are useful. *Don't confuse the "average" value of a variable (the mean) with its "typical" value, which we might describe by the median.*

APPLY YOUR KNOWLEDGE

2.2 **Deep-sea sediments.** Phytopigments are a marker of the amount of organic matter that settles in sediments. Phytopigment concentrations in deep-sea sediments collected worldwide showed a very strong right-skew.[5] Of these two summary statistics, 0.015 and 0.009 grams of phytopigment per square meter of bottom surface, which one is the mean and which one is the median? Explain your reasoning.

2.3 **Food poisoning.** In Exercise 2.1 you calculated the mean incubation period for 18 cases of food poisoning. Find the median for the 18 incubation periods. Compare the mean and median for these data. What general fact does your comparison illustrate? (Refer to the stemplot you made in Exercise 2.1.)

2.4 **Glucose levels.** People with diabetes must monitor and control their blood glucose level. The goal is to maintain "fasting plasma glucose" levels between about 90 and 130 milligrams per deciliter (mg/dl). Here are the fasting plasma glucose levels for 18 diabetics enrolled in a diabetes control class, five months after the end of the class.

141	158	112	153	134	95	96	78	148
172	200	271	103	172	359	145	147	255

The data set has one high outlier. Calculate the mean and median glucose level with and without the outlier. How does the outlier affect each measure of center? What general fact about the mean and median does your result illustrate?

Measuring spread: the quartiles

The mean and median provide two different measures of the center of a distribution. But a measure of center alone can be misleading. The mean and median survival times of guinea pigs after inoculation in Example 2.4 are 141.9 and 102.5 days, respectively. The mean is higher because the distribution of survival times is skewed to the right. But the median and mean don't tell the whole story. Ten percent of the 72 guinea pigs survived less than 2 months while a handful survived for over a year. We are interested in the *spread* or *variability* of survival times as well as their center. *The simplest useful numerical description of a distribution requires both a measure of center and a measure of spread.*

One way to measure spread is to give the smallest and largest observations. For example, the guinea pig survival times range from 43 days to 598 days. These single observations show the full spread of the data, but they may be outliers. We can improve our description of spread by also looking at the spread of the middle half of the data. The *quartiles* mark out the middle half. Count up the ordered list of observations, starting from the smallest. The *first quartile* lies one-quarter of the way up the list. The *third quartile* lies three-quarters of the way up the list. In other words, the first quartile is larger than 25% of the observations, and the third quartile is larger than 75% of the observations. The second quartile is the median, which is larger than 50% of the observations. That is the idea of quartiles. We need a rule to make the idea exact. The rule for calculating the quartiles uses the rule for the median.

THE QUARTILES Q_1 AND Q_3

To calculate the **quartiles:**

1. Arrange the observations in increasing order and locate the median M in the ordered list of observations.

2. The **first quartile Q_1** is the median of the observations whose position in the ordered list is to the left of the location of the overall median.

3. The **third quartile Q_3** is the median of the observations whose position in the ordered list is to the right of the location of the overall median.

Here are examples that show how the rules for the quartiles work for both odd and even numbers of observations.

EXAMPLE 2.5 Finding the quartiles: odd n ————————

Our sample of 15 Aleppo pine needles (Example 2.2), arranged in increasing order, is:

$$7.2 \quad 7.6 \quad 8.5 \quad 8.5 \quad 8.7 \quad 9.0 \quad 9.0 \quad \mathbf{9.3} \quad 9.4 \quad 9.4$$
$$10.2 \quad 10.9 \quad 11.3 \quad 12.1 \quad 12.8$$

We have an odd number of observations, so the median is the middle one, the bold **9.3** in the list. The first quartile is the median of the 7 observations to the left of the median. This is the 4th of these 7 observations, so $Q_1 = 8.5$ cm. If you want, you can use the recipe for the location of the median with $n = 7$:

$$\text{location of } Q_1 = \frac{n+1}{2} = \frac{7+1}{2} = 4$$

The third quartile is the median of the 7 observations to the right of the median, $Q_3 = 10.9$ cm.

The quartiles are resistant. For example, Q_3 would still be 10.9 if the largest needle length were 50 cm rather than 12.8 cm.

EXAMPLE 2.6 Finding the quartiles: even n ————————

Here are the 18 Torrey pine needles (Example 2.3), arranged in increasing order:

$$21.2 \quad 21.6 \quad 21.7 \quad 23.1 \quad 23.7 \quad 24.2 \quad 24.2 \quad 25.5 \quad 26.6 \quad * \quad 26.8$$
$$28.9 \quad 29.0 \quad 29.7 \quad 29.7 \quad 30.2 \quad 32.5 \quad 33.7 \quad 33.7$$

We have an even number of observations, so the median lies midway between the middle pair, the 9th and 10th in the list. Its value is M = 26.7 cm and its location is marked by an asterisk. The first quartile is the median of the first 9 observations, because these are the observations to the left of the location of the median. Check that $Q_1 = 23.7$ cm and $Q_3 = 29.7$ cm.

Be careful when, as in these examples, several observations take the same numerical value. Write down all of the observations and apply the rules just as if they all had distinct values.

The five-number summary and boxplots

The smallest and largest observations tell us little about the distribution as a whole, but they give information about the tails of the distribution that is missing if we know only Q_1, M, and Q_3. To get a quick summary of both center and spread, combine all five numbers.

THE FIVE-NUMBER SUMMARY

The **five-number summary** of a distribution consists of the smallest observation, the first quartile, the median, the third quartile, and the largest observation, written in order from smallest to largest. In symbols, the five-number summary is

$$\text{Minimum} \quad Q_1 \quad M \quad Q_3 \quad \text{Maximum}$$

These five numbers offer a reasonably complete description of center and spread. The five-number summaries from Examples 2.5 and 2.6 are

Aleppo pine:	7.2	8.5	9.3	10.9	12.8
Torrey pine:	21.2	23.7	26.7	29.7	33.7

The five-number summary of a distribution leads to a new graph, the *boxplot*. Figure 2.1 shows boxplots comparing needle lengths for the Aleppo pine and the Torrey pine species.

BOXPLOT

A **boxplot** is a graph of the five-number summary.

- A central box spans the quartiles Q_1 and Q_3.
- A line in the box marks the median M.
- Lines extend from the box out to the smallest and largest observations.

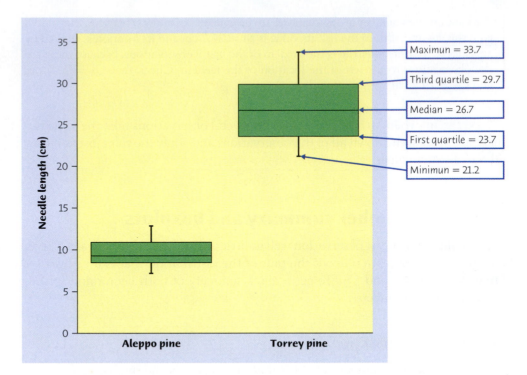

FIGURE 2.1 Boxplots comparing the lengths of 15 Aleppo pine needles and 18 Torrey pine needles.

Because boxplots show less detail than histograms or stemplots, they are best used for side-by-side comparison of more than one distribution, as in Figure 2.1. Be sure to include a numerical scale in the graph. When you look at a boxplot, first locate the median, which marks the center of the distribution. Then look at the spread. The box shows the spread of the middle half of the data, and the extremes (the smallest and largest observations) show the spread of the entire data set. We see from Figure 2.1 that needles are all longer in the Torrey pine set than in the Aleppo pine set. The median, both quartiles, and the maximum are all larger in the Torrey pine. Torrey pine needle lengths are also more variable, as shown by the spread of the box and the spread between the extremes.

Finally, the data for the Torrey pine are symmetrical but the data for the Aleppo pine are mildly right-skewed. In a symmetric distribution, the first and third quartiles are equally distant from the median. In contrast, in most distributions that are skewed to the right, the third quartile will be farther above the median than the first quartile is below it. The extremes behave the same way, but remember that they are just single observations and may say little about the distribution as a whole.

APPLY YOUR KNOWLEDGE

2.5   **Food poisoning.** In Exercise 2.1 you made a stemplot for the incubation period in 18 cases of food poisoning. Give the five-number summary of the distribution of incubation times in days. (The stemplot helps because it arranges the data in order, but make sure to use the exact numerical values provided.) Does the five-number summary reflect what you see in the stemplot? Remember that only a graph gives a clear picture of the shape of a distribution.

2.6   **Glucose levels.** Exercise 2.4 provides the fasting plasma glucose levels for 18 diabetics enrolled in a diabetes control class, five months after the end of the class. Here are the data for 16 diabetics who received individual instruction instead:[6]

128	195	188	158	227	198	163	164
159	128	283	226	223	221	220	160

128, 128, 158, 159, 160, 163, 164, 188, 195, 198, 220, 221, 223, 226, 227, 283 (handwritten, with Q1 and Q3 marked)

(a) Calculate the five-number summary for each of the two data sets.

(b) Make side-by-side boxplots comparing the two groups. What can you say from the graph about the differences between the two diabetes control training methods?

Spotting suspected outliers*

Figure 2.2 is a dotplot of the sizes (in cubic centimeters) of 11 acorns from oak trees found on the Pacific coast.[7] The five-number summary for this distribution is

$$0.4 \quad 1.6 \quad 4.1 \quad 6.0 \quad 17.1$$

How shall we describe the spread of this distribution? The largest observation is clearly an extreme that doesn't represent the majority of the data. The distance between the quartiles (the range of the center half of the data) is a more resistant measure of spread. This distance is called the *interquartile range*.

THE INTERQUARTILE RANGE *IQR*

The **interquartile range** *IQR* is the distance between the first and third quartiles,

$$IQR = Q_3 - Q_1$$

For our data on acorn sizes, $IQR = 6.0 - 1.6 = 4.4$ cm³. However, *no single numerical measure of spread, such as IQR, is very useful for describing skewed distributions*. The two sides of a skewed distribution have different spreads, so one number

CAUTION

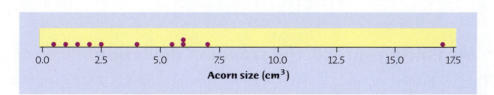

FIGURE 2.2 Dotplot of the sizes of 11 acorns.

Acorn size (cm³)

---

*This short section is optional.

can't summarize them. The interquartile range is mainly used as the basis for a rule of thumb for identifying suspected outliers.

THE 1.5 × *IQR* RULE FOR OUTLIERS

Call an observation a suspected outlier if it falls more than 1.5 × *IQR* above the third quartile or below the first quartile.

How extreme is too extreme?

In 1949, Mr. Hadlum sought divorce on the basis that his wife had given birth 349 days after his departure to war. He postulated that it was evidence of adultery because the average gestation time is 280 days. The judges ruled that such a long gestation time was unlikely but not impossible and denied his appeal. For comparison, while the average men's height in the United States is 1.76 m (5 ft 9 in), the *Guinness Book of World Records* lists the tallest man in recorded history as a 2.72 meter-tall American (8 ft 11 in).

EXAMPLE 2.7 Spotting suspected outliers

For the acorn sizes data, with five-number summary 0.4, 1.6, 4.1, 6.0, 17.1,

$$1.5 \times IQR = 1.5 \times 4.4 = 6.6$$

Any values not falling between

$$Q_1 - (1.5 \times IQR) = 1.6 - 6.6 = -5.0 \quad \text{and}$$
$$Q_3 + (1.5 \times IQR) = 6.0 + 6.6 = 12.6$$

are flagged as suspected outliers. Look again at the dotplot of the data in Figure 2.2: The only suspected outlier is the largest acorn, 17.1 cm^3. The 1.5 × *IQR* rule suggests that this high value is indeed an outlier.

The 1.5 × *IQR* rule is not a replacement for looking at the data. It is most useful when large volumes of data are scanned automatically. Some statistical software will draw boxplots with lines extending out only to the smallest and largest observations within the 1.5 × *IQR* limits and display the remaining extreme observations individually.

APPLY YOUR KNOWLEDGE

2.7 **Food poisoning.** In Exercise 2.1 you made a stemplot for the incubation period in 18 cases of food poisoning. Use the 1.5 × *IQR* rule to identify suspected outliers.

2.8 **Glucose levels.** In Exercise 2.6 you made boxplots comparing the group and individual training methods for diabetes control. Look at the raw data to see if there are unusually high or low values in either data set. Use the 1.5 × *IQR* rule to identify suspected outliers.

Measuring spread: the standard deviation

The five-number summary is not the most common numerical description of a distribution. That distinction belongs to the combination of the mean to measure center and the *standard deviation* to measure spread. The standard deviation and its close relative, the *variance*, measure spread by looking at how far the observations are from their mean.

THE STANDARD DEVIATION s

The **variance s^2** of a set of observations is an average of the squares of the deviations of the observations from their mean. In symbols, the variance of n observations $x_1, x_2, \ldots, x_n$ is

$$s^2 = \frac{(x_1 - \overline{x})^2 + (x_2 - \overline{x})^2 + \cdots + (x_n - \overline{x})^2}{n - 1}$$

or, more compactly,

$$s^2 = \frac{1}{n-1} \sum (x_i - \overline{x})^2$$

The **standard deviation s** is the square root of the variance s^2:

$$s = \sqrt{\frac{1}{n-1} \sum (x_i - \overline{x})^2}$$

In practice, use software or your calculator to obtain the standard deviation from keyed-in data. Doing an example step-by-step will help you understand how the variance and standard deviation work, however.

EXAMPLE 2.8   *Calculating the standard deviation*

A person's metabolic rate is the rate at which the body consumes energy. Metabolic rate is important in studies of weight gain, dieting, and exercise. Here are the metabolic rates of 7 men who took part in a study of dieting. The units are kilocalories (Cal) per 24 hours. These are the same calories used to describe the energy content of foods.

1792   1666   1362   1614   1460   1867   1439

The researchers reported $\overline{x}$ and s for these men. First find the mean:

$$\overline{x} = \frac{1792 + 1666 + 1362 + 1614 + 1460 + 1867 + 1439}{7}$$

$$= \frac{11,200}{7} = 1600 \text{ Cal}$$

Figure 2.3 displays the data as points above the number line, with their mean marked by an asterisk ($*$). The arrows mark two of the deviations from the mean. These deviations

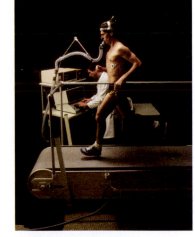

Tom Tracy Photography/Alamy

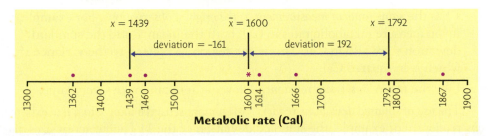

FIGURE 2.3 Metabolic rates for 7 men, with their mean ($*$) and the deviations of two observations from the mean.

show how spread out the data are about their mean. They are the starting point for calculating the variance and the standard deviation.

Observations x_i	Deviations $x_i - \overline{x}$		Squared deviations $(x_i - \overline{x})^2$	
1792	$1792 - 1600 =$	192	$192^2 =$	36,864
1666	$1666 - 1600 =$	66	$66^2 =$	4,356
1362	$1362 - 1600 =$	−238	$(-238)^2 =$	56,644
1614	$1614 - 1600 =$	14	$14^2 =$	196
1460	$1460 - 1600 =$	−140	$(-140)^2 =$	19,600
1867	$1867 - 1600 =$	267	$267^2 =$	71,289
1439	$1439 - 1600 =$	−161	$(-161)^2 =$	25,921
	sum $=$	0	sum $=$	214,870

The variance is the sum of the squared deviations divided by 1 less than the number of observations:

$$s^2 = \frac{214,870}{6} = 35,811.67$$

The standard deviation is the square root of the variance:

$$s = \sqrt{35,811.67} = 189.24 \text{ Cal}$$

Notice that the "average" in the variance s^2 divides the sum by 1 less than the number of observations, that is, $n - 1$ rather than n. The reason is that the deviations $x_i - \overline{x}$ always sum to exactly 0, so that knowing $n - 1$ of them determines the last one. Only $n - 1$ of the squared deviations can vary freely, and we average by dividing the total by $n - 1$. The number $n - 1$ is called the **degrees of freedom** of the variance or standard deviation. Many calculators offer a choice between dividing by n and dividing by $n - 1$, so be sure to use $n - 1$.

degrees of freedom

More important than the details of hand calculation are the properties that determine the usefulness of the standard deviation:

- s measures *spread about the mean* and should be used only when the mean is chosen as the measure of center.

- s is *always zero or greater than zero*. $s = 0$ only when there is no spread. This happens only when all observations have the same value. Otherwise, $s > 0$. As the observations become more spread out about their mean, s gets larger.

- s has the *same units of measurement* as the original observations. For example, if you measure metabolic rates in Cal, both the mean $\overline{x}$ and the standard deviation s are also in Cal. This is one reason to prefer s to the variance s^2, which is in squared Cal.

- Like the mean $\overline{x}$, s is *not resistant*. A few outliers can make s very large.

CAUTION

The use of squared deviations renders s even more sensitive than $\overline{x}$ to a few extreme observations. For example, the standard deviation of the 11 Pacific acorn sizes displayed in Figure 2.2 is 4.66 cm^3. If we omit the high outlier, the standard deviation drops to 2.40 cm^3.

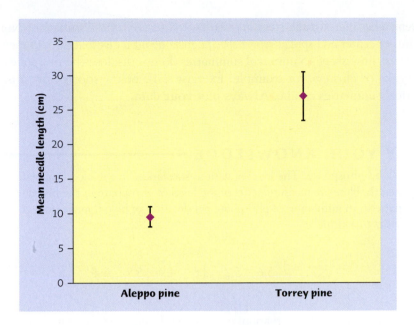

FIGURE 2.4 Same data as in Figure 2.1 now displayed as the mean plus and minus one standard deviation (error bars).

You may rightly feel that the importance of the standard deviation is not yet clear. We will see in later chapters that the standard deviation is the natural measure of spread for an important class of symmetric distributions, the Normal distributions.

Many scientific publications provide summary statistics for two or more groups in the form of one graph displaying each mean with **error bars** extending on either side to show the standard deviation in each group. Figure 2.4 is such a graph for the needle lengths collected for two species of pine trees, from Examples 2.2 and 2.3. *Always check the legend of such graphs, because error bars may represent some other measure also related to the spread of the data.*

error bars

Choosing measures of center and spread

We now have a choice between two descriptions of the center and spread of a distribution: the five-number summary, or $\overline{x}$ and s. Because $\overline{x}$ and s are sensitive to extreme observations, they can be misleading when a distribution is strongly skewed or has outliers. In fact, because the two sides of a skewed distribution have different spreads, no single number such as s describes the spread well. The five-number summary, with its two quartiles and two extremes, does a better job.

CHOOSING A SUMMARY

The five-number summary is usually better than the mean and standard deviation for describing a skewed distribution or a distribution with strong outliers. Use $\overline{x}$ and s only for reasonably symmetric distributions that are free of outliers.

Remember that a graph gives the best overall picture of a distribution. Numerical measures of center and spread report specific facts about a distribution, but they do not describe its entire shape. Numerical summaries do not disclose the presence of multiple peaks or clusters, for example. Exercise 2.10 below shows how misleading numerical summaries can be. **Always plot your data.**

APPLY YOUR KNOWLEDGE

2.9 **Blood phosphate.** The level of various substances in the blood influences our health. Here are measurements of the level of phosphate in the blood of a patient, in milligrams of phosphate per deciliter of blood, made on 6 consecutive visits to a clinic:

5.6   5.2   4.6   4.9   5.7   6.4

A graph of only 6 observations gives little information, so we proceed to compute the mean and standard deviation.

(a) Find the mean from its definition. That is, find the sum of the 6 observations and divide by 6.

(b) Find the standard deviation from its definition. That is, find the deviation of each observation from the mean, square the deviations, then obtain the variance and the standard deviation. Example 2.8 shows the method.

(c) Now enter the data into your calculator and use the mean and standard deviation buttons to obtain $\bar{x}$ and s. Do the results agree with your hand calculations?

2.10 $\bar{x}$ **and** s **are not enough.** The mean $\bar{x}$ and standard deviation s measure center and spread but are not a complete description of a distribution. Data sets with different shapes can have the same mean and standard deviation. To demonstrate this fact, use your calculator to find $\bar{x}$ and s for these two small data sets. Then make a stemplot of each and comment on the shape of each distribution.

Data A	9.14	8.14	8.74	8.77	9.26	8.10	6.13	3.10	9.13	7.26	4.74
Data B	6.58	5.76	7.71	8.84	8.47	7.04	5.25	5.56	7.91	6.89	12.50

2.11 **Choose a summary.** The shape of a distribution is a rough guide to whether the mean and standard deviation are a helpful summary of center and spread. For which of these distributions would $\bar{x}$ and s be useful? In each case, give a reason for your decision.

(a) Length of great white sharks, Figure 1.5 (page 12).

(b) Age onset for anorexia nervosa, Figure 1.8 (page 18).

(c) Number of fruit servings per day, Figure 1.13 (page 30).

(handwritten margin notes)

$\bar{x} = 7.50$
$s = 2.83$

A = 3.10, 4.74, 6.13, 7.26, 8.10, 8.14, 8.74, 8.77, 9.13, 9.14, 9.26

B = 6.25, 6.56, 5.76, 6.58, 6.89, 7.04, 7.71, 7.91, 8.47, 8.84, 12.50

$\bar{x} =$

$\bar{x}$ = mean
s = standard deviation

DISCUSSION: Dealing with outliers: recognition and treatment

When collecting data, we sometimes come across wild observations that clearly fall outside the overall pattern of data distribution. In this chapter, we have presented a way to identify suspected outliers (optional section) and discussed how outliers can affect numerical summaries. But what are these wild observations, and how do we deal with them? Entire books have been written on the topic. Here we describe three major types of wild observations and offer some general guidance on how to handle each type.

Human error in recording information

Perhaps the most famous example of an outlier caused by a data-recording error lies in the story of Popeye the Sailor. Created in 1929, Popeye is a friendly cartoon character who attains immediate superhuman strength whenever he eats iron-filled spinach. Indeed, an 1870 scientific publication reported that spinach had by far the highest iron content of any other green leafy vegetable, about 10 times more than lettuce or cabbage. This claim remained unquestioned until a 1937 study showed that spinach's iron content was similar to that of other leafy greens. It turns out that in the 1870 publication the iron content of spinach had a misplaced decimal point! The spinach value had been correctly identified as a wild observation, but for over half a century nobody had questioned its nature.

Errors in data recording are not that uncommon. Typos are an obvious concern. But the data themselves are not always clean. Surveys of individuals are particularly prone to errors. People may forget, lie, or simply misunderstand a question. In an online survey, undergraduate students enrolled in a biostatistics course were asked to record their heights in inches. Of the 149 numerical values submitted, two wild observations appeared as 5.3 and 6. These are obvious errors. Maybe the students had meant 5 feet 3 inches and 6 feet, respectively, or maybe the 5.3 value was a typo for 53 inches.

What should you do with wild observations that you have clearly identified as being errors in data entry? The obvious answer is that these values do not belong and should not be included with the whole data set. You might be able to correct the mistakes by checking your original records (notes, data tables, photos). Good scientific practice always includes keeping clear and extensive records of data and how they were obtained.

Human error in experimentation or data collection

Sloppy experimentation methods can lead to unexpected results. If you forgot to add bacteria to one of your petri dishes and find that nothing grew in it, that's a silly mistake. But some experimental blunders lead to more interesting results.

Fleming, for instance, had less-than-ideal lab techniques. A few of his petri dishes ended up being contaminated with a mold. Instead of simply throwing these away, he noticed a halo around the mold where no bacteria grew. He went on to cultivate the mold and discover its antibiotic properties, revolutionizing medicine and later earning a Nobel Prize. When Pasteur was studying chicken cholera, his assistant left some bacterial cultures out while he went on vacation. The dried-out cultures failed to kill inoculated chicken, as other cultures usually did. The assistant's first impulse was to discard the data, but Pasteur decided to take a closer look and explore the reason for this unexpected result. This work led him to understand the workings of the immune system and develop the vaccine.

Not all technical errors lead to fame or a Nobel Prize. Some mistakes are indeed not worth a second look. But sometimes the wild observations can be even more interesting than the original study. Again, you should always keep detailed notes of everything you do when gathering data, as it might help you identify mistakes and understand how they arose. Either way, these kinds of wild observations do not belong with the rest of your data (it would be like comparing apples and oranges). If you suspect an experimental error, the wild observation should be either ignored or studied separately.

Unexplainable but apparently legitimate wild observations
Most studies in the life sciences are conducted by collecting data about a small sample taken from the whole population of interest. Because of this, it can be difficult to determine whether a suspected outlier in a sample truly is a wild observation or just the consequence of studying only a small subset of the population. When you find a suspected outlier in a sample and have ruled out human error, you are faced with the challenging task of deciding what to do with it.

This is an important step because, for many statistical procedures, outliers are influential and can distort conclusions. Running the analysis with and then again without the outlier can help you determine whether the outlier affects your conclusions substantially or not. We will see in future chapters that some statistical methods are robust against mild outliers. That is, the method will be valid despite the presence of a mild outlier. Extreme outliers, however, are always a cause of concern. Many complex statistical approaches have been devised to deal with outliers. Most are well beyond the scope of this introductory textbook. Nonparametric tests, described in optional Chapter 27 on the CD, are just a few examples. For simple studies, however, deciding what to do with a suspected outlier typically boils down to deciding whether to include the wild observation with the rest of the data or not.

To some extent, deciding whether to include or exclude wild observations in your analysis depends on the purpose of your study. Are you more interested in describing a population in its entire extent or only in its most typical, general pattern? For example, before Pasteur invented the pasteurization method to prevent the spoilage of liquids, wine and beer producers often had some batches that developed a troublesome sour taste. If you were studying the performance of various wine production methods, you would want to include bad batches in your analysis because they too reflect the performance of the method. On the other hand, if you were studying the chemical composition of wines and beers to better understand their health effects, you would not include sour batches, because these would never be sold for consumption. Obviously, you should never discard wild observations simply because they lead to conclusions you do not like.

Lastly, and perhaps most importantly, whether you choose to keep or to discard them, all outliers should be disclosed. Outliers can be very influential or have a tremendous scientific relevance, and you should not act as if they did not exist. You should also explain why you chose to keep or discard them. Honesty and full disclosure are the foundation of a good scientific methodology.

Using technology

Although a "two-variables statistics" calculator will do the basic calculations we need in this book, more elaborate tools are helpful. Graphing calculators and computer software will do calculations and make graphs as you command, freeing you to concentrate on choosing the right methods and interpreting your results. Figure 2.5 displays output describing the 15 Aleppo pine needle lengths of Example 2.2. Can you find $\overline{x}$, s, and the five-number summary in each output? The big message of this section is: *Once you know what to look for, you can read output from any technological tool.*

The displays in Figure 2.5 come from the TI-83 (or TI-84) graphing calculator, three statistical programs, and the Microsoft Excel spreadsheet program. The statistical programs are CrunchIt! (available in the *E-Stat Pack* that accompanies this book), SPSS, and Minitab. Statistical programs allow us to choose what descriptive measures we want. Excel gives some things we don't need. Just ignore the extras. Excel's "Descriptive Statistics" menu item doesn't give the quartiles. We used the spreadsheet's separate quartile function to get Q_1 and Q_3.

EXAMPLE 2.9   What are the quartiles?

In Example 2.5, we saw that the quartiles of the Aleppo pine needle lengths are $Q_1 = 8.5$ and $Q_3 = 10.9$. Look at the output displays in Figure 2.5. The TI-83, Minitab, and CrunchIt! all agree with our work. Excel says that $Q_1 = 8.6$ and $Q_3 = 10.55$. What happened? *There are several rules for finding the quartiles. Some software packages use rules that give results that are different from ours for some sets of data.* Our rule is simplest for hand computation. SPSS provides the results from both methods. Results from the various rules are always close to each other, so *to describe data you should just use the answer your technology gives you.*

Organizing a statistical problem

Most of our examples and exercises have aimed at helping you learn basic tools (graphs and calculations) for describing and comparing distributions. You have also learned principles that guide use of these tools, such as "always start with a graph" and "look for the overall pattern and striking deviations from the pattern." The data you work with are not just numbers. They describe specific settings such

Texas Instruments TI-83

```
1-Var Stats
  x̄ = 9.593333333
  Σ x = 143.9
  Σ X² = 1415.79
  Sx = 1.588110587
  σx = 1.534260589
↓ n = 15
```

```
1-Var Stats
↑ n = 15
  minX = 7.2
  Q₁ = 8.5
  Med = 9.3
  Q₃ = 10.9
  maxX = 12.8
```

CrunchIt!

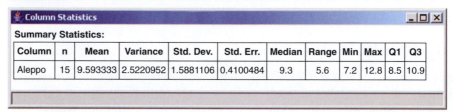

Column	n	Mean	Variance	Std. Dev.	Std. Err.	Median	Range	Min	Max	Q1	Q3
Aleppo	15	9.593333	2.5220952	1.5881106	0.4100484	9.3	5.6	7.2	12.8	8.5	10.9

Minitab

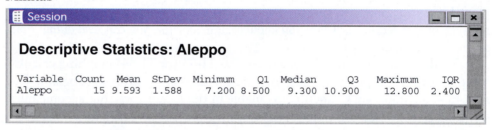

Descriptive Statistics: Aleppo

Variable	Count	Mean	StDev	Minimum	Q1	Median	Q3	Maximum	IQR
Aleppo	15	9.593	1.588	7.200	8.500	9.300	10.900	12.800	2.400

Excel

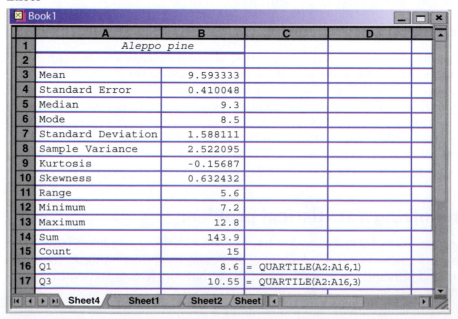

	A	B	C	D
1	*Aleppo pine*			
2				
3	Mean	9.593333		
4	Standard Error	0.410048		
5	Median	9.3		
6	Mode	8.5		
7	Standard Deviation	1.588111		
8	Sample Variance	2.522095		
9	Kurtosis	-0.15687		
10	Skewness	0.632432		
11	Range	5.6		
12	Minimum	7.2		
13	Maximum	12.8		
14	Sum	143.9		
15	Count	15		
16	Q1	8.6	= QUARTILE(A2:A16,1)	
17	Q3	10.55	= QUARTILE(A2:A16,3)	

SPSS

| Column Statistics | | | | | | | _ |□|×| |

Descriptive Statistics:

	N	Range	Minimum	Maximum	Mean	Std. Deviation
Aleppo	15	5.6	7.2	12.8	9.593	1.5881
Valid N (listwise)	15					

Percentiles

		Percentiles						
		5	10	25	50	75	90	95
Weighted Average (Definition 1)	Aleppo	7.200	7.440	8.500	9.300	10.900	12.380	
Turkey's Hinges	Aleppo			8.600	9.300	10.550		

FIGURE 2.5 Output from a graphing calculator, a spreadsheet program, and three software packages describing the Aleppo pine needle length data.

as water depth in the Everglades or needle length for two species of pine trees. Because data come from a specific setting, the final step in examining data is a conclusion for that setting. Water depth in the Everglades has a yearly cycle that reflects Florida's wet and dry seasons. Needle lengths are longer in Torrey pine trees than in Aleppo pine trees.

As you learn more statistical tools and principles, you will face more complex statistical problems. Although no framework accommodates all the varied issues arising in applying statistics to real settings, the following four-step thought process gives useful guidance. In particular, the first and last steps emphasize that statistical problems are tied to specific real-world settings and therefore involve more than doing calculations and making graphs.

ORGANIZING A STATISTICAL PROBLEM: THE FOUR-STEP PROCESS

STATE: What is the practical question, in the context of the real-world setting?

FORMULATE: What specific statistical operations does this problem call for?

SOLVE: Make the graphs and carry out the calculations needed for this problem.

CONCLUDE: Give your practical conclusion in the setting of the real-world problem.

To help you master the basics, many of our exercises will continue to tell you what to do—make a histogram, find the five-number summary, and so on. Real statistical problems don't come with detailed instructions. Especially in the later chapters of the book, you will meet some exercises that are more realistic. Use the four-step process as a guide to solving and reporting these problems. They are marked with the four-step icon, as the following example illustrates.

How many was that?

Good causes often breed bad statistics. An advocacy group claims, without much evidence, that 150,000 Americans suffer from the eating disorder anorexia nervosa. Soon someone misunderstands and says that 150,000 people *die* from anorexia nervosa each year. This wild number gets repeated in countless books and articles. It really is a wild number: only about 55,000 women aged 15 to 44 (the main group affected) die of *all causes* each year.

Art Wolfe/Getty Images

EXAMPLE 2.10   *Comparing tropical flowers*

STATE: Ethan Temeles of Amherst College, with his colleague W. John Kress, studied the relationship between varieties of the tropical flower *Heliconia* on the island of Dominica and the different species of hummingbirds that fertilize the flowers.[8] Over time, the researchers believe, the lengths of the flowers and the form of the hummingbirds' beaks have evolved to match each other. If that is true, flower varieties fertilized by different hummingbird species should have distinct distributions of length.

Table 2.1 gives length measurements (in millimeters) for samples of three varieties of *Heliconia*, each fertilized by a different species of hummingbird. Do the three varieties display distinct distributions of length? How do the mean lengths compare?

FORMULATE: Use graphs and numerical descriptions to describe and compare these three distributions of flower length.

SOLVE: We might use boxplots to compare the distributions, but stemplots preserve more detail and work well for data sets of these sizes. Figure 2.6 displays stemplots with the stems lined up for easy comparison. The lengths have been rounded to the nearest tenth of a millimeter. The *bihai* and red varieties have somewhat skewed distributions, so we might choose to compare the five-number summaries. But because the researchers plan to use $\bar{x}$ and s for further analysis, we instead calculate these measures:

Variety	Mean length	Standard deviation
bihai	47.60	1.213
red	39.71	1.799
yellow	36.18	0.975

CONCLUDE: The three varieties differ so much in flower length that there is little overlap among them. In particular, the flowers of *bihai* are longer than either red or yellow. The mean lengths are 47.6 mm for *H. bihai*, 39.7 mm for *H. caribaea* red, and 36.2 mm for *H. caribaea* yellow. If we can see a correspondence between the beaks of different hummingbird species and the flower lengths of different *Heliconia* varieties, the data will support the researchers' hypothesis.

TABLE 2.1	**Flower lengths (millimeters) for three *Heliconia* varieties**						
			H. bihai				
47.12	46.75	46.81	47.12	46.67	47.43	46.44	46.64
48.07	48.34	48.15	50.26	50.12	46.34	46.94	48.36
			H. caribaea red				
41.90	42.01	41.93	43.09	41.47	41.69	39.78	40.57
39.63	42.18	40.66	37.87	39.16	37.40	38.20	38.07
38.10	37.97	38.79	38.23	38.87	37.78	38.01	
			H. caribaea yellow				
36.78	37.02	36.52	36.11	36.03	35.45	38.13	37.10
35.17	36.82	36.66	35.68	36.03	34.57	34.63	

bihai		red		yellow	
34		34		34	6 6
35		35		35	2 5 7
36		36		36	0 0 1 5 7 8 8
37		37	4 8 9	37	0 1
38		38	0 0 1 1 2 2 8 9	38	1
39		39	2 6 8	39	
40		40	6 7	40	
41		41	5 7 9 9	41	
42		42	0 2	42	
43		43	1	43	
44		44		44	
45		45		45	
46	3 4 6 7 8 8 9	46		46	
47	1 1 4	47		47	
48	1 2 3 4	48		48	
49		49		49	
50	1 3	50		50	

FIGURE 2.6 Stemplots comparing the distributions of flower lengths from Table 2.1. The stems are whole millimeters and the leaves are tenths of a millimeter.

APPLY YOUR KNOWLEDGE

2.12 **Logging in the rain forest.** "Conservationists have despaired over destruction of tropical rainforest by logging, clearing, and burning." These words begin a report on a statistical study of the effects of logging in Borneo.[9] Researchers compared forest plots that had never been logged (Group 1) with similar plots nearby that had been logged 1 year earlier (Group 2) and 8 years earlier (Group 3). All plots were 0.1 hectare in area. Here are the counts of trees for plots in each group:

Group 1:	27	22	29	21	19	33	16	20	24	27	28	19
Group 2:	12	12	15	9	20	18	17	14	14	2	17	19
Group 3:	18	4	22	15	18	19	22	12	12			

To what extent has logging affected the count of trees? Follow the four-step process in reporting your work.

CHAPTER 2 SUMMARY

A numerical summary of a distribution should report at least its **center** and its **spread** or **variability.**

The **mean** $\bar{x}$ and the **median M** describe the center of a distribution in different ways. The mean is the arithmetic average of the observations, and the median is the midpoint of the values.

When you use the median to indicate the center of the distribution, describe its spread by giving the **quartiles.** The **first quartile** Q_1 has one-fourth of the observations below it, and the **third quartile** Q_3 has three-fourths of the observations below it.

The **five-number summary** consisting of the median, the quartiles, and the smallest and largest individual observations provides a quick overall description of a distribution. The median describes the center, and the quartiles and extremes show the spread.

Boxplots based on the five-number summary are useful for comparing several distributions. The box spans the quartiles and shows the spread of the central half of the distribution. The median is marked within the box. Lines extend from the box to the extremes and show the full spread of the data.

The **variance s^2** and especially its square root, the **standard deviation s,** are common measures of spread about the mean as center. The standard deviation s is zero when there is no spread and gets larger as the spread increases.

A **resistant measure** of any aspect of a distribution is relatively unaffected by changes in the numerical value of a small proportion of the total number of observations, no matter how large these changes are. The median and quartiles are resistant, but the mean and the standard deviation are not.

The mean and standard deviation are good descriptions for symmetric distributions without outliers. They are most useful for the Normal distributions, which we will describe in Chapter 11. The five-number summary is a better description for skewed distributions.

Numerical summaries do not fully describe the shape of a distribution. Always plot your data.

A statistical problem has a real-world setting. You can organize many problems using the four steps **state, formulate, solve,** and **conclude.**

CHECK YOUR SKILLS

2.13   Here are the IQ test scores of 10 randomly chosen fifth-grade students:

        145   139   126   122   125   130   96   110   118   118

The mean of these scores is

(a) 122.9.   (b) 123.4.   (c) 136.6.

2.14   The median of the 10 IQ test scores in Exercise 2.13 is

(a) 125.   (b) 123.5.   (c) 122.9.

2.15   The five-number summary of the 10 IQ scores in Exercise 2.13 is

(a) 96, 114, 125, 134.5, 145.

(b) 96, 118, 122.9, 130, 145.

(c) 96, 118, 123.5, 130, 145.

2.16   If a distribution is skewed to the right,

(a) the mean is less than the median.

(b) the mean and median are equal.

(c) the mean is greater than the median.

2.17   What percent of the observations in a distribution lie between the first quartile and the third quartile?

(a) 25%   (b) 50%   (c) 75%

2.18 To make a boxplot of a distribution, you must know

(a) all of the individual observations.

(b) the mean and the standard deviation.

(c) the five-number summary.

2.19 The standard deviation of the 10 IQ scores in Exercise 2.13 (use your calculator) is

(a) 13.23.   (b) 13.95.   (c) 194.6.

2.20 What are all the values that a standard deviation s can possibly take?

(a) $0 \leq s$ (b) $0 \leq s \leq 1$ (c) $-1 \leq s \leq 1$

2.21 You have data on the weights in grams of 5 baby pythons. The mean weight is 31.8 and the standard deviation of the weights is 2.39. The correct units for the standard deviation are

(a) no units—it's just a number.

(b) grams.

(c) grams squared.

2.22 Which of the following is least affected if an extreme high outlier is added to your data?

(a) the median   (b) the mean   (c) the standard deviation

CHAPTER 2 EXERCISES

2.23 **Florida lakes.** Alkalinity of lake waters is related to carbon content, which is mostly derived from ground and algal sources and plays an important role in aquatic ecosystems. Water analysis of 53 Florida lakes resulted in a mean alkalinity of 37.53 milliequivalents per liter and a median alkalinity of 19.60 meq/l.[10] What aspect, or aspects, of the data distribution might explain the difference between these two measures of center?

2.24 **Where are the doctors?** Table 1.2 (page 32) gives the number of medical doctors per 100,000 people in each state. Exercise 1.34 asked you to plot the data. The distribution is right-skewed with several high outliers.

(a) Do you expect the mean to be greater than the median, about equal to the median, or less than the median? Why? Calculate $\bar{x}$ and M and verify your expectation.

(b) The District of Columbia, at 683 MDs per 100,000 residents, is a high outlier. If you remove D.C. because it is a city rather than a state, do you expect $\bar{x}$ or M to change more? Why? Omitting D.C., calculate both measures for the 50 states and verify your expectation.

2.25 **Comparing tropical flowers.** An alternative presentation of the flower length data in Table 2.1 reports the five-number summary and uses boxplots to display the distributions. Do this. Do the boxplots fail to reveal any important information visible in the stemplots in Figure 2.6?

2.26 **Making resistance visible.** In the *Mean and Median* applet, place three observations on the line by clicking below it: two close together near the center of the line, and one somewhat to the right of these two.

APPLET

(a) Pull the single rightmost observation out to the right. (Place the cursor on the point, hold down a mouse button, and drag the point.) How does the mean behave? How does the median behave? Explain briefly why each measure acts as it does.

(b) Now drag the single rightmost point to the left as far as you can. What happens to the mean? What happens to the median as you drag this point past the other two (watch carefully)?

2.27 **How much fruit do adolescent girls eat?** Figure 1.13 (page 30) is a histogram of the number of servings of fruit per day claimed by 74 seventeen-year-old girls. With a little care, you can find the median and the quartiles from the histogram. What are these numbers? How did you find them?

2.28 **Weight of newborns.** Here is the distribution of the weight at birth for all babies born in the United States in 2002:[11]

Weight	Count	Weight	Count
Less than 500 grams	6,268	3,000 to 3,499 grams	1,521,884
500 to 999 grams	22,845	3,500 to 3,999 grams	1,125,959
1,000 to 1,499 grams	29,431	4,000 to 4,499 grams	314,182
1,500 to 1,999 grams	61,652	4,500 to 4,999 grams	48,606
2,000 to 2,499 grams	193,881	5,000 to 5,499 grams	5,396
2,500 to 2,999 grams	688,630		

(a) For comparison with other years and with other countries, we prefer a histogram of the *percents*, rather than the counts, in each weight class. Explain why.

(b) How many babies were there? Make a histogram of the distribution, using percents on the vertical scale.

(c) What are the positions of the median and quartiles in the ordered list of all birth weights? In which weight classes do the median and quartiles fall?

2.29 **Metabolic rate.** In Example 2.8 you examined the metabolic rate of 7 men. Here are the metabolic rates for 12 women from the same study:

995   1425   1396   1418   1502   1256   1189   913   1124   1052   1347   1204

(a) The most common methods for formal comparison of two groups use $\bar{x}$ and s to summarize the data. What kinds of distributions are best summarized by $\bar{x}$ and s?

(b) Make a summary graph comparing the metabolic rates of the 7 men and 12 women. What can you conclude about these two groups from your graph?

2.30 **Behavior of the median.** Place five observations on the line in the *Mean and Median* applet by clicking below it.

(a) Add one additional observation *without changing the median*. Where is your new point?

(b) Use the applet to convince yourself that when you add yet another observation (there are now seven in all), the median does not change no matter where you put the seventh point. Explain why this must be true.

2.31 **Which color attracts beetles best?** An experiment looked at the number of beetles trapped on sticky boards of identical size but various colors placed throughout a field of oats (there were six boards for each color). Here are the data:[12]

Board color	Insects trapped					
Blue	16	11	20	21	14	7
Green	37	32	20	29	37	32
White	21	12	14	17	13	20
Yellow	45	59	48	46	38	47

(a) Make a boxplot comparing the number of beetles trapped per board for each color (blue, green, white, and yellow).

(b) Briefly compare the four distributions. What are the major differences? Which color or colors seem to attract beetles best?

2.32 Guinea pig survival times. Example 1.6 (page 15) provides the numerical values for the guinea pig survival times plotted in Figure 1.6.[13] As often with survival times, this distribution is strongly skewed to the right.

(a) Which numerical summary would you choose for these data? Explain why. Does it show the expected right-skew?

(b) Calculate your chosen summary (Example 2.4 already provides the mean and median for this data set). How does it reflect the skewness of the distribution?

Dorling Kindersley/Getty Images

2.33 Never on Sunday: also in Canada? Exercise 1.4 (page 10) gives the average number of births in the United States on each day of the week during an entire year. The boxplots in Figure 2.7 are based on more detailed data from Toronto,

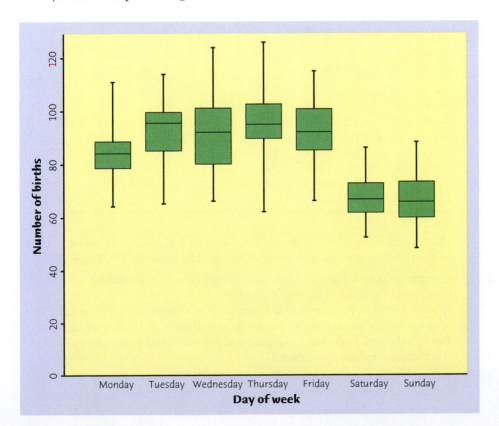

FIGURE 2.7 Boxplots of the distributions of number of births in Toronto, Canada, on each day of the week during a year, for Exercise 2.33.

Canada: the number of births on each of the 365 days in a year, grouped by day of the week.[14] Based on these plots, give a more detailed description of how births depend on the day of the week.

2.34   Fish sizes. Figure 1.7 (page 17) shows the body lengths of 56 perch caught in a lake in Finland.[15] Give a brief description of the important features of the distribution. Explain why no numerical summary would appropriately describe this distribution.

2.35   Ancient air. The composition of the earth's atmosphere may have changed over time. To try to discover the nature of the atmosphere long ago, we can examine the gas in bubbles inside ancient amber. Amber is tree resin that has hardened and been trapped in rocks. The gas in bubbles within amber should be a sample of the atmosphere at the time the amber was formed. Measurements on specimens of amber from the late Cretaceous era (75 to 95 million years ago) give these percents of nitrogen:[16]

David Sanger Photography/Alamy

> 63.4   65.0   64.4   63.3   54.8   64.5   60.8   49.1   51.0

(a) Make a stemplot (split the stems). Describe the distribution of nitrogen percents in amber specimens.

(b) Choose and calculate an appropriate numerical summary.

2.36   Does breast-feeding weaken bones? Breast-feeding mothers secrete calcium into their milk. Some of the calcium may come from their bones, so mothers may lose bone mineral. Researchers compared 47 breast-feeding women with 22 women of similar age who were neither pregnant nor lactating. They measured the percent change in the mineral content of the women's spines over three months. Here are the data:[17]

Breast-feeding women						Other women					
−4.7	−2.5	−4.9	−2.7	−0.8	−5.3	2.4	0.0	0.9	−0.2	1.0	1.7
−8.3	−2.1	−6.8	−4.3	2.2	−7.8	2.9	−0.6	1.1	−0.1	−0.4	0.3
−3.1	−1.0	−6.5	−1.8	−5.2	−5.7	1.2	−1.6	−0.1	−1.5	0.7	−0.4
−7.0	−2.2	−6.5	−1.0	−3.0	−3.6	2.2	−0.4	−2.2	−0.1		
−5.2	−2.0	−2.1	−5.6	−4.4	−3.3						
−4.0	−4.9	−4.7	−3.8	−5.9	−2.5						
−0.3	−6.2	−6.8	1.7	0.3	−2.3						
0.4	−5.3	0.2	−2.2	−5.1							

Do the data show distinctly greater bone mineral loss among the breast-feeding women? Follow the four-step process illustrated by Example 2.10 (page 56).

2.37   Compressing soil. Farmers know that driving heavy equipment on wet soil compresses the soil and injures future crops. Table 2.2 gives data on the "penetrability" of the same soil at three levels of compression.[18] Penetrability is a measure of how much resistance plant roots will meet when they try to grow through the soil. How does increasing compression affect penetrability? Follow the four-step outline in your work.

2.38   A standard deviation contest. This is a standard deviation contest. You must choose four numbers from the whole numbers 0 to 10, with repeats allowed.

TABLE 2.2	Penetrability of soil at three compression levels								
Soil Compression Level									
Compressed									
2.86	2.68	2.92	2.82	2.76	2.81	2.78	3.08	2.94	2.86
3.08	2.82	2.78	2.98	3.00	2.78	2.96	2.90	3.18	3.16
Intermediate									
3.13	3.38	3.10	3.40	3.38	3.14	3.18	3.26	2.96	3.02
3.54	3.36	3.18	3.12	3.86	2.92	3.46	3.44	3.62	4.26
Loose									
3.99	4.20	3.94	4.16	4.29	4.19	4.13	4.41	3.98	4.41
4.11	4.30	3.96	4.03	4.89	4.12	4.00	4.34	4.27	4.91

(a) Choose four numbers that have the smallest possible standard deviation.

(b) Choose four numbers that have the largest possible standard deviation.

(c) Is more than one choice possible in either (a) or (b)? Explain.

2.39   **Test your technology.** This exercise requires a calculator with a standard deviation button or statistical software on a computer. The observations

$$10,001 \quad 10,002 \quad 10,003$$

have mean $\bar{x} = 10,002$ and standard deviation $s = 1$. Adding a 0 in the center of each number, the next set becomes

$$100,001 \quad 100,002 \quad 100,003$$

The standard deviation remains $s = 1$ as more 0s are added. Use your calculator or software to find the standard deviation of these numbers, adding extra 0s until you get an incorrect answer. How soon did you go wrong? This demonstrates that calculators and software cannot handle an arbitrary number of digits correctly.

2.40   **You create the data.** Create a set of 5 positive numbers (repeats allowed) that have median 10 and mean 7. What thought process did you use to create your numbers?

2.41   **You create the data.** Give an example of a small set of data for which the mean is larger than the third quartile.

Exercises 2.42 to 2.44 make use of the optional material on the 1.5 × IQR rule for suspected outliers.

2.42   **Carbon dioxide emissions.** Table 1.3 (page 32) gives carbon dioxide (CO_2) emissions per person for countries with populations of at least 20 million. A stemplot or histogram shows that the distribution is strongly skewed to the right. The United States and several other countries appear to be high outliers.

(a) Give the five-number summary. Explain why this summary suggests that the distribution is right-skewed.

(b) Which countries are suspected outliers according to the $1.5 \times IQR$ rule? Make a stemplot of the data or look at your stemplot from Exercise 1.35. Do you agree with the rule's suggestions about which countries are and are not outliers?

2.43 **Older Americans.** The stemplot in Exercise 1.19 (page 26) displays the distribution of the percents of residents aged 65 and older in the 50 states. Stemplots help you find the five-number summary because they arrange the observations in increasing order.

(a) Give the five-number summary of this distribution.

(b) Does the $1.5 \times IQR$ rule identify Alaska and Florida as suspected outliers? Does it also flag any other states?

2.44 **Older Americans.** Refer to the previous exercise.

(a) Find the mean percent of residents aged 65 and older in the 50 states and compare it with the median. Explain your result.

(b) For this distribution, what numerical summary would be appropriate? Explain your reasoning.

Douglas Faulkner/Photo Researchers

Scatterplots and Correlation

A medical study finds that lung capacity decreases with the number of cigarettes smoked per day. The Department of Motor Vehicles warns that alcohol consumption reduces reflex time and that the effect is larger as more alcohol is consumed. These and many other statistical studies look at the *relationship between two variables*. Statistical relationships are overall tendencies, not ironclad rules. They allow individual exceptions. Although smokers on the average die younger than nonsmokers, some people live to 90 while smoking three packs a day.

To understand a statistical relationship between two variables, we measure both variables on the same individuals. Often, we must examine other variables as well. To conclude that smoking undermines lung capacity, for example, the researchers had to eliminate the effect of other variables such as each person's size and exercise habits. In this chapter we begin our study of relationships between variables. One of our main themes is that the relationship between two variables can be strongly influenced by other variables that are lurking in the background.

Explanatory and response variables

We think that blood alcohol content helps explain variations in reflex time and that smoking influences lung capacity and, eventually, life expectancy. In these relationships, the two variables play different roles: one explains or influences the other.

RESPONSE VARIABLE, EXPLANATORY VARIABLE

A **response variable** measures an outcome of a study. An **explanatory variable** may explain or influence changes in a response variable.

independent variable
dependent variable

You will often find explanatory variables called **independent variables,** and response variables called **dependent variables.** The idea behind this language is that the response variable depends on the explanatory variable. Because "independent" and "dependent" have other meanings in statistics that are unrelated to the explanatory-response distinction, we prefer to avoid those words in this textbook.

It is easiest to identify explanatory and response variables when we actually set values of one variable in order to see how it affects another variable.

EXAMPLE 3.1   Beer and blood alcohol

How does drinking beer affect the level of alcohol in our blood? The legal limit for driving is now 0.08% in all states. Student volunteers at Ohio State University drank different numbers of cans of beer. Thirty minutes later, a police officer measured their blood alcohol content. Number of beers consumed is the explanatory variable, and percent of alcohol in the blood is the response variable.

When we don't set the values of either variable but just observe both variables, there may or may not be explanatory and response variables. Whether there are depends on how we plan to use the data.

EXAMPLE 3.2   Height

The National Center for Health Statistics (NCHS) surveys the American population and collects, for each individual in the survey, information about body height, age, gender, and a long list of other attributes. The purpose of the NCHS survey is to document the characteristics of the American population. There are no explanatory or response variables in this context.

A pediatrician looks at the same data with an eye to using age and gender to discuss a child's growth. Now age and gender are explanatory variables, and height is the response variable.

In many studies, the goal is to show that changes in one or more explanatory variables actually *cause* changes in a response variable. But many explanatory-response relationships do not involve direct causation. The age and gender of a child can help predict the future height, but they certainly do not cause a particular height.

Most statistical studies examine data on more than one variable. Fortunately, statistical analysis of several-variable data builds on the tools we used to examine individual variables. The principles that guide our work also remain the same:

- Plot your data. Look for overall patterns and deviations from those patterns.
- Based on what your plot shows, choose numerical summaries for some aspects of the data.

APPLY YOUR KNOWLEDGE

3.1   Explanatory and response variables? In each of the following situations, is it more reasonable to simply explore the relationship between the two variables or to view one of the variables as an explanatory variable and the other as a response variable? In the latter case, which is the explanatory variable and which is the response variable?

(a) The typical amount of calories a person consumes per day and that person's percent of body fat.

(b) The weight in kilograms and height in centimeters of a person.

(c) Inches of rain in the growing season and the yield of corn in bushels per acre.

(d) A person's leg length and arm length in centimeters.

3.2   Coral reefs. How sensitive to changes in water temperature are coral reefs? To find out, measure the growth of corals in aquariums where the water temperature is controlled at different levels. Growth is measured by weighing the coral before and after the experiment. What are the explanatory and response variables? Are they categorical or quantitative?

3.3   Beer and blood alcohol. Example 3.1 describes a study in which college students drank different amounts of beer. The response variable was their blood alcohol content (BAC). BAC for the same amount of beer might depend on other facts about the students. Name two other variables that could influence BAC.

Displaying relationships: scatterplots

The most useful graph for displaying the relationship between two quantitative variables is a *scatterplot*.

EXAMPLE 3.3   An endangered species: the manatee

Manatees are large, herbivorous, aquatic mammals found primarily in the rivers and estuaries of Florida. This endangered species suffers from the cohabitation with human populations, and many manatees die each year from collisions with powerboats. Following our four-step process (page 55), let's look at the influence of the number of powerboats registered on manatee deaths from collisions with powerboats. We examine the relationship between the number of manatee deaths from powerboat collisions and the number of powerboats registered in any given year between 1977 and 2006, as displayed in Table 3.1.[1]

STATE: The number of powerboats registered in Florida varies from year to year. Does it help explain the differences from year to year in the number of manatee deaths from collisions with powerboats?

FORMULATE: Examine the relationship between powerboats registered and manatee deaths from collision. Choose the explanatory and response variables (if any). Make a

Douglas Faulkner/Photo Researchers

TABLE 3.1	Powerboat registrations (in thousands) and manatee deaths from powerboat collisions in Florida				
Year	Powerboats	Deaths	Year	Powerboats	Deaths
1977	447	13	1992	679	38
1978	460	21	1993	678	35
1979	481	24	1994	696	49
1980	498	16	1995	713	42
1981	513	24	1996	732	60
1982	512	20	1997	755	54
1983	526	15	1998	809	66
1984	559	34	1999	830	82
1985	585	33	2000	880	78
1986	614	33	2001	944	81
1987	645	39	2002	962	95
1988	675	43	2003	978	73
1989	711	50	2004	983	69
1990	719	47	2005	1010	79
1991	681	55	2006	1024	92

scatterplot to display the relationship between the variables. Interpret the plot to understand the relationship.

SOLVE (FIRST STEPS): We suspect that powerboats registered will help explain manatee deaths from collision. So powerboats registered is the explanatory variable, and manatee deaths from collision is the response variable. Time (year the data were gathered) is not a variable of interest here. We want to see how the variable manatee deaths changes when the variable powerboat registrations changes, so we put powerboat registrations (the explanatory variable, expressed in thousands) on the horizontal axis. Figure 3.1 is the scatterplot. Each point represents a single year. In 1997, for example, there were 755,000 powerboats registered and 54 manatee deaths due to powerboat collision. Find 755 on the x (horizontal) axis and 54 on the y (vertical) axis. The year 1997 appears as the point (755, 54) above 755 and to the right of 54, as shown in Figure 3.1.

Do big skulls house smart brains?

Nineteenth-century scientists thought that the volume of a human skull might be related to the intelligence of the skull's owner. Without imaging technology, it was difficult to measure a skull's volume accurately. Paul Broca, a professor of surgery, showed that filling an empty skull with small lead shot, then pouring out the shot and weighing it, gave quite accurate measurements of the skull's volume.

SCATTERPLOT

A **scatterplot** shows the relationship between two quantitative variables measured on the same individuals. The values of one variable appear on the horizontal axis, and the values of the other variable appear on the vertical axis. Each individual in the data appears as the point in the plot fixed by the values of both variables for that individual.

Always plot the explanatory variable, if there is one, on the horizontal axis (the x axis) of a scatterplot. As a reminder, we usually call the explanatory variable x and the response variable y. If there is no explanatory-response distinction, either variable can go on the horizontal axis.

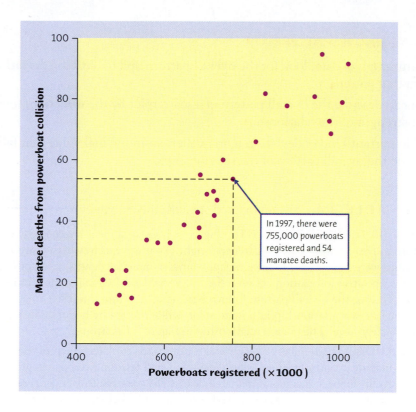

FIGURE 3.1 Scatterplot of the number of manatee deaths due to powerboat collisions in Florida each year against the number of powerboats registered (in thousands) that same year. The dotted lines intersect at the point (755, 54), the data for year 1997.

APPLY YOUR KNOWLEDGE

3.4   **Bird colonies.** One of nature's patterns connects the percent of adult birds in a colony that return from the previous year and the number of new adults that join the colony. Here are data for 13 colonies of sparrowhawks:[2]

Percent returning	New adults	Percent returning	New adults	Percent returning	New adults
74	5	62	15	46	18
66	6	52	16	60	19
81	8	45	17	46	20
52	11	62	18	38	20
73	12				

William S. Clark; Frank Lane Picture Agency/CORBIS

Plot the count of new birds (response) against the percent of returning birds (explanatory).

Interpreting scatterplots

To interpret a scatterplot, apply the strategies of data analysis learned in Chapters 1 and 2.

STEP

linear relationship

━━ **EXAMPLE 3.4**   *An endangered species: the manatee* ━━━━━━

SOLVE (INTERPRET THE PLOT):   Figure 3.1 shows a clear *direction:* The overall pattern moves up, from lower left to upper right. That is, years in which powerboat registrations were higher tend to have higher counts of manatee deaths from collision. We call this a positive association between the two variables. The *form* of the relationship is **linear.** That is, the overall pattern follows a straight line from lower left to upper right. The *strength* of a relationship in a scatterplot is determined by how closely the points follow a clear form. The overall relationship in Figure 3.1 is strong.

CONCLUDE: The number of powerboat registrations explains much of the variation among manatee deaths from collision. Years that had fewer powerboats registered also tended to have fewer accidental manatee deaths. To preserve this endangered species, restricting the number of powerboats registered might be helpful. However, the scatterplot in Figure 3.1 does not take into consideration other factors, such as speed limits, fines, or driver education, that might also be influential. Our conclusions are limited to the available data and what they say about the relationship between powerboat registrations and manatee accidents.

━━ **EXAMPLE 3.5**   *Counting carnivores* ━━━━━━━━━━━━━━━

Ecologists look at data to learn about nature's patterns. One pattern they have found relates the size of a carnivore (body mass in kilograms) to how many of those carnivores there are in an area. One way to measure of "how many" is to count carnivores per 10,000 kilograms of their prey in the area. Table 3.2 gives average data for 25 carnivore species.[3]

To see the pattern, plot carnivore abundance (response) against body mass (explanatory). Patterns involving sizes and counts are often simpler when we plot the logarithms of the data. Figure 3.2 does that—you can see that 1, 10, 100, and 1000 are equally spaced on the vertical scale.

TABLE 3.2	Size and abundance of carnivores				
Carnivore species	Body mass (kg)	Abundance	Carnivore species	Body mass (kg)	Abundance
Least weasel	0.14	1656.49	Eurasian lynx	20.0	0.46
Ermine	0.16	406.66	Wild dog	25.0	1.61
Small Indian mongoose	0.55	514.84	Dhole	25.0	0.81
Pine marten	1.3	31.84	Snow leopard	40.0	1.89
Kit fox	2.02	15.96	Wolf	46.0	0.62
Channel Island fox	2.16	145.94	Leopard	46.5	6.17
Arctic fox	3.19	21.63	Cheetah	50.0	2.29
Red fox	4.6	32.21	Puma	51.9	0.94
Bobcat	10.0	9.75	Spotted hyena	58.6	0.68
Canadian lynx	11.2	4.79	Lion	142.0	3.40
European badger	13.0	7.35	Tiger	181.0	0.33
Coyote	13.0	11.65	Polar bear	310.0	0.60
Ethiopian wolf	14.5	2.70			

This scatterplot shows a negative association. That is, bigger carnivores are less abundant. The form of the association between the log of carnivore body mass and the log of prey abundance is linear. The association is quite strong because the points don't deviate a great deal from the line. It is striking that animals from many different parts of the world should fit so simple a pattern.

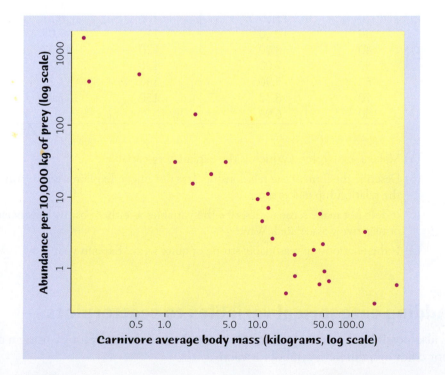

FIGURE 3.2 Scatterplot of the abundance of 25 species of carnivores against their body mass. Larger carnivores are less abundant. (Logarithmic scales are used for both variables.)

Of course, not all relationships have a simple form and a clear direction that we can describe as positive association or negative association. Exercise 3.6 gives an example that does not have a single direction.

APPLY YOUR KNOWLEDGE

3.5   **Bird colonies.** Describe the form, direction, and strength of the relationship between number of new sparrowhawks in a colony and percent of returning adults, as displayed in your plot from Exercise 3.4.

Ecological studies have shown that, for short-lived birds, the association between these variables is positive: Changes in weather and food supply drive the populations of new and returning birds up or down together. For long-lived territorial birds, on the other hand, the association is negative because returning birds claim their territories in the colony and don't leave room for new recruits. Which type of species is the sparrowhawk?

3.6   **Speed limits to preserve resources?** There are about half a billion cars worldwide,[4] depleting the planet's oil reserves and contributing to rising atmospheric carbon dioxide levels. Could changing driving habits influence fuel consumption? How does the fuel consumption of a car change as its speed increases? Here are data for a British Ford Escort. Speed is measured in kilometers per hour, and fuel consumption is measured in liters of gasoline used per 100 kilometers traveled.[5]

Speed (km/h)	Fuel used (liters/100 km)	Speed (km/h)	Fuel used (liters/100 km)
10	21.00	90	7.57
20	13.00	100	8.27
30	10.00	110	9.03
40	8.00	120	9.87
50	7.00	130	10.79
60	5.90	140	11.77
70	6.30	150	12.83
80	6.95		

(a) Make a scatterplot. (Which is the explanatory variable?)

(b) Describe the form of the relationship. It is not linear. Explain why the form of the relationship makes sense.

(c) It does not make sense to describe the variables as either positively associated or negatively associated. Why?

(d) Is the relationship reasonably strong or quite weak? Explain your answer.

Adding categorical variables to scatterplots

It is also possible to plot two relationships on the same scatterplot by using a different color or a different symbol for each.

CATEGORICAL VARIABLES IN SCATTERPLOTS

To add a categorical variable to a scatterplot, use a different plot color or symbol for each category.

EXAMPLE 3.6   The cost of reproduction in male fruit flies

Longevity in male fruit flies is positively associated with adult size. But do other factors, such as sexual activity, also matter? The cost of reproduction is well documented for the females of the species. A study looks at the association between longevity and adult size in male fruit flies kept under one of two conditions. One group is kept with sexually active females over their life span. The other group is cared for in the same way but kept with females that are not sexually active.[6]

Figure 3.3 shows the scatterplot of adult thorax length (indicative of body size, explanatory) versus longevity (response) for both conditions. The conditions are coded so that individual fruit flies from the sexually active group are represented by triangles, while the sexually inactive fruit flies from the second group are represented by circles. This coding introduces a third variable into the scatterplot. "Condition" is a categorical variable that has two values, identified by the two different plotting symbols.

The scatterplot is quite clear. Both groups show a positive, linear relationship between thorax length and longevity, as expected. However, the sexually active fruit flies have, on the whole, a lower longevity. That is, fruit flies of a given size tend to die sooner in the sexually active group than in the inactive group. In fact, only one male in the sexually active group outlived a male of the inactive group with similar thorax length.

Yoav Levy/Phototake

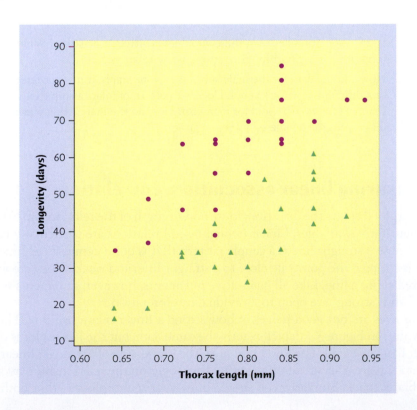

FIGURE 3.3 Cost of reproduction: scatterplot of thorax length (in millimeters) and longevity (in days) for sexually active (triangles) or inactive (circles) male fruit flies.

APPLY YOUR KNOWLEDGE ────────────────────────

3.7   **Do heavier people burn more energy?** Metabolic rate, the rate at which the body consumes energy, is important in studies of weight gain, dieting, and exercise. Here are data on the lean body mass and resting metabolic rate for 12 women and 7 men who are subjects in a study of dieting. Lean body mass, given in kilograms, is a person's weight leaving out all fat. Metabolic rate is measured in kilocalories (Cal) burned per 24 hours (the same calories used to describe the energy content of foods). Researchers believe that lean body mass is an important influence on metabolic rate.

Subject	Sex	Mass (kg)	Rate (Cal)	Subject	Sex	Mass (kg)	Rate (Cal)
1	M	62.0	1792	11	F	40.3	1189
2	M	62.9	1666	12	F	33.1	913
3	F	36.1	995	13	M	51.9	1460
4	F	54.6	1425	14	F	42.4	1124
5	F	48.5	1396	15	F	34.5	1052
6	F	42.0	1418	16	F	51.1	1347
7	M	47.4	1362	17	F	41.2	1204
8	F	50.6	1502	18	M	51.9	1867
9	F	42.0	1256	19	M	46.9	1439
10	M	48.7	1614				

(a) Make a scatterplot of the data for the female subjects. Which is the explanatory variable? Explain why the subject number is not part of the scatterplot.

(b) Is the association between these variables positive or negative? What is the form of the relationship? How strong is the relationship?

(c) Now add the data for the male subjects to your graph, using a different color or a different plotting symbol. Does the pattern of relationship that you observed for women hold for men also? How do the male subjects as a group differ from the female subjects as a group?

Measuring linear association: correlation

A scatterplot displays the direction, form, and strength of the relationship between two quantitative variables. Linear (straight-line) relations are particularly important because a straight line is a simple pattern that is quite common. A linear relation is strong if the points lie close to a straight line, and weak if they are widely scattered about a line. Like all qualitative judgments, however, statements such as "weak" and "strong" are open to individual interpretation.

Our eyes are not good judges of how strong a linear relationship is. The two scatterplots in Figure 3.4 depict exactly the same data, but the lower plot is drawn smaller in a large field. The lower plot gives the impression of a stronger linear relationship. Create your scatterplots so that they focus on the relationships, avoiding extra blank space. Similarly, you should draw your scatterplots so that both axes

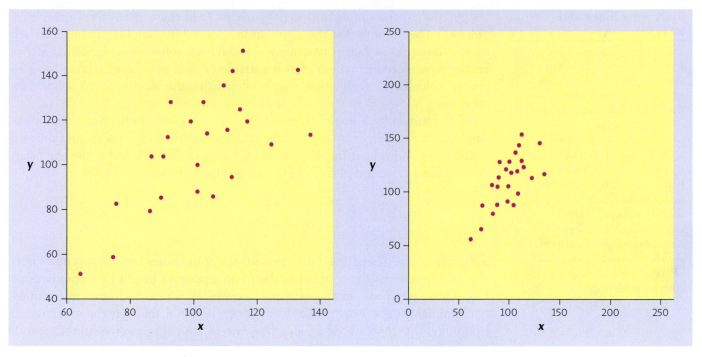

FIGURE 3.4 Two scatterplots of the same data. The straight-line pattern in the lower plot appears stronger because of the surrounding space.

are given the same emphasis, resulting in plots that are square rather than rectangular in shape. *When creating or studying a scatterplot, keep in mind that our eyes can be fooled by changing the plotting scales or the amount of white space around the cloud of points.*[7]

CAUTION

We need to follow our strategy for data analysis by using a numerical measure to supplement the graph. *Correlation* is the measure we use.

CORRELATION

The **correlation** measures the direction and strength of the linear relationship between two quantitative variables. Correlation is usually written as r.

Suppose that we have data on variables x and y for n individuals. The values for the first individual are x_1 and y_1, the values for the second individual are x_2 and y_2, and so on. The means and standard deviations of the two variables are $\overline{x}$ and s_x for the x-values, and $\overline{y}$ and s_y for the y-values. The correlation r between x and y is

$$r = \frac{1}{n-1} \sum \left(\frac{x_i - \overline{x}}{s_x} \right) \left(\frac{y_i - \overline{y}}{s_y} \right)$$

Body mass index and body fat

Obesity and, more specifically, body fat have been linked to cardiovascular disease, strokes, and type 2 diabetes. Methods that directly measure body fat content, like underwater weighing, dual-energy X-ray absorptiometry (DXA), or computerized tomography, are costly and require professional intervention. On the other hand, the body mass index (BMI) is calculated only from a person's height and weight. It has been shown to be strongly correlated with the results of these more complex methods.

As always, the summation sign $\sum$ means "add these terms for all the individuals." The formula for the correlation r is a bit complex. It helps us see what correlation is, but in practice you should use software or a calculator that finds r from keyed-in values of two variables x and y. The next exercise asks you to calculate a correlation step-by-step from the definition to solidify its meaning.

The formula for r begins by standardizing the observations. Suppose, for example, that x is height in centimeters and y is weight in kilograms and that we have height and weight measurements for n people. Then $\overline{x}$ and s_x are the mean and standard deviation of the n heights, both in centimeters. The value

$$\frac{x_i - \overline{x}}{s_x}$$

is the standardized height of the ith person. The standardized height says how many standard deviations above or below the mean a person's height lies. Standardized values have no units—in this example, they are no longer measured in centimeters. Standardize the weights also. Then, for each person, multiply the standardized height and the standardized weight. The correlation r is an average of these products for the n people.

APPLY YOUR KNOWLEDGE

3.8   **Coffee and deforestation.** Coffee is a leading export from several developing countries. When coffee prices are high, farmers often clear forest to plant more coffee trees. Here are data on prices paid to coffee growers in Indonesia and the rate of deforestation in a national park that lies in a coffee-producing region, for five years:[8]

Price (cents per pound)	Deforestation (percent)
29	0.49
40	1.59
54	1.69
55	1.82
72	3.10

(a) Make a scatterplot. Which is the explanatory variable? What kind of pattern does your plot show?

(b) Find the correlation r step-by-step. First find the mean and standard deviation of each variable. Then find the five standardized values for each variable. Finally, use the formula for r. Explain how your value for r matches your graph in (a).

(c) Enter these data into your calculator or software and use the correlation function to find r. Check that you get the same result as in (b), up to roundoff error.

Bill Ross/CORBIS

Facts about correlation

The formula for correlation helps us see that r is positive when there is a positive association between the variables. Height and weight, for example, have a positive association. People who are above average in height tend to also be above average in weight. For these people, the standardized height and the standardized weight tend to both be positive. People who are below average in height tend to also have below-average weight. For them, the standardized height and standardized weight tend to both be negative. Overall, the products in the formula for r are mostly positive and so r is positive. In the same way, we can see that r is negative when the association between x and y is negative. More detailed study of the formula gives more detailed properties of r. Here is what you need to know in order to interpret correlation.

1. *Correlation makes no distinction between explanatory and response variables.* It makes no difference which variable you call x and which you call y in calculating the correlation.

2. Because r uses the standardized values of the observations, r *does not change when we change the units of measurement of x, y, or both.* Measuring height in inches rather than centimeters and weight in pounds rather than kilograms does not change the correlation between height and weight. The correlation r itself has no unit of measurement; it is just a number.

3. *Positive r indicates positive association between the variables, and negative r indicates negative association.*

4. *The correlation r is always a number between -1 and 1.* Values of r near 0 indicate a very weak linear relationship. The strength of the linear relationship increases as r moves away from 0 toward either -1 or 1. Values of r close to -1 or 1 indicate that the points in a scatterplot lie close to a straight line. The extreme values $r = -1$ and $r = 1$ occur only in the case of a perfect linear relationship, when the points lie exactly along a straight line.

> **EXAMPLE 3.7** From scatterplot to correlation
>
> The scatterplots in Figure 3.5 illustrate how values of r closer to 1 or -1 correspond to stronger linear relationships. To make the meaning of r clearer, the standard deviations of both variables in these plots are equal, and the horizontal and vertical scales are the same. In general, it is not so easy to guess the value of r from the appearance of a scatterplot. Remember that changing the plotting scales in a scatterplot may mislead our eyes, but it does not change the correlation.
>
> The real data we have examined also illustrate how correlation measures the strength and direction of linear relationships. Figure 3.2 shows a strong negative linear relationship between the logarithms of body mass and abundance for carnivore species. The correlation is $r = -0.912$. Figure 3.3 shows a weaker but still quite strong positive association between thorax length and longevity in the group of sexually active fruit flies (displayed as triangles). The correlation is $r = 0.806$. Notice how the strength of the association depends on the absolute value of r (its numerical value irrespective of its sign).

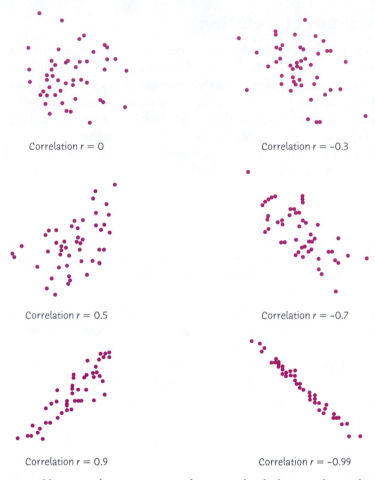

Correlation r = 0   Correlation r = –0.3

Correlation r = 0.5   Correlation r = –0.7

Correlation r = 0.9   Correlation r = –0.99

FIGURE 3.5 How correlation measures the strength of a linear relationship. Patterns closer to a straight line have correlations closer to 1 or −1.

Describing the relationship between two variables is a more complex task than describing the distribution of one variable. Here are some more facts about correlation, and cautions to keep in mind when you use r.

1. *Correlation requires that both variables be quantitative, so that it makes sense to do the arithmetic indicated by the formula for r.* We cannot calculate a correlation between the amount of fat in the diets of a group of people and their ethnicity, because ethnicity is a categorical variable.

2. Correlation measures the strength of only the linear relationship between two variables. *Correlation does not describe curved relationships between variables, no matter how strong they are.* Exercise 3.11 illustrates this important fact.

3. *Like the mean and standard deviation, the correlation is not resistant: r is strongly affected by a few outlying observations.* Use r with caution when outliers appear in the scatterplot. To explore how extreme observations can influence r, use the *Correlation and Regression* applet.

4. *Correlation calculated from averaged data is typically much stronger than correlation calculated from the raw individual data* because averaged values mask some of the individual-to-individual variations that make up scatter in the scatterplot. Exercise 3.35 illustrates this fact.

5. *Correlation is not a complete summary of two-variable data,* even after you have established that the relationship between the variables is linear. You should give the means and standard deviations of both x and y along with the correlation. Of course, these numerical summaries do not point out outliers or clusters in the scatterplot. Numerical summaries complement plots of data but don't replace them.

Lastly, we may come across data that have been collected over **time** instead of over a number of individuals. How do we handle these data?

time and scatterplots

Most often in biology, time is not used as an explanatory variable, even though time might be recorded during data collection. For instance, in Table 3.1, manatee deaths and powerboats registered are listed for each year between 1977 and 2006. It would be interesting to study the trends over time of each variable separately, say to find out how fast the number of powerboats registered in Florida has been increasing in the past few decades. However, that would not help us understand how powerboats affect manatee deaths. That is why we chose to build a scatterplot of the relationship between the number of powerboats registered and the number of manatee deaths from powerboat collisions in any given year. Both approaches provide very different kinds of information.

Some research fields, such as ecology and evolution, do include time in their correlation analyses as a means to understand patterns of change. While time in itself does not explain or cause the changes, it can help reveal the phenomenon of interest, such as global warming, habitat loss, or extinction rates. *Always examine carefully the question you are asking and match the proper analytic method to the question.*

APPLY YOUR KNOWLEDGE

3.9 **Changing the units.** Coffee is currently priced in dollars. If it were priced in euros, and the dollar prices in Exercise 3.8 were translated into the equivalent prices in euros, would the correlation between coffee price and percent deforestation change? Explain your answer.

3.10 **Changing the correlation.**

(a) Use your calculator to find the correlation between the percent of returning birds and the number of new birds from the data in Exercise 3.4.

(b) Make a scatterplot of the data with two new points added. Point A: 10% return, 25 new birds. Point B: 40% return, 5 new birds. Find two new correlations: for the original data plus Point A, and for the original data plus Point B.

(c) Explain in terms of what correlation measures why adding Point A makes the correlation stronger (closer to -1) and adding Point B makes the correlation weaker (closer to 0).

3.11   **Strong association but no correlation.** In Exercise 3.6 you made a scatterplot of the fuel consumption of a car for different driving speeds. Examine this scatterplot and describe the strength of the relationship. If you calculate the correlation based on the data provided in that exercise, you find $r = -0.17$ (check for yourself). Explain why the correlation is close to zero even though there is a strong relationship between speed and fuel consumption.

3.12   **Antarctica's ice core documents our past.** As part of a larger effort to document trends of global warming and its causes researchers took samples of ice core in the Antarctic. They were able to assess from them the year of each snow deposit as well as the concentrations of carbon dioxide and of methane in the air in each particular year. One data set provided information spanning 1978 to 1995.[9]

(a) To study graphically the evolution of carbon dioxide concentrations over time, what variables would you use for the horizontal and vertical axes? What point would you be trying to make?

(b) To study graphically how concentrations of carbon dioxide and methane are related in any given year, what variables would you use for the horizontal and vertical axes? What point would you be trying to make?

CHAPTER 3 SUMMARY

To study relationships between variables, we must measure the variables on the same group of individuals.

If we think that a variable x may explain or even cause changes in another variable y, we call x an **explanatory variable** (also called an independent variable) and y a **response variable** (also called a dependent variable).

A **scatterplot** displays the relationship between two quantitative variables measured on the same individuals. Mark values of one variable on the horizontal axis (x axis) and values of the other variable on the vertical axis (y axis). Plot each individual's data as a point on the graph. Always plot the explanatory variable, if there is one, on the x axis of a scatterplot.

Plot points with different colors or symbols to see the effect of a categorical variable in a scatterplot.

In examining a scatterplot, look for an overall pattern showing the **direction, form,** and **strength** of the relationship, and then for **outliers** or other deviations from this pattern.

Direction: If the relationship has a clear direction, we speak of either **positive association** (high values of the two variables tend to occur together) or **negative association** (high values of one variable tend to occur with low values of the other variable).

Form: Linear relationships, where the points show a straight-line pattern, are an important form of relationship between two variables. **Curved** relationships and **clusters** are other forms to watch for.

Strength: The **strength** of a relationship is determined by how close the points in the scatterplot lie to a simple form such as a line.

The **correlation** r measures the strength and direction of the linear association between two quantitative variables x and y. Although you can calculate a correlation for any scatterplot, r measures only straight-line relationships.

Correlation indicates the direction of a linear relationship by its sign: $r > 0$ for a positive association and $r < 0$ for a negative association. Correlation always satisfies $-1 \leq r \leq 1$ and indicates the strength of a relationship by how close it is to -1 or 1. Perfect correlation, $r = \pm 1$, occurs only when the points on a scatterplot lie exactly on a straight line.

Correlation ignores the distinction between explanatory and response variables. The value of r is not affected by changes in the unit of measurement of either variable. Correlation is not resistant, so outliers can greatly change the value of r.

CHECK YOUR SKILLS

3.13 You have data for many individuals on their walking speed and their heart rate after a 10-minute walk. When you make a scatterplot, the explanatory variable on the x axis

    (a) is walking speed.   (b) is heart rate.   (c) doesn't matter.

3.14 You have data for many individuals on their walking speed and their heart rate after a 10-minute walk. You expect to see

    (a) a positive association.

    (b) very little association.

    (c) a negative association.

3.15 Figure 3.6 is a scatterplot of the age (in months) at which a child begins to talk and the child's later score on a test of mental ability (Gesell Adaptive Score), for

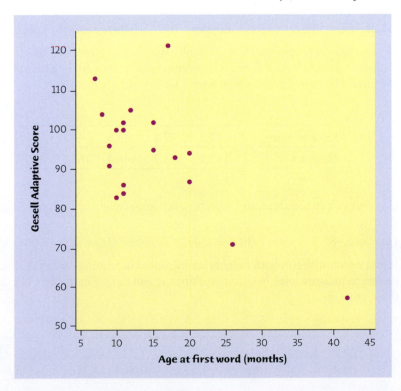

FIGURE 3.6 Scatterplot of Gesell Adaptive Score versus the age at first word for 21 children, for Exercise 3.15.

each of 21 children (2 sets of 2 children happen to have the same values).[10] One child clearly falls outside the overall pattern of the scatterplot and thus might be an outlier. Which child is it? The child with

(a) age 42 months; score 57.

(b) age 26 months; score 71.

(c) age 17 months; score 121.

3.16   To describe the relationship in Figure 3.6 you would say that

(a) it is a positive curved association.

(b) it is a negative linear association.

(c) it shows no association.

3.17   What are all the values that a correlation r can possibly take?

(a) $r \geq 0$ (b) $0 \leq r \leq 1$ (c) $-1 \leq r \leq 1$

3.18   If mothers were always 2 years younger than the fathers of their children, the correlation between the ages of mother and father would be

(a) 1.

(b) 0.5.

(c) Can't tell without seeing the data.

3.19   For a class project, you measure the weight in grams and the tail length in millimeters of a group of mice. The correlation is $r = 0.7$. You notice on the scatterplot one outlier of the relationship, falling clearly below the overall pattern of points. On closer inspection, that particular mouse looks somewhat sick and you decide to remove the outlier from the data set. With the outlier removed, you can expect the value of r to

(a) decrease.   (b) increase.   (c) stay the same.

3.20   Because elderly people may have difficulty standing to have their height measured, a study looked at the relationship between overall height and height to the knee. Here are data (in centimeters) for five elderly men:

Knee height x	57.7	47.4	43.5	44.8	55.2
Height y	192.1	153.3	146.4	162.7	169.1

Use your calculator: the correlation between knee height and overall height is about

(a) $r = 0.88$.   (b) $r = 0.09$.   (c) $r = 0.77$.

3.21   In the exercise above, both heights are measured in centimeters. A mad scientist prefers to measure knee height in millimeters and height in meters. The data in these units are

Knee height x	577	474	435	448	552
Height y	1.921	1.533	1.464	1.627	1.691

The correlation for the data using these units would

(a) be very close to the value calculated in the previous exercise.

(b) be exactly the same as in the previous exercise.

(c) be exactly 10 times smaller (1 cm = 10 mm; 1 cm = 1/100 m).

3.22   In Exercise 3.20 you calculated a correlation for knee height (x, in cm) and height (y, in cm). If, instead, you used height (in cm) as the x variable and knee height (in cm) as the y variable, the new correlation would

(a) have the inverse value (1 over).

(b) have the opposite value (a different sign).

(c) remain the same.

CHAPTER 3 EXERCISES

3.23   **Classifying fossils.** *Archaeopteryx* is an extinct beast having feathers like a bird but teeth and a long bony tail like a reptile. Only a few fossil specimens are known. Because these specimens differ greatly in size, some scientists think they belong to different species. We will examine some data. If the specimens belong to the same species and differ in size because some are younger than others, there should be a positive linear relationship between the lengths of a pair of bones from all individuals. An outlier from this relationship would suggest a different species. Here are data on the lengths in centimeters of the femur (a leg bone) and the humerus (a bone in the upper arm) for the five specimens that preserve both bones:[11]

Dr. John D. Cunningham/Visuals Unlimited

Femur	38	56	59	64	74
Humerus	41	63	70	72	84

(a) Make a scatterplot of femur and of humerus length. Do you think that all five specimens could come from the same species?

(b) Find the correlation r. How does it support your interpretation?

(c) Does it matter whether you use femur or humerus length for the x axis? Explain why.

3.24   **Is wine good for your heart?** There is some evidence that drinking moderate amounts of wine helps prevent heart attacks. Table 3.3 gives data on yearly wine consumption (liters of alcohol from drinking wine, per person) and yearly deaths from heart disease (deaths per 100,000 people) in 19 developed nations.[12]

(a) Make a scatterplot that shows how national wine consumption helps explain heart disease death rates.

(b) Describe the form of the relationship. Is there a linear pattern? How strong is the relationship?

(c) Is the direction of the association positive or negative? Explain in simple language what this says about wine and heart disease. Do you think these data give good evidence that drinking wine *causes* a reduction in heart disease deaths? Why?

TABLE 3.3	Wine consumption and heart attacks					
Country	Alcohol from wine	Heart disease deaths		Country	Alcohol from wine	Heart disease deaths
Australia	2.5	211		Netherlands	1.8	167
Austria	3.9	167		New Zealand	1.9	266
Belgium	2.9	131		Norway	0.8	227
Canada	2.4	191		Spain	6.5	86
Denmark	2.9	220		Sweden	1.6	207
Finland	0.8	297		Switzerland	5.8	115
France	9.1	71		United Kingdom	1.3	285
Iceland	0.8	211		United States	1.2	199
Ireland	0.7	300		West Germany	2.7	172
Italy	7.9	107				

3.25 **Classifying fossils.** Return to the fossil data from Exercise 3.23. If the scientists had recorded their data in inches instead of centimeters, how would the correlation be affected (1 cm = 0.3937 inches)? Explain your reasoning.

3.26 **Milk or soda?** The presence of soda vending machines in schools, under contracts with soft drink companies, is the subject of hot debate. Many see a link to childhood obesity as well as tooth decay and caffeine dependence. Has the soft drink industry changed our drinking habits? The Census Bureau reports U.S. per capita consumption of milk and carbonated soft drinks (in gallons per year) between 1980 and 2000:

Year	1980	1985	1990	1995	1998	1999	2000
Milk	27.6	26.7	25.7	23.9	23.0	22.9	23.3
Soda	35.1	35.7	46.2	47.4	47.9	49.7	49.3

(a) If you wanted to study the pattern of change over time in drinking habits for milk and soda, what graph would you create? Explain your reasoning.

(b) Make a graph showing the relationship between soda consumption and milk consumption in the same year. Which variable would you assign to be the explanatory variable, if any? Describe the relationship as revealed by your graph.

(c) If appropriate, calculate a numerical summary of the strength of the relationship between milk and soda consumption. What are your conclusions about American consumption of milk and soda?

3.27 **Correlation is not resistant.** Go to the *Correlation and Regression* applet. Click on the scatterplot to create a group of 10 points in the lower-left corner of the scatterplot with a strong straight-line pattern (correlation about 0.9).

(a) Add one point at the upper right that is in line with the first 10. How does the correlation change?

(b) Drag this last point down until it is opposite the group of 10 points. How small can you make the correlation? Can you make the correlation negative?

You see that a single outlier can greatly strengthen or weaken a correlation. Always plot your data to check for outlying points.

3.28 **How many corn plants are too many?** How much corn per acre should a farmer plant to obtain the highest yield? Too few plants will give a low yield. On the other hand, if there are too many plants, they will compete with each other for moisture and nutrients, and yields will fall. To find the best planting rate, plant at different rates on several plots of ground and measure the harvest. (Be sure to treat all the plots the same except for the planting rate.) Here are data from such an experiment:[13]

Plants per acre	Yield (bushels per acre)			
12,000	150.1	113.0	118.4	142.6
16,000	166.9	120.7	135.2	149.8
20,000	165.3	130.1	139.6	149.9
24,000	134.7	138.4	156.1	
28,000	119.0	150.5		

(a) Is yield or planting rate the explanatory variable?

(b) Make a scatterplot of yield and planting rate. Use a scale of yields from 100 to 200 bushels per acre so that the pattern will be clear. Notice how the data points along the x axis take 1 of only 5 possible values: the planting rates chosen by the experimenter. We will discuss in further chapters how this is one major difference between experiments and observational studies.

(c) Describe the overall pattern of the relationship. Is it linear? Is there a positive or negative association, or neither? Would calculating the correlation be a reasonable approach to describing this relationship?

3.29 **How many corn plants are too many?** The previous exercise gives data on corn yields for fields planted at different plant densities. Find the mean yield for each of the five planting rates. Plot each mean yield against its planting rate on the scatterplot you made in the previous exercise and connect these five points with lines. This combination of numerical description and graphing makes the relationship clearer. What planting rate would you recommend to a farmer whose conditions were similar to those in the experiment?

3.30 **Attracting beetles.** To detect the presence of harmful insects in farm fields, we can put up boards covered with a sticky material and examine the insects trapped on the boards. Which colors attract insects best? Experimenters placed boards of several colors at random locations in a field of oats (four colors and six boards of each color). Here are the counts of cereal leaf beetles trapped by each board:

Color	Insects trapped					
Blue	16	11	20	21	14	7
Green	37	32	20	29	37	32
White	21	12	14	17	13	20
Yellow	45	59	48	46	38	47

(a) Make a plot of the number of beetles trapped on each board against the board's color, arranging the colors to show blue first, then white, then green, and last yellow (space the four colors equally on the horizontal axis).

(b) Does it make sense to speak of a positive or negative association between beetles trapped and board color? Why?

(c) Based on your graph, which of the four colors seems to best attract cereal leaf beetles?

3.31 Attracting beetles, continued

(a) Using the data from the previous exercise, now make another plot of beetles trapped against board color, but this time arranging the colors in the order used in the table (alphabetical). Notice how different the two graphs look, although the same information is displayed. Explain why.

(b) What do you think a graph of the same data would look like if you arranged the colors in the reverse order of the previous exercise? Is one graph more acceptable than the others?

(c) Some statistical software packages do not allow the use of text for data and force you to use numerical codes instead (for example, blue = 1, green = 2, etc.). Point out the potential danger when creating graphs like the ones you made for this beetle problem.

3.32 Brain size and IQ score. Do people with larger brains have higher IQ scores? Figure 3.7 shows the result of a study conducted on 40 volunteers, 20 selected for

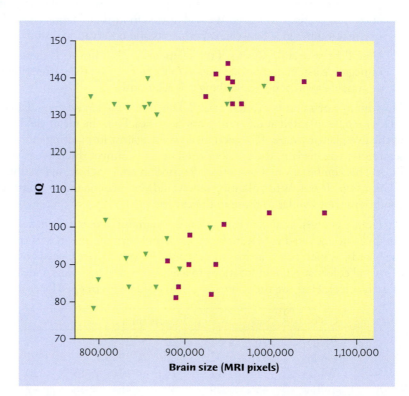

FIGURE 3.7 Scatterplot of the brain size (in MRI pixels) and IQ of 20 subjects selected for their high IQ (130 or more) and 20 selected for their low IQ (100 or less), for Exercise 3.32. Of the 40 subjects, 20 are women (triangles) and 20 are men (squares).

their high IQ (130 or more) and 20 selected for their low IQ (100 or less). The researchers' choice of studying high and low IQ scores only is reflected in the gap on the scatterplot for mid-range IQ scores. Brain size was measured by magnetic resonance imaging (MRI). The MRI count is the number of "pixels" the brain covered in the image. IQ was measured by the Wechsler test. Of the 40 volunteers, 20 were women (triangles) and 20 were men (squares).

(a) Men are larger than women on the average and have larger brains. How is this size effect visible in the plot? Guess the mean MRI count for men and women from Figure 3.7 to verify the difference.

(b) Comment on the nature and strength of the relationship between brain size and IQ for women and then for men. Explain why it makes sense to study men and women separately in this case.

(c) The study included only individuals with high IQ or low IQ. This is reflected on the scatterplot by the separate clusters of points. Explain why this makes interpreting the scatterplot particularly challenging.

3.33   **Primate species.** Some species tend to maintain an optimum population size that can be supported by available resources, while other species tend to maximize population growth. An evolutionary biologist postulates that primate species favor maintaining an optimum population size. Figure 3.8 shows the scatterplot

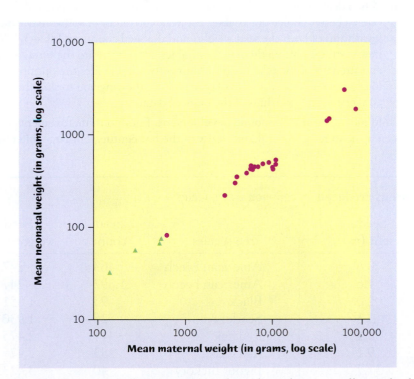

FIGURE 3.8 Scatterplot of average maternal weight and average offspring birth weight for 25 primate species (logarithmic scales), for Exercise 3.33. Circles represent species with only one offspring per birth, and triangles represent species with two or more offspring per litter.

of average maternal weight and average offspring birth weight for each of 25 primate species (monkeys and apes).[14] Both axes are shown in a logarithmic scale. Circles represent species with only one offspring per birth, and triangles represent species with two or more offspring per litter.

(a) What does the scatterplot show about the pattern of mother to offspring weight, when expressed in logarithm scales? Do you think that a scatterplot would show a stronger or a weaker relationship if it was made with the actual weights for individuals of each species instead of species averages?

(b) What does the scatterplot show about the difference between species with single and those with multiple offspring per litter? Rare multiple births and a large range of individual sizes is evidence against a survival strategy that maximizes population growth.

3.34 **Transforming data: counting seeds.** Table 3.4 gives data on the mean number of seeds produced in a year by several common tree species and the mean weight (in milligrams) of the seeds produced. (Some species appear twice because their seeds were counted in two locations.) We might expect that trees with heavy seeds produce fewer of them, but what is the form of the relationship?[15]

(a) Make a scatterplot showing how the weight of tree seeds helps explain how many seeds the tree produces. Describe the form, direction, and strength of the relationship.

(b) When dealing with sizes and counts, the logarithms of the original data are often a better choice of variable. Use your calculator or software to obtain the logarithms of both the seed weights and the seed counts in Table 3.4. Make a new scatterplot using these new variables. Now what are the form, direction, and strength of the relationship? Most software provides the option of using logarithm scales for the axes of scatterplots, allowing you to skip the conversion to logarithms of the original data.

3.35 **Child mortality and economic development.** Enormous improvements have been made in reducing child mortality in the last century, but large differences

TABLE 3.4 Count and weight of seeds produced by common tree species

Tree species	Seed count	Seed weight (mg)	Tree species	Seed count	Seed weight (mg)
Paper birch	27,239	0.6	American beech	463	247
Yellow birch	12,158	1.6	American beech	1,892	247
White spruce	7,202	2.0	Black oak	93	1,851
Engelmann spruce	3,671	3.3	Scarlet oak	525	1,930
Red spruce	5,051	3.4	Red oak	411	2,475
Tulip tree	13,509	9.1	Red oak	253	2,475
Ponderosa pine	2,667	37.7	Pignut hickory	40	3,423
White fir	5,196	40.0	White oak	184	3,669
Sugar maple	1,751	48.0	Chestnut oak	107	4,535
Sugar pine	1,159	216.0			

exist across the globe. We can ask whether there is a relationship between the economic development of countries, measured in per capita gross domestic product (GDP, in dollars per year, adjusted for local purchasing power), and child mortality (in percent children dying before five years of age). Here are data provided by the United Nations, arranged by geographic area for the year 2004 (the OECD is a group of 30 wealthy, democratic countries):[16]

Region	Child mortality	GDP per capita
Arab states	6.1	5,685
East Asia and the Pacific	3.9	5,100
Latin America and the Caribbean	3.2	7,404
South Asia	9.1	2,897
Sub-Saharan Africa	17.9	1,856
Central and Eastern Europe	2.4	7,939
OECD	0.6	30,181

(a) Figure 3.9(a) shows the scatterplot of these data, with each point labeled. Describe the relationship between GDP per capita and child mortality. Calculate the correlation.

(b) Figure 3.9(b) shows the scatterplot of child mortality against GDP per capita when the same data are broken down for each country making up the geographic region. The countries making up the South Asia region are color-coded to help show the correspondence between the two scatterplots. Describe this relationship and compare it with the one in Figure 3.9(a). The correlation for the relationship using individual countries is $r = -0.55$. How does it compare with the correlation you calculated in (a)? Explain why they are different.

3.36 Merlins breeding. Often the percent of an animal species in the wild that survive to breed again is lower following a successful breeding season. This is part of nature's self-regulation to keep population size stable. A study of merlins (small falcons) in northern Sweden observed the number of breeding pairs in an isolated area and the percent of males (banded for identification) who returned the next breeding season. Here are data for nine years:[17]

Breeding pairs	28	29	29	29	30	32	33	38	38
Percent return	82	83	70	61	69	58	43	50	47

Do the data support the theory that a smaller percent of birds survive following a successful breeding season? Use the four-step outline (page 55) in your answer.

3.37 Brighter sunlight? The brightness of sunlight at the earth's surface changes over time because the earth's atmosphere is more or less clear. Sunlight dimmed between 1960 and 1990. After 1990, air pollution dropped in industrial countries. Did sunlight brighten? Here are data from Boulder, Colorado,

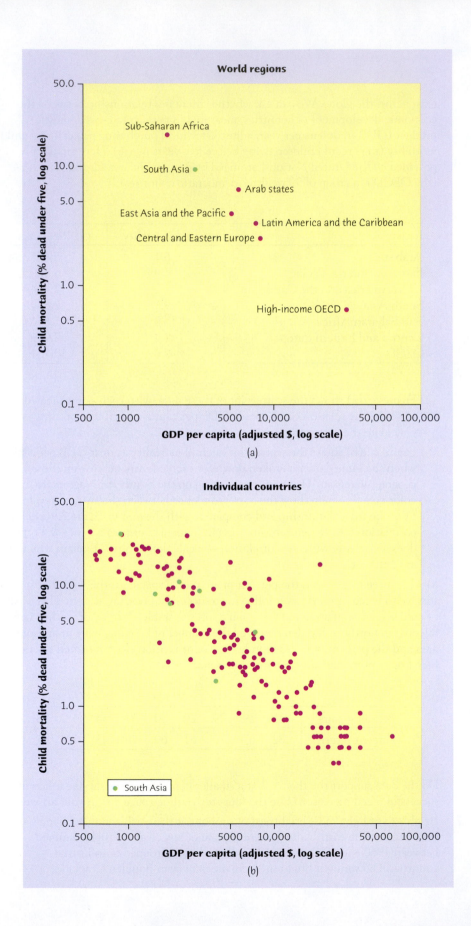

FIGURE 3.9 (a) Scatterplot of GDP per capita and child mortality of the main world regions, for Exercise 3.35. (b) Scatterplot of GDP per capita and child mortality when the same data are broken down for each country making up the regions. Countries making up South Asia are marked in green.

averaging over only clear days each year. (Other locations show similar trends.) The response variable is solar radiation in watts per square meter.[18]

Year	Sun	Year	Sun	Year	Sun
1992	243.2	1996	250.9	2000	251.7
1993	246.0	1997	250.9	2001	251.4
1994	248.0	1998	250.0	2002	250.9
1995	250.3	1999	248.9		

Follow the four-step outline (page 55) to answer the question posed.

3.38   Does social rejection hurt? We often describe our emotional reaction to social rejection as "pain." A clever study asked whether social rejection causes activity in areas of the brain that are known to be activated by physical pain. If it does, we really do experience social and physical pain in similar ways. Subjects were first included and then deliberately excluded from a social activity while changes in brain activity were measured. After each activity, the subjects filled out questionnaires that assessed how excluded they felt. Here are data for 13 subjects:[19]

Subject	Social distress	Change in brain activity	Subject	Social distress	Change in brain activity
1	1.26	−0.055	8	2.18	0.025
2	1.85	−0.040	9	2.58	0.027
3	1.10	−0.026	10	2.75	0.033
4	2.50	−0.017	11	2.75	0.064
5	2.17	−0.017	12	3.33	0.077
6	2.67	0.017	13	3.65	0.124
7	2.01	0.021			

The explanatory variable is "social distress" measured by each subject's questionnaire score after exclusion relative to the score after inclusion. (So values greater than 1 show the degree of distress caused by exclusion.) The response variable is change in activity in a region of the brain that is activated by physical pain. Discuss what the data show. Follow the four-step process (page 55) in your discussion.

3.39   Match the correlation. You are going to use the *Correlation and Regression* applet to make scatterplots with 10 points that have correlation close to 0.7. The lesson is that many patterns can have the same correlation. Always plot your data before you trust a correlation.

(a) Stop after adding the first 2 points. What is the value of the correlation? Why does it have this value?

(b) Make a lower-left to upper-right pattern of 10 points with correlation about $r = 0.7$. (You can drag points up or down to adjust r after you have 10 points.) Make a rough sketch of your scatterplot.

(c) Make another scatterplot with 9 points in a vertical stack at the left of the plot. Add 1 point far to the right, and move it until the correlation is close to 0.7. Make a rough sketch of your scatterplot.

(d) Make yet another scatterplot with 10 points in a curved pattern that starts at the lower left, rises to the right, then falls again at the far right. Adjust the points up or down until you have a quite smooth curve with correlation close to 0.7. Make a rough sketch of this scatterplot also.

James Leynse/CORBIS

Regression

Linear (straight-line) relationships between two quantitative variables are easy to understand and quite common. In Chapter 3, we found linear relationships in settings as varied as counting carnivores, manatee deaths, and fruit fly longevity. Correlation measures the direction and strength of these relationships. When a scatterplot shows a linear relationship, we would like to summarize the overall pattern by drawing a line on the scatterplot.

Regression lines

A *regression line* summarizes the relationship between two variables, but only in a specific setting: One of the variables helps explain or predict the other. That is, regression describes a relationship between an explanatory variable and a response variable.

> **REGRESSION LINE**
>
> A **regression line** is a straight line that describes how a response variable y changes as an explanatory variable x changes. We often use a regression line to predict the value of y for a given value of x.

EXAMPLE 4.1 Does fidgeting keep you slim?

Obesity is a growing problem around the world. Here, following our four-step process (page 55), is an account of a study that sheds some light on gaining weight.

Donna Day/Stone/Getty Images

STATE: Some people don't gain weight even when they overeat. Perhaps fidgeting and other "nonexercise activity" (NEA) explains why—some people may spontaneously increase nonexercise activity when fed more. Researchers deliberately overfed 16 healthy young adults for 8 weeks. They measured fat gain (in kilograms) and, as an explanatory variable, change in energy use (in kilocalories, or Calories, Cal) from activity other than deliberate exercise—fidgeting, daily living, and the like. Here are the data:[1]

NEA change (Cal)	−94	−57	−29	135	143	151	245	355
Fat gain (kg)	4.2	3.0	3.7	2.7	3.2	3.6	2.4	1.3
NEA change (Cal)	392	473	486	535	571	580	620	690
Fat gain (kg)	3.8	1.7	1.6	2.2	1.0	0.4	2.3	1.1

Do people with larger increases in NEA tend to gain less fat?

FORMULATE: Make a scatterplot of the data and examine the pattern. If it is linear, use correlation to measure its strength and draw a regression line on the scatterplot to predict fat gain from change in NEA.

SOLVE: Figure 4.1 is a scatterplot of these data. The plot shows a moderately strong negative linear association with no outliers. The correlation is $r = -0.7786$. The line on the plot is a regression line for predicting fat gain from change in NEA.

CONCLUDE: People with larger increases in nonexercise activity do indeed gain less fat. To add to this conclusion, we must study regression lines in more detail.

We can, however, already use the regression line to predict fat gain from NEA. Suppose that an individual's NEA increases by 400 Cal when she overeats. Go "up

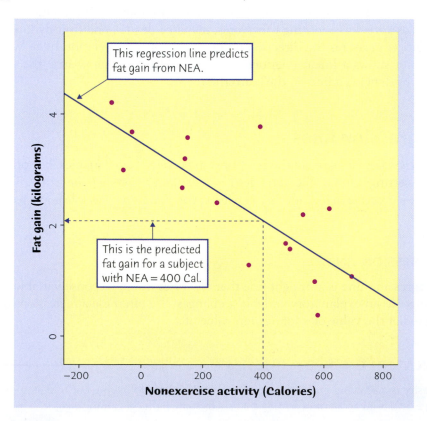

FIGURE 4.1 Fat gain after 8 weeks of overeating, plotted against increase in nonexercise activity over the same period.

and over" on the graph in Figure 4.1. From 400 Cal on the x axis, go up to the regression line and then over to the y axis. The graph shows that the predicted gain in fat is a bit more than 2 kilograms.

Many calculators and software programs will give you the equation of a regression line from keyed-in data. Understanding and using the line is more important than the details of where the equation comes from.

REVIEW OF STRAIGHT LINES

Suppose that y is a response variable (plotted on the vertical axis) and x is an explanatory variable (plotted on the horizontal axis). A straight line relating y to x has an equation of the form

$$y = a + bx$$

In this equation, b is the **slope,** the amount by which y changes when x increases by one unit. The number a is the **intercept,** the value of y when $x = 0$.

— **EXAMPLE 4.2**   Using a regression line ———————

Any straight line describing the NEA data has the form

$$\text{fat gain} = a + (b \times \text{NEA change})$$

The line in Figure 4.1 is the regression line with the equation

$$\text{fat gain} = 3.505 - (0.00344 \times \text{NEA change})$$

Be sure you understand the role of the two numbers in this equation:

- The slope $b = -0.00344$ tells us that fat gained goes down by 0.00344 kilogram for each added Calorie of NEA. The slope of a regression line is the *rate of change* in the response as the explanatory variable changes.

- The intercept, $a = 3.505$ kilograms, is the estimated fat gain if NEA does not change when a person overeats.

The slope of a regression line is an important numerical description of the relationship between the two variables. Although we need the value of the intercept to draw the line, this value is statistically meaningful only when, as in this example, the explanatory variable can actually take values close to zero.

The equation of the regression line makes it easy to predict fat gain. If a person's NEA increases by 400 Cal when she overeats, substitute $x = 400$ in the equation. The predicted fat gain is

$$\text{fat gain} = 3.505 - (0.00344 \times 400) = 2.13 \text{ kilograms}$$

To **plot the line** on the scatterplot, use the equation to find the predicted y for two values of x, one near each end of the range of x in the data. Plot each y above its x and draw the line through the two points.

plotting a line

The slope $b = -0.00344$ in Example 4.2 is small. This does *not* mean that change in NEA has little effect on fat gain. The size of the slope depends on the units in which we measure the two variables. In this example, the slope is the change in fat gain in kilograms when NEA increases by 1 Cal. There are

Regression toward the mean

To "regress" means to go backward. Why are statistical methods for predicting a response from an explanatory variable called "regression"? Sir Francis Galton (1822–1911), who was the first to apply regression to biological and psychological data, looked at examples such as the heights of children versus the heights of their parents. He found that the taller-than-average parents tended to have children who were also taller than average but not as tall as their parents. Galton called this fact "regression toward the mean," and the name came to be applied to the statistical method.

1000 grams in a kilogram. If we measured fat gain in grams, the slope would be 1000 times larger, $b = 3.44$. *You can't say how important a relationship is by looking at the size of the slope of the regression line.*

We have seen in the previous chapter that data can be transformed to reveal a linear relationship. Logarithm transformations, in particular, are quite frequent in biology to study problems ranging from population growth to animal physiology. A regression line can be computed using the transformed data, and prediction within the range of the data can be performed as with simple linear relationships. The difference is that we will need to transform the predicted y back into its original (nonlogarithmic) unit.

EXAMPLE 4.3 Using a regression line: transformed data

Humans have a large brain compared with other mammals, but they are also a relatively large mammal species. How does an animal's brain relate to its body size? Figure 4.2(a) shows the scatterplot of brain weight in grams against body weight in kilograms for 96 species of mammals.[2] Figure 4.2(b) shows the scatterplot for the same data after a logarithm transformation. The line is the least-squares regression line for predicting brain weight from body weight.

The regression line for the untransformed data in Figure 4.2(a) is clearly unsatisfactory because most mammals are so small compared to elephants. The correlation

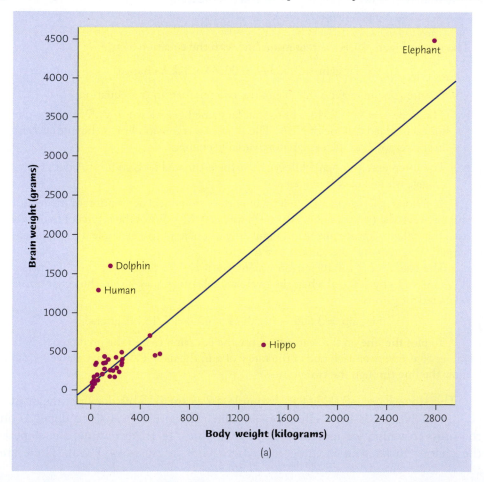

FIGURE 4.2(a) Scatterplot of body weight and brain weight for 96 species of mammals in the original numerical values of kilograms and grams.

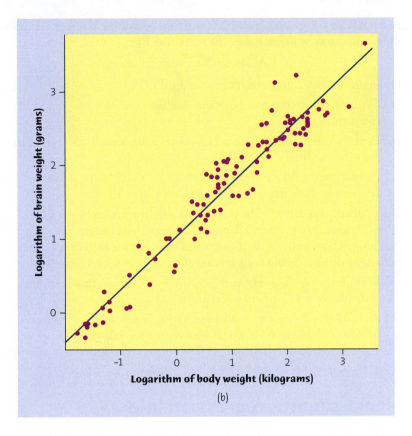

FIGURE 4.2(b) Scatterplot of the data in Figure 4.2(a) using the logarithm of brain weight and the logarithm of body weight.

between brain weight and body weight is $r = 0.86$, but this is misleading. If we remove the elephant data point, the correlation for the other 95 species falls to $r = 0.50$. Humans, dolphins, and hippos are also extreme outliers.

The scatterplot using transformed data in Figure 4.2(b) has no extreme outliers and is clearly linear with correlation $r = 0.96$. The least-squares regression for predicting the logarithm of brain weight from the logarithm of body weight has the equation

$$\log(\text{brain weight}) = 1.01 + [0.72 \times \log(\text{body weight})]$$

For a mammal species weighing on average 100 kg, the logarithm is $\log(100) = 2$, and the predicted logarithm of brain weight is

$$\log(\text{brain weight}) = 1.01 + (0.72 \times 2) = 2.45$$

To undo the logarithm transformation, remember that, for common logarithms with base 10, $y = 10^{\log(y)}$. Thus, the predicted brain weight, in grams, for the 100-kg species is

$$\text{brain weight} = 10^{2.45} = 282 \text{ g}$$

Verify on Figure 4.2 that this prediction makes sense on either scatterplot, although the scatterplot of transformed data is much easier to use.

APPLY YOUR KNOWLEDGE

4.1   **Don't drink and drive.** The Department of Motor Vehicles warns of the effect of drinking alcohol on one's blood alcohol content (BAC, in percent of volume) and behavior. At BAC of 0.8 or higher driving skills and response time are impaired and the risk of accidental death is greatly increased. One "drink" is defined as 12 ounces (oz) of beer or 4 oz of table wine (or 1.25 oz of 80-proof liquor). For a 160-pound man, BAC as a function of the number of drinks drunk in one hour can be expressed by the regression line

$$\text{BAC} = 0.023 \times \text{ number of drinks}$$

for predicting BAC from the number of drinks.[3]

(a) Say in words what the slope of this line tells you. Why do you think the intercept is zero?

(b) Find the predicted BAC for a 160-pound male who drank four 12-oz beers over the last hour.

(c) Draw a graph of the regression line for a range of drinks between 0 and 10. (Be sure to show the scales for the x and y axes.)

4.2   **Don't drink and drive.**   The impact of a single drink on blood alcohol content (BAC, in percent of volume) depends on gender and stature. For a woman weighing 120 pounds, the BAC increases by 0.038 for each drink taken in one hour. One drink is defined as a serving of 12 oz of beer or 4 oz of table wine.[4]

(a) What is the equation of the regression line for predicting BAC from number of drinks for a 120-pound woman?

(b) As of 2005, all states have adopted a BAC limit of 0.08 for drivers over 21 years old. What BAC would we expect for a 120-pound woman having 2 drinks in one hour? Would her BAC be over the legal limit, assuming she is over 21 years old? Would it be safe driving?

The least-squares regression line

In most cases, no line will pass exactly through all the points in a scatterplot. Different people will draw different lines by eye. We need a way to draw a regression line that doesn't depend on our guess as to where the line should go. Because we use the line to predict y from x, the prediction errors we make are errors in y, the vertical direction in the scatterplot. *A good regression line makes the vertical distances of the points from the line as small as possible*.

Figure 4.3 illustrates the idea. This plot shows three of the points from Figure 4.1, along with the line, on an expanded scale. The line passes above one of the points and below two of them. The three prediction errors appear as vertical line segments. For example, one subject had $x = -57$, a decrease of 57 Cal in NEA. The line predicts a fat gain of 3.7 kilograms, but the actual fat gain for this subject was 3.0 kilograms. The prediction error is

$$\text{error} = \text{observed response } - \text{ predicted response}$$
$$= 3.0 - 3.7 = -0.7 \text{ kilogram}$$

There are many ways to make the collection of vertical distances "as small as possible." The most common is the *least-squares* method.

LEAST-SQUARES REGRESSION LINE

The **least-squares regression line** of y on x is the line that makes the sum of the squares of the vertical distances of the data points from the line as small as possible.

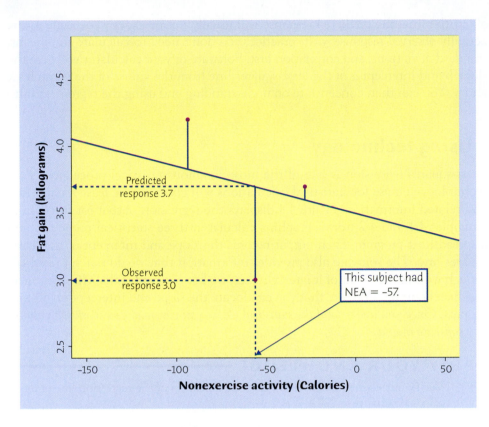

FIGURE 4.3 The least-squares idea (enlargement of the upper left corner of Figure 4.1). For each observation, find the vertical distance of each point on the scatterplot from a regression line. The least-squares regression line makes the sum of the squares of these distances as small as possible.

One reason for the popularity of the least-squares regression line is that the problem of finding the line has a simple answer. We can give the equation for the least-squares line in terms of the means and standard deviations of the two variables and the correlation between them.

EQUATION OF THE LEAST-SQUARES REGRESSION LINE

We have data on an explanatory variable x and a response variable y for n individuals. From the data, calculate the means $\overline{x}$ and $\overline{y}$ and the standard deviations s_x and s_y of the two variables and their correlation r. The least-squares regression line is the line

$$\hat{y} = a + bx$$

with **slope**

$$b = r\frac{s_y}{s_x}$$

and **intercept**

$$a = \overline{y} - b\overline{x}$$

We write $\hat{y}$ (read "y hat") in the equation of the regression line to emphasize that the line gives a *predicted* response $\hat{y}$ for any x. Because of the scatter of points about the line, the predicted response will usually not be exactly the same as the actually *observed* response y. In practice, you don't need to calculate the means, standard deviations, and correlation first. Software or your calculator will give the slope b and intercept a of the least-squares line from the values of the variables x and y. You can then concentrate on understanding and using the regression line.

Using technology

Least-squares regression is one of the most common statistical procedures. Any technology you use for statistical calculations will give you the least-squares line and related information. Figure 4.4 displays the regression output for the data of Examples 4.1 and 4.2 from a graphing calculator, three statistical programs, and a spreadsheet program. Each output records the slope and intercept of the least-squares line. The software also provides information that we do not yet need, although we will use much of it later. (In fact, we left out part of the Minitab, SPSS and Excel outputs.) Be sure that you can locate the slope and intercept on all four outputs. *Once you understand the statistical ideas, you can read and work with almost any software output.*

APPLY YOUR KNOWLEDGE ────────────────

4.3 **Verify our claims.** Example 4.2 gives the equation of the regression line of fat gain y on change in NEA x as

$$\hat{y} = 3.505 - 0.00344x$$

Enter the data from Example 4.1 into your software or calculator.

(a) Use the regression function to find the equation of the least-squares regression line.

(b) Also find the mean and standard deviation of both x and y and their correlation r. Calculate the slope b and intercept a of the regression line from these, using the facts in the box Equation of the Least-Squares Regression Line. Verify that in both part (a) and part (b) you get the equation in Example 4.2. (Results may differ slightly because of rounding off.)

4.4 **Bird colonies.** One of nature's patterns connects the percent of adult birds in a colony that return from the previous year and the number of new adults that join the colony. Here are data for 13 colonies of sparrowhawks:[5]

Percent return	74	66	81	52	73	62	52	45	62	46	60	46	38
New adults	5	6	8	11	12	15	16	17	18	18	19	20	20

As you saw in Exercise 3.4 (page 69), there is a linear relationship between the percent x of adult sparrowhawks that return to a colony from the previous year and the number y of new adult birds that join the colony.

(a) Find the correlation r for these data. The straight-line pattern is moderately strong.

(b) Find the least-squares regression line for predicting y from x. Make a scatterplot and draw your line on the plot.

Texas Instruments TI-83

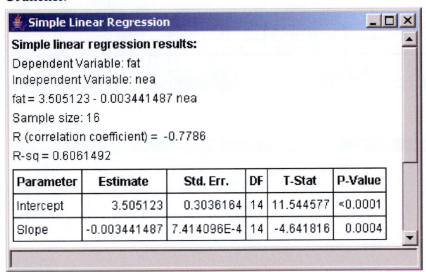

```
LinReg
 y=a+bx
 a=3.505122916
 b=-.003441487
 r²=.6061492049
 r=-.7785558457
```

CrunchIt!

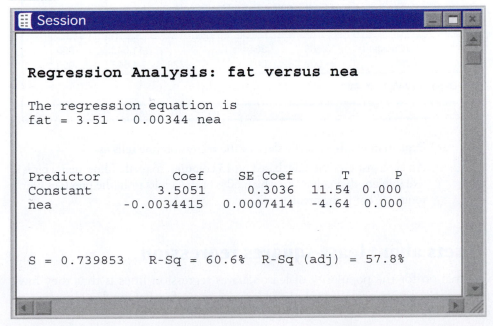

Minitab

Regression Analysis: fat versus nea

The regression equation is
fat = 3.51 - 0.00344 nea

Predictor	Coef	SE Coef	T	P
Constant	3.5051	0.3036	11.54	0.000
nea	-0.0034415	0.0007414	-4.64	0.000

S = 0.739853   R-Sq = 60.6%   R-Sq (adj) = 57.8%

FIGURE 4.4 Least-squares regression for the nonexercise activity data: output from a graphing calculator, three statistical programs, and a spreadsheet program (*continued*).

Microsoft Excel

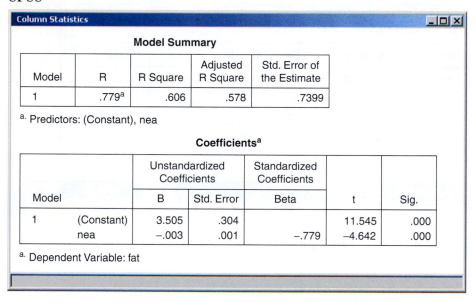

SPSS

Model Summary

Model	R	R Square	Adjusted R Square	Std. Error of the Estimate
1	.779[a]	.606	.578	.7399

a. Predictors: (Constant), nea

Coefficients[a]

Model		Unstandardized Coefficients		Standardized Coefficients	t	Sig.
		B	Std. Error	Beta		
1	(Constant)	3.505	.304		11.545	.000
	nea	−.003	.001	−.779	−4.642	.000

a. Dependent Variable: fat

FIGURE 4.4 (*continued*)

(c) Explain in words what the slope of the regression line tells us.

(d) An ecologist uses the line, based on 13 colonies, to predict how many birds will join another colony, to which 60% of the adults from the previous year return. What is the prediction?

Facts about least-squares regression

One reason for the popularity of least-squares regression lines is that they have many convenient special properties. Here are some facts about least-squares regression lines.

FACT 1. **The distinction between explanatory and response variables is essential in regression.** Least-squares regression makes the distances of the data points from the line small only in the y direction. If we reverse the roles of the two variables, we get a different least-squares regression line.

EXAMPLE 4.4   *Predicting fat, predicting NEA*

Figure 4.5 repeats the scatterplot of the nonexercise activity data in Figure 4.1, but with *two* least-squares regression lines. The solid line is the regression line for predicting fat gain from change in NEA. This is the line that appeared in Figure 4.1.

    We might also use the data on these 16 subjects to predict the change in NEA for another subject from that subject's fat gain when overfed for 8 weeks. Now the roles of the variables are reversed: fat gain is the explanatory variable and change in NEA is the response variable. The dashed line in Figure 4.5 is the least-squares line for predicting NEA change from fat gain. The two regression lines are not the same. *In the regression setting, you must know clearly which variable is explanatory.*

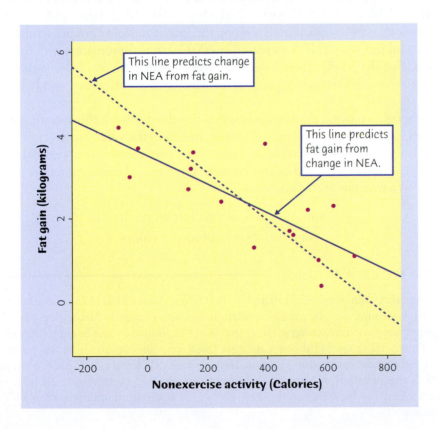

FIGURE 4.5 Two least-squares regression lines for the nonexercise activity data. The solid line predicts fat gain from change in nonexercise activity. The dashed line predicts change in nonexercise activity from fat gain.

FACT 2. There is a close connection between correlation and the slope of the least-squares line. The slope is

$$b = r \frac{s_y}{s_x}$$

This equation says that, along the regression line, **a change of one standard deviation in x corresponds to a change of r standard deviations in y.** When the variables are perfectly correlated ($r = 1$ or $r = -1$), the change in the predicted response $\hat{y}$ is the same (in standard deviation units) as the change in x. Otherwise, because $-1 \leq r \leq 1$, the change in $\hat{y}$ is less than the change in x. As the correlation grows less strong, the prediction $\hat{y}$ moves less in response to changes in x.

FACT 3. **The least-squares regression line always passes through the point $(\bar{x}, \bar{y})$** on the graph of y against x.

FACT 4. The correlation r describes the strength of a straight-line relationship. In the regression setting, this description takes a specific form: **The square of the correlation, r^2, is the fraction of the variation in the values of y that is explained by the least-squares regression of y on x.**

The idea is that when there is a linear relationship, some of the variation in y is accounted for by the fact that as x changes it pulls y along with it. Look again at Figure 4.1, the scatterplot of the NEA data. The variation in y appears as the spread of fat gains from 0.4 kg to 4.2 kg. Some of this variation is explained by the fact that x (change in NEA) varies from a loss of 94 Cal to a gain of 690 Cal. As x moves from -94 to 690, it pulls y along the line. You would predict a smaller fat gain for a subject whose NEA increased by 600 Cal than for someone with 0 change in NEA. But the straight-line tie of y to x doesn't explain *all* of the variation in y. The remaining variation appears as the scatter of points above and below the line.

Although we won't do the algebra, it is possible to break the variation in the observed values of y into two parts. One part measures the variation in $\hat{y}$ as x moves and pulls $\hat{y}$ with it along the regression line. The other measures the vertical scatter of the data points above and below the line. The squared correlation r^2 is the first of these as a fraction of the whole:

$$r^2 = \frac{\text{variation in } \hat{y} \text{ as } x \text{ pulls it along the line}}{\text{total variation in observed values of } y}$$

EXAMPLE 4.5 Using r^2

For the NEA data, $r = -0.7786$ and $r^2 = 0.6062$. About 61% of the variation in fat gained is accounted for by the linear relationship with change in NEA. The other 39% is individual variation among subjects that is not explained by the linear relationship.

Figure 3.2 (page 71) shows a stronger linear relationship in which the points are more tightly concentrated along a line. Here, $r = -0.9124$ and $r^2 = 0.8325$. More than 83% of the variation in carnivore abundance is explained by regression on body mass when taking the logarithm of the data. Only 17% is variation among species with the same mass.

When you report a regression, give r^2 as a measure of how successful the regression was in explaining the response. All of the outputs in Figure 4.4 include r^2, either in decimal form or as a percent. When you see a correlation, square it to get a better feel for the strength of the association. Perfect correlation ($r = -1$ or $r = 1$) means the points lie exactly on a line. Then $r^2 = 1$ and all of the variation in one variable is accounted for by the linear relationship with the other variable.

If $r = -0.7$ or $r = 0.7$, $r^2 = 0.49$ and about half the variation is accounted for by the linear relationship. In the r^2 scale, correlation ± 0.7 is about halfway between 0 and ± 1.

Facts 2, 3, and 4 are special properties of least-squares regression. They are not true for other methods of fitting a line to data.

APPLY YOUR KNOWLEDGE

4.5   Some regression math. Use the equation of the least-squares regression line (box on page 99) to show that these facts are true:

(a) The regression line for predicting y from x always passes through the point $(\overline{x}, \overline{y})$. That is, when $x = \overline{x}$, the equation gives $\hat{y} = \overline{y}$.

(b) Example 4.3 demonstrates that the regression line for predicting x from y differs from the line for predicting y from x. In fact, the slope of one of these lines is the reciprocal $(1/b)$ of the slope of the other.

4.6   Growing corn. Exercise 3.28 (page 85) gives data from an agricultural experiment. The purpose of the study was to see how the yield of corn changes as we change the planting rate (plants per acre).

(a) Make a scatterplot of the data. (Use a scale of yields from 100 to 200 bushels per acre.) Find the least-squares regression line for predicting yield from planting rate and add this line to your plot. Why should we *not* use the regression line for prediction in this setting?

(b) What is r^2? What does this value say about the success of the regression line in predicting yield?

(c) Even regression lines that make no practical sense obey Facts 2, 3, and 4. Use the equation of the regression line you found in (a) to show that when x is the mean planting rate, the predicted yield $\hat{y}$ is the mean of the observed yields.

4.7   How useful is regression? Figure 3.3 for Example 3.6 (page 73) displays the longevity and thorax length of two groups of male fruit flies, one sexually active and one sexually inactive. The correlation is $r = 0.806$ for the group of sexually active fruit flies. A biologist grows tomato plants under various conditions and records the plants' potassium and phosphorus levels.[6] The correlation is $r = -0.151$. Explain in simple language why knowing only these correlations enables you to say that prediction of longevity from thorax length for sexually active male fruit flies will be much more accurate than prediction of phosphorus level from potassium level in tomato plants.

Residuals

One of the first principles of data analysis is to look for an overall pattern and also for striking deviations from the pattern. A regression line describes the overall pattern of a linear relationship between an explanatory variable and a response variable. We see deviations from this pattern by looking at the scatter of the data points about the regression line. The vertical distances from the points to the least-squares regression line are as small as possible, in the sense that they have the smallest possible sum of squares. Because they represent "left-over" variation in the response after fitting the regression line, these distances are called *residuals*.

RESIDUALS

A **residual** is the difference between an observed value of the response variable and the value predicted by the regression line. That is, a residual is the prediction error that remains after we have chosen the regression line:

$$\text{residual} = \text{observed } y - \text{predicted } y$$
$$= y - \hat{y}$$

EXAMPLE 4.6 I feel your pain

"Empathy" means being able to understand what others feel. To see how the brain expresses empathy, researchers recruited 16 couples in their mid-20s who were married or had been dating for at least two years. They zapped the man's hand with an electrode while the woman watched and measured the activity in several parts of the woman's brain that would respond to her own pain. Brain activity was recorded as a fraction of the activity observed when the woman herself was zapped with the electrode. The women also completed a psychological test that measures empathy. Will women who are higher in empathy respond more strongly when their partner has a painful experience? Here are data for one brain region:[7]

Subject	1	2	3	4	5	6	7	8
Empathy score	38	53	41	55	56	61	62	48
Brain activity	−0.120	0.392	0.005	0.369	0.016	0.415	0.107	0.506
Subject	9	10	11	12	13	14	15	16
Empathy score	43	47	56	65	19	61	32	105
Brain activity	0.153	0.745	0.255	0.574	0.210	0.722	0.358	0.779

Figure 4.6 is a scatterplot, with empathy score as the explanatory variable x and brain activity as the response variable y. The plot shows a positive association. That is, women who are higher in empathy do indeed react more strongly to their partner's pain. The overall pattern is moderately linear, correlation $r = 0.515$.

The line on the plot is the least-squares regression line of brain activity on empathy score. Its equation is

$$\hat{y} = -0.0578 + 0.00761x$$

For Subject 1, with empathy score 38, we predict

$$\hat{y} = -0.0578 + (0.00761)(38) = 0.231$$

This subject's actual brain activity level was −0.120. The residual is

$$\text{residual} = \text{observed } y - \text{predicted } y$$
$$= -0.120 - 0.231 = -0.351$$

The residual is negative because the data point lies below the regression line. The dashed line segment in Figure 4.6 shows the size of the residual.

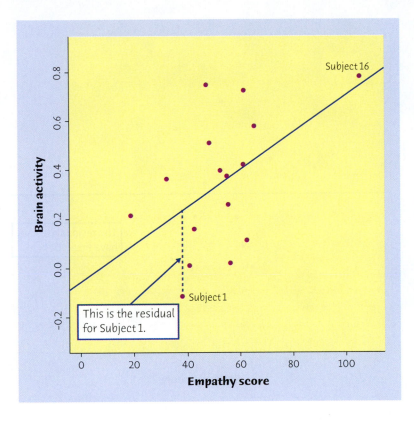

FIGURE 4.6 Scatterplot of activity in a region of the brain that responds to pain versus score on a test of empathy. Brain activity is measured as the subject watches her partner experience pain. The line is the least-squares regression line.

There is a residual for each data point. Finding the residuals is a bit unpleasant because you must first find the predicted response for every x. Software or a graphing calculator gives you the residuals all at once. Here are the 16 residuals for the empathy study data, from software:

```
residuals:
-0.3515 -0.2494 -0.3526 -0.3072 -0.1166 -0.1136  0.1231  0.1721
 0.0463  0.0080  0.0084  0.1983  0.4449  0.1369  0.3154  0.0374
```

Because the residuals show how far the data fall from our regression line, examining the residuals helps assess how well the line describes the data. Although residuals can be calculated from any curve fitted to the data, the residuals from the least-squares line have a special property: **The mean of the least-squares residuals is always zero.**

Compare the scatterplot in Figure 4.6 with the *residual plot* for the same data in Figure 4.7. The horizontal line at zero in Figure 4.7 helps orient us. This "residual = 0" line corresponds to the regression line in Figure 4.6.

RESIDUAL PLOTS

A **residual plot** is a scatterplot of the regression residuals against the explanatory variable. Residual plots help us assess how well a regression line fits the data.

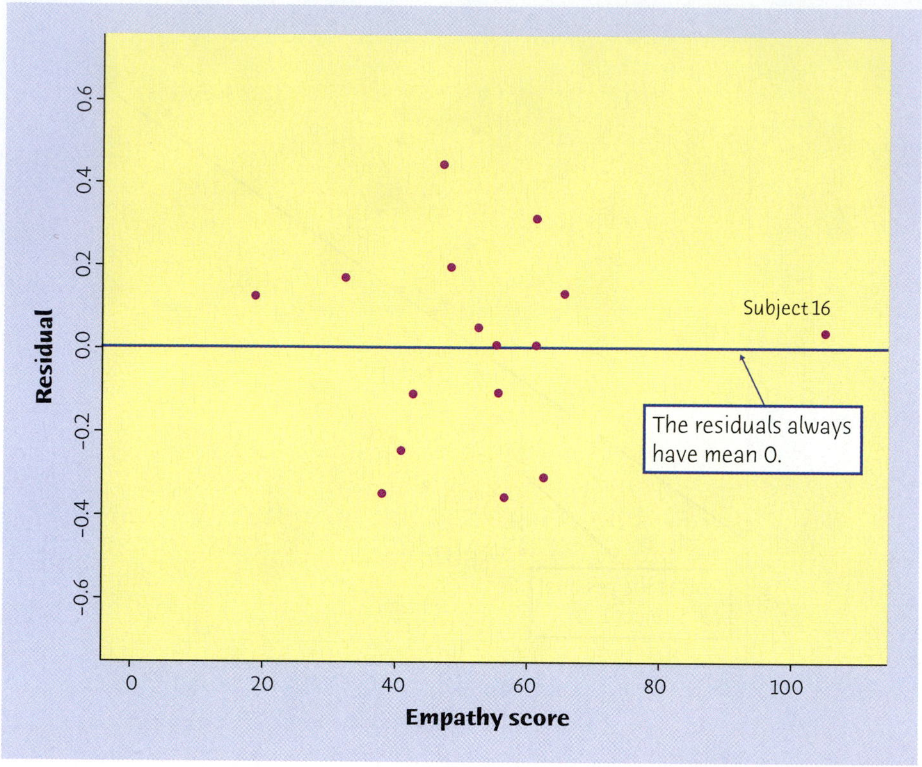

FIGURE 4.7 Residual plot for the data shown in Figure 4.6. The horizontal line at zero residual corresponds to the regression line in Figure 4.6.

A residual plot in effect turns the regression line horizontal. It magnifies the deviations of the points from the line and makes it easier to see unusual observations and patterns.

APPLY YOUR KNOWLEDGE

4.8   An endangered species: the manatee Figure 3.1 (page 69) shows the number of manatee deaths from collisions with powerboats y over 30 years with varying numbers of powerboats registered x (in thousands). The data are displayed in Table 3.1 (page 68). Software tells us that the equation of the least-squares regression line is

$$\hat{y} = -43.7 + 0.1301x$$

Using this line we can add the residuals to the original data:

Thousand powerboats	447	460	481	498	513	512	526	559	585	614
Manatee deaths	13	21	24	16	24	20	15	34	33	33
Residuals	−1.5	4.9	5.1	−5.1	1.0	−2.9	−9.7	5.0	0.6	−3.2
Thousand powerboats	645	675	711	719	681	679	678	696	713	732
Manatee deaths	39	43	50	47	55	38	35	49	42	60
Residuals	−1.2	−1.1	1.2	−2.8	10.1	−6.6	−9.5	2.2	−7.1	8.5
Thousand powerboats	755	809	830	880	944	962	978	983	1010	1024
Manatee deaths	54	66	82	78	81	95	73	69	79	92
Residuals	−0.5	4.4	17.7	7.2	1.9	13.5	−10.5	−15.2	−8.7	2.5

(a) Make a scatterplot of the observations and draw the regression line on your plot.

(b) Verify the value of the first residual, for $x = 447$. Verify that the residuals have sum zero (up to roundoff error).

(c) Make a plot of the residuals against the values of x. Draw a horizontal line at height zero on your plot. The residuals show the same pattern about this line as the data points show about the regression line in the scatterplot in (a).

(d) Would you use the regression line to predict y from x? Explain your answer.

Influential observations

Figures 4.6 and 4.7 show one unusual observation. Subject 16 is an outlier in the x direction, with an empathy score 40 points higher than any other subject. Because of its extreme position on the empathy scale, this point has a strong influence on the correlation. Dropping Subject 16 reduces the correlation from $r = 0.515$ to $r = 0.331$. You can see that this point extends the linear pattern in Figure 4.6 and so increases the correlation. We say that Subject 16 is *influential* for calculating the correlation.

INFLUENTIAL OBSERVATIONS

An observation is **influential** for a statistical calculation if removing it would markedly change the result of the calculation.

Points that are outliers in either the x or y direction of a scatterplot are often influential for the correlation. Points that are outliers in the x direction are often influential for the least-squares regression line.

EXAMPLE 4.7   *An influential observation?*

Subject 16 is influential for the correlation because removing it greatly reduces r. Is this observation also influential for the least-squares line? Figure 4.8 shows that it is not. The regression line calculated without Subject 16 (dashed) differs little from the line that uses all of the observations (solid). The reason that the outlier has little influence on the regression line is that it lies close to the dashed regression line calculated from the other observations.

To see why points that are outliers in the x direction are often influential, let's try an experiment. Pull Subject 16's point in the scatterplot straight down and watch the regression line. Figure 4.9 shows the result. The dashed line is the regression line with the outlier in its new, lower position. Because there are no other points with similar x-values, the line chases the outlier. An outlier in x pulls the least-squares line toward itself. If the outlier does not lie close to the line calculated from the other observations, it will be influential. You can use the *Correlation and Regression* applet to animate Figure 4.9 (see Exercise 4.42).

APPLET

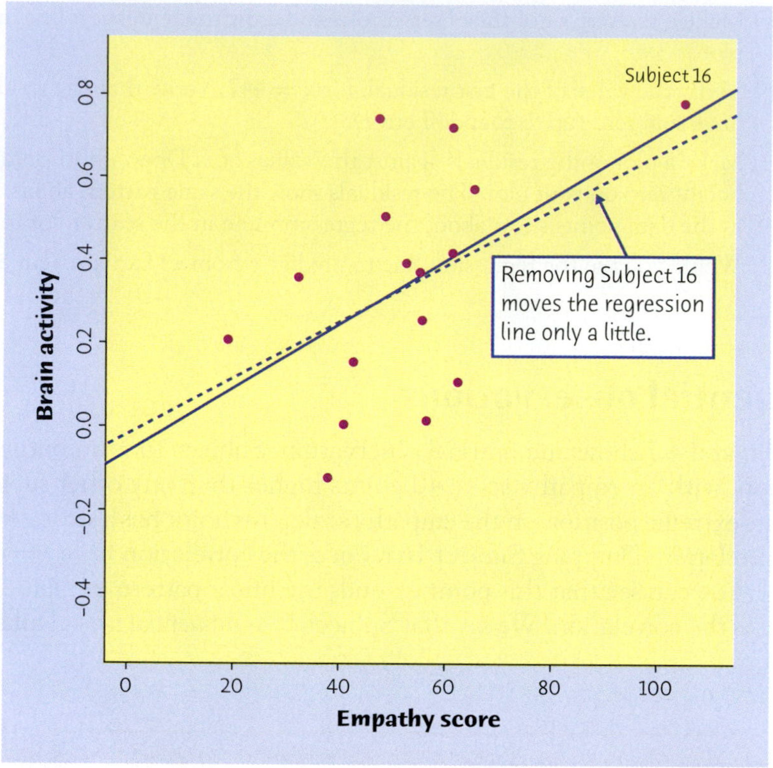

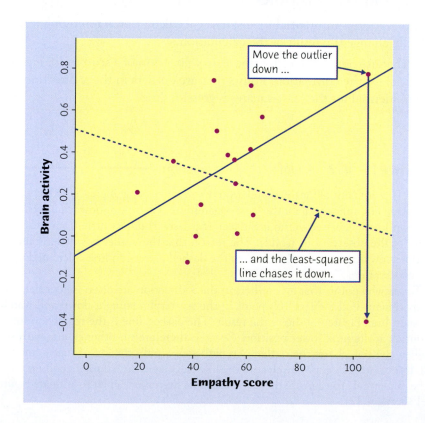

FIGURE 4.8 Subject 16 is an outlier in the *x* direction. The outlier is not influential for least-squares regression, because removing it moves the regression line only a little.

FIGURE 4.9 An outlier in the *x* direction pulls the least-squares line to itself because there are no other observations with similar values of *x* to hold the line in place. When the outlier moves down, the original regression line (solid) chases it down to the dashed line.

We did not need the distinction between outliers and influential observations in Chapter 2. When studying one variable by itself, an outlier that lies far above the other values pulls up the mean $\bar{x}$ toward it and is thus always influential. In the regression setting, however, not all outliers are influential.

APPLY YOUR KNOWLEDGE

4.9   **Bird colonies.** Return to the data of Exercise 4.4 on sparrowhawk colonies. We will use these data to illustrate influence.

(a) Make a scatterplot of the data suitable for predicting new adults from percent of returning adults. Then add two new points. Point A: 10% return, 15 new adults. Point B: 60% return, 28 new adults. In which direction is each new point an outlier?

(b) Add three least-squares regression lines to your plot: for the original 13 colonies, for the original colonies plus Point A, and for the original colonies plus Point B. Which new point is more influential for the regression line? Explain in simple language why each new point moves the line in the way your graph shows.

Cautions about correlation and regression

Correlation and regression are powerful tools for describing the relationship between two variables. When you use these tools, you must be aware of their limitations. You already know:

- *Correlation and regression lines describe only linear relationships.* You can do the calculations for any relationship between two quantitative variables, but the results are useful only if the scatterplot shows a linear pattern.

- *Correlation and least-squares regression lines are not resistant.* Always plot your data and look for observations that may be influential.

Here are two more things to keep in mind when you use correlation and regression.

Beware extrapolation. Suppose that you have data on a child's growth between 3 and 8 years of age. You find a strong linear relationship between age x and height y. If you fit a regression line to these data and use it to predict height at age 25 years, you will predict that the child will be 8 feet tall. Growth slows down and then stops at maturity, so extending the straight line to adult ages is foolish. *Few relationships are linear for all values of x. Don't make predictions far outside the range of x that actually appears in your data.*

EXTRAPOLATION

Extrapolation is the use of a regression line for prediction far outside the range of values of the explanatory variable x that you used to obtain the line. Such predictions are often not accurate.

Beware the lurking variable. Another caution is even more important: *the relationship between two variables can often be understood only by taking other variables into account. Lurking variables* can make a correlation or regression misleading.

> **LURKING VARIABLE**
>
> A **lurking variable** is a variable that is not among the explanatory or response variables in a study and yet may influence the interpretation of relationships among those variables.

You should always think about possible lurking variables before you draw conclusions based on correlation or regression.

Inversely, it makes no sense to study the relationship between any two variables just to see if something might come up. For instance, the worldwide AIDS epidemic and Internet use have both exploded over the past two decades, and thus you could expect to find a positive association between both variables. However, this association is purely coincidental and meaningless. Remember that statistics is the analysis of data in context.

EXAMPLE 4.8   *Nature, nurture, and lurking variables*

The Kalamazoo (Michigan) Symphony once advertised a "Mozart for Minors" program with this statement: "Question: Which students scored 51 points higher in verbal skills and 39 points higher in math? Answer: Students who had experience in music."[8]

The advertisement was worded in a way suggesting that children's brains are hardwired to respond to music developmentally. However, we could probably just as well answer "Students who played soccer." Why? Children with prosperous and well-educated parents are more likely than poorer children to have experience with music and also to play soccer. They are also likely to attend good schools, get good health care, and be encouraged to study hard. These advantages lead to high test scores. Family background is a lurking variable that explains, at least in part, why test scores are related to experience with music.

Attempts, such as this one, to tease out the relative influence of large networks of genes and complex environmental factors typically come up against numerous lurking variables that must be carefully evaluated before any conclusion can be formed.

APPLY YOUR KNOWLEDGE

4.10   **Monitoring a child's growth.** Marco's mom has measured his height in centimeters every few months between ages 4.5 and 7.5. The regression line of Marco's height (y) over time (x) has the equation:

$$\hat{y} = 80.46 + 6.25x$$

Use the equation to estimate how tall Marco was at age 6. Will Marco be taller than his mom by age 14 (she is 168 cm tall)? Discuss your answers.

4.11   **Blame the pharmacists?** Across American cities, the number of pharmacists and the number of deaths in a given year are positively associated. Should we limit the number of pharmacists allowed to work in our cities? What might influence the number of deaths in American cities? What might influence the number of pharmacists in American cities? What can you conclude from the positive association between pharmacists and deaths?

Association does not imply causation

Thinking about lurking variables leads to the most important caution about correlation and regression. When we study the relationship between two variables, we often hope to show that changes in the explanatory variable *cause* changes in the response variable. *A strong association between two variables is not enough to draw conclusions about cause and effect.* Sometimes an observed association really does reflect cause and effect. A household that heats with natural gas uses more gas in colder months because cold weather requires burning more gas to stay warm. In other cases, an association is explained by lurking variables, and the conclusion that x causes y is either wrong or not proved.

CAUTION

EXAMPLE 4.9   Does having more cars make you live longer?

A serious study once found that people with two cars live longer than people who own only one car.[9] Owning three cars is even better, and so on. There is a substantial positive correlation between number of cars x and length of life y.

The basic meaning of causation is that by changing x we can bring about a change in y. Could we lengthen our lives by buying more cars? No. The study used number of cars as a quick indicator of affluence. Well-off people tend to have more cars. They also tend to live longer, probably because they are better educated, take better care of themselves, and get better medical care. The cars have nothing to do with it. There is no cause-and-effect tie between number of cars and length of life.

Correlations such as that in Example 4.9 are sometimes called "nonsense correlations." The correlation is real. What is nonsense is the conclusion that changing one of the variables causes changes in the other. A lurking variable—such as personal affluence in Example 4.9—that influences both x and y can create a high correlation even though there is no direct connection between x and y.

ASSOCIATION DOES NOT IMPLY CAUSATION

An association between an explanatory variable x and a response variable y, even if it is very strong, is not by itself good evidence that changes in x actually cause changes in y.

EXAMPLE 4.10   Obesity in mothers and daughters

Obese parents tend to have obese children. But is obesity determined genetically? The results of a study of Mexican American girls aged 9 to 12 years are typical. The investigators measured body mass index (BMI), a measure of weight relative to height, for both the girls and their mothers. People with high BMI are overweight or obese. The correlation between the BMI of daughters and the BMI of their mothers was $r = 0.506$.[10]

Cereals to lower your cholesterol?

Many cereal packages and health foods claim that they may lower your cholesterol. Cheerios went a step further and sponsored a randomized, double-blind experiment, concluding from the study that "Cheerios is the only leading cold cereal clinically proven to help lower cholesterol in a low fat diet."

Body type is in part determined by heredity. Daughters inherit half their genes from their mothers. There is therefore the potential for a direct causal link between the BMI of mothers and daughters. But perhaps mothers who are overweight also set an example of little exercise, poor eating habits, and lots of television. Their daughters may pick up these habits, so the influence of heredity is mixed up with influences from the girls' environment. Both likely contribute to the mother-daughter correlation.

The lesson of Example 4.10 is more subtle than just "association does not imply causation." *Even when direct causation is present, it may not be the whole explanation for a correlation.* You must still worry about lurking variables. Careful statistical studies try to anticipate lurking variables and measure them. The mother-daughter study did measure TV viewing, exercise, and diet. Elaborate statistical analysis can remove the effects of these variables to come closer to the direct effect of mother's BMI on daughter's BMI. This remains a second-best approach to causation. The best way to get good evidence that *x* causes *y* is to do an **experiment** in which we change *x* and keep lurking variables under control. We will discuss experiments in Chapter 8.

experiment

When experiments cannot be done, explaining an observed association can be difficult and controversial. Many of the sharpest disputes in which statistics plays a role involve questions of causation that cannot be settled by experiment. Does using cell phones cause brain tumors? Are our extreme levels of carbon dioxide emissions causing global warming? Is human activity triggering a new, planet-wide mass extinction? All of these questions have become public issues. All concern associations among variables. And all have this in common: They try to pinpoint cause and effect in a setting involving complex relations among many interacting variables.

EXAMPLE 4.11 *Does smoking cause lung cancer?*

Despite the difficulties, it is sometimes possible to build a strong case for causation in the absence of experiments. The evidence that smoking causes lung cancer is about as strong as nonexperimental evidence can be.

Doctors had long observed that most lung cancer patients were smokers. Comparison of smokers and "similar" nonsmokers showed a very strong association between smoking and death from lung cancer. Could the association be explained by lurking variables? Might there be, for example, a genetic factor that predisposes people both to nicotine addiction and to lung cancer? Smoking and lung cancer would then be positively associated even if smoking had no direct effect on the lungs. How were these objections overcome?

James Leynse/CORBIS

Let's answer this question in general terms: What are the criteria for establishing causation when we cannot do an experiment?

* *The association is strong.* The association between smoking and lung cancer is very strong.

* *The association is consistent.* Many studies of different kinds of people in many countries link smoking to lung cancer. That reduces the chance that a lurking variable specific to one group or one study explains the association.

- *Higher doses are associated with stronger responses.* People who smoke more cigarettes per day or who smoke over a longer period get lung cancer more often. People who stop smoking reduce their risk.

- *The alleged cause precedes the effect in time.* Lung cancer develops after years of smoking. The number of men dying of lung cancer rose as smoking became more common, with a lag of about 30 years. Lung cancer kills more men than any other form of cancer. Lung cancer was rare among women until women began to smoke. Lung cancer in women rose along with smoking, again with a lag of about 30 years, and has now passed breast cancer as the leading cause of cancer death among women.

- *The alleged cause is plausible.* Experiments with animals show that tars from cigarette smoke do cause cancer.

Medical authorities do not hesitate to say that smoking causes lung cancer. The U.S. surgeon general has long stated that cigarette smoking is "the largest avoidable cause of death and disability in the United States."[11] The evidence for causation is overwhelming—but it is not as strong as the evidence provided by well-designed experiments.

APPLY YOUR KNOWLEDGE

4.12   The health and wealth of nations. There is a positive relationship between the wealth of nations (measured in dollars of income per capita adjusted for purchasing power) and the life expectancy at birth (measured in years) in these nations.[12] In part, this association reflects causation—wealth enables health through better nutrition and increased access to medical infrastructures. Suggest other explanations for the association between the wealth and health of nations. (Ask yourself how people's health can influence the economy.)

4.13   Predicting the risk of breast cancer. A large study of the detailed lifestyle of California Seventh-Day Adventists found that a woman's level of education is one of the better predictors of her risk of getting breast cancer later in life, with higher levels of education associated with higher risks of breast cancer.[13]

(a) Do you think that we should ask women to drop out of school in order to reduce the incidence of breast cancer? Is education a plausible candidate to be a direct cause of breast cancer? Explain why.

(b) Another variable associated with the risk of breast cancer is the age at which a woman has her first child, with later ages associated with higher cancer rates.[14] Is the age at first child a plausible cause of breast cancer? Explain why. How does knowing about the association between age at first child and breast cancer help us understand the association between education and breast cancer?

CHAPTER 4 SUMMARY

A **regression line** is a straight line that describes how a response variable *y* changes as an explanatory variable *x* changes. You can use a regression line to **predict** the value of *y* for any value of *x* by substituting this *x* into the equation of the line.

Cigarette warning labels

The history of cigarette warning labels illustrates how challenging establishing causation from evidence of association can be. Since the first surgeon general's report on smoking and health, there have been four successive warnings on packages sold in the United States: **1965**—"Caution: Cigarette Smoking May Be Hazardous to Your Health." **1967**—"Warning: Cigarette Smoking Is Dangerous to Health and May Cause Death from Cancer and Other Diseases." **1969**—"Warning: The Surgeon General Has Determined That Cigarette Smoking Is Dangerous to Your Health." **1984**—"SURGEON GENERAL'S WARNING: Smoking Causes Lung Cancer, Heart Disease, Emphysema, and May Complicate Pregnancy." The later is still in effect, along with three other warnings.

The **slope** b of a regression line $\hat{y} = a + bx$ is the rate at which the predicted response $\hat{y}$ changes along the line as the explanatory variable x changes. Specifically, b is the change in $\hat{y}$ when x increases by 1.

The **intercept** a of a regression line $\hat{y} = a + bx$ is the predicted response $\hat{y}$ when the explanatory variable $x = 0$. This prediction is of no statistical interest unless x can actually take values near 0.

The most common method of fitting a line to a scatterplot is least squares. The **least-squares regression line** is the straight line $\hat{y} = a + bx$ that minimizes the sum of the squares of the vertical distances of the observed points from the line.

The least-squares regression line of y on x is the line with slope $r s_y/s_x$ and intercept $a = \overline{y} - b\overline{x}$. This line always passes through the point $(\overline{x}, \overline{y})$.

Correlation and regression are closely connected. The correlation r is the slope of the least-squares regression line when we measure both x and y in standardized units. The **square of the correlation** r^2 is the fraction of the variation in one variable that is explained by least-squares regression on the other variable.

Correlation and regression must be **interpreted with caution. Plot the data** to be sure the relationship is roughly linear and to detect outliers and influential observations. A plot of the **residuals** makes these effects easier to see.

Look for **influential observations,** individual points that substantially change the correlation or the regression line. Outliers in the x direction are often influential for the regression line.

Avoid **extrapolation,** the use of a regression line for prediction for values of the explanatory variable far outside the range of the data from which the line was calculated.

Lurking variables may explain the relationship between the explanatory and response variables. Correlation and regression can be misleading if you ignore important lurking variables.

Most of all, be careful not to conclude that there is a cause-and-effect relationship between two variables just because they are strongly associated. **High correlation does not imply causation.** The best evidence that an association is due to causation comes from an **experiment** in which the explanatory variable is directly changed and other influences on the response are controlled.

CHECK YOUR SKILLS

4.14 Figure 4.10 is a scatterplot of tree frog mating call frequency (in notes per second) against outside temperature (in Celsius) for 18 American bird-voiced tree frogs.[15] The line is the least-squares regression line for predicting call frequency from temperature. If another frog of this species initiates a mating call when the outside temperature is 24 degrees Celsius, you predict the call frequency (in notes/s) to be close to

(a) 5.   (b) 7.   (c) 9.

4.15 The slope of the line in Figure 4.10 is closest to (in notes per second per degree Celsius)

(a) −0.5.   (b) 0.   (c) 0.5.

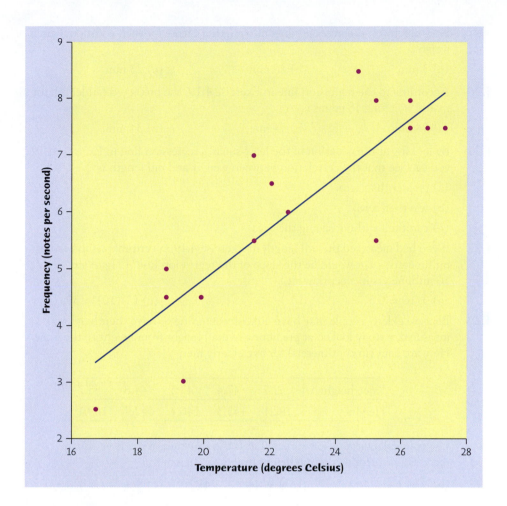

FIGURE 4.10 Outside temperature and frequency of mating calls for 16 tree frogs, for Exercise 4.14.

4.16 In the frog mating call data of Figure 4.10, temperature explains about 75% of the variability in call frequency. The correlation between call frequency and temperature is close to

(a) 0.5.   (b) 0.75.   (c) 0.85.

4.17 The metabolism of birds is very demanding and unusually cold winters are often accompanied by a reduction in the population sizes of many bird species. You want to use a regression line to predict bird mortality from the number of frost days in winter. Your best guess for the slope of this line is that it will be

(a) positive.   (b) negative.   (c) close to zero.

4.18 The points on a scatterplot lie close to the line whose equation is $y = 4 - 3x$. The slope of this line is

(a) 4.   (b) 3.   (c) −3.

4.19 For a class project, you measure the weight in grams and the tail length in millimeters (mm) of a group of mice. The equation of the least-squares line for predicting tail length from weight is

$$\text{predicted tail length} = 20 + 3 \times \text{weight}$$

How much (on the average) does tail length increase for each additional gram of weight?

(a) 3 mm   (b) 20 mm   (c) 23 mm

4.20   According to the regression line in Exercise 4.19, the predicted tail length for a mouse weighing 18 grams is

(a) 74 mm.   (b) 54 mm.   (c) 34 mm.

4.21   By looking at the equation of the least-squares regression line in Exercise 4.19, you can see that the correlation between weight and tail length is

(a) greater than zero.

(b) less than zero.

(c) can't tell without seeing the data.

4.22   If you had measured the tail length in Exercise 4.19 in centimeters instead of millimeters, what would be the slope of the regression line? (There are 10 millimeters in a centimeter.)

(a) $3/10 = 0.3$ (b) 3   (c) $(3)(10) = 30$

4.23   Because elderly people may have difficulty standing to have their heights measured, a study looked at predicting overall height from height to the knee. Here are data (in centimeters) for five elderly men:

Knee height x	57.7	47.4	43.5	44.8	55.2
Height y	192.1	153.3	146.4	162.7	169.1

Use your calculator or software: What is the equation of the least-squares regression line for predicting height from knee height?

(a) $\hat{y} = 2.4 + 44.1x$ (b) $\hat{y} = 44.1 + 2.4x$ (c) $\hat{y} = -2.5 + 0.32x$

CHAPTER 4 EXERCISES

Paul A. Souders/CORBIS

4.24   **Penguins diving.** A study of king penguins looked for a relationship between how deep the penguins dive to seek food and how long they stay underwater.[16] For all but the shallowest dives, there is a linear relationship that is different for different penguins. The study report gives a scatterplot for one penguin titled "The relation of dive duration (DD) to depth (D)." Duration DD is measured in minutes and depth D is in meters. The report then says, "The regression equation for this bird is: DD = 2.69 + 0.0138D."

(a) What is the slope of the regression line? Explain in specific language what this slope says about this penguin's dives.

(b) According to the regression line, how long does a typical dive to a depth of 200 meters last?

(c) The dives varied from 40 meters to 300 meters in depth. Plot the regression line from $x = 40$ to $x = 300$.

4.25   **Measuring water quality.** Biochemical oxygen demand (BOD) measures organic pollutants in water by measuring the amount of oxygen consumed by microorganisms that break down these compounds. BOD is hard to measure accurately. Total organic carbon (TOC) is easy to measure, so it is common to measure TOC and use regression to predict BOD. A typical regression equation for water entering a municipal treatment plant is[17]

$$BOD = -55.43 + 1.507\ TOC$$

Both BOD and TOC are measured in milligrams per liter of water.

(a) What does the slope of this line say about the relationship between BOD and TOC?

(b) What is the predicted BOD when TOC = 0? Values of BOD less than 0 are impossible. Why do you think the prediction gives an impossible value?

4.26 **Smoking in the United States.** The number of adult Americans who smoke has declined over the past few decades. Here are estimates from the Centers for Disease Control and Prevention for the percent of U.S. adults over 18 years of age who were smokers in any given year between 1965 and 2002:[18]

Year	1965	1974	1979	1983	1987	1990	1993	1997	2000	2002
Percent smokers	41.9	37.0	33.3	31.9	28.6	25.3	24.8	24.6	23.1	22.4

(a) Make a scatterplot of these data and find the least-squares regression line of percent adult smokers on year.

(b) According to the regression line, how much did the percent of adult smokers decline each year on the average during this period? What percent of the observed variation in percent adult smokers is accounted for by linear change over time?

(c) Use the regression equation to predict the percent of adult smokers in the United States by the middle of this century, 2050. Is this result reasonable? Why?

4.27 **Milk or soda?** Exercise 3.26 (page 84) provides the U.S. consumption per capita of milk and carbonated soft drinks (in gallons per year) between 1980 and 2000.

(a) Find the equation for the regression line (use soda for x and milk for y) and calculate r^2.

(b) What does the slope of the regression line tell you about the link between soda consumption and milk consumption (note that both variables are expressed in the same units)? What does r^2 tell you?

(c) The presence of soda vending machines in schools has been blamed for increased soda consumption. If removing soda machines from schools could bring soda consumption down to 37 gallons per year per person, what milk consumption per capita would you expect? And what if soda consumption could drop down to 25 gallons per person per year?

4.28 **Keeping water clean.** Keeping water supplies clean requires regular measurement of levels of pollutants. The measurements are indirect—a typical analysis involves forming a dye by a chemical reaction with the dissolved pollutant, then passing light through the solution and measuring its "absorbance." To calibrate such measurements, the laboratory measures known standard solutions and uses regression to relate absorbance and pollutant concentration. This is usually done every day. Here is one series of data on the absorbance for different levels of nitrates. Nitrates are measured in milligrams per liter of water.[19]

Nitrates	50	50	100	200	400	800	1200	1600	2000	2000
Absorbance	7.0	7.5	12.8	24.0	47.0	93.0	138.0	183.0	230.0	226.0

(a) Chemical theory says that these data should lie on a straight line. If the correlation is not at least 0.997, something went wrong and the calibration procedure is repeated. Plot the data and find the correlation. Must the calibration be done again?

(b) The calibration process sets the nitrate level and measures absorbance. Once established, the linear relationship will be used to estimate the nitrate level in water from a measurement of absorbance. What is the equation of the line used for estimation? What is the estimated nitrate level in a water specimen with absorbance of 40?

(c) Do you expect estimates of nitrate levels from absorbance to be quite accurate? Why?

4.29 **Counting carnivores.** Figure 3.2 (page 71) shows the relationship between carnivore body mass in kilograms (kg) and their abundance per 10,000 kg of prey, for 25 species of carnivores, as described in Example 3.5. The scatterplot uses logarithm scales. The relationship between log of body mass and log of abundance is linear and strong, with $r^2 = 0.83$. The least-squares regression gives

$$\log(\text{abundance}) = 1.95 - [1.05 \times \log(\text{body mass})]$$

Predict the abundance of a carnivore species whose individuals weigh about 1 kg. Make sure to transform the units both ways (remember that $y = 10^{\log(y)}$).

4.30 **The intriguing cerebellum.** The cerebellum is a highly convoluted brain structure sitting underneath the cerebral hemispheres. This intriguing structure is thought to facilitate the acquisition and use of sensory data by the rest of the brain, particularly the motor areas. Studies suggest that the cerebellum may scale up with the size of animals' bodies and brains, whereas other parts of the brain are clearly represented differentially across species. Here are data on 15 mammal species showing the weights in grams of their bodies and cerebellums, along with the logarithm transformed values:[20]

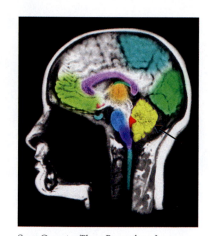

Scott Camazine/Photo Researchers, Inc.

Species	Body	Cerebellum	Log(body)	Log(cerebellum)
Mouse	58	0.09	1.76	−1.05
Bat	30	0.09	1.48	−1.05
Flying fox	130	0.30	2.11	−0.52
Pigeon	500	0.40	2.70	−0.40
Guinea pig	485	0.90	2.69	−0.05
Squirrel	350	1.50	2.54	0.18
Chinchilla	500	1.70	2.70	0.23
Rabbit	1,800	1.90	3.26	0.28
Hare	3,000	2.30	3.48	0.36
Cat	3,500	5.30	3.54	0.72
Dog	3,500	6.00	3.54	0.78
Macaque	6,000	7.80	3.78	0.89
Sheep	25,000	21.50	4.40	1.33
Bovine	300,000	35.70	5.48	1.55
Human	60,000	142.00	4.78	2.15

(a) Make a scatterplot of cerebellum weight (y) against body weight (x) and another scatterplot using the transformed values. Describe and compare both scatterplots. What did you learn about the relationship between cerebellum and body weight from these graphs?

(b) Find the equation for the regression line using the log-transformed values and add it to the scatterplot. In the logarithm scales, what proportion of the variation in cerebellum weight can be explained by the variation in body weight?

(c) Predict the cerebellum weight in grams of a species weighing 100,000 g (100 kg or about 220 lb). Make sure to transform the grams into log units first and then back into grams (remember that $y = 10^{\log(y)}$).

4.31 Predicting tree height. Measuring tree height is not an easy task. How well does trunk diameter predict tree height? A survey of 958 live trees in an old-growth forest in Canada provides us with the following information: the mean tree height is 15.6 m with standard deviation 13.4 m; the mean diameter, measured at "breast height" (1.3 m aboveground), is 23.4 cm with standard deviation 23.5 cm. The correlation between the height and diameter is very high: $r = 0.96$.[21]

(a) What is the slope of the regression line to predict tree height from trunk diameter? Draw a graph of this regression line.

(b) The tree diameters ranged from 1 to 101 cm. Predict the height of a tree in this area that is 50 cm wide in diameter at breast height. Use r^2 to argue the reliability of this prediction.

4.32 What's in your breakfast cereals? The EESEE data set on breakfast cereals examined the nutritional content of 77 brands.

(a) As expected, sugar content is positively correlated with calories per serving and explains 32% of the variations in calories per serving. What is the numerical value of the correlation between sugar content and calories per serving?

(b) Fiber content explains 9% of the calories per serving in this data set. What is the numerical value of the correlation between fiber content and calories per serving? What information are you missing to be certain of the value of the correlation?

4.33 Does social rejection hurt? Exercise 3.38 (page 91) gives data from a study that shows that social exclusion causes "real pain." That is, activity in an area of the brain that responds to physical pain goes up as distress from social exclusion goes up. A scatterplot shows a moderately strong linear relationship. Figure 4.11 shows regression output from software for these data.

(a) What is the equation of the least-squares regression line for predicting brain activity from social distress score? Use the equation to predict brain activity for social distress score 2.0.

(b) What percent of the variation in brain activity among these subjects is explained by the straight-line relationship with social distress score?

4.34 Merlins breeding. Exercise 3.36 (page 89) gives data on the number of breeding pairs of merlins in an isolated area in each of nine years and the percent of males who returned the next year. The data show that the percent returning is lower after successful breeding seasons and that the relationship is roughly linear. Figure 4.12 shows software regression output for these data.

Russell Burden/Index Stock Imagery/PictureQuest

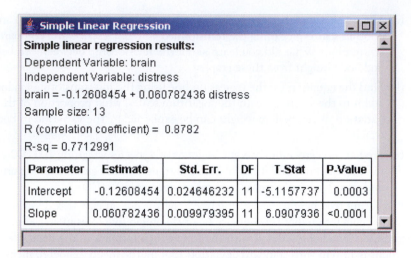

FIGURE 4.11 CrunchIt! regression output for a study of the effects of social rejection on brain activity, for Exercise 4.33.

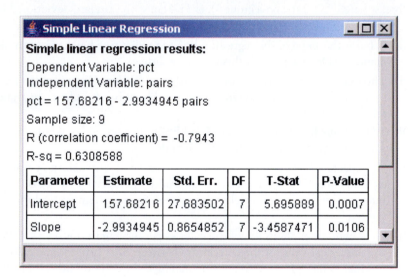

FIGURE 4.12 CrunchIt! regression output for a study of how breeding success affects survival in birds, for Exercise 4.34.

(a) What is the equation of the least-squares regression line for predicting the percent of males that return from the number of breeding pairs? Use the equation to predict the percent of returning males after a season with 30 breeding pairs.

(b) What percent of the year-to-year variation in percent of returning males is explained by the straight-line relationship with number of breeding pairs the previous year?

4.35   **Always plot your data!** Table 4.1 presents four sets of data prepared by the statistician Frank Anscombe to illustrate the dangers of calculating without first plotting the data.[22]

(a) Without making scatterplots, find the correlation and the least-squares regression line for all four data sets. What do you notice? Use the regression line to predict y for $x = 10$.

(b) Make a scatterplot for each of the data sets and, add the regression line to each plot.

TABLE 4.1   Four data sets for exploring correlation and regression

Data Set A

x	10	8	13	9	11	14	6	4	12	7	5
y	8.04	6.95	7.58	8.81	8.33	9.96	7.24	4.26	10.84	4.82	5.68

Data Set B

x	10	8	13	9	11	14	6	4	12	7	5
y	9.14	8.14	8.74	8.77	9.26	8.10	6.13	3.10	9.13	7.26	4.74

Data Set C

x	10	8	13	9	11	14	6	4	12	7	5
y	7.46	6.77	12.74	7.11	7.81	8.84	6.08	5.39	8.15	6.42	5.73

Data Set D

x	8	8	8	8	8	8	8	8	8	8	19
y	6.58	5.76	7.71	8.84	8.47	7.04	5.25	5.56	7.91	6.89	12.50

(c) In which of the four cases would you be willing to use the regression line to describe the dependence of y on x? Explain your answer in each case.

4.36 **Toxicology of aspartame.** The National Toxicology Program evaluates the toxicology of substances suspected of carcinogenicity (causing cancer). One such study evaluated the impact of aspartame added to the diet on the survival rates of mice. For each aspartame concentration studied (in parts per million, ppm), 30 mice were fed and monitored for 40 weeks and the group's survival rate recorded as the percent mice still alive at the end of the 40 weeks. Here are the results of this study:[23]

Concentration (ppm)	0	3125	6250	12,500	25,000	50,000
Survival rate (%)	67	73	57	70	73	60

(a) Make a scatterplot of mice survival rate (response) against aspartame concentration (explanatory) and add the regression line. Describe the direction, form, and strength of the relationship. What percent of the variation in mice survival rate can be explained by variations in aspartame concentrations?

(b) Make another scatterplot, this time ignoring the last data point (an aspartame concentration of 50,000 ppm). Add the regression line. What happened to the regression line? What does this say about the data point corresponding to the aspartame concentration of 50,000 ppm?

(c) Make another scatterplot, this time with all the data points except the one for an aspartame concentration of 6250 ppm. Add the regression line. What happened to the regression line? What does this say about the data point you just removed?

4.37 **Drilling into the past.** Drilling down beneath a lake in Alaska yields chemical evidence of past changes in climate. Biological silicon, left by the skeletons of single-celled creatures called diatoms, is a measure of the abundance of life in the lake. A rather complex variable based on the ratio of certain isotopes relative to ocean water gives an indirect measure of moisture, mostly from snow. As we drill down, we look further into the past. Here are data from 2300 to 12,000 years ago:[24]

Isotopes (%)	Silicon (mg/g)	Isotopes (%)	Silicon (mg/g)	Isotopes (%)	Silicon (mg/g)
−19.90	97	−20.71	154	−21.63	224
−19.84	106	−20.80	265	−21.63	237
−19.46	118	−20.86	267	−21.19	188
−20.20	141	−21.28	296	−19.37	337

(a) Make a scatterplot of silicon (response) against isotopes (explanatory). Ignoring the outlier, describe the direction, form, and strength of the relationship. The researchers say that this and relationships among other variables they measured are evidence for cyclic changes in climate that are linked to changes in the sun's activity.

(b) The researchers single out one point: "The open circle in the plot is an outlier that was excluded in the correlation analysis." Circle this outlier on your graph. What is the correlation with and without this point? The point strongly influences the correlation. Explain why the outlier moves r in the direction revealed by your calculations.

4.38 **Managing diabetes.** People with diabetes must manage their blood sugar levels carefully. They measure their fasting plasma glucose (FPG) several times a day with a glucose meter. Another measurement, made at regular medical checkups, is called HbA. This is roughly the percent of red blood cells that have a glucose molecule attached. It measures average exposure to glucose over a period of several months. Table 4.2 gives data on both HbA and FPG for 18 diabetics five months after they had completed a diabetes education class.[25]

(a) Make a scatterplot with HbA as the explanatory variable. There is a positive linear relationship, but it is surprisingly weak.

(b) Subject 15 is an outlier in the y direction. Subject 18 is an outlier in the x direction. Find the correlation for all 18 subjects, for all except Subject 15,

TABLE 4.2 Two measures of glucose level in diabetics

Subject	HbA (%)	FPG (mg/ml)	Subject	HbA (%)	FPG (mg/ml)	Subject	HbA (%)	FPG (mg/ml)
1	6.1	141	7	7.5	96	13	10.6	103
2	6.3	158	8	7.7	78	14	10.7	172
3	6.4	112	9	7.9	148	15	10.7	359
4	6.8	153	10	8.7	172	16	11.2	145
5	7.0	134	11	9.4	200	17	13.7	147
6	7.1	95	12	10.4	271	18	19.3	255

and for all except Subject 18. Are either or both of these subjects influential for the correlation? Explain in simple language why *r* changes in opposite directions when we remove each of these points.

4.39   **Drilling into the past, continued.** Is the outlier in Exercise 4.37 also strongly influential for the regression line? Calculate and draw on your graph two regression lines, and discuss what you see. Explain why adding the outlier moves the regression line in the direction shown on your graph.

4.40   **Managing diabetes, continued.** Add three regression lines for predicting FPG from HbA to your scatterplot from Exercise 4.38: for all 18 subjects, for all except Subject 15, and for all except Subject 18. Is either Subject 15 or Subject 18 strongly influential for the least-squares line? Explain in simple language what features of the scatterplot explain the degree of influence.

4.41   **Climate change and ecosystems.** Many studies have documented a clear pattern of climate change toward warmer temperatures globally. What has been the subject of controversies, however, is the role of human activity and the impact of climate change on ecosystems. A survey of the North Sea over a 25-year period (1977 to 2001) shows that, as sea winter temperatures have warmed up, fish populations have moved farther north. Here are the data for mean winter bottom temperature (in degrees Celsius) and mean northern latitude (in degrees) of anglerfish populations each year over the 25-year study period:[26]

Temperature	Latitude	Temperature	Latitude	Temperature	Latitude
6.26	57.20	6.52	57.72	7.09	58.07
6.26	57.96	6.68	57.83	7.13	58.49
6.27	57.65	6.76	57.87	7.15	58.28
6.31	57.59	6.78	57.48	7.29	58.49
6.34	58.01	6.89	58.13	7.34	58.01
6.32	59.06	6.90	58.52	7.57	58.57
6.37	56.85	6.93	58.48	7.65	58.90
6.39	56.87	6.98	57.89		
6.42	57.43	7.02	58.71		

(a) Make a scatterplot of the relationship between latitude of anglerfish distribution (*y*) and temperature (*x*). Describe the shape and strength of that relationship.

(b) Give the equation for the regression line to predict latitude from temperature and the value of the correlation. Add the regression line to your scatterplot.

(c) The scatterplot shows one outlier of the relationship. Take that point out of the data set and calculate the new regression line and correlation. Add the new regression line to the original scatterplot, making sure to use a different color or line thickness.

(d) How do the regression line and correlation omitting the outlier compare with the original ones in (b)? Use your findings to discuss how influential the outlier is.

4.42   **Influence in regression.** The *Correlation and Regression* applet allows you to animate Figure 4.9. Click to create a group of 10 points in the lower-left corner of the scatterplot with a strong straight-line pattern (correlation about 0.9). Click the "Show least-squares line" box to display the regression line.

APPLET

(a) Add one point at the upper right that is far from the other 10 points but exactly on the regression line. Why does this outlier have no effect on the line even though it changes the correlation?

(b) Now use the mouse to drag this last point straight down. You see that one end of the least-squares line chases this single point, while the other end remains near the middle of the original group of 10. What makes the last point so influential?

4.43 **How residuals behave.** Return to the merlin data of Exercise 3.36 (page 89). Figure 4.12 shows basic regression output.

(a) Use the regression equation from the output to obtain the residuals step-by-step. That is, find the predicted percent $\hat{y}$ of returning males for each number of breeding pairs, then find the residuals $y - \hat{y}$.

(b) The residuals are the part of the response left over after the straight-line tie to the explanatory variable is removed. Find the correlation between the residuals and the explanatory variable. Your result should not be a surprise.

4.44 **Using residuals.** Make a residual plot (residual against explanatory variable) for the merlin regression of the previous exercise. Use a y scale from -20 to 20 or wider to better see the pattern. Add a horizontal line at $y = 0$, the mean of the residuals.

(a) Describe the pattern if we ignore the two years with $x = 38$. Do the $x = 38$ years fit this pattern?

(b) Return to the original data. Make a scatterplot with two least-squares lines: with all nine years and without the two $x = 38$ years. Although the original regression in Figure 4.12 seemed satisfactory, the two $x = 38$ years are influential. We would like more data for years with x greater than 33.

4.45 **Do artificial sweeteners cause weight gain?** People who use artificial sweeteners in place of sugar tend to be heavier than people who use sugar. Does this mean that artificial sweeteners cause weight gain? Give a more plausible explanation for this association.

4.46 **How's your self-esteem?** People who do well tend to feel good about themselves. Perhaps helping people feel good about themselves will help them do better in school and life. Raising self-esteem became for a time a goal in many schools. California even created a state commission to advance the cause, hoping that it would help reduce crime, drug use, and teen pregnancy (the commission was ended in 1995). Can you think of explanations for the association between high self-esteem and good school performance other than "Self-esteem causes better work in school"?

4.47 **Are big hospitals bad for you?** A study shows that there is a positive correlation between the size of a hospital (measured by its number of beds x) and the median number of days y that patients remain in the hospital. Does this mean that you can shorten a hospital stay by choosing a small hospital? Why?

4.48 **Smoking and lung cancer.** One way to assess the link between smoking and lung cancer is to ask whether groups of individuals who tend to smoke more than the population average also experience higher rates of death from lung cancer than the general population. The data set below was obtained for men in 25 occupational groups in England (for example, farmers, sales workers). For each group, smoking extent and mortality from lung cancer are represented in the form of an index. An index value of 100 corresponds exactly to the national average, while index values above 100 represent group values above the national average.[27]

Smoking	77	137	117	94	116	102	111	93	88	102	91	104	107
Mortality	84	116	123	128	155	101	118	113	104	88	104	129	86
Smoking	112	113	110	125	133	115	105	87	91	100	76	66	
Mortality	96	144	139	113	146	128	115	79	85	120	60	51	

Analyze these data to uncover the nature and strength of the effect of smoking on mortality from lung cancer. Follow the four-step process (page 55) in reporting your work.

4.49 **Insect reproduction.** Reproduction is very taxing on an individual's physiology. What is the impact of better nutrition on reproduction? Insects are an easy model with which to answer this question because they produce large numbers of eggs and reach maturity very fast. Female fruit flies were kept in closed glass vials and fed a diet containing varying amounts (milligrams per vial) of yeast, a good source of proteins for fruit flies. The average number of eggs produced per female was then computed for each vial. Here are the data:[28]

Yeast	0	0	0	0	0	1	1	1	1	1
Eggs	9.85	6.25	10.05	5.2	7.15	21.35	25.95	21.1	14.94	20.6
Yeast	3	3	3	3	3	7	7	7	7	7
Eggs	36.75	58.05	67.05	42.85	49.2	63.8	86.45	82.6	73.35	85.5

Analyze these data to see if they support the notion that better nutrition (here, more yeast) improves the reproductive capacity of fruit flies. Follow the four-step process (page 55) in reporting your work.

4.50 **Beavers and beetles.** Ecologists sometimes find rather strange relationships in our environment. For example, do beavers benefit beetles? Researchers laid out 23 circular plots, each 4 meters in diameter, in an area where beavers were cutting down cottonwood trees. In each plot, they counted the number of stumps from trees cut by beavers and the number of clusters of beetle larvae. Ecologists think that the new sprouts from stumps are more tender than other cottonwood growth, so that beetles prefer them. If so, more stumps should produce more beetle larvae. Here are the data:[29]

Stumps	2	2	1	3	3	4	3	1	2	5	1	3
Beetle larvae	10	30	12	24	36	40	43	11	27	56	18	40
Stumps	2	1	2	2	1	1	4	1	2	1	4	
Beetle larvae	25	8	21	14	16	6	54	9	13	14	50	

Analyze these data to see if they support the "beavers benefit beetles" idea. Follow the four-step process (page 55) in reporting your work.

4.51   **Is regression useful?** In Exercise 3.39 (page 91) you used the *Correlation and Regression* applet to create three scatterplots having correlation about $r = 0.7$ between the horizontal variable x and the vertical variable y. Create three similar scatterplots again, and click the "Show least-squares line" box to display the regression lines. Correlation $r = 0.7$ is considered reasonably strong in many areas of work. Because there is a reasonably strong correlation, we might use a regression line to predict y from x. In which of your three scatterplots does it make sense to use a straight line for prediction?

4.52   **Guessing a regression line.** In the *Correlation and Regression* applet, click on the scatterplot to create a group of 15 to 20 points from lower left to upper right with a clear positive straight-line pattern (correlation around 0.7). Click the "Draw line" button and use the mouse (right-click and drag) to draw a line through the middle of the cloud of points from lower left to upper right. Note the "thermometer" above the plot. The red portion is the sum of the squared vertical distances from the points in the plot to the least-squares line. The green portion is the "extra" sum of squares for your line—it shows by how much your line misses the smallest possible sum of squares.

(a) You drew a line by eye through the middle of the pattern. Yet the right-hand part of the bar is probably almost entirely green. What does that tell you?

(b) Now click the "Show least-squares line" box. Is the slope of the least-squares line smaller (the new line is less steep) or larger (line is steeper) than that of your line? If you repeat this exercise several times, you will consistently get the same result. The least-squares line minimizes the *vertical* distances of the points from the line. It is *not* the line through the "middle" of the cloud of points. This is one reason why it is hard to draw a good regression line by eye.

Marco Secchi/Alamy

Two-Way Tables*

We have concentrated on relationships in which at least the response variable is quantitative. Now we will describe relationships between two categorical variables. Some variables—such as sex, species, and color—are categorical by nature. Other categorical variables are created by grouping values of a quantitative variable into classes—like age groups, for example. Published data often appear in grouped form to save space. To analyze categorical data, we use the *counts* or *percents* of individuals that fall into various categories.

EXAMPLE 5.1 Benefits of cranberry juice

Table 5.1 presents the results of a study of the recurrence of urinary tract infections (UTIs) in women just cured from a UTI with antibiotics and then taking one of three

---

*This material is important in statistics, but it is needed later in this book only for the section on conditional probabilities and the chapter on the chi-square test. You may omit it if you do not plan to cover these topics or delay reading it until you reach them.

TABLE 5.1   Urinary tract infection (UTI) recurrence by treatment

| Treatment | Outcome | | Total |
	Recurrence	No recurrence	
Cranberry juice	8	42	50
Lactobacillus drink	19	30	49
Neither drink	18	32	50
Total	45	104	149

two-way table

row and column variables

preventive treatments over a six-month period (cranberry juice daily, lactobacillus drink daily, or neither drink).[1] Both variables, treatment and outcome, are grouped into categories. (Outcome is categorical here because 1 or more UTI recurrences over the six-month period make up a single category, "Recurrence.") This is a **two-way table** because it describes the relationship between these two categorical variables. Treatment is the **row variable** because each row in the table describes women given one of three treatments. Outcome is the **column variable** because each column describes one possible outcome as classified in the study. The entries in the table are the counts of study participants in each treatment-by-outcome class.

Marginal distributions

How can we best grasp the information contained in Table 5.1? First, *look at the distribution of each variable separately.* The distribution of a categorical variable says how often each outcome occurred. The "Total" column at the right of the table contains the totals for each of the rows. These row totals give the distribution of treatment (the row variable) among women: 50 women drank cranberry juice daily, 49 women took a lactobacillus drink daily, and 50 abstained from both cranberry juice and lactobacillus beverages over the six months of the study (the researchers who designed this study had decided on 50 women for each group but one woman dropped out of the lactobacillus group for personal reasons). In the same way, the "Total" row at the bottom of the table gives the distribution of outcomes. The bottom row reveals that nearly a third of the women in the study had a recurrence of urinary tract infection during the six-month period.

If the row and column totals are missing, the first thing to do in studying a two-way table is to calculate them. The distributions of treatment alone and outcome alone are called **marginal distributions** because they appear at the right and bottom margins of the two-way table.

marginal distribution

Percents are often more informative than counts. We can display the marginal distribution of outcomes in terms of percents by dividing each column total by the table total and converting to a percent.

EXAMPLE 5.2 *Calculating a marginal distribution*

The percent of study participants who had no UTI recurrence is

$$\frac{\text{total No recurrence}}{\text{table total}} = \frac{104}{149} = 0.698 = 69.8\%$$

The percent of study participants who had a recurrence is

$$\frac{\text{total Recurrence}}{\text{table total}} = \frac{45}{149} = 0.302 = 30.2\%$$

The total is 100% because every woman in the study belongs to one, and only one, of the two outcome categories.

Each marginal distribution from a two-way table is a distribution for a single categorical variable. As we saw in Chapter 1, we can use a bar graph or a pie chart to display such a distribution. Figure 5.1 is a bar graph of the distribution of outcomes for the study participants.

In working with two-way tables, you must calculate lots of percents. Here's a tip to help decide what fraction gives the percent you want. Ask, "What group represents the total that I want a percent of?" The count for that group is the denominator of the fraction that leads to the percent. In Example 5.2, we want a percent of "study participants," so the count of study participants (the table total) is the denominator.

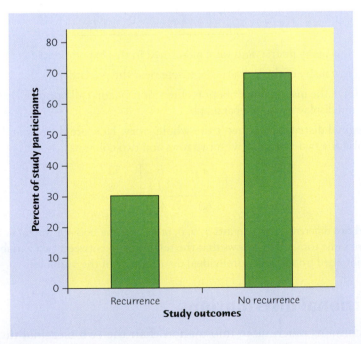

FIGURE 5.1 A bar graph of the distribution of outcomes for the UTI study. This is one of the marginal distributions for Table 5.1.

APPLY YOUR KNOWLEDGE

5.1   **Risks of playing soccer.** A study in Sweden compared former elite soccer players, people who had played soccer but not at the elite level, and people of the same age who did not play soccer. Here is a two-way table that classifies these subjects by whether or not they had arthritis of the hip or knee by their midfifties:[2]

	Elite	Non-elite	Did not play
Arthritis	10	9	24
No arthritis	61	206	548

(a) How many people do these data describe?

(b) How many of these people have arthritis of the hip or knee?

(c) Give the marginal distribution of participation in soccer, both as counts and as percents.

Jeff Gynane/Alamy

5.2   **Altruism in prairie dogs.** Prairie dogs are social rodents with sophisticated calls capable of identifying various kinds of predators to alert their kin. The warning calls are reminiscent of dogs barking, hence the species' name. Raising the alarm when a predator is spotted is the group's only defense mechanism but it places the caller at a higher risk of predation. A study examined whether proximity to the predator (a stuffed badger controlled remotely by an experimenter) influences the likelihood of a prairie dog raising the alarm. The table below shows the count of prairie dogs raising the alarm or not when they are either near the predator or far from it:[3]

	Alarm call	No call
Near	29	80
Far	55	78

(a) How many prairie dogs were monitored in this experiment?

(b) How many of these prairie dogs were near the predator?

(c) Give the marginal distribution of reaction (alarm call or no call) as percents and display them in a bar graph.

5.3   **Marginal distributions aren't the whole story.** Here are the row and column totals for a two-way table with two rows and two columns:

a	b	50
c	d	50
60	40	100

Find *two different* sets of counts a, b, c, and d for the body of the table that give these same totals. This shows that the relationship between two variables cannot be obtained from the two individual distributions of the variables.

Conditional distributions

Table 5.1 contains much more information than the two marginal distributions of treatment alone and outcome alone. The nature of the relationship between treatment and outcome in the study cannot be deduced from the separate distributions but requires the full table. *Relationships between categorical variables are described by calculating appropriate percents from the counts given.* We use percents because counts are often hard to compare. For example, there are 50 women in the cranberry group but only 49 in the lactobacillus group. This is only a minute difference but a two-way table can have widely different group sizes, in which case comparing counts across groups would obviously be misleading. When we compare the percents of UTI recurrence and of no recurrence in several treatment groups, we are comparing **conditional distributions.**

conditional distribution

MARGINAL AND CONDITIONAL DISTRIBUTIONS

The **marginal distribution** of one of the categorical variables in a two-way table of counts is the distribution of values of that variable among all individuals described by the table.

A **conditional distribution** of a variable is the distribution of values of that variable among only individuals who have a given value of the other variable. There is a separate conditional distribution for each value of the other variable.

EXAMPLE 5.3   Conditional distribution of outcome, given a particular treatment

If we know that a woman participating in the study drank cranberry juice daily for six months, we need only look at the "Cranberry juice" row in the two-way table, highlighted in Table 5.2. To find the distribution of outcome among only women in this treatment group, divide each count in the row by the row total, which is 50. The conditional distribution of outcome, *given* that a woman has been taking daily cranberry juice, is

	Recurrence	No recurrence
Percent of cranberry juice group	16.0	84.0

Smiling faces

Women smile more than men. The same data that produce this fact allow us to link smiling to other variables in two-way tables. For example, add as the second variable whether or not the person thinks they are being observed. If yes, that's when women smile more. If no, there's no difference between women and men. Or take the second variable to be the person's occupation or social role. Within each social category, there is very little difference in smiling between women and men.

TABLE 5.2   **Urinary tract infection recurrence by treatment: the daily cranberry juice group**

Treatment	Outcome		Total
	Recurrence	No recurrence	
Cranberry juice	**8**	**42**	**50**
Lactobacillus drink	19	30	49
Neither drink	18	32	50
Total	45	104	149

The two percents add to 100% because all women in the cranberry juice group had either a recurrence or no recurrence. We use the term "conditional" because these percents describe only study participants who satisfy the condition that they drank cranberry juice daily for six months.

EXAMPLE 5.4   Outcome from drinking cranberry juice

Let's follow the four-step process (page 55), starting with a practical medical question.

STATE: Cranberries were used by Native Americans to treat a number of ailments, particularly those related to the urinary tract. Could a daily intake of cranberry juice help prevent the recurrence of urinary tract infections (UTIs) in women who had just been treated with antibiotics for a UTI?

FORMULATE: Calculate and compare the conditional distributions of outcome for participants receiving the various treatments.

SOLVE: Comparing conditional distributions reveals the nature of the association between the outcome and treatment of study participants. Look at each row in Table 5.1 (that is, at each treatment group) in turn. Find the numbers of no recurrence and of recurrence as percents of each row total. Here are the four conditional distributions of outcome for given treatment groups:

	Recurrence	No recurrence
Percent of cranberry juice group	16.0	84.0
Percent of lactobacillus drink group	38.8	61.2
Percent of neither drink group	36.0	64.0

Because the variable "outcome" has just two values, comparing conditional distributions simply amounts to comparing the percents of no recurrence in the three treatment groups. The bar graph in Figure 5.2 compares the percents of no recurrence in the three

Medicinal properties of cranberries

The American cranberry (*Vaccinium macrocarpon*) is a fruit native to North America. It was used medicinally by Native Americans for the treatment of bladder and kidney ailments. Before the advent of antibiotics, the cranberry was still a popular treatment for urinary tract infections. Recent pharmacological studies have uncovered that cranberry compounds prevent bacteria from binding to the membrane of host cells.

Marco Secchi/Alamy

FIGURE 5.2 Bar graph comparing the percent of women with no UTI recurrence in the three treatment groups. In each treatment group, some women experienced no further UTIs, but that percent is highest among women assigned to drink cranberry juice daily.

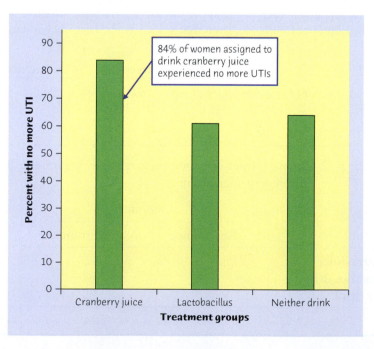

treatment groups. The heights of the bars do not add to 100% because they are not parts of a whole. Each bar describes a different treatment group.

CONCLUDE: The cranberry juice preventive treatment has the highest percent of women remaining free of recurring UTIs over the six-month period. Taking a lactobacillus drink and abstaining from either drink both resulted in a lower percent of women remaining free of UTIs. Native Americans might indeed have discovered an effective medicinal plant in the cranberry.

Remember that there are two sets of conditional distributions for any two-way table. Examples 5.3 and 5.4 looked at the conditional distributions of outcome for different treatments. We could also examine the conditional distributions of treatment for the two outcomes.

EXAMPLE 5.5 *Conditional distribution of treatment, given outcome*

What is the distribution of treatments among the study participants who experienced no UTI recurrence? Information about this outcome appears in the "No recurrence" column. Look only at this column, which is highlighted in Table 5.3. To find the conditional distribution of treatment, divide the count of participants experiencing no recurrence in each treatment group by the column total, which is 104. Here is the distribution:

Percent of participants who took daily		
cranberry juice	lactobacillus drink	neither drink
40.4	28.8	30.8

Looking only at the "Recurrence" column in the two-way table gives the following conditional distribution of treatment:

Percent of participants who took daily		
cranberry juice	lactobacillus drink	neither drink
17.8	42.2	40.0

Each set of percents adds to 100% because each conditional distribution includes all study participants with a given outcome. Comparing these two conditional distributions

---

TABLE 5.3	**Study participants by treatment and outcome: the no recurrence group**		
	Outcome		
Treatment	Recurrence	No recurrence	Total
Cranberry juice	8	42	50
Lactobacillus drink	19	30	49
Neither drink	18	32	50
Total	45	104	149

shows the relationship between treatment and outcome in another form. Participants who remained free of UTI recurrence were more likely to have drunk cranberry juice daily and less likely to have taken a lactobacillus drink daily.

Software will do these calculations for you once you understand how to enter and label the data for your particular program. Most programs allow you to choose which conditional distributions you want to compare. The output in Figure 5.3 compares the three conditional distributions of outcome, given treatment, and also the marginal distribution of outcome for all study participants. The row percents in the first two columns agree with the results in Example 5.4.

No single graph (such as a scatterplot) portrays the form of the relationship between categorical variables. No single numerical measure (such as the correlation) summarizes the strength of the association. Bar graphs are flexible enough to be helpful, but you must think about what comparisons you want to display. For numerical measures, we rely on appropriate percents. You must decide which percents you need. Here is a hint: *If there is an explanatory-response relationship, compare the conditional distributions of the response variable for the separate values of the explanatory variable.* In our UTI example, treatment is clearly explanatory, and the researchers hoped that the two preventive treatments would influence UTI outcome among the study participants. Thus, it makes sense to compare the conditional distributions of outcome among participants following the different treatments, as in Example 5.4.

APPLY YOUR KNOWLEDGE

5.4   **Check your calculations.** Starting with Table 5.1, show the calculations to find the conditional distribution of no recurrence among study participants following different treatments. Your results should agree with the "Percent of" study participants display in Example 5.4.

SPSS

Column Statistics					
Treatment* Outcome Crosstabulation					
			Outcome		
			Recurrence	No recurrence	Total
Treatment	Cranberry juice	Count	8	42	50
		% within treatment	16.0%	84.0%	100.0%
	Lactobacillus drink	Count	19	30	49
		% within treatment	38.8%	61.2%	100.0%
	Neither drink	Count	18	32	50
		% within treatment	36.0%	64.0%	100.0%
Total		Count	45	104	149
		% within treatment	30.2%	69.8%	100.0%

FIGURE 5.3 SPSS output of the two-way table of UTI outcomes for three different treatments, along with each entry as a percent of its row total. The percents in each row give the conditional distribution of UTI outcome for one treatment, and the Total row gives the marginal distribution of UTI outcome for the whole study (all treatments combined).

5.5   **Altruism in prairie dogs.** The two-way table in Exercise 5.2 describes a study of warning call behavior in prairie dogs exposed to different predation risks as defined by the level of proximity to a predator.

(a) Predator proximity is the explanatory variable in this study. Find the two conditional distributions of call behavior, one for the near-predator condition and one for the far-predator condition.

(b) Based on your calculations, describe the differences in call behavior when a predator is near or far with a graph and in words.

5.6   **Risks of playing soccer.** The two-way table in Exercise 5.1 describes a study of arthritis of the hip or knee among people with different levels of experience playing soccer. We suspect that more serious soccer players have more arthritis later in life. Do the data confirm this suspicion? Follow the four-step process, as illustrated in Example 5.4.

Simpson's paradox

As is the case with quantitative variables, the effects of lurking variables can change or even reverse relationships between two categorical variables. Here is an example that demonstrates the surprises that can await the unsuspecting user of data.

EXAMPLE 5.6   Do medical helicopters save lives?

Accident victims are sometimes taken by helicopter from the accident scene to a hospital. Helicopters save time. Do they also save lives? Let's compare the percents of accident victims who die with helicopter evacuation and with the usual transport to a hospital by road. Here are hypothetical (yet realistic) data that illustrate a practical difficulty:[4]

	Helicopter	Road
Victim died	64	260
Victim survived	136	840
Total	200	1100

We see that 32% (64 out of 200) of helicopter patients died, compared with only 24% (260 out of 1100) of the others. That seems discouraging. The explanation is that the helicopter is sent mostly to serious accidents, so that the victims transported by helicopter are more often seriously injured. They are more likely to die with or without helicopter evacuation. Here are the same data broken down by the seriousness of the accident:

Serious Accidents

	Helicopter	Road
Died	48	60
Survived	52	40
Total	100	100

Less Serious Accidents

	Helicopter	Road
Died	16	200
Survived	84	800
Total	100	1000

Inspect these tables to convince yourself that they describe the same 1300 accident victims as the original two-way table. For example, 200 (100 + 100) were moved by helicopter, and 64 (48 + 16) of these died.

Among victims of serious accidents, the helicopter saves 52% (52 out of 100) compared with 40% for road transport. If we look only at less serious accidents, 84% of those transported by helicopter survive, versus 80% of those transported by road. Both groups of victims have a higher survival rate when evacuated by helicopter.

Simpson's paradox

At first, it seems paradoxical that the helicopter does better for both groups of victims but worse when all victims are lumped together. This is called **Simpson's paradox.** Examining the data closely provides the explanation for this paradox. Half the helicopter transport patients are from serious accidents, compared with only 100 of the 1100 road transport patients. So helicopters carry patients who are more likely to die. The seriousness of the accident was a *lurking variable* that, until we uncovered it, made the relationship between survival and mode of transport to a hospital hard to interpret.

SIMPSON'S PARADOX

An association or comparison that holds for all of several groups can reverse direction when the data are combined to form a single group. This reversal is called **Simpson's paradox.**

Do left-handers die early?

Yes, said a study of 1000 deaths in California. Left-handed people died at an average age of 66 years; right-handers, at 75 years. Should left-handed people fear an early death? No—the lurking variable has struck again, resulting in Simpson's paradox. Older people grew up in an era when many natural left-handers were forced to use their right hands. So left-handers are more common among the young. When we look at deaths, the left-handers who die are younger on the average because left-handers in general are younger. Mystery solved.

The lurking variable in Simpson's paradox is categorical. That is, it breaks the individuals into groups, as when accident victims are classified as injured in a "serious accident" or a "less serious accident." Simpson's paradox is just an extreme form of the fact that observed associations can be misleading when there are lurking variables.

Many medical studies examine categorical variables such as "dead/still alive" or "condition improved/not improved." Because pooling data from nonhomogeneous groups can lead to Simpson's paradox, data gathered in different studies are not pooled together even if the studies looked at the same set of variables. Instead, a more complex method called *meta-analysis*, which compares the conclusions reached by each study, has been developed and is often found in the biomedical literature. This method is beyond the scope of our book.

APPLY YOUR KNOWLEDGE

5.7   **Kidney stones.** A study compared the success rates of two different procedures for removing kidney stones: open surgery and percutaneous nephrolithotomy (PCNL), a minimally invasive technique. Here are the number of procedures that were successful or not at getting rid of patients' kidney stones, by type of procedure. A separate table is given for patients with small kidney stones and for patients with large stones:[5]

Small Stones		
	Open surgery	PCNL
Success	81	234
Failure	6	36

Large Stones		
	Open surgery	PCNL
Success	192	55
Failure	71	25

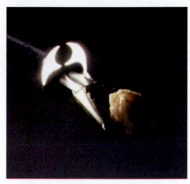

Alexander Tsiaras/Photo Researchers, Inc.

(a) Find the percent of kidney stones, combining the data for small and large stones, that were successfully removed for each of the two medical procedures. Which procedure had the higher overall success rate?

(b) What percent of all small kidney stones were successfully removed? What percent of all large kidney stones were successfully removed? Which type of kidney stones appears to be easier to treat?

(c) Now find the percent of successful procedures of each type for small kidney stones only. Do the same for large kidney stones. PCNL performed worse *for both* small and large kidney stones, yet it did better overall. That sounds impossible. Explain carefully, referring to the data, how this paradox can happen.

5.8   **Smoking and staying alive.** In the mid-1970s, a medical study contacted randomly chosen people in a district in England. Here are data on the 1314 women contacted who were either current smokers or who had never smoked. The tables classify these women by their smoking status and age at the time of the survey and whether they were still alive 20 years later.[6]

Age 18 to 44		
	Smoker	Not
Dead	19	13
Alive	269	327

Age 45 to 64		
	Smoker	Not
Dead	78	52
Alive	167	147

Age 65+		
	Smoker	Not
Dead	42	165
Alive	7	28

(a) Make from these data a two-way table of smoking (yes or no) by dead or alive. What percent of the smokers stayed alive for 20 years? What percent of the nonsmokers survived? It seems surprising that a higher percent of smokers stayed alive.

(b) The age of the women at the time of the study is a lurking variable. Show that, within each of the three age groups in the data, a higher percent of nonsmokers remained alive 20 years later. Explain why this is an example of Simpson's paradox.

(c) The study authors give this explanation: "Few of the older women (over 65 at the original survey) were smokers, but many of them had died by the time of follow-up." Compare the percent of smokers in the three age groups to verify the explanation.

CHAPTER 5 SUMMARY

A **two-way table** of counts organizes data about two categorical variables. Levels of the **row variable** label the rows that run across the table, and levels of the **column variable** label the columns that run down the table. Two-way tables are often used to summarize large amounts of information by grouping outcomes into categories.

The **row totals** and **column totals** in a two-way table give the **marginal distributions** of the two individual variables. It is clearer to present these distributions as percents of the table total. Marginal distributions tell us nothing about the relationship between the variables.

There are two sets of **conditional distributions** for a two-way table: the distributions of the row variable for each fixed value or level of the column variable and the distributions of the column variable for each fixed value or level of the row variable. Comparing one set of conditional distributions is one way to describe the association between the row and the column variables.

To find the **conditional distribution** of the row variable for one specific level of the column variable, look only at that one column in the table. Find each entry in the column as a percent of the column total.

Bar graphs are a flexible means of presenting categorical data. There is no single best way to describe an association between two categorical variables.

A comparison between two variables that holds for each individual value of a third variable can be changed or even reversed when the data for all values of the third variable are combined. This is **Simpson's paradox.** Simpson's paradox is an example of the effect of lurking variables on an observed association.

CHECK YOUR SKILLS

A study was designed to assess the effect of echinacea extracts on the rates of infection of individuals voluntarily exposed to the common-cold rhinovirus. Some subjects took an echinacea extract daily starting 7 days before viral exposure until the end of the study. Another group of subjects received the extract from the time of viral exposure onward only. A last group received just an inactive solution ("placebo group"). Table 5.4 is a two-way table of the diagnosed clinical colds among infected subjects in each treatment group.[7] Exercises 5.9 to 5.17 are based on this table.

5.9   How many infected subjects have been diagnosed with a cold?
(a) 219   (b) 351   (c) need more information

5.10   How many individuals are described by this table?
(a) 219   (b) 351   (c) need more information

5.11   The percent of infected subjects diagnosed with a cold was
(a) about 22%.   (b) about 38%.   (c) about 62%.

5.12   Your percent from the previous exercise is part of

(a) the marginal distribution of outcome (cold diagnosed).

Plantography/Alamy

TABLE 5.4 Echinacea and cold symptoms		
	Outcome	
Treatment	Cold	No cold
7 days before viral exposure and onward	73	59
From time of viral exposure onward	88	43
Placebo (no echinacea)	58	30

(b) the marginal distribution of treatment.

(c) the conditional distribution of outcome, given treatment.

5.13 What percent of the infected subjects in the placebo group have been diagnosed with a cold?
(a) about 58%   (b) about 62%   (c) about 66%

5.14 Your percent from the previous exercise is part of

(a) the marginal distribution of chance of outcome.

(b) the conditional distribution of outcome, given treatment.

(c) the conditional distribution of treatment, given outcome.

5.15 What percent of those diagnosed with a cold were in the placebo group?
(a) about 26%   (b) about 58%   (c) about 66%

5.16 Your percent from the previous exercise is part of

(a) the marginal distribution of chance of outcome.

(b) the conditional distribution of outcome, given treatment.

(c) the conditional distribution of treatment, given outcome.

5.17 A bar graph showing the conditional distribution of treatment, given that the infected subjects have been diagnosed with a cold, has
(a) 2 bars.   (b) 3 bars.   (c) 6 bars.

5.18 To help consumers make informed decisions about health care, the government releases data about patient outcomes in hospitals. A large regional hospital and a small private hospital both serve your community. The regional hospital receives a lot of patients in critical condition because of its state-of-the-art emergency room and diverse set of medical specialties. The private hospital specializes in scheduled surgeries and admits few patients in critical condition. The counts of patients who survived surgery or did not is provided for both hospitals, for all surgeries performed in the previous year. The regional hospital had the higher surgery survival rate (the percent of patients still alive 6 weeks after surgery) for both patients admitted in critical condition and patients coming in for scheduled surgery. Yet the private hospital had the higher overall survival rate when considering both types of surgery patients together. This finding is

(a) not possible: If the regional hospital had higher survival rates for each type of patient separately, then it must also have had a higher overall survival rate when both types of patient are combined.

(b) an example of Simpson's paradox: the regional hospital performed better with each type of patient but it did worse overall because it took in a lot more patients in critical condition, who come in with a lower chance of survival due to their condition.

(c) due to comparing two conditional distributions that should not be compared.

CHAPTER 5 EXERCISES

Sex ratios in geese. *Most species have a similar proportion of males and females born, maximizing the chances of genetic mix from both sexes. However, some disparities have been found in the conditions in which a male or a female is born. A field biologist surveyed the eggs of wild lesser snow geese, from egg production to hatching. Table 5.5 presents data from 29 clutches, each containing 4 eggs numbered 1 through 4 to reflect*

TABLE 5.5	Egg order and gosling sex				
	Order in which egg is laid				
Sex	1	2	3	4	Total
Male	17	16	7	5	45
Female	10	9	17	14	50
Total	27	25	24	19	95

the order in which they were laid. When an egg hatched, the sex of the gosling was assessed. A few eggs in the 29 clutches did not hatch, explaining why the column totals in Table 5.5 are not equal to 29.[8] Exercises 5.19 to 5.22 are based on these data.

5.19 **Marginal distributions.** Give (in percents) the two marginal distributions, for sex and for egg order.

5.20 **Percents.** What percent of male goslings are first eggs? What percent of first eggs are male?

5.21 **Conditional distribution.** Give (in percents) the conditional distribution of egg order among males. Should your percents add to 100% (up to roundoff error)?

5.22 **Sex and egg order.** One way to see the relationship is to look at which goslings were first eggs.

(a) There are 17 male first eggs and only 10 female first eggs. Explain why these counts by themselves don't describe the relationship between sex and egg order.

(b) Find the percent of goslings of each sex among the first eggs. Then find the percent of goslings of each sex among the second eggs, among the third eggs, and among the fourth eggs. What do these percents say about the relationship?

5.23 **May I have some chocolate?** Some people say that eating chocolate gives them a headache. A study looked at the role of chocolate as a headache trigger in 128 women suffering from chronic headache. Half the women ate a chocolate preparation and the other half ate the same preparation made with carob. Carob is an extract from the pods of the carob tree that tastes somewhat like sweetened cocoa but contains no caffeine or other psychoactive substances. The women did not know which preparation they were given and the two preparations were not discernible by taste. The table below presents the number of women in each test group who did or did not develop a headache after eating the preparation:[9]

FoodPix/Picturequest

	Outcome		
Preparation	No headache	Headache	Total
Chocolate	53	11	64
Carob	38	26	64
Total	91	37	128

(a) Find the marginal distribution of outcome. What does this distribution tell us?

(b) How does the outcome of women in the study differ depending on what preparation they ate? Use conditional distributions as a basis for your answer. What can you conclude about the role of chocolate in triggering headaches in women with chronic headaches?

5.24   **Helping cocaine addicts.** Cocaine addiction is hard to break. Addicts need cocaine to feel any pleasure, so perhaps giving them an antidepressant drug will help. An experiment assigned 72 chronic cocaine users to take either an antidepressant drug called desipramine, lithium, or a placebo. (Lithium is a standard drug to treat cocaine addiction. A placebo is a dummy drug, used so that the effect of being in the study but not taking any drug can be seen.) One-third of the subjects, chosen at random, received each drug. Here are the results after three years:[10]

	Desipramine	Lithium	Placebo
Relapse	10	18	20
No relapse	14	6	4
Total	24	24	24

(a) Compare the effectiveness of the three treatments in preventing relapse. Use percents and draw a bar graph.

(b) Do you think that this study gives good evidence that desipramine actually *causes* a reduction in relapses?

5.25   **Fast food for baby.** The Gerber company sponsored a large survey of the eating habits of American infants and toddlers in 2003. Among the many questions parents were asked was whether their child had eaten fried potatoes on one given day. Here are the data broken down by children's age range:[11]

	Ate fried potatoes	Did not
9–11 months	61	618
15–18 months	62	246
19–24 months	82	234

(a) Calculate the conditional distribution of children who ate fried potatoes for each age range. Briefly describe your findings. (The study also found that fried potatoes were actually the most common cooked vegetable in the diet of the 19- to 24-month-old toddlers, way ahead of green beans and peas!)

(b) Do you think that the association between age and diet found by this study is evidence that age actually *causes* a change in diet?

5.26   **Deaths.** Here is a two-way table of number of deaths in the United States in three age groups from selected causes in 2003. The entries are counts of deaths.[12] Because many deaths are due to other causes, the entries don't add to the "Total deaths" count. The total deaths in the three age groups are very different, so it is important to use percents rather than counts in comparing the age groups.

	15–24 years	25–44 years	45–64 years
Accidents	14,966	27,844	23,669
AIDS	171	6,879	5,917
Cancer	1,628	19,041	144,936
Heart diseases	1,083	16,283	101,713
Homicide	5,148	7,367	2,756
Suicide	3,921	11,251	10,057
Total deaths	33,022	128,924	437,058

The causes listed include the top three causes of death in each age group. For each age group, give the top three causes and the percent of deaths due to each. Use your results to explain briefly how the leading causes of death change as people get older.

5.27 **Type 1 and Type 2 diabetes.** Type 1 diabetes is usually first diagnosed in children or young adults and is due to the immune system interfering with the pancreas's ability to make insulin. Type 2 diabetes is by far the more common form of diabetes and can develop at any age but is especially common among older individuals; it arises typically in response to an excessively rich diet and limited physical activity, inducing insulin resistance in the body. Records from a diabetes clinic in the United Kingdom show the status (still alive or dead) of long-term diabetic patients as a function of their diabetes type:[13]

Patients 40 or Younger

	Type 1	Type 2
Alive	129	15
Dead	1	0
Total	130	15

Patients Older Than 40

	Type 1	Type 2
Alive	124	311
Dead	104	218
Total	228	529

(a) Compare the survival rates (percents still alive) for Type 1 and for Type 2 diabetes among the younger patients. Do the same for the older patients. What have you learned from these percents?

(b) Combine the data into a single two-way table of patient status ("alive" or "dead") by diabetes type (1 or 2). Now calculate the survival rates for Type 1 and for Type 2 diabetes among all patients together. Which type of diabetes shows the higher survival rate?

(c) This is an example of Simpson's paradox. What is the lurking variable here? Explain in simple language how the paradox can happen.

5.28 **Obesity and health.** Recent studies have shown that earlier reports underestimated the health risks associated with being overweight. The error was due to overlooking lurking variables. In particular, smoking tends both to reduce weight and to lead to earlier death. Illustrate Simpson's paradox by a simplified version of this situation. That is, make up two-way tables of overweight (yes or no) by early death (yes or no) separately for smokers and nonsmokers such that

• overweight smokers and overweight nonsmokers both tend to die earlier than those not overweight,

- but when smokers and nonsmokers are combined into a two-way table of overweight by early death, persons who are not overweight tend to die earlier.

5.29   **Python eggs.** How is the hatching of water python eggs influenced by the temperature of the snake's nest? Researchers assigned newly laid eggs to one of three temperatures: hot, neutral, or cold. Hot duplicates the warmth provided by the mother python. Neutral and cold are cooler, as when the mother is absent. Here are the data on the number of eggs and the number that hatched:[14]

	Cold	Neutral	Hot
Number of eggs	27	56	104
Number hatched	16	38	75

(a) Notice that this is not a two-way table! Explain why and how you could use the data to create a two-way table.

(b) The researchers anticipated that eggs would hatch less well at cooler temperatures. Do the data support that anticipation? Follow the four-step process (page 55) in your answer.

5.30   **Do angry people have more heart disease?** People who get angry easily tend to have more heart disease. That's the conclusion of a study that followed a random sample of 12,986 people from three locations for about four years. All subjects were free of heart disease at the beginning of the study. The subjects took the Spielberger Trait Anger Scale test, which measures how prone a person is to sudden anger. Here are data for the 8474 people in the sample who had normal blood pressure.[15] CHD stands for "coronary heart disease." This includes people who had heart attacks and those who needed medical treatment for heart disease.

	Low anger	Moderate anger	High anger	Total
CHD	53	110	27	190
No CHD	3057	4621	606	8284
Total	3110	4731	633	8474

Do these data support the study's conclusion about the relationship between anger and heart disease? Follow the four-step process (page 55) in your answer.

5.31   **Characteristics of a forest.** Forests are complex, evolving ecosystems. For instance, pioneer tree species can be displaced by successional species better adapted to the changing environment. Ecologists mapped a large Canadian forest plot dominated by Douglas fir with an understory of western hemlock and western red cedar. The two-way table below records all the trees in the plot by species and by life stage. The distinction between live and sapling trees is made for live trees taller or shorter than 1.3 meter respectively.[16]

	Dead	Live	Sapling	Total
Western red cedar	48	214	154	416
Douglas fir	326	324	2	652
Western hemlock	474	420	88	982
Total	848	958	244	2050

What can you tell from these data about the current composition and the evolution of this forest? Which tree species appears to be taking over and becoming dominant? Follow the four-step process (page 55) to guide your answer.

Gallo Images–Anthony Bannister/Getty Images

Exploring Data: Part I Review

Data analysis is the art of describing data using graphs and numerical summaries. The purpose of data analysis is to help us see and understand the most important features of a set of data. Chapter 1 commented on graphs to display distributions: pie charts and bar graphs for categorical variables, histograms and stemplots for quantitative variables. In addition, time plots show how a quantitative variable changes over time. Chapter 2 presented numerical tools for describing the center and spread of the distribution of one variable.

The first STATISTICS IN SUMMARY figure on the next page organizes the big ideas for exploring a quantitative variable. Plot your data, and then describe their center and spread using either the mean and standard deviation or the five-number summary. The question marks at the last stage remind us that the usefulness of numerical summaries depends on what we find when we examine graphs of our data. No short summary does justice to irregular shapes or to data with several distinct clusters.

Chapters 3 and 4 applied the same ideas to relationships between two quantitative variables. The second STATISTICS IN SUMMARY figure retraces the big ideas, with details that fit the new setting. Always begin by making graphs of your data. In the case of a scatterplot, we have learned a numerical summary only for data that show a roughly linear pattern on the scatterplot. The summary is then

147

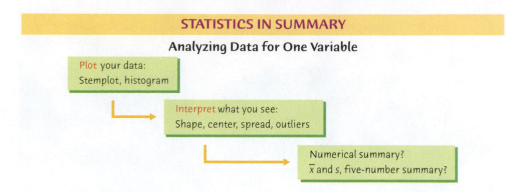

the means and standard deviations of the two variables and their correlation. A regression line drawn on the plot gives a compact description of the overall pattern that we can use for prediction. The question marks at the last two stages remind us that correlation and regression describe only straight-line relationships. Chapter 5 shows how to understand relationships between two categorical variables; comparing well-chosen percents is the key.

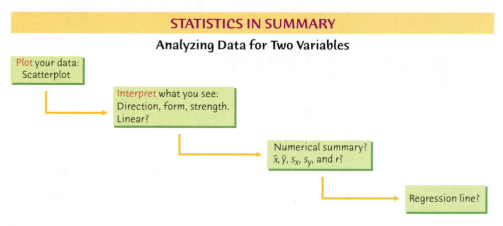

You can organize your work in any open-ended data analysis setting by following the four-step State, Formulate, Solve, and Conclude process first introduced in Chapter 2. After we have mastered the extra background needed for statistical inference, this process will also guide practical work on inference later in the book.

PART I SUMMARY

Here are the most important skills you should have acquired from reading Chapters 1 to 5.

A. DATA

1. Identify the individuals and variables in a set of data.

2. Identify each variable as categorical or quantitative. Identify the units in which each quantitative variable is measured.

3. Identify the explanatory (independent) and response (dependent) variables in situations where one variable explains or influences another.

B. DISPLAYING DISTRIBUTIONS

1. Recognize when a pie chart can and cannot be used.

2. Make a bar graph of the distribution of a categorical variable, or in general to compare related quantities.

3. Interpret pie charts and bar graphs.

4. Make a time plot of a quantitative variable over time. Recognize patterns such as trends and cycles in time plots.

5. Make a histogram of the distribution of a quantitative variable.

6. Make a stemplot of the distribution of a small set of observations. Round leaves or split stems as needed to make an effective stemplot.

C. DESCRIBING DISTRIBUTIONS (QUANTITATIVE VARIABLE)

1. Look for the overall pattern and for major deviations from the pattern.

2. Assess from a histogram or stemplot whether the shape of a distribution is roughly symmetric, distinctly skewed, or neither. Assess whether the distribution has one or more major peaks.

3. Describe the overall pattern by giving numerical measures of center and spread in addition to a verbal description of shape.

4. Decide which measures of center and spread are more appropriate: the mean and standard deviation (especially for symmetric distributions) or the five-number summary (especially for skewed distributions).

5. Recognize outliers and give plausible explanations for them.

D. NUMERICAL SUMMARIES OF DISTRIBUTIONS

1. Find the median M and the quartiles Q_1 and Q_3 for a set of observations.

2. Find the five-number summary and draw a boxplot; assess center, spread, symmetry, and skewness from a boxplot.

3. Find the mean $\bar{x}$ and the standard deviation s for a set of observations.

4. Understand that the median is more resistant than the mean. Recognize that skewness in a distribution moves the mean away from the median toward the long tail.

5. Know the basic properties of the standard deviation: $s \geq 0$ always; $s = 0$ only when all observations are identical and increases as the spread increases; s has the same units as the original measurements; s is pulled strongly up by outliers or skewness.

E. SCATTERPLOTS AND CORRELATION

1. Make a scatterplot to display the relationship between two quantitative variables measured on the same subjects. Place the explanatory variable (if any) on the horizontal scale of the plot.

2. Add a categorical variable to a scatterplot by using a different plotting symbol or color.

3. Describe the direction, form, and strength of the overall pattern of a scatterplot. In particular, recognize positive or negative association and linear (straight-line) patterns. Recognize outliers in a scatterplot.

4. Judge whether it is appropriate to use correlation to describe the relationship between two quantitative variables. Find the correlation r.

5. Know the basic properties of correlation: r measures the direction and strength of only straight-line relationships; r is always a number between -1 and 1; $r = \pm 1$ only for perfect straight-line relationships; r moves away from 0 toward ± 1 as the straight-line relationship gets stronger.

F. REGRESSION LINES

1. Understand that regression requires an explanatory variable and a response variable. Use a calculator or software to find the least-squares regression line of a response variable y on an explanatory variable x from data.

2. Explain what the slope b and the intercept a mean in the equation $\hat{y} = a + bx$ of a regression line.

3. Draw a graph of a regression line when you are given its equation.

4. Use a regression line to predict y for a given x. Recognize extrapolation and be aware of its dangers.

5. Find the slope and intercept of the least-squares regression line from the means and standard deviations of x and y and their correlation.

6. Use r^2, the square of the correlation, to describe how much of the variation in one variable can be accounted for by a straight-line relationship with another variable.

7. Recognize outliers and potentially influential observations from a scatterplot with the regression line drawn on it.

8. Calculate the residuals and plot them against the explanatory variable x. Recognize that a residual plot magnifies the pattern of the scatterplot of y versus x.

G. CAUTIONS ABOUT CORRELATION AND REGRESSION

1. Understand that both r and the least-squares regression line can be strongly influenced by a few extreme observations.

2. Recognize possible lurking variables that may explain the observed association between two variables x and y.

3. Understand that even a strong correlation does not mean that there is a cause-and-effect relationship between x and y.

4. Give plausible explanations for an observed association between two variables: direct cause and effect, the influence of lurking variables, or both.

H. CATEGORICAL DATA (Optional)

1. From a two-way table of counts, find the marginal distributions of both variables by obtaining the row sums and column sums.

2. Express any distribution in percents by dividing the category counts by their total.

3. Describe the relationship between two categorical variables by computing and comparing percents. Often this involves comparing the conditional distributions of one variable for the different categories of the other variable.

4. Recognize Simpson's paradox and be able to explain it.

REVIEW EXERCISES

Review exercises help you solidify the basic ideas and skills in Chapters 1 to 5.

6.1 **Describing colleges.** Popular magazines rank colleges and universities on their "academic quality" in serving undergraduate students. Give one categorical variable and two quantitative variables that you would like to see measured for each college if you were choosing where to study.

6.2 **Growing icicles.** Japanese researchers measured the growth of icicles in a cold chamber under various conditions of temperature, wind, and water flow.[1] Table 6.1 contains data produced under two sets of conditions. In both cases, there was no wind and the temperature was set at $-11°C$. Water flowed over the icicle at a higher rate (29.6 milligrams per second) in Run 8905 and at a slower rate (11.9 mg/s) in Run 8903.

  (a) Make a scatterplot of the length of the icicle in centimeters versus time in minutes, using separate symbols for the two runs.

  (b) What does your plot show about the pattern of growth of icicles? What does it show about the effect of changing the rate of water flow on icicle growth?

TABLE 6.1 **Growth of icicles over time**

| \<Run 8903\> | | | | \<Run 8905\> | | | |
Time (min)	Length (cm)	Time (min)	Length (cm)	Time (min)	Length (cm)	Time (min)	Length (cm)
10	0.6	130	18.1	10	0.3	130	10.4
20	1.8	140	19.9	20	0.6	140	11.0
30	2.9	150	21.0	30	1.0	150	11.9
40	4.0	160	23.4	40	1.3	160	12.7
50	5.0	170	24.7	50	3.2	170	13.9
60	6.1	180	27.8	60	4.0	180	14.6
70	7.9			70	5.3	190	15.8
80	10.1			80	6.0	200	16.2
90	10.9			90	6.9	210	17.9
100	12.7			100	7.8	220	18.8
110	14.4			110	8.3	230	19.9
120	16.6			120	9.6	240	21.1

6.3 **More on growing icicles.** Let's now examine Run 8903 only.

(a) How can you tell from a calculation, without drawing a scatterplot, that the pattern of growth is very close to a straight line?

(b) What is the equation of the least-squares regression line for predicting an icicle's length from time in minutes under these conditions?

(c) Predict the length of an icicle after one full day. This prediction can't be trusted. Why not?

6.4 **Weights are skewed.** The heights of people of the same sex and similar ages have a reasonably symmetric distribution. Weights, on the other hand, do not. The weights of women aged 20 to 29 have mean 141.7 pounds and median 133.2 pounds. The first and third quartiles are 118.3 pounds and 157.3 pounds. What can you say about the shape of the weight distribution? Why?

6.5 **Remember what you ate.** How well do people remember their past diet? Data are available for 91 people who were asked about their diet when they were 18 years old. Researchers asked them at about age 55 to describe their eating habits at age 18. For each subject, the researchers calculated the correlation between actual intakes of many foods at age 18 and the intakes the subjects now remember. The median of the 91 correlations was $r = 0.217$. The authors say, "We conclude that memory of food intake in the distant past is fair to poor."[2] Explain why $r = 0.217$ points to this conclusion.

6.6 **Cicadas as fertilizer?** Every 17 years, swarms of cicadas emerge from the ground in the eastern United States, live for about six weeks, and then die. (There are several "broods," so we experience cicada eruptions more often than every 17 years.) There are so many cicadas that their dead bodies can serve as fertilizer and increase plant growth. In an experiment, a researcher added 10 cicadas under some plants in a natural plot of American bellflowers in a forest, leaving other plants undisturbed. One of the response variables was the size of seeds produced by the plants. Here are data (seed mass in milligrams) for 39 cicada plants and 33 undisturbed (control) plants:[3]

Alastair Shay; Papilio/CORBIS

Cicada plants				Control plants			
0.237	0.277	0.241	0.142	0.212	0.188	0.263	0.253
0.109	0.209	0.238	0.277	0.261	0.265	0.135	0.170
0.261	0.227	0.171	0.235	0.203	0.241	0.257	0.155
0.276	0.234	0.255	0.296	0.215	0.285	0.198	0.266
0.239	0.266	0.296	0.217	0.178	0.244	0.190	0.212
0.238	0.210	0.295	0.193	0.290	0.253	0.249	0.253
0.218	0.263	0.305	0.257	0.268	0.190	0.196	0.220
0.351	0.245	0.226	0.276	0.246	0.145	0.247	0.140
0.317	0.310	0.223	0.229	0.241			
0.192	0.201	0.211					

Do the data support the idea that dead cicadas can serve as fertilizer? Follow the four-step process (page 55) in your work.

6.7 **More about cicadas.** Let's examine the distribution of seed mass for plants in the cicada group in more detail.

(a) Make a stemplot. Is the overall shape roughly symmetric or clearly skewed? There are both low and high observations that we might call outliers.

(b) Find the mean and standard deviation of the seed masses. Then remove both the smallest and the largest mass and find the mean and standard deviation of the remaining 37 seeds. Why does removing these two observations reduce s? Why does it have little effect on $\overline{x}$?

6.8 **Outliers? (optional)** In the previous exercise, you noticed that the smallest and largest observations might be called outliers. Are either of these observations suspected outliers by the $1.5 \times IQR$ rule (page 46)?

6.9 **Where does the water go?** Here are data on the amounts of water withdrawn from natural sources, including rivers, lakes, and wells, in 2000. The units are millions of gallons per day.[4]

Use	Water withdrawn
Public water supplies	43,300
Domestic water supplies	3,590
Irrigation	137,000
Industry	19,780
Power plant cooling	195,500
Fish farming	3,700

Make a bar graph to present these data. For clarity, order the bars by amount of water used. The total water withdrawn is about 408,000 million gallons per day. About how much is withdrawn for uses not mentioned above?

6.10 **A big toe problem.** Hallux abducto valgus (call it HAV) is a deformation of the big toe that is not common in youth and often requires surgery. Doctors used X-rays to measure the angle (in degrees) of deformity in 38 consecutive patients under the age of 21 who came to a medical center for surgery to correct HAV.[5] The angle is a measure of the seriousness of the deformity. The data appear in Table 6.2 as "HAV angle." Make a graph and give a numerical description of this distribution. Are there any outliers? Write a brief discussion of the shape, center,

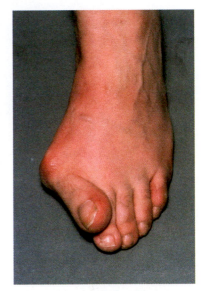

Wellcome Trust Medical Library/Custom Medical Stock Photo

TABLE 6.2	Angle of deformity (degrees) for two types of foot deformity				
HAV angle	MA angle	HAV angle	MA angle	HAV angle	MA angle
28	18	21	15	16	10
32	16	17	16	30	12
25	22	16	10	30	10
34	17	21	7	20	10
38	33	23	11	50	12
26	10	14	15	25	25
25	18	32	12	26	30
18	13	25	16	28	22
30	19	21	16	31	24
26	10	22	18	38	20
28	17	20	10	32	37
13	14	18	15	21	23
20	20	26	16		

and spread of the angle of deformity among young patients needing surgery for this condition.

6.11 **More on a big toe problem.** The HAV angle data in the previous exercise contain one high outlier. Calculate the median, the mean, and the standard deviation for the full data set and also for the 37 observations remaining when you remove the outlier. How strongly does the outlier affect each of these measures?

6.12 **Predicting foot problems.** Metatarsus adductus (call it MA) is a turning in of the front part of the foot that is common in adolescents and usually corrects itself. Table 6.2 gives the severity of MA ("MA angle") as well. Doctors speculate that the severity of MA can help predict the severity of HAV.

(a) Make a scatterplot of the data. (Which is the explanatory variable?)

(b) Describe the form, direction, and strength of the relationship between MA angle and HAV angle. Are there any clear outliers in your graph?

(c) Do you think the data confirm the doctors' speculation? Why or why not?

6.13 **Predicting foot problems, continued.**

(a) Find the equation of the least-squares regression line for predicting HAV angle from MA angle. Add this line to the scatterplot you made in the previous exercise.

(b) A new patient has MA angle 25 degrees. What do you predict this patient's HAV angle to be?

(c) Does knowing MA angle allow doctors to predict HAV angle accurately? Explain your answer from the scatterplot, and then calculate a numerical measure to support your finding.

6.14 **Data on mice.** For a biology project, you measure the tail length (centimeters) and weight (grams) of 12 mice of the same variety. What units of measurement do each of the following have?

(a) The mean length of the tails.

(b) The first quartile of the tail lengths.

(c) The standard deviation of the tail lengths.

(d) The correlation between tail length and weight.

Soap in the shower. *From Rex Boggs in Australia comes an unusual data set: before showering in the morning, he weighed the bar of soap in his shower stall. The weight goes down as the soap is used. The data appear in Table 6.3 (weights in grams). Notice*

TABLE 6.3	Weight (grams) of a bar of soap used to shower				
Day	Weight	Day	Weight	Day	Weight
1	124	8	84	16	27
2	121	9	78	18	16
5	103	10	71	19	12
6	96	12	58	20	8
7	90	13	50	21	6

that Mr. Boggs forgot to weigh the soap on some days. Exercises 6.15 to 6.17 are based on the soap data set.

6.15   Scatterplot. Plot the weight of the bar of soap against day. Is the overall pattern roughly linear? Based on your scatterplot, is the correlation between day and weight close to 1, positive but not close to 1, close to 0, negative but not close to -1, or close to -1? Explain your answer.

6.16   Regression. Find the equation of the least-squares regression line for predicting soap weight from day.

(a) What is the equation? Explain what it tells us about the rate at which the soap lost weight.

(b) Mr. Boggs did not measure the weight of the soap on Day 4. Use the regression equation to predict that weight.

(c) Draw the regression line on your scatterplot from the previous exercise.

6.17   Prediction? Use the regression equation in the previous exercise to predict the weight of the soap after 30 days. Why is it clear that your answer makes no sense? What's wrong with using the regression line to predict weight after 30 days?

SUPPLEMENTARY EXERCISES

Supplementary exercises apply the skills you have learned in ways that require more thought or more elaborate use of technology.

6.18   Change in the Serengeti. Long-term records from the Serengeti National Park in Tanzania show interesting ecological relationships. When wildebeest are more abundant, they graze the grass more heavily, so there are fewer fires and more trees grow. Lions feed more successfully when there are more trees, so the lion population increases. Here are data on one part of this cycle, wildebeest abundance (in thousands of animals) and the percent of the grass area that burned in the same year:[6]

Gallo Images–Anthony Bannister/Getty Images

Wildebeest (1000s)	Percent burned	Wildebeest (1000s)	Percent burned	Wildebeest (1000s)	Percent burned
396	56	360	88	1147	32
476	50	444	88	1173	31
698	25	524	75	1178	24
1049	16	622	60	1253	24
1178	7	600	56	1249	53
1200	5	902	45		
1302	7	1440	21		

To what extent do these data support the claim that more wildebeest reduce the percent of grasslands that burn? How rapidly does the burned area decrease as the number of wildebeest increases? Include a graph and suitable calculations. Follow the four-step process (page 55) in your answer.

6.19   Prey attract predators. Here is one way that nature regulates the size of animal populations: High population density attracts predators, who remove a higher proportion of the population than when the density of the prey is low. One study looked at kelp perch and their common predator, the kelp bass. The researcher set up four large circular pens on the sandy ocean bottom in southern California. He chose young perch at random from a large group and placed 10, 20, 40, and 60 perch in the four pens. Then he dropped the nets protecting the pens, allowing bass to swarm in, and counted the perch left after 2 hours. Here are data on the proportions of perch eaten in four repetitions of this setup:[7]

Perch	Proportion killed			
10	0.0	0.1	0.3	0.3
20	0.2	0.3	0.3	0.6
40	0.075	0.3	0.6	0.725
60	0.517	0.55	0.7	0.817

Do the data support the principle that "more prey attract more predators, who drive down the number of prey"? Follow the four-step process (page 55) in your answer.

6.20   **Extrapolation.** Your work in Exercise 6.18 no doubt included a regression line. Use the equation of this line to illustrate the danger of extrapolation, taking advantage of the fact that the percent of grasslands burned cannot be less than zero.

Falling through the ice. *The Nenana Ice Classic is an annual contest to guess the exact time in the spring thaw when a tripod erected on the frozen Tanana River near Nenana, Alaska, will fall through the ice. The 2006 jackpot prize was $308,000. The contest has been run since 1917. Table 6.4 gives simplified data that record only the date on which the tripod fell each year. The earliest date so far is April 20. To make the data easier to use, the table gives the date each year in days starting with April 20. That*

2006 Bill Watkins/AlaskaStock.com

TABLE 6.4   Days from April 20 for the Tanana River tripod to fall

Year	Day	Year	Day	Year	Day	Year	Day	Year	Day	Year	Day
1917	11	1932	12	1947	14	1962	23	1977	17	1992	25
1918	22	1933	19	1948	24	1963	16	1978	11	1993	4
1919	14	1934	11	1949	25	1964	31	1979	11	1994	10
1920	22	1935	26	1950	17	1965	18	1980	10	1995	7
1921	22	1936	11	1951	11	1966	19	1981	11	1996	16
1922	23	1937	23	1952	23	1967	15	1982	21	1997	11
1923	20	1938	17	1953	10	1968	19	1983	10	1998	1
1924	22	1939	10	1954	17	1969	9	1984	20	1999	10
1925	16	1940	1	1955	20	1970	15	1985	23	2000	12
1926	7	1941	14	1956	12	1971	19	1986	19	2001	19
1927	23	1942	11	1957	16	1972	21	1987	16	2002	18
1928	17	1943	9	1958	10	1973	15	1988	8	2003	10
1929	16	1944	15	1959	19	1974	17	1989	12	2004	5
1930	19	1945	27	1960	13	1975	21	1990	5	2005	9
1931	21	1946	16	1961	16	1976	13	1991	12	2006	13

is, April 20 is 1, April 21 is 2, and so on. You will need software or a graphing calculator to analyze these data in Exercises 6.21 to 6.23.[8]

6.21   **When does the ice break up?** We have 89 years of data on the date of ice breakup on the Tanana River. Describe the distribution of the breakup date with both a graph or graphs and appropriate numerical summaries. What is the median date (month and day) for ice breakup?

6.22   **Global warming?** Because of the high stakes, the falling of the tripod has been carefully observed for many years. If the date the tripod falls has been getting earlier, that may be evidence for the effects of global warming.

(a) Make a time plot of the date the tripod falls against year.

(b) There is a great deal of year-to-year variation. Fitting a regression line to the data may help us see the trend. Fit the least-squares line and add it to your time plot. What do you conclude?

(c) There is much variation about the line. Give a numerical description of how much of the year-to-year variation in ice breakup time is accounted for by the time trend represented by the regression line.

6.23   **More on global warming.** Side-by-side boxplots offer a different look at the data. Group the data into periods of roughly equal length: 1917 to 1939, 1940 to 1959, 1960 to 1979, and 1980 to 2006. Make boxplots to compare ice breakup dates in these four time periods. Write a brief description of what the plots show.

6.24   **Save the eagles.** The pesticide DDT was especially threatening to bald eagles. Here are data on the productivity of the eagle population in northwestern Ontario, Canada.[9] The eagles nest in an area free of DDT but migrate south and eat prey contaminated with the pesticide. DDT was banned at the end of 1972. The researcher observed every nesting area he could reach every year between 1966 and 1981. He measured productivity by the count of young eagles per nesting area.

ImageState/Alamy

Year	Count	Year	Count	Year	Count	Year	Count
1966	1.26	1970	0.54	1974	0.46	1978	0.82
1967	0.73	1971	0.60	1975	0.77	1979	0.98
1968	0.89	1972	0.54	1976	0.86	1980	0.93
1969	0.84	1973	0.78	1977	0.96	1981	1.12

(a) Make a time plot of the data. Does the plot support the claim that banning DDT helped save the eagles?

(b) It appears that the overall pattern might be described by *two* straight lines. Find the least-squares line for 1966 to 1972 (pre-ban) and also the least-squares line for 1975 to 1981 (allowing a few years for DDT to leave the environment after the ban). Draw these lines on your plot. Would you use the second line to predict young per nesting area in the several years after 1981?

6.25   **Thin monkeys, fat monkeys.** Animals and people that take in more energy than they expend will get fatter. Here are data on 12 rhesus monkeys: 6 lean monkeys (4% to 9% body fat) and 6 obese monkeys (13% to 44% body fat). The data report the energy expended in 24 hours (kilojoules per minute) and the lean body mass (kilograms, leaving out fat) for each monkey.[10]

Lean		Obese	
Mass	Energy	Mass	Energy
6.6	1.17	7.9	0.93
7.8	1.02	9.4	1.39
8.9	1.46	10.7	1.19
9.8	1.68	12.2	1.49
9.7	1.06	12.1	1.29
9.3	1.16	10.8	1.31

(a) What is the mean lean body mass of the lean monkeys? Of the obese monkeys? Because animals with higher lean mass usually expend more energy, we can't directly compare energy expended.

(b) Instead, look at how energy expended is related to body mass. Make a scatterplot of energy versus mass, using different plot symbols for lean and obese monkeys. Then add to the plot two regression lines, one for lean monkeys and one for obese monkeys. What do these lines suggest about the monkeys?

6.26   **Weeds among the corn.** Lamb's-quarter is a common weed that interferes with the growth of corn. An agriculture researcher planted corn at the same rate in 16 small plots of ground, then weeded the plots by hand to allow a fixed number of lamb's-quarter plants to grow in each meter of corn row. No other weeds were allowed to grow. Here are the yields of corn (bushels per acre) in each of the plots:[11]

Weeds per meter	Corn yield	Weeds per meter	Corn yield	Weeds per meter	Corn yield	Weeds per meter	Corn yield
0	166.7	1	166.2	3	158.6	9	162.8
0	172.2	1	157.3	3	176.4	9	142.4
0	165.0	1	166.7	3	153.1	9	162.8
0	176.9	1	161.1	3	156.0	9	162.4

(a) What are the explanatory and response variables in this experiment?

(b) Make side-by-side stemplots of the yields, after rounding to the nearest bushel. Give the median yield for each group (using the unrounded data). What do you conclude about the effect of this weed on corn yield?

6.27   **Weeds among the corn, continued.** We can also use regression to analyze the data on weeds and corn yield. The advantage of regression over the side-by-side comparison in the previous exercise is that we can use the fitted line to draw conclusions for counts of weeds other than the ones the researcher actually used.

(a) Make a scatterplot of corn yield against weeds per meter. Find the least-squares regression line and add it to your plot. What does the slope of the fitted line tell us about the effect of lamb's-quarter on corn yield?

(b) Predict the yield for corn grown under these conditions with 6 lamb's-quarter plants per meter of row.

6.28   **Influenza patterns.** According to the U.S. Centers for Disease Control and Prevention (CDC), millions of Americans get the flu every year, with an estimated 36,000 dying from complications. The two main viral types of infectious influenza are Type A and Type B, with Type A producing generally more severe symptoms. The CDC closely monitors reported flu cases and publishes weekly data aggregates from patient samples sent for laboratory analysis. Here are the yearly percents of laboratory-analyzed influenza isolates identified by the CDC as Type A Subtype H1 (H1N1), Type A Subtype H3 (H3N2), or Type B:[12]

Flu season	H1N1	H3N2	B	Flu season	H1N1	H3N2	B
1976–77	0.0	29.0	71.0	1991–92	18.0	82.0	0.0
1977–78	28.0	72.0	0.0	1992–93	3.0	26.0	71.0
1978–79	98.0	0.0	2.0	1993–94	1.0	98.2	0.8
1979–80	2.0	1.0	97.0	1994–95	1.0	77.0	22.0
1980–81	23.0	77.0	0.0	1995–96	54.0	38.0	8.0
1981–82	24.0	1.0	75.0	1996–97	0.0	81.0	19.0
1982–83	10.0	79.0	11.0	1997–98	0.2	99.6	0.2
1983–84	50.0	5.0	45.0	1998–99	0.8	76.2	23.0
1984–85	0.0	97.0	3.0	1999–2000	1.0	98.8	0.2
1985–86	0.0	24.0	76.0	2000–2001	55.7	2.3	42.0
1986–87	99.3	0.2	0.5	2001–2	1.7	85.3	13.0
1987–88	8.0	75.0	17.0	2002–3	43.0	14.0	43.0
1988–89	37.0	5.0	58.0	2003–4	0.1	98.9	1.0
1989–90	1.3	98.3	0.4	2004–5	0.2	75.2	24.6
1990–91	6.0	8.0	86.0	2005–6	6.1	74.8	19.1

(a) Make a time plot for the H1N1 strain. Describe the trend.

(b) Make a time plot for the H3N2 strain. Describe the trend.

(c) Make a time plot for the Type B strain. Describe the trend.

6.29   **Influenza patterns, continued.** Going back to the previous exercise, now plot all three strains on the same time plot. What new pattern emerges? What did you learn with this last plot that was not obvious from the single graphs in the previous exercise?

6.30   **Manatee deaths: trends.** Example 3.3 (page 67) describes a data set from Florida relating manatee deaths from collisions with powerboats and the number of powerboats registered in a given year. The data, available in Table 3.1, appear in a scatterplot in Figure 3.1. That scatterplot shows a clear, strong, positive linear relationship between the number of manatee deaths from collisions with powerboats and the number of powerboats registered in a given year. We also want to study the trends over time for these two variables. Create and interpret the corresponding two time plots. What do you conclude?

EESEE CASE STUDIES ————————————————

The Electronic Encyclopedia of Statistical Examples and Exercises (EESEE) is available on the text CD and Web site. These more elaborate stories, with data, provide settings for longer case studies. Here are some suggestions for EESEE stories that apply the ideas you have learned in Chapters 1 to 5.

6.31   **Is Old Faithful Faithful?** Write a response to Questions 1 and 3 for this case study. (Describing a distribution, scatterplots, and regression.)

6.32   **Checkmating and Reading Skills.** Write a report based on Question 1 in this case study. (Describing a distribution.)

6.33   **Counting Calories.** Respond to Questions 1, 4, and 6 for this case study. (Describing and comparing distributions.)

6.34   **Mercury in Florida's Bass.** Respond to Question 5. (Scatterplots, form of relationships. By the way, "homoscedastic" means that the scatter of points about the overall pattern is roughly the same from one side of the scatterplot to the other.)

6.35   **Brain Size and Intelligence.** Write a response to Question 3. (Scatterplots, correlation, and lurking variables.)

6.36   **Acorn Size and Oak Tree Range.** Write a report based on Questions 1 and 2. (Scatterplots, correlation, and regression.)

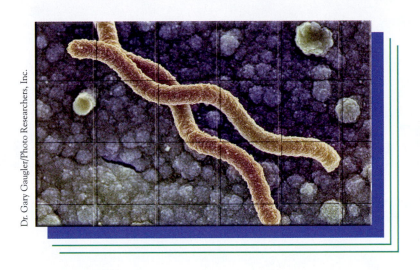

Dr. Gary Gaugler/Photo Researchers, Inc.

PART II

PART II: From Exploration to Inference

The purpose of statistics is to gain understanding from data. We can seek understanding in different ways, depending on the circumstances. We have studied one approach to data, *exploratory data analysis*, in some detail. Now we move from data analysis toward *statistical inference*. Both types of reasoning are essential to effective work with data. Here is a brief sketch of the differences between them.

EXPLORATORY DATA ANALYSIS	STATISTICAL INFERENCE
Purpose is unrestricted exploration of the data, searching for interesting patterns.	Purpose is to answer specific questions, posed before the data were produced.
Conclusions apply only to the individuals and circumstances for which we have data in hand.	Conclusions apply to a larger group of individuals or a broader class of circumstances.
Conclusions are informal, based on what we see in the data.	Conclusions are formal, backed by a statement of our confidence in them.

Our journey toward inference begins in Chapters 7 and 8 with *data production*, statistical ideas for producing data to answer specific questions. Chapters 9, 11, and 13 (and the optional Chapters 10 and 12) concern *probability*, the language of formal statistical conclusions. Finally, Chapters 14 and 15 present the core concepts of *inference*. Chapter 16 reviews the material of Chapters 7 to 15, along with more comprehensive exercises. In practice, data analysis and inference cooperate. Successful inference requires good data production, data analysis to ensure that the data are regular enough, and the language of probability to state conclusions.

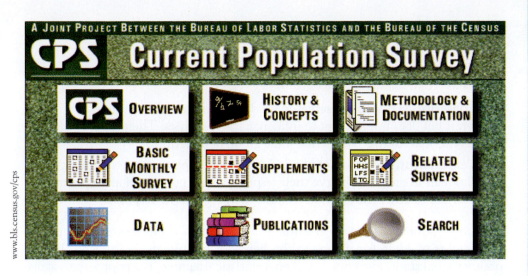

Samples and Observational Studies

Statistics, the science of data, provides ideas and tools that we can use in many settings. Raw data should always be carefully examined for trends and deviations from these trends using the tools of exploratory data analysis. But how do we *produce data* that can help us answer interesting scientific questions?

Suppose our question is "What percent of adult Americans use complementary and alternative medicine (CAM)?" To answer the question, we ask individuals aged 20 and older whether they have used any form of CAM in the past 12 months. We can't realistically ask every adult in America, so we put the question to a *sample* chosen to represent the entire adult *population*. How shall we choose such a sample? In this chapter we examine how to choose appropriate samples for observational studies.

Observation versus experiment

We collect data in order to understand phenomena or answer scientific questions. There are two distinct settings for collecting data. One setting is the *observational study*, in which the subject of our scientific inquiry is disturbed as little as possible by the act of gathering information. Observational studies can answer questions such as "What is the range of gestation times in the prairie dog?" or "What is the proportion of Americans who are overweight?" The other setting is that of the *experiment*. In doing an experiment, we actively impose some *treatment* in order to observe the *response*. Experiments can answer questions such as "Does aspirin reduce the chance of a heart attack?" or "Does ginkgo biloba enhance memory?" Experiments and observational studies provide useful data only when properly designed. The distinction between experiments and observational studies is one of the most important ideas in statistics.

OBSERVATION VERSUS EXPERIMENT

An **observational study** observes individuals and measures variables of interest but does not attempt to influence the responses. The purpose of an observational study is to describe some group or situation.

An **experiment,** on the other hand, deliberately imposes some treatment on individuals in order to observe their responses. The purpose of an experiment is to study whether the treatment causes a change in the response.

Observational studies are essential sources of data about topics ranging from the heights of adult Americans to the behavior of animals in the wild. But an observational study, even one based on a statistical sample, is a poor way to gauge the effect of an intervention. To see the response to a change, we must actually impose the change. When our goal is to understand cause and effect, experiments are the only source of fully convincing data.

EXAMPLE 7.1 The ups and downs of hormone replacement

Should women take hormones such as estrogen after menopause, when natural production of these hormones ends? In 1992, several major medical organizations said "Yes." In particular, women who took hormones seemed to reduce their risk of a heart attack by 35% to 50%. The risks of taking hormones appeared small compared with the benefits.

The evidence in favor of hormone replacement came from a number of observational studies that compared women who were taking hormones on their own accord with others who were not. But women who elect to take hormones are likely to be quite different from women who do not: They may be better informed and see doctors more often, and they may do many other things to maintain their health. So, is it surprising that they have fewer heart attacks?

Experiments don't let women decide what to do. The Women's Health Initiative (WHI) trial, sponsored by the National Institutes of Health, assigned women to either hormone replacement or dummy pills that look and taste the same as the hormone pills. The assignment is done by a coin toss, so that all kinds of women are equally likely to get either treatment. By 2002, the WHI trial published its first results, which indicated that women who took hormones had a *higher* incidence of cardiovascular disease and breast cancer. Taking hormones after menopause quickly fell out of favor. This first WHI study, however, had focused on older women, with an average age of 63 years. A follow-up WHI trial of women in their 50s was published in 2007, showing that younger women taking hormone therapy had lower levels of calcium deposits in their arteries, which may lower their risk of heart disease.[1]

When we simply observe women, the effects of actually taking hormones are *confounded* with (mixed up with) the characteristics of women who choose to take hormones.

CONFOUNDING

Two variables (explanatory variables or lurking variables) are **confounded** when their effects on a response variable cannot be distinguished from each other.

Observational studies of the effect of one variable on another often fail because the explanatory variable is confounded with lurking variables. We will see in the next chapter that well-designed experiments take steps to defeat confounding.

EXAMPLE 7.2   Wine, beer, or spirits?

Moderate use of alcohol is associated with better health. Observational studies suggest that drinking wine rather than beer or spirits confers added health benefits. But people who prefer wine are different from those who drink mainly beer or stronger stuff. Moderate wine drinkers as a group are richer and better educated. They eat more fruit and vegetables and less fried food. Their diets contain less fat and less cholesterol. They are less likely to smoke. The explanatory variable (What type of alcoholic beverage do you drink most often?) is confounded with many lurking variables (education, wealth, diet, and so on). A large study therefore concludes: "The apparent health benefits of wine compared with other alcoholic beverages, as described by others, may be a result of confounding by dietary habits and other lifestyle factors."[2] Figure 7.1 shows the confounding in picture form.

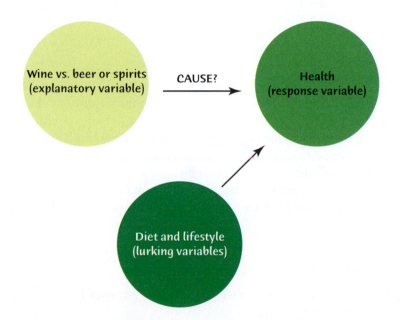

FIGURE 7.1 Confounding, for example 7.2: We can't distinguish the effects of what people drink from the effects of their overall diet and lifestyle.

APPLY YOUR KNOWLEDGE

7.1   **Brain asymmetry.** A study aims to evaluate the percent of adults whose right cerebral hemisphere is larger than their left hemisphere. A group of individuals aged 20 and older representative of the American adult population is asked to come to a brain-imaging facility to have their brains scanned. Each individual's scan is then analyzed to identify whether the right or left cerebral hemisphere is larger. Is this an observational study or an experiment? Explain your answer.

7.2   **Cell phones and brain cancer.** A study of cell phones and the risk of brain cancer looked at a group of 469 people who had brain cancer. The investigators matched each cancer patient with a person of the same sex, age, and race who did

AB/Getty Images

not have brain cancer, and then asked about use of cell phones. Result: "Our data suggest that use of handheld cellular telephones is not associated with risk of brain cancer."[3] Is this an observational study or an experiment? Why? What are the explanatory and response variables?

7.3   **Global warming.** Instrumental records show an increase in global mean surface air and ocean temperatures on the order of 1°F (0.5°C) over the 20th century. This trend is also evident in the reduced extent of snow cover, the accelerated rate of rise of sea level, and the increasingly earlier arrival and breeding times of migratory birds. This global warming coincides with dramatic changes in atmospheric gas composition due to human activity such as the burning of fossil fuels. In 2001 the Committee on the Science of Climate Change stated: "Because of the large and still uncertain level of natural variability inherent in the climate record … , a causal linkage between the buildup of greenhouse gases in the atmosphere and the observed climate changes during the 20th century cannot be unequivocally established."[4]

(a) Is the evidence for global warming experimental or observational?

(b) What explanatory variable is thought to influence climate change? What lurking variable is confounded with that explanatory variable?

Sampling

To help prepare next year's flu vaccine, the Centers for Disease Control and Prevention want to know the DNA sequences of the flu viruses circulating around the world. The Census Bureau wants to know what percent of children 16 and younger have received all recommended immunizations. In both cases, we want to gather information about a very large group of individuals. Time, cost, and practicality forbid collecting data about each flu case in the world or each American child's immunization records. So we gather information about only part of the group, a *sample,* in order to draw conclusions about the whole, the *population* of interest.

POPULATION, SAMPLE, SAMPLING DESIGN

The **population** in a statistical study is the entire group of individuals about which we want information.

A **sample** is a part of the population from which we actually collect information. We use a sample to draw conclusions about the entire population.

A **sampling design** describes exactly how to choose a sample from the population.

We often draw conclusions about a whole on the basis of a sample. Everyone has sipped a spoonful of soup and judged the entire bowl on the basis of that taste. When your doctor is concerned about your white blood cell count, she takes a "blood sample" to see how low your white cell count really is. But the bowl of soup and your blood are quite homogeneous, so that a single spoonful represents the whole bowl and a small vial of your blood represents all the blood in your body at the time of extraction. These are relatively easy sampling designs.

Choosing a representative sample from a large and varied population is not so simple. The first step in a proper sample design is to say exactly *what population* we want to describe. The second step is to say exactly *what we want to measure*—that is, to give exact definitions of our variables.

EXAMPLE 7.3 The Current Population Survey

www.bls.census.gov/cps

The Current Population Survey (CPS) is an important sample survey conducted by the U.S. government. The CPS contacts about 60,000 households each month. It produces the monthly unemployment rate and various demographic characteristics. Supplemental questionnaires on a variety of topics are added at regular intervals. They are designed to update other government surveys, such as the yearly National Health Interview Survey.

One question of interest to the government is the proportion of current U.S. smokers who attempted to quit smoking within the previous 12 months. The first step in producing data for this question is to specify the population we want to describe. Which age groups will we include? Will we include illegal aliens or people in prisons? The CPS defines its population as all U.S. residents (whether citizens or not) 15 years of age and over who are civilians and are not in an institution such as a prison. Persons from all 50 states who are age 15 and over are considered, representing both genders and all ethnic groups.

The second step involves choosing what precisely we want to measure: What does it mean to be a "current smoker"? Should we include individuals who smoke only occasionally? What constitutes an "attempt to quit smoking"? The CPS defines current smokers as individuals who currently smoke either every day or some days. If you are a current smoker, the interviewer then goes on to ask about quitting attempts. A quitting attempt is defined as an attempt to quit smoking during the past year that lasted for 24 hours or more.

EXAMPLE 7.4 Studying fruit flies

Evolutionary researchers want to study the impact of different diets on the fecundity of fruit flies.[5] The first step in producing data for this question is to specify the population we want to describe. The researchers are interested in the relationship between diet and fecundity rather than in describing the fruit fly species. Therefore, the population actually sampled could be all wild fruit flies in the world or any subpopulation of fruit flies not markedly different from wild fruit flies. What are researchers' options for obtaining lab animals? They can capture specimens in the wild themselves. In the case of common lab animals or highly regulated species, they can order specimens from breeding facilities. The researchers in this study chose to obtain fruit flies from a Massachusetts breeding facility. The advantage is that their population of interest is similar to that used by other researchers in their field who also ordered fruit flies from this colony. The disadvantage is that their fruit flies will not be representative of the population of wild fruit flies.

The second step involves choosing what we want to measure: How do we quantify "fecundity"? Fecundity here is defined as the female ability to lay eggs. To keep track of the tiny flying creatures, the experimenters kept them in closed vials containing a food supply representative of the diets studied. The number of eggs laid per female was then counted by examining the content of each vial.

These two examples bring up an important issue. *Not all statistical studies use samples that are drawn directly from the entire population of interest.* Experiments, in particular, rarely do because of practical and ethical considerations. Researchers in the fruit fly experiment could not realistically have captured fruit flies from all over the world. The consequence is that conclusions from the experiment cannot directly be extended to all fruit flies. Whether the results apply to a larger population of fruit flies (the entire Massachusetts colony or even the entire population of wild fruit flies) is a biological argument, not a statistical one.

EXAMPLE 7.5 *Gender and clinical research*

Researchers and drug manufacturers have historically avoided testing new drugs on women. Yet these same drugs have been routinely prescribed regardless of gender. Of course, an untested drug could potentially be harmful to the fetus if a woman were to become pregnant. However, this concern can easily be alleviated by requiring all women enrolled in clinical trials to use contraceptives.

The major reasons for omitting women from clinical research are variations in a woman's menstrual cycle and hormonal changes over time with menopause or oral contraceptive use. Response to treatment might be influenced by these hormonal fluctuations, confounding the results. Or the treatment itself might interfere with concurrent use of oral contraceptives. Hormonal fluctuations during the menstrual cycle or the use of an oral contraceptive have indeed been shown to affect how drugs are metabolized. This is true, for example, of such common drugs as acetaminophen, aspirin, diazepam, and even caffeine. Some drugs, like the antibiotic rifampin, also reduce the effectiveness of oral contraceptives.[6] The question is: Is this a good reason to exclude women from clinical trials?

Absolutely not! We know that samples are representative of only the population from which they are taken. Clinical trials based entirely on men ensure drug effectiveness and safety for men alone. Not including women in clinical research does not protect them; it only delays the assessment of efficacy and side effects until the drug is made available to the public.

Since the mid-1990s, the Food and Drug Administration (FDA) has required that clinical trials for drugs targeting both men and women include both genders, in proportions reflecting the actual proportions of men and women in the target population.

Sampling designs

There are a variety of possible sampling designs, but not all are appropriate for statistical purposes.

Poor sampling designs

convenience sample The easiest—but not the best—sampling design just chooses individuals close at hand. This is called a **convenience sample** and it often produces unrepresentative data. If we are interested in finding out what percent of adults take vitamin and mineral supplements, for example, we might go to a shopping mall and ask people we meet. However, a shopping mall sample will almost surely overrepresent middle-class and retired people and underrepresent the poor. This will happen almost every time we take such a sample. That is, it is a systematic error caused by a bad sampling design, not just bad luck on one sample. This is *bias:* The outcomes of shopping mall surveys will repeatedly miss the truth about the population in the same ways.

> **BIAS**
>
> The design of a statistical study is **biased** if it systematically favors certain outcomes.

Another biased sampling design, the **voluntary response sample,** lets individuals choose whether to participate or not. All write-in, call-in, and online polls are examples of voluntary response samples. These types of samples are biased because people with strong opinions are most likely to respond. The problem is that *people who take the trouble to respond to an open invitation are usually not representative of any clearly defined population.* Unfortunately, convenience and voluntary response samples have become very popular with the media because they are cheap and often (misleadingly) sensational.

voluntary response sample

EXAMPLE 7.6 Online polls: trust the lawyers instead …

Every day, the online news network `CNN.com` posts a new survey question; voting is open to anyone surfing the Web and interested in participating. You may select "Yes" or "No," typically, and view the QuickVote tally as it stands. Beyond entertainment, what is the value of such online polling?

To answer this, we need to think of who might make up the sample of individuals who actually participated in the poll and what larger population they might represent. No one says it better than the lawyers here. The fine print below the QuickVote results reads: "QuickVote is not scientific and reflects the opinions of only those Internet users who have chosen to participate. The results cannot be assumed to represent the opinions of Internet users in general, nor the public as a whole." So trust the lawyers: There is no larger, clearly defined population here.

APPLY YOUR KNOWLEDGE

7.4 **Sampling members.** A health care provider wants to rate the satisfaction of its members with physical therapy. A questionnaire is mailed to 800 members of the health plan selected at random from the list of members who were prescribed physical therapy in the past 12 months. Only 212 questionnaires are returned.

(a) What is the population in this study? Be careful: About what group does the health care provider *want information?*

(b) What is the sample? Be careful: From what group does the provider *actually obtain information?*

7.5   Live births. Every year, the state of California publishes the state's vital statistics. Among these are the total live births, a figure based on actual birth certificates and hospital records. There were 544,685 live births in California in 2004.[7] Is this number derived from a sample of the California population or does it represent the actual California population?

7.6   Sampling on campus. Your college wants to gather students' perceptions of the nutritional quality of campus food. It isn't practical to contact all students.

(a) Give an example of a way to choose a sample of students that is poor practice because it depends on voluntary response.

(b) Give an example of a bad way to choose a sample that doesn't use voluntary response.

Simple random samples

In a voluntary response sample, people choose whether to respond. In a convenience sample, the interviewer makes the choice. In both cases, personal choice produces bias. The statistician's remedy is to allow impersonal chance to choose the sample. A sample chosen by chance allows neither favoritism by the sampler nor self-selection by respondents. Choosing a sample by chance attacks bias by giving all individuals an equal chance to be chosen. Rich and poor, young and old, black and white, all have the same chance to be in the sample.

probability sampling   A sample chosen by chance is called a probability sample. The most common **probability sampling** designs are simple random sampling, stratified random sampling, and multistage random sampling. The simplest way to use chance to select a sample is to place names in a hat (the population) and draw out a handful (the sample). This is the idea of *simple random sampling*.

> **SIMPLE RANDOM SAMPLE**
>
> A **simple random sample (SRS)** of size n consists of n individuals from the population chosen in such a way that every set of n individuals has an equal chance to be the sample actually selected.

An SRS not only gives each individual an equal chance to be chosen but also gives every possible sample an equal chance to be chosen. When you think of an SRS, picture drawing names from a hat to remind yourself that an SRS doesn't favor any part of the population. That's why an SRS is a better method of choosing samples than convenience or voluntary response sampling.

But writing names on slips of paper and drawing them from a hat is slow and inconvenient. That's especially true if, like the Current Population Survey, we must draw a sample of size 60,000. In practice, most people use software. The *Simple Random Sample* applet makes choosing an SRS very fast. If you don't use the applet or other software, you can randomize by using a *table of random digits*.

RANDOM DIGITS

A **table of random digits** is a long string of the digits 0, 1, 2, 3, 4, 5, 6, 7, 8, 9 with these two properties:

1. Each entry in the table is equally likely to be any of the 10 digits, 0 through 9.
2. The entries are independent of each other. That is, knowledge of one part of the table gives no information about any other part.

Table A at the back of the book is a table of random digits. Using this method to select an SRS is antiquated in the computer era, but it helps us to understand the concept of random sampling from random digits, a concept on which software and applets rely. Table A begins with the digits 19223950340575628713. To make the table easier to read, the digits appear in groups of five and in numbered rows. The groups and rows have no meaning—the table is just a long list of randomly chosen digits. There are two steps in using the random digits table to choose a simple random sample.

USING TABLE A TO CHOOSE AN SRS

Step 1. LABEL.   Give each member of the population a numerical label of the *same length*.

Step 2. SELECT.   To choose an SRS, read from Table A successive groups of digits of the length you used as labels. Your sample contains the individuals whose labels you find in the table.

You can label up to 100 items with two digits: 01, 02, ..., 99, 00. Up to 1000 items can be labeled with three digits, and so on. Always use the shortest labels that will cover your population. As standard practice, we recommend that you begin with label 1 (or 01 or 001, as needed). Reading groups of digits from the table gives all individuals the same chance of being selected because all labels of the same length have the same chance of being found in the table. For example, any pair of digits in the table is equally likely to be any of the 100 possible labels 01, 02, ..., 99, 00. Ignore any group of digits that was not used as a label or that duplicates a label already in the sample. You can read digits from Table A in any order—across a row, down a column, and so on—because the table has no order. As standard practice, we recommend reading across rows.

EXAMPLE 7.7   *Sampling California native habitat*

You are interested in California's endangered plant species and want to survey four randomly chosen native California habitats. Table 7.1 is a list of California's 34 native habitats provided by the California Native Plant Society.[8]

TABLE 7.1	California's 34 native habitats		
01	Coastal dunes	18	Bogs and fens
02	Desert dunes	19	Marshes and swamps
03	Inland dunes	20	Riparian forest
04	Coastal bluff scrub	21	Riparian woodland
05	Sonoran desert scrub	22	Riparian scrub
06	Mojave desert scrub	23	Cismontane woodland
07	Coastal scrub	24	Pinyon and juniper woodland
08	Great Basin scrub	25	Joshua tree woodland
09	Chenopod scrub	26	Sonoran thorn woodland
10	Chaparral	27	Broad-leaved upland forest
11	Coastal prairie	28	North Coast conifer forest
12	Valley and foothill grasslands	29	Closed-cone conifer forest
13	Great Basin grasslands	30	Lower montane conifer forest
14	Vernal pools	31	Upper montane conifer forest
15	Meadows	32	Subalpine conifer forest
16	Playas	33	Alpine boulder and rock field
17	Pebble or pavement plain	34	Alpine dwarf scrub

Fernando Bengoechea/Beateworks/Corbis

APPLET

Step 1. LABEL. Because two digits are needed to label the 34 habitats, all labels will have two digits. We have added labels 01 to 34 in the list of habitats. Always say how you labeled the members of the population. (To sample from the 1774 quadrants making up California, you would label the quadrants 0001, 0002, …, 1773, 1774, for example.)

Step 2. SELECT. To use the *Simple Random Sample* applet, just enter 34 in the "Population = 1 to" box and 4 in the "Select a sample of size" box, click "Reset," and click "Sample." Figure 7.2 shows the result of one sample.

To use Table A, read two-digit groups until you have chosen 4 habitats. Starting at line 130 (any line will do), we find

69051   64817   87174   09517   84534   06489   87201   97245

Because the labels are two digits long, read successive two-digit groups from the table:

69 05 16 48 17 87 17 40 95 17 84 53 40 64 89 87 20 19 72 45

Ignore groups not used as labels, like the initial 69. Also ignore any repeated labels, like the second 17 in this row, because you can't choose the same habitat twice. Your sample contains the habitats labeled 05, 16, 17, and 20, which have been underlined in the row above. These habitats are Sonoran desert scrub, playas, pebble or pavement plain, and riparian forest.

We can trust results from an SRS because it uses impersonal chance to avoid bias. Online polls and mall interviews also produce samples. But we can't trust

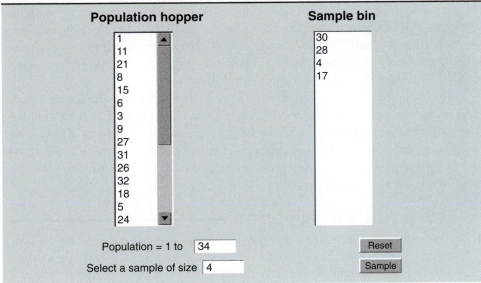

FIGURE 7.2 The *Simple Random Sample* applet used to choose an SRS of size $n = 4$ from a population of size 34, for example 7.7.

results from these samples, because they are chosen in ways that invite bias. *The first question to ask about any sample is whether it was chosen at random.*

EXAMPLE 7.8 Do you avoid soda?

A Gallup Poll on the American diet asked subjects about their attitudes toward various foods. The press release mentioned "the increasing proportion of Americans who say they try to avoid 'soda or pop' (51%, up from 41% in 2002)." Can we trust that 51%? Ask first how Gallup selected its sample. Later in the press release we read this: "These results are based on telephone interviews with a randomly selected national sample of 1,005 adults, aged 18 and older, conducted July 8–11, 2004."[9]

This is a good start toward gaining our confidence. Gallup tells us what population it has in mind (people at least 18 years old who live anywhere in the United States). We know that the sample from this population was of size 1005 and, most important, that it was chosen at random. There is more to say, but we have at least heard the comforting words "randomly selected."

Other probability sampling designs

The use of chance to select a sample is the essential principle of statistical sampling. Some probability sampling designs (such as an SRS) give each member of the population an equal chance to be selected. This may not be true in more elaborate sampling designs.

For example, it is common to sample important groups, often called strata, within the population separately, and then combine these samples. This is the idea

Computer-assisted interviewing

The days of the interviewer with a clipboard are past. Interviewers now read questions from a computer screen and use the keyboard to enter responses. The computer skips irrelevant items—once a woman says that she has no children, further questions about her children never appear. The computer can even present questions in random order to avoid bias due to always following the same order. Software keeps records of who has responded and prepares a file of data from the responses. The tedious process of transferring responses from paper to computer, once a source of errors, has disappeared.

stratified samples

multistage samples

of a **stratified** random sample. Large clinical trials often sample men and women separately or different ethnic groups separately in order to ensure inclusion of both majority and minority groups. Nationwide or statewide governmental surveys often use **multistage** samples for reasons of practicality. These samples typically involve choosing SRSs within SRSs. For example, an SRS of counties is first chosen; for each chosen county, an SRS of schools is chosen; then, within each chosen school, an SRS of students is drawn. This is more practical than directly selecting an SRS of students from all over the United States.

Most large-scale sample surveys use stratified, multistage samples that combine SRSs from each stage of the sampling process. Analysis of data from sampling designs more complex than an SRS takes us beyond basic statistics and will not be treated further in this book. But the SRS is the building block of more elaborate designs, and analysis of other probability designs differs more in complexity of detail than in fundamental concepts.

APPLY YOUR KNOWLEDGE

7.7   Sampling the forest. To gather data on a 1200-acre pine forest in Louisiana, the U.S. Forest Service laid a grid of 1410 equally spaced circular plots over a map of the forest. A ground survey visited a sample of 10% of these plots.[10]

(a) How would you label the plots?

(b) Use Table A, beginning at line 105, to choose the first 3 plots in an SRS of 141 plots.

7.8   A stratified sample. A club has 30 student members and 10 faculty members. The students are

Abel	Fisher	Huber	Miranda	Reinmann
Carson	Ghosh	Jimenez	Moskowitz	Santos
Chen	Griswold	Jones	Neyman	Shaw
David	Hein	Kim	O'Brien	Thompson
Deming	Hernandez	Klotz	Pearl	Utts
Elashoff	Holland	Liu	Potter	Varga

The faculty members are

Andrews	Fernandez	Kim	Moore	West
Besicovitch	Gupta	Lightman	Vicario	Yang

The club can send 4 students and 2 faculty members to a convention. It decides to choose those who will go by random selection. Use software or Table A to choose a stratified random sample of 4 students and 2 faculty members.

7.9   The National Health Interview Survey. The Centers for Disease Control and Prevention conduct the yearly National Health Interview Survey (NHIS). Here is a description of their methodology: "To achieve sampling efficiency and to keep survey operations manageable, cost-effective, and timely, the NHIS survey planners used multistage sampling techniques to select the sample of persons and households for the NHIS. These multistage methods partition the target universe into several nested levels of strata and clusters. The NHIS target universe is defined as all dwelling units in the U.S. that contain members of the civilian noninstitutionalized population." What is the population of interest? Explain how a multistage sample is different from a simple random sample and what they have in common.

Sample surveys

A sample survey is an observational study that relies on a random sample drawn from the entire population. Sample surveys have a wide array of applications. The Current Population Survey is a comprehensive survey of households from all parts of the United States (over 100 million households in all). Opinion polls are sample surveys that cover all sorts of topics and typically use voter registries or telephone numbers to select their samples. In epidemiology, sample surveys are used to establish the incidence (rate of new cases per year) and the prevalence (rate of all cases at one point in time) of various medical conditions and diseases.

Random selection eliminates bias in the choice of a sample from a list of the population. When the population consists of human beings, however, accurate information from a sample requires *more* than a good sampling design.

Typical problems with sample surveys are undercoverage and nonresponse. **Undercoverage** occurs when some groups in the population are left out of the process of choosing the sample; for example, homeless individuals are left out of all government surveys that select sample households. **Nonresponse** is a particularly serious source of bias in most sample surveys and occurs when a selected individual cannot be contacted or refuses to participate. Nonresponse to sample surveys often reaches 50% or more, even with careful planning and several callbacks. If the people contacted differ from those not responding, this creates very substantial bias.

undercoverage

nonresponse

> ### New York, New York
> New York City, they say, is bigger, richer, faster, ruder. Maybe there's something to that. The sample survey firm Zogby International says that as a national average it takes 5 telephone calls to reach a live person. When calling to New York, it takes 12 calls. Survey firms assign their best interviewers to make calls to New York and often pay them bonuses to cope with the stress.

EXAMPLE 7.9 How bad is nonresponse?

The Current Population Survey has the lowest nonresponse rate of any poll we know: Only about 6% or 7% of the households in the sample don't respond. People are more likely to respond to a government survey, and the CPS contacts its sample in person before doing later interviews by phone.

Opinion polls by news media and opinion-polling firms don't state their rates of nonresponse. That itself is a bad sign. The Pew Research Center imitated a careful telephone survey and published the results: Out of 2879 households called, 1658 were never at home, refused, or would not finish the interview. That's a nonresponse rate of 58%.[11]

The behavior of the respondent or of the interviewer can cause **response bias** in sample results. Drinking a lot of alcohol or having many sexual partners, for example, typically have negative connotations, and many interviewees understate their response. Responses may also differ if the interviewer is male or female, or from one ethnic group or another. Answers to questions that require recalling past events are often inaccurate because of faulty memory. For example, many people "telescope" events in the past, bringing them forward in memory to more recent time periods. "Have you visited a dentist in the last 6 months?" will often draw a "Yes" from someone who last visited a dentist 8 months ago.[12] Careful training

response bias

of interviewers and careful supervision to avoid variation among the interviewers can reduce response bias. Good interviewing technique is another aspect of a well-done sample survey.

wording effects

The **wording of questions** is the most important influence on the answers given to a sample survey. Confusing or leading questions can introduce strong bias, and even minor changes in wording can change a survey's outcome. For example, only 13% of Americans surveyed think we are spending too much on "assistance to the poor," but 44% think we are spending too much on "welfare."[13]

You can't trust the results of a sample survey until you have read the exact questions asked. The amount of nonresponse and the date of the survey are also important. A trustworthy survey must go beyond good sampling design.

He said, she said

If you need data about body weight, should you ask the individuals in your sample or should you take actual measurements? When *asked* their weight, almost all women say they weigh less than they really do. Heavier men also underreport their weight—but lighter men claim to weigh more than the scale shows. We leave you to ponder the psychology of the two sexes. Just remember that "say so" is no substitute for measuring.

APPLY YOUR KNOWLEDGE

7.10   Question wording. Comment on each of the following as a potential sample survey question. Is the question clear? Is it slanted toward a desired response?

(a) "It is estimated that disposable diapers account for less than 2% of the trash in today's landfills. In contrast, beverage containers, third-class mail, and yard wastes are estimated to account for about 21% of the trash in landfills. Given this, in your opinion, would it be fair to ban disposable diapers?"

(b) "Given the current trend of more home runs and more injuries in baseball today, do you think that steroid use should continue to be banned even though it is not enforced?"

(c) "In view of escalating environmental degradation and incipient resource depletion, would you favor economic incentives for recycling of resource-intensive consumer goods?"

7.11   In-line skaters. A study of injuries to in-line skaters used data from the National Electronic Injury Surveillance System, which collects data from a random sample of hospital emergency rooms. The researchers interviewed 161 people who came to emergency rooms with injuries from in-line skating. The interviews found that 53 people were wearing wrist guards and 6 of these had wrist injuries. Of the 108 who did not wear wrist guards, 45 had wrist injuries.[14]

(a) State carefully what population is sampled in this survey and what the sample size is. Should you draw conclusions from this study about all in-line skaters?

(b) What potential bias might affect the study results?

Comparative observational studies

Some studies may aim to compare different populations or to compare individuals within a population exposed to different conditions. Sample surveys can be used to compare subgroups within one population. For example, a study compared hearing impairment of blue-eyed and brown-eyed Dalmatian dogs by taking one random sample of 5333 Dalmatians. They found a substantially higher rate of hearing impairment among the blue-eyed dogs, suggesting a genetic link between lack of pigmentation (blue eyes) and hearing impairment in Dalmatians.

Case-control studies

Some biological traits, however, are too rare to study effectively with a sample survey. For instance, a trait found in about one in a thousand individuals would require a random sample of tens of thousands to observe maybe about 10 individuals with the trait of interest. It would be much simpler to select two separate random samples. In epidemiology, this design is often called *case-control*: A random sample of individuals with a condition (the cases) is compared with a random sample of individuals without the condition (the controls).

CASE-CONTROL STUDY

In a **case-control** observational study, case-subjects are selected based on defined outcomes, and a control group of subjects is selected separately to serve as a baseline with which the case group is compared.

EXAMPLE 7.10   Aflatoxicosis in Kenya

Aflatoxins are toxic compounds secreted by a fungus found in damaged crops. Aflatoxicosis is a rare but very severe poisoning that results from ingestion of aflatoxins in contaminated food. Kenya experienced an outbreak of aflatoxicosis in 2004, resulting in over three hundred cases of liver failure. The Kenya Ministry of Health suspected that improper maize (corn) storage was at least in part responsible for the outbreak. Forty case-patients and 80 healthy controls were asked how they had stored and prepared their maize.[15]

The case-patients were randomly selected from a list of individuals admitted to a hospital during the 2004 outbreak for unexplained acute jaundice. Selection of the control individuals was carefully thought out so that they would be as similar to the case-patients as possible, in a relevant way. A preliminary descriptive analysis of the outbreak data suggested that soil, microclimate, and farming practices might be influential factors, but not age or gender. Therefore, for each case-patient, the researchers selected at random two individuals from the patient's village who did not have a history of jaundice symptoms. Selecting individuals living in the same village ensured that the control subjects shared similar soil, microclimate, and farming practices with the case-patients. Because age or gender did not appear to matter, the controls were simply randomly selected from the village.

This case-control study started with two samples of individuals selected for a specific difference and looked for exposure factors in the subjects' past that differed between the two groups. One conclusion from the study was that improper maize storage was associated with a higher rate of aflatoxicosis and that the amount of maize consumed did not seem to matter.

Looking back into the past is called a **retrospective** approach. Memories are not always accurate, though, and recall bias may be a confounding factor. For instance, in Example 7.10, the individuals who were hospitalized for acute liver failure might have had different recollections from the control subjects, possibly because the issue mattered more to them.

retrospective

Case-control studies are efficient ways to approach rare outcomes and often give fast results, within the usual limitations of observational studies. The main challenge is in selecting a random sample of control individuals as similar to the case-subjects as possible. The aflatoxicosis study had very carefully selected controls, allowing the researchers to find plausible causes for the aflatoxin epidemic.

historical controls

Not all comparative observational studies are so careful about the selection of their control groups. **Historical-control** designs are case-control studies that utilize existing data from previous studies to make up their control groups. While this is particularly convenient and cost-effective, the design creates many confounding variables because the two groups are bound to differ substantially in several ways. However, when approval by ethical committees is difficult to obtain (for example, to study children), historical-control designs may be the only tool available for exploratory studies.

Cohort studies

Instead of searching the past to uncover possible causal factors, *cohort* studies enlist a homogeneous group of fairly similar individuals and keep track of them over a long period of time. These are **prospective** studies, which record at regular intervals all sorts of relevant information about the study participants. When the study is over, individuals who have developed a condition are compared with the remaining, unaffected individuals. Unlike case-control studies, cohort studies are very costly and better suited to investigating common outcomes such as Type 2 diabetes or heart disease.

prospective

COHORT STUDY

In a **cohort** study, subjects sharing a common demographic characteristic are enrolled and observed at regular intervals over an extended period of time.

EXAMPLE 7.11   The Nurses' Health Study

The Nurses' Health Study is one of the largest prospective observational studies designed to examine factors that may affect major chronic diseases in women. Since 1976 the study has followed a cohort of over 100,000 registered nurses with the idea that nurses would be able to respond accurately to technically worded medical questionnaires. Every two years, enrolled nurses receive a questionnaire about diseases and health-related topics such as diet and lifestyle. The response rates to the questionnaires are about 90% for each two-year cycle.

In the 2007 newsletter, study investigators reported their findings on age-related memory loss. About 20,000 women aged 70 and older had completed telephone interviews every two years to assess their memory with a set of cognitive tests. One of the study findings was that the more women walked during their late 50s and 60s, the better their memory was at age 70 and older. Although these women were asked to take a cognitive test, the study design was still observational because all women sampled took the test and the investigator did not randomly assign different walking regimens. Instead, they observed that women who had walked more during their late 50s and 60s ended up with the better memory scores at age 70 and older.

Cohort studies accumulate enormous amounts of detailed information and can examine the compounded effect of various factors over time. However, they also take a long time to complete and lose subjects over time, creating a potentially confounding effect. This is especially true when studying older individuals, who may die before the end of the study. For example, it is possible that the women with the greatest memory loss also had poor overall health and died younger and were therefore not included in the analysis.

Overall, though, cohort studies are less prone to confounding than case-control designs because cohorts start with one homogeneous group. This has other important advantages. Cohorts can provide information about the relative health risks of different subgroups. They also support incidence calculations, unlike case-control studies which select subjects based on an existing disease status.

However, like all observational studies, cohort designs cannot establish whether observed differences between groups can be attributed to the groups' differentiating feature or to confounding variables. For example, maybe the women with the better health were both more capable of walking and also less prone to memory loss. Therefore, we cannot unambiguously conclude that walking has a protective effect against memory loss.

Observational studies play an important role in the building of scientific knowledge. But they can easily be plagued by bias and confounding. Experiments, on the other hand, provide an opportunity for manipulating the environment and controlling confounding variables. In the next chapter we study ways in which experiments can be designed to avoid bias and confounding.

APPLY YOUR KNOWLEDGE

7.12   **Cell phones and brain cancer.** Exercise 7.2 describes a study of the link between cell phone use and brain cancer. What type of observational study is it? What population or populations are represented?

7.13   **Alcohol and heart attacks.** Many studies have found that people who drink alcohol in moderation have lower risk of heart attacks than either nondrinkers or heavy drinkers. Does alcohol consumption also improve survival after a heart attack? One study followed 1913 people who were hospitalized after severe heart attacks. In the year before their heart attacks, 47% of these people did not drink, 36% drank moderately, and 17% drank heavily. After four years, fewer of the moderate drinkers had died.[16] What type of observational study is this? What population or populations are represented in the study? What are the explanatory and response variables?

CHAPTER 7 SUMMARY

We can produce data intended to answer specific questions by **observational studies** or **experiments.** Experiments, unlike observational studies, actively impose some treatment on the subjects of the experiment.

Observational studies often fail to show that changes in an explanatory variable actually cause changes in a response variable because the explanatory variable is **confounded** with lurking variables. Variables are confounded when their effects on a response can't be distinguished from each other.

A practical way to study variables of interest is to select a **sample** from the **population** of all individuals about which we desire information. We base conclusions about the population on data from the sample.

The **design** of a sample describes the method used to select the sample from the population. **Probability sampling** designs use chance to select a sample.

The basic probability sample is a **simple random sample (SRS).** An SRS gives every possible sample of a given size the same chance to be chosen.

Choose an SRS by labeling the members of the population and using a **table of random digits** to select the sample. Software can automate this process.

Stratified random samples and **multistage samples** are more complex forms of probability sampling that are also commonly used.

Failure to use probability sampling often results in **bias,** or systematic errors in the way the sample represents the population. **Voluntary response samples,** in which the respondents choose themselves, are particularly prone to large bias.

Sample surveys are observational studies of a random sample drawn from the entire population. **Undercoverage** and **nonresponse** are two major challenges that can result in substantial bias. In addition, **response bias** and **poorly worded questions** can create misleading conclusions.

Case-control studies select two random samples based on a known characteristic and examine **retrospectively** what might have led to the difference. Finding controls that closely resemble the case-subjects is the most challenging part of this observational design.

In contrast, **cohort** studies enroll subjects and follow them **prospectively** over time, with data collected at regular intervals. These observational studies are very powerful but also expensive and may not be adequate to study rare outcomes.

CHECK YOUR SKILLS

7.14 A 2005 NBC News Poll asked, "Which do you think is more likely to actually be the explanation for the origin of human life on Earth: evolution or the biblical account of creation?" The poll surveyed 800 randomly selected adults nationwide. In all, 264 of the 800 responded "evolution." The sample in this setting is

(a) all 225 million adults in the United States.

(b) the 800 people interviewed.

(c) the 264 people who chose "evolution."

7.15 With online polls you simply click on a response to become part of the sample. The Excite poll question for December 13, 2005, was "Are you for or against opening the Arctic National Wildlife Refuge to oil drilling?" There were a total of 16,796 votes, with 40% (6812 votes) against and 55% (9319 votes) for opening the refuge for oil drilling. You can conclude that

(a) Americans favor opening the Arctic wildlife refuge for oil drilling.

(b) the poll uses voluntary response, so the results tell us little about the population.

(c) the sample is too small to draw any conclusion.

7.16   The population of interest in the Excite online poll described in the previous exercise is

(a) all Americans with access to the Internet.

(b) all Internet users who visit the Excite Poll Web site.

(c) all Internet users who chose to participate in this particular online poll.

7.17   You want to select an SRS of 25 of the 130 U.S. estuaries to study the extent to which they are polluted. How would you label this population in order to use Table A?

(a) 001, 002, 003, ..., 129, 130

(b) 000, 001, 002, ..., 129, 130

(c) 1, 2, ..., 129, 130

7.18   Based on your choice in the previous exercise, if you used line 104 of Table A, the first three members of your SRS would be

(a) 113, 074, 001.

(b) 113, 074, 011.

(c) 113, 074, 118.

7.19   A sample of households in a community is selected at random from the telephone directory. In this community, 4% of households have no telephone and another 35% have unlisted telephone numbers. The sample will certainly suffer from

(a) nonresponse.

(b) undercoverage.

(c) false responses.

7.20   A University of Pennsylvania survey of 1345 adults nationwide conducted in August 2004 asked, "Do you favor or oppose federal funding of research on diseases like Alzheimer's using stem cells taken from human embryos?" and found 64% in favor and 28% opposed. A *Newsweek* Poll of 1004 registered voters nationwide conducted in October 2004 asked, "Do you favor or oppose using federal tax dollars to fund medical research using stem cells obtained from human embryos?" and found 50% in favor and 36% opposed. The substantially different responses found by these two surveys can be explained by

(a) nonresponse.

(b) false responses.

(c) wording effect.

7.21   In the previous question, the substantially different responses found by the two surveys may also be explained by

(a) a difference in target population.

(b) respondent bias.

(c) interviewer bias.

7.22   A study surveyed from 1986 to 2000 a large number of male health professionals aged 40 to 75. Each respondent filled a lifestyle questionnaire every two years. The 22,086 subjects who had reported having good erectile function in 1986 were included in an analysis of risk factors in the development of erectile dysfunction. The analysis revealed that "obesity and smoking were positively associated, and physical activity was inversely associated with the risk of erectile dysfunction developing.[17]" This is

(a) a prospective observational study.

(b) a retrospective observational study.

(c) an experiment.

7.23   How strong is the evidence cited in the previous exercise that physical activity may lower the risk of erectile dysfunction?

(a) Quite strong because it comes from an experiment.

(b) Quite strong because it comes from a large random sample.

(c) Weak, because physical activity is confounded with many other variables.

CHAPTER 7 EXERCISES

In all exercises asking for an SRS, you may use Table A, the Simple Random Sample applet, or other software.

7.24   **Safety of anesthetics.** The National Halothane Study was a major investigation of the safety of anesthetics used in surgery. Records of over 850,000 operations performed in 34 major hospitals showed the following death rates for four common anesthetics:[18]

Anesthetic	A	B	C	D
Death rate	1.7%	1.7%	3.4%	1.9%

There is a clear association between the anesthetic used and the death rate of patients. Anesthetic C appears dangerous.

(a) Explain why we call the National Halothane Study an observational study rather than an experiment, even though it compared the results of using different anesthetics in actual surgery.

(b) When the study looked at other variables that are confounded with a doctor's choice of anesthetic, it found that Anesthetic C was not causing extra deaths. Suggest important lurking variables that are confounded with what anesthetic a patient receives.

7.25   **Medical marijuana.** A 2005 Mason-Dixon Poll asked, "Do you think adults should be allowed to legally use marijuana for medical purposes if their doctor recommends it, or do you think that marijuana should remain illegal even for medical purposes?" The poll surveyed 732 randomly selected registered voters nationwide. In all, 476 of the 732 were in favor of legalizing marijuana for medical purposes.[19]

(a) What population is the poll targeting?

(b) What sampling design was used? What is the sample size?

7.26   **Human origins.** An Excite online poll of November 2005 based on 12,748 votes found that 33% favored creationism and 31% favored evolution in

Emilio Ereza/Alamy

explaining human origins. National random samples taken that year showed 50% to 60% favoring creationism and 15% to 35% favoring evolution to explain human origins. (The results varied a bit depending on the exact question asked.) Explain briefly to someone who knows no statistics why the random samples report public opinion more reliably than the online poll.

7.27 Monitoring chimpanzee behavior. Jane Goodall dedicated years of her life to the study of chimpanzee behavior in their natural habitat of East Africa. Initially, the chimps would flee at the sight of her, and she had to observe them from a distance with binoculars. But she persisted until the animals eventually got accustomed enough to her to ignore her.[20]

(a) What kind of bias was she trying to minimize?

(b) Explain why this lengthy step is particularly important when observing animal behavior.

7.28 Random digits. Which of the following statements are true of a table of random digits, and which are false? Briefly explain your answers.

(a) There are exactly four 0s in each row of 40 digits.

(b) Each pair of digits has chance 1/100 of being 00.

(c) The digits 00000 can never appear as a group, because this pattern is not random.

7.29 Do you trust the Internet? You want to ask a sample of college students the question "How much do you trust information about health that you find on the Internet—a great deal, somewhat, not much, or not at all?" You try out this and other questions on a pilot group of 10 students chosen from your class. The class members are

Anderson	Deng	Glaus	Nguyen	Samuels
Arroyo	De Ramos	Helling	Palmiero	Shen
Batista	Drasin	Husain	Percival	Tse
Bell	Eckstein	Johnson	Prince	Velasco
Burke	Fernandez	Kim	Puri	Wallace
Cabrera	Fullmer	Molina	Richards	Washburn
Calloway	Gandhi	Morgan	Rider	Zabidi
Delluci	Garcia	Murphy	Rodriguez	Zhao

Choose an SRS of 10 students. If you use Table A, start at line 117.

7.30 Telephone area codes. There are approximately 371 active telephone area codes covering Canada, the United States, and parts of the Caribbean. (More are created regularly.) You want to choose an SRS of 25 of these area codes for a study of available telephone numbers. Label the codes 001 to 371 and use the *Simple Random Sample* applet or other software to choose your sample. (If you use Table A, start at line 129 and choose only the first 5 codes in the sample.)

APPLET

7.31 Sampling Amazon forests. Stratified samples are widely used to study large areas of forest. Based on satellite images, a forest area in the Amazon basin is divided into 14 types. Foresters studied the four most commercially valuable types: alluvial climax forests of quality levels 1, 2, and 3, and mature secondary forest. They divided the area of each type into large parcels, chose parcels of each type at random, and counted tree species in a 20- by 25-meter rectangle randomly

Wolfgang Kaehler/CORBIS

placed within each parcel selected. Here is some detail:

Forest type	Total parcels	Sample size
Climax 1	36	4
Climax 2	72	7
Climax 3	31	3
Secondary	42	4

The researchers chose the stratified sample of 18 parcels described in the table. Explain how a stratified random sample is different from a simple random sample. Why do you think the researchers preferred to use a stratified design?

7.32   **Seat belt use.** A study in El Paso, Texas, looked at seat belt use by drivers. Drivers were observed at randomly chosen convenience stores. After they left their cars, they were invited to answer questions about seat belt use. In all, 75% said they always used seat belts, yet only 61.5% were wearing seat belts when they pulled into the store parking lots.[21] Explain the reason for the bias observed in responses to the survey. Do you expect bias in the same direction in most surveys about seat belt use?

7.33   **Running red lights.** A survey of drivers started with a list of 5024 licensed drivers randomly selected from all over the United States. The investigators then chose an SRS of 880 of these drivers to answer questions about their driving habits.

(a) How would you assign labels to the 5024 drivers? Use Table A, starting at line 104, to choose the first 5 drivers in the sample.

(b) One question asked was, "Recalling the last ten traffic lights you drove through, how many of them were red when you entered the intersections?" Of the 880 respondents, 171 admitted that at least one light had been red. A practical problem with this survey is that people may not give truthful answers. What is the likely direction of the bias: Do you think more or fewer than 171 of the 880 respondents really ran a red light? Why?

7.34   **Random digit dialing.** The list of individuals from which a sample is actually selected is called the *sampling frame*. Ideally, the frame should list every individual in the population, but in practice this is often difficult. A frame that leaves out part of the population is a common source of undercoverage.

(a) Suppose that a sample of households in a community is selected at random from the telephone directory. What households are omitted from this frame? What types of people do you think are likely to live in these households? These people will probably be underrepresented in the sample.

(b) It is usual in telephone surveys to use random digit dialing equipment that selects the last four digits of a telephone number at random after being given the exchange (the first three digits). Which of the households you mentioned in your answer to (a) will be included in the sampling frame by random digit dialing?

7.35   **Wording survey questions.** Comment on each of the following as a potential sample survey question. Is the question clear? Is it slanted toward a desired response?

(a) "Some cell phone users have developed brain cancer. Should all cell phones come with a warning label explaining the danger of using cell phones?"

(b) "Do you agree that a national system of health insurance should be favored because it would provide health insurance for everyone and would reduce administrative costs?"

(c) "In view of the negative externalities in parent labor force participation and pediatric evidence associating increased group size with morbidity of children in day care, do you support government subsidies for day care programs?"

7.36 **Opinion on stem cell research.** An ABCNEWS/Beliefnet Poll was conducted by telephone in June 2001 among a random national sample of 1022 adults to evaluate the public's opinion about stem cell research. The online report states that 58% of Americans support stem cell research, while 30% oppose it, but the report does not provide the actual number of respondents for or against stem cell research. What likely statistical flaw does this lack of information cover up?

7.37 **Write your own bad questions.** Write your own examples of bad sample survey questions.

(a) Write a biased question designed to get one answer rather than another. Then rewrite it to avoid that bias.

(b) Write a question that is confusing, so that it is hard to answer. Then rewrite it in an improved form.

7.38 **Bone loss by nursing mothers.** Breast-feeding mothers secrete calcium into their milk. Some of the calcium may come from their bones, resulting in reduced bone mineral density. Researchers compared 47 breast-feeding women with 22 women of similar age who were neither pregnant nor lactating. They measured the percent change in the mineral content of the women's spines over three months.[22]

(a) What type of observational study is this? What two populations did the investigators want to compare?

(b) Give one possible confounding variable for the relationship between breast-feeding and change in bone mineral content.

7.39 **The Mediterranean diet and cancer.** Cancer of the colon and rectum is less common in the Mediterranean region than in other Western countries. The Mediterranean diet contains little animal fat and lots of olive oil. Italian researchers compared 1953 patients with colon or rectal cancer with a control group of 4154 patients admitted to the same hospitals for unrelated reasons. They estimated consumption of various foods from a detailed interview, then classified the patients according to their consumption of olive oil (low, medium, or high).[23]

(a) What type of observational study is this? Explain your answer.

(b) The investigators reported that fewer than 4% of subjects in each group refused to participate. Why is this important information?

(c) What would be a possible confounding variable for this study?

7.40 **Brain size and autism.** Is autism marked by different brain growth patterns in early life? One study looked at the brain sizes of 30 autistic boys and 12 nonautistic boys who had all received an MRI scan while toddlers. The autistic boys had received the MRI as part of a diagnostic for behavioral concerns, and the nonautistic boys had been participants in a previous quantitative MRI study reporting age-related changes in head and brain features.[24]

(a) What type of observational study is this? Explain your answer.

(b) The researcher honestly disclosed the limitations of the study, an important part of the ethical conduct of research. Suggest some potential confounding variables for this study.

7.41 **Red wine and prostate cancer.** In 1986, over 50,000 American male health professionals 40 to 75 years of age were enrolled in the Health Professionals Follow-up Study (HPFS). At the time of enrollment and again every two years, the subjects reported their average consumption of red wine, white wine, beer, and liquor, and prostate cancer diagnoses were recorded. Investigators found no clear evidence of association between amount of red wine consumed and risk of prostate cancer.[25]

(a) What type of observational study is this? Explain your answer.

(b) By 2002, the HPFS had recorded 3348 cases of prostate cancer. How is this information relevant when interpreting the study results?

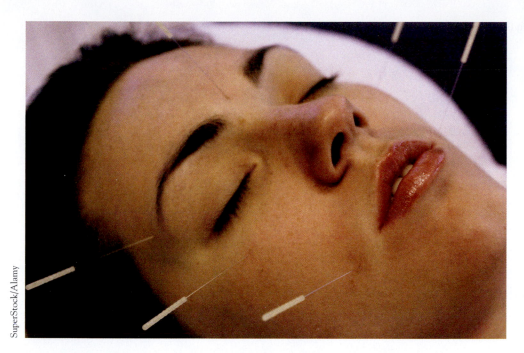

SuperStock/Alamy

Designing Experiments

How do we *produce data* that can help us answer interesting scientific questions? In the previous chapter we looked at ways to select samples from a population of interest and their applications for observational studies.

But what if we asked, "Is a new cholesterol-lowering drug more effective than the standard-treatment drug?" To answer this question, we would give the new drug to a group of high-cholesterol patients and measure the decrease in cholesterol level after one month of treatment. However, that wouldn't be enough to answer the question completely. We would also need to give the standard drug to another group of high-cholesterol patients and compare their results with those of the patients given the new drug. In this chapter we study how to design experiments.

Designing experiments

SUBJECTS, FACTORS, TREATMENTS

The **individuals** studied in an experiment are often called **subjects,** particularly when they are people.

The explanatory variables in an experiment are often called **factors.**

A **treatment** is any specific experimental condition applied to the subjects. If an experiment has more than one factor, a treatment is a combination of specific values of each factor.

EXAMPLE 8.1   Protective effect of the Mediterranean diet

Mediterranean populations are known to suffer comparatively less from heart disease and from some types of cancer. Their diet high in natural fibers, vitamins, and oleic acid has been offered as an explanation for this phenomenon.

To test this hypothesis, an experiment assigned 605 survivors of a first heart attack to follow for 5 years either a Mediterranean-type diet or a diet close to the American Heart Association (AHA) "prudent" diet (30% or less of caloric intake from fats, low cholesterol intake). The patients were assessed regularly, and a variety of outcomes were recorded, from plasma fatty acid levels to occurrences of cancer or heart attack, to death (cardiac, cancer, or other). The assignment to one diet or the other was random, and the two groups did not differ substantially in demographics. Possible confounders such as smoking and activity levels were closely monitored and included in the statistical analysis but were not restricted or imposed.[1]

This experiment compares two *treatments*: Mediterranean diet and AHA prudent diet. There is a single *factor*, "diet." The *subjects* are survivors of a first heart attack, and there are many *response variables* recorded over 5 years, such as cancer and heart attack.

EXAMPLE 8.2   Effect of corn variety and protein content on chick growth

New varieties of corn with altered amino acid content may have higher nutritional value than standard corn, which is low in the amino acid lysine. An experiment compares two new varieties, called opaque-2 and floury-2, with normal corn. The researchers mix corn-soybean meal diets using each type of corn at each of three protein levels: 12% protein, 16% protein, and 20% protein. They feed each diet to 10 one-day-old male chicks and record their weight gains after 21 days. The weight gain of the chicks is a measure of the nutritional value of their diet.[2]

This experiment has two *factors*: corn variety, with 3 values (opaque-2, floury-2, and normal), and protein level, with 3 values (12%, 16%, and 20%). The 9 combinations of both factors form 9 *treatments*. Figure 8.1 shows the layout of the treatments. The *individuals* are one-day-old male chicks, and the *response variable* is the weight gain after 21 days on the treatment diet.

FIGURE 8.1 The treatments in the experimental design of Example 8.2. Combinations of values of the two factors form nine treatments.

> Chicks assigned to Treatment 3 are fed the opaque-2 corn in a diet containing 20% protein.

		Factor B Protein		
		12%	16%	20%
	Opaque-2	1	2	3
Factor A Corn variety	Floury-2	4	5	6
	Normal	7	8	9

Examples 8.1 and 8.2 illustrate the advantages of experiments over observational studies. In an experiment, we can study the effects of the specific treatments we are interested in. By assigning subjects to treatments, we can avoid confounding. If, for example, we simply compare heart attack survivors who decided to follow a Mediterranean diet with those who decided to follow the AHA prudent diet, we may find that one group is richer or more social or more physically active. The experiment of Example 8.1 avoids that. Moreover, we can control the environment of the subjects to hold constant factors that are not directly of interest to us, such as the gender and age of the chicks in Example 8.2.

Another advantage of experiments is that we can study the combined effects of several factors simultaneously. The interaction of several factors can produce effects that could not be predicted from looking at the effect of each factor alone. Perhaps diets higher in proteins produce heavier chicks, and a corn variety richer in amino acids also produces heavier chicks, but if we feed the chicks both higher protein amounts and higher amino acid amounts (amino acids are the building blocks of proteins), chicks may not grow any better than they do with either higher protein levels or higher amino acid levels alone. The two-factor experiment in Example 8.2 will help us find out.

Experiments are the preferred method for examining the effect of one variable on another. By imposing the specific treatment of interest and controlling other influences, we can pin down cause and effect. Statistical designs are often essential for effective experiments, just as they are for sampling surveys and observational studies. To see why, let's start with an example of a bad design.

What's news?

Randomized comparative experiments provide the best evidence for medical advances. Do newspapers care? Maybe not. University researchers looked at 1192 articles in medical journals, of which 7% were turned into stories by the two newspapers examined. Of the journal articles, 37% concerned observational studies and 25% described randomized experiments. Among the articles publicized by the newspapers, 58% were observational studies and only 6% were randomized experiments. Conclusion: The newspapers want exciting stories, whether or not the evidence is good.

EXAMPLE 8.3 *An uncontrolled experiment*

Gastric freezing was introduced in the 1960s by a prominent surgeon as a way to relieve ulcer pain. Patients would swallow a deflated balloon which was later filled with a refrigerated liquid. The idea was that the refrigerant would cool the stomach lining, reduce acid production, and reduce pain. The procedures performed showed that gastric freezing did reduce ulcer pain, and the treatment was recommended based on this evidence.[3]

This experiment has a very simple design. A group of subjects (the patients) were exposed to a treatment (gastric freezing), and the outcome (pain reduction) was observed. Here is the design:

$$\text{Subjects} \longrightarrow \text{Gastric freezing} \longrightarrow \text{Pain reduction}$$

Some critical pieces of information are missing from this experiment. What would have happened if we had left the patients alone? Many medical conditions improve spontaneously over time. What if patients genuinely felt better but did so because the care and attention they received made them feel better overall, not because of the cooling of their stomach lining?

None of these questions can be addressed by the simple experiment. They remind us that the effect of gastric freezing is confounded with the effect of both spontaneous improvement and the response to care and attention. Figure 8.2 shows the confounding in picture form. Because of confounding, subject improvement cannot be unequivocally attributed to gastric freezing.

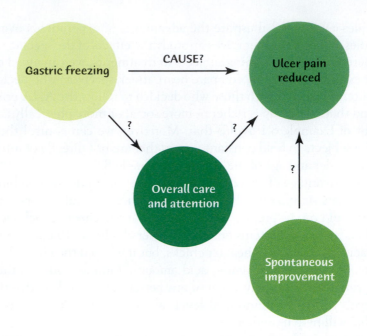

FIGURE 8.2 Confounding. We can't distinguish the effect of the treatment from the effects of lurking variables in Example 8.3.

Many laboratory experiments use a design like that of Example 8.3:

Subjects $\longrightarrow$ Treatment $\longrightarrow$ Measure response

In the controlled environment of the laboratory, simple designs may work well. However, *field experiments and experiments with human subjects are exposed to more variable conditions and deal with more variable individuals. A simple design often yields worthless results because of confounding with lurking variables.*

APPLY YOUR KNOWLEDGE

8.1   **Growing in the shade.** Ability to grow in shade may help pines found in the dry forests of Arizona resist drought. How well do these pines grow in shade? Investigators planted pine seedlings in a greenhouse in either full light, light reduced to 25% of normal by shade cloth, or light reduced to 5% of normal. At the end of the study, they dried the young trees and weighed them. What are the individuals, the treatments, and the response variable in this experiment?

8.2   **Improving adolescents' health habits.** Most American adolescents don't eat well and don't exercise enough. Can middle schools increase physical activity among their students? Can they persuade students to eat better? Investigators designed a "physical activity intervention" to increase activity in physical education classes and during leisure periods throughout the school day. They also designed a "nutrition intervention" that improved school lunches and offered ideas for healthy home-packed lunches. Each participating school was randomly assigned to one of the interventions, both interventions, or no intervention. The investigators observed physical activity and lunchtime consumption of fat. Identify the individuals, the factors, and the response variables in this experiment. Use a diagram like that in Figure 8.1 to display the treatments.

8.3   **Oral contraceptive use.** Oral contraceptives are known to have many side effects and interactions with other health risks. One study looked at 155 premenopausal women with blood clots in the deep veins of the legs and compared them with 169 premenopausal women randomly selected from the general population. The percent of women currently using oral contraceptives was found to be 70% among women with diagnosed vein blood clots but only 38% among women from the general population.

(a) Is this an experiment or an observational study?

(b) Explain why the researchers chose to compare the women with blood clots to women from the general population.

Randomized comparative experiments

The remedy for the confounding in Example 8.3 is to do a *comparative experiment* in which some patients undergo the gastric-freezing treatment and other, similar patients undergo a sham procedure or no procedure at all. Comparing patients undergoing gastric freezing with untreated patients allows us to control for the effect of natural improvement. However, that alone would not tell us if a lower pain rate in the *experimental group* (gastric freezing) would be due to the benefits of the medical procedure itself or to the psychological impact of undergoing any kind of treatment—the *placebo effect*. Only by comparing the gastric-freezing procedure to a sham procedure would we be able to conclude that gastric freezing itself is responsible for any observed improvement in patient condition.

Most well-designed experiments compare two or more treatments. Part of the design of an experiment is a description of the factors (explanatory variables) and the layout of the treatments, with comparison as the leading principle.

EXPERIMENTAL GROUP, CONTROL, PLACEBO

An **experimental group** is a group of individuals receiving a treatment whose effect we seek to understand.

A **control** is a treatment meant to serve as a baseline with which the experimental group is compared.

A **placebo** is a control treatment that is fake (for example, taking a sugar pill) but otherwise indistinguishable from the treatment in the experimental group.

EXAMPLE 8.4   A controlled experiment

The simple experiment outlined in Example 8.3 was followed up many years later with a randomized, placebo-controlled experiment in which patients undergoing gastric freezing were compared with similar patients undergoing the same procedure except that the liquid pumped into the balloon was not refrigerated. This second group was a placebo group. The comparative study found that the condition of 28 of the 82 gastric-freezing patients improved, while 30 of the 78 patients in the placebo group improved.

This comparative experiment demonstrated that cooling the stomach lining was not more effective in reducing ulcer pain than a sham procedure. The treatment was then abandoned.

The *placebo effect* is documented in human experimentation only, in contexts as diverse as the study of asthma or high blood pressure. The effect is particularly strong when the response variable is pain. The mechanisms of this mind-body effect are unknown but have a biological foundation. We know that the nervous system interacts with the immune system and that depression, for instance, suppresses the immune response.

Like experiments, observational studies often compare different groups and may even include a control group that does not exhibit the characteristic of interest. For example, a study could follow individuals over time and record diagnoses of lung cancer, comparing a sample of cigarette-smoking adults and a sample of nonsmoking adults. However, in observational studies the groups compared may differ by more characteristics than the ones used to distinguish the groups. For example, the cigarette smokers may also be poorer overall or more prone to other forms of addictions, rendering them more susceptible to lung cancer. Controls in observational studies are very informative but they are not sufficient to draw conclusions about a causal effect.

Thus, comparison alone isn't enough to produce results we can trust. If the treatments are given to groups that differ markedly when the experiment begins, bias will result. In Example 8.4, if we select all the patients most likely to improve to make up the gastric-freezing treatment group, we will see better results in the treatment group even if the procedure itself makes no contribution to the improvement. Personal choice would bias our results in the same way that volunteers bias the results of online opinion polls. The solution to the problem of bias is the same for experiments and for survey samples: use impersonal chance to select the groups.

Placebo for baby

The most famous and maybe most powerful placebo in the world is the one parents use on their little ones for every small misfortune. Different techniques exist but they all work equally well: a kiss, a hug, or even a Band-Aid. The effect gradually disappears, however, as children realize that sometimes they can get better without help and that at times more advanced treatment is provided.

RANDOMIZED COMPARATIVE EXPERIMENT

An experiment that uses both comparison of two or more treatments and chance assignment of subjects to treatments is a **randomized comparative experiment.**

EXAMPLE 8.5   May I have some chocolate?

Can chocolate help trigger headaches? A study recruited 128 women suffering from chronic headache and assigned them randomly to one of two groups: 64 women ate a chocolate preparation and 64 ate the same preparation made with carob, a natural extract with similar taste but without the caffeine and other psychoactive substances of chocolate.[4] Figure 8.3 outlines the design in graphical form.

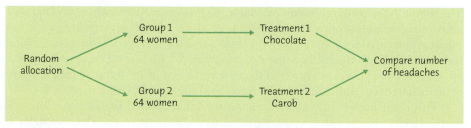

FIGURE 8.3 Outline of a randomized comparative experiment to compare the effect of carob and chocolate on headaches, for Example 8.5.

The selection procedure is exactly the same as it is for sampling: label and select. **Step 1. Label** the 128 women 001 to 128. **Step 2. Select.** Go to the table of random digits and read successive three-digit groups. The first 64 labels encountered select the chocolate group. As usual, ignore repeated labels and groups of digits not used as labels. For example, if you begin at line 102 in Table A, the first two women chosen are those labeled 099 and 019. Software such as the *Simple Random Sample* applet makes it particularly easy to choose treatment groups at random.

The design in Figure 8.3 is *comparative* because it compares two treatments (the two preparations). It is *randomized* because the subjects are assigned to the treatments by chance. This "flowchart" outline presents all the essentials: randomization, the sizes of the groups and which treatment they receive, and the response variable. There are, as we will see later, statistical reasons for generally using treatment groups about equal in size. We call designs like that in Figure 8.3 *completely randomized.*

COMPLETELY RANDOMIZED DESIGN

In a **completely randomized** experimental design, all the subjects are allocated at random among all the treatments.

Completely randomized designs can compare any number of treatments. They can also have more than one factor. The chick growth experiment of Example 8.2 has two factors: the corn variety and the protein level making up the chicks' diet. Their combinations form the nine treatments outlined in Figure 8.1. A completely randomized design assigns subjects at random to these nine treatments. Once the layout of treatments is set, the randomization needed for a completely randomized design is tedious but straightforward.

Randomized comparative experiments are designed to give good evidence that differences in the treatments actually *cause* the differences we see in the response. The logic is as follows:

- Random assignment of subjects forms groups that should be similar in all respects before the treatments are applied. Exercise 8.51 uses the *Simple Random Sample* applet to demonstrate this.

- Comparative design ensures that influences other than the experimental treatments operate equally on all groups.

- Therefore, differences in average response must be due either to the treatments or to the play of chance in the random assignment of subjects to the treatments.

This last statement deserves more thought. Even with an excellent experimental design, we can never say that *any* difference between two groups is entirely due to the treatments applied. Variables, by definition, vary from individual to individual, and therefore, one group of individuals will almost certainly give a different result from another group of individuals even when all conditions are the same. In Example 8.5, chance assigned the women to the chocolate group or the carob group, and this creates a chance difference between the groups. We would not trust an experiment with just one woman in each group. The results would depend too much on which group got lucky and received a woman with less frequent headaches, for example. If we assign many subjects to each group, however, the effects of chance will average out and there will be little difference in the *average* responses in the two groups unless the treatments themselves cause a difference. "Use enough subjects to reduce chance variation" is the third big idea of statistical design of experiments.

PRINCIPLES OF EXPERIMENTAL DESIGN

The basic principles of statistical design of experiments are

1. **Control** the effects of lurking variables on the response, most simply by comparing two or more treatments.

2. **Randomize**—use impersonal chance to assign subjects to treatments.

3. **Use enough subjects** in each group to reduce chance variation in the results.

We hope to see a difference in the responses so large that it is unlikely to happen just because of chance variation. We can use the laws of probability, which give a mathematical description of chance behavior, to learn if the treatment effects are larger than we would expect to see if only chance were operating. If they are, we call them *statistically significant*.

STATISTICAL SIGNIFICANCE

An observed effect so large that it would rarely occur by chance is called **statistically significant.**

If we observe statistically significant differences among the groups in a randomized comparative experiment, we have good evidence that the treatments actually caused these differences. You will often see the phrase "statistically significant" in scientific reports. The great advantage of randomized comparative experiments is that they can produce data that give good evidence for a cause-and-effect relationship between the explanatory and response variables. We know that in general a strong, *observed* association does not imply causation. A statistically significant association in data from a well-designed *experiment* does imply causation.

APPLY YOUR KNOWLEDGE ───────────────

8.4　**Can tea prevent cataracts?** Eye cataracts are responsible for over 40% of blindness around the world. Can drinking tea regularly slow the growth of cataracts? We can't experiment on people, so we use rats as subjects. Researchers injected 18 young rats with a substance that causes cataracts. One group of rats also received black tea extract; a second group received green tea extract; and a third got a placebo, a substance with no effect on the body. The response variable was the growth of cataracts over the next six weeks. Both tea extracts did slow cataract growth.[5]

(a) Following the model of Figure 8.3, outline the design of this experiment.

(b) The *Simple Random Sample* applet allows you to randomly assign subjects to more than two groups. Use the applet to choose an SRS of 6 of the 18 rats to form the first group. Which rats are in this group? The population hopper now contains the 12 remaining rats, in scrambled order. Click "Sample" again to choose an SRS of 6 of these to make up the second group. Which rats were chosen? The 6 rats remaining in the population hopper form the third group.

8.5 Benefit of acupuncture. Researchers investigated whether acupuncture can help prevent migraine attacks. Patients suffering from migraines were randomly assigned to either acupuncture, sham acupuncture (needles inserted in nontraditional anatomical points), or a waiting list for acupuncture. The treatments were administered weekly for 8 weeks and patients recorded headache days in diaries.

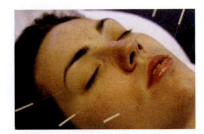

SuperStock/Alamy

(a) Following the model of Figure 8.3, outline the design of this experiment. Explain what purposes the sham acupuncture and the waiting-list groups serve.

(b) This experiment recruited a total of 302 patients. Explain step-by-step how you would assign the patients to each treatment.

8.6 Effectiveness of sport drinks. The Gatorade Sport Science Institute prepared a report on the scientific evidence supporting the benefits of sport drinks for soccer players. Here are some of the findings (creatine is a supplement added to sport drinks):

> *Creatine supplementation was tested in 14–19-year-old male soccer players from the 1st Yugoslav Junior League (Ostojic, 2004). Before and after 7 days of supplementation with either creatine or placebo, 10 players in each group performed a timed soccer-dribbling test around cones set 3 m apart, a sprint-power test that lasted about 3 s, a vertical-jump test, and a shuttle endurance run that lasted about 11 min. In contrast to two [other] studies of soccer skills, the authors correctly reported overall effects of creatine versus placebo, not simply pretest-posttest results within groups. The creatine group was significantly better than placebo for the dribbling test, the sprint test, and the vertical jump.*[6]

(a) A friend thinks that "significantly" in this article has its plain English meaning, roughly "I think this is important." Explain in simple language what "significantly better" tells us here.

(b) Your friend is puzzled by the comment on other studies not using a placebo. Explain what a placebo is and what using a placebo group in this study accomplishes.

Matched pairs and block designs

Completely randomized comparative designs are the simplest statistical designs for experiments. They illustrate clearly the principles of control, randomization, and adequate number of subjects. However, completely randomized designs are often inferior to more elaborate statistical designs. In particular, matching the subjects in various ways can produce more precise results than simple randomization.

One common design that combines matching with randomization is the *matched pairs design*. A matched pairs design compares just two treatments. Choose pairs of subjects that are as closely matched as possible. These can be individuals of the same sex, age, weight, and stature, for example, or they can be genetically related individuals such as twins or animals born in the same litter. Use chance to decide which subject in a pair gets the first treatment. The other subject in that pair gets the other treatment. That is, the random assignment of subjects to treatments is done within each matched pair, not for all subjects at once. Sometimes each "pair" in a matched pairs design consists of just one subject, who gets both treatments one after the other. In that case, each subject serves as his or her (or its) own control. *The order of the treatments can influence the subject's response, so we randomize the order for each subject.*

MATCHED PAIRS

A **matched pairs design** compares exactly two treatments, either by using a series of individuals that are closely matched two by two or by using each individual twice.

Matched pairs designs require that the assignment of the two treatments within each "pair" be **randomized** to avoid a systematic bias.

EXAMPLE 8.6 Does nature heal best?

Differences of electric potential occur naturally from point to point on a body's skin. Is the natural electric field strength best for helping wounds to heal? If so, changing the field will slow healing. Researchers compare the healing rates (in micrometers of skin growth per hour) after a small razor cut is made and either it is left to heal naturally or a change of electric potential is imposed.[7] Let's compare two designs for this experiment. The researchers have 14 anesthetized newts available for this experiment.

In a *completely randomized design*, all 14 newts would be assigned at random, 7 to heal naturally and 7 to heal while the skin electric potential is disturbed. In the *matched pairs design* that was actually used, all newts received two small razor cuts, one on each forelimb. One limb was left to heal naturally and the other received the electric field disturbance. The side receiving the disturbance was assigned at random, in case healing differed on the right and left sides.

Some newts spontaneously heal faster than others. The completely randomized design relies on chance to distribute the faster-healing newts roughly evenly between the two groups. The matched pairs design compares each newt's healing rate with and without the electric potential disturbance. This makes it easier to see the effects of disturbing the skin's natural electric potential.

Matched pairs designs use the principles of comparison of treatments and randomization. However, the randomization is not complete—we do not randomly assign all the subjects at once to the two treatments. Instead, we randomize only within each matched pair. This allows matching to reduce the effect of variation among the subjects. *Being able to distinguish a matched pairs design from a completely randomized design is critical because the design of an experiment impacts the statistical analysis and conclusions.*

A related design is the *block design,* in which *several* individuals sharing the same characteristic are pooled, forming a *block.*

BLOCK DESIGN

A **block** is a group of individuals that are known before the experiment to be similar in some way that is expected to affect the response to the treatments.

In a **block design,** the random assignment of individuals to treatments is carried out separately within each block.

Block designs can have blocks of any size. A block design combines the idea of creating equivalent treatment groups by matching with the principle of forming treatment groups at random. Blocks are another form of *control.* They control the effects of some outside variables by bringing those variables into the experiment to form the blocks. Here are some typical examples of block designs.

EXAMPLE 8.7 Gender differences

The progress of a type of cancer differs in women and men. A clinical experiment to compare three therapies for this cancer therefore treats sex as a blocking variable. Two separate randomizations are done, one assigning the female subjects to the treatments and the other assigning the male subjects. Figure 8.4 outlines the design of this experiment. Note that there is no randomization involved in making up the blocks. They are groups of subjects who differ in some way (gender in this case) that is apparent before the experiment begins.

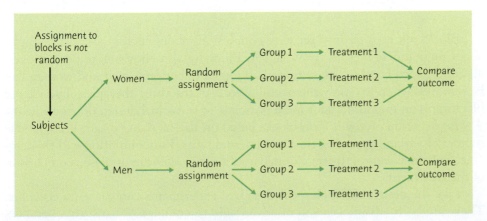

FIGURE 8.4 Outline of a block design comparing three cancer therapies for men and women, for Example 8.7.

EXAMPLE 8.8   Soil differences

The soil type and fertility of farmland differ by location. Because of this, experiments on plant growth often divide the available planting area into smaller fields making up the blocks. A test of the effect of tillage type (two types) and pesticide application (three application schedules) on soybean yields used small fields as blocks. Each block was further divided into six plots, and the six treatments were randomly assigned to plots separately within each block, as illustrated in Figure 8.5.

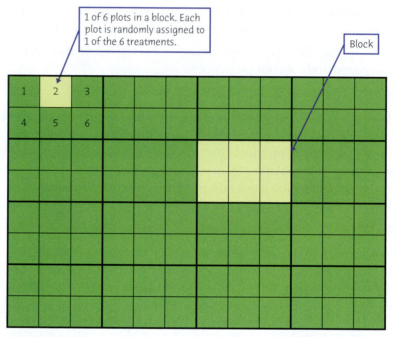

FIGURE 8.5   Outline of a block design, for Example 8.8. The blocks are land areas subdivided into six planting plots. The treatments are six combinations of tillage type and pesticide application.

Blocks allow us to draw separate conclusions about each block, for example, about men and women in Example 8.7. Blocking also allows more precise overall conclusions because the systematic differences between men and women can be removed when we study the overall effects of the three therapies. The idea of blocking is an important additional principle of statistical design of experiments. A wise experimenter will form blocks based on the most important unavoidable sources of variability among the experimental units. Randomization will then average out the effects of the remaining variation and allow an unbiased comparison of the treatments.

Like the design of samples, the design of experiments can get quite complex. There are both advantages and disadvantages to complex designs. In a complex design, the causal effect of one variable on another might be better isolated from the influences of other factors. On the other hand, the data collection, analysis, and interpretation are likely to also become much more complex. We will focus on simple completely randomized, matched pairs, and block designs.

APPLY YOUR KNOWLEDGE

8.7   **Cricket mating songs.** Male crickets sing to attract females, but singing also increases their risk of being found by a predator. What important function could the singing serve? The singing may provide females with insight into the health of male crickets. Or the singing may indicate a male's age, demonstrating the male's ability to survive. A researcher compares the female mating choices for healthy male crickets and for similar crickets that have been infected with intestinal parasites. Young male crickets 7 to 12 days old and older male crickets 15 days old or more are available for the experiment.

(a) Outline the design of this experiment, as in Figure 8.4. What type of design is it?

(b) Explain how you would carry out the randomization if you had 20 young crickets and 20 older crickets. (You need not actually do the randomization.)

8.8   **Comparing hand strength.** Is the right hand generally stronger than the left in right-handed people? You can crudely measure hand strength by placing a bathroom scale on a shelf with the end protruding, then squeezing the scale between the thumb below and the four fingers above it. The reading of the scale shows the force exerted.

(a) Describe the design of a matched pairs experiment to compare the strength of the right and left hands, using 10 right-handed people as subjects.

(b) Explain how you would carry out the randomization. (You need not actually do the randomization.)

8.9   **Alcohol in men and women.** Is drinking strong coffee more efficient in lowering blood alcohol content (BAC) than simply waiting after alcohol consumption? We should take into consideration that body fat content differs substantially between men and women overall, affecting how alcohol is absorbed and metabolized. Thirty subjects, 18 men and 12 women, agree to drink alcohol until their BAC reaches 0.08. The treatment is applied at that point, and the BAC is measured two hours later.

(a) Describe the design of a completely randomized comparative experiment to learn the effect of drinking coffee on BAC.

(b) Describe the design of a matched pairs experiment using the same 30 subjects.

(c) Describe the design of a block experiment considering men and women separately.

Cautions about experimentation

The logic of a randomized comparative experiment depends on our ability to treat all the subjects identically in every way except for the actual treatments being compared. Good experiments therefore require careful attention to details.

Scratch my furry ears

Rats and rabbits that are specially bred to be uniform in their inherited characteristics are the subjects in many experiments. Animals, like people, are quite sensitive to how they are treated. This can create opportunities for hidden bias. For example, human affection can change the cholesterol level of rabbits. Choose some rabbits at random and regularly remove them from their cages to have their heads scratched by friendly people. Leave other rabbits unloved. All the rabbits eat the same diet, but the rabbits that receive affection have lower cholesterol.

— **EXAMPLE 8.9** May I have some chocolate? —

The experiment on the effects of chocolate on headaches (Example 8.5) is a typical human experiment. All of the subjects received the same medical attention, and they all had tasted both preparations, one containing chocolate and the other carob (the placebo) before the experiment began. The investigators had asked a group of healthy subjects to taste both preparations ahead of time to verify that the two preparations were indistinguishable. The use of a control group ensured that the effect of chocolate would not be confounded with the placebo effect. This is particularly important because some subjects believed that chocolate triggers headaches. That belief alone might have made them anxious about eating chocolate, resulting in a headache. In addition, the study was *double-blind*: The subjects didn't know whether they were taking chocolate or a placebo, and neither did the investigators who gave them the preparation. The double-blind method avoids unconscious bias by, for example, a doctor who is convinced that chocolate triggers headaches.

DOUBLE-BLIND EXPERIMENTS

In a **double-blind** experiment, neither the subjects nor the people who interact with them know which treatment each subject is receiving.

Double-blinding is obviously necessary whenever the investigator evaluates the experimental outcome. In clinical trials, how a doctor interacts with a patient can also influence the patient's perception of his or her prognosis and thus create or reinforce a placebo effect. Double-blinding controls the influence of these human interactions by preventing potential bias. In many medical studies, only the statistician who does the randomization knows which treatment each patient is receiving.

Experimenter or subject bias is a very real problem that can plague both experiments and observational studies. For example, personal bias can skew our perceptions and data collection in field observations.

— **EXAMPLE 8.10** Humans and chimpanzees —

In the 1960s and 1970s, Jane Goodall defied scientific convention by naming the chimpanzees she observed over long periods of time instead of using a customary numerical ID. Her careful observations revealed displays of emotion and personalities in the chimpanzees, and she also documented the chimpanzees' creation and use of tools. The general belief at the time was that these were the very characteristics defining humanity alone.[8]

Using names can be construed as being anthropomorphic and therefore biased. But perhaps using numbers is just as biased because it reinforces a deep belief in human superiority and uniqueness.

Observational studies cannot address the problem of bias in a definitive manner. Experiments, on the other hand, provide the ability to control many aspects of the environment and data collection, reducing bias to a minimum. *Always be very cautious in your interpretation of an experiment if the likely sources of bias have not all been carefully controlled for.*

In practice, however, optimal controls are not always achievable. There are instances when the use of a placebo is unethical. And blinding might not always be practical. Imagine the complexity of keeping track of active and inactive pills for a number of different treatments under double-blind conditions involving hundreds of subjects for several years. Blinding is not even an option if patients are able to differentiate the treatment from the sham, as in the example below.

EXAMPLE 8.11   No placebo for marijuana ————————————

A study of the effects of marijuana recruited young men who used marijuana. Some were randomly assigned to smoke marijuana cigarettes, while others were given placebo cigarettes. This failed: The control group recognized that their cigarettes were phony and complained loudly. It may be quite common for blindness to fail because the subjects can tell which treatment they are receiving.[9]

Many—perhaps most—experiments have some weaknesses in detail. The environment of an experiment can influence the outcomes in unexpected ways. Although experiments are the gold standard for evidence of cause and effect, really convincing evidence usually requires that the study be **replicated** successfully by its investigators as well as by independent investigators from different locations. All researchers are required to keep detailed records of their experiments and must provide these records if other teams have been unsuccessful at replicating the published results.

replication

EXAMPLE 8.12   Environmental influence ————————————

To study genetic influence on behavior, experimenters "knock out" a gene in one group of mice and compare their behavior with that of a control group of normal mice. The results of these experiments often don't agree as well as hoped, so investigators did exactly the same experiment with the same genetic strain of mice in Oregon, Alberta (Canada), and New York. Many results were very different. It appears that small differences in the lab environments have big effects on the behavior of the mice. Remember this the next time you read that our genes control our behavior.[10]

The most serious potential weakness of experiments is **lack of realism.** The subjects or treatments or setting of an experiment may not realistically duplicate the conditions we really want to study. Carcinogenicity, for example, is often tested in extreme situations unlikely to reflect realistic human use.

lack of realism

EXAMPLE 8.13   Carcinogenicity of saccharin ————————————

The carcinogenicity of a product is its propensity to induce cancer when used, ingested, or simply present in the environment. Saccharin has been used as an artificial sweetener for almost a century. Because of experimental evidence of carcinogenicity in rats, in 1981 saccharin was listed in the *U.S. Report on Carcinogens* as reasonably anticipated to be a human carcinogen.

However, closer examination of the observed urinary bladder cancers in rats indicate that the biological mechanism of this cancer is specific to the rat urinary system and related to saccharin consumption at concentrations unlikely to represent realistic human consumption. Saccharin was officially delisted from the *Report on Carcinogens* in 2005 with the conclusion that "the factors thought to contribute to tumor induction by sodium saccharin in rats would not be expected to occur in humans."[11]

Lack of realism can limit our ability to apply the conclusions of an experiment to the settings of greatest interest. Most experimenters want to generalize their conclusions to some setting wider than that of the actual experiment. Example 7.5 (page 168) discussed how women were historically excluded from many clinical trials. A disturbing consequence is that medications were approved for use in women without knowing whether they would work at the dosage prescribed or what complications might be expected in this half of the population. It has only been 10 years since the Food and Drug Administration started requiring both male and female participation in clinical trials of drugs applicable to both sexes.

Statistical analysis of an experiment cannot tell us how far the results will generalize. The ability to extend conclusions beyond the experimental setting depends on a thorough discussion of the merits of the experiment and the representativeness of the subjects used. Nevertheless, a randomized comparative experiment, because of its ability to give convincing evidence for causation, is one of the most important ideas in statistics.

APPLY YOUR KNOWLEDGE

8.10   **Dealing with pain.** Health care providers are giving more attention to relieving the pain of cancer patients. An article in the journal *Cancer* surveyed a number of studies and concluded that controlled-release (CR) morphine tablets, which release the painkiller gradually over time, are more effective than giving standard morphine when the patient needs it. The "methods" section of the article begins: "Only those published studies that were controlled (i.e., randomized, double blind, and comparative), repeated-dose studies with CR morphine tablets in cancer pain patients were considered for this review."[12] Explain the terms in parentheses to someone who knows nothing about medical trials, and describe their purpose.

8.11   **Does meditation reduce anxiety?** An experiment that claimed to show that meditation reduces anxiety proceeded as follows. The experimenter interviewed the subjects and rated their level of anxiety. Then the subjects were randomly assigned to two groups. The experimenter taught one group how to meditate and they meditated daily for a month. The other group was simply told to relax more. At the end of the month, the experimenter interviewed all the subjects again and rated their anxiety level. The meditation group now had less anxiety. Psychologists said that the results were suspect because the ratings were not blind. Explain what this means and how lack of blindness could bias the reported results.

Ethics in experimentation

We have dedicated the bulk of this chapter to what kinds of experiments *can* be done and discussing their respective scientific merits. We will now ask what kinds of experiments *should* or *should not* be done. This critically important aspect of *ethics*   experimental design is the domain of **ethics.**

Ethical issues are particularly important in biological experimentation because investigators deal with live forms. From humans to animals and ecosystems, experimentation needs to be thought out not only to produce statistically powerful results but also to be *responsible*.

Experimental ethics is about all the questions we need to ask before, during, and after experimentation. But, unlike design, ethics doesn't always have clear-cut answers. Therefore, our objective here won't be to give you all the right answers but rather to help you ask all the right questions.

Ethics of human experimentation

Human experimentation undoubtedly lends itself to particular ethical scrutiny. All experiments involving humans raise ethical issues, even the mildest ones. Many experiments involving babies and young children have, for instance, helped us understand the nature and importance of attachment between a parent and a child. Yet we might legitimately ask what was the benefit to each subject and what negative impact, if any, the experiment might have had on them. These same questions become paramount in the area of medical experimentation where benefits and risks are clear and potentially great. Here are some questions we should ask ourselves:

- *Are all experiments acceptable in the name of science, knowledge, or the public's greater good?* Both medical ethics and international human rights standards say that "the interests of the subject must always prevail over the interests of science and society." The quoted words are from the 1964 Helsinki Declaration of the World Medical Association, the most respected international standard. The most outrageous examples of unethical experiments are those that ignore the interests of the subjects. Chief among them is the Tuskegee syphilis study discussed further below.

- *Is it morally right to include a placebo in a study design when there are safe and efficient treatments already available?* Because "the interests of the subject must always prevail," medical treatments can be tested in clinical trials only when there is reason to hope that they will help the patients who are subjects in the trials. Future benefits aren't enough to justify experiments with human subjects. So, why is it ethical to give a control group of patients a placebo? Well, we know that placebos often work. What is more, placebos have no harmful side effects other than withholding the benefits of a proven treatment. When we don't know whether the treatment works better than a placebo, it is ethical to try both and compare them. However, clinical trials involving a placebo are often modified half way through because the data already collected show that the treatment does better than the placebo. From then on, subjects who had received the placebo are given the better treatment.

- *Should we stop an experiment if we find that one of the treatments shows early signs of adverse effects? And what if one of the treatments shows early signs of clear superiority?* Again, "the interests of the subject must always prevail." Disregard of information gained in the course of experimentation or follow-up studies has been the basis of many private and class action lawsuits against pharmaceutical companies and is fueling the current debate over the prescription of VIOXX, for instance.

- *Is it ever acceptable to leave subjects in the dark about the purpose and expected results of an experiment?* This question goes hand in hand with the first question on what experiments are morally acceptable. Subjects should be able to judge for themselves the acceptability of their participation. That's why all human experiments are now required to obtain the subject's *informed consent* in writing. But can subjects truly grasp the purpose and consequences of an experiment? No amount of information can turn someone into a fully trained medical doctor. In fact, when information becomes too detailed and technical, it can have the opposite effect of simply being ignored, just as legal fine prints often are. And is complete informed consent achievable without compromising the experiment? Patients are informed that they may get a placebo, but they are not told whether they are getting it or not. That information would compromise the purpose of the placebo, but it would also likely affect the decision to participate.

- *Should subjects be rewarded for participation?* This is a tricky question related to informed consent and the freedom to choose participation. The poor and uninsured could easily become the primary targets for the riskiest experiments. In fact, it was once common to test new vaccines on prison inmates who gave their consent in return for good-behavior credit. The law now forbids medical experiments in prisons. And, of course, we must also learn to think globally in a world with disappearing frontiers: let us not repeat our Tuskegee studies in less privileged parts of the world simply because human rights have not yet fully matured there.

© CORBIS SYGMA

EXAMPLE 8.14 The Tuskegee syphilis study

The Tuskegee syphilis study is an extreme example of investigators following their own interests and ignoring the well-being of their subjects. In the 1930s, syphilis was common among black men in the rural South, a group that had almost no access to medical care. The Public Health Service recruited hundreds of poor black sharecroppers in order to observe how syphilis progressed when no treatment was given. The subjects were not aware that they were participants in an experiment and were actively deceived about it for decades. Beginning in 1943, penicillin became available to treat syphilis. The study subjects were not treated. In fact, the Public Health Service prevented any treatment until word leaked out and forced an end to the study in the 1970s. The horrifying details of the study and its context and legacy are discussed in greater detail below.

A 1996 review said about the study, "It has come to symbolize racism in medicine, ethical misconduct in human research, paternalism by physicians, and government abuse of vulnerable people." In 1997, President Clinton formally apologized to the surviving participants in a White House ceremony.[13]

The Tuskegee experiment was not the only unethical study or experiment ever conducted, not by a long shot. But its scale and horror were the catalyst for ethical reform of human experimentation in the United States. By law, all studies funded by the federal government must now obey the following principles:

- The organization that carries out the study must have an *institutional review board* that reviews all planned studies in advance in order to protect the subjects from possible harm.
- All individuals who are subjects in a study must give their *informed consent* in writing before data are collected. Subjects must be informed about the nature of a study and any risk of harm it may bring.
- All data on individuals must be kept *confidential*. Only statistical summaries for groups of subjects may be made public. Any breach of confidentiality is a serious violation of data ethics. The best practice is to separate the identity of the subjects from the rest of the data at once.

Human experimentation is not a simple one-sided problem, though. While it is important to question the necessity and ethicality of human experimentation, we also must consider the implications of not performing carefully designed experiments. There are many unfortunate examples of pharmaceutical products or surgical methods that have been widely used without this important preliminary stage. The result is that the whole population becomes unsuspecting test subjects of an experiment that has no guideline, no safeguard, and no way to reach informative conclusions about the benefits and dangers of the procedure.

Lobotomy, for instance, which is the severance of the frontal lobes from the rest of the brain, became a popular method in the 1950s and 1960s for dealing with psychotic individuals. It was performed on consenting and somewhat-informed individuals or with the approval of family members wishing to stop undesirable behavior (even on children). After thousands of lobotomies in the United States alone, it became clear that the procedure did provide improvement for some patients, but others showed no improvement at all and the rest suffered irreversible, lifelong psychological damage.[14]

This is not an isolated case. Surgical practice, especially, has evolved largely based on the ideas and practices of individual surgeons. Because surgery is by definition invasive, there has historically been a reluctance to perform randomized controlled experiments in this field out of concern for the subjects. However, with any new procedure, a placebo, or sham, operation may in fact prove as beneficial or more beneficial as the procedure tested and have many fewer or milder side effects. Gastric freezing for the treatment of stomach ulcers (described in Example 8.3) and mammary artery ligation for the treatment of angina (described in Exercise 15.2 page 387), are two medical procedures that were widely used before randomized comparative experiments showed that they performed no better than placebos and the procedures were abandoned. Likewise, the use of routine episiotomy to speed up child delivery and avoid natural tearing is now questioned by multiple studies.[15] Fortunately, these drastic practice reversals have led to a greater awareness of the necessity for careful experimentation to legitimate the use of medical and surgical procedures in humans.

DISCUSSION: The Tuskegee syphilis study

"For 40 years, the U.S. Public Health Service has conducted a study in which human guinea pigs, not given proper treatment, have died of syphilis and its side effects. The study was conducted to determine from autopsies what the disease does to the human body." The *Washington Evening Star* published these lines in a front-page article in the summer of 1972. The article would end one of the most controversial human experiments ever in America and result in drastic changes to federal regulations related to scientific and medical experimentation requiring human subjects.

The Public Health Service and the Tuskegee Institute began a study on syphilis in 1932. Nearly 400 poor black men with syphilis from Macon County, Alabama, were enrolled in the study. They were never told they had syphilis and were never treated for it even after penicillin was known to be a highly effective cure by 1943 and widely available by 1947. They were never told that they were now part of a medical experiment.

Syphilis

Syphilis is a sexually transmitted disease caused by the spirochete bacterium *Treponema pallidum*. The pathology progresses in stages, starting with hard ulcers primarily around the genitalia and followed by a transient, widespread rash. Without treatment the third stage of syphilis may develop, damaging the nervous system, eyes, heart, blood vessels, liver, bones, and joints, causing blindness, madness, and paralysis.

Untreated men can transmit the disease to their female partners, who, in turn, can transmit it to future offspring. About 40% of pregnancies where the mother has untreated syphilis result in stillbirth (baby dead at birth). If the baby survives birth, it then has a 40% to 70% chance of being infected with syphilis. If left untreated, infected newborns may later become retarded, have seizures, or die because of the disease.

How could this happen?

How could a study designed to deny treatment to uninformed subjects have come to be? The historical context can help us understand the logical journey.

At the end of the nineteenth century, philanthropists dreamed of black economic development, and the Tuskegee Institute was founded to help develop local schools, businesses, and agriculture. It was later expanded to fight the devastating impact of a syphilis pandemic that affected nearly a third of the African American population in the South. At that time, diagnosis and the currently known treatment were offered to participants.

When the stock market crashed in 1929, funds for the endeavor were cut. In 1932, the Tuskegee philanthropic effort was recycled into a study of the impact of untreated syphilis. In exchange for participating in the study, the men were given free medical exams, free meals, and free burial insurance. The study was originally intended to run for six months, but it lasted 40 years, until a whistle-blower turned to the press.

Acknowledging misconduct

The Tuskegee study was well known in the scientific community, and several papers were published in peer-reviewed journals. The first concerns over ethical aspects of the study were raised in 1968, but the study was terminated only after the outrage generated by the 1972 news article. It then took 25 more years for the government to voice a formal apology. Here is an excerpt from President Clinton's 1997 formal apology on behalf of the nation (the full transcript can be read at `http://clinton4.nara.gov/textonly/New/Remarks/Fri/19970516-898.html`):

> *The United States government did something that was wrong—deeply, profoundly, morally wrong. It was an outrage to our commitment to integrity and equality for all our citizens.*
>
> *To the survivors, to the wives and family members, the children and the grandchildren, I say what you know: No power on Earth can give you back the lives lost, the pain suffered, the years of internal torment and anguish. What was done cannot be undone. But we can end the silence. We can stop turning our heads away. We can look at you in the eye and finally say on behalf of the American people, what the United States government did was shameful, and I am sorry.*

How Tuskegee changed the ethics of human experimentation

The revelation of the Tuskegee syphilis study was the spark that initiated profound and long-lasting reforms of the ethical code of conduct for research requiring the use of human subjects in America. In 1974, the National Research Act was signed into law, creating the National Commission for the Protection of Human Subjects of Biomedical and Behavioral Research.

As a consequence, federally supported studies using human subjects now must be reviewed by institutional review boards, which decide whether study protocols meet ethical standards, and researchers must get voluntary informed consent from all study participants.

Complementing this approach, emphasis has also been placed on training researchers to be aware of the ethical issues related to human experimentation. For instance, the National Institutes of Health now require that all prebaccalaureate, predoctoral, and postdoctoral trainees supported by institutional training grants receive instruction in the responsible conduct of research.

Ethics of nonhuman experimentation

Experimentation on animals, ecosystems, cells, or embryos also raise many ethics questions. But standards for experiments involving these subjects are much more variable. Federal agencies regulate the use of animals for laboratory experiments, but the guidelines and scrutiny applied are widely different for different species. For example, primate experiments in the United States are closely monitored, in drastic contrast with research on insects. Rats are the typical animal model used for testing suspected carcinogenicity. Such experiments obviously do not have the animals' well-being in mind. But rats are considered pests routinely exterminated by farmers and house owners, and we consider the information gained from carcinogenicity studies important for the well-being and safety of entire human populations. Rat experiments used to test beauty products, on the other hand, have been much more controversial. Animal-rights activists advocate giving all animals (or at least species with some level of self-consciousness) the same ethical consideration as humans. The debate over animal experimentation is still far from being settled, and this is exemplified by the various laws, regulations, and practices enforced in different countries.

An alternative to actual experimentation with tremendous potential is to run "experiments" on the computer instead. With computers getting so powerful and fast and with our knowledge of the world ever deepening, we can *simulate* everything we know about a system and the forces that regulate it. The system can then be modified and observed, just as we would apply a treatment and observe the outcome in the real world.

Research on plants and ecosystems raises ethical issues too. The creation of genetically modified organisms (GMOs), for example, has been extremely controversial because of its long-term consequences.

EXAMPLE 8.15 Genetically modified organisms (GMOs)

The most common GMOs are crop plants modified to resist pests, disease, or herbicide, improve flavor, or increase shelf life. They range from soya, wheat, and corn to potatoes, peas, tomatoes, and even tobacco, among many others, with the United States leading their creation and use.

Improving crops through selective breeding has been a trademark of human evolution since we transitioned away from a hunter-gatherer species. So why are GMOs controversial?

One important issue is that the modifications we have engineered can be passed on to wild plants or other crops. Weeds could then become herbicide resistant, and pests could evolve resistance to all our current pesticides. The ethical concern is thus about the long-term implications of creating GMOs. This is not a hypothetical concern. Recent studies have shown that this has already happened, such as the contamination of Mexican traditional maize (corn) varieties by neighboring U.S. transgenic corn crops.

A completely different ethical concern is the appropriation of plant species. A company that creates a GMO owns its patent and all rights to exploitation. Some protect their assets by selling seeds with a genetic modification that renders the crops unable to produce fertile seeds. Farmers are then forced to keep buying seeds from the producer, creating potentially extremely dangerous monopolies.[16]

Finally, even the status of cells has been a subject of ethical controversy when these cells have the potential to become human beings. Stem cell research and its applications have been debated all over the world with varying results. Within the United States, the debate has changed over time. In 2001 President Bush changed the U.S. policy on human embryonic stem cell research by allowing federal funds to be used for research on existing stem cell lines.[17] In 2004, California passed its stem cell initiative allowing for $3 billion in funding over 10 years for stem cell research and research facilities in California.[18]

The answers to questions of ethics in experimentation are rarely easy. But perhaps as important as the decisions we make is keeping the debate alive and open to public discussion.

APPLY YOUR KNOWLEDGE

8.12   **Informed consent.** Long ago, doctors drew a blood specimen from you as part of treating minor anemia. Unknown to you, the sample was stored. Now researchers plan to used stored samples from you and many other people to look for genetic factors that may influence anemia. It is no longer possible to ask your consent. Modern technology can read your entire genetic makeup from the blood sample.

(a) Do you think it violates the principle of informed consent to use your blood sample if your name is on it but you were not told that it might be saved and studied later?

(b) Suppose that your identity is not attached. The blood sample is known only to come from (say) "a 20-year-old white female being treated for anemia." Is it now OK to use the sample for research?

(c) Perhaps we should use biological materials such as blood samples only from patients who have agreed to allow the material to be stored for later use in research. It isn't possible to say in advance what kind of research, so this falls short of the usual standard for informed consent. Is it nonetheless acceptable, given complete confidentiality and the fact that using the sample can't physically harm the patient?

8.13   **What is ethical?** Effective drugs for treating AIDS are very expensive, so most African nations cannot afford to give them to large numbers of people. Yet AIDS is more common in parts of Africa than anywhere else. Several clinical trials are looking at ways to prevent pregnant mothers infected with HIV from passing the infection to their unborn children, a major source of HIV infections in Africa. Some people say these trials are unethical because they do not give effective AIDS drugs to their subjects (the mothers), as would be required in rich nations. Others reply that the trials are looking for treatments that can work in the real world in Africa and that they promise benefits at least to the children of their subjects. What do you think?

CHAPTER 8 SUMMARY

In an experiment, we impose one or more **treatments** on individuals, often called **subjects.** Each treatment is a combination of values of the explanatory variables, which we call **factors.**

The **design** of an experiment describes the choice of treatments and the manner in which the subjects are assigned to the treatments.

The basic principles of statistical design of experiments are **control** and **randomization** to combat bias and **using enough subjects** to reduce chance variation.

The simplest form of control is **comparison.** Experiments should compare two or more treatments in order to avoid **confounding** of the effect of a treatment with other influences, such as lurking variables.

Randomization uses chance to assign subjects to the treatments. Randomization creates treatment groups that are similar (except for chance variation) before the treatments are applied. Randomization and comparison together prevent **bias,** or systematic favoritism, in experiments.

You can carry out randomization by using software or by giving numerical labels to the subjects and using a **table of random digits** to choose treatment groups.

Applying each treatment to many subjects reduces the role of chance variation and makes the experiment more sensitive to differences among the treatments.

Good experiments require attention to detail as well as good statistical design. Many behavioral and medical experiments are **double-blind.** Some give a **placebo** to a control group. **Replicating** an experiment helps ensure that the results were not merely due to unusual conditions. **Lack of realism** in an experiment can prevent us from generalizing its results.

In addition to comparison, a second form of control is to restrict randomization by forming **blocks** of individuals that are similar in some way that is important to the response. Randomization is then carried out separately within each block.

Matched pairs are a common form of blocking for comparing just two treatments. In some matched pairs designs, each subject receives both treatments in a random order. In others, the subjects are matched in pairs as closely as possible, and each subject in a pair receives one of the treatments.

In the end, **ethics** must be the overarching principle guiding experimentation. No matter what experiments could be done and what we might gain from them, we need to ask ourselves if the end justifies the means and to be open to a continuing debate.

CHECK YOUR SKILLS

8.14   A study of cell phones and the risk of brain cancer looked at a group of 469 people who have brain cancer. The investigators matched each cancer patient with a person of the same sex, age, and race who did not have brain cancer, and then asked about use of cell phones. This is

(a) an observational study.

(b) an uncontrolled experiment.

(c) a randomized comparative experiment.

8.15   What electrical changes occur in muscles as they get tired? Student subjects hold their arms above their shoulders until they can no longer do so. Meanwhile, the electrical activity in their arm muscles is measured. This is

(a) an observational study.

(b) an uncontrolled experiment.

(c) a randomized comparative experiment.

8.16   Can changing the diet reduce high blood pressure? Vegetarian diets and low-salt diets are both promising. Men with high blood pressure are assigned at random to four diets: (1) normal diet with unrestricted salt; (2) vegetarian with unrestricted salt; (3) normal with restricted salt; and (4) vegetarian with restricted salt. This experiment has

(a) one factor, the choice of diet.

(b) two factors, normal/vegetarian diet and unrestricted/restricted salt.

(c) four factors, the four diets being compared.

8.17   An important response variable in the experiment described in Exercise 8.16 is

(a) the amount of salt in the subject's diet.

(b) which of the four diets a subject is assigned to.

(c) change in blood pressure after 8 weeks on the assigned diet.

8.18   In the experiment described in Exercise 8.16, the 240 subjects are labeled 001 to 240. Software assigns an SRS of 60 subjects to Diet 1, then an SRS of 60 of the remaining 180 to Diet 2, then an SRS of 60 of the remaining 120 to Diet 3. The 60 who are left get Diet 4. This is a

(a) completely randomized design.

(b) block design, with four blocks.

(c) matched pairs design.

8.19   The Food and Drug Administration should be reluctant to use the results of the experiment described in Exercise 8.16 for its guideline on diets for Americans with high blood pressure because

(a) the study used a matched pairs design instead of a completely randomized design.

(b) results from men with high blood pressure may not generalize to the population of all Americans with high blood pressure.

(c) the study did not use a placebo and was not double-blind.

8.20   A medical experiment compares an antidepression medicine with a placebo for relief of chronic headaches. There are 36 headache patients available to serve as subjects. To choose 18 patients to receive the medicine, you would

(a) assign labels 01 to 36 and use Table A to choose 18.

(b) assign labels 01 to 18, because only 18 need to be chosen.

(c) assign the first 18 who signed up for the experiment to get the medicine.

8.21   The Community Intervention Trial for Smoking Cessation asked whether a community-wide advertising campaign would reduce smoking. The researchers located 11 pairs of communities for which the members of each pair were similar in location, size, economic status, and so on. One community in each pair participated in the advertising campaign and the other did not. This is

(a) an observational study.

(b) a matched pairs experiment.

(c) a completely randomized experiment.

8.22   To decide which community in each pair in the previous exercise should get the advertising campaign, it is best to

(a) toss a coin.

(b) choose the community that will help pay for the campaign.

(c) choose the community with a mayor who will participate.

8.23   An experimental design randomly assigns volunteer arthritis patients suffering from chronic pain to take either a new pain medication, aspirin (the traditional nonsteroidal anti-inflammatory treatment), or a placebo. An institutional review board discusses whether the experiment should include a placebo. This is

(a) an ethical issue.

(b) an issue of lack of replication.

(c) an issue of lack of realism.

CHAPTER 8 EXERCISES

In all exercises that require randomization, you may use Table A, the Simple Random Sample applet, or other software.

8.24 **Wine, beer, or spirits?** Example 7.2 (page 165) describes a study that compared three groups of people: the first group drinks mostly wine, the second drinks mostly beer, and the third drinks mostly spirits. This study is comparative, but it is not an experiment. Why not?

8.25 **Treating breast cancer.** The most common treatment for breast cancer discovered in its early stages was once removal of the breast. It is now usual to remove only the tumor and nearby lymph nodes, followed by radiation. To study whether these treatments differ in their effectiveness, a medical team examines the records of 25 large hospitals and compares the survival times after surgery of all women who have had either treatment.

(a) What are the explanatory and response variables?

(b) Explain carefully why this study is not an experiment.

(c) Explain why confounding will prevent this study from discovering which treatment is more effective. (The current treatment was in fact recommended after several large randomized comparative experiments.)

8.26 **Wine, beer, or spirits?** You have recruited 300 adults aged 45 to 65 who are willing to follow your orders about alcohol consumption over the next five years. You want to compare the effects on heart disease of moderate drinking of just wine, just beer, or just spirits. Outline the design of a completely randomized experiment to do this. (No such experiment has been done because subjects aren't willing to have their drinking regulated for years, and because there might be a risk of subjects developing alcohol addiction.)

8.27 **Marijuana and work.** How does smoking marijuana affect willingness to work? Canadian researchers persuaded young adult men who used marijuana to live for 98 days in a "planned environment." The men earned money by weaving belts. They used their earnings to pay for meals and other consumption and could keep any money left over. One group smoked two potent marijuana cigarettes every evening. The other group smoked two weak marijuana cigarettes. All subjects could buy more cigarettes but were given strong or weak cigarettes, depending on their group. Did the weak and strong groups differ in work output and earnings?[19]

(a) Outline the design of this experiment with 20 subjects available.

(b) Here are the names of the 20 subjects. Use software or Table A at line 131 to carry out the randomization your design requires.

Abate	Dubois	Gutierrez	Lucero	Rosen
Afifi	Engel	Huang	McNeill	Thompson
Brown	Fluharty	Iselin	Morse	Travers
Cheng	Gerson	Kaplan	Quinones	Ullmann

8.28   **Growing in the shade.** You have 45 pine seedlings available for the experiment described in Exercise 8.1. Outline the design of this experiment. Use software or Table A to randomly assign seedlings to the three treatment groups.

8.29   **Exercise and heart attacks.** Does regular exercise reduce the risk of a heart attack? Here are two ways to study this question. Explain clearly why the second design will produce more trustworthy data.

1. A researcher finds 2000 men over 40 who exercise regularly and have not had heart attacks. She matches each with a similar man who does not exercise regularly, and she follows both groups for 5 years.

2. Another researcher finds 4000 men over 40 who have not had heart attacks and are willing to participate in a study. She assigns 2000 of the men to a regular program of supervised exercise. The other 2000 continue their usual habits. The researcher follows both groups for 5 years.

8.30   **Improving adolescents' health habits.** Twenty-four public middle schools agree to participate in the experiment described in Exercise 8.2. Use a diagram to outline a completely randomized design for this experiment. Do the randomization required to assign schools to treatments. If you use the *Simple Random Sample* applet or other software, choose all four treatment groups. If you use Table A, start at line 105 and choose only the first two groups.

8.31   **Relieving headaches.** Doctors identify "chronic tension-type headaches" as headaches that occur almost daily for at least six months. Can antidepressant medications or stress management training reduce the number and severity of these headaches? Are both together more effective than either alone?

(a) Use a diagram like Figure 8.1 to display the treatments in a design with two factors: medication yes or no and stress management yes or no. Then outline the design of a completely randomized experiment to compare these treatments. Forty subjects are available.

(b) The headache sufferers named below have agreed to participate in the study. Randomly assign the subjects to the treatments. If you use the *Simple Random Sample* applet or other software, assign all the subjects. If you use Table A, start at line 130 and assign subjects to only the first treatment group.

Abbott	Decker	Herrera	Lucero	Richter
Abdalla	Devlin	Hersch	Masters	Riley
Alawi	Engel	Hurwitz	Morgan	Samuels
Broden	Fuentes	Irwin	Nelson	Smith
Chai	Garrett	Jiang	Nho	Suarez
Chuang	Gill	Kelley	Ortiz	Upasani
Cordoba	Glover	Kim	Ramdas	Wilson
Custer	Hammond	Landers	Reed	Xiang

8.32   **Frappuccino Light?** Here's the opening of a press release from June 2004: "Starbucks Corp. on Monday said it would roll out a line of blended coffee drinks intended to tap into the growing popularity of reduced-calorie and reduced-fat menu choices for Americans." You wonder if Starbucks customers like the new "Mocha Frappuccino Light" as well as regular Frappuccino coffee.

(a) Describe a matched pairs design to answer this question. Be sure to include proper blinding of your subjects.

(b) You have 20 regular Starbucks customers on hand. Use the *Simple Random Sample* applet or Table A at line 141 to do the randomization that your design requires.

8.33 **Growing trees faster.** The concentration of carbon dioxide (CO_2) in the atmosphere is increasing rapidly due to our use of fossil fuels. Because green plants use CO_2 to fuel photosynthesis, more CO_2 may cause trees to grow faster. An elaborate apparatus allows researchers to pipe extra CO_2 to a 30-meter circle of forest. We want to compare the growth in base area of trees in treated and untreated areas to see if extra CO_2 does in fact increase growth. We can afford to treat three circular areas.[20]

(a) Describe the design of a completely randomized experiment using 6 well-separated 30-meter circular areas in a pine forest. Sketch the circles and carry out the randomization your design calls for.

(b) Areas within the forest may differ in soil fertility. Describe a matched pairs design using three pairs of circles that will reduce the extra variation due to different fertility. Sketch the circles and carry out the randomization your design calls for.

8.34 **Athletes taking oxygen.** We often see players on the sidelines of a football game inhaling oxygen. Their coaches think this will speed their recovery. We might measure recovery from intense exercise as follows: Have a football player run 100 yards three times in quick succession. Then allow three minutes to rest before running 100 yards again. Time the final run. Because players vary greatly in speed, you plan a matched pairs experiment using 25 football players as subjects. Discuss the design of such an experiment to investigate the effect of inhaling oxygen during the rest period.

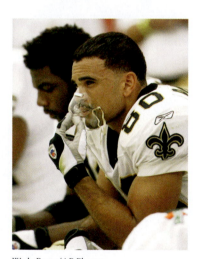

Wade Payne/AP Photos

8.35 **Protecting ultramarathon runners.** An ultramarathon, as you might guess, is a footrace longer than the 26.2 miles of a marathon. Runners commonly develop respiratory infections after an ultramarathon. Will taking 600 milligrams of vitamin C daily reduce these infections? Researchers randomly assigned ultramarathon runners to receive either vitamin C or a placebo. Separately, they also randomly assigned these treatments in a group of nonrunners the same age as the runners. All subjects were watched for 14 days after the big race to see if infections developed.[21]

(a) What is the name for this experimental design?

(b) Use a diagram to outline the design.

8.36 **Reducing spine fractures.** Fractures of the spine are common and serious among women with advanced osteoporosis (low mineral density in the bones). Can taking strontium renelate help? A large medical trial assigned 1649 women to take either strontium renelate or a placebo each day. All of the subjects had osteoporosis and had suffered at least one fracture. All were taking calcium supplements and receiving standard medical care. The response variables were measurements of bone density and counts of new fractures over three years. The subjects were treated at 10 medical centers in 10 different countries.[22] Outline a block design for this experiment, with the medical centers as blocks. Explain why this is the proper design.

8.37   Does ginkgo improve memory? The law allows marketers of herbs and other natural substances to make health claims that are not supported by evidence. Brands of ginkgo extract claim to "improve memory and concentration." A randomized comparative experiment found no evidence for such effects.[23] The subjects were 230 healthy people over 60 years old. They were randomly assigned to ginkgo or a placebo pill (a dummy pill that looks and tastes the same). All the subjects took a battery of tests for learning and memory before treatment started and again after six weeks.

(a) Following the model of Figure 8.3, outline the design of this experiment.

(b) Use the *Simple Random Sample* applet, other software, or Table A to assign half the subjects to the ginkgo group. If you use software, report the first 20 members of the ginkgo group (in the applet's "Sample bin") and the first 20 members of the placebo group (those left in the "Population hopper"). If you use Table A, start at line 103 and choose only the first 5 members of the ginkgo group.

Blickwinkel/Alamy

8.38   Wine, beer, or spirits? Women as a group develop heart disease much later than men. We can improve the completely randomized design of Exercise 8.26 by using women and men as blocks. Your 300 subjects include 120 women and 180 men. Outline a block design for comparing wine, beer, and spirits. Be sure to say how many subjects you will put in each group in your design.

8.39   Prayer and meditation. You read in a magazine that "nonphysical treatments such as meditation and prayer have been shown to be effective in controlled scientific studies for such ailments as high blood pressure, insomnia, ulcers, and asthma." Explain in simple language what the article means by "controlled scientific studies." Why can such studies in principle provide good evidence that, for example, meditation is an effective treatment for high blood pressure?

8.40   Quick randomizing. Here's a quick and easy way to randomize. You have 100 subjects, 50 women and 50 men. Toss a coin. If it's heads, assign the men to the treatment and the women to the control group. If the coin comes up tails, assign the women to treatment and the men to control. This gives every individual subject a 50-50 chance of being assigned to treatment or control. Why isn't this a good way to randomly assign subjects to treatment groups?

8.41   Explaining medical research. Observational studies had suggested that vitamin E reduces the risk of heart disease. Careful experiments showed that vitamin E has no effect, at least for women. According to the *Journal of the American Medical Association*:

> *Thus, vitamin E enters the category of therapies that were promising in epidemiologic and observational studies but failed to deliver in adequately powered randomized controlled trials. As in other studies, the "healthy user" bias must be considered, ie, the healthy lifestyle behaviors that characterize individuals who care enough about their health to take various supplements are actually responsible for the better health, but this is minimized with the rigorous trial design.*[24]

A friend who knows no statistics asks you to explain this.

(a) What is the difference between observational studies and experiments?

(b) What is a "randomized controlled trial?" (We'll discuss "adequately powered" in Chapter 15.)

(c) How does "healthy user bias" explain how people who take vitamin E supplements have better health in observational studies but not in controlled experiments?

8.42 **Do antioxidants prevent cancer?** People who eat lots of fruits and vegetables have lower rates of colon cancer than those who eat little of these foods. Fruits and vegetables are rich in "antioxidants" such as vitamins A, C, and E. Will taking antioxidants help prevent colon cancer? A medical experiment studied this question with 864 people who were at risk of colon cancer. The subjects were divided into four groups: daily beta-carotene, daily vitamins C and E, all three vitamins every day, or daily placebo. After four years, the researchers were surprised to find no significant difference in colon cancer among the groups.[25]

(a) What are the explanatory and response variables in this experiment?

(b) Outline the design of the experiment. Use your judgment in choosing the group sizes.

(c) The study was double-blind. What does this mean?

(d) What does "no significant difference" mean in describing the outcome of the study?

(e) Suggest some lurking variables that could explain why people who eat lots of fruits and vegetables have lower rates of colon cancer. The experiment suggests that these variables, rather than the antioxidants, may be responsible for the observed benefits of fruits and vegetables.

8.43 **An herb for depression?** Does the herb Saint-John's-wort relieve major depression? A study examined this issue. Here are some excerpts from the published report:[26]

> *Design: Randomized, double-blind, placebo-controlled clinical trial. . . . Participants . . . were randomly assigned to receive either Saint-John's-wort extract (n = 98) or placebo (n = 102). . . . The primary outcome measure was the rate of change in the Hamilton Rating Scale for Depression over the treatment period.*

(a) Use the information provided to draw a diagram outlining the design of this study.

(b) The report concluded that Saint-John's-wort is no more effective than a placebo at relieving symptoms of depression. Based on the information quoted, would you be inclined to trust this conclusion? Explain your answer.

superclic/Alamy

8.44 **Growing tomato plants.** An experiment measures the amounts of various nutrients in tomato plants subjected to three levels of nitrogen fertilization. To avoid confounding by the soil properties, the planting area was divided into 12 blocks within which subplots were randomly assigned to the 3 treatments. Draw an outline of this block design as in Figure 8.5.

8.45 **Reading a medical journal.** The article in the *New England Journal of Medicine* that presents the final results of the Physicians' Health Study begins with these words: "The Physicians' Health Study is a randomized, double-blind, placebo-controlled trial designed to determine whether low-dose aspirin (325 mg every other day) decreases cardiovascular mortality and whether beta carotene reduces the incidence of cancer."[27] Doctors are expected to understand this. Explain to a doctor who knows no statistics what "randomized," "double-blind," and "placebo-controlled" mean.

8.46   **Calcium and blood pressure.** Some medical researchers suspect that added calcium in the diet reduces blood pressure. You have available 40 men with high blood pressure who are willing to serve as subjects.

(a) Outline an appropriate design for the experiment, taking the placebo effect into account.

(b) The researchers conclude that "the blood pressure of the calcium group was significantly lower than that of the placebo group." "Significant" in this conclusion means statistically significant. Explain what "statistically significant" means in the context of this experiment, as if you were speaking to a doctor who knows no statistics.

8.47   **Sickle-cell disease.** Sickle-cell disease is an inherited disorder of the red blood cells that in the United States affects mostly blacks. It can cause severe pain and many complications. Can the drug hydroxyurea reduce the severe pain caused by sickle-cell disease? A study by the National Institutes of Health gave the drug to 150 sickle-cell sufferers and a placebo (a dummy medication) to another 150. The researchers then counted the episodes of pain reported by each subject.

(a) Outline the design of this experiment.

(b) It might seem humane to give all the subjects hydroxyurea in the hope that it will help them. Explain clearly why this would not provide information about the effectiveness of the drug.

(c) In fact, the experiment was stopped ahead of schedule because the hydroxyurea group had only half as many pain episodes as the control group. Explain why it would no longer be ethical to proceed with this study design once evidence of the effectiveness of hydroxyurea has been uncovered.

8.48   **What is minimal risk?** Federal regulations say that "minimal risk" means the risks are no greater than "those ordinarily encountered in daily life or during the performance of routine physical or psychological examinations or tests." You are a member of your college's institutional review board. You must decide whether several research proposals qualify for lighter review because they involve only minimal risk to subjects. Which of these do you think qualifies as "minimal risk"?

(a) Draw a drop of blood by pricking a finger in order to measure blood sugar.

(b) Draw blood from the arm for a full set of blood tests.

(c) Insert a tube that remains in the arm, so that blood can be drawn regularly.

8.49   **World ethics.** Researchers from Yale, working with medical teams in Tanzania, wanted to know how common infection with the AIDS virus is among pregnant women in that African country. To do this, they planned to test blood samples drawn from pregnant women. Yale's institutional review board insisted that the researchers get the informed consent of each woman and tell her the results of the test. This is the usual procedure in developed nations. The Tanzanian government did not want to tell the women why blood was drawn or tell them the test results. The government feared panic if many people turned out to have an incurable disease for which the country's medical system could not provide care. The study was canceled. Do you think that Yale was right to apply its usual standards for protecting subjects?

8.50 **Institutional review boards.** Your college or university has an institutional review board that screens all studies that use human subjects. Get a copy of the document that describes this board (you can probably find it online).

(a) According to this document, what are the duties of the board?

(b) How are members of the board chosen? How many members are not scientists? How many members are not employees of the college? Do these members have some special expertise, or are they simply members of the "general public"?

8.51 **Randomization avoids bias.** Suppose that you have 50 subjects available, 25 women and 25 men, to test reaction time to a visual cue in the presence or absence of a noise distraction. Let's label the women from 1 to 25 and the men from 26 to 50. We hope that randomization will distribute these subjects roughly equally between the noise and no-noise groups. Use the *Simple Random Sample* applet to take 20 samples of size 25 from the 50 subjects. (Be sure to click "Reset" after each sample.) Record the counts of women (labels 1 to 25) in each of your 20 samples. You see that there is considerable chance variation but no systematic bias in favor of one or the other group in assigning the women. Larger samples from a larger population will on the average do an even better job of creating two similar groups.

In this chapter we cover...

The idea of probability

Probability models

Probability rules

Discrete probability models

Continuous probability models

Random variables

Personal probability*

Risk and odds*

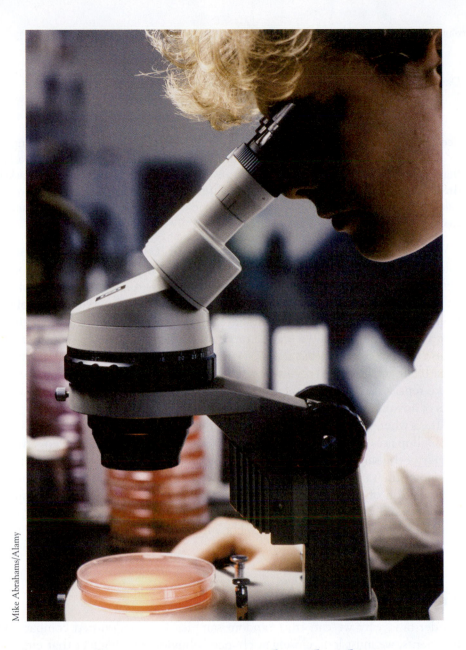

Mike Abrahams/Alamy

Introducing Probability

Why is probability, the mathematics of chance behavior, needed to understand statistics, the science of data? Let's look at a typical sample survey.

┌─ **EXAMPLE 9.1**   Do you believe in natural evolution? ─────────

What proportion of all adults believe in natural (Darwinian) evolution? We don't know, but we do have results from a Gallup Poll. In September 2005, Gallup took a random sample of 1005 adults. The poll found that 121 of the people in the sample believed in

Darwinian evolution. The proportion who believed in Darwinian evolution was

$$\text{sample proportion} = \frac{121}{1005} = 0.12 \quad \text{(that is, 12\%)}$$

Because all adults had the same chance to be among the chosen 1005, it seems reasonable to use this 12% as an estimate of the unknown proportion in the population. It's a *fact* that 12% of the sample believe in Darwinian evolution—we know because Gallup asked them. We don't know what percent of all adults believe in Darwinian evolution, but we *estimate* that about 12% do. This is a basic move in statistics: Use a result from a sample to estimate something about a population.

What if Gallup took a second random sample of 1005 adults? The new sample would have different people in it. It is almost certain that there would not be exactly 121 responses for Darwinian evolution. That is, Gallup's estimate of the proportion of adults who believe in Darwinian evolution will vary from sample to sample. Could it happen that one random sample finds that 12% of adults believe in Darwinian evolution and a second random sample finds that 32% do? *Random samples eliminate bias from the act of choosing a sample, but they can still be wrong because of the variability that results when we choose at random.* If the variation when we take repeat samples from the same population is too great, we can't trust the results of any one sample.

This is where we need facts about probability to make progress in statistics. Because Gallup uses chance to choose its samples, the laws of probability govern the behavior of the samples. Gallup says that the probability is 0.95 that an estimate from one of their samples comes within ±3 percentage points of the truth about the population of all adults. The first step toward understanding this statement is to understand what "probability 0.95" means. Our purpose in this chapter is to understand the language of probability, but without going into the mathematics of probability theory.

The idea of probability

To understand why we can trust random samples and randomized comparative experiments, we must look closely at chance behavior. The big fact that emerges is this: **Chance behavior is unpredictable in the short run but has a regular and predictable pattern in the long run.**

Toss a coin, or choose an SRS. The result can't be predicted in advance, because the result will vary when you toss the coin or choose the sample repeatedly. But there is still a regular pattern in the results, a pattern that emerges clearly only after many repetitions. This remarkable fact is the basis for the idea of probability.

> **EXAMPLE 9.2**   *Coin tossing*

When you toss a coin, there are only two possible outcomes, heads or tails. Figure 9.1 shows the results of tossing a coin 5000 times on two separate occasions. For each number of tosses from 1 to 5000, we have plotted the proportion of those tosses that gave a head. Trial A (solid blue line) begins tail, head, tail, tail. You can see that the proportion of heads for Trial A starts at 0 on the first toss, rises to 0.5 when the second toss gives a

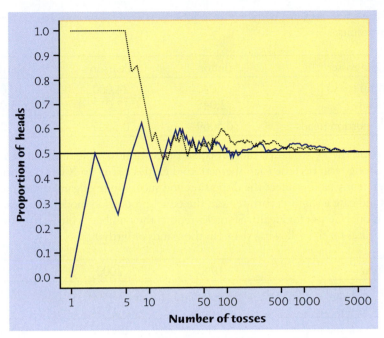

FIGURE 9.1 The proportion of tosses of a coin that give a head changes as we make more tosses. Eventually, however, the proportion approaches 0.5, the probability of a head. This figure shows the results of two trials of 5000 tosses each.

head, then falls to 0.33 and 0.25 as we get two more tails. Trial B, on the other hand, starts with five straight heads, so the proportion of heads is 1 until the sixth toss.

The proportion of tosses that produce heads is quite variable at first. Trial A starts low and Trial B starts high. As we make more and more tosses, however, the proportion of heads for both trials gets close to 0.5 and stays there. If we made yet a third trial at tossing the coin a great many times, the proportion of heads would again settle down to 0.5 in the long run. This is the intuitive idea of probability. Probability 0.5 means "occurs half the time in a very large number of trials." The probability 0.5 appears as a horizontal line on the graph.

We might suspect that a coin has probability 0.5 of coming up heads just because the coin has two sides. But we can't be sure. The coin might be unbalanced. In fact, spinning a penny or nickel on a flat surface, rather than tossing the coin, doesn't give heads probability 0.5. The idea of probability is empirical. That is, it is based on observation rather than theorizing. Probability describes what happens in very many trials, and we must actually observe many trials to pin down a probability. In the case of tossing a coin, some diligent people have in fact made thousands of tosses.

EXAMPLE 9.3 Theory and practice

How does practice agree with theory? When it comes to coin tossing, we have a few historical examples of very large trial numbers. The French naturalist Count Buffon (1707–1788) tossed a coin 4040 times. Around 1900, the English statistician Karl Pearson heroically tossed a coin 24,000 times. While imprisoned by the Germans during World

War II, the South African mathematician John Kerrich tossed a coin 10,000 times. Here are their results:

Coin tosser	Buffon	Pearson	Kerrich
Total tosses	4,040	24,000	10,000
Heads	2,048	12,012	5,067
Proportion heads	0.5069	0.5005	0.5067

Genetic theory would predict an equal proportion of male and female newborns over the long term because each spermatozoid carries either an X or a Y chromosome. However, many factors affect the dispersion and success rate of gametes, and birth certificates indicate a slight departure from the equal-proportion model. For example, the U.S. National Center for Health Statistics reports the number of births by gender in the United States for the following years (in thousands of births):[1]

Year	1990	2000	2002
Males	2129	2077	2058
Females	2029	1982	1964
Proportion males	0.5120	0.5117	0.5117

RANDOMNESS AND PROBABILITY

We call a phenomenon **random** if individual outcomes are uncertain but there is nonetheless a regular distribution of outcomes in a large number of repetitions.

The **probability** of any outcome of a random phenomenon is the proportion of times the outcome would occur in a very long series of repetitions.

That some things are random is an observed fact about the world. The outcome of a coin toss, the time between emissions of particles by a radioactive source, and the sexes of the next litter of lab rats are all random. So is the outcome of a random sample or a randomized experiment. Probability theory is the branch of mathematics that describes random behavior. Of course, we can never observe a probability exactly. We could always continue tossing the coin, for example. Mathematical probability is an idealization based on imagining what would happen in an indefinitely long series of trials.

The best way to understand randomness is to observe random behavior, as in Figure 9.1. You can do this with physical devices like coins, but computer simulations (imitations) of random behavior allow faster exploration. The *Probability* applet is a computer simulation that animates Figure 9.1. It allows you to choose the probability of a head and simulate any number of tosses of a coin with that probability. Experience shows that the proportion of heads gradually settles down close to the probability. Equally important, it also shows that *the proportion in a small or moderate number of tosses can be far from the probability. Probability describes only what happens in the long run.*

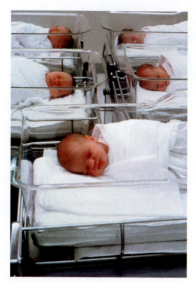

Randy Duchaine/CORBIS

APPLET

CAUTION

Computer simulations like the *Probability* applet start with given probabilities and imitate random behavior, but we can estimate a real-world probability only by actually observing many trials. Nonetheless, computer simulations are very useful because we need long runs of trials. In situations such as coin tossing, the proportion of an outcome often requires several hundred trials to settle down to the probability of that outcome. Shorter runs give only rough estimates of a probability.

APPLY YOUR KNOWLEDGE

9.1   Hemophilia. Hemophilia refers to a group of rare hereditary disorders of blood coagulation. Because the disorder is caused by defective genes on the X chromosome, hemophilia affects primarily men. According to the Centers for Disease Control and Prevention, the prevalence of hemophilia (that is, the number in the population with hemophilia at any given time) among American males is 13 in 100,000. Explain carefully what this means. In particular, explain why it does *not* mean that if you obtain the medical records of 100,000 males, exactly 13 will be diagnosed with hemophilia.

9.2   Random digits. The table of random digits (Table A) was produced by a random mechanism that gives each digit probability 0.1 of being a 0.

(a) What proportion of the first 50 digits in the table are 0s? This proportion is an estimate, based on 50 repetitions, of the true probability, which in this case is known to be 0.1.

(b) The *Probability* applet can imitate random digits. Set the probability of heads in the applet to 0.1. Check "Show true probability" to show this value on the graph. A head stands for a 0 in the random digit table, and a tail stands for any other digit. Simulate 200 digits (40 at a time—don't click "Reset"). If you kept going forever, presumably you would get 10% heads. What was the result of your 200 tosses?

9.3   Probability says ... Probability is a measure of how likely an event is to occur. Match one of the probabilities that follow with each statement of likelihood given. (The probability is usually a more exact measure of likelihood than is the verbal statement.)

$$0 \quad 0.01 \quad 0.3 \quad 0.6 \quad 0.99 \quad 1$$

(a) This event is impossible. It can never occur.

(b) This event is certain. It will occur on every trial.

(c) This event is very unlikely, but it will occur once in a while in a long sequence of trials.

(d) This event will occur more often than not.

Probability models

Gamblers have known for centuries that the fall of coins, cards, and dice displays clear patterns in the long run. After all, the first formal studies of probabilities were aimed at understanding various games of chance. The idea of probability rests on the observed fact that the average result of many thousands of chance outcomes can be known with near certainty. How can we give a mathematical description of long-run regularity?

To see how to proceed, think first about a very simple random phenomenon, the birth of one child. When the child is conceived, we cannot know the outcome in advance. What do we know? We are willing to say that the outcome will be either male or female. We believe that each of these outcomes has probability 1/2. This description of a child's birth has two parts:

- A list of possible outcomes.
- A probability for each outcome.

Such a description is the basis for all probability models. Here is the basic vocabulary we use.

PROBABILITY MODELS

The **sample space S** of a random phenomenon is the set of all possible outcomes.

An **event** is an outcome or a set of outcomes of a random phenomenon. That is, an event is a subset of the sample space.

A **probability model** is a mathematical description of a random phenomenon consisting of two parts: a sample space S and a way of assigning probabilities to events.

A sample space S can be very simple or very complex. When one child is born, there are only two outcomes, male and female. The sample space is S = {M, F}. When Gallup records the body weights of a random sample of 1000 adults, the sample space contains all possible weights of 1000 of the 225 million adults in the country.

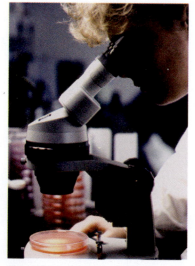

Mike Abrahams/Alamy

EXAMPLE 9.4   Blood types

Your blood type greatly impacts, for instance, the kind of blood transfusion or organ transplant you can safely get. There are 8 different blood types based on the presence or absence of certain molecules on the surface of red blood cells. A person's blood type is given as a combination of a group (O, A, B, or AB) and a Rhesus factor (+ or −). The combinations make up the sample space S:

$$S = \{ O+, O-, A+, A-, B+, B-, AB+, AB- \}$$

How can we assign probabilities to this sample space? First of all, these 8 blood types are represented differently in different ethnic groups. Within a given ethnic group, we can use the blood types' frequencies in that group to assign their respective probabilities. The American Red Cross reports that, among Asian Americans, there are 39% blood type O+, 1% O−, 27% A+, 0.5% A−, 25% B+, 0.4% B−, 7% AB+, 0.1% AB−.[2] Because 39% of all Asian Americans have blood type O+, the probability that a randomly chosen Asian American has blood type O+ is 39%, or 0.39. We can thus construct the complete probability model for blood types among Asian Americans:

Blood type	O+	O−	A+	A−	B+	B−	AB+	AB−
Probability	0.39	0.01	0.27	0.005	0.25	0.004	0.07	0.001

In Example 9.4, we used the known frequencies of blood types among Asian Americans to construct the probability model. In some cases, we can use known properties of the random phenomenon (for example, physical properties, genetic laws) to compute the probabilities of the outcomes in the sample space. Here is an example.

		First child	
		Girl	Boy
Second child	Girl	GG	BG
	Boy	GB	BB

FIGURE 9.2 The four possible outcomes in gender sequence for couples having two children, for Example 9.5. If we assume that male and female newborns are equally likely, all four of these outcomes have the same probability.

EXAMPLE 9.5   A boy or a girl?

Young couples often discuss how many children they would like to have. One couple wants to have two children. There are four possible outcomes when we examine the gender of the two children in order (first child, second child). Figure 9.2 displays these four outcomes.

Parents often care more about how many boys or girls they could have than about the particular order. The sample space for the number of girls a couple with two children could have is

$$S = \{0, 1, 2\}$$

What are the probabilities for this sample space? For each newborn, the probabilities that it will be a girl or a boy are approximately equal. We also know that the gender of the first child does not influence the gender of the second child. Therefore, all four outcomes in Figure 9.2 will be *equally likely*. That is, each of the four outcomes will, in the long run, occur for one-fourth of all couples with two children. So each outcome in Figure 9.2 has probability 1/4. However, the three possible outcomes in our sample space are *not* equally likely, because there are two ways to have exactly one girl and only one way to have no girl at all. So "no girl" has probability 1/4 but "one girl" has probability 2/4 (2 outcomes from Figure 9.2). Here is the complete probability model:

Number of girls	0	1	2
Probability	1/4	2/4	1/4

We built the probability model in Example 9.5 by assuming equal probability of both genders at birth. This model is reasonably accurate. However, we have seen in Example 9.3 that national data point to a slight imbalance between the genders at birth. So, in reality, all four outcomes in Figure 9.2 are not exactly equally likely.

APPLY YOUR KNOWLEDGE

9.4   **Sample space.** Choose a student at random from a large statistics class. Describe a sample space S for each of the following. (In some cases you may have some freedom in specifying S.)

(a) Ask whether the student is male or female.

(b) Ask how tall the student is.

(c) Ask what the student's blood type is.

(d) Ask how many times a day the student brushes his or her teeth.

(e) Ask how long it has been since the student's last flu or cold.

9.5   **More on boys and girls.** A couple wants to have three children. Assume that the probabilities of a newborn being male or female are the same and that the gender of one child does not influence the gender of another child.

(a) There are 8 possible arrangements of girls and boys. What is the sample space for having three children (gender of the first, second, and third child)? All 8 arrangements are equally likely.

(b) The future parents are wondering how many boys they might get if they have three children. Give a probability model (sample space and probabilities of outcomes) for the number of boys. Follow the method of Example 9.5.

Probability rules

Examples 9.4 and 9.5 describe pretty simple random phenomena. However, we don't always have a probability model available to answer our questions. We can make progress by listing some facts that must be true for *any* assignment of probabilities. These facts follow from the idea of probability as "the long-run proportion of repetitions on which an event occurs."

1. **Any probability is a number between 0 and 1.** Any proportion is a number between 0 and 1, so any probability is also a number between 0 and 1. An event with probability 0 never occurs, and an event with probability 1 occurs on every trial. An event with probability 0.5 occurs in half the trials in the long run.

2. **All possible outcomes together must have probability 1.** Because some outcome must occur on every trial, the sum of the probabilities for all possible outcomes must be exactly 1.

3. **If two events have no outcomes in common, the probability that one or the other occurs is the sum of their individual probabilities.** If one event occurs in 40% of all trials, a different event occurs in 25% of all trials, and the two can never occur together, then one or the other occurs on 65% of all trials because 40% + 25% = 65%.

4. **The probability that an event does not occur is 1 minus the probability that the event does occur.** If an event occurs in (say) 70% of all trials, it fails to occur in the other 30%. The probability that an event occurs and the probability that it does not occur always add to 100%, or 1.

We can use mathematical notation to state Facts 1 to 4 more concisely. Capital letters near the beginning of the alphabet denote events. If A is any event, we write its probability as $P(A)$. Here are our probability facts in formal language. As you apply these rules, remember that they are just another form of intuitively true facts about long-run proportions.

PROBABILITY RULES

Rule 1. The probability $P(A)$ of any event A satisfies $0 \leq P(A) \leq 1$.

Rule 2. If S is the sample space in a probability model, then $P(S) = 1$.

Rule 3. Two events A and B are **disjoint** if they have no outcomes in common and so can never occur together. If A and B are disjoint,

$$P(A \text{ or } B) = P(A) + P(B)$$

This is the **addition rule for disjoint events.**

Rule 4. For any event A,

$$P(A \text{ does not occur}) = 1 - P(A)$$

The addition rule extends to more than two events that are disjoint in the sense that no two have any outcomes in common. If events A, B, and C are disjoint, the probability that one of these events occurs is $P(A) + P(B) + P(C)$.

What is the chance of rain?

We have all checked the weather forecast to see if it would rain on our weekend plans. A report says that the chance of rain in your city tomorrow is 20%. What does that mean and how is this probability calculated? The National Weather Service keeps a historical database of daily weather conditions such as temperature, pressure, and humidity. The chance of rain on a given day is calculated as the percent of days in the database with similar weather conditions that had rain. So, a 20% chance of rain tomorrow means that it has rained in only 20% of days with similar weather conditions.

EXAMPLE 9.6   Using the probability rules

We already used the addition rule, without calling it by that name, to find the probabilities in Example 9.5. The event "one girl" contains the two disjoint outcomes displayed in Example 9.5, so the addition rule (Rule 3) says that its probability is

$$P(\text{one girl}) = P(\text{GB}) + P(\text{BG})$$
$$= \frac{1}{4} + \frac{1}{4}$$
$$= \frac{2}{4} = 0.5$$

Check that the probabilities in Example 9.5, found using the addition rule, are all between 0 and 1 and add to exactly 1. That is, this probability model obeys Rules 1 and 2.

What is the probability that a couple with two children would not have two girls? By Rule 4,

$$P(\text{couple does not have two girls}) = 1 - P(\text{has two girls})$$
$$= 1 - 0.25 = 0.75$$

APPLY YOUR KNOWLEDGE

9.6   **What's your weight?** According to the U.S. Census Bureau, 24% of American males 18 years of age or older are obese, 43% are overweight, 32% have a healthy weight, and 1% are underweight.[3]

(a) Does this assignment of probabilities to adult American males satisfy Rules 1 and 2?

(b) What percent of adult American males have a weight higher than what is considered healthy?

(c) The Census Bureau reports that the percent of adult American females with a weight higher than what is considered healthy is 50%. Does the assignment of probabilities for males and females over a healthy weight satisfy Rules 1 and 2?

9.7 **Rabies in Florida.** Rabies is a viral disease of mammals transmitted through the bite of a rabid animal. The virus infects the central nervous system, causing encephalopathy and ultimately death. The Florida Department of Health reports the distribution of documented cases of rabies for all of 2004:[4]

Species	Raccoon	Fox	Bat	Other
Probability	0.70	0.15	0.10	?

(a) What probability should replace "?" in the distribution?

(b) What is the probability that a reported case of rabies is not a raccoon?

(c) What is the probability that a reported case of rabies is either a bat or a fox?

Discrete probability models

Examples 9.4, 9.5, and 9.6 illustrate one way to assign probabilities to events: assign a probability to every individual outcome, and then add these probabilities, if necessary, to find the probability of any event. This idea works well when there are only a finite (fixed and limited) number of outcomes.

DISCRETE PROBABILITY MODEL

A probability model with a sample space made up of a list of individual outcomes[5] is called **discrete.**

To assign probabilities in a discrete model, list the probabilities of all the individual outcomes. These probabilities must be numbers between 0 and 1 and must have a sum of 1. The probability of any event is the sum of the probabilities of the outcomes making up the event.

EXAMPLE 9.7 Hearing impairment in Dalmatians

Pure dog breeds are often highly inbred, leading to high numbers of congenital defects. A study examined hearing impairment in 5333 Dalmatians.[6] Call the number of ears impaired (deaf) in a randomly chosen Dalmatian X for short. The researchers found the following probability model for X:

X	0	1	2
Probability	0.70	0.22	0.08

Check that the probabilities of the outcomes sum to exactly 1. This is therefore a legitimate discrete probability model.

The probability that a randomly chosen Dalmatian has some hearing impairment is the probability that X is equal to or greater than 1:

$$P(X \geq 1) = P(X = 1) + P(X = 2)$$
$$= 0.22 + 0.08 = 0.30$$

Almost a third of Dalmatians are deaf in one or both ears. This is a very high proportion that may be explained in part by the fact that breeders cannot detect partial deafness behaviorally. The study suggested giving the dogs a hearing test before considering them for breeding.

Note that the probability that X is equal to or greater than 1 is not the same as the probability that X is strictly greater than 1. The latter probability is

CAUTION

$$P(X > 1) = P(X = 2) = 0.08$$

The outcome $X = 1$ is included in "equal to or greater than" and is not included in "strictly greater than."

APPLY YOUR KNOWLEDGE

9.8 **Breakfast cereals.** The EESEE story "Nutrition and Breakfast Cereals" provides the nutritional content of 77 brands of packaged breakfast cereals. Let's call X the fat content in grams per serving. The researchers found the following probability model for X:

X	0	1	2	3	4	5
Probability	0.35	0.39	0.18	0.07	0.00	0.01

Consider the events

$$A = \{\text{fat content is 4 g or greater}\}$$
$$B = \{\text{fat content is 5 g or less}\}$$

(a) What outcomes make up the event A? What is $P(A)$?

(b) What outcomes make up the event B? What is $P(B)$?

(c) What outcomes make up the event "A or B"? What is $P(A \text{ or } B)$? Why is this probability not equal to $P(A) + P(B)$?

9.9 **Watching TV.** Some argue that millions of Americans are so hooked on television that their viewing habits fit the criteria for substance abuse as defined in the official psychiatric manual.[7] Choose a young person (aged 19 to 25) at random and ask, "In the past seven days, how many days did you watch television?" Call the response X for short. Here is a probability model for the response:[8]

Days X	0	1	2	3	4	5	6	7
Probability	0.04	0.03	0.06	0.08	0.09	0.08	0.05	0.57

(a) Verify that this is a legitimate discrete probability model.

(b) Describe the event $X < 7$ in words. What is $P(X < 7)$?

(c) Express the event "watched TV at least once" in terms of X. What is the probability of this event?

Continuous probability models

When we use the table of random digits to select a digit between 0 and 9, the discrete probability model assigns probability 1/10 to each of the 10 possible outcomes. Suppose that we want to choose a number at random between 0 and 1, allowing *any* number between 0 and 1 as the outcome. Software random number generators will do this. The sample space is now an entire interval of numbers:

$$S = \{\text{all numbers between 0 and 1}\}$$

Call the outcome of the random number generator Y for short. How can we assign probabilities to such events as $\{0.3 \leq Y \leq 0.7\}$? As in the case of selecting a random digit, we would like all possible outcomes to be equally likely. But we cannot assign probabilities to each individual value of Y and then add them, because there are infinitely many possible values.

We use a new way of assigning probabilities directly to events—as *areas under a density curve*. Density curves are models for continuous distributions. Any density curve has area exactly 1 underneath it, corresponding to total probability 1.

DENSITY CURVE

A **density curve** is a curve that

- is always on or above the horizontal axis, and

- has area exactly 1 underneath it.

A density curve describes the overall pattern of a distribution. The area under the curve and above any range of values on the horizontal axis is the proportion of all observations that fall in that range.

In Chapter 1 we described continuous distributions with histograms. Sometimes the overall pattern of a large number of observations is so regular that we can describe it by a smooth curve.

EXAMPLE 9.8   From histogram to density curve

Figure 9.3 is a histogram of the heights of a large government sample of women 40 to 49 years of age.[9] Overall, the distribution of heights is quite regular. The histogram is symmetric, and both tails fall off smoothly from a single center peak. There are no large gaps or obvious outliers. The smooth curve drawn over the histogram is a good description of the overall pattern of the data. Our eyes respond to the *areas* of the bars in a histogram. The bar areas represent the percents of the observations. Figure 9.4(a)

Jon Feingersh/zefa/Corbis

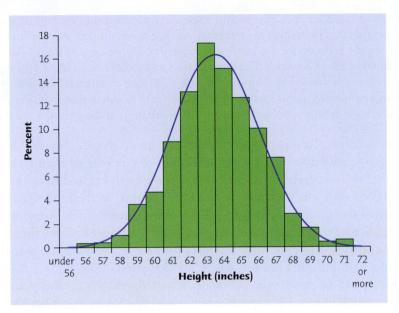

FIGURE 9.3 Histogram of the heights in inches of women aged 40 to 49 in the United States, for Example 9.8. The smooth curve shows the overall shape of the distribution.

is a copy of Figure 9.3 with the leftmost bars shaded. The area of the shaded bars represents the women 62 inches tall or less. They make up 30.7% of all women in the sample—this is the cumulative percent, the sum of all bars for 62 inches and below.

Now look at the curve drawn through the bars. In Figure 9.4(b), the area under the curve to the left of 62 inches is shaded. The smooth curve we use to model the

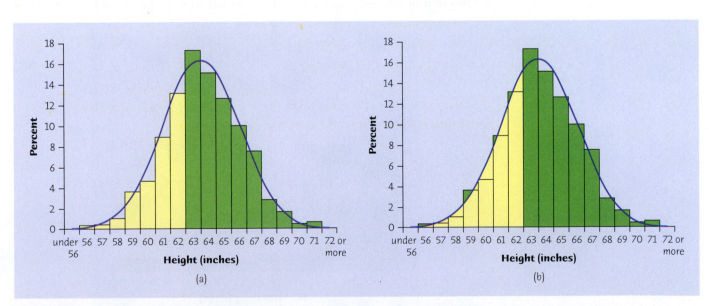

FIGURE 9.4 (a) The proportion of women 62 inches tall or less in the sample is 0.307. (b) The proportion of women 62 inches tall or less calculated from the density curve is 0.316. The density curve is a good approximation to the distribution of the data.

histogram distribution is chosen with the specific constraint that *the total area under the curve is exactly 1*. The total area represents 100%, that is, all the observations. We can then interpret areas under the curve as proportions of the observations. The curve is now a *density curve*. The shaded area under the density curve in Figure 9.4(b) represents the proportion of women in their 40s who are 62 inches or shorter. This area is 31.6%, less than 1 percentage point away from the actual 30.7%. Areas under the density curve give quite good approximations to the actual distribution of the sampled women.

Density curves, like distributions, come in many shapes. Figure 9.5 shows a strongly skewed distribution, the survival times of guinea pigs from Example 1.6 (page 16). The histogram and density curve were both created from the data by software. Both show the overall shape and the "bumps" in the long right tail. The density curve shows a higher single peak as a main feature of the distribution. The histogram divides the observations near the peak between two bars, thus reducing the height of the peak. A density curve is often a good description of the overall pattern of a distribution. Outliers, which are deviations from the overall pattern, are not described by the curve. *Of course, no set of real data is exactly described by a density curve. The curve is a model, an idealized description that is easy to use and accurate enough for practical purposes.* Conceptually, a density curve is similar to a regression line: We use a least-squares regression line to model an observed linear trend and to make predictions about similar individuals in the population.

Measures of center and spread apply to density curves as well as to actual sets of observations. Areas under a density curve represent proportions of the total number of observations. The median is the point with half the observations on either side. Thus, *the median of a density curve is the equal-areas point*. The mean of a set of observations is their arithmetic average. If we think of observations as a series of different weights strung out along a thin rod, the mean is the point at

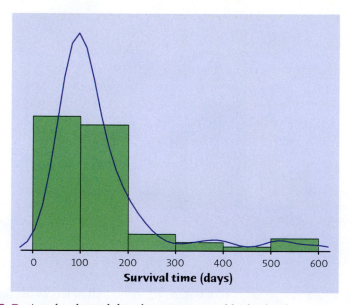

FIGURE 9.5 A right-skewed distribution, pictured by both a histogram and a density curve, representing the survival times in days of guinea pigs infected with a pathogen.

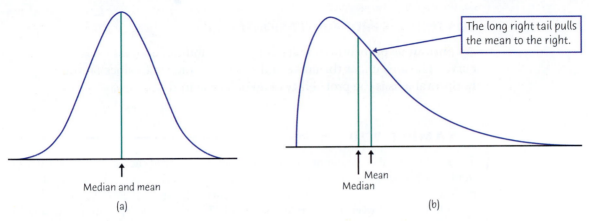

The long right tail pulls
the mean to the right.

Median and mean

(a)

Mean
Median

(b)

FIGURE 9.6 The mean and median of (a) a symmetrical density curve and (b) a
skewed density curve.

which the rod would balance. This fact is also true of density curves. *The mean is
the point at which the curve would balance if made of solid material.*

Because density curves are idealized patterns, a symmetric density curve is exactly symmetric. The mean and median of a symmetric density curve are therefore
equal, as shown on Figure 9.6(a). The mean of a skewed distribution is pulled toward the long tail more than is the median, as shown on Figure 9.6(b) (finding the
mean and standard deviation of a density curve is beyond the scope of this textbook). Density curves represent the whole population. Their mean and standard
deviation are expressed with Greek letters to distinguish them from the mean $\bar{x}$
and standard deviation s computed from actual sample observations. The usual
notation for the **mean of a density curve** is μ (the Greek letter mu). We write the
standard deviation of a density curve as σ (the Greek letter sigma).

mean μ
standard deviation σ

How do we use density curves to assign probabilities over continuous intervals? There is a direct relationship between the representation (or proportion) of
a given type of individual in a population and the probability that one individual randomly selected from the population will be of that given type. Just as with
discrete probabilities, we define probabilities over continuous intervals by the frequency of relevant individuals in the population—in this case, all individuals from
the population that belong to the desired interval.

EXAMPLE 9.9   From population distribution to probability distribution

In Example 9.8, we found that the proportion of women in their 40s who are 62 inches
or shorter is 31.6%. This is the area under the density curve for heights of 62 inches or
less, as shown in Figure 9.4(b).

Let's now ask: What is the probability that a randomly chosen woman in her 40s has
a height of 62 inches or less? Because the selection is random, this probability depends
on the frequency of women in the population who are 62 inches tall or shorter, and that
frequency is 31.6%. Therefore, the probability that a randomly selected woman in her
40s would measure 62 inches or less is 0.316. This is the area under the density curve for
heights 62 inches or less.

> **CONTINUOUS PROBABILITY MODEL**
>
> A **continuous probability model** assigns probabilities as areas under a density curve. The area under the curve and above any range of values on the horizontal axis is the probability of an outcome in that range.

EXAMPLE 9.10   *Random numbers*

The random number generator will spread its output uniformly across the entire interval from 0 to 1 as we allow it to generate a long sequence of numbers. The results of many trials are represented by the uniform density curve shown in Figure 9.7. This density curve has height 1 over the interval from 0 to 1. The area under the curve is 1, and the probability of any event is the area under the curve and above the event in question.

As Figure 9.7(a) illustrates, the probability that the random number generator produces a number between 0.3 and 0.7 is

$$P(0.3 \leq Y \leq 0.7) = 0.4$$

because the area under the density curve and above the interval from 0.3 to 0.7 is 0.4. The height of the curve is 1 and the area of a rectangle is the product of height and length, so the probability of any interval of outcomes is just the length of the interval. Similarly,

$$P(Y \leq 0.5) = 0.5$$
$$P(Y > 0.8) = 0.2$$
$$P(Y \leq 0.5 \text{ or } Y > 0.8) = 0.7$$

The last event consists of two nonoverlapping intervals, so the total area above the event is found by adding two areas, as illustrated by Figure 9.7(b). This assignment of probabilities obeys all of our rules for probability.

The probability model for a continuous random variable assigns probabilities to intervals of outcomes rather than to individual outcomes. In fact, *all continuous probability models assign probability 0 to any individual outcome. Only intervals of values have positive probability.* To see that this is true, consider a specific outcome such

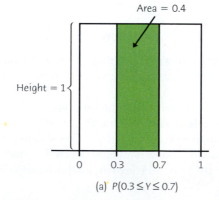
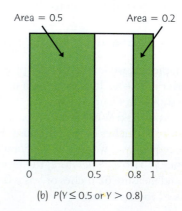

FIGURE 9.7 Probability as area under a density curve, for Example 9.10. This uniform density curve spreads probability evenly between 0 and 1.

as $P(Y = 0.8)$ in Example 9.10. The probability of any interval is the same as its length. The point 0.8 has no length, so its probability is 0.

We can use any density curve to assign probabilities. The density curves that are most familiar to us are the *Normal curves.* We will describe these important distributions in a following chapter.

APPLY YOUR KNOWLEDGE

9.10 Sketch density curves. Sketch density curves that describe distributions with the following shapes:

(a) Symmetric, but with two peaks (that is, two strong clusters of observations).

(b) Single peak and skewed to the left.

9.11 Random numbers. Let X be a random number between 0 and 1 produced by the random number generator described in Example 9.10 and Figure 9.7. Find the following probabilities:

(a) $P(X \leq 0.4)$

(b) $P(X < 0.4)$

(c) $P(0.3 \leq X \leq 0.5)$

(d) $P(X < 0.3 \ or \ X > 0.5)$

9.12 TV viewing among high school students. TV viewing promotes obesity because of physical inactivity and high exposure to commercials for foods with high fat and sugar content. In the two examples provided here, decide whether the probability model is discrete or continuous. Explain your reasoning.

(a) According to the National Center for Health Statistics, there is a 38% chance that a high school student watches at least 3 hours of television on an average school day.[10] This nationwide government survey recorded the amount of time high school students watched TV on an average school day.

(b) A large sample survey of school-age children by the Kaiser Family Foundation found a 42% chance that the TV is on most of the time at home.[11] The survey asked whether, at home, the TV was on most of the time, some of the time, a little bit of the time, or never.

Random variables

Examples 9.7 and 9.10 use a shorthand notation that is often convenient. In Example 9.7, we let X stand for the result of choosing a Dalmatian at random and assessing its hearing impairment. We know that X would take a different value if we made another random choice. Because its value changes from one random choice to another, we call the hearing impairment X a *random variable*.

RANDOM VARIABLE

A **random variable** is a variable whose value is a numerical outcome of a random phenomenon.

The **probability distribution** of a random variable X tells us what values X can take and how to assign probabilities to those values.

There are two main types of random variables, corresponding to two types of probability models: *discrete* and *continuous*.

EXAMPLE 9.11 Discrete and continuous random variables

discrete random variable

The hearing impairment X in Example 9.7 is a random variable whose possible values are the whole numbers {0, 1, 2}. The distribution of X assigns a probability to each of these outcomes. Random variables that have a countable (typically finite) list of possible outcomes are called **discrete.**

Compare the random number generator Y in Example 9.10. The values of Y fill the entire interval of numbers between 0 and 1. The probability distribution of Y is given by its density curve, shown in Figure 9.7. Random variables that can take on any value in an interval, with probabilities given as areas under a density curve, are called **continuous.**

continuous random variable

APPLY YOUR KNOWLEDGE

9.13　**Discrete or continuous random variable?** Indicate in the following examples whether the random variable X is discrete or continuous. Explain your reasoning.

(a) X is the number of days last week that a randomly chosen child exercised for at least one hour.

(b) X is the amount of time in hours that a randomly chosen child spends watching television today.

(c) X is the number of television sets in a randomly chosen household.

9.14　**Discrete or continuous random variable?** Indicate in the following examples whether the random variable X is discrete or continuous. Explain your reasoning.

(a) X is the number of petals on a randomly chosen daisy.

(b) X is the stem length in centimeters of a randomly chosen daisy.

(c) X is the number of daisies found in a randomly chosen grassy area 1 square meter in size.

(d) X is the average number of petals per daisy computed from all the daisies found in a randomly chosen grassy area 1 square meter in size.

Personal probability*

We began our discussion of probability with one idea: The probability of an outcome of a random phenomenon is the proportion of times that outcome would occur in a very long series of repetitions. This idea ties probability to actual outcomes. It allows us, for example, to estimate probabilities by simulating random phenomena. Yet we often meet another, quite different, idea of probability.

EXAMPLE 9.12 Intelligent life in the universe

Joe reads an article discussing the Search for Extraterrestrial Intelligence (SETI) project. We ask Joe, "What's the chance that we will find evidence of extraterrestrial intelligence in this century?" Joe responds, "Oh, about 1%."

---

*This section is optional.

Does Joe assign probability 0.01 to humans' finding extraterrestrial intelligence this century? The outcome of our search is certainly unpredictable, but we can't reasonably ask what would happen in many repetitions. This century will happen only once and will differ from all other centuries in many ways, especially technology. If probability measures "what would happen if we did this many times," Joe's 0.01 is not a probability. The frequentist definition of probability is based on data about many repetitions of the same random phenomenon. Joe is giving us something else, his personal judgment.

Although Joe's 0.01 isn't a probability in the frequentist sense, it gives useful information about Joe's opinion. Closer to home, a government asking, "How likely is it that building a new nuclear power plant will pay off within five years?" can't employ an idea of probability based on many repetitions of the same thing. The opinions of science and business advisers are nonetheless useful information, and these opinions can be expressed in the language of probability. These are *personal probabilities*.

PERSONAL PROBABILITY

A **personal probability** of an outcome is a number between 0 and 1 that expresses an individual's judgment of how likely the outcome is.

Rachel's opinion about the finding of extraterrestrial intelligence may differ from Joe's, and the opinions of several advisers about the new power plant may differ. Personal probabilities are indeed personal: They vary from person to person. Moreover, a personal probability can't be called right or wrong. If we say, "In the long run, this coin will come up heads 60% of the time," we can find out if we are right by actually tossing the coin several thousand times. If Joe says, "I think there is a 1% chance of finding extraterrestrial intelligence this century," that's just Joe's opinion.

Why think of personal probabilities as probabilities? Because any set of personal probabilities that makes sense obeys the same basic Rules 1 to 4 that describe any legitimate assignment of probabilities to events. If Joe thinks there's a 1% chance that we find extraterrestrial intelligence this century, he must also think that there's a 99% chance that we won't. There is just one set of rules of probability, even though we now have two interpretations of what probability means.

APPLY YOUR KNOWLEDGE

9.15 Will you have an accident? The probability that a randomly chosen adult will be involved in a car accident in the next year is about 0.2. This is based on the proportion of millions of drivers who have accidents. "Accident" includes things like crumpling a fender in your own driveway, not just highway accidents.

(a) What do you think is your own probability of being in an accident in the next year? This is a personal probability.

(b) Give some reasons why your personal probability might be a more accurate prediction of your "true chance" of having an accident than the probability for a random driver.

(c) Almost everyone says their personal probability is lower than the random driver probability. Why do you think this is true?

Risk and odds*

Random events can be described in a variety of ways. The most common descriptors of random events in the life sciences are probability, risk, and odds. So far in this chapter we have introduced the idea of probability and fundamental probability rules. We now briefly describe what risk and odds are, and how they relate to the notion of probability. Chapter 20 will further discuss the applications of risk and odds in clinical research when comparing two groups.

risk   Risk means different things in different fields. In statistics, **risk** corresponds to the probability of an undesirable event such as death, disease, or side effects. This term is particularly common in the health sciences, which aim to assess risk and find ways to reduce it. The risk of a given adverse event, like its probability, is defined by the frequency of that adverse event in a population or sample of interest.

Odds are a somewhat less intuitive concept, with a foundation in gambling (although mathematical odds are not to be confused with betting odds which typ-
odds   ically reflect payoffs offered by bookies to winning bets). An **odds** is a ratio of two probabilities where the numerator represents the probability of an event and the denominator the complementary probability of that event not occurring. As usual, each probability can be obtained from the frequency of the event in a population or sample of interest. Unlike risk and probability, odds can take any positive value. The odds of an event can be expressed as the numerical value of the ratio or as a ratio of two integers with no common denominator.

RISK AND ODDS

The **risk** of an *undesirable* outcome of a random phenomenon is the probability of that undesirable outcome.

The **odds** of any outcome of a random phenomenon is the ratio of the probability of that outcome over the probability of that outcome not occurring.

That is, if an outcome A has probability p of occurring, then

$$\text{risk}(A) = p$$
$$\text{odds}(A) = p/(1 - p)$$

---

*This section is optional.

┌─ **EXAMPLE 9.13**   Blood clots in immobilized patients ─────────

Patients immobilized for a substantial amount of time can develop deep vein thrombosis (DVT), a blood clot in a leg or pelvis vein. DVT can have serious adverse health effects and can be difficult to diagnose. On its Web site, drug manufacturer Pfizer reports the outcome of a study looking at the effectiveness of the drug Fragmin (dalteparin) in preventing DVT in immobilized patients. Of the 1518 randomly chosen immobilized patients given Fragmin, 42 experienced a complication from DVT.[12]

The proportion of patients experiencing DVT complications is $42/1518 = 0.0277$, or 2.77%. We can use this information to compute the risk and odds of experiencing DVT complications for immobilized patients treated with Fragmin:

$$\text{risk} = 0.0277, \text{ or } 2.77\%$$

$$\text{odds} = 0.0277/(1 - 0.0277) = 42/1476 = 7/246 = 0.0285$$

The odds of experiencing DVT complications among immobilized patients given Fragmin are 7:246, or about 1:35. That is, for every patient experiencing a DVT complication, about 35 do not experience a DVT complication.

The numerical values for the risk and odds in this example are very close, 0.277 and 0.0285. In general, when the sample size is very large and the undesirable event not very frequent, risk and odds give similar numerical values. In other situations, risk and odds can be very different.

┌─ **EXAMPLE 9.14**   Sickle-cell anemia ─────────────────────────

Sickle-cell anemia is a serious, inherited blood disease affecting the shape of red blood cells. Individuals with both genes causing the defect suffer pain from blocked arteries and can have their lives shortened from organ damage. Individuals carrying only one copy of the defective gene ("sickle-cell trait") are generally healthy but may pass on the gene to their offspring. An estimated two million Americans carry the sickle-cell trait.[13]

If a couple learns from blood tests that they both carry the sickle-cell trait, genetic laws of inheritance tell us that there is a 25% chance that they could conceive a child who will suffer from sickle-cell anemia. That is, the risk of conceiving a child who will suffer from sickle-cell anemia is 0.25, or 25%. The odds of this are

$$\text{odds} = 0.25/(1 - 0.25) = 0.333, \text{ or } 1:3$$

In this second example, the risk and the odds of conceiving a child who will suffer from sickle-cell anemia have quite different numerical values, 0.25 and 0.33. Always make sure that you understand how risk and odds relate to probability when you read reports about these concepts.

APPLY YOUR KNOWLEDGE ─────────────────────────────

9.16   Blood clots in immobilized patients, continued. The Fragmin study from Example 9.13 compared patients treated with Fragmin with patients given a placebo in a randomized, double-blind design. Of the 1473 immobilized patients given a placebo, 73 experienced a complication from DVT.

(a) Compute the proportion of patients given a placebo who experienced a complication from DVT. What are the risk and the odds of experiencing a complication from DVT when an immobilized patient is given a placebo? How do these values compare?

(b) Compare your results with those of Example 9.13. What do you conclude? We will learn in Chapter 20 how to formally compare risks or odds for two groups.

9.17 **Breast cancer.** The National Cancer Institute (NCI) compiles U.S. epidemiology data for a number of different cancers (the Surveillance, Epidemiology, and End Results Program, or SEER). Among women, breast cancer is one of the most common and deadliest types of cancer.

(a) Based on SEER data, the NCI estimates that a woman in her 30s has a 0.43% chance of being diagnosed with breast cancer.[14] Compute the risk and the odds of being diagnosed with breast cancer for a randomly chosen woman in her 30s.

(b) The risk of breast cancer increases with age. The NCI estimates that a woman in her 60s has a 3.65% chance of being diagnosed with breast cancer. Compute the risk and the odds of being diagnosed with breast cancer for a randomly chosen woman in her 60s.

(c) The NCI also reports a lifetime risk of breast cancer. This is the probability that a woman born in the United States today will develop breast cancer at some time in her life, based on current rates of breast cancer. The most recent estimate is 12.7%. Give the probability, risk, and odds of developing breast cancer over a lifetime for a randomly chosen woman born in the United States today.

(d) Compare the lifetime values you got in (c) with the values by decade you found in (a) and (b). Which way of communicating risk would you say is more informative (lifetime or by decade)?

CHAPTER 9 SUMMARY

A **random phenomenon** has outcomes that we cannot predict but that nonetheless have a regular distribution in very many repetitions.

The **probability** of an event is the proportion of times the event occurs in many repeated trials of a random phenomenon.

A **probability model** for a random phenomenon consists of a sample space S and an assignment of probabilities P.

The **sample space** S is the set of all possible outcomes of the random phenomenon. Sets of outcomes are called **events.** P assigns a number $P(A)$ to an event A as its probability.

Any assignment of probability must obey the rules that state the basic properties of probability:

1. $0 \le P(A) \le 1$ for any event A.
2. $P(S) = 1$.
3. **Addition rule:** Events A and B are **disjoint** if they have no outcomes in common. If A and B are disjoint, then $P(A \text{ or } B) = P(A) + P(B)$.
4. For any event A, $P(A \text{ does not occur}) = 1 - P(A)$.

When a sample space S contains a finite number of possible values, a **discrete probability model** assigns each of these values a probability between 0 and 1 such that the sum of all the probabilities is exactly 1. The probability of any event is the sum of the probabilities of all the values that make up the event.

A sample space can contain all values in some interval of numbers. A **continuous probability model** assigns probabilities as areas under a density curve. A **density curve** has total area 1 underneath it. The probability of any event is the area under the curve and above the values on the horizontal axis that make up the event.

A density curve is an idealized description of the overall pattern of a distribution that smooths out the irregularities in the actual data. We write the **mean of a density curve** as μ and the **standard deviation of a density curve** as σ to distinguish them from the mean $\bar{x}$ and standard deviation s of the actual data.

A **random variable** is a variable taking numerical values determined by the outcome of a random phenomenon. The **probability distribution** of a random variable X tells us what the possible values of X are and how probabilities are assigned to those values.

A random variable X and its distribution can be **discrete** or **continuous.** A **discrete random variable** has finitely many possible values. Its distribution gives the probability of each value. A **continuous random variable** takes all values in some interval of numbers. A density curve describes the probability distribution of a continuous random variable.

Personal probabilities also follow the rules of probability, although they represent an individual's judgment of how likely an outcome is rather than its frequency in a long run of trials.

A probability can also be described as a **risk** when it applies to an undesirable outcome. An **odds** is the ratio of the probability of an outcome over the probability of that outcome not occurring.

CHECK YOUR SKILLS

9.18 You read that in native Hawaiians, the probability of having blood type AB is 1/100 (1 in 100). This means that

(a) if you pick 100 Hawaiians randomly, the fraction of them having blood type AB will be very close to 1/100.

(b) if you pick 100 Hawaiians randomly, exactly 1 of them will have blood type AB.

(c) if you pick 10,000 Hawaiians randomly, exactly 100 of them will have blood type AB.

9.19 A cat is about to have 6 kittens. The sample space for counting the number of female kittens she has is

(a) S = any number between 0 and 1.

(b) S = whole numbers 0 to 6.

(c) S = all sequences of 6 males or females by order of birth, like FMMFFF.

Here is the probability model for the blood type of a randomly chosen person in the United States. Exercises 9.20 to 9.23 use this information.

Blood type	O	A	B	AB
Probability	0.45	0.40	0.11	?

9.20   This probability model is
(a) continuous.   (b) discrete.   (c) discretely continuous.

9.21   The probability that a randomly chosen American has type AB blood must be
(a) any number between 0 and 1.   (b) 0.04.   (c) 0.4.

9.22   Maria has type B blood. She can safely receive blood transfusions from people with blood types O and B. What is the probability that a randomly chosen American can donate blood to Maria?
(a) 0.11   (b) 0.44   (c) 0.56

9.23   What is the probability that a randomly chosen American does not have type O blood?
(a) 0.55   (b) 0.45   (c) 0.04

9.24   In a table of random digits such as Table A, each digit is equally likely to be any of 0, 1, 2, 3, 4, 5, 6, 7, 8, or 9. What is the probability that a digit in the table is a 0?
(a) 1/9   (b) 1/10   (c) 9/10

9.25   In a table of random digits such as Table A, each digit is equally likely to be any of 0, 1, 2, 3, 4, 5, 6, 7, 8, or 9. What is the probability that a digit in the table is 7 or greater?
(a) 7/10   (b) 4/10   (c) 3/10

9.26   A survey looked at the eating habits of adolescent girls. Let X be the number of fruit servings per day. Here is the probability model:

Number of fruit servings X	0	1	2	3	4	5	6	7	8
Probability	0.20	0.15	0.20	0.15	0.11	0.07	0.04	0.04	0.04

What percent of adolescent girls eat the recommended 3 or more fruit servings per day?
(a) 15%   (b) 45%   (c) 55%

9.27   In the previous exercise, if we picked an adolescent girl at random the probability that X is equal to 1 or less is
(a) 0.15   (b) 0.20   (c) 0.35

CHAPTER 9 EXERCISES

9.28   **Taking your pulse.** The throbbing of your arteries is a consequence of the heartbeat. A normal resting pulse rate for a healthy adult ranges from 60 to 90 beats per minute (BPM). While at rest, take your pulse for 30 seconds (for example, at the neck or at the wrist) and record whether your pulse was higher than 75 BPM or not. Repeat the experiment until you have 20 measurements. How many times did you get a pulse above 75 BPM? What is your approximate probability of having a pulse above 75 BPM?

9.29   **Land in Canada.** Statistics Canada says that the land area of Canada is 9,094,000 square kilometers. Of this land, 4,176,000 square kilometers are forested. Choose a square kilometer of land in Canada at random.

(a) What is the probability that the area you choose is forested?

(b) What is the probability that it is not forested?

9.30   **Sample space.** A randomly chosen subject arrives for a study of exercise and fitness. Describe a sample space for each of the following. (In some cases, you may have some freedom in your choice of S.)

(a) You record the gender of the subjects.

(b) After 10 minutes on an exercise bicycle, you ask the subject to rate his or her effort on the Rate of Perceived Exertion (RPE) scale. RPE ranges in whole-number steps from 6 (no exertion at all) to 20 (maximal exertion).

(c) You measure VO2, the maximum volume of oxygen consumed per minute during exercise. VO2 is generally between 2.5 liters per minute and 6 liters per minute.

(d) You measure the maximum heart rate (beats per minute).

9.31   **More sample spaces.** In each of the following situations, describe a sample space S for the random phenomenon. In some cases, you have some freedom in your choice of S.

(a) A seed is planted in the ground. It either germinates or fails to grow.

(b) A patient with a usually fatal form of cancer is given a new treatment. The response variable is the length of time that the patient lives after treatment.

(c) A nutrition researcher feeds a new diet to a young male white rat. The response variable is the weight (in grams) that the rat gains in 8 weeks.

(d) Buy a hot dog and record how many calories it has.

9.32   **Probability models?** In each of the following situations, state whether or not the given assignment of probabilities to individual outcomes is legitimate, that is, satisfies the rules of probability. If not, give specific reasons for your answer.

(a) Among Navajo Native Americans, the distribution of blood type is[15]
$P(A) = 0.27$, $P(B) = 0.00$, $P(AB) = 0.00$, $P(O) = 0.73$.

(b) Choose a cat at random from an animal shelter and record sex and hair type:
$P(\text{female short-haired}) = 0.56$, $P(\text{female long-haired}) = 0.24$,
$P(\text{male short-haired}) = 0.44$, $P(\text{male long-haired}) = 0.16$.

(c) A hospital tests every newborn for phenylketonuria (PKU). PKU is one of the most common genetic disorders, easily treated but resulting in brain damage if undiagnosed. The incidence of PKU varies around the world:[16]
$P(\text{U.S.}) = 1/10{,}000$, $P(\text{Ireland}) = 1/4500$, $P(\text{Finland}) = 1/100{,}000$,
$P(\text{rest of the world}) = 1/15{,}000$.

9.33   **Causes of death.** Government data assign a single cause for each death that occurs in the United States. The data show that the probability is 0.45 that a randomly chosen death was due to cardiovascular (mainly heart) disease, and 0.22 that it was due to cancer.

(a) What is the sample space for the probability model of major causes of death based on the information you are given?

(b) What is the probability that a death was due either to cardiovascular disease or to cancer? What is the probability that the death was due to some other cause?

Solid
bone matrix

Weakened
bone matrix

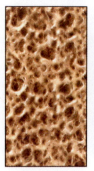 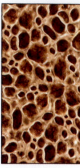

9.34 **Osteoporosis.** Choose a postmenopausal woman at random. The probability is 0.40 that the woman chosen has osteopenia (low bone density) and 0.07 that she has osteoporosis (pathologically low bone mineral density).[17]

(a) What must be the probability that a randomly chosen postmenopausal woman has a healthy bone density?

(b) What is the probability that a randomly chosen postmenopausal woman has some problem with low bone density (osteopenia or osteoporosis)?

9.35 **Hepatitis C.** The Centers for Disease Control and Prevention provide the breakdown of the sources of infection leading to hepatitis C in Americans. Here are the probabilities of each infection source for a randomly chosen individual with hepatitis C:[18]

Source of infection	Probability
Intravenous drug use	0.60
Unprotected sex	0.15
Transfusion (before screening)	0.10
Unknown or other	0.11
Occupational	?

(a) What is the probability that a person with hepatitis C was infected in the course of his or her professional occupation?

(b) What is the probability that a person with hepatitis C was infected through a known risky behavior (intravenous drugs or unprotected sex)?

9.36 **Contraception.** A 2002 large survey of women of reproductive age (15 to 44 years old) looked at contraceptive use. Of the 38,109 women using contraceptives, here are the various contraceptive methods used at the time of the survey:[19]

Method	Percent
Methods involving the woman only	
Female sterilization	27.0
Pill or hormonal patches	37.0
Diaphragm, IUD, sponge, and other female methods	3.2
Methods involving the man only	
Male sterilization	9.2
Male withdrawal	4.0
Methods involving both partners	
Male condoms	18.0
Periodic abstinence	1.6

(a) What is the probability that a woman using a contraceptive method relies on a method requiring the woman only?

(b) What is the probability that a woman using a contraceptive method relies on a method requiring the man only?

(c) What is the probability that a woman using a contraceptive method relies on a method that involves her male partner in some way?

9.37 **Taste buds.** Two wine tasters rate each wine they taste on a scale of 1 to 5. From data on their ratings of a large number of wines, we obtain the following probabilities for both tasters' ratings of a randomly chosen wine:

		Taster 2			
Taster 1	1	2	3	4	5
1	0.03	0.02	0.01	0.00	0.00
2	0.02	0.08	0.05	0.02	0.01
3	0.01	0.05	0.25	0.05	0.01
4	0.00	0.02	0.05	0.20	0.02
5	0.00	0.01	0.01	0.02	0.06

(a) Why is this a legitimate assignment of probabilities to outcomes?

(b) What is the probability that the tasters agree when rating a wine?

(c) What is the probability that Taster 1 rates a wine higher than 3? What is the probability that Taster 2 rates a wine higher than 3?

9.38 **How big are farms?** Choose an American farm at random and measure its size in acres. Here are the probabilities that the farm chosen falls in several acreage categories:

Acres	< 10	10–49	50–99	100–179	180–499	500–999	1000–1999	≥ 2000
Probability	0.09	0.20	0.15	0.16	0.22	0.09	0.05	0.04

Let A be the event that the farm is less than 50 acres in size, and let B be the event that it is 500 acres or more.

(a) Find $P(A)$ and $P(B)$.

(b) Describe the event "A does not occur" in words and find its probability by Rule 4.

(c) Describe the event "A or B" in words and find its probability by the addition rule.

9.39 **Dentist appointment.** A large governmental survey of dental care among 2- to 17-year-old children found the following distribution of times since last visit to the dentist:[20]

Time since last visit	Probability
6 months or less	0.57
More than 6 months but no more than 1 year	0.18
More than 1 year but no more than 2 years	0.08
More than 2 years but no more than 5 years	0.03
More than 5 years	?

(a) What must be the probability that a child has not seen a dentist in more than 5 years?

(b) What is the probability that a randomly chosen child has not seen a dentist within the last 6 months?

(c) What is the probability that the child has seen a dentist within the last year?

Joe McDonald/CORBIS

9.40   Eye color in hawks. An observational study examined iris coloration in a population of North American Cooper's hawks. Iris colors range from yellow to red and are numerically coded to reflect pigmentation intensity.[21] Among female hawks, the given eye color X has the following distribution:

Eye color (X)	Yellow 1	Light orange 2	Orange 3	Dark orange 4	Red 5
Probability	0.05	0.29	0.47	0.18	0.01

(a) Is the random variable X discrete or continuous? Why?

(b) Write the event "eye color at least 4" in terms of X. What is the probability of this event?

(c) Describe the event $X \leq 2$ in words. What is its probability? What is the probability that $X < 2$?

9.41   Eye color in hawks again. The study in the previous exercise also gave the distribution of eye color X among male hawks:

Eye color (X)	Yellow 1	Light orange 2	Orange 3	Dark orange 4	Red 5
Probability	0.00	0.11	0.40	0.32	0.17

(a) Write the event "eye color less than 2 or more than 4" in terms of X. What is the probability of this event?

(b) Describe the event $2 < X \leq 4$ in words. What is its probability?

9.42   Understanding density curves. Remember that it is areas under a density curve, not the height of the curve, that give proportions in a distribution. To illustrate this, sketch a density curve that has its peak at 0 on the horizontal axis but has greater area within 0.25 on either side of 1 than within 0.25 on either side of 0.

9.43   Random numbers. Many random number generators allow users to specify the range of the random numbers to be produced. Suppose that you specify that the random number Y can take any value between 0 and 2. Then the density curve of the outcomes has constant height between 0 and 2 and height 0 elsewhere.

(a) Is the random variable Y discrete or continuous? Why?

(b) What is the height of the density curve between 0 and 2? Draw a graph of the density curve.

(c) Use your graph from (b) and the fact that probability is area under the curve to find $P(Y \leq 1)$.

9.44   More random numbers. Find these probabilities as areas under the density curve that you sketched in the previous exercise.

(a) $P(0.5 < Y < 1.3)$

(b) $P(Y \geq 0.8)$

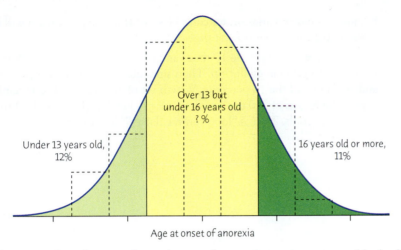

FIGURE 9.8 Distribution of age of onset of anorexia nervosa, pictured by both a histogram and a density curve, for Exercise 9.45.

9.45   **Anorexia.** Exercise 1.8 (page 18) introduced a Canadian study of the age of onset of anorexia nervosa—an eating disorder—for 691 diagnosed adolescent girls. Figure 9.8 shows the density curve (solid line) summarizing the histogram of the data (dotted line). The density curve has been chopped into three parts shown in different shades.

(a) If we call X the age at onset of anorexia, describe the event $13 \leq X < 16$ in words. What is its probability?

(b) What is the probability that the age at onset for a randomly chosen adolescent girl is some time after the 13th birthday? Write this event in terms of X.

9.46   **Hospital births.** The proportion of baby boys born is about 0.5 in the general population. Use the *Probability* applet or software to simulate 100 consecutive births at a local hospital, with probability 0.5 of a boy in each birth. (In most software, the key phrase to look for is "Bernoulli trials." This is the technical term for independent trials with Yes/No outcomes. Our outcomes here are "Boy" and "Girl.")

(a) What percent of the 100 births are boys?

(b) Examine the sequence of boys and girls. How long was the longest run of boys? Of girls? (Sequences of random outcomes often show runs longer than our intuition thinks likely.)

9.47   **Simulating an opinion poll.** Example 9.1 described a 2005 Gallup Poll on evolution, which found that 12% of the Americans sampled believe in Darwinian evolution. Suppose that this is exactly true. Choosing an American at random then has probability 0.12 of getting one who believes in Darwinian evolution. Use the *Probability* applet or your statistical software to simulate choosing many people at random. (In most software, the key phrase to look for is "Bernoulli trials." This is the technical term for independent trials with Yes/No outcomes. Our outcomes here are "believes in Darwinian evolution" or not.)

(a) Simulate drawing 20 people, then 80 people, then 320 people. What proportion believes in Darwinian evolution in each case? We expect (but because of chance variation we can't be sure) that the proportion will be closer to 0.12 in longer runs of trials.

(b) Simulate drawing 20 people 10 times and record the percent in each trial who believe in Darwinian evolution. Then simulate drawing 320 people 10 times and again record the 10 percents. Which set of 10 results is less variable? We expect the results of 320 trials to be more predictable (less variable) than the results of 20 trials. That is "long-run regularity" showing itself.

John Sylvester/Alamy

General Rules of Probability*

The mathematics of probability can provide models to describe the genetic makeup of populations, the spread of epidemics or rumors, and risk factors for a given disorder. Although we are interested in probability because of its usefulness in statistics, the mathematics of chance is important in many fields of study. Our study of probability in Chapter 9 concentrated on basic ideas and facts. Now we look at some details. With more probability at our command, we can model more complex random phenomena.

Relationships among several events

In the previous chapter we have described the probability that *one or the other* of two events A and B occurs in the special situation when A and B cannot occur together. Now we will describe the probability that *both* events A and B occur.

You may find it helpful to draw a picture to display relationships among several events. A picture like Figure 10.1 that shows the sample space S as a rectangular area and events as areas within S is called a **Venn diagram.** The events A and B in Figure 10.1 are disjoint because they do not overlap. The Venn diagram in

Venn diagram

---

*This chapter gives more details about probability that are not needed to read the rest of the book. These concepts, however, are particularly important in medicine and clinical research.

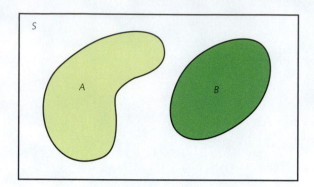

FIGURE 10.1 Venn diagram showing disjoint events A and B.

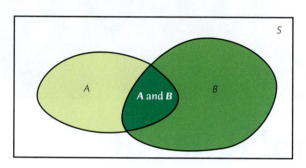

FIGURE 10.2 Venn diagram showing events A and B that are not disjoint. The event {A and B} consists of outcomes common to A and B.

Figure 10.2 illustrates two events that are not disjoint. The event {A and B} appears as the overlapping area that is common to both A and B.

Suppose that you want to study the next two single births at a local hospital. You are counting girls, so two events of interest are

$$A = \text{first baby is a girl}$$
$$B = \text{second baby is a girl}$$

The events A and B are not disjoint. They occur together whenever both babies are girls. We want to find the probability of the event {A and B} that *both* babies are girls.

Genetic laws and the U.S. statistics described at the beginning of Chapter 9 make us willing to assign probability 1/2 to a girl when a baby is born. So

$$P(A) = 0.5$$
$$P(B) = 0.5$$

What is $P(A \text{ and } B)$? Common sense says that it is 1/4. The first baby will be a girl half the time and then the second will be a girl half the time, so both babies will be girls on $1/2 \times 1/2 = 1/4$ of all births in the long run. This reasoning assumes that the second baby still has probability 1/2 of being a girl after the first one born was a girl. This is true—the birth outcome of one couple is not influenced by the birth outcome of another couple. We say that the events "girl on the first birth" and "girl on the second birth" are independent. **Independence** means that the outcome of the first event cannot influence the outcome of the second event.

independence

EXAMPLE 10.1   Independent or not?

Because a coin has no memory and most coin tossers cannot influence the fall of the coin, it is safe to assume that successive coin tosses are independent. For a balanced coin this means that after we see the outcome of the first toss, we still assign probability 1/2 to heads on the second toss.

On the other hand, the colors of successive cards dealt from the same deck are not independent. A standard 52-card deck contains 26 red and 26 black cards. For the first card dealt from a shuffled deck, the probability of a red card is $26/52 = 0.50$ (equally likely outcomes). Once we see that the first card is red, we know that there are only 25 reds among the remaining 51 cards. The probability that the second card is red is therefore only $25/51 = 0.49$. Knowing the outcome of the first deal changes the probability for the second.

If a nurse measures your height twice, it is reasonable to assume that the two results are independent observations. Each records your actual height plus a measurement error, and the size of the error in the first result does not influence the instrument that makes the second reading. But if you take an IQ test or other mental test twice in succession, the two test scores are not independent. The learning that occurs on the first attempt influences your second attempt.

> **We want a boy**
>
> Misunderstanding independence can be disastrous. "Dear Abby" once published a letter from a mother of eight girls. She and her husband had planned a family of four children. When all four were girls, they kept trying for a boy. After seven girls, her doctor assured her that "the law of averages was in our favor 100 to 1." Unfortunately, having children is like tossing coins. Having eight girls in a row is highly unlikely, but once you have seven girls it is not at all unlikely that the next child will be a girl—and she was.

MULTIPLICATION RULE FOR INDEPENDENT EVENTS

Two events A and B are **independent** if knowing that one occurs does not change the probability that the other occurs. If A and B are independent,

$$P(A \text{ and } B) = P(A)P(B)$$

The multiplication rule also extends to collections of more than two events, provided that all are independent. Independence of events A, B, and C means that no information about any one or any two can change the probability of the remaining events. Independence is often assumed in setting up a probability model when the events we are describing seem to have no connection.

If two events A and B are independent, the event that A does not occur is also independent of B, and so on. For example, 45% of all Americans have blood type O.[1] If the Census Bureau interviews two individuals chosen independently, the probability that the first is type O and the second is not type O is $(0.45)(0.55) = 0.2475$.

EXAMPLE 10.2   Rapid HIV testing

STATE: Many people who come to clinics to be tested for HIV, the virus that causes AIDS, don't come back to learn the test results. Clinics now use "rapid HIV tests" that give a result while the client waits. In a clinic in Malawi, for example, use of rapid tests increased the percent of clients who learned their test results from 69% to 99.7%.

The tradeoff for fast results is that rapid tests are less accurate than slower laboratory tests. Applied to people who have no HIV antibodies, one rapid test has a probability of about 0.004 of producing a false positive (that is, of falsely indicating that antibodies are present).[2] If a clinic tests 200 people who are free of HIV antibodies, what is the chance that at least one false positive will occur?

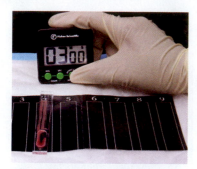

CDC/Cheryl Tryon; Stacy Howard

FORMULATE: We have 200 individuals not infected with HIV taking a test detecting the presence of HIV antibodies. It is reasonable to assume that the test results for different individuals are independent. We have 200 independent events, each with probability 0.004 of a false positive result. What is the probability that at least one of these events occurs?

SOLVE: The probability of a negative result for any one person not infected with HIV is $1 - 0.004 = 0.996$. The probability of at least one false positive among the 200 people tested is therefore

$$P \,(\text{at least one positive}) = 1 - P \,(\text{no positives})$$
$$= 1 - P \,(200 \text{ negatives})$$
$$= 1 - 0.996^{200}$$
$$= 1 - 0.4486 = 0.5514$$

CONCLUDE: The probability is greater than 1/2 that at least one of the 200 people will test positive for HIV, even though no one has the virus.

The multiplication rule $P \,(A \text{ and } B) = P \,(A)P \,(B)$ holds if A and B are independent, but not otherwise. The addition rule $P \,(A \text{ or } B) = P \,(A) + P \,(B)$ discussed in Chapter 9 holds if A and B are disjoint, but not otherwise. Resist the temptation to use these simple rules when the circumstances that justify them are not present. You must also be careful not to confuse disjointness and independence. If A and B are disjoint, then the fact that A occurs tells us that B cannot occur—look again at Figure 10.1. So disjoint events are not independent. Unlike disjointness, we cannot picture independence in a Venn diagram, because it involves the probabilities of the events rather than just the outcomes that make up the events.

APPLY YOUR KNOWLEDGE ————————

10.1   **Unintended pregnancies.** Pharmaceutical companies advertise for the birth control pill an annual efficacy of 99.5% in preventing pregnancy. However, under typical use the real efficacy is only about 95%. That is, 5% of women taking the pill for a year will experience an unplanned pregnancy that year. The difference between these two rates is that the real world is not perfect: for example, a woman might get sick or forget to take the pill one day, or she might be prescribed antibiotics which interfere with hormonal metabolism.[3] If a sexually active woman takes the pill for the four years she is in college, what is the chance that she will become pregnant at least once? Give your answer using first the theoretical efficacy of the pill and then the real efficacy of the pill. Compare both answers.

10.2   **Unintended pregnancies, continued.** Condoms have a failure rate of about 14%. That is, if a woman uses condoms as a means of contraception for one year, she has a 14% chance of becoming pregnant that year. Condom failure is independent of failure of the pill in preventing pregnancy.[4]

(a) If a woman chooses to use both the pill and condoms as a means of contraception for a year, what is the chance that she will become pregnant that year? Use the real pill's efficacy (95%), not its theoretical efficacy.

(b) If a woman uses both condoms and the pill for the four years she is in college, what is the chance that she will become pregnant at least once?

10.3   Spinal cord injuries. The state of Florida reports that 75% of all patients first diagnosed with a spinal cord injury are males.[5]

(a) What is the probability that the next 2 patients diagnosed with a spinal cord injury in Florida will be males?

(b) Of the next five patients diagnosed with a spinal cord injury in Florida, what is the probability that at least 1 will be female? What is the probability that at least 1 will be male?

10.4   Prescriptions. The Centers for Disease Control and Prevention reported for 2002–2003 that 46.6% of Americans 55 to 64 years of age visiting a physician were prescribed a blood glucose regulator and that 52.8% were prescribed a cholesterol-lowering drug. Nonetheless, we can't conclude that because $(0.466)(0.528) = 0.245$ about 24.5% of Americans 55 to 64 years old are prescribed both a blood glucose regulator and a cholesterol-lowering drug. Why not?

Conditional probability

When 2 events A and B are independent, the outcome of the first event cannot influence the outcome of the second event. This is true of 2 independent coin tosses or 2 independent measurements of your blood pressure. However, not all events are independent. Sometimes the probability we assign to an event can change if we know that some other event is true or has occurred. This idea is the key to many applications of probability.

EXAMPLE 10.3   Characteristics of a forest

Forests are complex, evolving ecosystems. For instance, pioneer tree species can be displaced by successional species better adapted to the changing environment. Ecologists mapped a large Canadian forest plot dominated by pioneer Douglas fir with an understory of the invading successional species western hemlock and western red cedar. The two-way table below records the distribution of all 2050 trees in the plot by species and by life stage. The distinction between live and sapling trees is made for live trees taller and shorter than 1.3 meters, respectively.[6]

John Sylvester/Alamy

	Dead	Live	Sapling	Total
Western red cedar (RC)	0.02	0.10	0.08	0.20
Douglas fir (DF)	0.16	0.16	0.00	0.32
Western hemlock (WH)	0.23	0.21	0.04	0.48
Total	0.41	0.47	0.12	

The "Total" row and column are obtained from the probabilities in the body of the table by the addition rule. For example, the probability that a randomly selected tree is a western hemlock (WH) is

$$P(\text{WH}) = P(\text{WH and dead}) + P(\text{WH and live}) + P(\text{WH and sapling})$$

$$= 0.23 + 0.21 + 0.04 = 0.48$$

Now we are told that the tree selected is a sapling. That is, it is one of the 12% in the "Sapling" column of the table. The probability that a tree is a western hemlock, *given the information that it is a sapling*, is the proportion of western hemlock in the "Sapling" column,

$$P(\text{WH} \mid \text{sapling}) = \frac{0.04}{0.12} = 0.33$$

conditional probability

This is a **conditional probability**. You can read the bar | as "given the information that."

Although 48% of the trees in this forest are western hemlock, only 33% of sapling trees are western hemlock. It's common sense that knowing that one event (the tree is a sapling) occurs often changes the probability of another event (the tree is a western hemlock). The example also shows how we should define conditional probability. The idea of a conditional probability $P(B \mid A)$ of one event B given that another event A occurs is the proportion *of all occurrences of A for which B also occurs.*

CONDITIONAL PROBABILITY

When $P(A) > 0$, the **conditional probability** of B given A is

$$P(B \mid A) = \frac{P(A \text{ and } B)}{P(A)}$$

The conditional probability $P(B \mid A)$ makes no sense if the event A can never occur, so we require that $P(A) > 0$ whenever we talk about $P(B \mid A)$. *Be sure to keep in mind the distinct roles of the events A and B in* $P(B \mid A)$. Event A represents the information we are given, and B is the event whose probability we are calculating. Here is an example that emphasizes this distinction.

EXAMPLE 10.4   Characteristics of a forest

What is the conditional probability that a randomly chosen tree is a sapling, *given the information that it is a western hemlock?* Using the definition of conditional probability,

$$P(\text{sapling} \mid \text{western hemlock}) = \frac{P(\text{sapling and western hemlock})}{P(\text{western hemlock})}$$

$$= \frac{0.04}{0.48} = 0.08$$

Only 8% of western hemlock trees are saplings.

Be careful not to confuse the two different conditional probabilities

$$P(\text{western hemlock} \mid \text{sapling}) = 0.35$$
$$P(\text{sapling} \mid \text{western hemlock}) = 0.13$$

The first answers the question, "What proportion of sapling trees are western hemlock?" The second answers "What proportion of western hemlock trees are sapling?"

If you have covered optional Chapter 5 on two-way tables, you will immediately notice how similar conditional probabilities are to the conditional distributions in two-way tables.

Personal probability, defined in Chapter 9, is a special case of conditional probability. Different persons may assign different personal probabilities to a given event because their knowledge and beliefs are different. We also often update the probability we give to a particular event when new information is provided to us (or at least we should; failure to do so is known as "stubbornness"!).

APPLY YOUR KNOWLEDGE

10.5   Composition of a forest. What do the data in Example 10.3 say about the current composition and the evolution of this Canadian forest? Compute the conditional probabilities that a randomly chosen tree is dead given information about its species, for each of the three tree species in this forest. What do these say about the past of this forest?

10.6   Composition of a forest, continued. Using the data provided in Example 10.4, compute the conditional probabilities that a randomly chosen tree is a sapling given information about its species, for each of the three tree species. What does this say about the likely future of this forest? Which tree species appears to be taking over and becoming dominant?

10.7   Cancer and metastases. About 30% of women with early breast cancer will experience metastases, a spread of cancer from its initial location. A new genetic test was developed to help identify those women with early breast cancer who will later develop metastases. A study of women with early breast cancer who took this new genetic test finds that 27% had a positive test and later developed metastases.[7] What is the new test's ability to identify women who will experience metastases? That is, what is the probability that a woman gets a positive test given that she later develops metastases?

10.8   Alcohol dependency. A study of the US clinical population found that 22.5% are diagnosed with a mental disorder, 13.5% are diagnosed with an alcohol-related disorder, and 5% are diagnosed with both disorders.[8]

(a) What is the probability that someone from the clinical population is diagnosed with a mental disorder, knowing that the person is diagnosed with an alcohol-related disorder?

(b) What is the probability that someone from the clinical population is diagnosed with an alcohol-related disorder, knowing that the person is diagnosed with a mental disorder?

General probability rules

We know that if A and B are disjoint events, then $P(A \text{ or } B) = P(A) + P(B)$. If events A and B are *not* disjoint, they can occur together. The probability that one or the other occurs is then *less* than the sum of their probabilities. As Figure 10.3 illustrates, outcomes common to both are counted twice when we add probabilities, so we must subtract this probability once. Here is the addition rule for any two events, disjoint or not.

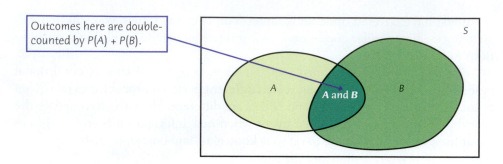

Outcomes here are double-counted by $P(A) + P(B)$.

FIGURE 10.3 The general addition rule: $P(A \text{ or } B) = P(A) + P(B) - P(A \text{ and } B)$ for any two events A and B.

GENERAL ADDITION RULE FOR ANY TWO EVENTS

For any two events A and B,

$$P(A \text{ or } B) = P(A) + P(B) - P(A \text{ and } B)$$

If A and B are disjoint, the event $\{A \text{ and } B\}$ that both occur contains no outcomes and therefore has probability 0. So the general addition rule includes Rule 3, the addition rule for disjoint events.

Petra Wegner/Alamy

> ### EXAMPLE 10.5 *Hearing impairment in dalmatians*
>
> Congenital sensorineural deafness is the most common form of deafness in dogs and is often associated with congenital pigmentation deficiencies. A study of hearing impairment in dogs examined over 5000 dalmatians for both hearing impairment and iris color. Hearing "impaired" was defined as deafness in either one or both ears. Dogs with one or both irises blue (a trait due to low iris pigmentation) were labeled "blue" eyed.
>
> The study found that 28% of the dalmatians were hearing impaired, 11% were blue eyed, and 5% were hearing impaired and blue eyed.[9] Choose a dalmatian at random. Then
>
> $$P(\text{blue or impaired}) = P(\text{blue}) + P(\text{impaired}) - P(\text{blue impaired})$$
> $$= 0.11 + 0.28 - 0.05 = 0.34$$
>
> That is, 34% of dalmatians were either blue eyed or hearing impaired. A dalmatian is a traditional unimpaired, brown-eyed dog if it is *neither* hearing impaired nor blue eyed. So
>
> $$P(\text{unimpaired brown}) = 1 - 0.34 = 0.66$$

Venn diagrams are a great help in finding such probabilities, because you can just think of adding and subtracting areas. Look carefully at Figure 10.4, which shows some events and their probabilities for Example 10.5. What is the probability that a randomly chosen dalmatian is a hearing-impaired, brown-eyed dog? The Venn diagram shows that this is the probability that the dog is hearing impaired minus the probability that it is a hearing-impaired, blue-eyed dog, $0.28 - 0.05 = 0.23$. The four probabilities that appear in the figure add to 1 because they refer to four disjoint events that make up the entire sample space.

The definition of conditional probability reminds us that in principle all probabilities, including conditional probabilities, can be found from the assignment of

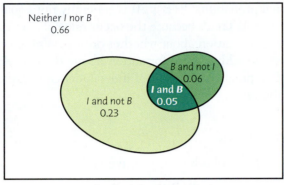

I = Dalmatian is hearing impaired
B = Dalmatian is blue eyed

FIGURE 10.4 Venn diagram and probabilities for hearing impairment and eye color in dalmatian dogs, for Example 10.5.

probabilities to events that make up a random phenomenon. More often, however, conditional probabilities are part of the information given to us in a probability model. The definition of conditional probability then turns into a rule for finding the probability that both of two events occur.

GENERAL MULTIPLICATION RULE FOR ANY TWO EVENTS

The probability that both of two events A and B happen together can be found by

$$P(A \text{ and } B) = P(A)P(B \mid A)$$

Here $P(B \mid A)$ is the conditional probability that B occurs, given the information that A occurs.

In words, this rule says that for both of two events to occur, first one must occur and then, given that the first event has occurred, the second must occur. This is often just common sense expressed in the language of probability, as the following example illustrates.

EXAMPLE 10.6 Blood types

An individual's blood type is described by the ABO system and the Rhesus factor. In the American population 16% of individuals have a negative Rhesus factor (Rh−), and 43.75% of those with Rh− are blood type O. What percent of Americans are Rh− *and* have the blood type O?

Use the general multiplication rule:

$$P(\text{Rh}-) = 0.16$$

$$P(\text{O} \mid \text{Rh}-) = 0.4375$$

$$P(\text{Rh}- \text{ and } \text{O}) = P(\text{Rh}-) \times P(\text{O} \mid \text{Rh}-)$$

$$= (0.16)(0.4375) = 0.07$$

That is, 7% of Americans have the blood type "O−".

You should think your way through this: if 16% of Americans are Rh− and 43.75% *of these* are blood type O, then 43.75% of 16% are Rh− and O.

The conditional probability $P(B \mid A)$ is generally not equal to the unconditional probability $P(B)$. That's because the occurrence of event A generally gives us some additional information about whether or not event B occurs. If knowing that A occurs gives no additional information about B, then A and B are independent events. The precise definition of independence is expressed in terms of conditional probability.

INDEPENDENT EVENTS

Two events A and B that both have positive probability are **independent** if

$$P(B \mid A) = P(B)$$

Because of the multiplication rule, this also implies that

$$P(A \text{ and } B) = P(A) * P(B)$$

You can use either formula to verify independence.

We now see that the multiplication rule for independent events, $P(A \text{ and } B) = P(A)P(B)$, is a special case of the general multiplication rule, $P(A \text{ and } B) = P(A)P(B \mid A)$, just as the addition rule for disjoint events is a special case of the general addition rule.

The multiplication rule extends to the probability that all of several events occur. The key is to condition each event on the occurrence of *all* of the preceding events. For example, we have for three events A, B, and C that

$$P(A \text{ and } B \text{ and } C) = P(A)P(B \mid A)P(C \mid A \text{ and } B)$$

The special multiplication rule $P(A \text{ and } B) = P(A) * P(B)$ applies only to independent events; you cannot use it if events are not independent. Here is a distressing example of misuse of the multiplication rule.

EXAMPLE 10.7 Sudden infant death syndrome

Sudden infant death syndrome (SIDS) causes babies to die suddenly (often in their cribs) with no explanation. Deaths from SIDS have been greatly reduced by placing babies on their backs, but as yet no cause is known.

When more than one SIDS death occurs in a family, the parents are sometimes accused. One "expert witness" popular with prosecutors in England told juries that there is only a 1 in 73 million chance that two children in the same family could have died naturally. Here's his calculation: The rate of SIDS in a nonsmoking middle-class family is 1 in 8500. So the probability of two deaths is

$$\frac{1}{8500} \times \frac{1}{8500} = \frac{1}{72,250,000}$$

Several women were convicted of murder on this basis, without any direct evidence that they harmed their children.

As the Royal Statistical Society said, this reasoning is nonsense. It assumes that SIDS deaths in the same family are independent events. The cause of SIDS is unknown:

"There may well be unknown genetic or environmental factors that predispose families to SIDS, so that a second case within the family becomes much more likely."[10] The British government decided to review the cases of 258 parents convicted of murdering their babies.

Independence is often part of the information given to us in a probability model. However, it can also be an important property that we want to identify. Lack of independence in biology can suggest, for example, a shared mechanism of action or transmission, genetic linkage, or interaction between two factors.

EXAMPLE 10.8 Dalmatians again

Example 10.5 described the frequencies of hearing impairment and eye color in dalmatians. We know that there is a 28% chance that a randomly chosen dalmatian is hearing impaired, an 11% chance that it is blue eyed, and a 5% chance that it is hearing impaired *and* blue eyed. We want to know if the two traits, hearing impairment and eye color, are independent or linked in dalmatians.

We know that

$$P(\text{blue and impaired}) = 0.05$$

If both events are independent, we should find the same value by using the multiplication rule for independent events:

$$P(\text{blue}) \times P(\text{impaired}) = 0.11 \times 0.28 = 0.03$$

The frequency of hearing-impaired, blue-eyed dalmatians is larger than we would predict if the two traits were independent (0.05 instead of 0.03). Therefore, hearing impairment and blue eye pigmentation in dalmatians are not independent. Eye color is a well-known genetic trait. The lack of independence suggests that hearing impairment may also be genetically inherited in at least some dogs (hearing loss could result from a microbial infection in some of the dogs) or that the lack of dark eye pigmentation may also render the dogs more susceptible to hearing loss.

APPLY YOUR KNOWLEDGE

10.9 **At the gym.** Suppose that 10% of adults belong to health clubs and 40% of these health club members go to the club at least twice a week. What percent of all adults go to a health club at least twice a week? Write the information given in terms of probabilities and use the general multiplication rule.

10.10 **Sickle-cell and malaria.** Sickle-cell anemia is a hereditary medical condition affecting red blood cells that is thought to protect against malaria, a debilitating parasitic infection of the liver and blood. That would explain why the sickle-cell trait is found in people who originally came from Africa, where malaria is widespread. A study in Africa tested 543 children for the sickle-cell trait and also for malaria infection. In all, 25% of the children had sickle-cell and 6.6% of the children had both sickle-cell and malaria. Overall, 34.6% of the children had malaria.[11]

(a) Make a Venn diagram with the information provided.

(b) What is the probability that a given child has neither malaria nor sickle-cell? What is the probability that a given child has either malaria or sickle-cell?

10.11 Sickle-cell and malaria, continued. Write the information in the previous exercise in terms of probabilities and use them to answer the following questions:

(a) What is the probability that a child has malaria given that the child has the sickle-cell trait? What is the probability that a child has malaria given that the child doesn't have sickle-cell?

(b) Are the events sickle-cell trait and malaria independent? What might that tell you about the relationship between sickle-cell and malaria?

10.12 Alcohol dependency, continued. Exercise 10.8 describes a study of the U.S. clinical population which found that 22.5% are diagnosed with a mental disorder, 13.5% are diagnosed with an alcohol-related disorder, and 5% are diagnosed with both disorders. Make a Venn diagram and use it to answer these questions:

(a) What percent of the U.S. clinical population is diagnosed with a mental disorder but not with an alcohol-related disorder?

(b) What percent has neither a mental disorder nor an alcohol-related disorder?

Tree diagrams

People often find it difficult to understand formal probability calculations, especially when they involve conditional probabilities. The discussion (page 267) on interpreting medical test results illustrates this problem. There are, fortunately, some graphical representations of probabilities that can greatly help guide calculations.

If you have covered optional Chapter 5 on two-way tables, you will have noticed the similarity between conditional probabilities and conditional distributions calculated from a two-way table. It can indeed be easier to arrange the problem in a 2-way table before calculating conditional probabilities.

However, some problems can have too many stages to be easily represented in a two-way table. We can still display these problems graphically by using *tree diagrams*. Here is an example.

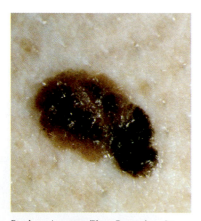

Biophoto Associates/Photo Researchers, Inc.

── **EXAMPLE 10.9** *Skin cancer in men and women* ────────────

STATE: Intense or repetitive sun exposure can lead to skin cancer. Patterns of sun exposure, however, differ between men and women. A study of cutaneous malignant myeloma in the Italian population found that 15% of skin cancers are located on the head and neck, another 41% on the trunk, and the remaining 44% on the limbs. Moreover, 44% of the individuals with a skin cancer on the head are men, as are 63% of those with a skin cancer on the trunk and only 20% of those with a skin cancer on the limbs.[12] What percent of all individuals with skin cancer are women?

FORMULATE: To use the tools of probability, restate the percents as probabilities. If we choose an individual with skin cancer at random,

$$P(\text{head}) = 0.15$$
$$P(\text{trunk}) = 0.41$$
$$P(\text{limbs}) = 0.44$$

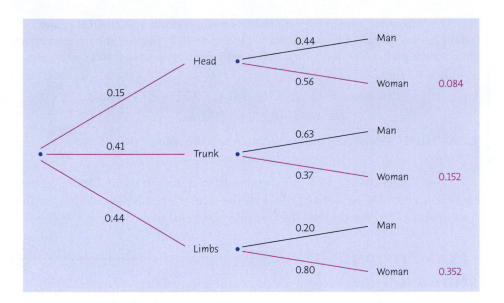

FIGURE 10.5 Tree diagram of skin cancer for Example 10.9. The two stages are the location on the body of the cancer and the gender of the cancer patient.

These three probabilities add to 1 because every individual with skin cancer is in one of these three groups. The percents in these groups that are men are *conditional* probabilities:

$$P(\text{man} \mid \text{head}) = 0.44$$

$$P(\text{man} \mid \text{trunk}) = 0.63$$

$$P(\text{man} \mid \text{limbs}) = 0.20$$

We want to find the unconditional probability $P(\text{woman})$.

SOLVE: The **tree diagram** in Figure 10.5 organizes this information. Each segment in the tree is one stage of the problem. Each complete branch shows a path through the two stages. The probability written on each segment is the conditional probability that an individual with skin cancer will follow that segment, given that he or she has reached the node from which it branches.

    Starting at the left, an individual with skin cancer falls into one of the three cancer location groups. The probabilities of these groups mark the leftmost branches in the tree. Look at the limbs group, the bottom branch. The two segments going out from the "limbs" branch point carry the conditional probabilities

$$P(\text{man} \mid \text{limbs}) = 0.20$$

$$P(\text{woman} \mid \text{limbs}) = 0.80$$

The full tree shows the probabilities for all three cancer location groups.

    Now use the multiplication rule. The probability that a randomly chosen individual with skin cancer is a woman with cancer on the limbs is

$$P(\text{limbs and woman}) = P(\text{limbs})P(\text{woman} \mid \text{limbs})$$

$$= (0.44)(0.80) = 0.352$$

This probability appears at the end of the bottommost branch. You see that the probability of any complete branch in the tree is the product of the probabilities of the segments in that branch.

tree diagram

There are three disjoint paths to female gender, starting with the three cancer location groups. These paths are colored red in Figure 10.5. Because the three paths are disjoint, the probability that an individual with skin cancer is a woman is the sum of their probabilities,

$$P(\text{woman}) = (0.15)(0.56) + (0.41)(0.37) + (0.44)(0.80)$$
$$= 0.084 + 0.152 + 0.352 = 0.588$$

CONCLUDE: About 59% of all individuals with skin cancer are women.

It takes longer to explain a tree diagram than it does to use it. Once you have understood a problem well enough to draw the tree, the rest is easy. Here is another question about skin cancer that the tree diagram helps us answer.

EXAMPLE 10.10 Skin cancer in men and women

STATE: What percent of women with skin cancer have the cancer on the limbs?

FORMULATE: In probability language, we want the conditional probability

$$P(\text{limbs} \mid \text{woman}) = \frac{P(\text{limbs and woman})}{P(\text{woman})}$$

SOLVE: Look again at the tree diagram. $P(\text{woman})$ is the overall outcome. $P(\text{limbs and woman})$ is the result of following a branch of the tree diagram. So

$$P(\text{limbs} \mid \text{woman}) = \frac{0.352}{0.588} = 0.599$$

CONCLUDE: Sixty percent of skin cancers in women are located on the limbs. Compare this conditional probability with the original information (unconditional) that 44% of skin cancers are located on the limbs. Knowing that a person is a woman increases the probability that her skin cancer is located on the limbs.

Notice how $P(\text{limbs} \mid \text{woman})$ and $P(\text{woman} \mid \text{limbs})$ are two very different probabilities. The conditional probability $P(\text{woman} \mid \text{limbs})$ is displayed on the tree and is complementary to $P(\text{man} \mid \text{limbs})$—that is, they add up to 1. The conditional probability $P(\text{limbs} \mid \text{woman})$, on the other hand, is calculated from other probabilities and is complementary to $P(\text{trunk} \mid \text{woman})$ and $P(\text{head} \mid \text{woman})$.

Examples 10.9 and 10.10 illustrate a common setting for tree diagrams. Some outcome (such as the person's gender) has several sources (such as the three cancer location groups). Starting from

- the probability of each source
- and the conditional probability of the outcome given each source,

the tree diagram leads to the overall probability of the outcome. Example 10.9 does this. You can then use the probability of the outcome and the definition of conditional probability to find the conditional probability of one of the sources given that the outcome occurred. Example 10.10 shows how.

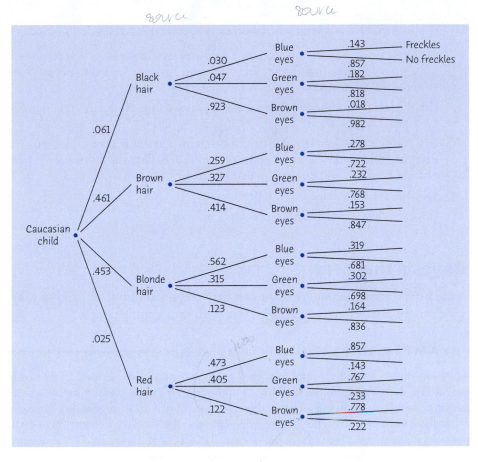

FIGURE 10.6 Tree diagram of the features of children of Caucasian descent in Germany for Exercise 10.13. The three stages are hair color, eye color, and whether or not the child has freckles.

APPLY YOUR KNOWLEDGE

10.13 Eye color, hair color, and freckles. A large study of children of Caucasian descent in Germany looked at the effect of eye color, hair color and freckles on the reported extent of burning from sun exposure.[13] The population's distribution of hair color, eye color, and freckles was as shown in the tree diagram of Figure 10.6. Find the following probabilities and describe them in plain English:

(a) P(blue eyes | red hair), P(blue eyes and red hair).

(b) P(freckles | red hair and blue eyes), P(freckles and red hair and blue eyes).

10.14 Testing for HIV. Enzyme immunoassay tests are used to screen blood specimens for the presence of antibodies to HIV, the virus that causes AIDS. Antibodies indicate the presence of the virus. The test is quite accurate but is not always correct. Here are approximate probabilities of positive and negative test results when the blood tested does and does not actually contain antibodies to HIV:[14]

	Test result	
	+	−
Antibodies present	0.9985	0.0015
Antibodies absent	0.0060	0.9940

HIV screening

In 2006, the Centers for Disease Control and Prevention (CDC) announced a new recommendation that voluntary HIV screening be offered to all patients aged 13 to 64 as part of routine medical care. This is a sharp departure from previous guidelines recommending that only high-risk individuals be tested. Why such a drastic change? The CDC estimates that about 25% of all HIV-positive Americans are not aware of their serologic status. This is thought to be one of the primary reasons why new HIV infections remain high in the United States, with about 40,000 new cases every year.

Suppose that 1% of a large population carries antibodies to HIV in its blood.

(a) Draw a tree diagram for selecting a person from this population (outcomes: antibodies present or absent) and for testing his or her blood (outcomes: test positive or negative).

(b) What is the probability that the test is positive for a randomly chosen person from this population?

10.15   Eye color, hair color, and freckles, continued. Continue your work from Exercise 10.13 using the tree diagram of Figure 10.6. Find the following probabilities and describe them in plain English:

(a) P (red hair), P (blue eyes), P (freckles).

(b) P (freckles and red hair), P (freckles | red hair).

Bayes's theorem

In some situations, we might know the conditional probability $P(B \mid A)$ but would be much more interested in $P(A \mid B)$. Here is an important example.

EXAMPLE 10.11   Diagnostic tests in medicine

sensitivity

specificity

positive predictive value

The performance of a diagnostic test for a given disease can be defined by two properties. The first one is the test's ability to appropriately give a positive result when a person tested has the disease. This is the test's **sensitivity** or, in probability language, P (positive test | disease). The second property is the test's ability to come up negative when a person tested doesn't have the disease. This is the test's **specificity**, P(negative test | no disease). These two values are determined experimentally.

Information about the sensitivity of a medical test can easily be found in the Internet age. However, when a patient is given a diagnostic test, what we really want to know is the probability that a positive test result truly reflects the presence of the disease. This is the conditional probability of having the disease knowing that you received a positive test result, P(disease | positive test). It is called the **positive predictive value**, or *PPV*, of a test.

Many patients—and indeed, even some physicians[15]—assume that if you get a positive test result, then you must have the disease. That's not true. No test is perfectly accurate, and human error in interpreting or communicating the results can also occur. Figure 10.7 shows all possible outcomes of a diagnostic test, displayed in a two-way table (a) and in a tree diagram (b). The information contained in both displays is the same, but you might feel that one is more intuitive for you than the other. It is clear in both displays that

- there are 4 possible outcomes to a diagnostic test
- and a positive test may be obtained for subjects who have the disease (these results are called "true positives") or for subjects who do not have the disease ("false positives").

Therefore, the probability that a subject whose test result is positive indeed has the disease, P(disease | positive test), depends on what proportion of all positive

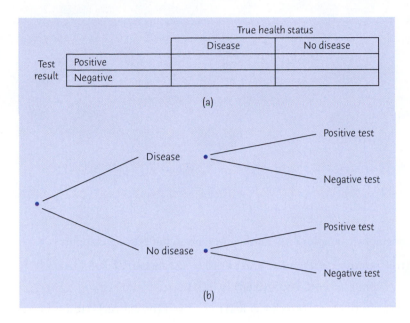

(a)

(b)

FIGURE 10.7 All possible outcomes of a diagnostic test, represented either (a) with a two-way table or (b) with a tree diagram.

tests (both true positives and false positives) can be expected to be true positives. And this depends on both the properties of the diagnostic test (the test's sensitivity and specificity) and how common or rare the disease is in the population of interest (estimated from large-scale epidemiological studies).

EXAMPLE 10.12   Mammography screening

STATE: Breast cancer occurs most frequently among older women. Of all age groups, women in their 60s have the highest rate of breast cancer. The National Cancer Institute (NCI) compiles U.S. epidemiology data for a number of different cancers. The NCI estimates that 3.65% of women in their 60s get breast cancer.[16]

Mammograms are X-ray images of the breast used to detect breast cancer. A mammogram can typically identify correctly 85% of cancer cases (sensitivity = 85%) and 95% of cases without cancer (specificity = 95%).[17] If a woman in her 60s gets a positive mammogram, what is the probability that she indeed has breast cancer?

FORMULATE: We know the following probabilities:

$$P(\text{cancer}) = 0.0365$$

$$P(\text{no cancer}) = 0.9635$$

$$P(\text{test+} \mid \text{cancer}) = 0.85 \text{ (sensitivity)}$$

$$P(\text{test−} \mid \text{cancer}) = 0.15$$

$$P(\text{test+} \mid \text{no cancer}) = 0.05$$

$$P(\text{test−} \mid \text{no cancer}) = 0.95 \text{ (specificity)}$$

What we want to know is

$$P(\text{cancer} \mid \text{test+}) = ???$$

SOLVE: We can create a tree diagram like that of Figure 10.7 to visualize the problem. Figure 10.8 is such a tree diagram populated with the probabilities from the

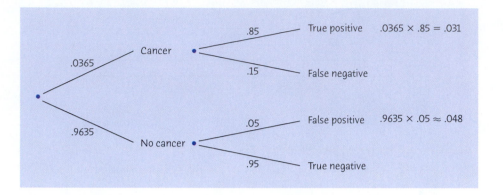

FIGURE 10.8 Tree diagram for a mammogram given to a woman in her 60s, for Example 10.12.

mammography example. For a random woman in her 60s, we compute:

$$P(\text{true positive}) = P(\text{cancer and test+}) = P(\text{cancer})P(\text{test+ | cancer})$$
$$= (0.0365)(0.85) = 0.031$$

$$P(\text{false positive}) = P(\text{no cancer and test+}) = P(\text{no cancer})P(\text{test+ | no cancer})$$
$$= (0.9635)(0.05) = 0.048$$

The positive predictive value is simply the proportion of true positives among all positive test results for women in their 60s:

$$PPV = P(\text{cancer | test+}) = \frac{P(\text{true positive})}{P(\text{true positive}) + P(\text{false positive})}$$

$$= \frac{0.031}{0.031 + 0.048} = 0.392$$

CONCLUDE: If a randomly selected woman in her 60s gets a positive mammogram, there is a 39% chance that she indeed has breast cancer. It also means that there is a 61% chance that she does *not* have breast cancer. This important information should be clear to the physician and communicated clearly to the patient.

What we have done in Example 10.12 is state that the probability we want (*PPV*) represents the frequency of one outcome (the true positives) in a series of complementary and mutually exclusive outcomes (all positive test results: true positives and false positives). Using a tree diagram made it easy visually to understand why. Our approach can be summarized formally with conditional probabilities into an equation known as Bayes's theorem:

BAYES'S THEOREM

Suppose that A_1, A_2, …, A_k are disjoint events whose probabilities are not 0 and add to exactly 1. That is, any outcome has to be exactly in one of these events. Then if B is any other event whose probability is not 0 or 1,

$$P(A_i \mid B) = \frac{P(B \mid A_i)P(A_i)}{P(B \mid A_1)P(A_1) + P(B \mid A_2)P(A_2) + \cdots + P(A_k)P(B \mid A_k)}$$

The numerator in Bayes's theorem is always one of the terms in the sum that make up the denominator. It has far reaching implications in probability because it lets us express a probability as a function of prior knowledge (the A_i's). Let's apply Bayes's theorem to the breast cancer example.

EXAMPLE 10.13 Mammography screening: computations from Bayes's theorem

We can express all possible breast cancer outcomes as either cancer or no cancer. These events are disjoint (that is, mutually exclusive) and complementary; therefore their probabilities must add up to 1. Using Bayes's theorem, we find:

$$P(\text{cancer} \mid \text{test}+) = \frac{P(\text{test}+ \mid \text{cancer})P(\text{cancer})}{P(\text{test}+ \mid \text{cancer})P(\text{cancer}) + P(\text{test}+ \mid \text{no cancer})P(\text{no cancer})}$$

$$= \frac{(0.85)(0.0365)}{(0.85)(0.0365) + (0.05)(0.9635)}$$

$$= \frac{0.031}{0.031 + 0.048} = 0.392$$

Bayes's theorem is extremely powerful and extensively used in more advanced statistics. The importance of Bayes's theorem justifies its inclusion in an introductory textbook. However, many struggle with its formal use. It is far better to think your way through such problems rather than memorize these formal expressions. The discussion topic below illustrates this in the context of medical diagnostic tests.

APPLY YOUR KNOWLEDGE

10.16 False HIV positives. Continue your work from Exercise 10.14.

(a) What is the probability that a person has the antibody, given that the test is positive? Explain in your own words what this means.

(b) Identify the test's sensitivity, specificity, and positive predictive value.

DISCUSSION: Making sense of conditional probabilities in diagnostic tests

Conditional probabilities are often grossly misinterpreted. Yet they are very important, particularly in medicine, where understanding test results requires understanding these probabilities. Many studies have documented that most physicians understand and communicate the true meaning of test results incorrectly. Here is a revealing quote from one such study:[18]

"For instance, doctors with an average of 14 years of professional experience were asked to imagine using the Haemoccult test to screen for colorectal cancer. The prevalence [rate] of cancer was 0.3%, the sensitivity of the test was 50%, and the

false positive rate was 3%. The doctors were asked: What is the probability that someone who tests positive actually has colorectal cancer? The correct answer is about 5%. However, the doctors' answers ranged from 1% to 99%, with about half of them estimating the probability as 50% (the sensitivity) or 47% (sensitivity minus false positive rate). If patients knew about this degree of variability and statistical innumeracy they would be justly alarmed."

The positive predictive value, PPV, describes the probability that a person with a positive test result actually has a given disease, $P(\text{disease} \mid \text{positive test})$. You can check that PPV in this study is indeed only

$$\frac{(0.003)(0.5)}{(0.003)(0.5) + (0.997)(0.03)} = 0.048, \text{ or about } 5\%$$

The discrepancy between a perceived (incorrect) PPV and its actual value is particularly large for screening tests, because they are typically administered to a large population with a relatively low rate of disease. When a disease is rare in the population screened, the percentage of false positives tends to be large, even when sensitivity and specificity are high. The consequence is a low PPV. The merits of screening are obvious, as early detection often plays a key role in patient prognosis. However, the emotional, medical, and financial consequences of misinterpreted test results cannot be ignored. The U.S. Food and Drug Administration, for instance, has refused approval of over-the-counter HIV screening tests for years because of concerns with public reaction to false positives.[19]

Beyond debating the pros and cons of various diagnostic tests, we should ask how such misunderstanding can happen. Physicians are highly trained professionals, and studies show that most do know the definitions of sensitivity and specificity. The challenge seems to be interpreting conditional probabilities, particularly confusing $P(A \mid B)$ with $P(B \mid A)$. In the study cited, many physicians confused the PPV, $P(\text{disease} \mid \text{positive test})$, with the test's sensitivity, $P(\text{positive test} \mid \text{disease})$. Probability concepts also tend to be more abstract than natural frequencies. Studies show that reframing information in a simpler, more natural format using a concrete example increases dramatically the proportion of physicians who correctly identify the PPV. The colorectal cancer case, for example, can be reframed as follows:

Ten thousand patients take the Haemoccult test. Of these 10,000 patients, we expect that about 30 actually have colorectal cancer and the remaining 9,970 do not. Of the 30 patients with colorectal cancer, about 15 (50%) can be expected to receive a positive test result. Of the 9,970 patients without colorectal cancer, about 299 (3%) can also be expected to receive a positive test result. What proportion of patients who receive a positive test result can be expected to actually have colorectal cancer?

The PPV seems much more intuitive now as the proportion of positive tests that are true positives: $\text{PPV} = 15/(15 + 299) = 0.048$, or about 5%. Researchers indeed advocate providing information about diagnostic tests in terms of natural frequencies in addition to giving the usual sensitivity and specificity conditional probabilities. Rather than being expected to do elaborate probability calculations on their own, physicians, as well as patients, should be given clear information about the predictive value of a test. This would improve medical practices and support informed patient decisions.

Until the general public understands these statistical concepts, we will continue to pay a high medical and financial price. Here is a most unusual and interesting illustration. In 2006, a Texas judge uncovered an extensive legal scam in which screening tests were used to fuel massive class action asbestos and silicone lawsuits. Instead of representing workers who sought them because of existing health problems, some lawyers had advertised widely to the general public and indiscriminately tested very large numbers of people, including some with existing diagnoses for unrelated lung diseases. This resulted obviously in large numbers of false positives ... and lots of money for the lawyers and screening companies. The many workers who tested positive received a meager check in the mail and were never informed that a positive test could be a false positive. Interestingly, the scheme was uncovered by a judge who had been a nurse earlier in her career. Judge Janis Jack was the first person, in years of such litigation, to ever ask how the plaintiffs had been screened.[20]

CHAPTER 10 SUMMARY

Events A and B are **disjoint** if they have no outcomes in common. In that case, $P(A \text{ or } B) = P(A) + P(B)$.

The **conditional probability** $P(B \mid A)$ of an event B given an event A is defined by

$$P(B \mid A) = \frac{P(A \text{ and } B)}{P(A)}$$

when $P(A) > 0$. In practice, we most often find conditional probabilities from directly available information rather than from the definition.

Events A and B are **independent** if knowing that one event occurs does not change the probability we would assign to the other event; that is, $P(B \mid A) = P(B)$. In that case, $P(A \text{ and } B) = P(A)P(B)$.

Any assignment of probability obeys these general rules:

Addition rule: If events $A, B, C, \ldots$ are all **disjoint** in pairs, then

$$P(\text{at least one of these events occurs}) = P(A) + P(B) + P(C) + \cdots$$

Multiplication rule: If events $A, B, C, \ldots$ are **independent,** then

$$P(\text{all of the events occur}) = P(A)P(B)P(C)\cdots$$

General addition rule: For any two events A and B,

$$P(A \text{ or } B) = P(A) + P(B) - P(A \text{ and } B)$$

General multiplication rule: For any two events A and B,

$$P(A \text{ and } B) = P(A)P(B \mid A)$$

Bayes's theorem states that, if a series of disjoint and complementary events A_1 through A_k have non-zero probabilities, we can express the conditional probability of event A_i knowing that event B occurred (when event B has probability not equal to zero or 1) as

$$P(A_i \mid B) = \frac{P(B \mid A_i)P(A_i)}{P(B \mid A_1)P(A_1) + P(B \mid A_2)P(A_2) + \cdots + P(A_k)P(B \mid A_k)}$$

CHECK YOUR SKILLS

10.17 The weather forecast for the weekend is a 50% chance of rain for Saturday and a 50% chance of rain for Sunday. If we assume that consecutive days are independent events, the probability that it rains over the weekend (either Saturday or Sunday) is

(a) 0.75.   (b) 1.0.   (c) unknown given the information provided.

10.18 In fact, weather on consecutive days depends on similar atmospheric conditions and therefore consecutive days are not independent events. With the weather forecast of the previous exercise, the probability that it rains over the weekend (either Saturday or Sunday) is

(a) 0.75.   (b) 1.0.   (c) unknown given the information provided.

10.19 An athlete suspected of having used steroids is given two tests that operate independently of each other. Test A has probability 0.9 of being positive if steroids have been used. Test B has probability 0.8 of being positive if steroids have been used. What is the probability that *neither* test is positive if steroids have been used?

(a) 0.72.   (b) 0.38.   (c) 0.02.

Exercises 10.20 to 10.23 are based on the following table.

Government data give the following counts of violent deaths in a recent year among people 20 to 24 years of age by sex and cause of death:

	Female	Male
Accidents	1818	6457
Homicide	457	2870
Suicide	345	2152

10.20 Choose a violent death in this age group at random. The probability that the victim was male is about

(a) 0.81.   (b) 0.78.   (c) 0.19.

10.21 The conditional probability that the victim was male, given that the death was accidental, is about

(a) 0.81.   (b) 0.78.   (c) 0.56.

10.22 The conditional probability that the death was accidental, given that the victim was male, is about

(a) 0.81.   (b) 0.78.   (c) 0.56.

10.23 Let A be the event that a victim of violent death was a woman and B the event that the death was a suicide. The proportion of suicides among violent deaths of women is expressed in probability notation as

(a) $P(A \text{ and } B)$.   (b) $P(A \mid B)$.   (c) $P(B \mid A)$.

10.24 A survey reports that 3.7% of French people from 15 to 19 years of age who smoke daily have also used another psychoactive substance (e.g., cocaine, heroin, crack), compared with 0.3% among those who do not smoke daily.[21] If A is the event "smokes daily" and B is the event "uses another psychoactive substance," then the probability 0.037 is

(a) $P(A \text{ and } B)$.   (b) $P(A \mid B)$.   (c) $P(B \mid A)$.

10.25 Of people who died in the United States in a recent year, 86% were white, 12% were black, and 2% were Asian. (This ignores a small number of deaths among other races.) Diabetes caused 2.8% of deaths among whites, 4.4% among blacks, and 3.5% among Asians. The probability that a randomly chosen death is a white person who died of diabetes is about

(a) 0.107.   (b) 0.030.   (c) 0.024.

10.26 Using the information in the previous exercise, the probability that a randomly chosen death was due to diabetes is about

(a) 0.107.   (b) 0.030.   (c) 0.024.

CHAPTER 10 EXERCISES

10.27 **Blood types.** All human blood can be "ABO-typed" as O, A, B, or AB, but the distribution of the types varies a bit among groups of people. Here is the distribution of blood types for a randomly chosen person in China and in the United States:

Blood type	O	A	B	AB
China probability	0.35	0.27	0.26	0.12
U.S. probability	0.45	0.40	0.11	0.04

Choose an American and a Chinese at random, independently of each other. What is the probability that both have type O blood? What is the probability that both have the same blood type?

10.28 **The Rhesus factor.** Human blood is typed as O, A, B, or AB and also as Rh-positive or Rh-negative. ABO type and Rh-factor type are independent because they are governed by different genes. In the American population, 84% of people are Rh-positive. Use the information about ABO type in the previous exercise to give the probability distribution of blood type (ABO and Rh) for a randomly chosen American.

10.29 **Universal blood donors.** People with type O-negative blood are universal donors. That is, any patient can receive a transfusion of O-negative blood. Only 7.2% of the American population has O-negative blood. If 10 people appear at random to give blood, what is the probability that at least one of them is a universal donor?

10.30 **Lost Internet sites.** Internet sites often vanish or move, so that references to them can't be followed. In fact, 13% of Internet sites referenced in major scientific journals are lost within 2 years after publication.[22] If a paper contains seven Internet references, what is the probability that all seven are still good 2 years later? What specific assumptions did you make in order to calculate this probability?

10.31   **Tasting phenylthiocarbamide.** Phenylthiocarbamide, or PTC, is a molecule that tastes very bitter to those with the ability to taste it. However, in about 30% of the world population a recessive allele causes a chemosensory deficit for PTC.[23] A researcher administers a simple PTC tasting test to 10 randomly chosen subjects. What is the probability that none of these 10 subjects is PTC chemodeficient? What is the probability that at least one subject is chemodeficient for PTC?

10.32   **Polydactyly.** Polydactyly is a fairly common congenital abnormality in which a baby is born with one or more extra fingers or toes. It is reported in about one child in every 500.[24] A young obstetrician celebrates her first 100 deliveries. What is the probability that the obstetrician has delivered no child with polydactyly? What is the probability that she has delivered at least one child with polydactyly?

10.33   **Arthritis and sport.** A Swedish study examined cases of arthritis of the hip or knee among former soccer players in their mid-50s.[25]

(a) Among former elite soccer players, the probability of having arthritis was 0.14. What is the probability that at least 1 player in an elite soccer team of 11 players will have arthritis in his mid-50s? What assumption are you making about arthritis in different soccer players?

(b) The probability of having arthritis was only 0.04 among former non-elite soccer players. What is the probability that at least 1 player in a non-elite soccer team of 11 players will have arthritis in his mid-50s?

10.34   **Older women.** Government data show that 6% of the American population are at least 75 years of age and that about 52% of Americans are women. Explain why it is wrong to conclude that because $(0.06)(0.52) = 0.0312$ about 3% of the population are women ages 75 or over.

10.35   **Hair color and freckles.** A large study of German children of Caucasian descent found that 46% have brown hair and that 25% have freckles.[26] Nonetheless, we can't conclude that because $(0.46)(0.25) = 0.115$ about 11.5% of German children of Caucasian descent are brown-haired with freckles. Why not?

10.36   **A probability teaser.** Suppose (as is roughly correct) that each child born is equally likely to be a boy or a girl and that the sexes of successive children are independent. If we let BG mean that the older of two children is a boy and the younger child is a girl, then each of the combinations BB, BG, GB, GG has probability 0.25. Ashley and Brianna each have two children.

(a) You know that at least one of Ashley's children is a boy. What is the conditional probability that she has two boys?

(b) You know that Brianna's older child is a boy. What is the conditional probability that she has two boys?

10.37   **Tendon surgery.** You have torn a tendon and are facing surgery to repair it. The surgeon explains the risks to you: Infection occurs in 3% of such operations, the repair fails in 14%, and infection and failure occur together in 1%. What percent of these operations succeed and are free of infection?

10.38 Julie's job prospects. Julie is graduating from college. She has studied biology, chemistry, and computing and hopes to use her science background in crime investigation. Late one night she thinks about some jobs for which she has applied. Let A, B, and C be the events that Julie is offered a job by

A = the Connecticut Office of the Chief Medical Examiner
B = the New Jersey Division of Criminal Justice
C = the federal Disaster Mortuary Operations Response Team

Julie writes down her personal probabilities for being offered these jobs:

$P(A) = 0.6$ $P(B) = 0.4$ $P(C) = 0.2$
$P(A \text{ and } B) = 0.1$ $P(A \text{ and } C) = 0.05$ $P(B \text{ and } C) = 0.05$
$P(A \text{ and } B \text{ and } C) = 0$

Make a Venn diagram of the events A, B, and C. As in Figure 10.4, mark the probabilities of every intersection involving these events and their complements. Use this diagram for Exercises 10.39 to 10.41.

10.39 What is the probability that Julie is offered at least one of the three jobs?

10.40 What is the probability that Julie is offered both the Connecticut and New Jersey jobs but not the federal job?

10.41 If Julie is offered the federal job, what is the conditional probability that she is also offered the New Jersey job? If Julie is offered the New Jersey job, what is the conditional probability that she is also offered the federal job?

10.42 Continuous probability. Choose a point at random in a square with sides $0 \le x \le 1$ and $0 \le y \le 1$. This means that the probability that the point falls in any region within the square is equal to the area of that region. Let X be the x coordinate and Y the y coordinate of the point chosen. Find the conditional probability $P(Y < 1/2 \mid Y > X)$. (Hint: Draw a diagram of the square and the events $Y < 1/2$ and $Y > X$.)

10.43 The geometric distributions. Osteoporosis is a disease defined as very low bone density resulting in a high risk of fracture, hospitalization, and immobilization. Seven percent of postmenopausal women suffer from osteoporosis.[27] A physician administers routine checkups. Visits by postmenopausal women are independent. We are interested in how long we must wait to get the first postmenopausal woman with osteoporosis.

(a) Considering only patient visits by postmenopausal women, the probability of a patient with osteoporosis on the first visit is 7%. What is the probability that the first patient does not have osteoporosis and the second patient does?

(b) What is the probability that the first two patients do not have osteoporosis and the third one does? This is the probability that the first patient with osteoporosis occurs on the third visit.

(c) Now you see the pattern. What is the probability that the first patient with osteoporosis occurs on the fourth visit? On the fifth visit? Give the general result: What is the probability that the first patient with osteoporosis occurs on the kth visit?

Comment: The distribution of the number of trials to the first "success" (in our example, a patient with osteoporosis) is called a **geometric distribution.** In this problem you have found geometric distribution probabilities when the probability of a success on each trial is 7%. The same idea works for any probability of success.

geometric distribution

10.44 Hepatitis C and HIV coinfection. According the 2003 report of the U.S. Health Resources and Services Administration, coinfection of HIV and hepatitis C virus (HCV) is on the rise. Only about 2% of Americans have HCV, but 25% of Americans with HIV have HCV. And about 10% of Americans with HCV also have HIV.[28]

(a) Write the information provided in terms of probabilities. Are the two viral infections independent? Explain your reasoning.

(b) What is the probability that a randomly chosen American has both HIV and HCV?

10.45 Cystic fibrosis. Cystic fibrosis is a hereditary lung disorder that often results in death. It can be inherited only if both parents are carriers of an abnormal gene. In 1989, the CF gene that is abnormal in carriers of cystic fibrosis was identified. The probability that a randomly chosen person of European ancestry carries an abnormal CF gene is 1/25. (The probability is less in other ethnic groups.) The CF20m test detects most but not all harmful mutations of the CF gene. The test is positive for 90% of people who are carriers. It is (ignoring human error) never positive for people who are not carriers.[29] Jason tests positive. What is the probability that he is a carrier?

10.46 Smoking by students and their parents. How are the smoking habits of students related to their parents' smoking? Here are data from a large survey of students in eight Arizona high schools:[30]

	Student smokes	Student does not smoke
Both parents smoke	0.07	0.26
One parent smokes	0.08	0.34
Neither parent smokes	0.03	0.22

(a) Verify that this is a legitimate probability model. What is the overall probability that a student smokes?

(b) What is the conditional probability that a student smokes, knowing that both parents smoke? What is the conditional probability that a student smokes, knowing that neither parent smokes? What did you learn from the conditional probabilities that was not apparent in the overall probability you computed in (a)?

10.47 Fundraising by telephone. Tree diagrams can organize problems having more than two stages. Figure 10.9 shows probabilities for a charity calling potential donors by telephone.[31] Each person called is either a recent donor, a past donor, or a new prospect. At the next stage, the person called either does or does not pledge to contribute, with conditional probabilities that depend on the class the person belongs to. Finally, those who make a pledge either do or don't actually make a contribution.

(a) What percent of calls result in a contribution?

(b) What percent of those who contribute are recent donors?

10.48 Teenage pregnancies. The U.S. National Center for Health Statistics reports for 2002 the distribution of teenage pregnancies by age range and ethnicity of the mother:[32]

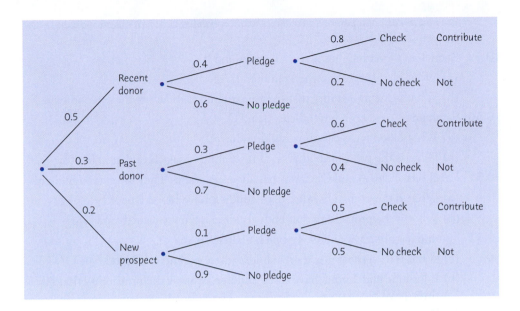

FIGURE 10.9 Tree diagram of fundraising by telephone for Exercise 10.47. The three stages are the type of prospect called, whether or not the person makes a pledge, and whether or not a person who pledges actually makes a contribution.

	15 to 17 years old	18 to 19 years old
White	0.23	0.49
Black	0.09	0.15
Other	0.01	0.03

(a) Verify that this is a legitimate probability model. Make a probability tree of teenage pregnancies.

(b) Overall, what percent of teenage pregnancies are in young mothers 15 to 17 years old? What is the conditional probability that a teenage pregnancy is in a young mother 15 to 17 years old, given that she is classified as "white"? What is the conditional probability that a teenage pregnancy is in a young mother 15 to 17 years old, given that she is classified as "black"?

Mendelian inheritance. *Some traits of plants and animals depend on inheritance of a single gene. This is called Mendelian inheritance, after Gregor Mendel (1822–1884). Exercises 10.49 to 10.52 are based on the following information about Mendelian inheritance of blood type.*

Each of us has an ABO blood type, which describes whether two characteristics called A and B are present. Every human being has two blood-type alleles (gene forms), one inherited from our mother and one from our father. Each of these alleles can be A, B, or O. Which two we inherit determines our blood type. Here is a table that shows what our blood type is for each combination of two alleles:

Alleles inherited	Blood type
A and A	A
A and B	AB
A and O	A
B and B	B
B and O	B
O and O	O

We inherit each of a parent's two alleles with probability 0.5. We inherit independently from our mother and father.

10.49 Rachel and Jonathan both have alleles A and B.

(a) What blood types can their children have?

(b) What is the probability that their next child has each of these blood types?

10.50 Sarah and David both have alleles B and O.

(a) What blood types can their children have?

(b) What is the probability that their next child has each of these blood types?

10.51 Isabel has alleles A and O. Carlos has alleles A and B. They have two children.

(a) What is the probability that both children have blood type A?

(b) What is the probability that both children have the same blood type?

10.52 Jasmine has alleles A and O. Tyrone has alleles B and O.

(a) What is the probability that a child of these parents has blood type O?

(b) If Jasmine and Tyrone have three children, what is the probability that all three have blood type O?

(c) What is the probability that the first child has blood type O and the next two do not?

10.53 **Tay-Sachs disease.** Tay-Sachs disease (TS) is a fatal, recessive genetic disorder in which harmful quantities of a fatty waste substance accumulate in the brain, slowly progressing to death by age five. Tay-Sachs carrier status is common in individuals of eastern European Jewish descent, with a carrier frequency of about 1 in 27. In the general population, on the other hand, the carrier rate is about 1 in 250.[33]

(a) Give the probability that both parents are carriers in each of the following conditions: both mom and dad are of European Jewish descent; mom is of European Jewish descent but dad isn't; neither mom or dad are of European Jewish descent.

(b) If both parents are carriers, the probability is 0.25 that their child has TS. Otherwise, the probability is zero. Draw the probability tree for the children of parents who are both of European Jewish descent. Use the tree to find the probability of a child with TS given that both parents are of European Jewish descent.

(c) What would be the probability that a child has TS given that neither parent is of European Jewish descent?

10.54 **Albinism.** The gene for albinism in humans is recessive. That is, carriers of this gene have probability 0.5 of passing it to a child, and the child is albino only if both parents pass the albinism gene. Parents pass their genes independently of each other. If both parents carry the albinism gene but are not albino, what is the probability that their first child is albino? If they have two children (who inherit independently of each other), what is the probability that both are albino? That neither is albino?

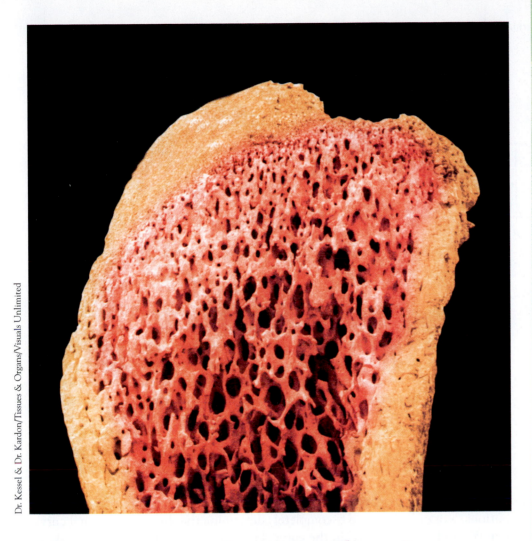

Dr. Kessel & Dr. Kardon/Tissues & Organs/Visuals Unlimited

The Normal Distributions

Of all the continuous probability distributions, none is more widely used than the Normal distributions. Normal distributions are extremely important in inference, as we will see in the next chapters. They are also a good mathematical model for many biological variables, such as blood pressure, bone mineral density, and the height of plants or animals.

Normal distributions

Chapter 9 introduced the idea of density curves to describe continuous probability distributions. One particularly important class of density curves has already appeared in Figures 9.4 and 9.8. These density curves are symmetric, single-peaked, and bell-shaped. They are called *Normal curves*, and they describe *Normal distributions*. Normal distributions play a large role in statistics, but they are rather special

277

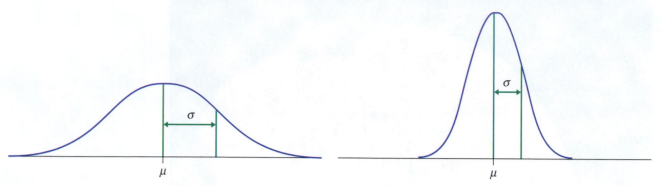

FIGURE 11.1 Two Normal curves, showing the mean μ and standard deviation σ.

and not at all "normal" in the sense of being usual or average. We capitalize Normal to remind you that these curves are special. All Normal distributions have the same overall shape. *The exact density curve for a particular Normal distribution is described by giving its mean μ and its standard deviation σ.* We use the Greek letters μ and σ to distinguish them from the mean $\bar{x}$ and standard deviation s of sample data. The mean is located at the center of the symmetric curve and is the same as the median. Changing μ without changing σ moves the Normal curve along the horizontal axis without changing its spread. The standard deviation σ controls the spread of a Normal curve. Figure 11.1 shows two Normal curves with different values of σ. The curve with the larger standard deviation is more spread out.

The standard deviation σ is the natural measure of spread for Normal distributions. Not only do μ and σ completely determine the shape of a Normal curve, but we can locate σ by eye on the curve. Here's how. Imagine that you are skiing down a mountain that has the shape of a Normal curve. At first, you descend at an ever-steeper angle as you go out from the peak:

Fortunately, before you find yourself going straight down, the slope begins to grow flatter rather than steeper as you go out and down:

The points at which this change of curvature takes place are located at distance σ on either side of the mean μ. You can feel the change as you run a pencil along a Normal curve, and so find the standard deviation. Remember that μ *and σ alone do not specify the shape of most distributions* and that the shape of density curves in general does not reveal σ. These are special properties of Normal distributions.

> **NORMAL DISTRIBUTIONS**
>
> A **Normal distribution** is described by a Normal density curve. Any particular Normal distribution is completely specified by two numbers: its mean and its standard deviation.
>
> The mean of a Normal distribution is at the center of the symmetric Normal curve. The standard deviation is the distance from the center to the change-of-curvature points on either side.

Why are the Normal distributions important in statistics? Here are three reasons. First, Normal distributions are good descriptions for some distributions of *real data*. Distributions that are often close to Normal include scores on tests taken by many people (such as IQ tests and SAT exams), repeated careful measurements of the same quantity, and characteristics of biological populations (such as lengths of crickets and yields of corn). Second, Normal distributions are good approximations to the results of many kinds of *chance outcomes*, such as the proportion of boys over many hospital births. Third, we will see that many *statistical inference* procedures based on Normal distributions work well for other roughly symmetric distributions.

Students and professionals often make the error of assuming that their variable is Normally distributed without first verifying this assumption by plotting the data. However, many data sets do not follow a Normal distribution. Distributions of body weights or survival times, for example, are skewed to the right and so are not Normal. Non-Normal data not only are common but are sometimes more interesting than their Normal counterparts.

The 68–95–99.7 rule

Although there are many Normal curves (with different values of μ and σ), they all have common properties. In particular, all Normal distributions obey the following rule.

> **THE 68–95–99.7 RULE**
>
> In the Normal distribution with mean μ and standard deviation σ:
> - Approximately **68%** of the observations fall within σ of the mean μ.
> - Approximately **95%** of the observations fall within 2σ of μ.
> - Approximately **99.7%** of the observations fall within 3σ of μ.

Figure 11.2 illustrates the 68–95–99.7 rule. By remembering these three numbers, you can think about Normal distributions without constantly making detailed calculations.

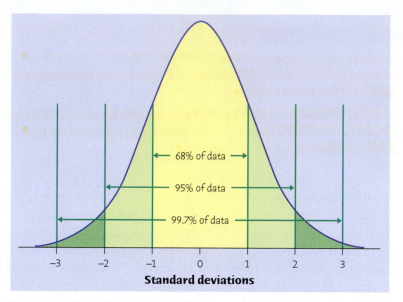

FIGURE 11.2 The 68–95–99.7 rule for Normal distributions.

Jupiterimages/Thinkstock/Alamy

EXAMPLE 11.1 Heights of young women

The distribution of heights of young women aged 18 to 24 is approximately Normal with mean $\mu = 64.5$ inches and standard deviation $\sigma = 2.5$ inches. Figure 11.3 applies the 68–95–99.7 rule to this distribution.

The 95 part of the 68–95–99.7 rule says that the middle 95% of young women are approximately between

$$\mu - 2\sigma = 64.5 - (2)(2.5) = 64.5 - 5 = 59.5$$

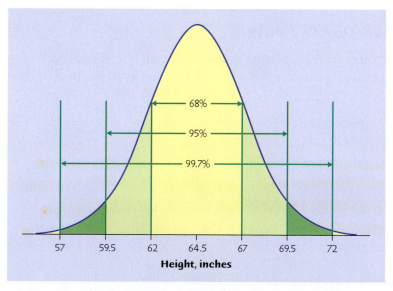

FIGURE 11.3 The 68–95–99.7 rule applied to the distribution of heights among young women aged 18 to 24, with $\mu = 64.5$ inches and $\sigma = 2.5$ inches.

and

$$\mu + 2\sigma = 64.5 + (2)(2.5) = 64.5 + 5 = 69.5$$

that is, between 59.5 inches and 69.5 inches tall.

The other 5% of young women have heights outside this range. Because the Normal distributions are symmetric, half of these women are on the tall side. So the tallest 2.5% of young women are taller than 69.5 inches. Rephrased in probability terms, this means that there is probability approximately 0.025 that a randomly selected woman is taller than 69.5 inches.

EXAMPLE 11.2 *Heights of young women*

Look again at Figure 11.3. A height of 62 inches is one standard deviation below the mean. What is the probability that a woman is taller than 62 inches? Find the answer by adding areas in the figure. Here is the calculation in pictures:

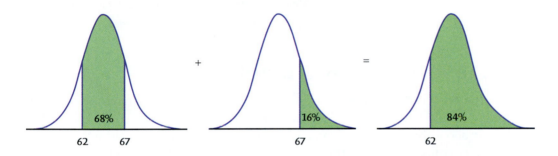

percent between 62 and 67	+	percent above 67	=	percent above 62
68%	+	16%	=	84%

Be sure you see where the 16% came from: 32% of heights are outside the range 62 to 67 inches, and half of these are above 67 inches.

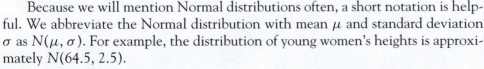*The 68–95–99.7 rule describes distributions that are exactly Normal. Real data are never "exactly" Normal, however.* Look back at Example 9.8 (page 230) on the heights of women in their 40s. The example and Figure 9.4 show that calculations using the histogram of raw data are quite close to calculations using a Normal density curve. Differences are due, in part, to the fact that someone's height is usually reported only to the nearest quarter of an inch. A height is reported as 59.0 or 59.25, but not as 59.1436. We can use a Normal distribution because it's a good approximation and because we know that heights are actually continuous—people do not grow by steps of a quarter of an inch!

Because we will mention Normal distributions often, a short notation is helpful. We abbreviate the Normal distribution with mean μ and standard deviation σ as $N(\mu, \sigma)$. For example, the distribution of young women's heights is approximately $N(64.5, 2.5)$.

APPLY YOUR KNOWLEDGE

11.1   **Men's bladders.** The distribution of bladder volume in men is approximately Normal with mean 550 ml and standard deviation 100 ml.[1] Draw a Normal curve on which this mean and standard deviation are correctly located. (*Hint:* Draw the curve first, locate the points where the curvature changes, and then mark the horizontal axis.)

11.2   **Men's bladders, continued.** The distribution of bladder volume in men is approximately Normal with mean 550 ml and standard deviation 100 ml. Use the 68–95–99.7 rule to answer the following questions. (Use the sketch you made in the previous exercise.)

(a) Between what volumes do the middle 95% of men's bladders fall?

(b) What percent of men's bladders have a volume larger than 650 ml?

11.3   **Women's bladders.** The distribution of bladder volume in women is approximately Normal with mean 400 ml and standard deviation 75 ml. Use the 68–95–99.7 rule to answer the following questions.

(a) Between what values do almost all (99.7%) of women's bladder volumes fall?

(b) How small are the smallest 2.5% of all bladders among women?

The standard Normal distribution

As the 68–95–99.7 rule suggests, all Normal distributions share many common properties. In fact, all Normal distributions are the same if we measure in units of size σ about the mean μ as center. Changing to these units is called *standardizing*. To standardize a value, subtract the mean of the distribution and then divide by the standard deviation.

STANDARDIZING AND z-SCORES

If x is an observation from a distribution that has mean μ and standard deviation σ, the **standardized value** of x is

$$z = \frac{x - \mu}{\sigma}$$

A standardized value is often called a **z-score.**

A z-score tells us how many standard deviations the original observation falls away from the mean, and in which direction. Observations larger than the mean are positive when standardized, and observations smaller than the mean are negative.

EXAMPLE 11.3   *Standardizing women's heights*

The heights of young women age 18 to 24 are approximately Normal with $\mu = 64.5$ inches and $\sigma = 2.5$ inches. The standardized height is

$$z = \frac{\text{height} - 64.5}{2.5}$$

A woman's standardized height is the number of standard deviations by which her height differs from the mean height of all young women. A woman 70 inches tall, for example, has standardized height

$$z = \frac{70 - 64.5}{2.5} = 2.2$$

or 2.2 standard deviations above the mean. Similarly, a woman 5 feet (60 inches) tall has standardized height

$$z = \frac{60 - 64.5}{2.5} = -1.8$$

or 1.8 standard deviations less than the mean height.

We often standardize observations from symmetric distributions to express them in a common scale. We might, for example, measure the height of a boy at two different ages by calculating each z-score. The standardized heights tell us where the child stands in the distribution for his age group, and whether he might be growing more slowly than other children his age.

If the variable we standardize has a Normal distribution, standardizing does more than give a common scale. It makes all Normal distributions into a single distribution, and this distribution is still Normal. Standardizing a variable that has any Normal distribution produces a new variable that has the *standard Normal distribution*.

STANDARD NORMAL DISTRIBUTION

The **standard Normal distribution** is the Normal distribution $N(0, 1)$ with mean 0 and standard deviation 1.

If a variable x has any Normal distribution $N(\mu, \sigma)$ with mean μ and standard deviation σ, then the standardized variable

$$z = \frac{x - \mu}{\sigma}$$

has the standard Normal distribution.

APPLY YOUR KNOWLEDGE

11.4   Men's and women's heights. The heights of women aged 20 to 29 are approximately Normal with mean 64 inches and standard deviation 2.7 inches. Men the same age have mean height 69.3 inches with standard deviation 2.8 inches. What are the z-scores for a woman 6 feet tall and for a man 6 feet tall? Say in simple language what information the z-scores give that the actual heights do not.

11.5   Growth chart. Boys' growth charts indicate that the heights of 5-year-old boys are approximately Normal with mean 43 inches and standard deviation 1.8 inches, whereas those of 7-year-old boys are approximately Normal with mean 48 inches and standard deviation 2.1 inches. John measured 44 inches on his fifth birthday, and now, on his seventh birthday, he measures 49 inches. Find John's standardized scores for both ages and compare them. Has he been growing faster or more slowly than the general population of boys his age?

Finding Normal probabilities

Areas under a Normal curve represent proportions (frequencies) of observations from that Normal distribution. Therefore, they also represent probabilities of randomly selecting an individual from that Normal distribution. There is no direct formula for areas under a Normal curve. Calculations use either software that calculates areas or a table of areas. Tables and most software calculate one kind of area, *cumulative probabilities*.

CUMULATIVE PROBABILITIES

The **cumulative probability** for a value x in a distribution is the proportion of observations in the distribution that lie at or below x.

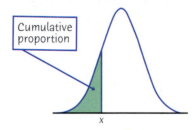

The key to calculating Normal probabilities is to match the area you want with areas that represent cumulative probabilities. *If you make a sketch of the area you want, you will almost never go wrong.* Find areas for cumulative proportions either from software or (with an extra step) from a table. The following example shows the method in a picture.

EXAMPLE 11.4   Length of human pregnancies

The length of human pregnancies from conception to birth varies according to a distribution that is approximately Normal with mean 266 days and standard deviation 16 days. Babies born substantially before term must be given special medical attention. What percent of babies are born after 8 months (240 days) or more of gestation from conception?

Here is the calculation in a picture: The proportion of babies born with 240 days of gestation or more is the area under the curve to the right of 240. That's the total area under the curve (which is always 1) minus the cumulative proportion up to 240.

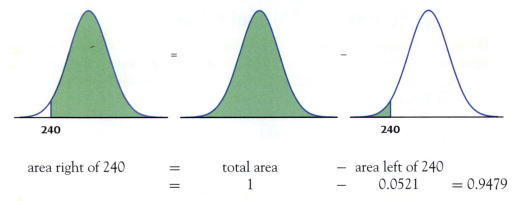

area right of 240	=	total area	−	area left of 240	
	=	1	−	0.0521	= 0.9479

About 95% of all babies born in the United States have had 240 days or more of gestation since conception.

There is *no* area under a smooth curve and exactly over the point 240. Consequently, the area to the right of 240 (the proportion of babies born at > 240 days) is the same as the area at or to the right of this point (the proportion of babies born at ≥ 240 days).

Actual birth records may show babies born 240 days after conception. However, a child born "exactly 239.8 days" after conception would be labeled as born 240 days after conception because of rounding and lack of precision in estimating time of conception. That the proportion of babies born exactly 240 days after conception is 0 for a Normal distribution is a consequence of the idealized smoothing of Normal distributions for data. Under a Normal distribution, the proportion of babies born exactly 239.8 days after conception is also 0. ==Be sure to distinguish the everyday language we use to describe a variable—such as pregnancy length—and a mathematical description like the Normal distribution.==

To find the numerical value 0.0521 of the cumulative proportion in Example 11.4 using software, plug in mean 266 and standard deviation 16 and ask for the cumulative proportion for 240. Software often uses terms such as "cumulative distribution" or "cumulative probability." Here, for example, is Minitab's output:

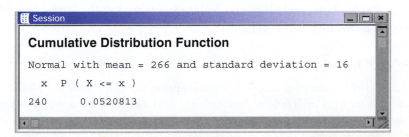

Cumulative Distribution Function

```
Normal with mean = 266 and standard deviation = 16

  x   P ( X <= x )
240     0.0520813
```

The *P* in the output stands for "probability," but we can read it as "proportion of the observations." CrunchIt! and the *Normal Curve* applet are even handier, because they draw pictures as well as finding areas. If you are not using software, you can find cumulative proportions for Normal curves from a table. That requires an extra step.

Using the standard Normal table*

The extra step in finding cumulative proportions from a table is that we must first standardize to express the problem in the standard scale of z-scores. This allows us to get by with just one table, a table of *standard Normal cumulative proportions*.

Table B in the back of the book gives cumulative proportions for the standard Normal distribution. The pictures at the top of the table remind us that the entries are cumulative proportions, areas under the curve to the left of a value z.

EXAMPLE 11.5   The standard Normal table

What proportion of observations of a standard Normal variable z take values less than 1.47?

Solution: To find the area to the left of 1.47, locate 1.4 in the left-hand column of Table B, and then locate the remaining digit 7 as .07 in the top row. The entry opposite 1.4 and under .07 is 0.9292. This is the cumulative proportion we seek. Figure 11.4 illustrates this area.

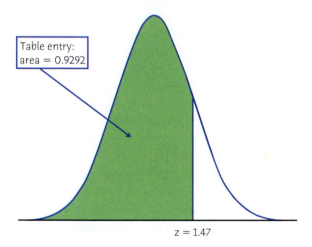

Table entry:
area = 0.9292

z = 1.47

FIGURE 11.4 The area under a standard Normal curve to the left of the point $z = 1.47$ is 0.9292, for Example 11.5. Table B gives areas under the standard Normal curve.

Now that you see how Table B works, let's redo Example 11.4 using the table. We can break it down into three steps.

EXAMPLE 11.6   Length of human pregnancies

The lengths of human pregnancies from conception to birth follow the Normal distribution with mean $\mu = 266 \, days$ and standard deviation $\sigma = 16 \, days$. What percent of pregnancies are at least 240 days long?

1. *Draw a picture.* The picture is exactly as in Example 11.4.
2. *Standardize.* Call pregnancy length x. Subtract the mean, and then divide by the standard deviation to transform the problem about x into a problem about a

---

*This section is unnecessary if you will always use software for Normal distribution calculations.

standard Normal z:

$$x \geq 240$$

$$\frac{x - 266}{16} \geq \frac{240 - 266}{16}$$

$$z \geq -1.625$$

3. *Use the table*. The picture says that we want the cumulative proportion for $x = 240$. Step 2 says this is the same as the cumulative proportion for $z = -1.625$, or approximately $z = -1.63$. The Table B entry for $z = -1.63$ says that this cumulative proportion is 0.0516. The area to the right of -1.63 is therefore $1 - 0.0516 = 0.9484$, or about 95%.

The area from the table in Example 11.6 (0.9484) is slightly less accurate than the area from software in Example 11.4 (0.9479), because we must round z to two decimal places when we use Table B. The difference is rarely important in practice. Here's the method in outline form.

FINDING NORMAL PROBABILITIES WITH TABLE B

1. **State the problem** in terms of the observed variable x. **Draw a picture** that shows the proportion you want in terms of cumulative proportions.
2. **Standardize** x to restate the problem in terms of a standard Normal variable z.
3. **Use Table B** and the fact that the total area under the curve is 1 to find the required area under the standard Normal curve.

EXAMPLE 11.7   *Heights of young women*

The heights of young women are approximately Normal with $\mu = 64.5$ inches and $\sigma = 2.5$ inches. What is the probability that a randomly selected woman measures between 60 and 68 inches?

1. *State the problem and draw a picture*. Call the height x. The variable x has the $N(64.5, 2.5)$ distribution. The probability that a randomly selected young woman measures between 60 and 68 inches reflects the proportion of women measuring between 60 and 68 inches in the population of young women. What is this proportion? Here is the picture:

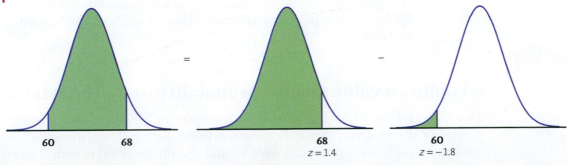

2. *Standardize*. Subtract the mean, and then divide by the standard deviation to turn x into a standard Normal z:

$$60 \leq x < 68$$

$$\frac{60 - 64.5}{2.5} \leq \frac{x - 64.5}{2.5} < \frac{68 - 64.5}{2.5}$$

$$-1.8 \leq z < 1.4$$

3. *Use the table*. Follow the picture (we added the z-scores to the picture label to help you):

$$\text{area between } -1.8 \text{ and } 1.4 = (\text{area left of } 1.4) - (\text{area left of } -1.8)$$

$$= 0.9192 - 0.0359 = 0.8833$$

The probability that a randomly selected young woman measures between 60 and 68 inches is about 0.88, or 88%.

Sometimes we encounter a value of z more extreme than those appearing in Table B. For example, the area to the left of $z = -4$ is not given directly in the table. The z-values in Table B leave only an area of 0.0002 in each tail unaccounted for. For practical purposes, we can act as if there is zero area outside the range of Table B.

APPLY YOUR KNOWLEDGE

11.6 Use the Normal table. Use Table B to find the proportion of observations from a standard Normal distribution that satisfies each of the following statements. In each case, sketch a standard Normal curve and shade the area under the curve that is the answer to the question.

(a) $z < 2.85$

(b) $z > 2.85$

(c) $z > -1.66$

(d) $-1.66 < z < 2.85$

11.7 Men's bladders. The distribution of bladder volume in men is approximately Normal with mean $\mu = 550$ ml and standard deviation $\sigma = 100$ ml.

(a) What proportion of male bladders are larger than 500 ml?

(b) What proportion of male bladders are between 500 and 600 ml?

11.8 Women's bladders. Women's bladders, on average, are smaller than those of men. The distribution of bladder volume in women is approximately $N(400, 75)$ in milliliters. Find the proportion of women's bladders with a volume between 500 and 600 ml.

Finding a value given a probability or a proportion

Examples 11.4 and 11.6 illustrate the use of software or Table B to find what proportion of the observations satisfy some condition, such as "pregnancies of 240 days or longer." We may instead want to find the observed value with a given proportion of the observations above or below it. Statistical software will do this directly.

EXAMPLE 11.8   Find the top 10% using software

The lengths of human pregnancies in days from conception to birth follow approximately the $N(266, 16)$ distribution. How long are the longest 10% of pregnancies?

We want to find the pregnancy length x with area 0.1 to its *right* under the Normal curve with mean $\mu = 266$ and standard deviation $\sigma = 16$. That's the same as finding the pregnancy length x with area 0.9 to its *left*. Most software will tell you x when you plug in mean 266, standard deviation 16, and cumulative probability 0.9. Figure 11.5 gives both Minitab's and Excel's output.

Both software programs give approximately $x = 286.505$. So pregnancies longer than 287 days are in the top 10%. (Round up because pregnancy durations are expressed only in whole numbers.) That's three weeks or more longer than the mean pregnancy length of 266 days.

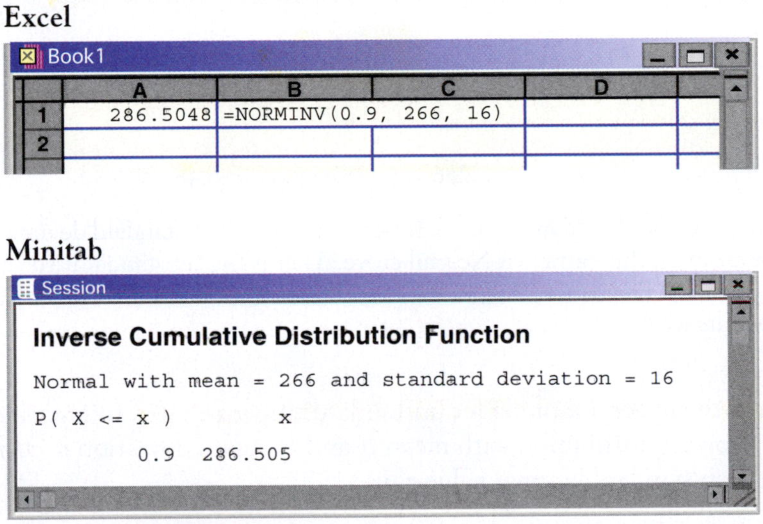

Excel

	A	B	C	D
1	286.5048	=NORMINV(0.9, 266, 16)		
2				

Minitab

Session

Inverse Cumulative Distribution Function

Normal with mean = 266 and standard deviation = 16

P(X <= x)   x

0.9   286.505

FIGURE 11.5 Software output for Example 11.8.

Without software, use Table B backward. Find the given proportion in the body of the table and then read the corresponding z from the left column and top row. There are again three steps.

EXAMPLE 11.9   Find the top 10% using Table B

The lengths of human pregnancies in days from conception to birth follow approximately the $N(266, 16)$ distribution. How long are the longest 10% of pregnancies?

1. *State the problem and draw a picture.* This step is exactly as in Example 11.8. The picture is Figure 11.6.

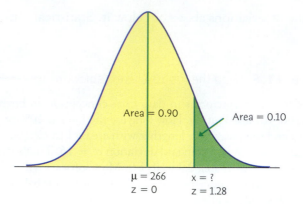

FIGURE 11.6 Locating the point on a Normal curve with area 0.1 to its right, for Example 11.9.

2. *Use the table.* Look in the body of Table B for the entry closest to 0.9. It is 0.8997. This is the entry corresponding to $z = 1.28$. So $z = 1.28$ is the standardized value with approximately area 0.9 to its left.

3. *Unstandardize* to transform z back to the original x scale. We know that the standardized value of the unknown x is $z = 1.28$. So x itself satisfies

$$\frac{x - 266}{16} = 1.$$

Solving this equation for x gives

$$x = 266 + (1.28)(16) = 286.48$$

This equation should make sense: It says that x lies 1.28 standard deviations above the mean on this particular Normal curve. That is the "unstandardized" meaning of $z = 1.28$. Pregnancies longer than 287 days are among the longest 10% of pregnancies.

Here is the general formula for unstandardizing a z-score. To find the value x from the Normal distribution with mean μ and standard deviation σ corresponding to a given standard Normal value z, use

$$x = \mu + z\sigma$$

EXAMPLE 11.10   Find the third quartile

High levels of cholesterol in the blood increase the risk of heart disease. For 14-year-old boys, the distribution of blood cholesterol is approximately Normal with mean $\mu = 170$ milligrams of cholesterol per deciliter of blood (mg/dl) and standard deviation $\sigma = 30$ mg/dl.[2] What is the third quartile of the distribution of blood cholesterol?

1. *State the problem and draw a picture.* Call the cholesterol level x. The variable x has the N(170, 30) distribution. The third quartile is the value with 75% of the distribution to its left. Figure 11.7 is the picture.

2. *Use the table.* Look in the body of Table B for the entry closest to 0.75. It is 0.7486. This is the entry corresponding to $z = 0.67$. So $z = 0.67$ is the standardized value with area 0.75 to its left.

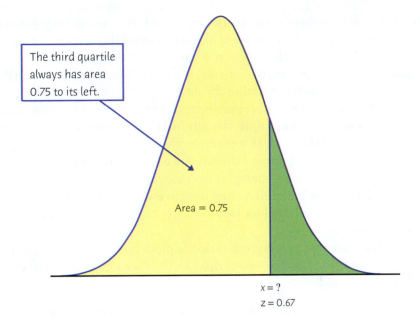

The third quartile
always has area
0.75 to its left.

Area = 0.75

x = ?
z = 0.67

FIGURE 11.7 Locating the
third quartile of a Normal curve,
for Example 11.10.

3. *Unstandardize*. The cholesterol level corresponding to $z = 0.67$ is

$$x = \mu + z\sigma$$
$$= 170 + (0.67)(30) = 190.1$$

The third quartile of blood cholesterol levels in 14-year-old boys is about 190 mg/dl.

APPLY YOUR KNOWLEDGE

11.9   Table B. Use Table B to find the value z of a standard Normal variable that satisfies each of the following conditions. (Use the value of z from Table B that comes closest to satisfying the condition.) In each case, sketch a standard Normal curve with your value of z marked on the axis.

(a) The point z with 25% of the observations falling below it.

(b) The point z with 40% of the observations falling above it.

11.10 IQ test scores. Scores on the Wechsler Adult Intelligence Scale are approximately Normally distributed with $\mu = 100$ and $\sigma = 15$.

(a) What scores fall in the lowest 25% of the distribution?

(b) How high a score is needed to be in the highest 5%?

Normal quantile plots*

The Normal distributions provide good descriptions of some distributions of real data, such as IQ tests or individual heights. The distributions of some other common variables are skewed and therefore distinctly non-Normal. Examples include

*This section is optional for a quarter-based course.

The bell curve?

Does the distribution of human intelligence follow the "bell curve" of a Normal distribution? Scores on IQ tests do roughly follow a Normal distribution. That is because a test score is calculated from a person's answers in a way that is designed to produce a Normal distribution in the population. To conclude that *intelligence* follows a bell curve, we must agree that the test scores directly measure intelligence. Many psychologists don't think there is one human characteristic that we can call "intelligence" and that can be measured by a single test score.

the survival times of cancer patients after treatment, the amount of a given toxin in the blood among a random sample of individuals, and the average size or weight of mammalian species. While experience can suggest whether or not a Normal distribution is plausible in a particular case, it is risky to assume that a distribution is Normal without actually inspecting the data.

A histogram or stemplot can reveal distinctly non-Normal features of a distribution, such as outliers (for example, the survival time of guinea pigs in Figure 1.6, page 16), pronounced skewness (for example, the number of fruit servings per day in Figure 1.13, page 29), or gaps and clusters (for example, the body length of perch in Figure 1.7, page 17). If the stemplot or histogram appears roughly symmetric and unimodal, however, we need a more sensitive way to judge the adequacy of a Normal model. The most useful tool for assessing Normality is another graph, the

Normal quantile plot   **Normal quantile plot.**

Here is the basic idea of a Normal quantile plot. The graphs produced by software use more sophisticated versions of this idea. It is not practical to make Normal quantile plots by hand.

1. Arrange the observed data values from smallest to largest. Record what percentile of the data each value occupies. For example, the smallest observation in a set of 20 is at the 5% point, the second smallest is at the 10% point, and so on.

2. Do Normal distribution calculations to find the z-scores at these percentiles. For example, $z = -1.645$ is the 5% point of the standard Normal distribution, and $z = -1.282$ is the 10% point.

3. Plot each data point x against the corresponding z. If the data distribution is close to standard Normal, the plotted points will lie close to the 45-degree line $x = z$. If the data distribution is close to any Normal distribution, the plotted points will lie close to some straight line.

Any Normally distributed data set will produce a straight line on a Normal quantile plot, because the data distribution and the z distribution are both Normal and their relationship is thus linear. If the data are not Normally distributed, the data and the z distribution are unrelated and a Normal quantile plot will not be linear.

USE OF NORMAL QUANTILE PLOTS

If the points on a Normal quantile plot lie close to a straight line, the plot indicates that the data are Normal. Systematic deviations from a straight line indicate a non-Normal distribution. Outliers appear as points that are far away from the overall pattern of the plot.

EXAMPLE 11.11   *Shark body lengths*

Figure 11.8 is a Normal quantile plot of the body lengths of sharks from Example 1.4 (page 11). Lay a transparent straightedge over the center of the plot to see that most

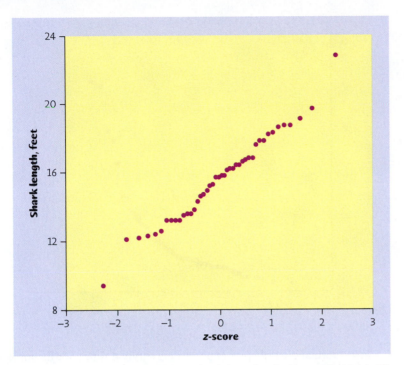

FIGURE 11.8 Normal quantile plot of shark lengths, for Example 11.11. This distribution is approximately Normal except for a low outlier and a high outlier.

of the points lie close to a straight line. A Normal distribution describes these points quite well. The low and high outliers at 9.4 and 22.8 feet lie, respectively, below and above the line formed by the center of the data. Compare Figure 11.8 with the histogram of these data in Figure 1.5(b) (page 13).

EXAMPLE 11.12   Guinea pig survival times

Figure 11.9 is a Normal quantile plot of the guinea pig survival times from Example 1.6 (page 16). The data x are plotted vertically against the corresponding standard Normal z-score plotted horizontally. The z-score scale extends from -3 to 3 because almost all of a standard Normal curve lies between these values.

The points on the plot clearly do not follow a linear pattern. If you draw a line through the leftmost points, which correspond to the smaller observations, you find that the larger observations fall systematically above this line. This deviation in the Normal quantile plot indicates a right skewness. That is, the right-of-center observations have larger values than in a Normal distribution. The histogram in Figure 1.6 (page 16) also clearly shows that the distribution of guinea pig survival times is strongly skewed to the right.

In a right-skewed distribution, the largest observations fall distinctly above a line drawn through the main body of points. Similarly, left skewness is evident when the smallest observations fall below the line.

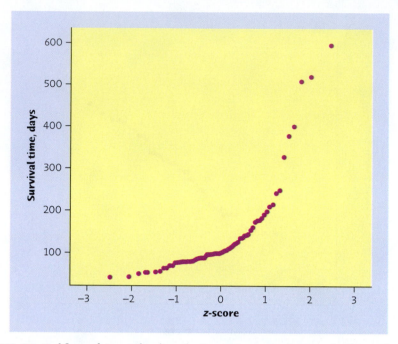

FIGURE 11.9 Normal quantile plot of guinea pig survival times, for Example 11.12. This distribution is skewed to the right.

EXAMPLE 11.13   Acid rain

Exercise 1.33 (page 30) provided data on the acidity (pH) of 105 samples of rain-water. Figure 11.10 shows side by side the histogram of these data with a Normal curve fitted over it (a) and a Normal quantile plot of the same data (b). Lay a transparent straight-edge over the center of the plot to see that most of the points lie close to a straight line. A Normal distribution describes these points quite well. Notice also that the majority of the points on the Normal quantile plot have z-scores between -1 and 1 and few have z-scores close to 3 or -3, reflecting the 68–95–99.7 property of Normal distributions.

As Exercise 1.33 (page 30) illustrated, histograms don't settle the question of approximate Normality because their particular shape depends on the choice of classes. The Normal quantile plot makes it clear that a Normal distribution is a good description for these data—there are only minor wiggles in a generally straight-line pattern.

As Figure 11.10 illustrates, real data almost always show some departure from the theoretical Normal model. *When you examine a Normal quantile plot, look for shapes that show clear departures from Normality. Don't overreact to minor wiggles in the plot.* When we discuss statistical methods that are based on the Normal model, we will pay attention to the sensitivity of each method to departures from Normality. Many common methods work well as long as the data are approximately Normal and outliers are not present.

APPLY YOUR KNOWLEDGE

11.11 Heart rate. Figure 11.11 is a Normal quantile plot of the heart rates of 200 male runners after a 6-minute treadmill run.[3] The distribution is close to Normal. How

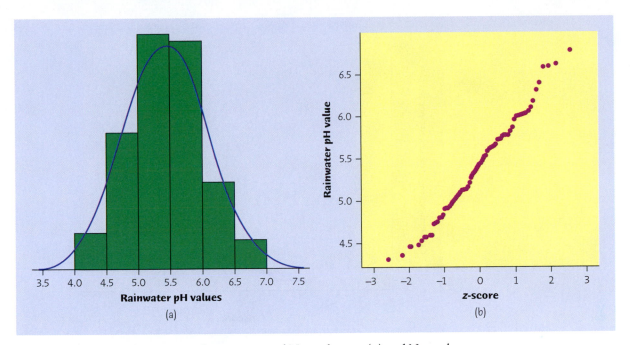

FIGURE 11.10 Histogram with superimposed Normal curve (a) and Normal quantile plot (b) of rainwater pH values, for Example 11.13. This distribution is roughly Normal.

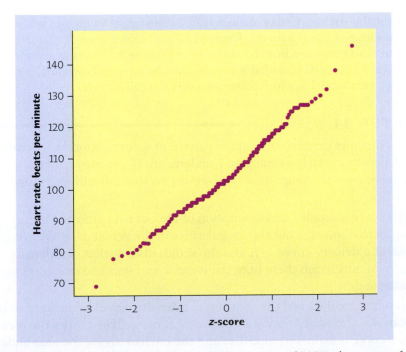

FIGURE 11.11 Normal quantile plot of the heart rates of 200 male runners, for Exercise 11.11.

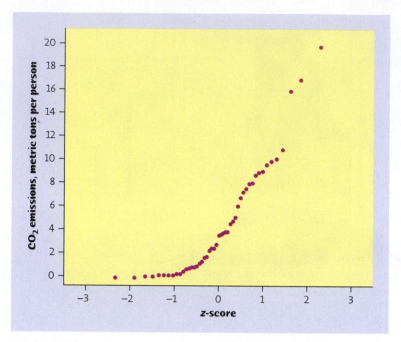

FIGURE 11.12 Normal quantile plot of CO_2 emissions in 48 countries, for Exercise 11.12.

can you see this? Describe the nature of the small deviations from Normality that are visible in the plot.

11.12 **Global warming.** Carbon dioxide (CO_2) emissions are argued to be the main gas source of global warming. Figure 11.12 is a Normal quantile plot of CO_2 emissions per person in 48 countries, from Exercise 1.35 with data displayed in Table 1.3 (page 32). In what way is this distribution non-Normal? Comparing the plot with Table 1.3, which countries would you call outliers?

CHAPTER 11 SUMMARY

We can sometimes describe the overall pattern of a distribution by a **density curve.** A density curve has total area 1 underneath it. An area under a density curve gives the proportion of observations that fall within a range of values.

A density curve is an idealized description of the overall pattern of a distribution that smooths out the irregularities in the actual data. We write the **mean of a density curve** as μ and the **standard deviation of a density curve** as σ to distinguish them from the mean $\bar{x}$ and standard deviation s of the actual sample data.

The mean, the median, and the quartiles of a density curve can be located by eye. The **mean** μ is the balance point of the curve. The **median** divides the area under the curve in half. The **quartiles** and the median divide the area under the curve into quarters. The **standard deviation** σ cannot be located by eye on most density curves.

The mean and median are equal for symmetric density curves. The mean of a skewed curve is located farther toward the long tail than is the median.

The **Normal distributions** are described by a special family of bell-shaped, symmetric density curves, called **Normal curves.** The mean μ and standard deviation σ completely specify a Normal distribution $N(\mu, \sigma)$. The mean is the center of the curve, and σ is the distance from μ to the change-of-curvature points on either side.

To **standardize** any observation x, subtract the mean of the distribution and then divide by the standard deviation. The resulting **z-score**

$$z = \frac{x - \mu}{\sigma}$$

says how many standard deviations x lies away from the distribution mean.

All Normal distributions are the same when measurements are transformed to the standardized scale. In particular, all Normal distributions satisfy the **68–95–99.7 rule,** which describes what percent of observations lie within one, two, and three standard deviations of the mean.

If x has the $N(\mu, \sigma)$ distribution, then the **standardized variable** $z = (x - \mu)/\sigma$ has the **standard Normal distribution** $N(0, 1)$ with mean 0 and standard deviation 1. Table B gives the **cumulative proportions** of standard Normal observations that are less than z for many values of z. By standardizing, we can use Table B for any Normal distribution.

The adequacy of a Normal model for describing a distribution of data is best assessed by a **Normal quantile plot,** which is available in most statistical software packages and graphing calculators. A pattern on such a plot that deviates substantially from a straight line indicates that the data are not Normal.

CHECK YOUR SKILLS

11.13 Which of these variables is least likely to have a Normal distribution?

(a) Survival time after diagnosis with fatal disease.

(b) Lengths of 50 newly hatched pythons.

(c) Heights of 100 white pine trees in a forest.

11.14 To completely specify the shape of a Normal distribution, you must give

(a) the mean and the standard deviation.

(b) the five-number summary.

(c) the mean and the median.

The respiratory rate per minute in newborns varies according to a distribution that is approximately Normal with mean 50 and standard deviation 5.[4] Exercises 11.15 to 11.17 use this information.

11.15 Approximately 95% of all newborn respiratory rates per minute are between

(a) 45 and 55.

(b) 40 and 60.

(c) 35 and 65.

11.16 The probability that a randomly chosen newborn has a respiratory rate of 55 per minute or more is approximately

(a) 0.16. (b) 0.68. (c) 0.84.

11.17 The probability that a randomly chosen newborn has a respiratory rate per minute between 40 and 55 is approximately

(a) 0.498. (b) 0.815. (c) 0.997.

11.18 The proportion of observations from a standard Normal distribution that take values less than 1.15 is about

(a) 0.1251. (b) 0.8531. (c) 0.8749.

11.19 The proportion of observations from a standard Normal distribution that take values larger than −0.75 is about

(a) 0.2266. (b) 0.7734. (c) 0.8023.

The scale of scores on an IQ test is approximately Normal with mean 100 and standard deviation 15. Exercises 11.20 to 11.22 use this information.

11.20 The organization MENSA, which calls itself "the high IQ society," requires an IQ score of 130 or higher for membership. What percent of adults would qualify for membership?

(a) 95%. (b) 5%. (c) 2.5%.

11.21 Corinne scored 118 on her IQ test. Her z-score is about

(a) 1.2. (b) 7.87. (c) 18.

11.22 Corinne scored 118 on her IQ test. She scored higher than what percent of all adults?

(a) About 12%

(b) About 88%

(c) About 98%

CHAPTER 11 EXERCISES

11.23 Actual IQ test scores. Here are the IQ test scores of 31 seventh-grade girls in a Midwest school district:[5]

114	100	104	89	102	91	114	114	103	105	
108	130	120	132	111	128	118	119	86	72	
111	103	74	112	107	103	98	96	112	112	93

(a) We expect IQ scores to be approximately Normal. Make a stemplot to check that there are no major departures from Normality.

(b) Nonetheless, proportions calculated from a Normal distribution are not always very accurate for small numbers of observations. Find the mean $\bar{x}$ and standard deviation s for these IQ scores. What proportions of the scores are within one standard deviation and within two standard deviations of the mean? What would these proportions be in an exactly Normal distribution?

11.24 IQ test scores. The Wechsler Adult Intelligence Scale (WAIS) is the most common "IQ test." The scale of scores is set separately for each age group and is approximately Normal with mean 100 and standard deviation 15. According to the 68–95–99.7 rule, about what percent of people have WAIS scores

(a) above 100?   (b) above 145?   (c) below 85?

11.25 Low IQ test scores. Scores on the Wechsler Adult Intelligence Scale (WAIS) are approximately Normal with mean 100 and standard deviation 15. People with WAIS scores below 70 are considered mentally retarded when, for example, applying for Social Security disability benefits. According to the 68–95–99.7 rule, about what percent of adults are retarded by this criterion?

11.26 Pregnancy length in horses. Bigger mammals tend to carry their young longer before birth. The length of horse pregnancies from conception to birth varies according to a roughly Normal distribution with mean 336 days and standard deviation 3 days. Use the 68–95–99.7 rule to answer the following questions.

(a) Almost all (99.7%) horse pregnancies fall within what range of lengths?

(b) What percent of horse pregnancies are longer than 339 days?

11.27 Using Table B. Use Table B to find the proportion of observations from a standard Normal distribution that falls in each of the following regions. In each case, sketch a standard Normal curve and shade the area representing the region.

(a) $z \le -2.25$.   (b) $z \ge -2.25$.
(c) $z > 1.77$.   (d) $-2.25 < z < 1.77$.

11.28 Using Table B. Use Table B to find the proportion of observations from a standard Normal distribution that satisfies each of the following statements. In each case, sketch a standard Normal curve and shade the area under the curve that is the answer to the question.

(a) $z < 1.85$.   (b) $z > 1.85$.
(c) $z > -0.66$.   (d) $-0.66 < z < 1.85$.

11.29 Using Table B. Use Table B to find the standardized value z that satisfies each of the following conditions (report the value of z that comes closest to satisfying the condition). In each case, sketch a standard Normal curve with your value of z marked on the axis.

(a) The probability is 0.8 that a randomly selected observation falls below z.

(b) The probability is 0.35 that a randomly selected observation falls above z.

11.30 Using Table B. Find the z value that satisfies each of the following conditions (report the value of z that comes closest to satisfying the condition). In each case, sketch a standard Normal curve with your value of z marked on the axis.

(a) 20% of the observations fall below z.

(b) 30% of the observations fall above z.

(c) 30% of the observations fall below z.

11.31 Grading on the curve. Grades in large classes are often approximately Normally distributed, and therefore many large college classes "grade on a bell curve." A common practice is to give 16% A grades, 34% B grades, 34% C grades and 16% D and F grades. Assuming a Normal distribution of grades, what are the z-scores for these letter grades?

11.32 **Hemoglobin.** The distribution of hemoglobin in g/dl of blood is approximately $N(14, 1)$ in women and approximately $N(16, 1)$ in men.[6]

(a) What percent of women have more hemoglobin than the men's mean hemoglobin?

(b) What percent of men have less hemoglobin than the women's mean hemoglobin?

(c) What percent of men have more hemoglobin than the women's mean hemoglobin?

11.33 **Osteoporosis.** Osteoporosis is a condition in which the bones become brittle due to loss of minerals. To diagnose osteoporosis, an elaborate apparatus measures bone mineral density (BMD). BMD is usually reported in standardized form. The standardization is based on a population of healthy young adults. The World Health Organization (WHO) criterion for osteoporosis is a BMD 2.5 or more standard deviations *below* the mean for young adults. BMD measurements in a population of people similar in age and sex roughly follow a Normal distribution.[7]

(a) What percent of healthy young adults have osteoporosis by the WHO criterion?

(b) Women aged 70 to 79 are, of course, not young adults. The mean BMD in this age is about −2 on the standard scale for young adults. Suppose that the standard deviation is the same as for young adults. What percent of this older population has osteoporosis?

11.34 **Osteopenia.** The previous exercise described the criteria for diagnosing osteoporosis. Likewise, osteopenia is low bone mineral density, defined by the WHO as a BMD between 1 and 2.5 standard deviations *below* the mean of young adults.

(a) What percent of healthy young adults have osteopenia by the WHO criterion?

(b) The mean BMD among women aged 70 to 79 is about −2 on the standard scale for young adults. Suppose that the standard deviation is the same as for young adults. What percent of this older population has osteopenia?

11.35 **Cholesterol in young women.** Too much cholesterol in the blood increases the risk of heart disease. Young women are generally less afflicted with high cholesterol than other groups. The cholesterol levels for women aged 20 to 34 follow an approximately Normal distribution with mean 185 milligrams per deciliter (mg/dl) and standard deviation 39 mg/dl.[8]

(a) Cholesterol levels above 240 mg/dl demand medical attention. What percent of young women have levels above 240 mg/dl?

(b) Levels above 200 mg/dl are considered elevated and place the person at some risk of cardiovascular disease. What percent of young women have blood cholesterol between 200 and 240 mg/dl?

11.36 **Cholesterol in middle-aged men.** Middle-aged men are more susceptible to high cholesterol than the young women of the previous exercise. The blood cholesterol levels of men aged 55 to 64 are approximately Normal with mean 222 mg/dl and standard deviation 37 mg/dl.

(a) What percent of these men have high cholesterol (levels above 240 mg/dl)?

(b) What percent have elevated cholesterol (between 200 and 240 mg/dl)?

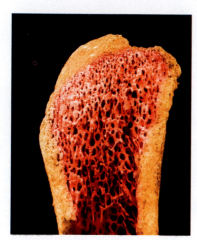

Dr. Kessel & Dr. Kardon/Tissues & Organs/Visuals Unlimited

Is your cholesterol too high?

High blood cholesterol increases your risk of heart disease and stroke. The higher your total cholesterol, the greater your risk, but there is no sharp cutoff point between a healthy and a pathological level. Historically, total cholesterol levels above 240 mg/dl have been considered a serious risk factor. In a shift toward early detection and prevention, a new "elevated cholesterol" label for the 200 to 240 mg/dl range was created to identify individuals who may be at some risk of cardiovascular disease. A study estimates that this resulted in over 40 million new candidates for pharmacological treatment.

11.37 Pregnancy length in humans. The length of human pregnancies from conception to birth varies according to a distribution that is approximately Normal with mean 266 days and standard deviation 16 days.

(a) What percent of pregnancies last less than 270 days (about 9 months)?

(b) What percent of pregnancies last between 240 and 270 days (roughly between 8 months and 9 months)?

(c) How long do the longest 20% of pregnancies last?

11.38 Pregnancy length in humans, continued. The quartiles of any distribution are the values with cumulative proportions 0.25 and 0.75.

(a) What are the quartiles of the standard Normal distribution?

(b) Using your numerical values from (a), write an equation that gives the quartiles of the $N(\mu, \sigma)$ distribution in terms of μ and σ.

(c) The length of human pregnancies from conception to birth varies according to a distribution that is approximately Normal with mean 266 days and standard deviation 16 days. Apply your result from (b): What are the quartiles of the distribution of lengths of human pregnancies?

11.39 Are we getting smarter? When the Stanford-Binet "IQ test" came into use in 1932, it was adjusted so that scores for each age group of children followed roughly the Normal distribution with mean $\mu = 100$ and standard deviation $\sigma = 15$. The test is readjusted from time to time to keep the mean at 100. If present-day American children took the 1932 Stanford-Binet test, their mean score would be about 120. The reasons for the increase in IQ over time are not known but probably include better childhood nutrition and more experience in taking tests.[9]

(a) IQ scores above 130 are often called "very superior." What percent of children had very superior scores in 1932?

(b) If present-day children took the 1932 test, what percent would have very superior scores? (Assume that the standard deviation $\sigma = 15$ does not change.)

The Normal Curve applet allows you to do Normal calculations quickly. It is somewhat limited by the number of pixels available for use, so it can't hit every value exactly. In the exercises below, use the closest available values. In each case, make a sketch of the curve from the applet marked with the values you used to answer the questions.

11.40 How accurate is 68–95–99.7? The 68–95–99.7 rule for Normal distributions is a useful approximation. To see how accurate the rule is, drag one flag across the other so that the applet shows the area under the curve between the two flags.

(a) Place the flags one standard deviation on either side of the mean. What is the area between these two values? What does the 68–95–99.7 rule say this area is?

(b) Repeat for locations two and three standard deviations on either side of the mean. Again compare the 68–95–99.7 rule with the area given by the applet.

11.41 Where are the quartiles? How many standard deviations above and below the mean do the quartiles of any Normal distribution lie? (Use the standard Normal distribution to answer this question.)

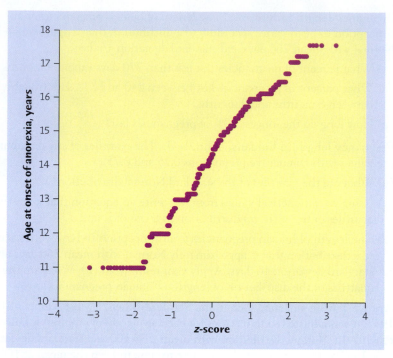

FIGURE 11.13 Normal quantile plot of age at onset of anorexia nervosa in 691 anorexic teenage girls, for Exercise 11.42.

Digital Vision/Getty Images

granularity

Exercises 11.42 to 11.47 make use of the optional material on Normal quantile plots.

11.42   Onset of anorexia nervosa. Figure 11.13 shows the Normal quantile plot for data on the onset of anorexia in 691 teenage girls from Exercise 1.8. Examine the Normal quantile plot for deviations from Normality. Explain your conclusions and compare this plot with the histogram of the same data shown in Figure 1.8 (page 18).

11.43   Acorn sizes. Figure 11.14 shows the Normal quantile plot for the volumes of 28 acorns from the Atlantic area.[10] Examine the plot for deviations from Normality and explain your conclusions.

11.44   Fruit flies. Figure 11.15 shows the Normal quantile plot for data on the thorax lengths of 49 fruit flies.[11] Examine the plot for deviations from Normality and explain your conclusions. Because of limited precision in measurements, many observations happen to have the same numerical value. This gives a steplike appearance to the Normal quantile plot. This phenomenon is called **granularity.** It is not an important feature and can be ignored in your interpretation.

11.45   Logging in the rainforest. Exercise 2.12 (page 57) described a study of the effects of logging on tree counts in the Borneo rainforest.

(a) Make your own Normal quantile plot of the data in Group 3 for plots of forest that were logged 8 years earlier: First, sort and rank the data from smallest to largest. Second, calculate z-scores for the ranks. Third, make a scatterplot of number of trees per plot against their calculated z-scores.

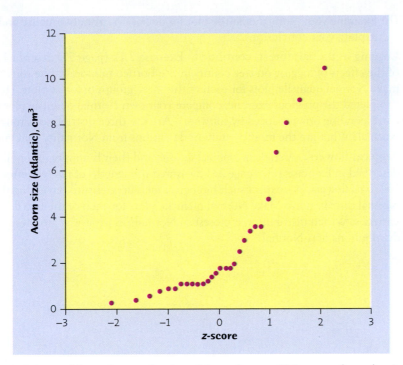

FIGURE 11.14 Normal quantile plot of the volumes of 28 acorns from the Atlantic region, for Exercise 11.43.

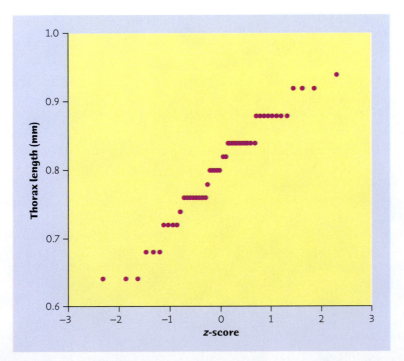

FIGURE 11.15 Normal quantile plot of the thorax lengths of 49 male fruit flies, for Exercise 11.44.

(b) Examine your Normal quantile plot for deviations from Normality and explain your conclusions.

11.46 Logging in the rainforest, continued. Exercise 2.12 (page 57) described a study of the effects of logging on tree counts in the Borneo rainforest. Use software to make Normal quantile plots for each of the three groups of forest plots. (If you completed the previous exercise, compare your own Normal quantile plot to the corresponding one produced by software). Are the three distributions roughly Normal? What are the most prominent deviations from Normality that you see?

11.47 Tropical flowers. A study of tropical flowers and their hummingbird pollinators described in Exercise 1.37 (page 33) measured the lengths of two varieties of *Heliconia* flowers. We expect such biological measurements to have roughly Normal distributions. Make Normal quantile plots for each of the two flower varieties. Which distribution is closest to Normal? In what way are these distributions non-Normal?

David Young-Wolff/Photo Edit

Discrete Probability Distributions*

An elementary school administers eye exams to 800 students. How many students have perfect vision? A new treatment for pancreatic cancer is tried on 250 patients. How many survive for five years? You plant 10 dogwood trees. How many live through the winter? In all these situations, we want a probability model for a *count* of successful outcomes.

The binomial setting and binomial distributions

The distribution of a count depends on how the data are produced. Here is a common situation.

---

*This more advanced chapter concerns a special topic in probability. This material is not needed to read the rest of the book.

> **THE BINOMIAL SETTING**
>
> 1. There are a fixed number n of observations.
> 2. The n observations are all **independent.** That is, knowing the result of one observation does not change the probabilities we assign to other observations.
> 3. Each observation falls into one of just two categories, which for convenience we call "success" and "failure."
> 4. The probability of a success, call it p, is the same for each observation.

Think of an obstetrician overseeing n single-birth deliveries as an example of the binomial setting. Each single-birth delivery is either a baby girl or a baby boy. Knowing the outcome of one birth doesn't change the probability of a girl on any other birth, so the births are independent. If we call a girl a "success" (this is an arbitrary choice), then p is the probability of a girl and remains the same over all the births overseen by that obstetrician. The number of girls we count is a discrete random variable X. The distribution of X is called a *binomial distribution*.

> **BINOMIAL DISTRIBUTION**
>
> The count X of successes in the binomial setting has the **binomial distribution** with parameters n and p. The parameter n is the number of observations, and p is the probability of a success on any one observation. The possible values of X are the whole numbers from 0 to n.

The binomial distributions are an important class of probability distributions. *Pay attention to the binomial setting, because not all counts have binomial distributions.*

EXAMPLE 12.1 Inheriting blood type

Genetics says that children receive genes from their parents independently. Each child of one particular pair of parents has probability 0.25 of having blood type O. If these parents have 5 children, the number who have blood type O is the count X of successes in 5 independent observations with probability 0.25 of a success on each observation. So X has the binomial distribution with $n = 5$ and $p = 0.25$.

EXAMPLE 12.2 Selecting volunteer subjects

A researcher has access to 40 healthy volunteers, 20 men and 20 women. The researcher selects 10 of them at random to participate in an experiment on the effect of sport drinks taken during endurance exercise. There are 10 observations, and each is either a man or a woman. But the observations are *not* independent. If the first volunteer selected is a man, the second one is now more likely to be a woman because there are more women (20) than men (19) left in the pool of volunteers. The count X of men randomly selected in this situation does *not* have a binomial distribution.

Binomial distributions in statistical sampling

The binomial distributions are important in statistics when we wish to make inferences about the proportion p of "successes" in a population. Here is a typical example.

EXAMPLE 12.3 *Selecting an SRS*

A pharmaceutical company inspects an SRS of 10 empty plastic containers from a shipment of 10,000. The containers are examined for traces of benzene, a common chemical solvent but also a known human carcinogen. Suppose that (unknown to the company) 10% of the containers in the shipment contain traces of benzene. Count the number X of containers contaminated with benzene in the sample.

This is not quite a binomial setting. Just as selecting one man in Example 12.2 changes the makeup of the remaining pool of volunteers, removing one plastic container changes the proportion of contaminated containers remaining in the shipment. So the probability that the second container chosen is contaminated changes when we know that the first is contaminated once one container is removed from the shipment. But removing one container from a shipment of 10,000 changes the makeup of the remaining 9999 very little. In practice, the distribution of X is very close to the binomial distribution with $n = 10$ and $p = 0.1$.

Allan H. Shoemake/Taxi/Getty Images

Example 12.3 shows how we can use the binomial distributions in the statistical setting of selecting an SRS. When the population is much larger than the sample, a count of successes in an SRS of size n has approximately the binomial distribution with n equal to the sample size and p equal to the proportion of successes in the population.

SAMPLING DISTRIBUTION OF A COUNT

Choose an SRS of size n from a population with proportion p of successes. When the population is much larger than the sample, the count X of successes in the sample has approximately the binomial distribution with parameters n and p.

APPLY YOUR KNOWLEDGE

In each of Exercises 12.1 to 12.3, X is a count. Does X have a binomial distribution? Give your reasons in each case.

12.1 **Random digit dialing.** When an opinion poll calls residential telephone numbers at random, only 20% of the calls reach a live person. You watch the random dialing machine make 15 calls. X is the number that reach a live person.

12.2 **A boy or a girl?** A couple really wants to have a boy. They decide that they will keep having children until they have their first boy. Successive births are independent and each has about probability 0.5 of being a boy or a girl. X is the number of the pregnancy that finally succeeds at bringing a baby boy to the couple.

12.3 **Medical marijuana.** Opinion polls find that 15% of Americans think that "marijuana should remain illegal even for medical purposes."[1] X is the number who think that marijuana should remain illegal even for medical purposes in an SRS of 500 adults.

Binomial probabilities

We can find a formula for the probability that a binomial random variable takes any value by adding probabilities for the different ways of getting exactly that many successes in n observations. Here is an example that illustrates the idea.

EXAMPLE 12.4   Inheriting blood type

Each child born to a particular set of parents has probability 0.25 of having blood type O. If these parents have 5 children, what is the probability that exactly 2 of them have type O blood?

The count of children with type O blood is a binomial random variable X with $n = 5$ children and probability $p = 0.25$ that a child has type O blood (as explained in Example 12.1). We want $P(X = 2)$.

Because the method doesn't depend on the specific example, let's use "S" for success (has type O blood) and "F" for failure (does not have type O blood) for short. Do the work in two steps.

Step 1. Find the probability that a specific 2 of the 5 tries, say the first and the third, give successes. This is the outcome SFSFF. Because tries are independent (the blood type of one child doesn't change the probability that the next child has blood type O), the multiplication rule for independent events applies. The probability we want is

$$P(\text{SFSFF}) = P(S)P(F)P(S)P(F)P(F)$$
$$= (0.25)(0.75)(0.25)(0.75)(0.75)$$
$$= (0.25)^2(0.75)^3$$

Step 2. Observe that *any one arrangement* of 2 S's and 3 F's has this same probability. This is true because we multiply together 0.25 twice and 0.75 three times whenever we have 2 S's and 3 F's. The probability that $X = 2$ is the probability of getting 2 S's and 3 F's in any arrangement whatsoever. Here are all the possible arrangements:

SSFFF   SFSFF   SFFSF   SFFFS   FSSFF
FSFSF   FSFFS   FFSSF   FFSFS   FFFSS

There are 10 of them, all with the same probability. The overall probability of 2 successes is therefore

$$P(X = 2) = 10(0.25)^2(0.75)^3 = 0.2637$$

The pattern of this calculation works for any binomial probability. To use it, we must count the number of arrangements of k successes in n observations. We use the following fact to do the counting without actually listing all the arrangements.

BINOMIAL COEFFICIENT

The number of ways of arranging k successes among n observations is given by the **binomial coefficient**

$$\binom{n}{k} = \frac{n!}{k!\,(n-k)!}$$

for $k = 0, 1, 2, \ldots, n$.

The formula for binomial coefficients uses the **factorial** notation. For any positive whole number n, its factorial $n!$ is

$$n! = n \times (n-1) \times (n-2) \times \cdots \times 3 \times 2 \times 1$$

Also, $0! = 1$.

The larger of the two factorials in the denominator of a binomial coefficient will cancel much of the $n!$ in the numerator. For example, the binomial coefficient we need for Example 12.4 is

$$
\begin{aligned}
\binom{5}{2} &= \frac{5!}{2!\,3!} \\
&= \frac{(5)(4)(3)(2)(1)}{(2)(1) \times (3)(2)(1)} \\
&= \frac{(5)(4)}{(2)(1)} = \frac{20}{2} = 10
\end{aligned}
$$

The notation $\binom{n}{k}$ is not related to the fraction $\frac{n}{k}$. A helpful way to remember its meaning is to read it as "binomial coefficient n choose k." Binomial coefficients have many uses, but we are interested in them only as an aid to finding binomial probabilities. The binomial coefficient $\binom{n}{k}$ counts the number of different ways in which k successes can be arranged among n observations. The binomial probability $P(X = k)$ is this count multiplied by the probability of any one specific arrangement of the k successes. Here is the result we seek.

factorial

CAUTION

Unusual performances

When a baseball player hits .300, people applaud or suspect performance enhancing drugs. A .300 hitter gets a hit in 30% of times at bat. Could a .300 year just be luck? Typical Major Leaguers bat about 500 times a season and hit about .260. A hitter's successive tries seem to be independent, so we have a binomial setting. From this model, the probability of hitting .300 is about 0.025. Out of 100 run-of-the-mill Major League hitters, two or three each year will bat .300 because they were lucky.

BINOMIAL PROBABILITY

If X has the binomial distribution with n observations and probability p of success on each observation, the possible values of X are 0, 1, 2, ..., n. If k is any one of these values,

$$P(X = k) = \binom{n}{k} p^k (1 - p)^{n-k}$$

EXAMPLE 12.5 Inspecting pharmaceutical containers

The number X of containers contaminated with benzene in the sample in Example 12.3 has approximately the binomial distribution with $n = 10$ and $p = 0.1$.

The probability that the sample contains no more than 1 contaminated container is

$$P(X \leq 1) = P(X = 1) + P(X = 0)$$
$$= \binom{10}{1} (0.1)^1 (0.9)^9 + \binom{10}{0} (0.1)^0 (0.9)^{10}$$
$$= \frac{10!}{1! \, 9!}(0.1)(0.3874) + \frac{10!}{0! \, 10!}(1)(0.3487)$$
$$= (10)(0.1)(0.3874) + (1)(1)(0.3487)$$
$$= 0.3874 + 0.3487 = 0.7361$$

This calculation uses the facts that $0! = 1$ and that $a^0 = 1$ for any number a other than 0. We see that about 74% of all samples will contain no more than 1 contaminated container. In fact, 35% of the samples will contain no contaminated containers. A sample of size 10 cannot be trusted to alert the pharmaceutical company to the presence of unacceptable containers in the shipment.

Using technology

The binomial probability formula is awkward to use, particularly for the probabilities of events that contain many outcomes. You can find tables of binomial probabilities $P(X = k)$ and cumulative probabilities $P(X \leq k)$ for selected values of n and p.

The most efficient way to do binomial calculations is to use technology. Figure 12.1 shows output for the calculation in Example 12.5 from a graphing calculator, two statistical programs, and a spreadsheet program. We asked all four to give cumulative probabilities. The TI-83, CrunchIt!, and Minitab have menu entries for binomial cumulative probabilities. Excel has no menu entry, but the worksheet function BINOMDIST is available. All of the outputs agree with the result 0.7361 of Example 12.5.

APPLY YOUR KNOWLEDGE

12.4 Random digit dialing. When a survey calls residential telephone numbers at random, only 20% of the calls reach a live person. You watch the random dialing machine make 15 calls.

CrunchIt!

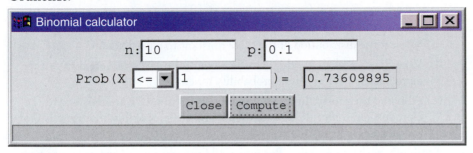

Texas Instruments TI-83 Plus

```
binomcdf(10,0.1,
1)
              .7361
```

Minitab

Session

Cumulative Distribution Function

Binomial with n = 10 and p = 0.100000

x	p(x <= x)
0.00	0.3487
1.00	0.7361
2.00	0.9298
3.00	0.9872

SPSS

Untitled - SPSS Data Editor

	X	CumBinomial
1	0	.3487
2	1	.7361
3	2	.9298
4	3	.9872
5	4	.9984
6		

Microsoft Excel

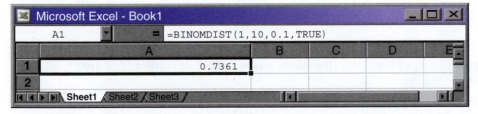

FIGURE 12.1 The binomial probability $P(X \leq 1)$ for Example 12.5: output from a graphing calculator, two statistical programs, and a spreadsheet program.

(a) What is the probability that exactly 3 calls reach a person?

(b) What is the probability that 3 or fewer calls reach a person?

12.5 Antibiotic resistance. Antibiotic resistance occurs when disease-causing microbes become resistant to antibiotic drug therapy. Because this resistance is typically genetic and transferred to the next generation of microbes, it is a very serious public health problem. According to the CDC, 7% of gonorrhea cases tested in 2004 were resistant to the antibiotic ciprofloxacin.[2] A physician prescribed ciprofloxacin for the treatment of 10 cases of gonorrhea during one week in 2004.

(a) What is the distribution of the cases resistant to ciprofloxacin?

(b) What is the probability that exactly 1 out of the 10 cases were resistant to ciprofloxacin? What is that probability for exactly 2 out of 10?

(c) Also find the probability that 1 or more out of the 10 were resistant to ciprofloxacin. (*Hint:* It is easier to first find the probability that exactly 0 of the 10 cases were resistant.)

12.6   Antibiotic resistance again. Because antibiotic resistance is passed on to new generations of microbes, the problem is an increasing concern. In 2000, for instance, only 0.5% of gonorrhea cases tested were resistant to ciprofloxacin.[3] Let's now think of a physician who prescribed ciprofloxacin for the treatment of 10 cases of gonorrhea during one week in 2000.

(a) What is the distribution of the cases resistant to ciprofloxacin?

(b) What is the probability that exactly 1 out of the 10 cases were resistant to ciprofloxacin? What is that probability for exactly 2 out of 10?

(c) Also find the probability that 1 or more out of the 10 were resistant to ciprofloxacin. (*Hint:* It is easier to first find the probability that exactly 0 of the 10 cases were resistant.)

Binomial mean and standard deviation

If a count X has the binomial distribution based on n observations with probability p of success, what is its mean μ? That is, in very many repetitions of the binomial setting, what will be the average count of successes? We can guess the answer. If a psychic correctly guesses a pattern in a test 60% of the time, the mean number of correct guesses in 10 tries should be 60% of 10, or 6. In general, the mean of a binomial distribution should be $\mu = np$. Here are the facts.

> **BINOMIAL MEAN AND STANDARD DEVIATION**
>
> If a count X has the binomial distribution with number of observations n and probability of success p, the **mean** and **standard deviation** of X are
>
> $$\mu = np$$
> $$\sigma = \sqrt{np(1-p)}$$

Remember that these short formulas are good only for binomial distributions. They can't be used for other distributions.

┌─ **EXAMPLE 12.6** Inspecting pharmaceutical containers ───────

Continuing Example 12.5, the count X of containers contaminated with benzene is binomial with $n = 10$ and $p = 0.1$. The histogram in Figure 12.2 displays this probability distribution. (Because probabilities are long-run proportions, using probabilities as the

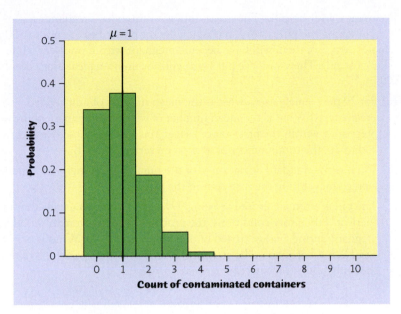

FIGURE 12.2 Probability histogram for the binomial distribution with $n = 10$ and $p = 0.1$, for Example 12.6.

heights of the bars shows what the distribution of X would be in very many repetitions.) The distribution is strongly skewed. Although X can take any whole-number value from 0 to 10, the probabilities of values larger than 5 are so small that they do not appear in the histogram.

The mean and standard deviation of the binomial distribution in Figure 12.2 are

$$\mu = np$$
$$= (10)(0.1) = 1$$
$$\sigma = \sqrt{np(1-p)}$$
$$= \sqrt{(10)(0.1)(0.9)} = \sqrt{0.9} = 0.9487$$

The mean is marked on the probability histogram in Figure 12.2.

APPLY YOUR KNOWLEDGE

12.7 Antibiotic resistance

(a) What is the mean number of gonorrhea cases that are resistant to the antibiotic ciprofloxacin out of 10 cases, as described in Exercise 12.5 for the year 2004?

(b) What is the standard deviation σ of the count of antibiotic-resistant cases?

(c) In the year 2000, the probability that a gonorrhea case was resistant to ciprofloxacin was only $p = 0.005$, as described in Exercise 12.6. How does this new p affect the standard deviation? What is the standard deviation for $p = 0.005$? What does your work show about the behavior of the standard deviation of a binomial distribution as the probability of a success gets closer to 0?

12.8   **Pregnancy and the pill.** The actual rate of unintended pregnancy for women taking the pill is 5%. That is, if a doctor prescribes the pill to 100 new female users, 5 of them will have become pregnant within one year while taking the pill. Therefore, the pill's true efficacy under typical usage conditions is 95%.[4]

(a) For 20 new female users, what is the mean number of unintended pregnancies? What is the mean number of women not experiencing any pregnancy within the first year of usage? You see that these two means must add to 20, the total number of women prescribed the pill.

(b) What is the standard deviation σ of the number of women not experiencing a pregnancy during the first year on the pill?

(c) The pill is often advertised as being 99% effective. This means that under "perfect" laboratory conditions, the probability that a woman would become pregnant during a usage period of one year is $p = 0.99$. What is σ in this case? What happens to the standard deviation of a binomial distribution as the probability of a "success" gets close to 1?

The Normal approximation to binomial distributions

The formula for binomial probabilities becomes awkward as the number of observations n increases. You can use software or a graphing calculator to handle many problems for which the formula is not practical. If technology does not rescue you, there is another alternative: *As the number of observations* n *gets larger, the binomial distribution gets close to a Normal distribution.* When n is large, we can use Normal probability calculations to approximate binomial probabilities.

David Young-Wolff/Photo Edit

EXAMPLE 12.7 Overweight Americans

Nearly 60% of American adults are either overweight or obese, according to the U.S. National Center for Health Statistics. Suppose that we take a random sample of 2500 adults. What is the probability that 1520 or more of the sample are overweight or obese?

Because there are almost 225 million adults, we can take the weights of 2500 randomly chosen adults to be independent. So the number in our sample who are either overweight or obese is a random variable X having the binomial distribution with $n = 2500$ and $p = 0.6$. To find the probability that at least 1520 of the people in the sample are overweight or obese, we must add the binomial probabilities of all outcomes from $X = 1520$ to $X = 2500$. This isn't practical. Here are two ways to do this problem.

1. Use technology. The CrunchIt! output is shown in Figure 12.3. The result is

$$P(X \geq 1520) = 0.2131$$

CrunchIt!

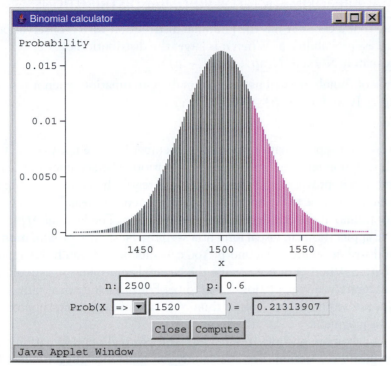

FIGURE 12.3 CrunchIt! probability distribution for the binomial model $n = 2500$, $p = 0.6$, displayed graphically. The height of each bar represents the probability for X when it takes a value on the horizontal axis. Notice how the shape of this binomial probability distribution closely resembles a Normal curve.

This answer is exactly correct to four decimal places. Note that some programs will not even perform such intensive calculations. Excel, for example, gives an error message instead.

2. Use the Normal approximation to the binomial distribution. Look again at the CrunchIt! output in Figure 12.3. Above the numerical result is the actual binomial probability distribution for X, with events $X \geq 1520$ in red. The shape of this distribution closely resembles the Normal distribution with the same mean and standard deviation as the binomial variable X:

$$\mu = np = (2500)(0.6) = 1500$$

$$\sigma = \sqrt{np(1-p)} = \sqrt{(2500)(0.6)(0.4)} = 24.49$$

EXAMPLE 12.8 Normal calculation of a binomial probability

If we act as though the count X has the $N(1500, 24.49)$ distribution, here is the probability we want, using Table B:

$$P(X \geq 1520) = P\left(\frac{X - 1500}{24.49} \geq \frac{1520 - 1500}{24.49}\right)$$

$$= P(Z \geq 0.82)$$

$$= 1 - 0.7939 = 0.2061$$

The Normal approximation 0.2061 differs from the software result 0.2131 by only 0.007.

> ### NORMAL APPROXIMATION FOR BINOMIAL DISTRIBUTIONS
>
> Suppose that a count X has the binomial distribution with n observations and success probability p. When n is large, the distribution of X is approximately Normal, $N(np, \sqrt{np(1 - p)})$.
>
> As a rule of thumb, we will use the Normal approximation when n is so large that $np \geq 10$ and $n(1 - p) \geq 10$.

The Normal approximation is easy to remember because it says that X is Normal, with its binomial mean and standard deviation. The accuracy of the Normal approximation improves as the sample size n increases. It is most accurate for any fixed n when p is close to 0.5 and least accurate when p is near 0 or 1. This is why the rule of thumb in the box depends on p as well as n. The *Normal Approximation to Binomial* applet shows in visual form how well the Normal approximation fits the binomial distribution for any n and p. You can slide n and watch the approximation get better. Whether or not you use the Normal approximation should depend on how accurate your calculations need to be. For most statistical purposes great accuracy is not required. Our "rule of thumb" for use of the Normal approximation reflects this judgment.

continuity correction

For slightly more accurate results with the Normal approximation, you can use a **continuity correction.** Because counts can take only integer values but the Normal distribution can take any real value, the proper continuous equivalent to a count is the interval around it with size 1. For example, the continuous equivalent to a 1520 count is the interval between 1519.5 and 1520.5. Therefore, in Example 12.8, we would get a slightly more accurate Normal approximation of the probability that 1520 or more individuals in the sample are overweight or obese by finding $P(X \geq 1519.5)$. This gives probability 0.2119, which is closer still to the binomial software output. The continuity correction is especially helpful when the sample size is small.

APPLY YOUR KNOWLEDGE

12.9 **Bone density.** Forty percent of postmenopausal women have low bone density (osteopenia), placing them at risk for osteoporosis with ensuing spontaneous fractures. Osteoporosis is estimated to cost $14 billion per year in medical expenses alone, and yet it can be prevented if treated early enough. A physicians group has 248 postmenopausal primary patients and has diagnosed 82 with low bone density.

(a) What is the probability that only 82 or fewer of the 248 patients have low bone density? Use the Normal approximation to find this probability.

(b) If you have access to technology, find the probability in (a) using the binomial probability model. How does it compare to the Normal approximation?

(c) Based on your answer, what would you suggest at the next physicians meeting?

12.10 **Checking for survey errors.** One way of checking the effect of undercoverage, nonresponse, and other sources of error in a sample survey is to compare the sample with known facts about the population. About 12% of American adults are black. The number X of blacks in a random sample of 1500 adults should therefore vary with the binomial ($n = 1500$, $p = 0.12$) distribution.

(a) What are the mean and standard deviation of X?

(b) Use the Normal approximation to find the probability that the sample will contain between 165 and 195 blacks. Be sure to check that you can safely use the approximation.

The Poisson distributions

Binomial distributions are not the only type of discrete distribution for whole-number counts. Exercises 10.43 (page 273) and 12.43 show examples of geometric distributions, and Exercise 12.45 walks you through an example of a multinomial distribution. These distributions are beyond the scope of an introductory textbook and will not be discussed beyond these exercises.

Poisson distributions are yet another class of discrete distributions[5] and have important applications in the life sciences. They describe the count of events that occur at random over space or time. For example, a Poisson distribution would be a good model for the count of chocolate chips in cookies made from the same well-mixed batch or for the number of times you walk up the stairs on any workday.

POISSON DISTRIBUTION

A **Poisson distribution** describes the count X of occurrences of a defined event in fixed, finite intervals of time or space when

1. occurrences are all **independent** (that is, knowing that one event has occurred does not change the probability that another event may occur),

2. and the probability of an occurrence is the same over all possible intervals.

— **EXAMPLE 12.9** *Counting daisies* —

On a lazy summer afternoon, you lay on an unkept lawn adorned with daisies growing here and there. In an inquisitive spirit, you divide the lawn into quadrants of 1 square foot and count the number of daisies in each quadrant. The number of daisies per quadrant is the random variable X. If the daisies present in each quadrant are random, independent events, then the distribution of X follows approximately a Poisson distribution.

Figure 12.4 illustrates this concept via a simulation. The large rectangle represents a lawn divided into 50 quadrants of equal size. Each pink circle is a daisy on the simulated lawn. The daisies were randomly "planted" according to a Poisson distribution giving an overall density of 2.1 daisies per quadrant. The first 10 quadrants on the upper row, starting from the left, have respectively 2, 3, 2, 1, 3, 3, 0, 2, 2, and 2 daisies each. Notice how the overall pattern of daisies on the lawn appears to indeed be entirely random.

Julie Habel/CORBIS

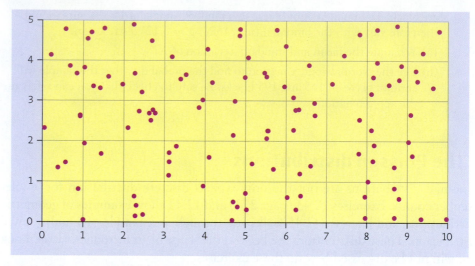

FIGURE 12.4 Graphical display of a simulated lawn planted with daisies (pink circles), for Example 12.9. Daisy location on the rectangular lawn follows a Poisson distribution with mean 2.1 daisies per quadrant.

The Poisson distributions can be used to predict what counts we would expect for a random event occurring within an interval. For instance, a medical group can model the count of non-life-threatening emergency room visits on weekend days and decide whether to open a walk-in clinic and how to staff it.

Because Poisson distributions describe random, independent events, it can also be interesting to find out that a random variable substantially deviates from that model. For instance, when the count of meningitis cases in a county exceeds what would be reasonably expected under a Poisson distribution, we can conclude that the meningitis cases are not randomly and independently distributed but suggest instead an epidemic outbreak.

POISSON PROBABILITY

If X has the Poisson distribution with mean number of occurrences per interval μ, the possible values of X are 0, 1, 2, …. If k is any one of these values,

$$P(X = k) = \frac{e^{-\mu}\mu^k}{k!}$$

The mean and variance of the Poisson distribution are both equal to μ, the mean number of occurrences per interval. The distribution's standard deviation σ is equal to $\sqrt{\mu}$.

Because the mean and variance of a Poisson distribution are equal, when the mean number of occurrences is large, the variance is also large and the distribution looks very flat and wide. Therefore, Poisson distributions are typically used to describe rarely occurring, random phenomena.

EXAMPLE 12.10 *Monitoring mumps outbreaks*

Mumps is an acute viral infection that is generally mild and, in about 20% of infected individuals, even asymptomatic. However, complications can arise with long-lasting consequences such as meningitis/encephalitis, sterility, spontaneous abortion, or deafness. Mandatory vaccinations of infants have largely prevented mumps epidemics in the United States, which has experienced an average of only 265 mumps cases per year since 1996. In Iowa alone, the average monthly number of reported cases is about 0.1 over that same period.[6]

Assuming that cases of mumps are random and independent, the number X of monthly mumps cases in Iowa has approximately the Poisson distribution with $\mu = 0.1$.

The probability that in a given month there is no more than 1 mumps case is

$$P(X \le 1) = P(X = 0) + P(X = 1)$$
$$= \frac{e^{-0.1} \times 0.1^0}{0!} + \frac{e^{-0.1} \times 0.1^1}{1!}$$
$$= \frac{0.9048 \times 1}{1} + \frac{0.9048 \times 0.1}{1}$$
$$= 0.9048 + 0.0905 = 0.9953$$

We would expect no more than one case of mumps per month with probability 99.53%. That is, we would consider seeing two or more mumps cases in a month a very unlikely event (probability 1 minus 0.9953) if cases are random and independent.

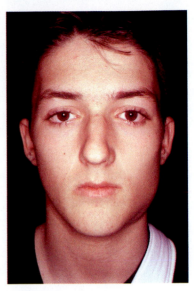

ISM/Phototake

Using technology

The Poisson probability formula is simple to use for any value of X but can be more cumbersome for cumulative probabilities. Again, using technology is very useful. Figure 12.5 shows output for the calculation in Example 12.10 from the

Minitab

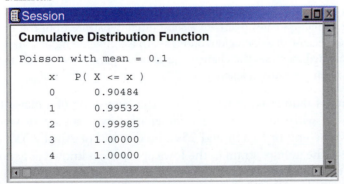

Cumulative Distribution Function

Poisson with mean = 0.1

x	P(X <= x)
0	0.90484
1	0.99532
2	0.99985
3	1.00000
4	1.00000

Texas Instruments TI-83 Plus

```
Poissoncdf (0.1,1
)
                 .995321159
```

Mumps in perspective

According to the Centers for Disease Control and Prevention, before approval of the mumps vaccine in 1967, nearly everyone in the United States experienced mumps, with about 200,000 new cases reported every year. The mumps vaccine has a reported effectiveness of about 95%. That is, some individuals will not be helped by the vaccine because they lack the genetic ability to produce antibodies against that particular viral strain.

FIGURE 12.5 The Poisson probability $P(X \le 1)$, for Example 12.10: output from a statistical program and a graphing calculator.

TI-83 graphing calculator and from the statistical program Minitab. Both outputs agree with the result 0.9953 of Example 12.10.

EXAMPLE 12.11   Monitoring mumps outbreaks

Mumps cases, like all cases of rare diseases, must be reported to the CDC and are published in the *Morbidity and Mortality Weekly Report*. In January 2006, Iowa reported 4 cases of mumps. What is the probability of getting 4 or more cases of mumps under the model used in Example 12.10?

The probability that in a given month there are 4 or more mumps cases is

$$P(X \geq 4) = 1 - P(X \leq 3)$$

$$= 1 - 0.999996 = 0.000004$$

We found $P(X \leq 3)$ with the spreadsheet program Excel, using the cumulative Poisson distribution with parameter $\mu = 0.1$. Excel's function and output are displayed in Figure 12.6.

Microsoft Excel

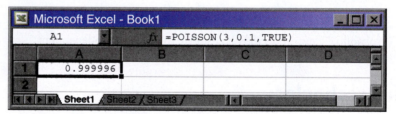

FIGURE 12.6 The Poisson probability $P(X \leq 3)$, for Example 12.11: output from a spreadsheet program.

Assuming that cases of mumps are random and independent, we would expect to almost never see 4 or more cases of mumps in Iowa in any given month. The unusually high count of mumps cases points to a substantial departure from the Poisson model, for instance, because of a contagious outbreak. In a contagion model, knowing that one person is infected increases the chance that another person also is infected. Therefore, contagious events are not independent.

These first 4 mumps cases were indeed the beginning of a major mumps outbreak in Iowa, with an increasing number of new cases every week, totaling 219 cases by the end of March and 2597 by the end of April 2006, spread over several states. The mumps strain in the Iowa epidemic is known as genotype G, the same strain that has been responsible for other epidemic outbreaks in the United States and the United Kingdom, even in vaccinated individuals.[7] The CDC tracks all reports of rare diseases in order to understand their origin and prevent their spread and future recurrence.

Plant and animal distributions can also be compared to a Poisson distribution to determine whether the events are indeed random and independent. In Example 12.9, you looked at the distribution of daisies in lawn quadrants. Plants disseminated by animals or by wind would be expected to have geographical distributions that can be approximated by a Poisson model. *Remember that biological data should not be expected to exactly match a mathematical model. This is a common misconception. Rather, people have found mathematical models that are reasonable representations for various biological features.* Other plants disseminate asexually, for example, by underground rhizomes capable of sprouting new clones (for example, lily of the valley). Such plants would be expected to be more clustered than predicted by a Poisson model.

APPLY YOUR KNOWLEDGE

12.11 Polydactyly. A hospital delivers an average of 268 children per month. In the United States, one in every 500 babies is born with polydactyly, and cases are random and independent. Polydactyly is a fairly common congenital abnormality in which a baby is born with one or more extra fingers or toes. Let X be the count of babies born with polydactyly in a month at that hospital.

  (a) What values could X take? Notice how there is no clear limit to this range.

  (b) What distribution does X follow, approximately?

  (c) Give the mean and standard deviation of X.

12.12 Typhoid fever. The CDC receives reports of 7.7 cases of typhoid fever per week on average from all over the United States, although most cases were acquired while traveling internationally. Typhoid fever is a life-threatening bacterial illness caused by *Salmonella typhi* and transmitted by ingestion of contaminated food or drink.[8]

  (a) What distribution would best represent the distribution of weekly counts of typhoid fever?

  (b) What are the mean and standard deviation of this distribution?

12.13 Polydactyly. In the setting of Exercise 12.11,

  (a) what is the probability that no child is born with polydactyly in a given month at that hospital?

  (b) what is the probability that exactly one child is born with polydactyly in a given month?

  (c) what is the probability that more than one child is born with polydactyly in a given month?

12.14 Typhoid fever. In the setting of Exercise 12.12,

  (a) use technology to find the cumulative probabilities for X equals 0, 1, 2, 3, to 15.

  (b) what is the probability that the CDC receives reports of more than 15 cases next month? If this were to happen, what would you conclude?

CHAPTER 12 SUMMARY

A count X of successes has a **binomial distribution** in the **binomial setting:** there are n observations; the observations are independent of each other; each observation results in a success or a failure; and each observation has the same probability p of a success.

The binomial distribution with n observations and probability p of success gives a good approximation to the sampling distribution of the count of successes in an SRS of size n from a large population containing proportion p of successes.

If X has the binomial distribution with parameters n and p, the possible values of X are the whole numbers $0, 1, 2, \ldots, n$. The **binomial probability** that X takes any value is

$$P(X = k) = \binom{n}{k} p^k (1 - p)^{n-k}$$

Binomial probabilities in practice are best found using software.

The **binomial coefficient**

$$\binom{n}{k} = \frac{n!}{k!\,(n - k)!}$$

counts the number of ways k successes can be arranged among n observations. Here the **factorial $n!$** is

$$n! = n \times (n - 1) \times (n - 2) \times \cdots \times 3 \times 2 \times 1$$

for positive whole numbers n, and $0! = 1$.

The **mean** and **standard deviation** of a binomial count X are

$$\mu = np$$
$$\sigma = \sqrt{np(1 - p)}$$

The **Normal approximation** to the binomial distribution says that if X is a count having the binomial distribution with parameters n and p, then when n is large, X is approximately $N(np, \sqrt{np(1 - p)}\,)$. Use this approximation only when $np \geq 10$ and $n(1 - p) \geq 10$.

The **Poisson distribution** with mean μ describes the count X of occurrences of random, independent events in fixed intervals of time or space when the probability of an event occurring is the same over all possible intervals.

If X has the Poisson distribution with mean number of occurrences per interval μ, the possible values of X are $0, 1, 2, \ldots$. The **Poisson probability** that X takes any value is

$$P(X = k) = \frac{e^{-\mu}\mu^k}{k!}$$

The **mean** and **standard deviation** of a Poisson count X are

$$\mu = \text{mean number of occurrences per interval}$$
$$\sigma = \sqrt{\mu}$$

CHECK YOUR SKILLS

12.15 Joe reads that 1 out of 4 eggs contains salmonella bacteria. So he never uses more than 3 eggs in cooking. If eggs do or don't contain salmonella independently of each other, the number of contaminated eggs when Joe uses 3 chosen at random has the distribution

(a) binomial with $n = 4$ and $p = 1/4$.

(b) binomial with $n = 3$ and $p = 1/4$.

(c) binomial with $n = 3$ and $p = 1/3$.

12.16 In the previous exercise, the probability that at least 1 of Joe's 3 eggs contains salmonella is about

(a) 0.68.   (b) 0.58.   (c) 0.30.

12.17 In a group of 10 college students, 4 are biology majors. You choose 3 of the 10 students at random and ask their major. The distribution of the number of biology majors you choose is

(a) binomial with $n = 10$ and $p = 0.4$.

(b) binomial with $n = 3$ and $p = 0.4$.

(c) not binomial.

About 5% of emergency room visits (excluding births) in South Carolina are for respiratory and chest symptoms, making up the top reason for emergency room visits.[9] Assume that visits to the emergency room are independent. Exercises 12.18 to 12.20 use this setting.

12.18 If 2 of 10 emergency room visits are because of respiratory and chest symptoms, in how many ways can you arrange the sequence of emergency room visits due to such symptoms and due to other symptoms?

(a) $\binom{10}{2} = 45$ (b) $\binom{10}{2} = 5$ (c) $\binom{8}{2} = 28$

12.19 Of 10 emergency room visits in South Carolina, the probability that the first 2 are because of respiratory and chest symptoms but not the remaining 8 is about

(a) 0.2.   (b) 0.05.   (c) 0.0017.

12.20 Of 10 emergency room visits in South Carolina, the probability that 2 are because of respiratory symptoms is about

(a) 0.2.   (b) 0.0746.   (c) 0.0017.

12.21 Each entry in a table of random digits like Table A has probability 0.1 of being a 0, and digits are independent of each other. Ten lines in the table contain 400 digits. The count of 0s in these lines is approximately Normal with

(a) mean 40 and standard deviation 36.

(b) mean 40 and standard deviation 6.

(c) mean 36 and standard deviation 6.

12.22 The mean number of daily surgeries at a local hospital is 6.2. Assuming that surgeries are random, independent events, the count of daily surgeries follows approximately

(a) a binomial distribution with mean 6.2 and standard deviation 3.8.

(b) a Poisson distribution with mean 6.2 and standard deviation 6.2.

(c) a Poisson distribution with mean 6.2 and standard deviation 2.49.

12.23 The mean number of daily surgeries at a local hospital is 6.2. Assuming that surgeries are random, independent events, the probability that there would be only 2 or fewer surgeries in a given day is approximately,

(a) 0.015. (b) 0.039. (c) 0.054.

12.24 Which of the following would you expect to follow approximately a Poisson distribution?

(a) The weekly count of toddlers with a runny nose at a day care center.

(b) The count of adults with a disability visiting the post office daily.

(c) The count of children with color blindness in a classroom of 20.

CHAPTER 12 EXERCISES

12.25 **Binomial setting?** In each situation below, is it reasonable to use a binomial distribution for the random variable X? Give reasons for your answer in each case.

(a) A manufacturer of medical catheters randomly selects 8 catheters, one from each hour's production for a detailed quality inspection. One variable recorded is the count X of catheters with an unacceptable diameter (too small or too large).

(b) A sample survey asks 100 persons chosen at random from the adult residents of a large city whether they oppose the legalization of medical marijuana; X is the number of persons who say "Yes."

(c) A pediatrician sees 24 unrelated children on one winter day. X is the number of patients who came because of cold or flu symptoms.

12.26 **Binomial setting?** Your pulse reflects the rate at which your heart beats. A healthy resting pulse rate is between 60 and 100 beats per minute in adults. In which of the following two settings is a binomial distribution more likely to be at least approximately correct? Explain your answer.

(a) You take your pulse 10 times throughout the day at random intervals over a 24-hour period regardless of what you are doing (you are very dedicated and even set the clock for some measurement sessions in the middle of the night). X is the number of measurements that are below 100 beats per minute.

(b) A nurse instructs an adult patient to take his resting pulse 10 times over the next few days. A resting pulse requires resting for at least 10 minutes before taking the pulse. X is the number of measurements that are below 100 beats per minute.

12.27 **Competitive sports.** During the 2006 Turin Winter Olympics, a survey of Team USA found that 70.8% of the athletes had previously experienced one or more bone fractures.[10] Let's assume that this finding is representative of all athletes involved in winter competitive sports. You interview a random sample of 12 athletes involved in winter competitive sports.

(a) What is the distribution of the number who have experienced some bone fracture?

(b) What is the probability that exactly 8 of the 12 have experienced some bone fracture? If you have software, also find the probability that at least 8 of the 12 have experienced some bone fracture.

12.28 Universal donors. People with type O-negative blood are universal donors whose blood can be safely given to anyone. Only 7.2% of the population have O-negative blood. A blood center is visited by 20 donors in an afternoon. What is the probability that there are at least 2 universal donors among them?

12.29 Testing ESP. In a test for ESP (extrasensory perception), a subject is told that cards the experimenter can see but that the subject cannot contain either a star, a circle, a wave, or a square. As the experimenter looks at each of 20 cards in turn, the subject names the shape on the card. A subject who is just guessing has probability 0.25 of guessing correctly on each card.

(a) The count of correct guesses in 20 cards has a binomial distribution. What are n and p?

(b) What is the mean number of correct guesses in many repetitions of the experiment?

(c) What is the probability of exactly 5 correct guesses?

12.30 Malaria epidemics. Malaria is a debilitating and potentially deadly parasitic infection that is mostly preventable and curable, yet many developing countries lack the necessary resources. The West African country Guinea has the highest rate of malaria in the world, with 75% of its population infected.[11] A physician from Doctors without Borders sees 6 patients in a given hour. Assuming that patient visits are independent, what is the probability that all of them are infected? What is the probability that all but 1 of these 6 patients are infected with malaria?

12.31 Malaria epidemics. In the previous exercise, X is the count of patients who are infected with malaria among the 6 patients the doctor saw.

(a) X has a binomial distribution. What are n and p?

(b) What are the possible values that X can take?

(c) Find the probability of each value of X. Draw a probability histogram for the distribution of X. (See Figure 12.2 for an example of a probability histogram.)

(d) What are the mean and standard deviation of this distribution? Mark the location of the mean on your histogram.

12.32 False-positives in testing for HIV. A rapid test for the presence in the blood of antibodies to HIV, the virus that causes AIDS, gives a positive result with probability about 0.004 when a person who is free of HIV antibodies is tested. A clinic tests 1000 people who are all free of HIV antibodies.

(a) What is the distribution of the number of positive tests?

(b) What is the mean number of positive tests?

(c) You cannot safely use the Normal approximation for this distribution. Explain why.

12.33 TV viewing and children. TV viewing promotes obesity because of physical inactivity and high exposure to commercials for foods with high fat and sugar content. A large sample survey of school-age children by the Kaiser Family Foundation found a 42% chance that the TV is on most of the time at home.

(a) You choose 8 households with children at random. What is the probability that none of the chosen households has the TV on most of the time? What is the probability that only 1 household has the TV on most of the time? What is the probability that at least 2 households have the TV on most of the time?

(b) A polling agency surveys a random sample of 806 households with children. What are the mean and standard deviation of the number of households in this large sample that have the TV on most of the time? What is the probability that at least 320 households have the TV on most of the time (use technology)?

(c) In which of (a) and (b) would it be safe to use a Normal approximation?

12.34 Multiple-choice tests. Here is a simple probability model for multiple-choice tests. Suppose that each student has probability p of correctly answering a question chosen at random from a universe of possible questions. (A strong student has a higher p than a weak student.) Answers to different questions are independent. Jodi is a good student for whom $p = 0.75$.

(a) Use the Normal approximation to find the probability that Jodi scores 70% or lower on a 100-question test. What is the probability that Jodi scores 65% or lower on this 100-question test?

(b) If the test contains 250 questions, what is the probability that Jodi scores 70% or lower?

12.35 Multiple-choice tests again. Follow the model described in the previous exercise. Jay never came to class, never did the homework, and did not study for the exam. For each question, four choices are offered, and thus Jay has probability $p = 0.25$ of guessing correctly by chance.

(a) Use the Normal approximation to find the probability that Jay scores 50% or more on a 20-question test.

(b) If the test contains 100 questions, what is the probability that Jay will score 50% or higher?

12.36 Survey demographics. According to the Census Bureau, 12.4% of American adults (aged 18 and over) are Hispanics (data for 2004). An opinion poll plans to contact an SRS of 1200 adults.

(a) What is the mean number of Hispanics in such samples? What is the standard deviation?

(b) According to the 68–95–99.7 rule, what range will include the counts of Hispanics in 95% of all such samples?

(c) How large a sample is required to make the mean number of Hispanics at least 200?

12.37 Leaking gas tanks. Leakage from underground gasoline tanks at service stations can damage the environment. It is estimated that 25% of these tanks leak. You examine 15 tanks chosen at random, independently of each other.

(a) What is the mean number of leaking tanks in such samples of 15?

(b) What is the probability that 10 or more of the 15 tanks leak?

(c) Now you do a larger study, examining a random sample of 1000 tanks nationally. What is the probability that at least 275 of these tanks are leaking?

12.38 Genetics. According to genetic theory, the blossom color in the second generation of a certain cross of sweet peas should be red or white in a 3:1 ratio. That is, each plant has probability 3/4 of having red blossoms, and the blossom colors of separate plants are independent.

(a) What is the probability that exactly 6 out of 8 of these plants have red blossoms?

(b) What is the mean number of red-blossomed plants when 80 plants of this type are grown from seeds?

(c) What is the probability of obtaining at least 50 red-blossomed plants when 80 plants are grown from seeds?

12.39 Inheriting blood type. If the parents in Example 12.4 have 5 children, the number who have type O blood is a random variable X that has the binomial distribution with $n = 5$ and $p = 0.25$.

(a) What are the possible values of X?

(b) Find the probability of each value of X. Draw a histogram to display this distribution. (Because probabilities are long-run proportions, a histogram with the probabilities as the heights of the bars shows what the distribution of X would be in very many repetitions.)

12.40 Inheriting blood type. What are the mean and standard deviation of the number of children with type O blood in the previous exercise? Mark the location of the mean on the probability histogram you made in that exercise.

12.41 Sex ratio at birth. Example 9.3 (page 221) gave the counts of male and female babies born in the United States over the past decade. In 2002, for instance, there were 1,964,000 baby girls and 2,058,000 baby boys born. We often assume that the probability of having a boy or a girl is 0.5 because of the chromosomal determination of sex in humans. Is there reason to think that the sex ratio in the United States favors more baby boys? To answer this question, find the probability that an equal chance of getting a boy would give 2,058,000 boys or more in 4,022,000 births. What do you conclude?

12.42 Inspecting pharmaceutical containers. Example 12.5 discusses the count of pharmaceutical containers contaminated with benzene in inspection samples of size 10. The count has the binomial distribution with $n = 10$ and $p = 0.1$. Set these values for the number of tosses and probability of heads in the *Probability* applet. In the example, we calculated that the probability of getting a sample with exactly 1 contaminated container is 0.3874. Of course, when we inspect only a few lots of 10 containers, the proportion of samples with exactly 1 contaminated container will differ from this probability. Click "Toss" and "Reset" repeatedly to simulate inspecting a total of 20 lots of 10 containers each. Record the number of contaminated containers (the count of heads) in each of these 20 samples. What proportion of the 20 lots had exactly 1 contaminated container? Remember that probability tells us only what happens in the long run.

APPLET

12.43 The geometric distributions. Osteoporosis is a disease defined as very low bone density, resulting in a high risk of fracture, hospitalization, and immobilization. Seven percent of postmenopausal women suffer from osteoporosis.[12] A physician administers routine checkups. Visits by postmenopausal women are independent.

We are interested in how long we must wait to get the first postmenopausal woman with osteoporosis.

(a) Considering only patient visits by postmenopausal women, the probability of a patient with osteoporosis on the first visit is 7%. What is the probability that the first patient does not have osteoporosis and the second patient does?

(b) What is the probability that the first two patients do not have osteoporosis and the third one does? This is the probability that the first patient with osteoporosis occurs on the third visit.

(c) Now you see the pattern. What is the probability that the first patient with osteoporosis occurs on the fourth visit? On the fifth visit? Give the general result: What is the probability that the first patient with osteoporosis occurs on the kth visit? (*Comment:* The distribution of the number of trials to the first "success" [in our example, a patient with osteoporosis] is called a **geometric distribution.** In this problem you have found geometric distribution probabilities when the probability of a success on each trial is 7%. The same idea works for any probability of success.)

geometric distribution

12.44 **Sex selection.** A couple decides to continue to have children until their first girl is born; X is the total number of children the couple has. Does X have a binomial distribution?

12.45 **The multinomial distributions.** The National Center for Health Statistics estimates that 3% of adult American women are underweight, 47% have a healthy weight, and 50% are overweight (including obese individuals).[13] A gynecologist sees 3 patients for a routine exam.

(a) What is the probability that the first one has a healthy weight and the next two are overweight?

(b) What is the probability that the second patient has a healthy weight and the other two are overweight? What is the probability that the first two are overweight and the last one has a healthy weight?

(c) Using the results from (a) and (b) give the probability that, out of any three patients, one has a healthy weight and two are overweight. (*Comment:* We have 3 independent trials, each with 3 possible outcomes: underweight, healthy weight, overweight. Counts of the three outcomes follow a **multinomial distribution.** Multinomial distributions can represent any number of possible outcomes, generalizing the binomial setting. Binomial probabilities are a special case of the multinomial model when only two outcomes are possible.)

multinomial distribution

12.46 **The common cold.** During the cold season, a pediatrician sees an average of 4.5 children with common-cold symptoms per day. The pediatrician's patients are unrelated, and so we can assume that patients with colds represent random, independent events. Let X be the count of children with common-cold symptoms per day. What distribution does X follow, approximately? Give the mean and standard deviation of X.

12.47 **The common cold.** In the setting of the previous exercise,

(a) what is the probability that the pediatrician sees no child with common-cold symptoms on a given workday?

(b) what is the probability that the pediatrician sees exactly one child with common-cold symptoms in a day?

(c) what is the probability that the pediatrician sees two or more children with common-cold symptoms in a day?

12.48 Blood serum donations. Blood type AB is somewhat rare in the United States (4% of the general population) and is particularly useful for plasma transfusions because this plasma can be accepted by individuals with all blood types. The plasma is the liquid part of the blood once all cells are removed. You count the number X of AB donors that come to a fixed Red Cross location per day. This count has the Poisson distribution with parameter μ being the mean number of AB donors per day.

(a) A given blood donation center sees an average of 2.1 AB donors per day. Find the Poisson probabilities for $X = 0, 1, 2, 3, 4, 5$.

(b) What is the probability that this particular location would get more than 5 AB donors in one day?

12.49 Planning for hospital admissions. A private hospital has an average of 3.2 hospital admissions a day from its emergency room. X is the count of hospital admissions from the emergency room in a day. In order to plan for the number of beds that should be kept available, you compute the probabilities $P(X \geq 5)$, $P(X \geq 6)$, and $P(X \geq 7)$. What do you conclude?

12.50 Salmonellosis outbreak. Improperly handled or undercooked poultry and eggs are the most frequent cause of salmonella food poisoning. Salmonellosis can be deadly, although most individuals recover on their own. Cases must be reported to the CDC, where they are closely monitored. The average number of salmonellosis cases per month in South Dakota is 1.67.

(a) If X is the monthly count of salmonellosis cases in South Dakota, what distribution does X follow, approximately? Give the mean and standard deviation of X.

(b) Give the probabilities $P(X = 0)$, $P(X \leq 1)$, $P(X \leq 2)$, $P(X \leq 3)$, and $P(X \leq 4)$. What is the probability that there would be more than 4 cases of salmonellosis in South Dakota in a given month?

(c) In September 1994, 14 cases of salmonellosis in South Dakota were reported to the CDC. What is the probability that 14 or more cases would arise in one month in South Dakota? What does this unusual report suggest? If you worked at the CDC at that time, what do you think would be your next action?

12.51 Salmonellosis outbreak, continued. Following on the previous example, the CDC reports that the average number of salmonellosis cases per month in Wisconsin is 15.58.

(a) If X is the monthly count of salmonellosis cases in Wisconsin, what distribution does X follow, approximately? Give the mean and standard deviation of X.

(b) Use technology to find the probabilities $P(X = 0)$, $P(X \leq 5)$, $P(X \leq 15)$, and $P(X \leq 25)$. What is the probability that there would be more than 25 cases of salmonellosis in Wisconsin in a given month?

(c) As with South Dakota, September 1994 had an unusually high number of salmonella cases, with 48 cases that month. What is the probability that 48 or more cases would arise in one month in Wisconsin? What do you make of the fact that a salmonellosis outbreak occurred in two states at the same time? In fact, the salmonellosis outbreak was traced back to nationally distributed ice cream products from one factory. The factory was shut down until cleared, and its contaminated production were recalled, thus preventing many more salmonellosis cases.

Enigma/Alamy

Sampling Distributions

Obesity has become the unrelenting focus of health debates. But how do we know that it is a problem and that it is worsening? In the National Health and Nutrition Examination Survey (NHANES), a large representative sample of Americans are contacted every few years. In 1994, the average weight of American adult males aged 20 to 74 was $\bar{x} = 182.4$ pounds, based on a sample of 6,860 individuals. Less than 10 years later, a 2002 NHANES random sample of 3,791 produced an average weight of $\bar{x} = 191.0$ pounds. The value 191.0 describes the sample taken in 2002, but we use it to estimate the mean weight of all American adult males in 2002. This is an example of statistical inference: We use information from a sample to infer something about a wider population.

Because the results of random samples and randomized comparative experiments include an element of chance, we can't guarantee that our inferences are correct. What we can guarantee is that our methods usually give correct answers. We will see that the reasoning of statistical inference rests on asking, "How often would this method give a correct answer if I used it very many times?" If our data come from random sampling or randomized comparative experiments, the laws of probability answer the question "What would happen if we did this many times?" This chapter presents some facts about probability that help answer this question.

Parameters and statistics

As we begin to use sample data to draw conclusions about a wider population, we must take care to keep straight whether a number describes a sample or a population. Here is the vocabulary we use.

> **PARAMETER, STATISTIC**
>
> A **parameter** is a number that describes the population. In statistical practice, the value of a parameter is not known, because we cannot examine the entire population.
>
> A **statistic** is a number that can be computed from the sample data without making use of any unknown parameters. In practice, we often use a statistic to estimate an unknown parameter.

- **EXAMPLE 13.1** Obesity

The mean weight of the sample of American adult males contacted by the 2002 NHANES was $\overline{x} = 191.0$ pounds. The number 191.0 is a *statistic*, because it describes this one NHANES sample. The population that the survey wants to draw conclusions about is all American adult males between 20 and 74 years of age. The *parameter* of interest is the mean weight of all of these individuals. We don't know the value of this parameter.

population mean μ
sample mean $\overline{x}$

Remember: **S**tatistics come from **s**amples, and **p**arameters come from **p**opulations. The distinction between population and sample is essential, and the notation we use must reflect this distinction. We write μ (the Greek letter mu) for the **mean of a population.** This is a fixed parameter that is unknown when we use a sample for inference. The **mean of the sample** is the familiar $\overline{x}$, the average of the observations in the sample. This is a statistic that would almost certainly take a different value if we chose another sample from the same population. The sample mean $\overline{x}$ from a sample or an experiment is an estimate of the mean μ of the underlying population.

APPLY YOUR KNOWLEDGE

13.1 **Effects of caffeine.** How does caffeine affect our bodies? In a matched-pairs experiment, subjects pushed a button as quickly as they could after taking a caffeine pill and also after taking a placebo pill. The mean pushes per minute were **283** for the placebo and **311** for caffeine. Is each of the boldface numbers a parameter or a statistic?

13.2 **Lab rats.** Extensive scientific research on nerve conduction has led to the conclusion that the mean refractory period in a healthy rat is **1.3 millisecond** (ms). An experiment designed to study the effect of the pesticide DDT on nerve conduction finds that the mean refractory period of 4 rats poisoned with DDT is **1.75 ms.** Is each boldface number a parameter or a statistic?

Statistical estimation and the law of large numbers

Statistical inference uses sample data to draw conclusions about the entire population. *Because good samples are chosen randomly, statistics such as $\overline{x}$ are random variables.* We can describe the behavior of a sample statistic by a probability model that answers the question "What would happen if we did this many times?" Here is an example that will lead us toward the probability ideas most important for statistical inference.

EXAMPLE 13.2   Does this wine smell bad?

Enigma/Alamy

Sulfur compounds such as dimethyl sulfide (DMS) are sometimes present in wine. DMS causes "off-odors" in wine, so winemakers want to know the odor threshold, the lowest concentration of DMS that the human nose can detect. Different people have different thresholds, so we start by asking about the mean threshold μ in the population of all adults. The number μ is a parameter that describes this population.

To estimate μ, we present tasters with both natural wine and the same wine spiked with DMS at different concentrations to find the lowest concentration at which they identify the spiked wine. Here are the odor thresholds (measured in micrograms of DMS per liter of wine) for 10 randomly chosen subjects:

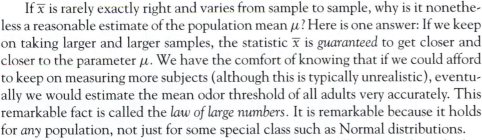

| 28 | 40 | 28 | 33 | 20 | 31 | 29 | 27 | 17 | 21 |

The mean threshold for these subjects is $\overline{x} = 27.4$. It seems reasonable to use the sample result $\overline{x} = 27.4$ to estimate the unknown μ. An SRS should fairly represent the population, so the mean $\overline{x}$ of the sample should be somewhere near the mean μ of the population. Of course, we don't expect $\overline{x}$ to be exactly equal to μ. We realize that if we choose another SRS, the luck of the draw will probably produce a different $\overline{x}$.

If $\overline{x}$ is rarely exactly right and varies from sample to sample, why is it nonetheless a reasonable estimate of the population mean μ? Here is one answer: If we keep on taking larger and larger samples, the statistic $\overline{x}$ is *guaranteed* to get closer and closer to the parameter μ. We have the comfort of knowing that if we could afford to keep on measuring more subjects (although this is typically unrealistic), eventually we would estimate the mean odor threshold of all adults very accurately. This remarkable fact is called the *law of large numbers*. It is remarkable because it holds for *any* population, not just for some special class such as Normal distributions.

The law of large numbers can be proved mathematically starting from the basic laws of probability. The behavior of $\overline{x}$ is similar to the idea of probability. In the long run, the *proportion* of outcomes taking any value gets close to the probability of that value, and the *average* outcome gets close to the population mean. Figure 9.1 (page 221) shows how proportions approach probability in a coin-tossing example. The *Law of Large Numbers* applet helps illustrate this concept using dice. You can use the applet to watch $\overline{x}$ change as you average more observations until it eventually settles down at the mean μ. Using software to imitate chance behavior is called **simulation**.

The law of large numbers is usually impractical for direct applications in the life sciences. Indeed, it is only *in the very long run* that the mean outcome is predictable, and studies in the life sciences rarely deal with such large numbers. However, the law of large numbers is the foundation of such businesses as gambling casinos and insurance companies. The winnings (or losses) of a gambler on a few plays are uncertain—that's why, unfortunately, gambling can be addictive. The house plays tens of thousands of times. So the house, unlike individual gamblers, can count on

simulation

The probability of dying

We can't predict whether a specific person will die next year. But if we observe millions of people, deaths are random. Life insurance is based on this fact. Each year, the proportion of men aged 25 to 34 who die is about 0.0021. We can use this as the probability that a given young man will die next year. For women in that age group, the probability of death is about 0.0007. An insurance company that sells many policies to people aged 25 to 34 will have to pay off on about 0.21% of the policies sold to men and about 0.07% of the policies sold to women.

the long-run regularity described by the law of large numbers. The average winnings of the house on tens of thousands of plays will be very close to the mean of the distribution of winnings, guaranteeing the house a profit. For all other problems, however, the law of large numbers is impractical. In the next section, we introduce another mathematical property of random samples that is much more useful in practice.

APPLY YOUR KNOWLEDGE

13.3   Insurance. The idea of insurance is that we all face risks that are unlikely but carry high costs. Workers' compensation insurance provides for the cost of medical care and rehabilitation for injured workers and other related expenses. In a given year, most workers will not have any work injury, some will have minor injuries, and a few will die or become permanently disabled. Insurance spreads the risk: We all pay a small amount, and the insurance policy pays a large amount to these few extreme claims. The National Academy of Social Insurance reports that the mean workers' compensation cost in a year is $\mu = \$439$ per covered worker. An insurance company plans to sell workers compensation insurance for $439 plus enough to cover its administrative costs and profit. Explain clearly why it would be unwise to sell only 12 policies. Then explain why selling thousands of such policies is a safe business.

13.4   Means in action. Use the *Law of Large Numbers* applet. The applet simulates successive rolls of a die and recomputes the average number of spots on the up face after each roll. Over the long run, the mean number of spots is 3.5.

(a) Type 20 in the "Rolls" window. Then click on the "Roll dice" button. What do you observe? Reset the applet and repeat this simulation three more times. How are these four simulations different? What do they have in common? You can expect a lot of variability for such small sample sizes.

(b) Now type 100 in the "Rolls" window and click on the "Roll dice" button three times so that you have a total of 300 rolls. What do you observe? Reset the applet and repeat this simulation one more time. How are these two simulations different? What do they have in common? Three hundred is not a huge number, but you can get an idea of how the law of large numbers works.

Sampling distributions

The law of large numbers assures us that if we measured enough subjects, the statistic $\overline{x}$ would eventually get very close to the unknown parameter μ. But our study in Example 13.2 had just 10 subjects. What can we say about $\overline{x}$ from 10 subjects as an estimate of μ? We ask: "What would happen if we took many samples of 10 subjects from this population?" Here's how to answer this question:

• Take a large number of samples of size 10 from the population.

• Calculate the sample mean $\overline{x}$ for each sample.

• Make a histogram of the values of $\overline{x}$.

• Examine the distribution displayed in the histogram for shape, center, and spread, as well as outliers or other deviations.

Students are sometimes tempted to take many samples to build a sampling distribution. However, sampling distributions represent all possible samples of a given size. They are important mathematical concepts and are not built experimentally. To illustrate the concept, we can imitate many samples by using software.

EXAMPLE 13.3   What would happen in many samples?

Extensive studies have found that the DMS odor threshold of adults follows roughly a Normal distribution with mean $\mu = 25$ micrograms per liter and standard deviation $\sigma = 7$ micrograms per liter. With this information, we can simulate many repetitions of Example 13.2 with different subjects drawn at random from the population.

Figure 13.1 illustrates the process of choosing many samples and finding the sample mean threshold $\bar{x}$ for each one. Follow the flow of the figure from the population at the left, to choosing an SRS and finding the $\bar{x}$ for this sample, to collecting together the $\bar{x}$'s from many samples. The first sample has $\bar{x} = 26.42$. The second sample contains a different 10 people, with $\bar{x} = 24.28$, and so on. The histogram at the right of the figure shows the distribution of the values of $\bar{x}$ from 1000 separate SRSs of size 10. This histogram displays the *sampling distribution* of the statistic $\bar{x}$.

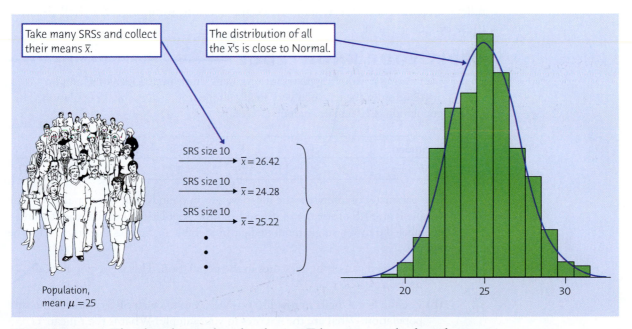

FIGURE 13.1 The idea of a sampling distribution: Take many samples from the same population, collect the $\bar{x}$ from all the samples, and display the distribution of $\bar{x}$. The histogram shows the results of 1000 samples.

SAMPLING DISTRIBUTION

The **sampling distribution** of a statistic is the distribution of values taken by the statistic in all possible samples of the same size from the same population.

Strictly speaking, the sampling distribution is the ideal pattern that would emerge if we looked at all possible samples of size 10 from our population. A distribution obtained from a fixed number of trials, like the 1000 trials in Figure 13.1, is only an approximation to the sampling distribution. One of the uses of probability theory in statistics is to obtain exact sampling distributions without simulation. The interpretation of a sampling distribution is the same, however, whether we obtain it by simulation or by the mathematics of probability.

We can use the tools of data analysis to describe any distribution. Let's apply those tools to Figure 13.1. What can we say about the shape, center, and spread of this distribution?

- **Shape:** It looks Normal! Detailed examination confirms that the distribution of $\bar{x}$ from many samples does have a distribution that is very close to Normal.

- **Center:** The mean of the 1000 $\bar{x}$'s is 24.95. That is, the distribution is centered very close to the population mean $\mu = 25$.

- **Spread:** The standard deviation of the 1000 $\bar{x}$'s is 2.217, notably smaller than the standard deviation $\sigma = 7$ of the population of individual subjects.

Although these results describe just one simulation of a sampling distribution, they reflect facts that are true whenever we use random sampling. We shall describe these properties next.

APPLY YOUR KNOWLEDGE

13.5 **Generating a sampling distribution.** Let's illustrate the idea of a sampling distribution in the case of a very small sample from a very small population. The population is 10 female students in a class:

Student	0	1	2	3	4	5	6	7	8	9
Weight	136	99	118	129	125	170	130	128	120	147

The parameter of interest is the mean weight in pounds μ in this population. The sample is an SRS of size $n = 4$ drawn from the population. Because the students are labeled 0 to 9, a single random digit from Table A chooses one student for the sample.

(a) Find the mean of the 10 scores in the population. This is the population mean μ.

(b) Use the first 4 digits in row 116 of Table A to draw an SRS of size 4 from this population. What are the 4 scores in your sample? What is their mean $\bar{x}$? This statistic is an estimate of μ.

(c) Repeat this process 9 more times, using the first 4 digits in rows 116 to 125 of Table A. Make a histogram of the 10 values of $\bar{x}$. You are constructing the sampling distribution of $\bar{x}$. Is the center of your histogram close to μ?

(d) This is a lengthy process, even for such a small population. Explain why it would be unrealistic to try building a sampling distribution by hand for a large or unknown population.

APPLET

13.6 **A sampling distribution.** We can use the *Simple Random Sample* applet to help grasp the idea of a sampling distribution. Form a population labeled 1 to 100. We will choose an SRS of 10 of these numbers. That is, in this exercise, the numbers themselves are the population, not just labels for 100 individuals. The mean of the whole numbers 1 to 100 is $\mu = 50.5$. This is the population mean.

(a) Use the applet to choose an SRS of size 10. Which 10 numbers were chosen? What is their mean? This is the sample mean $\bar{x}$.

(b) Although the population and its mean $\mu = 50.5$ remain fixed, the sample mean changes as we take more samples. Take another SRS of size 10. (Use the "Reset" button to return to the original population before taking the second sample.) What are the 10 numbers in your sample? What is their mean? This is another value of $\bar{x}$.

(c) Take 8 more SRSs from this same population and record their means. You now have 10 values of the sample mean $\bar{x}$ from 10 SRSs of the same size from the same population. Make a histogram of the 10 values and mark the population mean $\mu = 50.5$ on the horizontal axis. Are your 10 sample values roughly centered at the population value μ? (If you kept going forever, your $\bar{x}$-values would form the sampling distribution of the sample mean; the population mean μ would indeed be the center of this distribution.)

(d) Explain some of the obstacles to trying to build a sampling distribution through repeated samples, even with software.

The sampling distribution of $\bar{x}$

Figure 13.1 suggests that when we choose many SRSs from a population, the sampling distribution of the sample means is centered at the mean of the original population and less spread out than the distribution of individual observations. Although we won't elaborate, this can be demonstrated mathematically. Here are the facts.

> **MEAN AND STANDARD DEVIATION OF A SAMPLE MEAN**[1]
>
> Suppose that $\bar{x}$ is the mean of an SRS of size n drawn from a large population with mean μ and standard deviation σ. Then the sampling distribution of $\bar{x}$ has **mean μ** and **standard deviation $\sigma/\sqrt{n}$**.

These facts about the mean and the standard deviation of the sampling distribution of $\bar{x}$ are true for *any* population, not just for some special class such as Normal distributions. Both facts have important implications for statistical inference.

- The mean of the statistic $\bar{x}$ is always equal to the mean μ of the population. That is, the sampling distribution of $\bar{x}$ is centered at μ. In repeated sampling, $\bar{x}$ will sometimes fall above the true value of the parameter μ and sometimes below it, but there is no systematic tendency to overestimate or underestimate the parameter. This makes the idea of lack of bias in the sense of "no favoritism" more precise. Because the mean of $\bar{x}$ is equal to μ, we say that the statistic $\bar{x}$ is an **unbiased estimator** of the parameter μ.

unbiased estimator

- An unbiased estimator is "correct on average" over many samples. How close the estimator falls to the parameter in most samples is determined by the spread of the sampling distribution. If individual observations have standard deviation σ, then sample means $\bar{x}$ from samples of size n have standard deviation $\sigma/\sqrt{n}$. That is, **averages are less variable than individual observations.**

We have described the center and spread of the sampling distribution of a sample mean $\bar{x}$, but not its shape. The shape of the distribution of $\bar{x}$ depends on the shape of the population. Here is one important case: If measurements in the population follow a Normal distribution, then so does the sample mean.

> ### SAMPLING DISTRIBUTION OF A SAMPLE MEAN FOR A NORMAL POPULATION
>
> If individual observations have the $N(\mu, \sigma)$ distribution, then the sample mean $\bar{x}$ of an SRS of size n has the $N(\mu, \sigma/\sqrt{n})$ distribution.

— **EXAMPLE 13.4**   Population distribution versus sampling distribution —

population distribution

If we measure the DMS odor thresholds of individual adults, the values follow the Normal distribution with mean $\mu = 25$ micrograms per liter and standard deviation $\sigma = 7$ micrograms per liter. We call this the **population distribution,** because it shows how measurements vary within the population.

Take many SRSs of size 10 from this population and find the sample mean $\bar{x}$ for each sample, as in Figure 13.1. The *sampling distribution* describes how the values of $\bar{x}$ vary among samples. That sampling distribution is also Normal, with mean $\mu = 25$ and standard deviation

$$\frac{\sigma}{\sqrt{n}} = \frac{7}{\sqrt{10}} = 2.2136$$

Figure 13.2 contrasts these two Normal distributions. Both are centered at the population mean, but sample means are much less variable than individual observations.

Not only is the standard deviation of the distribution of $\bar{x}$ smaller than the standard deviation of individual observations, but it gets smaller as we take larger samples. **The results of large samples are less variable than the results of small samples.** If n is large, the standard deviation of $\bar{x}$ is small, and almost all samples will give values of $\bar{x}$ that lie very close to the true parameter μ. That is, the sample mean from a large sample can be trusted to estimate the population mean accurately. Unlike the law of large numbers, this property of sampling distributions has many important, direct applications in statistics. *However, the standard deviation of the sampling distribution gets smaller only at the rate $\sqrt{n}$. To cut the standard deviation of $\bar{x}$ by 10, we must take 100 times as many observations, not just 10 times as many, and large sample sizes are not always an option.*

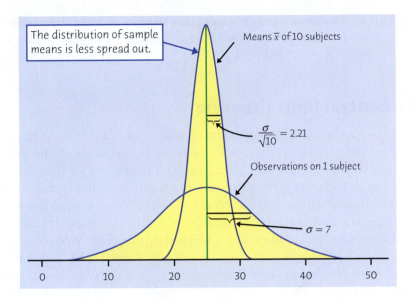

The distribution of sample means is less spread out.

Means $\overline{x}$ of 10 subjects

$\dfrac{\sigma}{\sqrt{10}} = 2.21$

Observations on 1 subject

$\sigma = 7$

FIGURE 13.2 The distribution of single observations compared with the distribution of the means $\overline{x}$ of 10 observations. Averages are less variable than individual observations.

APPLY YOUR KNOWLEDGE

13.7 **A sample of teens.** A study of the health of teenagers plans to measure the blood cholesterol level of an SRS of youth aged 13 to 16. The researchers will report the mean $\overline{x}$ from their sample as an estimate of the mean cholesterol level μ in this population.

   (a) Explain to someone who knows no statistics what it means to say that $\overline{x}$ is an "unbiased" estimator of μ.

   (b) The sample result $\overline{x}$ is an unbiased estimator of the true population μ no matter what size SRS the study uses. Explain to someone who knows no statistics why a large sample gives more trustworthy results than a small sample.

13.8 **Measurements in the lab.** Juan makes a measurement in a chemistry laboratory and records the result in his lab report. The standard deviation of lab measurements made by students is $\sigma = 10$ milligrams. Juan repeats the measurement 3 times and records the mean $\overline{x}$ of his 3 measurements.

   (a) What is the standard deviation of Juan's mean result? (That is, if Juan kept on making 3 measurements and averaging them, what would be the standard deviation of all his $\overline{x}$'s?)

   (b) How many times must Juan repeat the measurement to reduce the standard deviation of $\overline{x}$ to 5? Explain to someone who knows no statistics the advantage of reporting the average of several measurements rather than the result of a single measurement.

13.9 **Potassium in the blood.** Judy's doctor is concerned that she may suffer from hypokalemia (low potassium in the blood). There is variation both in the actual potassium level and in the blood test that measures the level. Judy's measured potassium level varies according to the Normal distribution with $\mu = 3.8$ and $\sigma = 0.2$. A patient is classified as hypokalemic if the potassium level is below 3.5.

   (a) If a single potassium measurement is made, what is the probability that Judy is diagnosed as hypokalemic?

Spencer Grant/Photo Edit

(b) If measurements are made instead on 4 separate days and the mean result is compared with the criterion 3.5, what is the probability that Judy is diagnosed as hypokalemic?

The central limit theorem

The facts about the mean and standard deviation of $\bar{x}$ are true no matter what the shape of the population distribution may be. But what about the *shape of the sampling distribution?* We saw in Figure 13.1 from Example 13.4 that the shape of the sampling distribution is Normal when the population distribution is Normal. But what if the population distribution is not Normal? It is a remarkable fact that as the sample size increases, the distribution of $\bar{x}$ changes shape: It looks less like that of the population and more like a Normal distribution. When the sample is large enough, the distribution of $\bar{x}$ is very close to Normal. This is true no matter what shape the population distribution has, as long as the population has a finite standard deviation σ. This famous fact of probability theory is called the *central limit theorem* and is extremely useful because of its wide applications.

CENTRAL LIMIT THEOREM

Draw an SRS of size n from any population with mean μ and finite standard deviation σ. When n is large, the sampling distribution of the sample mean $\bar{x}$ is approximately Normal:

$$\bar{x} \text{ is approximately } N\left(\mu, \frac{\sigma}{\sqrt{n}}\right)$$

The central limit theorem allows us to use Normal probability calculations to answer questions about sample means from many observations even when the population distribution is not Normal.

EXAMPLE 13.5   *The central limit theorem in action*

The *Central Limit Theorem* applet allows you to watch the central limit theorem in action. Figure 13.3 presents snapshots from the applet. Figure 13.3(a) shows the density curve of a single observation, that is, of the population. The distribution is strongly right-skewed, and the most probable outcomes are near 0. The mean μ of this distribution is 1, and its standard deviation σ is also 1. This particular distribution is called an *exponential distribution*. Exponential distributions are used as models for the lifetime in service of electronic components and for the time required to serve a customer or repair a machine.

Figures 13.3(b), (c), and (d) are the density curves of the sample means of 2, 10, and 25 observations from this population. As n increases, the shape becomes more Normal. The mean remains at $\mu = 1$, and the standard deviation decreases, taking the value $1/\sqrt{n}$. The density curve for 10 observations is still somewhat skewed to the right but already resembles a Normal curve having $\mu = 1$ and $\sigma = 1/\sqrt{10} = 0.32$. The density curve for $n = 25$ is yet more Normal. The contrast between the shapes of the population distribution and of the distribution of the mean of 10 or 25 observations is striking.

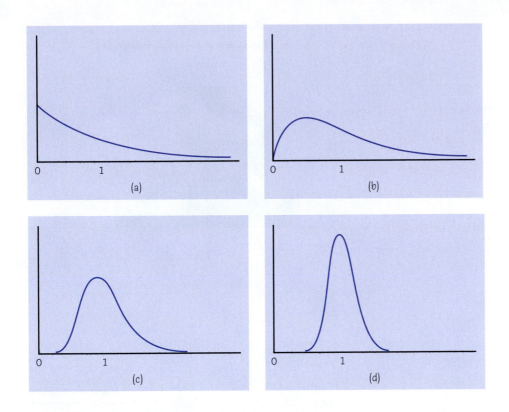

FIGURE 13.3 The central limit theorem in action: The distribution of sample means $\bar{x}$ from a strongly non-Normal population becomes more Normal as the sample size increases. (a) The distribution of one observation. (b) The distribution of $\bar{x}$ for 2 observations. (c) The distribution of $\bar{x}$ for 10 observations. (d) The distribution of $\bar{x}$ for 25 observations.

The central limit theorem applies to sampling distributions, not to the distribution of a single sample. Many students mistakenly believe that larger sample sizes yield more Normal sample histograms. This is not the case. If a population is skewed, for instance, chances are that a histogram of a random sample from that population will be skewed too. The central limit theorem describes only what happens to the distribution of the averages from repeated samples of a given size.

How large a sample size n *is needed for* $\bar{x}$ *to be close to Normal depends on the population distribution. More observations are required if the shape of the population distribution is far from Normal.* The following is always true, though.

THINKING ABOUT SAMPLE MEANS

Means of random samples are **less variable** than individual observations.

Means of random samples are **more Normal** than individual observations.

Mathematically proving the Central Limit Theorem is beyond the scope of an introductory statistics textbook. We will instead focus on its practical uses.

— EXAMPLE 13.6   *Practical use of the central limit theorem* ——

Figure 13.4 shows the histograms for three biological sample data sets. Each histogram describes one sample exactly. By observing the overall shape of the histogram we can guess whether the population might be approximately Normal, somewhat Normal, or

FIGURE 13.4 Histograms of three data sets for Example 13.6.
(a) The angle of big toe deformations in 38 patients.
(b) The number of daily fruit servings for 74 adolescent girls.
(c) The lengths of 56 perch from a Swedish lake.

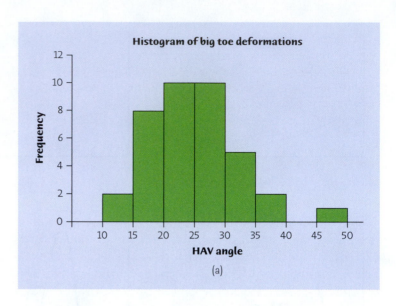

(a)

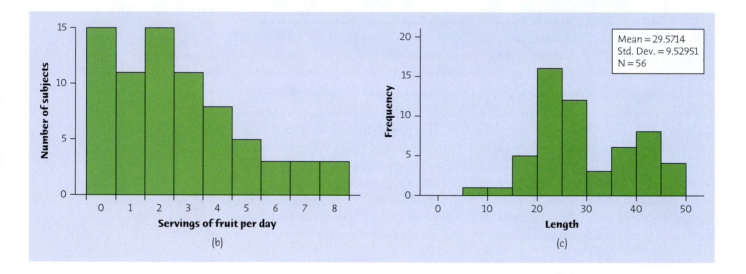

(b)

(c)

definitely not Normal. From this and the size of the sample shown we can deduce whether the corresponding sampling distribution of $\bar{x}$ might be approximately Normal or not.

Figure 13.4 (a) shows the angle of big toe deformations in 38 patients. The histogram is nearly Normal except for one mild outlier. Because the outlier is rather mild and the sample size is reasonably large, the sampling distribution of $\bar{x}$ should be approximately Normal.

Figure 13.4 (b) shows the number of daily fruit servings for 74 adolescent girls. The histogram is strongly right-skewed, suggesting that the population is not Normal but skewed. The sampling distribution for this variable should be Normal, nonetheless, because the sample size is very large.

Figure 13.4 (c) shows the lengths of 56 perch from a Swedish lake. The histogram is not particularly skewed and does not have any outlier. With a sample of 56 fish, the central limit theorem guarantees that the sampling distribution would be approximately Normal. However, it is clear from the histogram that this sample is not homogeneous and instead seems to be made up of two types of fish: some smaller fish with a peak at 20–25 inches and some larger fish with a peak at 40–45 inches. The histogram does not reveal what might be the reason for this bimodal distribution, and we would have to examine each fish closely to find it out. What is clear is that the sampling distribution would be centered on a value "μ" that would reflect neither of these two subpopulations of fish and therefore would be misleading. Always keep in mind that poorly designed studies create problems that mathematics cannot remedy.

Sometimes it may be useful to look at published data to make an educated guess about the distribution of your own variable and decide on an appropriate sample size. For example, if you wanted to conduct your own study on the number of daily fruit servings among teenagers in your school district, you could rely on a publication showing the histogram of Figure 13.4 (b) to guess that your own data would probably be skewed too and decide on using at least 40 teenagers in your own sample to take advantage of the central limit theorem.

More general versions of the central limit theorem say that the distribution of any sum or average of many small random quantities is close to Normal. This is true even if the quantities are correlated with each other (as long as they are not too highly correlated) and even if they have different distributions (as long as no one random quantity is so large that it dominates the others). The central limit theorem suggests why the Normal distributions are common models for observed data. Any variable that is a sum of many small influences (for example, the height of men or the height of women) will have an approximately Normal distribution.

APPLY YOUR KNOWLEDGE

13.10 What does the central limit theorem say? Asked what the central limit theorem says, a student replies, "As you take larger and larger samples from a population, the histogram of the sample values looks more and more Normal." Is the student right? Explain your answer.

13.11 Detecting gypsy moths. The gypsy moth is a serious threat to oak and aspen trees. A state agriculture department places traps throughout the state to detect the moths. When traps are checked periodically, the mean number of moths trapped is only 0.5, but some traps have several moths. The distribution of moth counts is discrete and strongly skewed, with standard deviation 0.7. What are the mean and standard deviation of the average number of moths per trap $\overline{x}$ in 50 traps? What is the shape of this sampling distribution?

Bruce Coleman/Alamy

13.12 More on insurance. An insurance company knows that in the entire population of millions of insured workers, the mean annual cost of workers' compensation claims is $\mu = \$439$ per insured worker and the standard deviation is $\sigma = \$20,000$. The distribution of losses is strongly right-skewed: Most policies have no loss, but a few have large losses, up to millions of dollars. If the company sells 40,000 policies, can it safely base its rates on the assumption that its average loss will be no greater than $500? No greater than $700? Follow the four-step process in your answer.

The sampling distribution of $\hat{p}$

count

sample proportion

An experiment finds that 6 of 20 birds exposed to an avian flu strain develop flu symptoms. The number of birds that develop flu symptoms is a random variable X. X is a **count** of the occurrences of some categorical outcome in a fixed number of observations. If the number of observations is n, then the **sample proportion** is

$$\hat{p} = \frac{\text{count of successes in sample}}{\text{size of sample}} = \frac{X}{n}$$

Like the sample average $\overline{x}$ when studying quantitative variables, sample counts and sample proportions are common statistics when dealing with categorical data. This section describes their sampling distributions.

EXAMPLE 13.7   Risky behavior in the age of AIDS

How common is behavior that puts people at risk of AIDS? The National AIDS Behavioral Surveys interviewed a random sample of 2673 adult heterosexuals. Of these, 170 had more than one sexual partner in the past year.[2]

The value 170 is the count X of adult heterosexuals in the sample who had multiple partners last year. The proportion of adult heterosexuals among the 2673 in the sample who had multiple partners last year is the sample proportion $\hat{p} = 170/2673 = 6.36\%$. The statistic 6.36% is representative of the population proportion p of all American adult heterosexuals who had multiple partners in the past year.

Categorical variables can take any of a finite number of possible outcomes. We may choose to call one such possible outcome a "success" and define all other possible outcomes as non-successes, or failures. (This is an arbitrary decision, not a moral judgment.) If you have covered Chapter 12, you know that the distribution of the count X of a random "success" event occurring within n observations is binomial when the probability p of that event is constant over all observations and successive observations are independent. You also learned that the binomial distribution of X can be approximated by the Normal distribution $N(np, \sqrt{np(1-p)})$ when n is so large that each the count of successes and the complementary count of non-successes in the sample are each at least 10.

A count of successes in itself is meaningful only in the context of the total number of observations and is therefore of limited use when comparing different studies. That's why we typically favor a more informative statistic, the sample proportion $\hat{p}$ of successes. How good is the statistic $\hat{p}$ as an estimate of the parameter p, the proportion of successes in the population? To find out, we ask, "What would happen if we took many samples?" The sampling distribution of $\hat{p}$ answers this question. Here are the facts.

> **SAMPLING DISTRIBUTION OF A SAMPLE PROPORTION**
>
> Choose an SRS of size n from a large population that contains population proportion p of successes. Let $\hat{p}$ be the **sample proportion** of successes,
>
> $$\hat{p} = \frac{\text{count of successes in the sample}}{n}$$
>
> Then:
>
> - The **mean** of the sampling distribution is p.
> - The **standard deviation** of the sampling distribution is $\sqrt{p(1-p)/n}$.
> - As the sample size increases, the sampling distribution of $\hat{p}$ becomes **approximately Normal.**

Figure 13.5 summarizes these facts in a form that helps you recall the big idea of a sampling distribution. The behavior of sample proportions $\hat{p}$ is similar to the behavior of sample means $\overline{x}$. When the sample size n is large, the sampling distribution is approximately Normal. The larger the sample, the more nearly Normal the distribution. The mean of the sampling distribution is the true value of the population proportion p. That is, $\hat{p}$ is an unbiased estimator of p. The standard deviation of $\hat{p}$ gets smaller as the sample size n gets larger, so that estimation is likely to be more accurate when the sample is larger. As is the case for $\overline{x}$, the standard deviation gets smaller only at the rate $\sqrt{n}$. We need 100 times as many observations to cut the standard deviation by 10.

You should not use the Normal approximation to the distribution of $\hat{p}$ when the sample size n is small. What is more, the formula given for the standard deviation of $\hat{p}$ is not accurate unless the population is much larger than the sample—say, at least 20 times larger.[3] We will give guidelines to help you decide when methods for inference based on this sampling distribution are trustworthy.

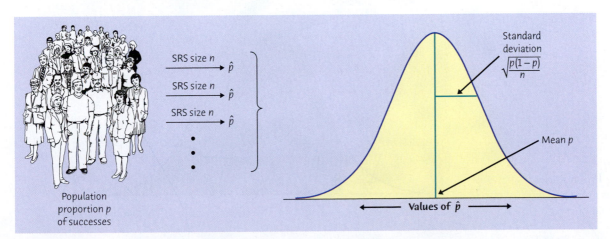

FIGURE 13.5 Select a large SRS from a population of which the proportion p are successes. The sampling distribution of the proportion $\hat{p}$ of successes in the sample is approximately Normal. The mean is p, and the standard deviation is $\sqrt{p(1-p)/n}$.

┌─ **EXAMPLE 13.8** Asking about risky behavior ─────────────────

Suppose that in fact 6% of all adult heterosexuals had more than one sexual partner in the past year (and would admit it when asked). The National AIDS Behavioral Surveys interviewed a random sample of 2673 people from this population. What is the probability that at least 5% of such a sample report having more than one partner?

If the sample size is $n = 2673$ and the population proportion is $p = 0.06$, the sample proportion $\hat{p}$ has mean 0.06 and standard deviation

$$\sqrt{\frac{p(1-p)}{n}} = \sqrt{\frac{(0.06)(0.94)}{2673}}$$

$$= \sqrt{0.0000211} = 0.00459$$

We want the probability that $\hat{p}$ is 0.05 or greater.

Standardize $\hat{p}$ by subtracting the mean 0.06 and dividing by the standard deviation 0.00459. This produces a new statistic that has approximately the standard Normal distribution. As usual, we call this statistic z:

$$z = \frac{\hat{p} - 0.06}{0.00459}$$

Figure 13.6 shows the probability we want as an area under the standard Normal curve.

$$P(\hat{p} \geq 0.05) = P\left(\frac{\hat{p} - 0.06}{0.00459} \geq \frac{0.05 - 0.06}{0.00459}\right)$$

$$= P(z \geq -2.18)$$

$$= 1 - 0.0146 = 0.9854$$

If we repeat the National AIDS Behavioral Surveys many times, more than 98% of all the samples will contain at least 5% of respondents who report having more than one sexual partner in a year. The sample described in Example 13.7 is one such sample.

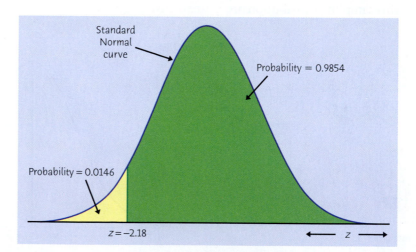

FIGURE 13.6 Probabilities in Example 13.8 as areas under the standard Normal curve.

The Normal approximation for the sampling distribution of $\hat{p}$ is least accurate when p is close to 0 or 1. You can see that if $p = 0$, any sample must contain only failures. That is, $\hat{p} = 0$ every time and there is no Normal distribution in sight. In just the same way, the approximation works poorly when p is close to 1. In practice, this means that we need larger n for values of p near 0 or 1.

APPLY YOUR KNOWLEDGE

13.13 Aging American population. Government data show that **6%** of the American population are at least 75 years of age. To test a random-digit dialing device, you use the device to call randomly chosen residential telephones in your county. Of the 150 members of the households contacted, **4%** are 75 years or older.

(a) Is each of the boldface numbers a parameter or a statistic?

(b) Assume that your county's population is similar to the American population in age distribution. What are the mean and standard deviation of the proportion $\hat{p}$ who are at least 75 years of age in samples of 150 respondents?

13.14 Student drinking. The College Alcohol Study interviewed an SRS of 14,941 college students about their drinking habits. Suppose that half of all college students "drink to get drunk" at least once in a while. That is, $p = 0.5$.

(a) What are the mean and standard deviation of the statistic $\hat{p}$, the sample proportion who drink to get drunk?

(b) Use the Normal approximation to find the probability that $\hat{p}$ is between 0.49 and 0.51.

13.15 Aging American population, continued. Refer back to Exercise 13.13. What is the probability that a random sample of 150 respondents would contain less than 4% of individuals at least 75 years of age?

13.16 Student drinking, continued. Suppose that half of all college students drink to get drunk at least once in a while. Exercise 13.14 asks the probability that the sample proportion $\hat{p}$ estimates $p = 0.5$ within ±1 percentage point ($\hat{p}$ between 0.49 and 0.51). Find this probability for SRSs of sizes 1000, 4000, and 16,000. What general fact do your results illustrate?

CHAPTER 13 SUMMARY

When we want information about the **population mean μ** for some quantitative variable, we often take an SRS and use the **sample mean $\overline{x}$** to estimate the unknown parameter μ.

The **law of large numbers** states that the actually observed mean outcome $\overline{x}$ must approach the mean μ of the population as the number of observations increases.

The **sampling distribution** of $\overline{x}$ describes how the statistic $\overline{x}$ varies in all possible SRSs of the same size from the same population.

The **mean** of the sampling distribution of $\overline{x}$ is the same as the population mean μ, and therefore $\overline{x}$ is an **unbiased estimator** of μ.

The **standard deviation** of the sampling distribution of $\overline{x}$ is $\sigma/\sqrt{n}$ for an SRS of size n if the population has standard deviation σ. That is, averages are less variable than individual observations.

If the population has a Normal distribution, so does $\overline{x}$.

The **central limit theorem** states that for large n the sampling distribution of $\overline{x}$ is approximately Normal for any population with finite standard deviation σ. That is, averages are more Normal than individual observations. We can use the

$N(\mu, \sigma/\sqrt{n})$ distribution to calculate approximate probabilities for events involving $\bar{x}$.

When we want information about the **population proportion** p for some categorical variable, we often take an SRS and use the **sample proportion** $\hat{p}$ to estimate the unknown parameter p.

The **sampling distribution of the proportion** $\hat{p}$ **of successes** in an SRS of size n from a large population containing proportion p of successes has mean p and standard deviation $\sqrt{p(1-p)/n}$.

The **mean** of the sampling distribution of $\hat{p}$ is p, and therefore $\hat{p}$ is an **unbiased estimator** of p.

When the sample size n is large, when both np and $n(1-p)$ are large enough, and when the population is at least 20 times larger than the sample, the sampling distribution of $\hat{p}$ has approximately the $N(p, \sqrt{np(1-p)/n})$ distribution. We can use this distribution to calculate approximate probabilities for events involving $\hat{p}$.

CHECK YOUR SKILLS

13.17 The National Center for Health Statistics interviewed all adult household members of a nationally representative sample of the civilian noninstitutionalized household population; **62%** of the interviewees had used some form of complementary or alternative medicine, including prayer, during the past 12 months. The boldface number is a

(a) sampling distribution. (b) parameter. (c) statistic.

13.18 Medical care and compensation costs for workers injured on the job vary a lot and are strongly skewed as a whole. The National Academy of Social Insurance reports that the mean workers' compensation claim cost in a year is $\mu = \$439$ per insured worker, with standard deviation $\sigma = \$20,000$. The law of large numbers says that

(a) an insurance company can get an average workers' compensation cost lower than the mean \$439 by insuring a large number of workers.

(b) as a company insures more and more workers chosen at random, the average claim cost gets ever closer to \$439.

(c) if a company insures a large number of workers chosen at random, that company's average claim cost will have approximately a Normal distribution.

13.19 A newborn baby has extremely low birth weight (ELBW) if it weighs less than 1000 grams. A study of the health of such children in later years examined a random sample of 219 children. Their mean weight at birth was $\bar{x} = 810$ grams. This sample mean is an *unbiased estimator* of the mean weight μ in the population of all ELBW babies. This means that

(a) in many samples from this population, the mean of the many values of $\bar{x}$ will be equal to μ.

(b) as we take larger and larger samples from this population, $\bar{x}$ will get closer and closer to μ.

(c) in many samples from this population, the many values of $\bar{x}$ will have a distribution that is close to Normal.

Use the following for Exercises 13.20 to 13.22. Cholesterol levels among 14-year-old boys are roughly Normal with mean 170 and standard deviation 30 mg/dl.

13.20 You choose an SRS of four 14-year-old boys and average their cholesterol level. If you do this many times, the mean of the average cholesterol levels you get will be close to

(a) 170.   (b) $170/4 = 42.5$.   (c) $170/\sqrt{4} = 85$.

13.21 You choose an SRS of four 14-year-old boys and average their cholesterol level. If you do this many times, the standard deviation of the average cholesterol levels you get will be close to

(a) 30.   (b) $4/\sqrt{30} = 0.73$.   (c) $30/\sqrt{4} = 15$.

13.22 In an SRS of four 14-year-old boys, the probability that the average cholesterol level is 200 mg/dl or more is close to

(a) 0.023.   (b) 0.159.   (c) 0.977.

13.23 The survival times of guinea pigs inoculated with an infectious viral strain vary from animal to animal. The distribution of survival times is strongly skewed to the right. The central limit theorem says that

(a) as we study more and more infected guinea pigs, their average survival time gets ever closer to the mean μ for all infected guinea pigs.

(b) the average survival time of a large number of infected guinea pigs has a distribution of the same shape (strongly skewed) as the distribution for individual infected guinea pigs.

(c) the average survival time of a large number of infected guinea pigs has a distribution that is close to Normal.

Use the following for Exercises 13.24 to 13.26. The proportion of baby boys born in the United States has historically been 0.512. You choose an SRS of 20 newborn babies and record the proportion of boys.

13.24 The number 0.512 is a

(a) sampling distribution.   (b) parameter.   (c) statistic.

13.25 If you take samples of 20 newborns many times, the standard deviation of the proportion of boys will be close to

(a) 0.512.

(b) $0.512 \times 0.488/\sqrt{20} = 0.056$.

(c) $\sqrt{0.512 \times 0.488/20} = 0.112$.

13.26 In an SRS of 20 newborn babies, the probability of getting 40% boys or less is close to

(a) 0.023.   (b) 0.159.   (c) 0.655.

CHAPTER 13 EXERCISES

13.27 Women's heights. A random sample of female college students has a mean height of **65** inches, which is greater than the **64**-inch mean height of all young women. Is each of the bold numbers a parameter or a statistic? Explain your answer.

13.28 Lightning strikes. The number of lightning strikes on a square kilometer of open ground in a year has mean 6 and standard deviation 2.4. (These values are typical of much of the United States.) The National Lightning Detection Network uses automatic sensors to watch for lightning in a sample of 10 randomly chosen square kilometers. What are the mean and standard deviation of $\bar{x}$, the mean number of strikes per square kilometer?

13.29 Heights of male students. Suppose that the distribution of heights of all male students on your campus is Normal with mean 70 inches and standard deviation 2.8 inches.

(a) If you choose one student at random, what is the probability that he is between 69 and 71 inches tall?

(b) What is the standard deviation of $\bar{x}$, the mean height for a sample of n male students? How large an SRS must you take to reduce the standard deviation of the sample mean to 0.5 inch?

(c) What is the probability that the mean height of the sample in (b) is between 69 and 71 inches?

13.30 Lightning strikes, continued. The number of lightning strikes on a square kilometer of open ground in a year has mean 6 and standard deviation 2.4. What is the probability that the average number of lightning strikes per square kilometer per year for 10 randomly chosen square kilometers is 3.6 or less?

13.31 Heights of male students, continued. Suppose that the distribution of heights of all male students on your campus is Normal with mean 70 inches and standard deviation 2.8 inches.

(a) What standard deviation must $\bar{x}$ have so that 99.7% of all samples give an $\bar{x}$ within one-half inch of μ? (Use the 68–95–99.7 rule.)

(b) How large an SRS do you need to reduce the standard deviation of $\bar{x}$ to the value you found in (a)?

13.32 Glucose testing. Shelia's doctor is concerned that she may suffer from gestational diabetes (high blood glucose levels during pregnancy). There is variation both in the actual glucose level and in the blood test that measures the level. A patient is classified as having gestational diabetes if her glucose level is above 140 milligrams per deciliter (mg/dl) one hour after a sugary drink. Shelia's measured glucose level one hour after the sugary drink varies according to the Normal distribution with $\mu = 125$ mg/dl and $\sigma = 10$ mg/dl.

(a) If a single glucose measurement is made, what is the probability that Shelia is diagnosed as having gestational diabetes?

(b) If measurements are made on 4 separate days and the mean result is compared with the criterion 140 mg/dl, what is the probability that Shelia is diagnosed as having gestational diabetes?

13.33 Pollutants in auto exhausts. The level of nitrogen oxides (NOx) in the exhaust of cars of a particular model varies Normally with mean 0.2 grams per mile (g/mi) and standard deviation 0.05 g/mi. Government regulations call for NOx emissions no higher than 0.3 g/mi.

(a) What is the probability that a single car of this model fails to meet the NOx requirement?

(b) A company has 25 cars of this model in its fleet. What is the probability that the average NOx level $\bar{x}$ of these cars is above the 0.3 g/mi limit?

13.34 Glucose testing, continued. Shelia's measured glucose level one hour after a sugary drink varies according to the Normal distribution with $\mu = 125$ mg/dl and $\sigma = 10$ mg/dl. Let's consider what could happen if we took 4 separate glucose level measurements from Shelia. What is the blood glucose level L such that the probability is only 0.05 that the average of 4 measurements is larger than L? (Hint: This requires a backward Normal calculation.)

13.35 Pollutants in auto exhausts, continued. The level of NOx in the exhaust of cars of a particular model varies Normally with mean 0.2 g/mi and standard deviation 0.05 g/mi. A company has 25 cars of this model in its fleet. Let's consider what could happen if we measured the exhaust NOx level of these 25 cars. What is the NOx level L such that the probability is only 0.01 that the average of 25 cars is larger than L? (Hint: This requires a backward Normal calculation.)

13.36 Blood alcohol content. The distribution of blood alcohol content in evening drivers is very skewed: Most drivers don't drink and drive, and some limit their drinking, while a few are intoxicated. If you wanted to make Normal calculations using the sampling distribution of the average blood alcohol content of evening drivers, would a random sample of 10 drivers be sufficient? 50 drivers? 200 drivers? Explain your answers.

13.37 Airline passengers get heavier. In response to the increasing weight of airline passengers, the Federal Aviation Administration in 2003 told airlines to assume that passengers average 190 pounds in the summer, including clothing and carry-on baggage. But passengers vary, and the FAA did not specify a standard deviation. A reasonable standard deviation is 35 pounds. Weights are not Normally distributed, especially when the population includes both men and women, but they are not very non-Normal. A commuter plane carries 19 passengers. What is the approximate probability that the total weight of the passengers exceeds 4000 pounds? (Hint: To apply the central limit theorem, restate the problem in terms of the mean weight.)

13.38 Extrasensory perception. In tests for extrasensory perception (ESP), an experimenter looks at cards that are hidden from the subject. Each card contains one of 5 symbols, and the subject must name the symbol on each card presented. A subject with no suspected psychic ability has, by chance alone, a 20% chance of correctly guessing the symbol on each card presented. Because the cards are independent, the overall proportion of success can thus be expected to be **20%**. You have no psychic ability and score **28%** correct guesses out of a 25-card deck. Is each of the bold values a parameter or a statistic? Explain your answer.

13.39 Sampling bias. One way of checking the effect of undercoverage, nonresponse, and other sources of error in a sample survey is to compare the sample with known demographic facts about the population. The 2000 Census found that 11.4%, or 23,772,494, of the 209,128,094 adults (age 18 and over) in the United States called themselves "Black or African American." Is the value 11.4% a parameter or a statistic? Explain your answer.

13.40 Students on diets. A sample survey interviews an SRS of 267 college women. Suppose (as is roughly true) that 70% of all college women have been on a diet within the past 12 months. What are the mean and standard deviation of the sampling distribution for samples of 267 women?

13.41 Sampling bias, continued. The 2000 Census found that 11.4% of adults in the United States called themselves "Black or African American."

(a) An opinion poll plans to interview 1500 adults at random. What are the mean and standard deviation of the sampling distribution for samples of this size?

(b) Use a Normal approximation to find the probability that such a sample will contain 11% or fewer blacks.

13.42 Students on diets, continued. A sample survey interviews an SRS of 267 college women. Suppose (as is roughly true) that 70% of all college women have been on a diet within the past 12 months. Use a Normal approximation to find the probability that 75% or more of the women in the sample have been on a diet.

13.43 Contraception and unintended pregnancies. Pharmaceutical companies advertise for the pill an efficacy of 99.5% in preventing pregnancy. However, under typical use the real efficacy is only about 95%. That is, 5% of women taking the pill for a year will experience an unplanned pregnancy that year.[4] A gynecologist looks back at a random sample of medical records from patients who had been prescribed the pill one year before.

(a) What are the mean and standard deviation of the distribution of sample proportions of women experiencing unplanned pregnancies?

(b) The gynecologist takes an SRS of 200 records and finds that 14 women had become pregnant within 1 year while taking the pill. How surprising is this finding? Give the probability of finding 7% or more pregnant women in the sample.

(c) How surprising would it be if the gynecologist had found 16 pregnant women (8%)? Twenty pregnant women (10%)?

Getty Images/Glowimages

Introduction to Inference

After we have selected a sample, we know the responses of the individuals in the sample. The usual reason for taking a sample is not to learn about the individuals in the sample but to *infer* from the sample data some conclusion about the wider population that the sample represents.

STATISTICAL INFERENCE

Statistical inference provides methods for drawing conclusions about a population from sample data.

Because a different sample might lead to different conclusions, we can't be certain that our conclusions are correct. Statistical inference uses the language of probability to say how trustworthy our conclusions are. This chapter introduces the two most common types of inference, *confidence intervals* for estimating the value of a population parameter and *tests of significance* to assess the evidence for or against a claim about a population. Both types of inference are based on the sampling distributions of statistics. That is, both report probabilities that state what would happen if we used the inference method many times.

This chapter presents the basic reasoning of statistical inference. To make the reasoning as clear as possible, we start with a setting that is too simple to be realistic, especially in the life sciences. Here is the setting for our work in this chapter.

> **SIMPLE CONDITIONS FOR INFERENCE ABOUT A MEAN**
>
> 1. We have an SRS from the population of interest. There is no nonresponse or other practical difficulty.
> 2. The variable we measure has a perfectly Normal distribution $N(\mu, \sigma)$ in the population.
> 3. We don't know the population mean μ. But we do know the population standard deviation σ.

The conditions that we have a perfect SRS, that the population is perfectly Normal, and that we know the population σ are all unrealistic. We will discuss some important practical issues when using inference in Chapter 15, and later chapters will show how inference deals with more realistic settings.

The reasoning of statistical estimation

The National Health and Nutrition Examination Survey (NHANES) reports the weights and heights of Americans every few years. Each survey is based on a nationwide probability sample of the U.S. civilian noninstitutionalized population. What can we learn from the survey about the height of eight-year-old boys in America?

David Young-Wolff/Photo Edit

EXAMPLE 14.1 How tall are children these days?

The most recent NHANES reports that the mean height of a sample of 217 eight-year-old boys was $\overline{x} = 132.5$ centimeters (cm). (That's 52.2 inches.) On the basis of this sample, we want to estimate the mean height μ in the population of over a million American eight-year-old boys.

To match the "simple conditions," we will treat the NHANES sample as a perfect SRS of all American eight-year-old boys and the height in this population as having an exactly Normal distribution with standard deviation $\sigma = 10$ cm.[1]

Here is the reasoning of statistical estimation in a nutshell.

1. To estimate the unknown population mean μ, use the mean $\overline{x} = 132.5$ of the random sample. We don't expect $\overline{x}$ to be exactly equal to μ, so we want to say how accurate this estimate is.

2. We know the sampling distribution of $\overline{x}$. In repeated samples, $\overline{x}$ has the Normal distribution with mean μ and standard deviation $\sigma/\sqrt{n}$. So the

average height $\bar{x}$ of 217 eight-year-old American boys has standard deviation

$$\frac{\sigma}{\sqrt{n}} = \frac{10}{\sqrt{217}} = 0.7 \text{ cm} \quad \text{(rounded off)}$$

3. The 95 part of the 68–95–99.7 rule for Normal distributions says that $\bar{x}$ is within 1.4 cm (that's two standard deviations) of the mean μ in 95% of all samples. That is, for 95% of all samples, 1.4 cm is the maximum distance separating $\bar{x}$ and μ. So if we estimate that μ lies somewhere in the interval from $\bar{x} - 1.4$ to $\bar{x} + 1.4$, we'll be right 95% of the times we take a sample. For this particular sample, this interval is

$$\bar{x} - 1.4 = 132.5 - 1.4 = 131.1 \text{ cm}$$

to

$$\bar{x} + 1.4 = 132.5 + 1.4 = 133.9 \text{ cm}$$

Because we got the interval 131.1 to 133.9 from a method that captures the population mean μ 95% of the time, we say that we are 95% *confident* that the mean height of all eight-year-old boys in the United States is some value in that interval, maybe as low as 131.1 cm or as high as 133.9 cm.

The big idea is that the sampling distribution of $\bar{x}$ tells us how close to μ the sample mean $\bar{x}$ is likely to be. A confidence interval just turns that information around to say how close to $\bar{x}$ the unknown population mean μ is likely to be.

EXAMPLE 14.2 *Statistical estimation in pictures*

Figures 14.1 and 14.2 illustrate the reasoning of estimation in graphical form. Figure 14.1 summarizes the idea of the sampling distribution. Starting with the population, imagine taking many SRSs of 217 eight-year-old American boys. The first sample has mean

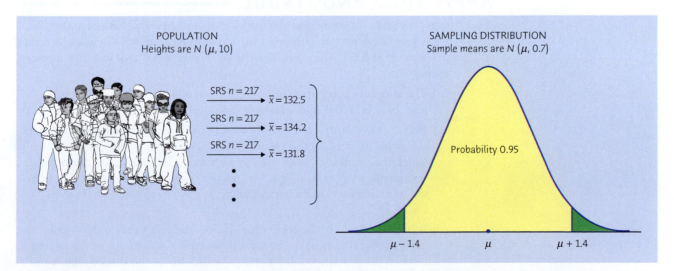

FIGURE 14.1 The sampling distribution of the mean height $\bar{x}$ of an SRS of 217 eight-year-old boys. In 95% of all samples, $\bar{x}$ lies within ± 1.4 of the unknown population mean μ.

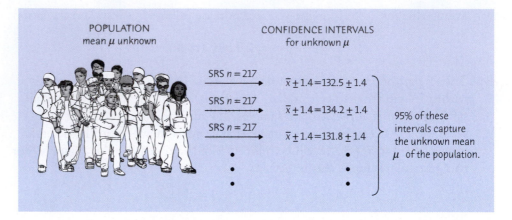

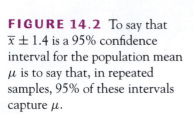

FIGURE 14.2 To say that $\bar{x} \pm 1.4$ is a 95% confidence interval for the population mean μ is to say that, in repeated samples, 95% of these intervals capture μ.

height $\bar{x} = 132.5$ cm, the second has mean $\bar{x} = 134.2$ cm, the third has mean $\bar{x} = 131.8$ cm, and so on. If we collect all these sample means and display their distribution, we get the Normal distribution with mean equal to the unknown μ and standard deviation 0.7 cm.

The 68–95–99.7 rule says that in 95% of all samples, the sample mean $\bar{x}$ lies within 1.4 cm (two standard deviations) of the population mean μ. Whenever this happens, the interval $\bar{x} \pm 1.4$ contains, or captures, μ.

Figure 14.2 summarizes the behavior of this interval. Starting with the population, imagine taking many SRSs of 217 eight-year-old American boys. The formula $\bar{x} \pm 1.4$ gives an interval based on each sample; 95% *of these intervals capture the unknown population mean* μ.

The interval of numbers between the values $\bar{x} \pm 1.4$ is called a 95% *confidence interval* for μ.

APPLY YOUR KNOWLEDGE

14.1 **Girls' heights.** Suppose that you measure the heights of an SRS of 400 eight-year-old American girls, a population with mean $\mu = 140$ cm and standard deviation $\sigma = 8$ cm. The mean $\bar{x}$ of the 400 scores will vary if you take repeated samples.

(a) The sampling distribution of $\bar{x}$ is approximately Normal. It has mean $\mu = 140$ cm. What is its standard deviation?

(b) Sketch the Normal curve that describes how $\bar{x}$ varies in many samples from this population. Mark the mean $\mu = 140$. According to the 68–95–99.7 rule, about 95% of all the values of $\bar{x}$ fall within _____ cm of the mean. What is the missing number? Call it m for "margin of error." Shade the region from the mean minus m to the mean plus m on the axis of your sketch, as in Figure 14.1.

(c) Whenever $\bar{x}$ falls in the region you shaded, the true value of the population mean, $\mu = 140$, lies in the interval between $\bar{x} - m$ and $\bar{x} + m$. Draw that interval below your sketch for one value of $\bar{x}$ inside the shaded region and one value of $\bar{x}$ outside the shaded region.

(d) In what percent of all samples will the confidence interval $\bar{x} \pm m$ capture the true mean $\mu = 140$?

Margin of error and confidence level

Our confidence interval for the mean height of all eight-year-old boys, based on the NHANES sample, is 132.5 ± 1.4. Most confidence intervals take the form

$$\text{estimate} \pm \text{margin of error}$$

The estimate ($\bar{x} = 132.5$ in our example) is the center of the interval. It is our guess for the value of the unknown parameter, based on the sample data. The *margin of error* (in our example, ± 1.4) shows how accurate we believe our guess is, based on the variability of the estimate. We have a 95% confidence interval because the interval catches the unknown parameter (μ in this case) with 95% probability. That is, we are 95% confident that the unknown parameter is one value within the confidence interval.

CONFIDENCE INTERVAL

A **level C confidence interval** for a parameter has two parts:

- An interval calculated from the data, usually of the form

$$\text{estimate} \pm \text{margin of error}$$

  where the estimate is a sample statistic and the **margin of error** represents the accuracy of our guess for the parameter.

- A **confidence level C,** which gives the probability that the interval will capture the true parameter value in repeated samples. That is, the confidence level is the success rate for the method.

Users can choose the confidence level, usually 90% or higher because we want to be quite sure of our conclusions. It may be disconcerting at first to rely on a method that is not 100% reliable. However, we must keep in mind that the only way to be 100% confident about the value of an unknown parameter would be to record the entire population; this is rarely an option, especially in the life sciences.

INTERPRETING A CONFIDENCE INTERVAL

The confidence level is the success rate of the method that produces the interval. We don't know whether the 95% confidence interval from a particular sample is one of the 95% that capture μ or one of the unlucky 5% that miss.

To say that we are **95% confident** that the unknown μ lies between 131.1 and 133.9 cm is shorthand for **"We got these numbers using a method that gives correct results 95% of the time."**

Figure 14.2 is one way to picture the idea of a 95% confidence interval. Figure 14.3 illustrates the idea in a different form. Study these figures carefully. If you understand what they say, you have mastered one of the big ideas of statistics. Figure 14.3 shows the result of drawing many SRSs from the same population and

Ranges are for statistics?

Many people like to think that statistical estimates are exact. The Nobel prize-winning economist Daniel McFadden tells a story of his time on the Council of Economic Advisers. Presented with a range of forecasts for economic growth, President Lyndon Johnson replied: "Ranges are for cattle; give me one number."

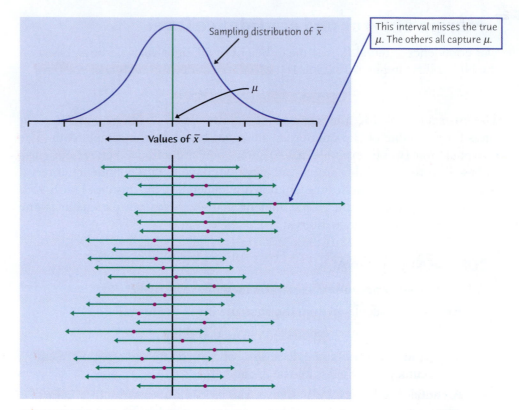

FIGURE 14.3 Twenty-five samples from the same population gave these 95% confidence intervals. In the long run, 95% of all samples give an interval that contains the population mean μ.

calculating a 95% confidence interval from each sample. The center of each interval is at $\overline{x}$ and therefore varies from sample to sample. The sampling distribution of $\overline{x}$ appears at the top of the figure to show the long-term pattern of this variation. The population mean μ is at the center of the sampling distribution. The 95% confidence intervals from 25 SRSs appear underneath. The center $\overline{x}$ of each interval is marked by a dot. The arrows on either side of the dot span the confidence interval. All except one of these 25 intervals contain the true value of μ. In a very large number of samples, 95% of the confidence intervals would contain μ. The *Confidence Interval* applet animates Figure 14.3. You can use the applet to watch confidence intervals from one sample after another capture or fail to capture the true parameter.

APPLY YOUR KNOWLEDGE

14.2   **Confidence intervals in action.** The idea of an 80% confidence interval is that the interval captures the true parameter value in 80% of all samples. That's not high enough confidence for practical use, but 80% hits and 20% misses make it easy to see how a confidence interval behaves in repeated samples from the same population. Go to the *Confidence Interval* applet.

(a) Set the confidence level to 80%. Click "Sample" to choose an SRS and calculate the confidence interval. Do this 10 times to simulate 10 SRSs with their 10 confidence intervals. How many of the 10 intervals captured the true mean μ? How many missed?

(b) You see that we can't predict whether the next sample will hit or miss. The confidence level, however, tells us what percent will hit in the long run. Reset the applet and click "Sample 50" to get the confidence intervals from 50 SRSs. How many hit? Keep clicking "Sample 50" and record the percent of hits among 100, 200, 300, 400, 500, 600, 700, 800, and 1000 SRSs. Even 1000 samples is not truly "the long run," but we expect the percent of hits in 1000 samples to be fairly close to the confidence level, 80%.

14.3 **Number skills of young men.** The National Assessment of Educational Progress (NAEP) gave a test of basic arithmetic and the ability to apply it in everyday life to a sample of 840 men 21 to 25 years of age.[2] Scores range from 0 to 500; for example, someone with a score of 325 can determine the price of a meal from a menu. The mean score for these 840 young men was $\bar{x} = 272$. Consider the NAEP sample as an SRS from a Normal population with standard deviation $\sigma = 60$.

(a) What distribution describes how the sample mean $\bar{x}$ will vary if we take many such samples?

(b) What is the 95% confidence interval for the population mean score μ based on this one sample?

(c) What does it mean to say that we have "95% confidence" in this interval?

Confidence intervals for the mean μ

In Example 14.1 we outlined the reasoning that leads to a 95% confidence interval for the unknown mean μ of a population. Now we will see how to reduce the reasoning to a formula.

To find a 95% confidence interval for the mean height of eight-year-old American boys, we first caught the central 95% of the Normal sampling distribution by going out two standard deviations in both directions from the mean. To find a level C confidence interval, we first catch the central area C under the Normal sampling distribution. Because all Normal distributions are the same in the standard scale, we can obtain everything we need from the standard Normal curve.

Figure 14.4 shows how the central area C under a standard Normal curve is marked off by two points z^* and $-z^*$. Numbers like z^* that mark off specified areas are called **critical values** of the standard Normal distribution. Values of z^* for many choices of C appear in the top row of Table C in the back of the book. This row is labeled z^*. Here are the entries for the most common confidence levels:

critical value

Confidence level C	90%	95%	99%
Critical value z^*	1.645	1.960	2.576

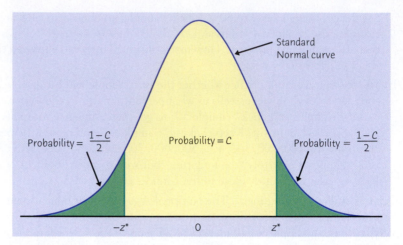

FIGURE 14.4 The critical value z^* is the number that catches central probability C under a standard Normal curve between $-z^*$ and z^*.

Notice that for C = 95% the table gives $z^* = 1.960$. This is a bit more precise than the approximate value $z^* = 2$ based on the 68–95–99.7 rule. You can of course use software to find critical values z^*, as well as the entire confidence interval.

Figure 14.4 shows that there is area C under the standard Normal curve between $-z^*$ and z^*. If we start at the sample mean $\overline{x}$ and go out z^* standard deviations, we get an interval that contains the population mean μ in a proportion C of all samples. This interval is

$$\text{from} \quad \overline{x} - z^* \frac{\sigma}{\sqrt{n}} \quad \text{to} \quad \overline{x} + z^* \frac{\sigma}{\sqrt{n}}$$

or

$$\overline{x} \pm z^* \frac{\sigma}{\sqrt{n}}$$

It is a level C confidence interval for μ.

CONFIDENCE INTERVAL FOR THE MEAN OF A NORMAL POPULATION

Draw an SRS of size n from a Normal population having unknown mean μ and known standard deviation σ. A level C **confidence interval for μ** is

$$\overline{x} \pm z^* \frac{\sigma}{\sqrt{n}}$$

The critical value z^* is illustrated in Figure 14.4 and found in Table C.

The steps in finding a confidence interval mirror the overall four-step process for organizing statistical problems.

> **CONFIDENCE INTERVALS: THE FOUR-STEP PROCESS**
>
> **STATE:** What is the practical question that requires estimating a parameter?
>
> **FORMULATE:** Identify the parameter and choose a level of confidence.
>
> **SOLVE:** Carry out the work in two phases:
>
> 1. **Check the conditions** for the interval you plan to use.
>
> 2. Calculate the **confidence interval.**
>
> **CONCLUDE:** Return to the practical question to describe your results in this setting.

EXAMPLE 14.3   Healing of skin wounds

STATE: Biologists studying the healing of skin wounds measured the rate at which new cells closed a razor cut made in the skin of an anesthetized newt. Here are data from 18 newts, measured in micrometers (millionths of a meter) per hour:[3]

$$
\begin{array}{ccccccccc}
29 & 27 & 34 & 40 & 22 & 28 & 14 & 35 & 26 \\
35 & 12 & 30 & 23 & 18 & 11 & 22 & 23 & 33
\end{array}
$$

This is one of several sets of measurements made under different conditions. We want to estimate the mean healing rate for comparison with rates under other conditions.

FORMULATE: We will estimate the mean rate μ for all newts of this species by giving a 95% confidence interval.

SOLVE: We should start by checking the conditions for inference. For this first example, we will find the interval, then discuss how statistical practice deals with conditions that are never perfectly satisfied.

   The mean of the sample is $\bar{x} = 25.67$. As part of the "simple conditions," suppose that from past experience with this species of newts we know that the standard deviation of healing rates is 8 micrometers per hour. For 95% confidence, the critical value is $z^* = 1.960$. A 95% confidence interval for μ is therefore

$$
\bar{x} \pm z^* \frac{\sigma}{\sqrt{n}} = 25.67 \pm 1.960 \frac{8}{\sqrt{18}}
$$
$$
= 25.67 \pm 3.70
$$
$$
= 21.97 \text{ to } 29.37
$$

CONCLUDE: We are 95% confident that the mean healing rate for all newts of this species is between 21.97 and 29.37 micrometers per hour.

David A. Northcott/CORBIS

   In practice, the first part of the *Solve* step is to check the conditions for inference. The "simple conditions" are:

1. **SRS:** We don't have an actual SRS from the population of all newts of this species. Scientists usually act as if a set of animal subjects is an SRS from their species or genetic type if there is nothing special about how the subjects were obtained. This study was a randomized comparative experiment in which these 18 newts were assigned at random from a larger group of newts to get one of the treatments being compared.

```
1 | 1 2 4
1 | 8
2 | 2 2 3 3
2 | 6 7 8 9
3 | 0 3 4
3 | 5 5
4 | 0
```

FIGURE 14.5 Stemplot of the healing rates in Example 14.3.

2. **Normal distribution:** The biologists expect from past experience that measurements like this on several animals of the same species under the same conditions will follow an approximately Normal distribution. We can't look at the population, but we can examine the sample. Figure 14.5 is a stemplot, with split stems. The shape is irregular, but there are no outliers or strong skewness. Shapes like this often occur in small samples from Normal populations, so we have no reason to doubt that the population distribution is Normal.

3. **Known σ:** It really is unrealistic to suppose that we know that $\sigma = 8$. We will see in Chapter 17 that it is easy to do away with the need to know σ.

As this discussion suggests, inference methods are often used when conditions like SRS and Normal population are not exactly satisfied. In this introductory chapter, we act as though the "simple conditions" are satisfied. In reality, wise use of inference requires judgment. Chapter 15 and the later chapters on each inference topic will give you a better basis for judgment.

APPLY YOUR KNOWLEDGE

14.4   Find a critical value. The critical value z^* for confidence level 97.5% is not in Table C. Use software or Table B of standard Normal probabilities to find z^*. Include in your answer a copy of Figure 14.4 with $C = 0.975$ that shows how much area is left in each tail when the central area is 0.975.

14.5   Pharmaceutical production. A manufacturer of pharmaceutical products analyzes each batch of a product to verify the concentration of the active ingredient. The chemical analysis is not perfectly precise. In fact, repeated measurements follow a Normal distribution with mean μ equal to the true concentration and standard deviation $\sigma = 0.0068$ grams per liter (g/l). Three analyses of one batch give concentrations 0.8403, 0.8363, and 0.8447 g/l. To estimate the true concentration, give a 95% confidence interval for μ. Follow the four-step process as illustrated in Example 14.3.

14.6   IQ test scores. Here are the IQ test scores of 31 seventh-grade girls in a midwest school district:[4]

114	100	104	89	102	91	114	114	103	105
108	130	120	132	111	128	118	119	86	72
111	103	74	112	107	103	98	96	112	112	93

(a) These 31 girls are an SRS of all seventh-grade girls in the school district. Suppose that the standard deviation of IQ scores in this population is known to be $\sigma = 15$. We expect the distribution of IQ scores to be close to Normal. Make a stemplot of the distribution of these 31 scores (split the stems) to verify that there are no major departures from Normality. You have now checked the "simple conditions" to the extent possible.

(b) Estimate the mean IQ score for all seventh-grade girls in the school district using a 99% confidence interval. Follow the four-step process as illustrated in Example 14.3. (Note that IQ scores are typically rounded to the nearest integer when provided. For inference's sake, we treat IQ scores as a continuous random variable.)

The reasoning of tests of significance

Confidence intervals are one of the two most common types of statistical inference. Use a confidence interval when your goal is to estimate a population parameter. The second common type of inference, called *tests of significance*, has a different goal: to assess the evidence provided by data about some claim concerning a population. Here is the reasoning of statistical tests in a nutshell.

EXAMPLE 14.4   I'm a psychic

I claim that I am a psychic and that I can correctly guess the color, red or black, of a card from a regular playing card deck 80% of the time. To test my claim, you ask me to guess the color of 20 cards chosen at random from several regular decks. I correctly guess only 8 of the 20 cards. "Aha!" you say. "Someone who makes 80% correct guesses would almost never make only 8 out of 20. So I don't believe your claim."

Your reasoning is based on asking what would happen if my claim were true and we repeated the sample of 20 card guesses many times—I would almost never correctly identify as few as 8 out of 20. This outcome is so unlikely that it gives strong evidence that my claim is not true.

You can say how strong the evidence against my claim is by giving the probability that I would correctly guess as few as 8 out of 20 cards if I really am right 80% in the long run. This probability is 0.0001. I would make as few as 8 correct guesses out of 20 card guesses only once in 10,000 tries in the long run if my claim to make 80% were true. The small probability convinces you that my claim is false.

The *Reasoning of a Statistical Test* applet animates Example 14.4. You can ask a person to guess the color of the cards until the data do (or don't) convince you that he guesses correctly less than 80% of the time. Significance tests use an elaborate vocabulary, but the basic idea is simple: *An outcome that would rarely happen if a claim were true is good evidence that the claim is not true.*

The reasoning of statistical tests, like that of confidence intervals, is based on asking what would happen if we repeated the sample or experiment many times. We will act as if the "simple conditions" listed on page 354 are true: We have a perfect SRS from an exactly Normal population with standard deviation σ known to us. Here is an example we will explore.

EXAMPLE 14.5   Sweetening colas

Diet colas use artificial sweeteners to avoid sugar. These sweeteners gradually lose their sweetness over time. Manufacturers therefore test new colas for loss of sweetness before marketing them. Trained tasters sip the cola along with drinks of standard sweetness and score the cola on a "sweetness score" of 1 to 10. The cola is then stored for a month at high temperature to imitate the effect of four months' storage at room temperature. Each taster scores the cola again after storage. This is a matched-pairs experiment. Our data are the differences (score before storage minus score after storage) in the tasters' scores. The bigger these differences, the greater the loss of sweetness.

Suppose we know that for any cola, the sweetness loss scores vary from taster to taster according to a Normal distribution with standard deviation $\sigma = 1$. The mean μ for all tasters measures loss of sweetness and is different for different colas.

Here are the sweetness losses for a new cola, as measured by 10 trained tasters:

$$2.0 \quad 0.4 \quad 0.7 \quad 2.0 \quad -0.4 \quad 2.2 \quad -1.3 \quad 1.2 \quad 1.1 \quad 2.3$$

Most are positive. That is, most tasters found a loss of sweetness. But the losses are small, and two tasters (the negative scores) thought the cola gained sweetness. The average sweetness loss is given by the sample mean $\bar{x} = 1.02$. Are these data good evidence that the cola lost sweetness in storage?

The reasoning is the same as in Example 14.4. We make a claim and ask if the data give evidence *against* it. We seek evidence that there *is* a sweetness loss, so the claim we test is that there *is not* a loss. In that case, the mean loss for the population of all trained testers would be $\mu = 0$.

•   If the claim that $\mu = 0$ is true, the sampling distribution of $\bar{x}$ from 10 tasters is Normal with mean $\mu = 0$ and standard deviation

$$\frac{\sigma}{\sqrt{n}} = \frac{1}{\sqrt{10}} = 0.316$$

Figure 14.6 shows this sampling distribution. We can judge whether any observed $\bar{x}$ is surprising by locating it on this distribution.

•   Suppose for a moment that the 10 tasters had mean loss $\bar{x} = 0.3$. It is clear from Figure 14.6 that an $\bar{x}$ this large could easily occur just by chance when the population mean is $\mu = 0$. That 10 tasters would find $\bar{x} = 0.3$ would not be evidence of a sweetness loss.

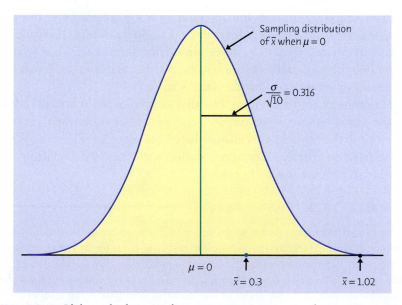

FIGURE 14.6 If the cola does not lose sweetness in storage, the mean sweetness loss $\bar{x}$ for 10 tasters will have this sampling distribution. The actual result for the new cola was $\bar{x} = 1.02$. That's so far out on the Normal curve that it is good evidence that this cola did lose sweetness. Hypothetically, if some cola had given a mean sweetness loss $\bar{x} = 0.3$, that result would be close enough to the center of the sampling distribution that it could easily have happened just by chance.

- In fact, the taste test really produced $\bar{x} = 1.02$. That's way out on the Normal curve in Figure 14.6—so far out that *an observed value this large would rarely occur just by chance if the true μ were 0*. This observed value is good evidence that the true μ is in fact greater than 0, that is, that the cola lost sweetness. The manufacturer must reformulate the cola and try again.

APPLY YOUR KNOWLEDGE

14.7 **Anemia.** Hemoglobin is a protein in red blood cells that carries oxygen from the lungs to body tissues. People with less than 12 grams of hemoglobin per deciliter of blood (g/dl) are anemic. A public health official in Jordan suspects that the mean μ for all children in Jordan is less than 12. He measures a sample of 50 children. Suppose that the "simple conditions" hold: The 50 children are an SRS from all Jordanian children and the hemoglobin level in this population follows a Normal distribution with standard deviation $\sigma = 1.6$ g/dl.

(a) We seek evidence *against* the claim that $\mu = 12$. What is the sampling distribution of $\bar{x}$ in many samples of size 50 if in fact $\mu = 12$? Make a sketch of the Normal curve for this distribution. (Sketch a Normal curve, and then mark the axis using what you know about locating the mean and standard deviation on a Normal curve.)

(b) The sample mean was $\bar{x} = 11.3$ g/dl. Mark this outcome on the sampling distribution. Also mark the outcome $\bar{x} = 11.8$ g/dl of a different study of 50 children in another country. Explain carefully from your sketch why one of these outcomes is good evidence that μ is lower than 12 and also why the other outcome is not good evidence for this conclusion.

14.8 **Arsenic contamination.** Arsenic is a compound naturally occurring in very low concentrations. Arsenic blood concentrations in healthy individuals are Normally distributed with mean $\mu = 3.2$ micrograms per deciliter (μg/dl) and standard deviation $\sigma = 1.5$ μg/dl.[5] Some areas are known to have naturally elevated concentrations of arsenic in the ground and water supplies. We take two SRSs of 25 adults residing in two different high-arsenic areas.

(a) We seek evidence *against* the claim that $\mu = 3.2$. What is the sampling distribution of the mean blood arsenic concentration $\bar{x}$ of a sample of 25 adults if the claim is true? Sketch the density curve of this distribution. (Sketch a Normal curve, and then mark the axis using what you know about locating the mean and standard deviation on a Normal curve.)

(b) Suppose that the data from the first sample give $\bar{x} = 3.35$. Mark this point on the axis of your sketch. Suppose that the data from the second sample give $\bar{x} = 3.75$ μg/dl. Mark this point on your sketch. Using your sketch, explain in simple language why one result is good evidence that the mean blood arsenic concentration of all adults in one high-arsenic area is greater than 3.2 μg/dl and why the outcome for the other high-arsenic area is not.

Stating hypotheses

A statistical test starts with a careful statement of the claims we want to compare. In Example 14.5, we saw that the taste test data are not plausible if the new cola loses no sweetness. Because the reasoning of tests looks for evidence *against* a claim, we start with the claim we seek evidence against, such as "no loss of sweetness."

NULL AND ALTERNATIVE HYPOTHESES

The claim tested by a statistical test is called the **null hypothesis.** The test is designed to assess the strength of the evidence *against* the null hypothesis. Usually the null hypothesis is a statement of "no effect" or "no difference."

The claim about the population that we are trying to find evidence *for* is the **alternative hypothesis.** The alternative hypothesis is **one-sided** if it states that a parameter is *larger than* or that it is *smaller than* the null hypothesis value. It is **two-sided** if it states that the parameter is *different from* the null value (it could be either smaller or larger).

We abbreviate the null hypothesis H_0 and the alternative hypothesis H_a. *Hypotheses always refer to a population, not to a particular outcome. Be sure to state H_0 and H_a in terms of population parameters.* Because H_a expresses the effect that we hope to find evidence *for*, it is sometimes easier to begin by stating H_a and then set up H_0 as the statement that the hoped-for effect is not present.

In Example 14.5, we are seeking evidence *for* loss in sweetness. The null hypothesis says "no loss" on the average in a large population of tasters. The alternative hypothesis says "there is a loss." So the hypotheses are

$$H_0: \mu = 0$$
$$H_a: \mu > 0$$

The alternative hypothesis is *one-sided* because we are interested only in whether the cola *lost* sweetness.

EXAMPLE 14.6   *Manufacturing aspirin tablets*

A pharmaceutical company manufactures aspirin tablets sold with the label "active ingredient: aspirin 325 mg." No production, even if machine-made, is ever perfect, and individual tablets do vary a little bit in actual aspirin content. A small amount of variation is acceptable provided that, on average, the whole production run has mean $\mu = 325$ mg.

The parameter of interest is the mean aspirin content per tablet μ. The null hypothesis says that the population of all aspirin tablets produced by this manufacturer has mean $\mu = 325$ mg, that is,

$$H_0: \mu = 325 \text{ mg}$$

Proper dosage of prescription medicine is important, and deviations from the expected dose distribution in either direction (too much or too little aspirin per tablet) would need to be identified and then, of course, remedied. The alternative hypothesis is therefore *two-sided*:

$$H_a: \mu \neq 325 \text{ mg}$$

Getty Images/Glowimages

*The hypotheses should express the hopes or suspicions we have **before** we see the data. It is cheating to first look at the data, and then frame hypotheses to fit what the data show.* Thus, the fact that a random sample of aspirin tablets in Example 14.6 gave a sample average $\overline{x}$ larger than the population mean μ should not influence our choice of H_a. If you do not have a specific direction firmly in mind in advance, you must use a two-sided alternative.

CAUTION

APPLY YOUR KNOWLEDGE

14.9   Anemia. State the null and alternative hypotheses for the anemia study described in Exercise 14.7.

14.10   Arsenic contamination. State the null and alternative hypotheses for the study of blood arsenic levels described in Exercise 14.8.

14.11   Women's heights. Young American women aged 18 to 24 have an average height of 64.5 inches. You wonder whether the mean height of female students at your university is different from the national average. You find that the mean height of a sample of 78 female students on campus is $\overline{x} = 63.1$ inches. What are your null and alternative hypotheses?

14.12   Stating hypotheses. In planning a study of the birth weights of babies whose mothers did not see a doctor before delivery, a researcher states the hypotheses as

$$H_0: \overline{x} = 1000 \text{ grams}$$
$$H_a: \overline{x} < 1000 \text{ grams}$$

What's wrong with this?

Honest hypotheses?

Chinese and Japanese, who consider the number 4 unlucky, die more often on the fourth day of the month than on other days. The authors of a study did a statistical test of the claim that the fourth day has more deaths than other days and found good evidence in favor of this claim. Can we trust this? Not if the authors looked at all days, picked the one with the most deaths, and then made "this day is different" the claim to be tested. A critic raised that issue, and the authors replied, "No, we had day 4 in mind in advance, so our test was legitimate."

P-value and statistical significance

The idea of making a claim (H_0) in order to evaluate evidence *against* it may seem odd at first. The approach is not unique to tests of significance, though. Think of a criminal trial. The defendant is "innocent until proven guilty." That is, the null hypothesis is innocence and the prosecution must try to provide convincing evidence against this hypothesis. That's exactly how statistical tests work, though in statistics we deal with evidence provided by data and use a probability to say how strong the evidence is.

The probability that measures the strength of the evidence against a null hypothesis is called a *P-value.* Statistical tests generally work like this:

TEST STATISTIC AND *P*-VALUE

A **test statistic** calculated from the sample data measures how far the data diverge from the null hypothesis H_0. Large values of the statistic show that the data are far from what we would expect if H_0 were true.

The probability, computed assuming that H_0 is true, that the test statistic would take a value as extreme as or more extreme than that actually observed is called the **P-value** of the test. The smaller the *P*-value, the stronger the evidence against H_0 provided by the data.

Small P-values are evidence against H_0, because they say that the observed result would be unlikely to occur if H_0 were true. Large P-values fail to provide evidence against H_0. Statistical software will give you the P-value of a test when you enter your null and alternative hypotheses and your data. So your most important task is to understand what a P-value says.

EXAMPLE 14.7   Sweetening colas: one-sided P-value

The study of sweetness loss in Example 14.5 tests the hypotheses

$$H_0: \mu = 0$$

$$H_a: \mu > 0$$

Because the alternative hypothesis says that $\mu > 0$, values of $\overline{x}$ greater than 0 favor H_a over H_0. The 10 tasters found mean sweetness loss $\overline{x} = 1.02$. *The P-value is the probability of getting an $\overline{x}$ at least as large as 1.02 when the null hypothesis is really true.*

The *P-Value of a Test of Significance* applet automates the work of finding P-values for samples of size 50 or smaller under the simple inference conditions. We enter the information for Example 14.5 into the applet: hypotheses, n, σ, and $\overline{x}$; then we click "Show P." Figure 14.7 shows the applet output. It displays the P-value as an area under a Normal curve and its corresponding value $P = 0.0006$. We would very rarely observe a mean sweetness loss of 1.02 or larger if H_0 were true. The small P-value provides strong evidence against H_0 and in favor of the alternative $H_a: \mu > 0$.

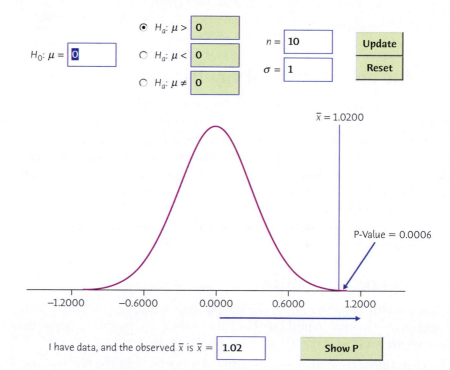

FIGURE 14.7 Output from the *P-Value of a Test of Significance* applet for Example 14.7. The P-value for the one-sided test, 0.0006, is the area under the Normal curve to the right of $\overline{x} = 1.02$.

The alternative hypothesis sets the direction that counts as evidence against H_0. In Example 14.7, only large positive values count because the alternative is one-sided on the high side. If the alternative is two-sided, both directions count.

EXAMPLE 14.8 Aspirin tablets: two-sided *P*-value

Suppose we know that the aspirin content of aspirin tablets in Example 14.6 follows a Normal distribution with standard deviation $\sigma = 5$ mg. If the manufacturing process is well calibrated, the mean aspirin content is $\mu = 325$ mg. This is our null hypothesis. The alternative hypothesis says simply "the mean is not 325 mg."

$$H_0: \mu = 325$$

$$H_a: \mu \neq 325$$

Data from a random sample of 10 aspirin tablets gives $\bar{x} = 326.9$ mg. *Because the alternative is two-sided, the P-value is the probability of getting an $\bar{x}$ at least as far from $\mu = 0$ in either direction as the observed $\bar{x} = 17$.*

Enter the information for this example into the *P-Value of a Test of Significance* applet and click "Show P." Figure 14.8 shows the applet output as well as the information we entered. The *P*-value is the sum of the two shaded areas under the Normal curve. It is $P = 0.2302$. Values as far from 325 as $\bar{x} = 326.9$ (in either direction) would happen 23% of the time if the true population mean were $\mu = 325$. An outcome that would occur so often if H_0 were true is not good evidence against H_0.

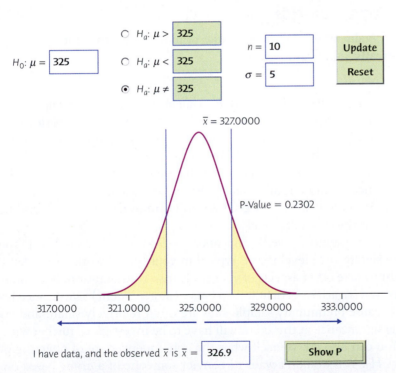

FIGURE 14.8 Output from the *P-Value of a Test of Significance* applet for Example 14.8. The *P*-value for the two-sided test, 0.2302, is twice the value of the area under the Normal curve to the right of $\bar{x} = 326.9$.

The conclusion of Example 14.8 is *not* that H_0 is true. The study looked for evidence against H_0: $\mu = 325$ and failed to find strong evidence. That is all we can say. Tests of significance assess the evidence *against* H_0. If the evidence is strong, we can confidently reject H_0 in favor of the alternative. *Failing to find evidence against H_0 means only that the data are consistent with H_0, not that we have clear evidence that H_0 is true.* The mean μ for the population of all aspirin tablets could, for instance, be a value very close but not exactly equal to 325 mg.

One key question is how small a P-value is needed to reject H_0. Very small P-values, as in Example 14.7, and large P-values, as in Example 14.8, are easy to interpret but there is a grey area in between. To avoid this problem, we can compare the test P-value with a fixed value that we regard as decisive. It amounts to announcing in advance how much evidence against H_0 we will insist on for *significance level* rejecting H_0. This decisive value of P is called the **significance level**. We write it as α, the Greek letter alpha.

If we choose $\alpha = 0.05$, we are requiring in order to reject H_0 that the data give evidence against H_0 so strong that it would happen no more than 5% of the time (1 time in 20 samples in the long run) if H_0 were true. If we choose $\alpha = 0.01$, we are insisting on stronger evidence against H_0, evidence so strong that it would appear only 1% of the time (1 time in 100 samples) if H_0 were in fact true.

STATISTICAL SIGNIFICANCE

If the P-value is as small as or smaller than α, we say that the data are **statistically significant at level α.**

"Significant" in the statistical sense does not mean "important." It means simply "not likely to happen just by chance because of random variations from sample to sample." The significance level α helps describe statistical results. Significance at level 0.01 is often expressed by the statement "The results were significant $(P < 0.01)$." Here P stands for the P-value. The actual P-value is more informative than a statement of significance, because it allows us to assess significance at any level we choose. For example, a result with $P = 0.03$ is significant at the $\alpha = 0.05$ level but is not significant at the $\alpha = 0.01$ level.

The most commonly used significance levels are 0.1, 0.05, and 0.01. You should choose a significance level that is typical in your line of work. However, you may also want to take other aspects of your study into consideration. For instance, preliminary investigations benefit from a larger (more lenient) significance level such as 0.1, because preliminary sample sizes are often relatively small, and any effect reaching significance at this point will have to be investigated further with a larger study. You can also ask yourself what the consequences are of rejecting or failing to reject H_0. You wouldn't want to convict someone of a crime based on biological data that has a 9% probability of occurring by chance alone. Inversely, you wouldn't want to set a strict significance level (for example, 0.01 or 0.001) for a study of the possible lethal side effect of a new drug.

APPLY YOUR KNOWLEDGE

14.13 Sweetening colas. Figure 14.6 for Example 14.5 compares the real cola taste test result of $\bar{x} = 1.02$ and a hypothetical result of $\bar{x} = 0.3$. A mean $\bar{x} = 1.02$ is far out on the Normal curve and so is good evidence against $H_0: \mu = 0$ ($P = 0.0006$.). A mean $\bar{x} = 0.3$ is not far enough out to convince us that the population mean is greater than 0. Follow Example 14.7 to find the test P-value when $\bar{x} = 0.3$. This P-value says, "A sample outcome this large or larger would often occur just by chance when the true mean is really 0."

14.14 Protecting ultramarathon runners. Exercise 8.35 (page 214) describes an experiment designed to learn whether taking vitamin C reduces respiratory infections among ultramarathon runners. The report of the study said:

> *Sixty-eight percent of the runners in the placebo group reported the development of symptoms of upper respiratory tract infection after the race; this was significantly more ($P < 0.01$) than that reported by the vitamin C–supplemented group (33%).*

(a) Explain to someone who knows no statistics why "significantly more" means there is good reason to think that vitamin C works.

(b) Now explain more exactly: What does $P < 0.01$ mean?

14.15 Anemia. Go back to the anemia study of Exercise 14.7.

(a) Enter the hypotheses and n, σ, and $\bar{x}$ for the Jordan study in the *P-Value of a Test of Significance* applet. What is the P-value for this study? Is this outcome statistically significant at the $\alpha = 0.05$ level? At the $\alpha = 0.01$ level?

(b) Now enter the values for the other study of anemia. What is the P-value for that study? Is this outcome statistically significant at the $\alpha = 0.05$ level? At the $\alpha = 0.01$ level?

(c) Explain briefly why these P-values tell us that one outcome is strong evidence against the null hypothesis and that the other outcome is not.

14.16 Arsenic contamination. Go back to the study of arsenic contamination in two high-arsenic areas in Exercise 14.8.

(a) Enter the hypotheses, n, σ, and $\bar{x}$ for the first study in the *P-Value of a Test of Significance* applet. What is the P-value for this study? Is this outcome statistically significant at the $\alpha = 0.05$ level? At the $\alpha = 0.01$ level?

(b) Now enter into the applet the mean blood arsenic level for the residents of the second high-arsenic area. What is the P-value for this study? Is it statistically significant at the $\alpha = 0.05$ level? At the $\alpha = 0.01$ level?

(c) Explain briefly why these P-values tell us that one outcome is strong evidence against the null hypothesis and that the other outcome is not.

Tests for a population mean

We have used tests for hypotheses about the mean μ of a population, under the "simple conditions," to introduce tests of significance. The big idea is the reasoning of a test: *Data that would rarely occur if the null hypothesis H_0 were true provide evidence that H_0 is not true.* The P-value gives us a probability to measure "would rarely occur." In practice, the steps in carrying out a significance test mirror the overall four-step process for organizing realistic statistical problems.

> ## TESTS OF SIGNIFICANCE: THE FOUR-STEP PROCESS
>
> **STATE:** What is the practical question that requires a statistical test?
>
> **FORMULATE:** Identify the parameter and state null and alternative hypotheses.
>
> **SOLVE:** Carry out the test in two phases:
>
> 1. **Check the conditions** for the test you plan to use.
> 2. Calculate the **test statistic** and find the **P-value,** or obtain the **P-value** using technology.
>
> **CONCLUDE:** Return to the practical question to describe your results in this setting.

Once you have stated your question, formulated hypotheses, and checked the conditions for your test, you or your software can find the P-value by following a rule. Here is the rule for the test we have used in our examples.

> ## z TEST FOR A POPULATION MEAN
>
> Draw an SRS of size n from a Normal population that has unknown mean μ and known standard deviation σ. To **test the null hypothesis that μ has a specified value**
>
> $$H_0: \mu = \mu_0$$
>
> use the **one-sample z test statistic**
>
> $$z = \frac{\bar{x} - \mu_0}{\sigma/\sqrt{n}}$$
>
> In terms of a variable Z having the standard Normal distribution, the P-value for a test of H_0 against
>
> $H_a: \mu > \mu_0$ is $P(Z \geq z)$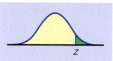
>
> $H_a: \mu < \mu_0$ is $P(Z \leq z)$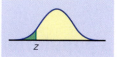
>
> $H_a: \mu \neq \mu_0$ is $2P(Z \geq |z|)$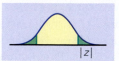
>
> Software will find both z and its P-value for you.

The test statistic measures how far the observed sample mean $\overline{x}$ deviates from the hypothesized population value μ_0. The measurement is in the familiar standard scale obtained by dividing by the standard deviation of $\overline{x}$. So we have a common scale for all z tests, and the 68–95–99.7 rule helps us see at once if $\overline{x}$ is far from μ_0. The pictures that illustrate the P-value look just like Figures 14.7 and 14.8, except that they are in the standard scale.

Know that many statistical software packages call a P-value obtained for a one-sided H_a a "one-sided" or **"one-tailed"** P-value. A P-value obtained for a two-sided H_a is then called a "two-sided" or **"two-tailed"** P-value. The logic for this naming convention is apparent if you sketch areas under the Normal curve, because a P-value computed for a two-sided H_a includes both "tails" of the Normal sampling distribution.

one-tailed
two-tailed

EXAMPLE 14.9 *Executives' blood pressures*

STATE: The National Center for Health Statistics reports that the systolic blood pressure for males 35 to 44 years of age has mean 128 and standard deviation 15. The medical director of a large company looks at the medical records of 72 male executives in this age group and finds that the mean systolic blood pressure in this sample is $\overline{x} = 126.07$. Is this evidence that the company's executives have a different mean blood pressure from the general population?

FORMULATE: The null hypothesis is "no difference" from the national mean $\mu_0 = 128$. The alternative is two-sided, because the medical director did not have a particular direction in mind before examining the data. So the hypotheses about the unknown mean μ of the executive population are

$$H_0: \mu = 128$$
$$H_a: \mu \neq 128$$

SOLVE: As part of the "simple conditions," suppose we know that executives' blood pressures follow a Normal distribution with standard deviation $\sigma = 15$. The one-sample z **test statistic** is

$$z = \frac{\overline{x} - \mu_0}{\sigma/\sqrt{n}} = \frac{126.07 - 128}{15/\sqrt{72}}$$

$$= -1.09$$

To help find a P-value, sketch the standard Normal curve and mark on it the observed value of z. Figure 14.9 shows that the P-value is the probability that a standard Normal variable Z takes a value at least 1.09 away from zero. From Table B or software, this probability is

$$P = 2P(Z \geq 1.09) = 2P(Z \leq -1.09) = (2)(0.1379) = 0.2758$$

CONCLUDE: More than 27% of the time, an SRS of size 72 from the general male population would have a mean blood pressure at least as far from 128 as that of the executive sample. The observed $\overline{x} = 126.07$ is therefore not good evidence that male executives differ from other men.

In this chapter we are acting as if the "simple conditions" stated on page 354 are true. In practice, you must verify these conditions.

1. **SRS:** The most important condition is that the 72 executives in the sample are an SRS from the population of all middle-aged male executives in the company. We should check this requirement by asking how the data were produced. If medical records are available only for executives with recent medical problems, for example, the data are of little value for our purpose because of the obvious health bias. It turns out that all executives are given a free annual medical exam, and that the medical director selected 72 exam results at random.

2. **Normal distribution:** We should also examine the distribution of the 72 observations to look for signs that the population distribution is not Normal.

3. **Known σ:** It really is unrealistic to suppose that we know that $\sigma = 15$. We will see in Chapter 17 that it is easy to do away with the need to know σ.

APPLY YOUR KNOWLEDGE

14.17 Water quality. An environmentalist group collects a liter of water from each of 45 random locations along a stream and measures the amount of dissolved oxygen in each specimen. The mean is 4.62 milligrams (mg). Is this strong evidence that the stream has a mean oxygen content of less than 5 mg per liter? (Suppose we know that dissolved oxygen varies among locations according to a Normal distribution with $\sigma = 0.92$ mg.) Follow the four-step process as illustrated in Example 14.9.

14.18 Reading a computer screen. Does the use of fancy type fonts slow down the reading of text on a computer screen? Adults can read four paragraphs of text in an average time of 22 seconds in the common Times New Roman font. Ask

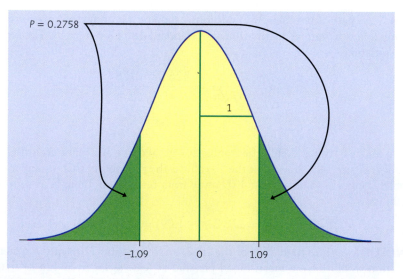

FIGURE 14.9 The P-value for the two-sided test in Example 14.9. The observed value of the test statistic is $z = -1.09$.

25 adults to read this text in the ornate font named Gigi. Here are their times:[6]

$$
\begin{array}{ccccccccc}
23.2 & 21.2 & 28.9 & 27.7 & 29.1 & 27.3 & 16.1 & 22.6 & 25.6 \\
34.2 & 23.9 & 26.8 & 20.5 & 34.3 & 21.4 & 32.6 & 26.2 & 34.1 \\
31.5 & 24.6 & 23.0 & 28.6 & 24.4 & 28.1 & 41.3
\end{array}
$$

Suppose that reading times are Normal with $\sigma = 6$ seconds. Is there good evidence that the mean reading time for Gigi is greater than 22 seconds? Follow the four-step process as illustrated in Example 14.9.

Tests from confidence intervals

Both tests and confidence intervals for a population mean μ start by using the sample mean $\bar{x}$ to estimate μ. Both rely on probabilities calculated from a Normal sampling distribution. In fact, a two-sided test at significance level α can be carried out from a confidence interval with confidence level $C = 1 - \alpha$.

CONFIDENCE INTERVALS AND TWO-SIDED TESTS

A level α two-sided significance test rejects a hypothesis $H_0: \mu = \mu_0$ exactly when the value μ_0 falls outside a level $1 - \alpha$ confidence interval for μ.

EXAMPLE 14.10   *Tests from a confidence interval*

In Example 14.9, a medical director found mean blood pressure $\bar{x} = 126.07$ for an SRS of 72 executives. Is this value significantly different from the national mean $\mu_0 = 128$ at the 10% significance level?

We can answer this question directly by a two-sided test or indirectly from a 90% confidence interval. The confidence interval is

$$
\bar{x} \pm z^* \frac{\sigma}{\sqrt{n}} = 126.07 \pm 1.645 \frac{15}{\sqrt{72}}
$$

$$
= 126.07 \pm 2.91
$$

$$
= 123.16 \text{ to } 128.98
$$

The hypothesized value $\mu_0 = 128$ falls *inside* this confidence interval, so we *cannot* reject

$$
H_0: \mu = 128
$$

at the 10% significance level. On the other hand, a two-sided test *can* reject

$$
H_0: \mu = 129
$$

at the 10% level, because 129 lies *outside* the confidence interval.

Statisticians argue over the relative merits of computing a confidence interval or a P-value. A test of significance is obviously particularly well suited when you have a clearly defined set of hypotheses. The P-value will say how strong the evidence against the null hypothesis is. However, if you reject H_0, you are left not knowing what kind of value the parameter might have. And if you fail to reject H_0, there are still other likely values for the parameter besides that defined by H_0. That is, the P-value gives no information about effect size (we will discuss this further in the next chapter). All things being equal, a confidence interval will provide more useful information than a P-value alone.

APPLY YOUR KNOWLEDGE

14.19 Test and confidence interval. The P-value for a two-sided test of the null hypothesis H_0: $\mu = 10$ is 0.06.

    (a) Does the 95% confidence interval include the value 10? Why?

    (b) Does the 90% confidence interval include the value 10? Why?

14.20 Executives' blood pressures. Examples 14.9 and 14.10 computed a P-value and a confidence interval from the data about executives' blood pressure. What can be concluded from either method? What can be concluded from the confidence interval but not from the test of significance? What can be concluded from the test of significance but not from the confidence interval?

14.21 Confidence interval and test. A 95% confidence interval for a population mean is 31.5 ± 3.5.

    (a) With a 2-sided alternative, can you reject the null hypothesis that $\mu = 34$ at the 5% significance level? Why?

    (b) With a 2-sided alternative, can you reject the null hypothesis that $\mu = 36$ at the 5% significance level? Why?

CHAPTER 14 SUMMARY

A **confidence interval** uses sample data to estimate an unknown population parameter with an indication of how accurate the estimate is and of how confident we are that the result is correct.

Any confidence interval has two parts: an interval calculated from the data and a confidence level C. The **interval** often has the form

$$\text{estimate} \pm \text{margin of error}$$

The **confidence level** is the success rate of the method that produces the interval. That is, C is the probability that the method will give a correct answer. If you use 95% confidence intervals often, in the long run 95% of your intervals will contain the true parameter value. You do not know whether a 95% confidence interval calculated from a particular set of data contains the true parameter value.

A level C **confidence interval for the mean μ** of a Normal population with known standard deviation σ, based on an SRS of size n, is given by

$$\bar{x} \pm z^* \frac{\sigma}{\sqrt{n}}$$

The **critical value** z^* is chosen so that the standard Normal curve has area C between $-z^*$ and z^*.

A **test of significance** assesses the evidence provided by data against a **null hypothesis** H_0 in favor of an **alternative hypothesis** H_a.

Hypotheses are always stated in terms of population parameters. Usually H_0 is a statement that no effect is present and H_a says that a parameter differs from its null value in a specific direction (**one-sided alternative**) or in either direction (**two-sided alternative**).

The essential reasoning of a significance test is as follows. Suppose for the sake of argument that the null hypothesis is true. If we repeated our data production many times, would we often get data as inconsistent with H_0 as the data we actually have? If the data are unlikely when H_0 is true, they provide evidence against H_0.

A test is based on a **test statistic** that measures how far the sample outcome is from the value stated by H_0.

The **P-value** of a test is the probability, computed supposing H_0 to be true, that the test statistic will take a value at least as extreme as that actually observed. Small P-values indicate strong evidence against H_0. To calculate a P-value we must know the sampling distribution of the test statistic when H_0 is true.

If the P-value is as small as or smaller than a specified value α, the data are **statistically significant** at significance level α.

Significance tests for the null hypothesis $H_0 : \mu = \mu_0$ concerning the unknown mean μ of a population are based on the **one-sample z test statistic**

$$z = \frac{\overline{x} - \mu_0}{\sigma/\sqrt{n}}$$

The z test assumes an SRS of size n from a Normal population with known population standard deviation σ. P-values can be obtained either with computations from the standard Normal distribution or by using technology (applet or software).

CHECK YOUR SKILLS

14.22 To give a 98% confidence interval for a population mean μ, you would use the critical value

(a) 1.960.   (b) 2.054.   (c) 2.326.

14.23 An opinion poll says that the result of their latest sample has a margin of error of plus or minus three percentage points. This means that

(a) we can be certain that the poll result is within 3 percentage points of the truth about the population.

(b) we could be certain that the poll result was within 3 percentage points of the truth if there were no nonresponse.

(c) the poll used a method that gives a result within three percentage points of the truth in 95% of all samples.

Use the following information for Exercises 14.24 through 14.26. A laboratory scale is known to have a standard deviation of $\sigma = 0.001$ gram in repeated weighings. Scale readings in repeated weighings are Normally distributed, with mean equal to the true weight of the specimen.

14.24 Three weighings of a specimen on this scale give 3.412, 3.416, and 3.414 g. A 95% confidence interval for the true weight is

(a) 3.414 ± 0.00113.   (b) 3.414 ± 0.00065.   (c) 3.414 ± 0.00196.

14.25 Suppose that we needed a 99% confidence level for this specimen. The margin of error would be

(a) about the same.   (b) larger.   (c) smaller.

14.26 Another specimen is weighed 8 times on this scale. The average weight is 4.1602 grams. A 99% confidence interval for the true weight of this specimen is

(a) 4.1602 ± 0.00032.   (b) 4.1602 ± 0.00069.   (c) 4.1602 ± 0.00091.

Use the following information for Exercises 14.27 through 14.29. The human average gestation time is 266 days from conception. A researcher suspects that proper nutrition plays an important role and that poor women with inadequate food intake would have shorter gestation times even when given vitamin supplements. A random sample of 20 poor women given vitamin supplements throughout the pregnancy has mean gestation time from conception $\overline{x} = 256$ days.

14.27 The null hypothesis for the researcher's test is

(a) $H_0: \mu = 266$.   (b) $H_0: \mu = 256$.   (c) $H_0: \mu < 266$.

14.28 The researcher's alternative hypothesis for the test is

(a) $H_a: \mu \neq 256$.   (b) $H_a: \mu < 266$.   (c) $H_a: \mu < 256$.

14.29 Human gestation times are approximately Normal with standard deviation $\sigma = 16$ days. The P-value for the researcher's test is

(a) more than 0.1.   (b) less than 0.01.   (c) less than 0.001.

14.30 You use software to run a test of significance. The program tells you that the P-value is 0.031. This result is

(a) not significant at the 5% level.

(b) significant at the 5% level but not at the 1% level.

(c) significant at the 1% level.

14.31 The average human gestation time is 266 days long, when counted from conception. A hospital gives a 90% confidence interval for the mean gestation time from conception among its patients. That interval is 264 ± 5 days. Is the mean gestation time in that hospital significantly different from 266 days?

(a) It is not significantly different at the 10% level and therefore is also not significantly different at the 5% level.

(b) It is not significantly different at the 10% level but might be significantly different at the 5% level.

(c) It is significantly different at the 10% level.

CHAPTER 14 EXERCISES

14.32 **Explaining confidence.** You calculate a 95% confidence interval of 27 ± 2 cm for the mean needle length of Torrey pine trees. You ask a friend to explain this result. He believes it means that "95% of all Torrey pine needles have lengths between 25 and 29 cm." Is he right? Explain your answer.

14.33 **Explaining confidence.** Here is an explanation from the Associated Press concerning one of its opinion polls. Explain briefly but clearly in what way this explanation is incorrect.

For a poll of 1,600 adults, the variation due to sampling error is no more than three percentage points either way. The error margin is said to be valid at the 95 percent confidence level. This means that, if the same questions were repeated in 20 polls, the results of at least 19 surveys would be within three percentage points of the results of this survey.

14.34 Explaining confidence. You ask another friend to explain your finding that a 95% confidence interval for the mean needle length of Torrey pines is 27 ± 2 cm. She thinks it means that "We can be 95% confident that the true mean needle length of Torrey pine trees is 27 cm." Is she right? Explain your answer.

14.35 Deer mice. Deer mice (*Peromyscus maniculatus*) are small rodents native to North America. Their body lengths (excluding tail) are known to vary approximately Normally with mean $\mu = 86$ mm and standard deviation $\sigma = 8$ mm.[7] Deer mice are found in diverse habitats and exhibit different adaptations to their environment. A random sample of 14 deer mice in a rich forest habitat gives an average body length of $\bar{x} = 91.1$ mm. Assume that the standard deviation σ of all deer mice in this area is also 8 mm.

(a) What is the standard deviation of the mean length $\bar{x}$?

(b) What critical value do you need to use in order to compute a 95% confidence interval for the mean μ?

(c) Give a 95% confidence interval for the mean body length of all deer mice in the forest habitat.

Inga Spence/Visuals Unlimited

14.36 More about deer mice. The 14 deer mice described in the previous exercise had average body length of $\bar{x} = 91.1$ mm. Assume that the standard deviation of body lengths in the population of all deer mice in the forest habitat is the same as the $\sigma = 8$ mm for the general deer mouse population.

(a) Following your approach in the previous exercise, now give a 90% confidence interval for the mean body length of all deer mice in the forest habitat.

(b) This confidence interval is shorter than your interval in the previous exercise, even though the intervals come from the same sample. Why does the second interval have a smaller margin of error?

14.37 Wood strength. Different species of trees have different biomechanical properties, affecting their commercial use. Wood density, for instance, affects the strength of produced wooden beams. Wood strength can be determined by the load (in pounds) needed to pull apart wood pieces of a given size. Here are data from students doing a laboratory exercise to measure the strength of Douglas fir pieces 4 inches long and 1.5 inches square:

33,190	31,860	32,590	26,520	33,280
32,320	33,020	32,030	30,460	32,700
23,040	30,930	32,720	33,650	32,340
24,050	30,170	31,300	28,730	31,920

(a) We are willing to regard the wood pieces prepared for the lab session as an SRS of all similar pieces of Douglas fir. Engineers also commonly assume that characteristics of materials vary Normally. Make a graph to show the shape of the distribution for these data. Does the Normality condition appear safe? Suppose, nonetheless, that the strength of pieces of wood like these follows a Normal distribution with standard deviation 3000 lb.

(b) Give a 90% confidence interval for the mean load required to pull the wood apart. Follow the four-step process in your work.

14.38 **This wine stinks.** Sulfur compounds cause "off-odors" in wine, so winemakers want to know the odor threshold, the lowest concentration of a compound that the human nose can detect. The odor threshold for dimethyl sulfide (DMS) in trained wine tasters is about 25 micrograms per liter of wine (μg/l). The untrained noses of consumers may be less sensitive, however. Here are the DMS odor thresholds for 10 untrained students:

31   31   43   36   23   34   32   30   20   24

(a) Assume that the standard deviation of the odor threshold for untrained noses is known to be $\sigma = 7$ μg/l. Briefly discuss the other two "simple conditions," using a stemplot to verify that the distribution is roughly symmetric with no outliers.

(b) Following the four-step process, give a 95% confidence interval for the mean DMS odor threshold among all students.

In all exercises that call for P-values, give the actual value if you use software or the P-value applet. Otherwise, calculate the test statistic and use Table B or Table C to give values between which P must fall.

14.39 **Fortified breakfast cereals.** The Food and Drug Administration recommends that breakfast cereals be fortified with folic acid. In a matched-pairs study, volunteers ate either fortified or unfortified cereal for some time, and then switched to the other cereal. The response variable is the difference in blood folic acid levels, fortified minus unfortified, computed for each subject. Does eating fortified cereal raise the level of folic acid in the blood? State H_0 and H_a for a test to answer this question. State carefully what the parameter μ in your hypotheses is.

14.40 **Cartons of eggs.** According to the U.S. Department of Agriculture (USDA), cartons of 12 large eggs should have a minimum weight of 24 oz; that's an average egg weight of 2 oz or more.[8] You suspect that a certain producer sells eggs of inferior caliber (lower weight) in its cartons labeled "large eggs." State H_0 and H_a for a test to assess your suspicion. State what the parameter μ in your hypotheses is.

14.41 **The wrong alternative.** One of your friends is comparing yeast fermentation rates when yeast are grown in a solution containing glucose (G) and in a solution containing starch(S). She starts with no expectations as to which feeding medium will produce the highest fermentation rate. After seeing the results of her experiment, in which the fermentation rate was higher for yeast fed glucose, she tests a one-sided alternative about the mean fermentation rates,

$$H_0: \mu_G = \mu_S$$
$$H_a: \mu_G > \mu_S$$

She finds $z = 2.1$ with one-sided P-value $P = 0.0179$.

(a) Explain why your friend should have used the two-sided alternative hypothesis.

(b) What is the correct two-sided P-value for $z = 2.1$?

14.42 Is this what P means? When asked to explain the meaning of "the *P*-value was 0.03," a student says, "This means there is only probability 0.03 that the null hypothesis is true." Is this an essentially correct explanation? Explain your answer.

14.43 Is this what significance means? Another student, when asked why statistical significance appears so often in research reports, says, "Because saying that results are significant tells us that they cannot easily be explained by chance variation alone." Do you think that this statement is essentially correct? Explain your answer.

14.44 Cicadas as fertilizer? Every 17 years, swarms of cicadas emerge from the ground in the eastern United States, live for about 6 weeks, and then die. There are so many cicadas that their dead bodies can serve as fertilizer. In an experiment, a researcher added cicadas under some plants in a natural plot of bellflowers on the forest floor, leaving other plants undisturbed. "In this experiment, cicada-supplemented bellflowers from a natural field population produced foliage with 12% greater nitrogen content relative to controls ($P = 0.031$)."[9] A colleague who knows no statistics says that an increase of 12% isn't a lot—maybe it's just an accident due to natural variation among the plants. Explain in simple language how "$P = 0.031$" answers this objection.

14.45 Forests and windstorms. Does the destruction of large trees in a windstorm change forests in any important way? Here is the conclusion of a study that found that the answer is no:

> We found surprisingly little divergence between treefall areas and adjacent control areas in the richness of woody plants ($P = 0.62$), in total stem densities ($P = 0.98$), or in population size or structure for any individual shrub or tree species.[10]

The two *P*-values refer to null hypotheses that say "no change" in measurements between treefall and control areas. Explain clearly why these values provide no evidence of change.

14.46 The wrong P. The report of a study of seat belt use by drivers says, "Hispanic drivers were not significantly more likely than White/non-Hispanic drivers to overreport safety belt use (27.4 vs. 21.1%, respectively; $z = 1.33$, $P > 1.0$)."[11] How do you know that the *P*-value given is incorrect? What is the correct one-sided *P*-value for test statistic $z = 1.33$?

14.47 Significance level: 5% versus 1%. Sketch the standard Normal curve for the z test statistic and mark off areas under the curve to show why a value of z that is significant at the 1% level in a one-sided test is always significant at the 5% level. If z is significant at the 5% level, what can you say about its significance at the 1% level?

14.48 This wine stinks. Exercise 14.38 gives the thresholds for DMS odor in wine for 10 untrained students. Assume that the odor threshold for untrained noses is Normally distributed with $\sigma = 7$ μg/l. Is there evidence that the mean threshold for untrained tasters is greater than 25 μg/l? Follow the four-step process, as illustrated in Example 14.9, in your answer.

14.49 Deer mice again. The sample of 14 forest deer mice in Exercise 14.35 had an average body length of $\bar{x} = 91.1$ mm. Do forest deer mice on the average differ significantly in body length from the general deer mice population? Assume that the standard deviation σ of all deer mice in this area is also 8 mm. Follow the four-step process, as illustrated in Example 14.9, in your answer.

14.50 IQ test scores. Exercise 14.6 gives the IQ test scores of 31 seventh-grade girls in a midwest school district. IQ scores follow a Normal distribution with standard deviation $\sigma = 15$. Treat these 31 girls as an SRS of all seventh-grade girls in this district. IQ scores in a broad population are supposed to have mean $\mu = 100$. Is there evidence that the mean in this district differs from 100? Follow the four-step process, as illustrated in Example 14.9, in your answer.

14.51 Significance tests and confidence intervals. In Exercise 14.35 you found the 95% confidence interval for the mean body length μ of deer mice in forest habitats. In Exercise 14.49 you used the same data to test the hypothesis that deer mice from forest habitats have significantly different body lengths from the general population of deer mice. Compare the two results and explain why the confidence interval is more informative than the test P-value.

14.52 Significance tests and confidence intervals. In Exercise 14.6 you calculated a 99% confidence interval for the mean IQ μ of seventh-grade girls in a midwest district. In Exercise 14.50 you used the same data to test the hypothesis that this population has a mean IQ significantly different from 100. Compare your two results. What did you learn from the confidence interval? What did you learn from the test P-value? Describe the relative merits of both methods.

14.53 Significance tests and confidence intervals. To assess the accuracy of a laboratory scale, a reference weight known to weigh exactly 10 g is weighed repeatedly. The scale readings are Normally distributed with standard deviation $\sigma = 0.0002$ g. The reference weight is weighed 5 times on that scale. The mean result is 10.0023 g.

(a) Do the 5 weighings give good evidence that the scale is not well calibrated (that is, its mean μ for weighing this weight is not 10 g)?

(b) Give a 95% confidence interval for the mean weight on this scale for all possible measurements of the reference weight. What do you conclude about the calibration of this scale?

(c) Compare your results for questions (a) and (b). Explain why the confidence interval is more informative than the test result.

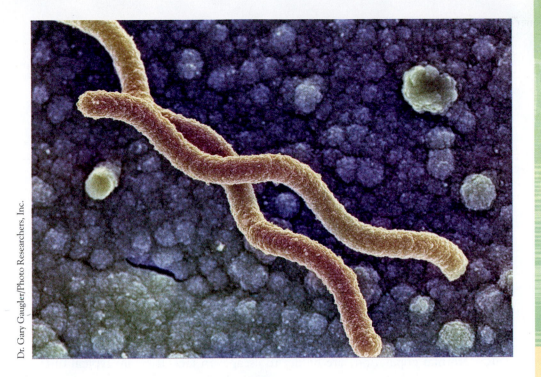

Dr. Gary Gaugler/Photo Researchers, Inc.

Inference in Practice

To this point, we have learned just two procedures for statistical inference. Both concern inference about the mean μ of a population when the "simple conditions" (page 354) are true: *The data are an SRS, the population has a Normal distribution, and we know the standard deviation σ of the population.* Under these conditions, a confidence interval for the mean μ is

$$\overline{x} \pm z^* \frac{\sigma}{\sqrt{n}}$$

To test a hypothesis H_0: $\mu = \mu_0$, we use the one-sample z statistic:

$$z = \frac{\overline{x} - \mu_0}{\sigma/\sqrt{n}}$$

We call these **z procedures** because they both start with the one-sample z statistic and use the standard Normal distribution.

z procedures

In later chapters we will modify these procedures for inference about a population mean to make them useful in practice. We will also introduce procedures for confidence intervals and tests in most of the settings we encountered in learning to explore data. There are libraries—both of books and of software—full of more elaborate statistical techniques. The reasoning of confidence intervals and tests is the same, no matter how elaborate the details of the procedure are.

There is a saying among statisticians that "mathematical theorems are true; statistical methods are effective when used with judgment." That the one-sample z statistic has the standard Normal distribution when the null hypothesis is true is a mathematical theorem. Effective use of statistical methods requires more than knowing such facts.

This chapter begins the process of helping you develop the judgment needed to use statistics in practice. That process will continue in examples and exercises through the rest of this book.

Conditions for inference in practice

Any confidence interval or significance test can be trusted only under specific conditions. It's up to you to understand these conditions and judge whether they fit your problem. With that in mind, let's look back at the "simple conditions" for the z confidence interval and test.

The final "simple condition," that we know the standard deviation σ of the population, is rarely satisfied in practice. The z procedures are therefore of little practical use, especially in the life sciences. Fortunately, it's easy to do without the "known σ" condition. Chapter 17 shows how. The first two "simple conditions" (SRS, Normal population) are harder to escape. In fact, they are typical of the conditions needed if we are to trust any statistical inference. As you plan inference, you should always ask, "Where did the data come from?" and you must often also ask, "What is the shape of the population distribution?" This is the point where knowing mathematical facts gives way to the need for judgment.

Where did the data come from? *The most important requirement for any inference procedure is that the data come from a process to which the laws of probability apply.* Inference is most reliable when the data come from a probability sample or a randomized comparative experiment. Probability samples use chance to choose respondents. Randomized comparative experiments use chance to assign subjects to treatments. The deliberate use of chance ensures that the laws of probability apply to the outcomes, and this in turn ensures that statistical inference makes sense.

> **WHERE THE DATA COME FROM MATTERS**
>
> When you use statistical inference, you are acting as if your data are a probability sample or come from a randomized experiment.

If your data don't come from a probability sample or a randomized comparative experiment, your conclusions may be challenged. To answer the challenge, you must usually rely on subject-matter knowledge, not on statistics. It is common to apply statistics to data that are not produced by random selection. When you see such a study, ask whether the data can be trusted as a basis for the conclusions of the study.

EXAMPLE 15.1   The neurobiologist and the sociologist

A neurobiologist is interested in how our visual perception can be fooled by optical illusions. Her subjects are students in Neuroscience 101 at her university. Most neurobiologists would agree that it's safe to treat the students as an SRS of all people with normal vision. There is nothing special about being a student that changes the neurobiology of visual perception.

A sociologist at the same university uses students in Sociology 101 to examine attitudes toward the use of human subjects in science. Students as a group are younger than the adult population as a whole. Even among young people, students as a group tend to come from more prosperous and better-educated homes. Even among students, this university isn't typical of all campuses. Even on this campus, students in a sociology course may have opinions that are quite different from those of engineering students or biology students. The sociologist can't reasonably act as if these students are a random sample from any interesting population.

Our first examples of inference, using the z procedures, act as if the data are an SRS from the population of interest. Let's look back at the examples in Chapter 14.

EXAMPLE 15.2 Is it really an SRS?

The NHANES survey that produced the height data for Example 14.1 uses a complex multistage sample design, so it's a bit of an oversimplification to treat the height data as coming from an SRS from the population of eight-year-old boys.[1] The overall effect of the NHANES sample is close to that of an SRS, but professional statisticians would use more complex inference procedures to match the more complex design of the sample.

The 18 newts for the skin-healing study in Example 14.3 were chosen from a laboratory population of newts to receive one of several treatments being compared in a randomized comparative experiment. Recall that each treatment group in a completely randomized experiment is an SRS of the available subjects. Scientists usually act as if the available animal subjects are an SRS from their species or genetic type if there is nothing special about where the subjects came from. We can treat these newts as an SRS from this species.

The cola taste test in Example 14.5 uses scores from 10 tasters. All were examined to be sure that they had no medical condition that might interfere with normal taste and then carefully trained to score sweetness against a set of standard drinks. We are willing to take their scores as an SRS from the population of trained tasters.

The medical director who examined executives' blood pressures in Example 14.9 actually chose an SRS from the medical records of all executives in this company.

These four examples are typical. One is an actual SRS and two are situations in which common practice would act as if the sample were an SRS. The NHANES data come from a more complex probability sampling design that necessitates advanced analytical procedures but we could assume an SRS to do a quick preliminary analysis. There is no simple rule for deciding when you can act as if a sample is an SRS. Here are a few more cautions.

- *Practical problems such as nonresponse in samples or dropouts from an experiment can hinder inference from even a well-designed study.* The NHANES survey has about an 80% response rate. This is much higher than opinion polls and most other national surveys, so by realistic standards NHANES data are quite trustworthy. (NHANES uses advanced methods to try to correct for nonresponse, but these methods work a lot better when response is high to start with.)

CAUTION

- *Different methods are needed for different designs.* The *z* procedures aren't correct for probability samples more complex than an SRS. Later chapters give methods for some other designs, but we won't discuss inference for really complex designs like that used by NHANES. Always be sure that you (or your statistical consultant) know how to carry out the inference your design calls for.

- *There is no cure for fundamental flaws like voluntary response surveys or uncontrolled experiments.* Look back at the bad examples in Chapters 7 and 8 and steel yourself to just ignore data from such studies.

What is the shape of the population distribution? Most statistical inference procedures require some conditions on the shape of the population distribution. Many of the most basic methods of inference are designed for Normal populations. That's the case for the *z* procedures and also for the more practical procedures for inference about means that we will see in Chapters 17 and 18. Fortunately, this condition is less essential than where the data come from. Here are a few practical guidelines.

- The *z* procedures and many other procedures designed for Normal distributions are based on Normality of the sample mean $\bar{x}$, not Normality of individual observations. The central limit theorem tells us that $\bar{x}$ is more Normal than the individual observations and that $\bar{x}$ becomes more Normal as we take more observations. In practice, the *z* procedures are reasonably accurate for any roughly symmetric distribution for samples of even moderate size. If the sample is large, $\bar{x}$ will be close to Normal even if individual measurements are strongly skewed.

- We rarely know the shape of the population distribution. In practice we rely on previous studies, common sense, and, most important, data analysis. When the data are chosen at random from a population, the shape of the data distribution mirrors the shape of the population distribution. By using descriptive data analysis tools, such as histograms, stemplots, or Normal quantile plots, we can guess whether the population is likely to be somewhat Normal or not. Remember, though, that *small samples have a lot of chance variation, so that Normality is hard to judge from just a few observations.*

- There is one important exception to the principle that the shape of the population is less critical than how the data were produced. *Any inference procedure based on sample statistics like the sample mean $\bar{x}$ that are not resistant to outliers can be strongly influenced by a few extreme observations.* In some cases you may be justified in removing the outlier. If not, you may want to run your analysis both with and without the outlier to see how influential that wild observation is. The discussion "Dealing with outliers" in Chapter 2 (pages 51–53) goes into more depth, describing different types of outliers and practical options for dealing with each type.

When outliers are present or the data suggest that the population is strongly non-Normal and sample size is somewhat small, Chapter 27 (available online and on the text CD) describes procedures that don't require Normality and are not sensitive to outliers.

APPLY YOUR KNOWLEDGE

15.1   **PBS takes a poll.** The PBS television program "NOVA Science Now" has a Web site for its feature programs. The section dedicated to stem cells asks Internet viewers to vote on a question after viewing arguments for and against the issue. As of September 2006, to the question "Should we allow cloning for stem cell research?" 78% have answered Yes and 16% No; 6% have remained undecided after considering the arguments.[2]

     (a) Would it be reasonable to calculate from these data a confidence interval for the percentage answering Yes in the American population? Explain your answer.

     (b) Would it be reasonable to calculate from these data a confidence interval for the percentage answering Yes among PBS viewers? Explain your answer.

15.2   **Mammary artery ligation.** Angina is the severe pain caused by inadequate blood supply to the heart. Perhaps we can relieve angina by tying off the mammary arteries to force the body to develop other routes to supply blood to the heart. Surgeons tried this procedure, called "mammary artery ligation." Patients reported a statistically significant reduction in angina pain. A randomized comparative experiment later showed that ligation was no more effective than a sham operation (placebo) procedure. Surgeons abandoned the procedure at once.[3] Explain how a study can have statistically significant results and nonetheless reach uninformative or even misleading conclusions.

15.3   **Sampling medical records.** A retrospective study published in the *British Medical Journal* examined the content quality of medical records by retrieving the clinical information of 298 consecutive patients' records from a general practice. The presence of notes related to tuberculin skin tests or BCG vaccination was chosen to represent good medical procedure in record keeping.[4] Why is it risky to regard these 298 patients as an SRS from the population of all patients at this general practice? Name some factors that might make these 298 consecutive patient records unrepresentative of all general practice medical records.

15.4   **Central limit theorem and exploratory data analysis.** Go back to the three histograms presented in Figure 13.4 (page 342) for Example 13.6. Decide for each case whether it would be appropriate to calculate a confidence interval for the population mean, assuming that we know the population standard deviation σ. Explain your decision.

15.5   **Shape of the population.** In Example 14.10 (page 375), you calculated a 95% confidence interval for the mean blood pressure of executives at a large company based on an SRS of 72 executives from that company, assuming that blood pressures follow a Normal distribution with standard deviation $\sigma = 15$. Would you be willing to calculate a 95% confidence interval for μ if blood pressures were strongly skewed? Explain your answer.

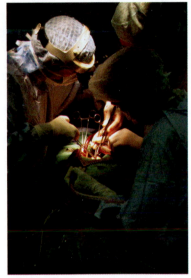

BSIP/Phototake

How confidence intervals behave

The z confidence interval $\bar{x} \pm z^{*}\sigma/\sqrt{n}$ for the mean of a Normal population illustrates several important properties that are shared by all confidence intervals in common use. The user chooses the confidence level, and the margin of error follows from this choice. We would like high confidence and also a small margin of error. High confidence says that our method almost always gives correct answers. A small margin of error says that we have pinned down the parameter quite precisely. The factors that influence the margin of error of the z confidence interval are typical of most confidence intervals.

How do we get a small margin of error? The margin of error for the z confidence interval is

$$\text{margin of error} = z^* \frac{\sigma}{\sqrt{n}}$$

This expression has z^* and σ in the numerator and $\sqrt{n}$ in the denominator. Therefore, the margin of error gets smaller when

- z^* gets smaller. A smaller z^* is the same as a lower confidence level C (look at Figure 14.4 again). *There is a trade-off between the confidence level and the margin of error. To obtain a smaller margin of error from the same data, you must be willing to accept lower confidence.*

- σ is smaller. The standard deviation σ measures the variation in the population. You can think of the variation among individuals in the population as noise that obscures the average value μ. It is easier to pin down μ when σ is small.

- n gets larger. Increasing the sample size n reduces the margin of error for any confidence level. Larger samples thus allow more precise estimates. However, *because* n *appears under a square root sign, we must take four times as many observations in order to cut the margin of error in half.*

EXAMPLE 15.3 *Confidence level and margin of error*

In Example 14.3 (page 361), biologists measured the rate of healing of the skin of 18 newts. The data gave $\overline{x} = 25.67$ micrometers per hour, and we know that $\sigma = 8$ micrometers per hour. The 95% confidence interval for the mean healing rate for all newts is

$$\overline{x} \pm z^* \frac{\sigma}{\sqrt{n}} = 25.67 \pm 1.960 \frac{8}{\sqrt{18}}$$

$$= 25.67 \pm 3.70$$

The 90% confidence interval based on the same data replaces the 95% critical value $z^* = 1.960$ with the 90% critical value $z^* = 1.645$. This interval is

$$\overline{x} \pm z^* \frac{\sigma}{\sqrt{n}} = 25.67 \pm 1.645 \frac{8}{\sqrt{18}}$$

$$= 25.67 \pm 3.10$$

Lower confidence results in a smaller margin of error, ±3.10 in place of ±3.70. In the same way, you can calculate that the margin of error for 99% confidence is larger, ±4.86. Figure 15.1 compares these three confidence intervals.

You can check that if we had a sample of only 9 newts, the margin of error for 95% confidence would increase from ±3.70 to ±5.23. Cutting the sample size in half does *not*

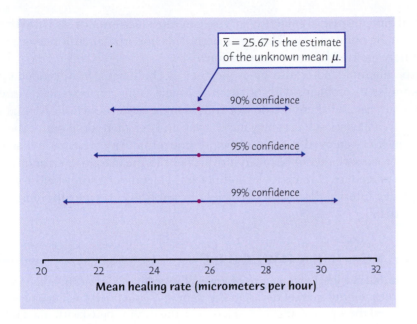

FIGURE 15.1 The lengths of three confidence intervals for Example 15.3. All three are centered at the estimate $\bar{x} = 25.67$. When the data and the sample size remain the same, higher confidence requires a larger margin of error.

double the margin of error, because the sample size n appears under a square root sign. Also keep in mind that a sample of a different size would most likely produce a different sample average and thus change both the center and the width of the 95% confidence interval.

The margin of error accounts only for sampling error. The most important caution about confidence intervals in general is a consequence of the use of a sampling distribution. A sampling distribution shows how a statistic such as $\bar{x}$ varies in repeated random sampling. This variation causes "random sampling error," because the statistic misses the true parameter by a random amount. No other source of variation or bias in the sample data influences the sampling distribution. So *the margin of error in a confidence interval ignores everything except the sample-to-sample variation due to choosing the sample randomly.*

> **THE MARGIN OF ERROR DOESN'T COVER ALL ERRORS**
>
> The margin of error in a confidence interval covers only random sampling errors.
>
> Practical difficulties such as undercoverage and nonresponse are often more serious than random sampling error. The margin of error does not take such difficulties into account.

Recall from Chapter 7 that national opinion polls often have response rates less than 50% and that even small changes in the wording of questions can strongly influence results. In such cases, the announced margin of error is probably unrealistically small. The Gallup organization always includes the following statement in its detailed survey results: "In addition to sampling error, question wording and practical difficulties in conducting surveys can introduce error or bias into the findings of public opinion polls." Unfortunately, this important cautionary statement is routinely omitted by news organizations.

Experimental studies also have their set of challenges, from producing random samples to problems with attrition. Clinical trials, for instance, can enroll only volunteers and therefore do not truly sample from the whole population of interest. In addition, subjects may move away and decide to drop out of the study, or they may be removed because of health concerns. Any systematic bias in the attrition process would not be taken into account in the margin of error of a simple inference procedure. Always look carefully at the details of a study before you trust a confidence interval. Peer-reviewed scientific publications typically address these issues in detail.

APPLY YOUR KNOWLEDGE

15.6 **Confidence level and margin of error.** Example 14.8 (page 369) described a quality control study of the aspirin content of aspirin tablets manufactured. We treated the aspirin content of tablets as Normally distributed with standard deviation $\sigma = 5$ mg.

(a) Give a 95% confidence interval for the mean aspirin content μ in this population. The actual sample, an SRS of 10 tablets, gave $\bar{x} = 326.9$ mg.

(b) Now give the 90% and 99% confidence intervals for μ.

(c) What are the margins of error for 90%, 95%, and 99% confidence? How does increasing the confidence level change the margin of error of a confidence interval?

15.7 **Sample size and margin of error.** In the previous exercise, you calculated a 95% confidence interval for the population mean aspirin content μ based on this sample.

(a) Suppose that the SRS had only 4 tablets. Obviously the sample average would be specific to that particular sample. What would be the margin of error of a 95% confidence interval for the population mean μ in this case?

(b) Suppose that the SRS had 25 tablets. What would be the margin of error of a 95% confidence interval for μ?

(c) Compare the margins of error for samples of size 4, 10, and 25. How does increasing the sample size change the margin of error of a confidence interval?

15.8 **Rating the environment.** A Gallup Poll asked the question, "How would you rate the overall quality of the environment in this country today—as excellent, good, only fair, or poor?" In all, 46% of the sample rated the environment as good or excellent. Gallup announced the poll's margin of error for 95% confidence as ±3 percentage points. Which of the following sources of error are included in the margin of error?

(a) The poll dialed telephone numbers at random and so missed all people without phones.

(b) Nonresponse—some people whose numbers were chosen never answered the phone in several calls or answered but refused to participate in the poll.

(c) There is chance variation in the random selection of telephone numbers.

How significance tests behave

Significance tests are widely used in reporting the results of research in many fields of applied science and in industry. New pharmaceutical products require significant evidence of effectiveness and safety. Courts inquire about statistical significance in hearing class action law suits (many of which are against pharmaceutical products or biohazards in the workplace). Marketers want to know whether a new ad campaign significantly outperforms the old one, and medical researchers want to know whether a new therapy performs significantly better. In all these uses, statistical significance is valued because it points to an effect that is unlikely to occur simply by chance. Here are some points to keep in mind when using or interpreting significance tests.

How small a P is convincing? The purpose of a test of significance is to describe the degree of evidence provided by the sample against the null hypothesis. The P-value does this. But how small a P-value is convincing evidence against the null hypothesis? This depends mainly on two circumstances:

- *How plausible is H_0?* If H_0 represents an assumption that the people you must convince have believed for years, strong evidence (small P) will be needed to persuade them.

- *What are the consequences of rejecting H_0?* If rejecting H_0 in favor of H_a means making an expensive changeover from one therapy to another, you need strong evidence that the new treatment will save or improve lives. On the other hand, preliminary studies do not require taking any action beyond deciding whether to pursue the topic or not. It makes sense to continue work in light of even weakly significant evidence.

These criteria are a bit subjective. Different people will often insist on different levels of significance. Giving the P-value allows each of us to decide individually if the evidence is sufficiently strong.

Users of statistics have often emphasized standard levels of significance, such as 10%, 5%, and 1%. This emphasis reflects the time when tables of critical values rather than software dominated statistical practice. The 5% level ($\alpha = 0.05$) is particularly common. However, *there is no sharp border between "significant" and "insignificant," only increasingly strong evidence as the P-value decreases. There is no practical distinction between the P-values 0.049 and 0.051. It makes no sense to treat $P \leq 0.05$ as a universal rule for what is significant.* This is a particularly important concept, considering that significance is affected by factors such as population variability and sample size.

Significance depends on sample size. A sample survey shows that significantly fewer students are heavy drinkers at colleges that ban alcohol on campus. "Significantly fewer" is not enough information to decide whether there is an important difference in drinking behavior at schools that ban alcohol. *How important an effect is depends on the size of the effect as well as on its statistical significance.* If the number of heavy drinkers is only 1% less at colleges that ban alcohol than at other colleges,

Should tests be banned?

Significance tests don't tell us how large or how important an effect is. Research in psychology has emphasized such tests, so much so that some think their weaknesses should ban them from use. The American Psychological Association asked a group of experts about this. They said: "Use anything that sheds light on your study. Use more data analysis and confidence intervals." But they also said: "The task force does not support any action that could be interpreted as banning the use of null hypothesis significance testing or P-values in psychological research and publication."

this is not an important effect even if it is statistically significant. In fact, the sample survey found that 38% of students at colleges that ban alcohol are "heavy episodic drinkers," compared with 48% at other colleges.[5] That difference is large enough to be important. (Of course, this observational study doesn't prove that an alcohol ban directly reduces drinking; it may be that colleges that ban alcohol attract more students who don't want to drink heavily.)

Such examples remind us to always look at the size of an effect (like 38% versus 48%) as well as its significance. They also raise a question: Can a tiny effect really be highly significant? Yes. The behavior of the z test statistic is typical. The statistic is

$$z = \frac{\overline{x} - \mu_0}{\sigma/\sqrt{n}}$$

The numerator of z measures how far the sample mean deviates from the hypothesized mean μ_0. Larger values of the numerator give stronger evidence against H_0: $\mu = \mu_0$. The denominator is the standard deviation of $\overline{x}$. It measures how much random variation we expect. There is less variation when the number of observations n is large. So z gets larger (more significant) when the estimated effect $\overline{x} - \mu_0$ gets larger *or* when the number of observations n gets larger. Significance depends both on the size of the effect we observe *and* on the size of the sample. Understanding this fact is essential to understanding significance tests.

> **SAMPLE SIZE AFFECTS STATISTICAL SIGNIFICANCE**
>
> Because large random samples have small chance variation, very small population effects can be highly significant if the sample is large.
>
> Because small random samples have a lot of chance variation, even large population effects can fail to be significant if the sample is small.
>
> Statistical significance does not tell us whether an effect is large enough to be important. That is, **statistical significance is not the same thing as practical significance.**

Keep in mind that statistical significance means "the sample showed an effect larger than would often occur just by chance." The extent of chance variation changes with the size of the sample, so sample size does matter. Exercises 15.9 and 15.10 demonstrate in detail how increasing the sample size drives down the P-value. Here is another example.

EXAMPLE 15.4 It's significant. Or not. So what?

We are testing the hypothesis of no correlation between two variables. This requires a different statistical test (which we will see in Chapter 23), but the reasoning is still the same. With 1000 observations, an observed correlation of only $r = 0.08$ is significant evidence at the $\alpha = 1\%$ level that the correlation in the population is not zero but positive. *The small P-value does not mean that there is a strong association, only that there is strong evidence of* some *association*. The true population correlation is probably quite close to the observed sample value, $r = 0.08$. We might well conclude that for practical purposes we can ignore the association between these variables, even though we are confident (at the 1% level) that the correlation is positive.

On the other hand, if we have only 10 observations, a correlation of $r = 0.5$ is not significantly greater than zero even at the 5% level. Small samples vary so much that a large r is needed if we are to be confident that we aren't just seeing chance variation at work. So a small sample will often fall short of significance even if the true population correlation is quite large.

Confidence intervals have the advantage over tests of hypotheses of providing an estimate of the effect size in the population. But they still don't tell whether an effect is practically important or not. Only your understanding of the subject matter can guide you in making that call.

Beware of multiple analyses. Statistical significance ought to mean that you have found an effect that you were looking for. The reasoning behind statistical significance works well if you decide what effect you are seeking, design a study to search for it, and use a test of significance to weigh the evidence you get. In other settings, significance may have little meaning.

─ EXAMPLE 15.5 *Cell phones and brain cancer* ─

Might the radiation from cell phones be harmful to users? Many studies have found little or no connection between using cell phones and various illnesses. Here is part of a news account of one study:

> *A hospital study that compared brain cancer patients and a similar group without brain cancer found no statistically significant association between cell phone use and a group of brain cancer types known as gliomas. But when 20 types of glioma were considered separately, an association was found between phone use and one rare form. Puzzlingly, however, this risk appeared to decrease rather than increase with greater mobile phone use.*[6]

Think for a moment: Suppose that the 20 null hypotheses (no association) for these 20 significance tests are all true. Then each test has a 5% chance of being significant at the 5% level. That's what $\alpha = 0.05$ means: Results this extreme occur 5% of the time just by chance when the null hypothesis is true. Because 5% is 1/20, we expect about 1 of 20 tests to give a significant result just by chance. That's what the study observed.

Jupiterimages/Comstock Premium/Alamy

Running one test and reaching the 5% level of significance is reasonably good evidence that you have found something. Running 20 tests and reaching that level only once is not. A related mistake happens when an experiment is performed under a number of different conditions but the experimenter decides to run a test of significance only on the most extreme results. This is still equivalent to running multiple analyses.

The caution about multiple analyses applies to confidence intervals as well. A single 95% confidence interval has probability 0.95 of capturing the true parameter each time you use it. The probability that all of 20 confidence intervals will capture their parameters is much less than 95%. If you think that multiple tests or intervals may have discovered an important effect, you need to gather new data to do inference about that specific effect.

APPLY YOUR KNOWLEDGE

15.9   Detecting acid rain. Emissions of sulfur dioxide by industry set off chemical changes in the atmosphere that result in "acid rain." The acidity (or basicity) of liquids is measured by pH on a scale of 0 to 14. Distilled water has pH 7.0, and lower pH values indicate acidity. Typical rain is somewhat acidic, so acid rain is defined as rainfall with a pH below 5.0. Suppose that pH measurements of rainfall on different days in a Canadian forest follow a Normal distribution with standard deviation $\sigma = 0.5$. A sample of n days finds that the mean pH is $\bar{x} = 4.8$. Is this good evidence that the mean pH μ for all rainy days is less than 5.0? The answer depends on the size of the sample.

(a) Use the *P-Value of a Test of Significance* applet. Enter H_0: $\mu = 5.0$, H_a: $\mu < 5.0$, $\sigma = 0.5$, and $\bar{x} = 4.8$. Then enter $n = 5$, $n = 15$, and $n = 40$ one after the other, clicking "Show P" each time to get the three P-values. What are they?

(b) Sketch the three Normal curves displayed by the applet, with $\bar{x} = 4.8$ marked on each curve. Explain why the P-value of the same result $\bar{x} = 4.8$ gets smaller (more significant) as the sample size increases.

15.10   Detecting acid rain, by hand. The previous exercise is very important to your understanding of tests of significance. If you don't use the applet, you should do the calculations by hand. (Your results may differ slightly from those obtained by the applet due to rounding error.) Find the P-value in each of the following hypothetical situations:

(a) We measure the acidity of rainfall on 5 days. The average pH is $\bar{x} = 4.8$.

(b) Use a larger sample of 15 days. The average pH is $\bar{x} = 4.8$.

(c) Finally, measure acidity for a sample of 40 days. The average pH is $\bar{x} = 4.8$.

15.11   Confidence intervals help. Give a 95% confidence interval for the mean pH μ in each part of the previous two exercises. The intervals, unlike the P-values, give a clear picture of what mean pH values are plausible for each sample.

15.12   Is it significant? A study examined dairy cows during pregnancy and after giving birth ("calving") and reported the following:

> *The effect of time was almost significant at the 5% level ($P = 0.058$), with the mean locomotion scores tending to be higher after calving than in late pregnancy.*[7]

(a) Why is it more informative to report the actual P-value than to simply reject or fail to reject H_0 at a chosen significance level?

(b) Explain what "almost significant at the 5% level" means.

15.13   Searching for ESP. A researcher looking for evidence of extrasensory perception (ESP) tests 500 subjects. Four of these subjects do significantly better ($P < 0.01$) than random guessing.

(a) Is it proper to conclude that these 4 people have ESP? Explain your answer.

(b) What should the researcher now do to test whether any of these 4 subjects have ESP?

15.14   Treating shortness of breath. Dyspnea, or shortness of breath, is a common complaint in patients with chronic obstructive pulmonary disease (COPD). It is often assessed by an FEV_1 test, measuring the forced expiratory volume in the first second. A study examined the effects of various treatments on perceived dyspnea in patients with advanced COPD. The researchers reported that for the 23 patients who had received 6 weeks of therapy with a long-acting bronchodilator, "there was a small, statistically insignificant, increase in FEV_1."[8]

(a) What does "statistically insignificant" mean?

(b) Why is it important that the effect was small in size as well as insignificant?

DISCUSSION: The scientific approach: hypotheses, study design, data collection, and statistical analysis

Citing a P-value is not the ultimate goal and focus point of a research study. In fact, a P-value obtained from a poor design means little, whether it is statistically significant or not. A test of significance is only part of the whole scientific approach. Good scientific methodology follows a four-pronged approach: hypotheses, study design, data collection, and statistical analysis. Statistical analysis can lead to valuable conclusions only when the rest of the approach is sound.

So how do we get there? Hypotheses arise in complicated ways from the results of past studies, preliminary data, theory, or even simply an educated guess. They must be specific enough to allow planning studies designed to confirm or refute them. Under appropriate circumstances, such as random sample selection or large enough samples, statistical inference can then be applied and sound conclusions reached. Here is an outstanding example of the scientific method.

Uncovering the cause of peptic ulcers[9]

It was once commonplace to assume that stress, an anxious personality, and even spicy food were causing peptic ulcers, painful life-long lesions of the stomach or duodenum for which only temporary relief was available. Several reports had been made of unusual, curved bacteria found in biopsies of patients suffering from peptic ulcers. These were undeniable observations, but at the time, most researchers and physicians believed that nothing could survive very long in the stomach because of its extremely acidic environment (pH between 2 and 4). Thus, the reports had been generally ignored, except by Australian researchers Robin Warren and Barry Marshall. They thought that it was worth exploring a possible link between the curved bacteria and peptic ulcers, however counterintuitive it might be. They had a working hypothesis.

How can you test the hypothesis that a given bacteria causes peptic ulcers? It would obviously not be ethical to infect people with the bacteria and see if they developed a peptic ulcer. Interestingly, that's exactly what Marshall did to himself later on, after he and Warren had developed a treatment against the bacteria. The only real research option at this stage, though, was to conduct an observational study. They examined biopsies of the stomach lining from a random sample of patients given an endoscopy referral for any number of problems along the upper digestive tract. Statistical analysis showed that the bacteria was found significantly more often in patients with peptic ulcers than in patients with other digestive disorders. This clearly established a link between the curved bacteria and peptic ulcers, an important preliminary step in determining a possible causal effect.

Establishing causality would require direct experimentation, however, and ethically, this would require offering possible treatments. If the bacteria caused the peptic ulcers, then a treatment killing them or inhibiting their growth should relieve the symptoms. This logic gave Warren and Marshall a second hypothesis to test. They therefore designed a double-blind experiment in which patients with an established diagnosis of peptic ulcer received one of several treatments or a placebo. They found that patients whose treatment completely eradicated the bacteria did significantly better over both the short term and the long term, with very few relapses even many years later. The experiment showed that the bacteria were indeed responsible, at least in part, for the peptic ulcers. It also showed that the majority of peptic ulcers could be successfully treated, thus preventing a lifetime of suffering in these patients.

How these curved bacteria, *Helicobacter pylori*, could survive in the acidic environment of the stomach and duodenum was still a mystery that needed to be solved to undeniably establish the bacterial role in peptic ulcers. Studying *H. pylori* showed that it produces urease, an enzyme that had already been found in the gastric mucosa of animals. Urease breaks down urea into carbon dioxide and ammonia, resulting in a higher (more basic) pH in solution. Marshall and colleagues hypothesized that *H. pylori* might use urease to survive in the extremely acidic gastric environment by neutralizing the pH locally. To test this hypothesis, they compared the survival in environments with various pH of urease-producing *H. pylori* and two other types of bacteria with no or limited capacity to produce urease. In the absence of urea, none of the three bacteria types survived in significant numbers at pH below 3. When urea was added to the solution, however, *H. pylori* survived in significant numbers at extreme pH as low as 1.5, whereas the other two types still couldn't survive below pH 3. This experiment showed that *H. pylori* possessed a mechanism enabling it to survive in the stomach long enough to have a lasting impact. It also paved the way for a new, fast, and noninvasive method to assess the presence of *H. pylori* in patients with gastric problems simply by having patients swallow a solution of carbon-labeled urea and monitoring signs of labeled carbon dioxide in the breath shortly afterward.

These are only three of the many studies Warren and Marshall conducted on the link between *H. pylori* and peptic ulcers over 2 decades. In 2005, they were awarded the Nobel Prize in Medicine "*for their discovery of the bacterium* Helicobacter pylori *and its role in gastritis and peptic ulcer disease.*" More can be found on the Nobel Prize Web site at `nobelprize.org/nobel_prizes/medicine/laureates/2005`.

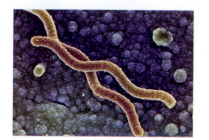

Dr. Gary Gaugler/Photo Researchers, Inc.

Planning studies: sample size for confidence intervals

A wise user of statistics never plans a sample or an experiment without at the same time planning the inference. The number of observations is a critical part of planning a study. Larger samples give smaller margins of error in confidence intervals and make significance tests better able to detect effects in the population. But taking observations costs both time and money. How many observations are enough? We will look at this question first for confidence intervals and then for tests.

You can arrange to have both high confidence and a small margin of error by taking enough observations. The margin of error of the confidence interval for the mean of a Normally distributed population is $m = z^*\sigma/\sqrt{n}$. To obtain a desired margin of error m, put in the value of z^* for your desired confidence level and solve for the sample size n. Here is the result.

SAMPLE SIZE FOR DESIRED MARGIN OF ERROR

The confidence interval for the mean of a Normal population will have a specified margin of error m when the sample size is

$$n = \left(\frac{z^*\sigma}{m}\right)^2$$

Notice that it is the size of the sample that determines the margin of error. The size of the population does not influence the sample size we need. (This is true as long as the population is much larger than the sample.)

EXAMPLE 15.6 How many observations?

Example 14.3 (page 361) reports a study of the healing rate of cuts in the skin of newts. We know that the population standard deviation is $\sigma = 8$ micrometers per hour. We want to estimate the mean healing rate μ for this species of newts within ± 3 micrometers per hour with 90% confidence. How many newts must we measure?

The desired margin of error is $m = 3$. For 90% confidence, Table C gives $z^* = 1.645$. We know that $\sigma = 8$. Therefore,

$$n = \left(\frac{z^*\sigma}{m}\right)^2 = \left(\frac{1.645 \times 8}{3}\right)^2 = 19.2$$

Because 19 newts will give a slightly larger margin of error than desired and 20 newts a slightly smaller margin of error, we must measure 20 newts. *Always round up to the next higher whole number when finding n.*

APPLY YOUR KNOWLEDGE

15.15 How tall are children these days? Example 14.1 (page 354) assumed that the heights of all American eight-year-old boys are Normally distributed with standard deviation $\sigma = 10$ cm. How large a sample would be needed to estimate the mean height μ in this population to within ± 1 cm with 95% confidence?

15.16 Estimating mean IQ. IQ scores are typically Normally distributed within a homogeneous population. How large a sample of schoolgirls in Exercise 14.6 (page 362) would be needed to estimate the mean IQ score μ within ± 5 points with 99% confidence, assuming that $\sigma = 15$ in this population?

15.17 Pharmaceutical production. Exercise 14.5 described the manufacturing process of a pharmaceutical product. Repeated measurements follow a Normal distribution with mean μ equal to the true product concentration and standard deviation $\sigma = 0.0068$ g/l.

(a) How many measurements would be needed to estimate the true concentration within ± 0.001 g/l with 95% confidence?

(b) How many measurements would be needed to estimate the true concentration within ± 0.001 g/l with 99% confidence? What is the implication of choosing a higher confidence level?

15.18 Arsenic blood level. Arsenic is a poisonous compound naturally present in soil and water at very low levels. Arsenic blood concentrations in healthy individuals are Normally distributed with standard deviation $\sigma = 1.5$ μg/dl.[10]

(a) To assess the mean arsenic blood level μ of a population living in a defined geographical area, how many individuals should have their blood tested so that we can estimate μ to within ± 0.5 μg/dl with 95% confidence?

(b) How many would be needed to estimate μ to within ± 0.1 μg/dl with 95% confidence? What is the implication of requiring a smaller margin of error?

Planning studies: the power of a statistical test*

How large a sample should we take when we plan to carry out a test of significance? We know that if our sample is too small, even large effects in the population will often fail to give statistically significant results. Here are the questions we must answer to decide how large a sample we must take.

Significance level. How much protection do we want against getting a significant result from our sample when there really is no effect in the population?

Effect size. How large an effect in the population is important in practice?

Power. How confident do we want to be that our study will detect an effect of the size we think is important?

The three boldface terms are statistical shorthand for three pieces of information. *Power* is a new idea.

EXAMPLE 15.7 Sweetening colas: planning a study

Let's illustrate typical answers to these questions in the example of testing a new cola for loss of sweetness in storage (Example 14.5, page 363). Ten trained tasters rated the sweetness on a 10-point scale before and after storage, so that we have each taster's judgment of loss of sweetness. From experience, we know that sweetness loss scores vary from taster to taster according to a Normal distribution with standard deviation about

---

* The remainder of this chapter presents more advanced material that is not needed to read the rest of the book. The idea of the power of a test is, however, important in practice.

$\sigma = 1$. To see if the taste test gives reason to think that the cola does lose sweetness, we will test

$$H_0: \mu = 0$$
$$H_a: \mu > 0$$

Are 10 tasters enough, or should we use more?

Significance level. Requiring significance at the 5% level is enough protection against declaring there is a loss in sweetness when in fact no change would be found if we could look at the entire population. This means that when there is no change in sweetness in the population, 1 out of 20 samples of tasters will wrongly find a significant loss.

Effect size. A mean sweetness loss of 0.8 point on the 10-point scale will be noticed by consumers and so is important in practice.

Power. We want to be 90% confident that our test will detect a mean loss of 0.8 point in the population of all tasters. We agreed to use significance at the 5% level as our standard for detecting an effect. So we want probability at least 0.9 that a test at the $\alpha = 0.05$ level will reject the null hypothesis $H_0: \mu = 0$ when the true population mean is $\mu = 0.8$.

The probability that the test successfully detects a sweetness loss of the specified size is the *power* of the test. You can think of tests with high power as being highly sensitive to deviations from the null hypothesis. In Example 15.7, we decided that we wanted power 90% when the truth about the population was that $\mu = 0.8$.

POWER

The **power** of a test against a specific alternative is the probability that the test will reject H_0 at a chosen significance level α when the specified alternative value of the parameter is true.

For most statistical tests, calculating power is a job for comprehensive statistical software. Finding it for the z test is easier, but we will nonetheless skip the details. The two following examples illustrate two approaches: an applet that shows the meaning of power, and statistical software. Exercise 15.22 walks you through the details of the calculations.

EXAMPLE 15.8 Finding power: use an applet

Finding the power of the z test is less challenging than most other power calculations, because it requires only a Normal distribution probability calculation. The *Power of a Test* applet does this and illustrates the calculation with Normal curves. Enter the information from Example 15.7 into the applet: hypotheses, significance level $\alpha = 0.05$, alternative value $\mu = 0.8$, and sample size $n = 10$. Click "Update." The applet output appears in Figure 15.2.

The power of the test against the specific alternative $\mu = 0.8$ is 0.808. That is, the test will reject H_0 about 81% of the time when this alternative is true. So 10 observations are too few to give power 90%.

Fish, fishermen, and power

Are the stocks of cod in the ocean off eastern Canada declining? Studies over many years failed to find significant evidence of a decline. These studies had low power—that is, they might fail to find a decline even if one were present. When it became clear that the cod were vanishing, quotas on fishing ravaged the economy in parts of Canada. If the earlier studies had had high power, they would likely have seen the decline. Quick action might have reduced the economic and environmental costs.

APPLET

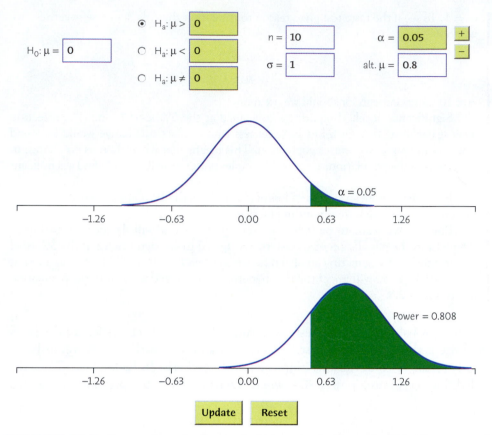

FIGURE 15.2 Output from the *Power of a Test* applet for Example 15.8. The power of the test $H_0: \mu = 0$ against the specific alternative $H_a: \mu = 0.8$ is 0.808.

The two Normal curves in Figure 15.2 show the sampling distribution of $\overline{x}$ under the null hypothesis $\mu = 0$ (top) and also under the specific alternative $\mu = 0.8$ (bottom). The curves have the same shape because σ does not change. The top curve is centered at $\mu = 0$ and the bottom curve at $\mu = 0.8$. The shaded region at the right of the top curve has area 0.05. It marks off values of $\overline{x}$ that are statistically significant at the $\alpha = 0.05$ level. The lower curve shows the probability of these same values when $\mu = 0.8$. This area is the power, 0.808.

The applet will find the power for any given sample size. It's more helpful in practice to turn the process around and learn what sample size we need to achieve a given power. Statistical software will do this, but it usually doesn't show the helpful Normal curves that are part of the applet's output.

EXAMPLE 15.9 Finding power: use software

We asked Minitab to find the number of observations needed for the one-sided z test to have power 0.9 against several specific alternatives at the 5% significance level when the population standard deviation is $\sigma = 1$. Here is the table that results:

Difference	Sample Size	Target Power	Actual Power
0.1	857	0.9	0.900184
0.2	215	0.9	0.901079
0.3	96	0.9	0.902259
0.4	54	0.9	0.902259
0.5	35	0.9	0.905440
0.6	24	0.9	0.902259
0.7	18	0.9	0.907414
0.8	14	0.9	0.911247
0.9	11	0.9	0.909895
1.0	9	0.9	0.912315

In this output, "Difference" is the difference between the null hypothesis value $\mu = 0$ and the alternative we want to detect. This is the effect size. The "Sample Size" column shows the smallest number of observations needed for power 0.9 against each effect size.

We see again that our earlier sample of 10 tasters is not large enough to be 90% confident of detecting (at the 5% significance level) an effect of size 0.8. If we want 90% power against effect size 0.8 we need at least 14 tasters. The actual power with 14 tasters is 0.911247.

Statistical software, unlike the applet, will do power calculations for most of the tests in this book.

The table in Example 15.9 makes it clear that smaller effects require larger samples to reach 90% power. Here is an overview of influences on the sample size needed.

- If you insist on a smaller significance level (such as 1% rather than 5%), you will need a larger sample. A smaller significance level requires stronger evidence to reject the null hypothesis.

- If you insist on higher power (such as 99% rather than 90%), you will need a larger sample. Higher power gives a better chance of detecting an effect when it is really there.

- At any significance level and desired power, a two-sided alternative requires a larger sample than a one-sided alternative.

- At any significance level and desired power, detecting a small effect requires a larger sample than detecting a large effect.

Planning a serious statistical study always requires an answer to the question "How large a sample do I need?" If you intend to test the hypothesis H_0: $\mu = \mu_0$ about the mean μ of a population, you need at least a rough idea of the size of the population standard deviation σ and of how big a deviation $\mu - \mu_0$ of the population mean from its hypothesized value you want to be able to detect. To effectively plan a new study, you should find the power for a range of sample sizes and effect sizes to get a full picture of how the test will behave.

More elaborate settings, such as a comparison of the mean effects of several treatments, require more elaborate advance information. You can leave the details to experts, but you should understand the idea of power and the factors that influence how large a sample you need.

Type I and Type II errors in significance tests. We can assess the performance of a test by giving two probabilities: the significance level α and the power for an alternative that we want to be able to detect. The significance level of a test is the probability of making the *wrong* decision when the null hypothesis is true. The power for a specific alternative is the probability of making the *right* decision when that alternative is true. We can just as well describe the test by giving the probability of a *wrong* decision under both conditions.

TYPE I AND TYPE II ERRORS

If we reject H_0 when in fact H_0 is true, this is a **Type I error.**

If we fail to reject H_0 when in fact H_a is true, this is a **Type II error,** also called β.

The **significance level** α of any fixed-level test is the probability of a Type I error.

The **power** of a test against any alternative is 1 minus the probability of a Type II error for that alternative: power $= 1 - \beta$.

The possibilities are summed up in Figure 15.3. If H_0 is true, our decision is correct if we fail to reject H_0 and is a Type I error if we reject H_0. If H_a is true, our decision is either correct or a Type II error. Only one error is possible at one time.

Figure 15.4 illustrates graphically how these probabilities are obtained for a significance test for a population mean with a fixed significance level α and a one-sided alternative. Figure 15.4(a) shows that when the null hypothesis is true, $\mu = \mu_0$ and the probability of making a Type I error is the chosen significance level α (dark green area). Figure 15.4(b) shows the two possible statistical outcomes when H_0 is not true but instead μ is equal to a specific value μ_a chosen to represent a practically important effect. All tests of significance are computed assuming that H_0 is true (the distribution in dotted line on Figure 15.4(b)); therefore we would correctly reject H_0 whenever results go beyond the rejection cutoff established in Figure 15.4(a). Correct decisions are represented by the dark green area, and that area is the power of the test when $\mu = \mu_a$. If we failed to reject H_0 when $\mu = \mu_a$

		Truth about the population	
		H_0 true	H_a true
Decision based on sample	Reject H_0	Type I error, α	Correct decision
	Fail to reject H_0	Correct decision	Type II error, β

FIGURE 15.3 The two types of error in testing hypotheses.

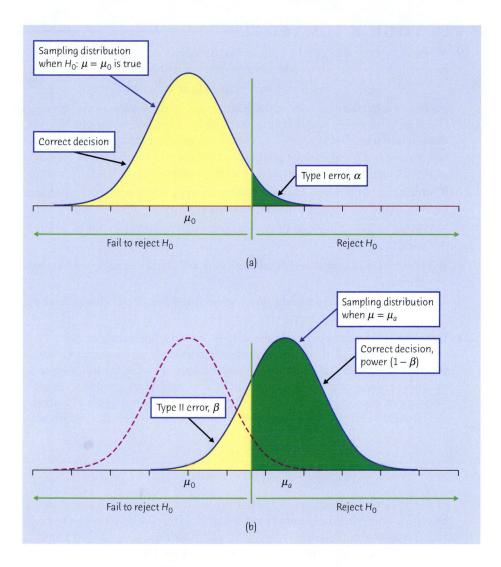

FIGURE 15.4 Graphical representation of Type I and Type II errors (a) when the null hypothesis $H_0: \mu = \mu_0$ is true and (b) when the alternative hypothesis $H_a: \mu = \mu_a$ is true. The value μ_a is chosen to represent a practically important effect.

for results below that cutoff, we would commit a Type II error. The probability of a Type II error is given by the light yellow area in Figure 15.4(b).

EXAMPLE 15.10   *Calculating error probabilities*

Because the probabilities of the two types of error are just a rewording of significance level and power, we can see from Figure 15.2 what the error probabilities are for the test in Example 15.7.

$$P(\text{Type I error}) = P(\text{reject } H_0 \text{ when in fact } \mu = 0)$$
$$= \text{significance level } \alpha = 0.05$$
$$P(\text{Type II error}) = P(\text{fail to reject } H_0 \text{ when in fact } \mu = 0.8)$$
$$= 1 - \text{power} = 1 - 0.808 = 0.192$$

The two Normal curves in Figure 15.2 are used to find the probabilities of a Type I error (top curve, $\mu = 0$) and of a Type II error (bottom curve, $\mu = 0.8$).

APPLY YOUR KNOWLEDGE

15.19 What is power? Exercise 14.17 described a study of water quality, testing the following hypotheses about the mean oxygen content of a stream:

$$H_0: \mu = 5 \text{ mg/l versus } H_a: \mu < 5 \text{ mg/l}$$

Forty-five water samples were taken from randomly chosen locations. We assume that dissolved oxygen varies among sampling locations according to a Normal distribution with $\sigma = 0.92$ mg/l. A statistician tells you that the power of the z test with $\alpha = 0.05$ against the alternative that the true mean oxygen content is $\mu = 4.75$ mg/l is 0.57. Explain in simple language what "power = 0.57" means.

15.20 Thinking about power. Answer these questions in the setting of the previous exercise about the oxygen content of a stream.

(a) To get higher power against the same alternative with the same α, what must we do?

(b) If we decide to use $\alpha = 0.10$ in place of $\alpha = 0.05$, does the power increase or decrease?

(c) If we shift our interest to the alternative $\mu = 4.5$ mg/l and change nothing else, does the power increase or decrease?

15.21 Detecting acid rain: power. Exercise 15.9 concerned detecting acid rain (rainfall with pH less than 5) from measurements made on a sample of n days for several sample sizes n. That exercise shows how the P-value for an observed sample mean $\bar{x}$ changes with n. It would be wise to do power calculations before deciding on the sample size. Suppose that pH measurements follow a Normal distribution with standard deviation $\sigma = 0.5$. You plan to test the hypotheses

$$H_0: \mu = 5$$
$$H_a: \mu < 5$$

APPLET

at the 5% level of significance. You want to use a test that will almost always reject H_0 when the true mean pH is 4.7. Use the *Power of a Test* applet to find the power against the alternative $\mu = 4.7$ for samples of size $n = 5$, $n = 15$, and $n = 40$. What happens to the power as the size of the sample increases? Which of these sample sizes are adequate for use in this setting?

15.22 Detecting acid rain: power by hand. Even though software is used in practice to calculate power, doing the work by hand in a few examples builds your understanding. Find the power of the test in the previous exercise for a sample of size $n = 15$ by following these steps.

(a) Write the z test statistic for a sample of size 15. What values of z lead to rejection of H_0 at the 5% significance level?

(b) Starting from your result in (a), what values of $\bar{x}$ lead to rejection of H_0?

(c) What is the probability of rejecting H_0 when $\mu = 4.7$? This probability is the power against this alternative.

15.23 Two types of error. Your company markets a computerized medical diagnostic program used to evaluate thousands of people. The program scans the results of routine medical tests (pulse rate, blood tests, etc.) and refers the case to a doctor if there is evidence of a medical problem. The program makes a decision about each person.

(a) What are the two hypotheses and the two types of error that the program can make? Describe the two types of error in terms of "false positive" and "false negative" test results.

(b) The program can be adjusted to decrease one error probability, at the cost of an increase in the other error probability. Which error probability would you choose to make smaller, and why? (This is a matter of judgment. There is no single correct answer.)

CHAPTER 15 SUMMARY

A specific confidence interval or test is correct only under specific conditions. The most important conditions concern the method used to produce the data. Other factors, such as the shape of the population distribution, may also be important.

Whenever you use statistical inference, you are acting as if your data are a probability sample or come from a randomized comparative experiment.

Always do data analysis before inference to detect outliers or other problems that would make inference untrustworthy.

Other things being equal, the **margin of error** of a confidence interval gets smaller as

- the confidence level C decreases,
- the population standard deviation σ decreases, and
- the sample size n increases.

The margin of error in a confidence interval accounts for only the chance variation due to random sampling. In practice, errors due to nonresponse or undercoverage are often more serious.

There is no universal rule for how small a P-value is convincing. Beware of placing too much weight on traditional significance levels, such as $\alpha = 0.05$.

Very small effects can be highly significant (small P) when a test is based on a large sample. A statistically significant effect need not be practically important. Plot the data to display the effect you are seeking, and use confidence intervals to estimate the actual values of parameters.

On the other hand, lack of significance does not imply that H_0 is true. Even a large effect can fail to be significant when a test is based on a small sample.

Many tests run at once will probably produce some significant results by chance alone, even if all the null hypotheses are true.

Smart experimenters plan their studies with inference in mind. In particular, it is good practice to find what sample size would allow successful inference.

The sample size required to obtain a confidence interval with specified margin of error m for a Normal mean is

$$n = \left(\frac{z^*\sigma}{m}\right)^2$$

where z^* is the critical value for the desired level of confidence. Always round n up when you use this formula.

The **power** of a significance test measures its ability to detect an alternative hypothesis. The power against a specific alternative is the probability that the test will reject H_0 when that alternative is true.

Increasing the size of the sample increases the power of a significance test. Using statistical applets or software, you can find the required sample to achieve a desired test power.

We can describe the performance of a test at fixed level α by giving the probabilities of two types of error. A **Type I error** occurs if we reject H_0 when it is in fact true. A **Type II error** occurs if we fail to reject H_0 when in fact H_a is true.

In a fixed-level α significance test, the significance level α is the probability of a Type I error, and the power against a specific alternative is 1 minus the probability of a Type II error for that alternative.

CHECK YOUR SKILLS

15.24 The most important condition for sound conclusions from statistical inference is usually

  (a) that the data can be thought of as a random sample from the population of interest.

  (b) that the population distribution is exactly Normal.

  (c) that no calculation errors are made in the confidence interval or test statistic.

15.25 The coach of a college basketball team records the resting pulse rates of the team members. A confidence interval for the mean resting pulse rate of all college-age adults based on these data is of little use because

  (a) the number of team members is small, so the margin of error will be large.

  (b) many of the students on the team will probably refuse to respond.

  (c) the college students on the basketball team can't be considered a random sample from the population.

15.26 You turn your Web browser to the online Excite Poll and view yesterday's poll results based on 10,282 responses. You should refuse to calculate any 95% confidence interval based on this sample because

  (a) yesterday's responses are meaningless today.

  (b) inference from a voluntary response sample can't be trusted.

  (c) the sample is too large.

15.27 Many sample surveys use well-designed random samples, but half or more of the original sample can't be contacted or refuse to take part. Any errors due to this nonresponse

  (a) have no effect on the accuracy of confidence intervals.

  (b) are included in the announced margin of error.

  (c) are in addition to the random variation accounted for by the announced margin of error.

15.28 An opinion poll reports that 60% of adults have tried to lose weight. It adds that the margin of error for 95% confidence is $\pm 3\%$. The true probability that such polls give results within $\pm 3\%$ of the truth is

  (a) 0.95, because the poll uses 95% confidence intervals.

(b) less than 0.95, because of nonresponse and other errors not included in the margin of error ±3%.

(c) only approximately 0.95, because the sampling distribution is only approximately Normal.

15.29 Here's a quote from a medical journal: "An uncontrolled experiment in 17 women found a significantly improved mean clinical symptom score after treatment. Methodologic flaws make it difficult to interpret the results of this study." The authors of this paper are skeptical about the significant improvement because

(a) there is no control group, so the improvement might be due to the placebo effect or to the fact that many medical conditions improve over time.

(b) the P-value given was $P = 0.03$, which is too large to be convincing.

(c) the response variable might not have an exactly Normal distribution in the population.

15.30 Vigorous exercise helps people live several years longer (on the average). Whether mild activities like slow walking extend life is not clear. Suppose that the added life expectancy from regular slow walking is just 2 months. A statistical test is more likely to find a significant increase in mean life if

(a) it is based on a very large random sample.

(b) it is based on a very small random sample.

(c) The size of the sample doesn't have any effect on the significance of the test.

15.31 A medical experiment compared the herb echinacea with a placebo for preventing colds. One response variable was "volume of nasal secretions" (if you have a cold, you blow your nose a lot). Take the average volume of nasal secretions in people without colds to be $\mu = 1$. An increase to $\mu = 3$ indicates a cold. The significance level of a test of H_0: $\mu = 1$ versus H_a: $\mu > 1$ is

(a) the probability that the test rejects H_0 when $\mu = 1$ is true.

(b) the probability that the test rejects H_0 when $\mu = 3$ is true.

(c) the probability that the test fails to reject H_0 when $\mu = 3$ is true.

15.32 (**Optional**) The power of the test in the previous exercise against the specific alternative $\mu = 3$ is

(a) the probability that the test rejects H_0 when $\mu = 1$ is true.

(b) the probability that the test rejects H_0 when $\mu = 3$ is true.

(c) the probability that the test fails to reject H_0 when $\mu = 3$ is true.

15.33 (**Optional**) A laboratory scale is known to have a standard deviation of $\sigma = 0.001$ g in repeated weighings. Scale readings in repeated weighings are Normally distributed with mean equal to the true weight of the specimen. How many times must you weigh a specimen on this scale in order to get a margin of error no larger than ±0.0005 with 95% confidence?

(a) 4 times   (b) 15 times   (c) 16 times

CHAPTER 15 EXERCISES

15.34 **Deer mice.** In Exercise 14.35 (page 379) you gave a confidence interval based on the body lengths of 14 deer mice (*Peromyscus maniculatus*) from a rich forest habitat. Before you can trust your results, you would like more information about the data. What facts would you most like to know?

15.35 Color blindness in Africa. An anthropologist suspects that color blindness is less common in societies that live by hunting and gathering than in settled agricultural societies. He tests a number of adults in two populations in Africa, one of each type. The proportion of color-blind people is significantly lower ($P < 0.05$) in the hunter-gatherer population. What additional information would you want to help you decide whether you accept the claim about color blindness?

15.36 Prescription drugs. A 2005 Gallup Poll based on telephone interviews with 1011 adults nationwide found that 52% of adult Americans take prescription drugs. Gallup says:

> For results based on the total sample of national adults, one can say with 95 percent confidence that the maximum error attributable to sampling and other random effects is plus or minus 3 percentage points.

Give one example of a source of error in the poll result that is *not* included in this margin of error.

15.37 Sensitive questions. The National AIDS Behavioral Surveys found that 170 individuals in its random sample of 2673 adult heterosexuals said they had multiple sexual partners in the past year. That's 6.36% of the sample. Why is this estimate likely to be biased? Does the margin of error of a 95% confidence interval for the proportion of all adults with multiple partners allow for this bias?

15.38 Saccharin no longer listed as a possible human carcinogen. In its 11th edition, the U.S. Report on Carcinogens removed saccharin from the list of possible human carcinogens. Earlier carcinogenicity experiments had found rare instances of bladder tumors among male rats under very specific circumstances. Female rats never exhibited this sensitivity to saccharin, and neither did male or female mice. The toxicity was later explained by the unique metabolic pathway of certain male rats only. Years of observational studies in humans failed to show any sign of carcinogenicity from the regular consumption of saccharin.[11]

(a) What was the assumption made when saccharin was first placed in the list of possible human carcinogens based on the rat data? Why do you think this assumption was made?

(b) What simple condition is not met when applying animal research to humans? Explain why confirmation of the unique physiology of certain male rats led to the removal of saccharin from the list of possible human carcinogens.

15.39 What is significance good for? Which of the following questions does a test of significance answer? Briefly explain your replies.

(a) Is the sample or experiment properly designed?

(b) Is the observed effect due to chance?

(c) Is the observed effect important?

15.40 When to use pacemakers. A medical panel prepared guidelines for when cardiac pacemakers should be implanted in patients with heart problems. The panel reviewed a large number of medical studies to judge the strength of the evidence supporting each recommendation. For each recommendation, they ranked the evidence as level A (strongest), B, or C (weakest). Here, in scrambled order, are the panel's descriptions of the three levels of evidence.[12] Which is A, which B, and which C? Explain your ranking.

Layne Kennedy/CORBIS

Evidence was ranked as level _____ when data were derived from a limited number of trials involving comparatively small numbers of patients or from well-designed data analysis of nonrandomized studies or observational data registries.

Evidence was ranked as level _____ if the data were derived from multiple randomized clinical trials involving a large number of individuals.

Evidence was ranked as level _____ when consensus of expert opinion was the primary source of recommendation.

15.41 **Sex ratios.** Example 9.3 (page 221) gives the counts of male and female live births in the United States. In 2002, there were 2,058,000 male and 1,964,000 female live births. It makes no sense to calculate a confidence interval for the percent of male births in the United States. Why not?

15.42 **Island life.** When human settlers bring new plants and animals to an island, they may drive out native plants and animals. A study of 220 oceanic islands far from other land counted "exotic" (introduced from outside) bird species and the number of bird species that have become extinct since Europeans arrived on the islands. The study report says, "Numbers of exotic bird species and native bird extinctions are also positively correlated ($r = 0.62$, $n = 220$ islands, $P < 0.01$)."[13]

(a) The hypotheses concern the correlation for all oceanic islands, the population from which these 220 islands are a sample. Call this population correlation ρ. The hypotheses tested are

$$H_0: \rho = 0 \text{ versus } H_a: \rho > 0$$

In simple language, explain what $P < 0.01$ tells us.

(b) Before drawing practical conclusions from a P-value, we must look at the sample size and at the size of the observed effect. If the sample is large, effects too small to be important may be statistically significant. Do you think that is the case here? Why?

15.43 **Effect of an outlier.** Examining data on the exploration time it takes rats to find food in a maze, you find one outlier. Will this outlier have a greater effect on a confidence interval for mean exploration time if your sample is small or large? Why?

15.44 **What distinguishes schizophrenics?** A group of psychologists once measured 77 variables on a sample of schizophrenic people and a sample of people who were not schizophrenic. They compared the two samples using 77 separate significance tests. Two of these tests were significant at the 5% level. Suppose that there is in fact no difference in any of the variables between people who are and people who are not schizophrenic, so that all 77 null hypotheses are true.

(a) What is the probability that one specific test shows a difference significant at the 5% level?

(b) Why is it not surprising that 2 of the 77 tests were significant at the 5% level?

15.45 **Why are larger samples better?** Statisticians prefer large samples. Describe briefly the effect of increasing the size of a sample (or the number of subjects in an experiment) on each of the following:

(a) The margin of error of a 95% confidence interval.

(b) The P-value of a test when H_0 is false and all facts about the population remain unchanged as n increases.

(c) (Optional) The power of a fixed-level α test, when α, the alternative hypothesis, and all facts about the population remain unchanged.

15.46 Controlling diabetes. The Diabetes Control and Complications Trial (DCCT) is the most comprehensive diabetes study, conducted on 1441 volunteers with type 1 diabetes (insulin-dependent diabetes mellitus) in the United States and Canada. The study randomly assigned these volunteers to either the conventional therapy or an intensive therapy designed to tightly control blood glucose levels. Medical complications from diabetes (such as neuropathy, eye disease, and kidney disease) were then compared between the two groups.[14]

(a) One of the study findings is that intensive therapy significantly reduced the occurrence of clinical neuropathy by 60 percent (95% confidence interval, 38 to 74 percent). Why is giving a confidence interval more informative than giving a statement of significance supported only by a P-value?

(b) In fact, the study found that neuropathy occurred in about 9% of the conventional care group but in only about 3% of the intensively treated group. What additional information is provided here?

(c) The study also reported that the higher cost of intensive therapy would be offset by the reduction in medical expenses related to long-term medical complications. What are the statistical significance, effect size, and practical conclusions of this study? Explain briefly how each tells a different aspect of the findings.

The following exercises concern the optional material on planning studies for successful inference, choosing a sample size and power calculations.

15.47 Aspirin tablets: How large a sample? Suppose we know that the aspirin content of aspirin tablets in Example 14.6 (page 366) follows a Normal distribution with standard deviation $\sigma = 5$ mg.

(a) You must verify the aspirin content of tablets produced in a day to within ± 1 mg with 99% confidence. How large a sample of aspirin tablets from the daily production do you need?

(b) If you needed to produce only a 95% confidence interval with a margin of error no larger than 1 mg, how large a sample of aspirin tablets from the daily production would you need instead?

15.48 Deer mice. Deer mice (*Peromyscus maniculatus*) are small rodents native to North America. Their body lengths (excluding tail) are known to vary approximately Normally, with mean $\mu = 86$ mm and standard deviation $\sigma = 8$ mm.[15] You decide to study the effect of vitamin supplements on deer mice body length, but first you want to figure out how many deer mice you would need for successful inference about the mean μ in the population of all deer mice that would ever receive vitamin supplements. How many deer mice would be needed to estimate μ to within ± 2 mm with 95% confidence?

15.49 Neural mechanism of valium. Diazepam is the generic name of Valium, a common antidepressant and sedative. A study investigated the molecule's neural mechanism by comparing the effect of diazepam on sleep in 7 knock-in mice in which the α_2-GABA$_A$ receptor is diazepam-insensitive and in 8 wild-type control mice. The study found that diazepam reduced sleep latency in both mutant and wild-type mice, but the two mice types did not differ significantly.

The authors attributed this lack of significance to a large interindividual variability, resulting in low power (20%) of the significance test.[16]

(a) Explain in simple language why tests having low power often fail to give evidence against a null hypothesis even when the null hypothesis is really false.

(b) Which aspects of this experiment most likely contributed to a low test power?

15.50 **Deer mice, continued.** Continuing on Exercise 15.48, you are thinking further about the design of your experiment.

(a) How many deer mice would be needed to estimate μ to within ± 3 mm with 95% confidence? When planning your study, what would be the most important factor in deciding that you should aim for a 2-mm versus a 3-mm margin of error?

(b) How should you design your study so that you can truly make a conclusion about the possible effect of vitamin supplements on mouse body length?

15.51 **Treating knee pain.** Arthroscopic surgery is a minimally invasive surgical procedure on a joint, most commonly performed for persistent knee pain. Several uncontrolled, retrospective studies of the procedure reported pain relief among patients following the surgery.

(a) What can you conclude from these studies? What do they say about the benefits of arthroscopic surgery?

(b) A recent double-blind, randomized, controlled experiment found that outcomes after arthroscopic surgery are no better than those after a sham or placebo procedure.[17] How do you interpret this result? What does it say about the benefits of arthroscopic surgery?

(c) The controlled experiment was "designed to have 90 percent power, with a two-sided type I error of 0.04, to detect a moderate effect size (0.55) between the placebo group and the combined arthroscopic-treatment groups." How is this discussion of power important in interpreting the results of the controlled experiment? What do you conclude about the benefits of arthroscopic surgery in treating knee pain?

15.52 **Sweetening colas: calculating power** The cola maker of Example 14.5 wants to test at the 5% significance level the following hypotheses:

$$H_0: \mu = 0 \text{ versus } H_a: \mu > 0$$

Ten taste scores were used for the significance test. The distribution of taste scores is assumed to be roughly Normal with standard deviation $\sigma = 1$. We want to calculate the power of this test when the true mean sweetness loss is $\mu = 0.8$.

(a) What values of the z statistic would lead us to reject H_0? To what values of $\bar{x}$ do they correspond? Use the inverse Normal calculations to figure it out.

(b) When the true mean sweetness loss is $\mu = 0.8$, how often would we reject H_0? That is, what is the probability of obtaining sample averages like the ones defined in (a) when $\mu = 0.8$? Use Table B to calculate that probability. This probability is the power of your test against the alternative $\mu = 0.8$, the probability of rejecting H_0 when the alternative $\mu = 0.8$ is true.

15.53 **Sweetening colas: power for various sample sizes** In the previous exercise, you just calculated by hand the power of the cola sweetness test performed in Example 14.5 when the true mean sweetness loss is $\mu = 0.8$.

(a) Use the *Power of a Test* applet to find the power against the alternative $\mu = 0.8$ as in the previous exercise. Check that your answers are similar.

(b) Now use the applet to calculate the power of the test against the alternative $\mu = 0.8$ for samples of size $n = 5$, $n = 15$, and $n = 40$. How does sample size affect power?

(c) Using a sample size of $n = 10$, find with the applet the power of the test against the alternatives $\mu = 0.5$, $\mu = 1.0$, and $\mu = 1.5$. How does effect size affect power?

15.54 Power of a two-sided test. Power calculations for two-sided tests follow the same outline used for one-sided tests. Example 14.9 (page 373) presents a test of

$$H_0: \mu = 128 \text{ versus } H_a: \mu \neq 128$$

at the 5% level of significance. The company medical director failed to find significant evidence that the mean blood pressure of a population of executives differed from the national mean $\mu = 128$. The data used were an SRS of size 72 from a population with standard deviation $\sigma = 15$. The medical director now wonders if the test used would detect an important difference if one were present. What would be the power of this test against the alternative $\mu = 134$?

(a) The test in Example 14.9 rejects H_0 when $|z| \geq 1.96$. The test statistic z is

$$z = \frac{\bar{x} - 128}{15/\sqrt{72}}$$

Write the rule for rejecting H_0 in terms of the values of $\bar{x}$. (Because the test is two-sided, it rejects H_0 when $\bar{x}$ is either too large or too small.)

(b) Now find the probability that $\bar{x}$ takes values that lead to rejecting H_0 if the true mean is $\mu = 134$. This probability is the power.

(c) What is the probability that this test makes a Type II error when $\mu = 134$?

15.55 Power of a two-sided test, continued. Let's now use technology to perform the power calculations for the two-sided test of the previous exercise.

(a) Use the *Power of a Test* applet to find the power of the two-sided test against the alternative $\mu = 134$. Make sure that it is similar to what you calculated in the previous exercise.

(b) Use the applet to calculate the power of the test against the alternative $\mu = 122$. Can the test be relied on to detect a mean that differs from 128 by 6?

(c) If the alternative were farther from H_0, say $\mu = 136$, would the power be higher or lower than the values calculated in (a) and (b)?

15.56 Error probabilities. You read that a statistical test at significance level $\alpha = 0.05$ has power 0.78. What are the probabilities of Type I and Type II errors for this test?

15.57 Island life. Exercise 15.42 describes a study that tested the null hypothesis that there is 0 correlation between the number of exotic bird species on an island and the number of native bird extinctions. Describe in words what it means to make a Type I and a Type II error in this setting.

15.58 Power. You read that a statistical test at the $\alpha = 0.01$ level has probability 0.14 of making a Type II error when a specific alternative is true. What is the power of the test against this alternative?

Peter Lilja/Getty Images

From Exploration to Inference: Part II Review

Designs for producing data are essential parts of statistics in practice. Figures 16.1 and 16.2 display the big ideas visually. Random sampling and randomized comparative experiments are perhaps the most important statistical inventions of the 20th century. Both were slow to gain acceptance, and you will still see many voluntary response samples and uncontrolled experiments. You should now understand good techniques for producing data and also why bad techniques often produce worthless data. The deliberate use of chance in producing data is a central idea in statistics. It not only reduces bias but allows use of the laws of probability to analyze data. Fortunately, we need only some basic facts about probability in order to understand statistical inference.

Statistical inference draws conclusions about a population on the basis of sample data and uses probability to indicate how reliable the conclusions are. A confidence interval estimates an unknown parameter. A significance test shows how strong the evidence is for some claim about a parameter.

The probabilities in both confidence intervals and tests tell us what would happen if we used the same method for the interval or test very many times.

- A confidence level is the success rate of the method for a confidence interval. This is the probability that the method actually produces an interval that contains the unknown parameter. A 95% confidence interval gives a correct result 95% of the time when we use it repeatedly. That is, when we produce

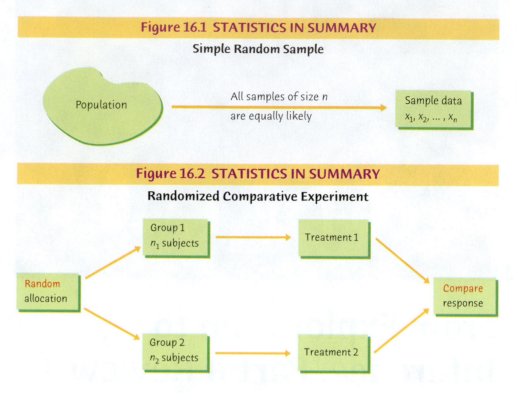

Figure 16.1 STATISTICS IN SUMMARY

Simple Random Sample

Figure 16.2 STATISTICS IN SUMMARY

Randomized Comparative Experiment

one 95% confidence interval, we do not know whether it gave a correct result or not, but we know that there is a 95% chance that it did.

- A P-value tells us how surprising the observed outcome would be if the null hypothesis were true. That is, P is the probability that the test would produce a result at least as extreme as the observed result if the null hypothesis really were true. Very surprising outcomes (small P-values) are good evidence that the null hypothesis is not true.

Figures 16.3 and 16.4 use the z procedures introduced in Chapter 14 to present in picture form the big ideas of confidence intervals and significance tests. These ideas are the foundation for the rest of this book. We will have much to say about many statistical methods and their use in practice. In every case, the basic reasoning of confidence intervals and significance tests remains the same.

PART II SUMMARY

Here are the most important skills you should have acquired from reading Chapters 7 through 15.

A. SAMPLING

1. Identify the population in a sampling situation.
2. Recognize bias due to voluntary response samples and other inferior sampling methods.
3. Use software or Table A of random digits to select a simple random sample (SRS) from a population.

Figure 16.3 STATISTICS IN SUMMARY

The Idea of a Confidence Interval

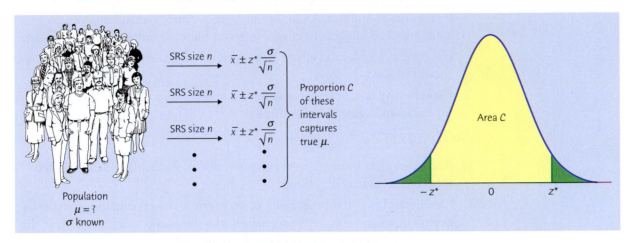

Figure 16.4 STATISTICS IN SUMMARY

The Idea of a Significance Test

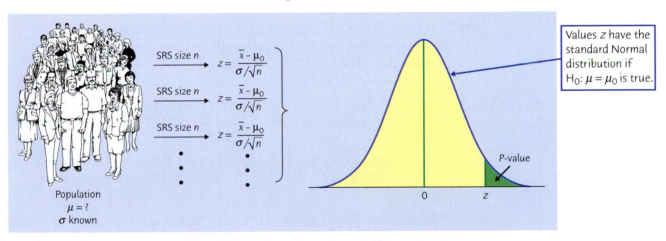

4. Recognize the presence of undercoverage and nonresponse as sources of error in a sample survey. Recognize the effect of the wording of questions on the responses.

5. Use random digits to select a stratified random sample from a population when the strata are identified.

B. EXPERIMENTS

1. Recognize whether a study is an observational study or an experiment.

2. Recognize bias due to confounding of explanatory variables with lurking variables in either an observational study or an experiment.

3. Identify the factors (explanatory variables), treatments, response variables, and individuals or subjects in an experiment.

4. Outline the design of a completely randomized experiment using a diagram like that in Figure 16.2. The diagram in a specific case should show the sizes of the groups, the specific treatments, and the response variable.

5. Use software or Table A of random digits to carry out the random assignment of subjects to groups in a completely randomized experiment.

6. Recognize the placebo effect. Recognize when the double-blind technique should be used.

7. Explain why randomized comparative experiments can give good evidence for cause-and-effect relationships.

C. PROBABILITY

1. Recognize that some phenomena are random. Probability describes the long-run regularity of random phenomena.

2. Understand that the probability of an event is the proportion of times the event occurs in very many repetitions of a random phenomenon. Use this idea of long-run proportion to think about probability.

3. Use basic probability rules to detect illegitimate assignments of probability: any probability must be a number between 0 and 1, and the total probability assigned to all possible outcomes must be 1.

4. Use basic probability rules to find the probabilities of events that are made up of other events. The probability that an event does not occur is 1 minus its probability. If two events are disjoint, the probability that one or the other occurs is the sum of their individual probabilities.

5. Find probabilities in a discrete probability model by adding the probabilities of their outcomes. Find probabilities in a continuous probability model as areas under a density curve.

6. Use the notation of random variables to make compact statements about random outcomes, such as $P(\overline{x} \leq 4) = 0.3$. Be able to interpret such statements.

D. GENERAL RULES OF PROBABILITY (Optional)

1. Use Venn diagrams to picture relationships among several events.

2. Use the general addition rule to find probabilities that involve overlapping events.

3. Understand the idea of independence. Judge when it is reasonable to assume independence as part of a probability model.

4. Use the multiplication rule for independent events to find the probability that all of several independent events occur.

5. Use the multiplication rule for independent events in combination with other probability rules to find the probabilities of complex events.

6. Understand the idea of conditional probability. Find conditional probabilities for individuals or outcomes chosen at random from a table of counts of possible outcomes.

7. Use the general multiplication rule to find $P(A \text{ and } B)$ from $P(A)$ and the conditional probability $P(B \mid A)$.

8. Use two-way tables or tree diagrams to organize several-stage probability models.

9. Understand the idea of Bayes's theorem. Use the formula or a tree diagram to find a conditional probability based on other, known probabilities.

E. DENSITY CURVES AND NORMAL DISTRIBUTIONS

1. Know that density curves represent probability distributions for continuous variables. The total area under a density curve is 1.

2. Know that any given area under a density curve represents both the proportion of all corresponding observations in a population and the probability of randomly selecting such observations from the population.

3. Approximately locate the median (equal-areas point) and the mean (balance point) on a density curve.

4. Recognize the shape of Normal curves and estimate by eye both the mean and the standard deviation from such a curve.

5. Use the 68–95–99.7 rule and symmetry to state what percent of the observations from a Normal distribution fall between two points when both points lie at the mean or one, two, or three standard deviations on either side of the mean.

6. Find the standardized value (z-score) of an observation. Interpret z-scores and understand that any Normal distribution becomes standard Normal $N(0, 1)$ when standardized.

7. Given that a variable has a Normal distribution with a stated mean μ and standard deviation σ, calculate the proportion/probability of values above a stated number, below a stated number, or between two stated numbers.

8. Given that a variable has a Normal distribution with a stated mean μ and standard deviation σ, calculate the point having a stated proportion/probability above it or below it.

F. DISCRETE DISTRIBUTIONS (Optional)

1. Recognize the binomial setting: a fixed number n of independent success-failure trials with the same probability p of success on each trial.

2. Recognize and use the binomial distribution of the count of successes in a binomial setting.

3. Use the binomial probability formula or software to find probabilities of events involving the count X of successes in a binomial setting for small values of n.

4. Find the mean and standard deviation of a binomial count X.

5. Recognize when you can use the Normal approximation to a binomial distribution. Use the Normal approximation to calculate probabilities that concern a binomial count X.

6. Recognize and use the Poisson distribution of the count of occurrences of a defined event in fixed, finite intervals of time or space.

7. Find the mean and standard deviation of a Poisson distribution.

8. Use the Poisson probability formula or software to find probabilities of events involving the count X of occurrences over a finite interval.

G. SAMPLING DISTRIBUTIONS

1. Identify parameters and statistics in a statistical study.

2. Recognize the fact of sampling variability: A statistic will take different values when you repeat a sample or experiment.

3. Interpret a sampling distribution as describing the values taken by a statistic in all possible repetitions of a sample or experiment under the same conditions.

4. Interpret the sampling distribution of a statistic as describing the probabilities of its possible values.

H. THE SAMPLING DISTRIBUTION OF A SAMPLE MEAN

1. Recognize when a problem involves the mean $\bar{x}$ of a sample. Understand that $\bar{x}$ estimates the mean μ of the population from which the sample is drawn.

2. Use the law of large numbers to describe the behavior of $\bar{x}$ as the size of the sample increases.

3. Find the mean and standard deviation of a sample mean $\bar{x}$ from an SRS of size n when the mean μ and standard deviation σ of the population are known.

4. Understand that $\bar{x}$ is an unbiased estimator of μ and that the variability of $\bar{x}$ about its mean μ gets smaller as the sample size increases.

5. Understand that $\bar{x}$ has approximately a Normal distribution when the sample is large (central limit theorem). Use this Normal distribution to calculate probabilities that concern $\bar{x}$.

I. THE SAMPLING DISTRIBUTION OF A SAMPLE PROPORTION

1. Recognize when a problem involves the proportion $\hat{p}$ of successes in a sample. Understand that $\hat{p}$ estimates the proportion p of successes in the population from which the sample is drawn.

2. Use the law of large numbers to describe the behavior of $\hat{p}$ as the size of the sample increases.

3. Find the mean and standard deviation of a sample proportion $\hat{p}$ from an SRS of size n when the population proportion p is known.

4. Understand that $\hat{p}$ is an unbiased estimator of p and that the variability of $\hat{p}$ about its mean p gets smaller as the sample size increases.

5. Understand that $\hat{p}$ has approximately a Normal distribution when the sample is large. Use this Normal distribution to calculate probabilities that concern $\hat{p}$.

J. CONFIDENCE INTERVALS

1. State in nontechnical language what is meant by "95% confidence" or other statements of confidence in statistical reports.

2. Know the four-step process (page 361) for any confidence interval.

3. Calculate a confidence interval for the mean μ of a Normal population with known standard deviation σ, using the formula $\bar{x} \pm z^* \sigma / \sqrt{n}$.

4. Understand how the margin of error of a confidence interval changes with the sample size and the level of confidence C.

5. Find the sample size required to obtain a confidence interval of specified margin of error m when the confidence level and other information are given.

6. Identify sources of error in a study that are *not* included in the margin of error of a confidence interval, such as undercoverage or nonresponse.

K. SIGNIFICANCE TESTS

1. State the null and alternative hypotheses in a testing situation when the parameter in question is a population mean μ.

2. Explain in nontechnical language the meaning of the P-value when you are given the numerical value of P for a test.

3. Know the four-step process (page 372) for any significance test.

4. Calculate the one-sample z test statistic and the P-value for both one-sided and two-sided tests about the mean μ of a Normal population.

5. Assess statistical significance at standard levels α, either by comparing P with α or by comparing z with standard Normal critical values.

6. Recognize that significance testing does not measure the size or importance of an effect. Explain why a small effect can be significant in a large sample and why a large effect can fail to be significant in a small sample.

7. Recognize that any inference procedure acts as if the data were properly produced. The z confidence interval and test require that the data be an SRS from the population.

REVIEW EXERCISES

Review exercises help you solidify the basic ideas and skills in Chapters 7 to 15.

16.1 Marijuana and driving. Questioning a sample of young people in New Zealand revealed a positive association between use of marijuana (cannabis) and traffic accidents caused by the members of the sample. Both cannabis use and accidents were measured by interviewing the young people themselves. The study report says, "It is unlikely that self reports of cannabis use and accident rates will be perfectly accurate."[1] Is the response bias likely to make the reported association stronger or weaker than the true association? Why?

Don't touch the plants

We know that confounding can distort inference. We don't always recognize how easy it is to confound data. Consider the innocent scientist who visits plants in the field once a week to measure their size. A study of six plant species found that one touch a week significantly increased leaf damage by insects in two species and significantly decreased damage in another species.

16.2 **California's endangered animals.** The California Department of Fish and Game publishes a list of the state's endangered animals. Here are the reptiles on the list:

Desert tortoise	Green sea turtle	Loggerhead sea turtle
Olive ridley sea turtle	Leatherback sea turtle	Barefoot banded gecko
Island night lizard	Alameda whipsnake	Coachella Valley fringe-toed lizard
Flat-tailed horned lizard	Southern rubber boa	
Giant garter snake	San Francisco garter snake	Blunt-nosed leopard lizard

Your class can't decide which 2 endangered reptiles to choose for special study, so you agree to choose an SRS from the list. Describe briefly three methods that could be used to select an SRS of two of these species.

16.3 **Elephants and bees.** Elephants sometimes damage crops in Africa. It turns out that elephants dislike bees. They recognize beehives in areas where they are common and avoid them. Can this be used to keep elephants away from trees? A group in Kenya placed active beehives in some trees and empty beehives in others. Will elephant damage be less in trees with hives? Will even empty hives keep elephants away?[2]

Peter Lilja/Getty Images

(a) Outline the design of an experiment to answer these questions using 72 acacia trees (be sure to also include a control group).

(b) Use software or the *Simple Random Sample* applet to choose the trees for the active-hive group, or Table A at line 137 to choose the first 4 trees in that group.

(c) What is the response variable in this experiment?

16.4 **Support groups for breast cancer.** Does participating in a support group extend the lives of women with breast cancer? There is no good evidence for this claim, but it was hard to carry out randomized comparative experiments because breast cancer patients believe that support groups help and want to be in one. When the first such experiment was finally completed, it showed that support groups have no effect on survival time. The experiment assigned 235 women with advanced breast cancer to two groups: 158 to "expressive group therapy" and 77 to a control group.[3]

(a) Outline the design of this experiment.

(b) Use software or the *Simple Random Sample* applet to choose the 77 members of the control group (list only the first 10), or use Table A at line 110 to choose the first 5 members of the control group.

16.5 **Informed consent.** The requirement that human subjects give their informed consent to participate in an experiment can greatly reduce the number of available subjects. For example, a study of new physical education (PE) curriculum incorporating health education asks the consent of parents for their children to be taught either the new or the standard PE curriculum. Many parents do not return the forms, so their children must continue to follow the standard curriculum. Why is it not correct to consider these children part of the control group along with children who are randomly assigned to the standard curriculum?

16.6 **Sample space.** A randomly chosen subject arrives for a study of exercise and fitness. Describe a sample space for each of the following. (In some cases, you may have some freedom in your choice of *S*.)

(a) The subject is either female or male.

(b) After 10 minutes on an exercise bicycle, you ask the subject to rate his or her effort on the Rate of Perceived Exertion (RPE) scale. The RPE scale ranges in whole-number steps from 6 (no exertion at all) to 20 (maximal exertion).

(c) You measure VO_2, the maximum volume of oxygen consumed per minute during exercise. VO_2 is generally between 2.5 and 6.1 liters per minute.

(d) You measure the maximum heart rate (beats per minute).

16.7 **How many in the house?** In government data, a household consists of all occupants of a dwelling unit. Here is the distribution of household size in the United States:

Number of persons	1	2	3	4	5	6	7
Probability	0.27	0.33	0.16	0.14	0.06	0.03	0.01

Choose an American household at random and let the random variable Y be the number of persons living in the household.

(a) Express "more than one person lives in this household" in terms of Y. What is the probability of this event?

(b) What is $P(2 < Y \leq 4)$?

(c) What is $P(Y \neq 2)$?

16.8 **The addition rule.** The addition rule for probabilities, $P(A \text{ or } B) = P(A) + P(B)$, is true only in some circumstances. Give (in words) an example of real-world events A and B for which this rule is not true.

16.9 **Weighing bean seeds.** Many biological measurements on the same species follow a Normal distribution quite closely. The weights of seeds of a variety of winged bean are approximately Normal with mean 525 milligrams (mg) and standard deviation 110 mg.

(a) What percent of seeds weigh more than 500 mg?

(b) If we discard the lightest 10% of all seeds, what is the smallest weight among the remaining seeds?

16.10 **An IQ test.** The Wechsler Adult Intelligence Scale (WAIS) is a common "IQ test" for adults. The distribution of WAIS scores for persons over 16 years of age is approximately Normal with mean 100 and standard deviation 15.

(a) What is the probability that a randomly chosen individual has a WAIS score of 105 or higher?

(b) What are the mean and standard deviation of the average WAIS score $\bar{x}$ for an SRS of 60 people?

(c) What is the probability that the average WAIS score of an SRS of 60 people is 105 or higher?

(d) Would your answers to any of (a), (b), or (c) be affected if the distribution of WAIS scores in the adult population were distinctly non-Normal?

16.11 **Distributions: means versus individuals.** The z confidence interval and test are based on the sampling distribution of the sample mean $\bar{x}$. Suppose that the distribution of the scores of young men on the National Assessment of

Educational Progress quantitative test is Normal with mean $\mu = 272$ and standard deviation $\sigma = 60$.

(a) You take an SRS of 100 young men. According to the 99.7 part of the 68–95–99.7 rule, what approximate range of scores do you expect to see in your sample?

(b) You look at many SRSs of size 100. About what range of sample mean scores $\bar{x}$ do you expect to see?

16.12 Distributions: larger samples. In the setting of the previous exercise, how many men must you sample to cut the range of values of $\bar{x}$ in half? This will also cut the margin of error of a confidence interval for μ in half. Do you expect the range of individual scores in the new sample to also be much less than that in a sample of size 100? Why?

16.13 Brains at work. When our brains store information, complicated chemical changes take place. In trying to understand these changes, researchers blocked some processes in brain cells taken from rats and compared these cells with a control group of normal cells. They say that "no differences were seen" between the two groups in four response variables. They give P-values 0.45, 0.83, 0.26, and 0.84 for these four comparisons.[4]

(a) Say clearly what P-value $P = 0.45$ says about the response that was observed.

(b) It isn't literally true that "no differences were seen." That is, the mean responses were not exactly alike in the two groups. Explain what the researchers mean when they give $P = 0.45$ and say "no difference was seen."

Use the following information for Exercises 16.14 to 16.17. The distribution of blood cholesterol level in the population of young men aged 20 to 34 years is close to Normal with standard deviation $\sigma = 41$ milligrams per deciliter (mg/dl).

16.14 Estimating blood cholesterol. You measure the blood cholesterol of 14 cross-country runners. The mean level is $\bar{x} = 172$ mg/dl. Assuming that σ is the same as in the general population, give a 90% confidence interval for the mean level μ among cross-country runners.

16.15 Testing blood cholesterol. The mean blood cholesterol level for all men aged 20 to 34 years is $\mu = 188$ mg/dl. We suspect that the mean for cross-country runners is lower.

(a) State hypotheses, find the test statistic, and give the P-value. Is the result significant at the $\alpha = 0.10$ level? At $\alpha = 0.05$? At $\alpha = 0.01$?

(b) Explain how you can use the confidence interval you computed in the previous exercise to test these hypotheses. What significance level does that 90% confidence interval yield when testing the hypotheses you stated in (a)? Remember that your test was one-sided.

16.16 Smaller margin of error. How large a sample is needed to cut the margin of error in Exercise 16.14 in half? How large a sample is needed to cut the margin of error to ±5 mg/dl?

16.17 More significant results. Suppose, hypothetically, that a larger sample of 56 cross-country runners yields the same mean level, $\bar{x} = 172$ mg/dl. Redo the test of hypotheses in Exercise 16.15. What is the P-value now? At which of the levels

$\alpha = 0.10$, $\alpha = 0.05$, and $\alpha = 0.01$ is the result significant? What general fact about significance tests does comparing your results here and in Exercise 16.15 illustrate?

16.18 Support groups for breast cancer, continued. Here are some of the results of the medical study described in Exercise 16.4. Women in the treatment group reported less pain ($P = 0.04$), but there was no significant difference between the groups in median survival time ($P = 0.72$). Explain carefully why $P = 0.04$ is evidence that the treatment *does* make a difference and why $P = 0.72$ means that there is no evidence that support groups prolong life.

SUPPLEMENTARY EXERCISES

Supplementary exercises apply the skills you have learned in ways that require more thought or more elaborate use of technology.

16.19 Gallup takes a poll. In August 2005, the Gallup organization asked a randomly selected national sample of 1004 adults whether they thought that "the federal government should—or should not—fund research that would use newly created stem cells obtained from human embryos." The poll was conducted through telephone interviews and found 56% answering "Yes (should)." Gallup added that "question wording and practical difficulties in conducting surveys can introduce error or bias into the findings of public opinion polls."[5]

(a) Is it reasonable to state that these results are representative of the American adult population? Explain your answer.

(b) Explain what "practical difficulties" were likely experienced during the conducting of this poll. What aspect of the "question wording" might bias responses? How might these affect the conclusions?

16.20 Sampling students. You want to investigate the attitudes of students at your school toward getting a yearly flu shot. You have a grant that will pay the costs of contacting about 500 students.

(a) Specify the exact population for your study. For example, will you include part-time students?

(b) Describe your sample design. Will you use a stratified sample?

(c) Briefly discuss the practical difficulties that you anticipate. For example, how will you contact the students in your sample?

16.21 The placebo effect. A survey of physicians found that some doctors give a placebo to a patient who complains of pain for which the physician can find no cause. If the patient's pain improves, these doctors conclude that it had no physical basis. The medical school researchers who conducted the survey claimed that these doctors do not understand the placebo effect. Why?

16.22 Alcohol and mortality. It appears that people who drink alcohol in moderation have lower death rates than either people who drink heavily or people who do not drink at all. The protection offered by moderate drinking is concentrated among people over 50 and on deaths from heart disease. The Nurses' Health Study played an essential role in establishing these facts for women. This part of the study followed 85,709 female nurses for 12 years, during which time 2658 of the subjects died. The nurses completed a questionnaire that described their diet, including their use of alcohol. They were reexamined every two years.

Conclusion: "As compared with nondrinkers and heavy drinkers, light-to-moderate drinkers had a significantly lower risk of death."[6]

(a) Was this study an experiment? Explain your answer.

(b) What does "significantly lower risk of death" mean in simple language?

(c) Suggest some lurking variables that might be confounded with how much a person drinks. The investigators used advanced statistical methods to adjust for many such variables before concluding that the moderate drinkers really have a lower risk of death.

16.23 Making french fries. Few people want to eat discolored french fries. Potatoes are kept refrigerated before being cut for french fries to prevent spoiling and preserve flavor. But immediate processing of cold potatoes causes discoloring due to complex chemical reactions. The potatoes must therefore be brought to room temperature before processing. Design an experiment in which tasters will rate the color and flavor of french fries prepared from several groups of potatoes. The potatoes will be fresh picked or stored for a month at room temperature or stored for a month refrigerated. They will then be sliced and cooked either immediately or after an hour at room temperature.

(a) What are the factors and their levels, the treatments, and the response variables?

(b) Describe and outline the design of this experiment.

(c) It is efficient to have each taster rate fries from all treatments. How will you use randomization in presenting fries to the tasters?

Use the following information for Exercises 16.24 to 16.26. The level of pesticides found in the blubber of whales is a measure of pollution of the oceans by runoff from land and can also be used to identify different populations of whales. A sample of 8 male minke whales in the West Greenland area of the North Atlantic found the mean concentration of the insecticide dieldrin to be $\overline{x} = 357$ nanograms per gram (ng/g) of blubber.[7] Suppose that the concentration in all such whales varies Normally with standard deviation $\sigma = 50$ ng/g.

16.24 Pesticides in whale blubber. Use a 95% confidence interval to estimate the mean dieldrin level in whale blubber. Follow the four-step process for confidence intervals (page 361) in your work.

16.25 Testing pesticide levels. The Food and Drug Administration regulates the amount of dieldrin in raw food. For some foods, no more than 100 ng/g is allowed. Is there good evidence that the mean concentration in whale blubber is above this level? Follow the four-step process for significance tests (page 372) in your work.

16.26 Other confidence levels. Give an 80% confidence interval and a 90% confidence interval for the mean concentration of dieldrin in the whale population. What general fact about confidence intervals do the margins of error of your three intervals illustrate?

16.27 Birth weight and IQ: estimation. Infants weighing less than 1500 grams at birth are classed as "very low birth weight." Low birth weight carries many risks. One study followed 113 male infants with very low birth weight to adulthood. At age 20, the mean IQ score for these men was $\overline{x} = 87.6$.[8] IQ scores vary Normally with standard deviation $\sigma = 15$. Give a 95% confidence interval for the mean IQ

Pete Atkinson/Getty Images

score at age 20 for all very-low-birth-weight males. Use the four-step process for confidence intervals (page 361) as a guide.

16.28 Birth weight and IQ: testing. IQ tests are scaled so that the mean score in a large population should be $\mu = 100$. We suspect that the very-low-birth-weight population has mean score less than 100. Does the study described in the previous exercise give good evidence that this is true? Use the four-step process for significance tests (page 372) as a guide.

16.29 Birth weight and IQ: causation? Very-low-birth-weight babies are more likely to be born to unmarried mothers and to mothers who did not complete high school.

(a) Explain why the study of Exercise 16.27 was not an experiment.

(b) Explain clearly why confounding prevents us from concluding that very low birth weight in itself reduces adult IQ.

16.30 Normal body temperature? Suppose that body temperature varies Normally with standard deviation 0.7 degree. Here are the daily average body temperatures (degrees Fahrenheit) for 20 healthy adults.[9]

```
 98.74   98.83   96.80   98.12   97.89   98.09   97.87   97.42   97.30   97.84
100.27   97.90   99.64   97.88   98.54   98.33   97.87   97.48   98.92   98.33
```

(a) Do these data give evidence that the mean body temperature for all healthy adults is not equal to the traditional 98.6 degrees? Follow the four-step process for significance tests (page 372).

(b) Estimate mean body temperature with 90% confidence. Follow the four-step process for confidence intervals (page 361).

16.31 The first kid has higher IQ. Does the birth order of a family's children influence their IQ scores? A careful study of 241,310 Norwegian 18- and 19-year-olds found that firstborn children scored 2.3 points higher on the average than second children in the same family. This difference was highly significant ($P < 0.001$). A commentator said,[10] "One puzzle highlighted by these latest findings is why certain other within-family studies have failed to show equally consistent results. Some of these previous null findings, which have all been obtained in much smaller samples, may be explained by inadequate statistical power."

(a) Explain in simple language why tests having low power often fail to give evidence against a null hypothesis even when the hypothesis is really false.

(b) A population mean IQ is typically 100. For comparison, the Norwegian study found an effect size of 2.3 points. Is that a large effect size?

(c) The study examined 18- and 19-year-olds. By definition, firstborn children are older than second children in the same family. What confounding variable likely plays a role in the findings of this study?

16.32 Low power? (Optional) It appears that eating oat bran lowers cholesterol slightly. At a time when oat bran was something of a fad, a paper in the *New England Journal of Medicine* found that it had no significant effect on cholesterol.[11] The paper reported a study with just 20 subjects. Letters to the journal denounced publication of a negative finding from a study with very low power. Explain why lack of significance in a study with low power gives no reason to accept the null hypothesis that oat bran has no effect.

16.33 Type I and Type II errors (Optional). Exercise 16.28 asks for a significance test of the null hypothesis that the mean IQ of very-low-birth-weight male babies is 100 against the alternative hypothesis that the mean is less than 100. State in words what it means to make a Type I error and a Type II error in this setting.

OPTIONAL EXERCISES

These exercises concern the optional material in Chapters 10 and 12.

16.34 Smoking and social class. As the dangers of smoking have become more widely known, clear class differences in smoking have emerged. British government statistics classify adult men by occupation as "managerial and professional" (30% of the population), "intermediate" (29%) or "routine and manual" (34%). A survey finds that 20% of men with managerial and professional occupations smoke, 29% of the intermediate group smoke, and 38% in routine and manual occupations smoke.[12]

(a) Use a tree diagram to find the percent of all adult British men who smoke.

(b) Find the percent of male smokers who have routine and manual occupations. (Start by expressing this as a conditional probability.) This information is used in planning anti-smoking campaigns.

16.35 Comparing wine tasters. Two wine tasters rate each wine they taste on a scale from 1 to 5. From data on their ratings of a large number of wines, we obtain the following probabilities for both tasters' ratings of a randomly chosen wine:

	Taster 2				
Taster 1	1	2	3	4	5
1	0.03	0.02	0.01	0.00	0.00
2	0.02	0.08	0.05	0.02	0.01
3	0.01	0.05	0.25	0.05	0.01
4	0.00	0.02	0.05	0.20	0.02
5	0.00	0.01	0.01	0.02	0.06

(a) What is the probability that Taster 1 rates a wine higher than Taster 2? What is the probability that Taster 2 rates a wine higher than Taster 1?

(b) If Taster 1's rating for a wine is 3, what is the conditional probability that Taster 2's rating is higher than 3?

16.36 Life tables. The National Center for Health Statistics produces a "life table" for the American population. For each year of age, the table gives the probability that a randomly chosen U.S. resident will die during that year of life. These are *conditional* probabilities, given that the person lived to the birthday that marks the beginning of the year. Here is an excerpt from the table:

Year of life	Probability of death
51	0.00439
52	0.00473
53	0.00512
54	0.00557
55	0.00610

What is the probability that a person who lives to age 50 (the beginning of the 51st year) will live to age 55?

16.37 Cystic fibrosis. Cystic fibrosis is a lung disorder that often results in death. It is inherited but can be inherited only if both parents are carriers of an abnormal gene. In 1989, the CF gene that is abnormal in carriers of cystic fibrosis was identified. The probability that a randomly chosen person of European ancestry carries an abnormal CF gene is 1/25. (The probability is less in other ethnic groups.) The CF20m test detects most but not all harmful mutations of the CF gene. The test is positive for 90% of people who are carriers. It is (ignoring human error) never positive for people who are not carriers. What is the probability that a randomly chosen person of European ancestry tests positive?

16.38 Teenage drivers. An insurance company has the following information about drivers aged 16 to 18: 20% are involved in accidents each year; 10% in this age group are A students; among those involved in an accident, 5% are A students.

(a) Let A be the event that a young driver is an A student and C the event that a young driver is involved in an accident this year. State the information given in terms of probabilities and conditional probabilities for the events A and C.

(b) What is the probability that a randomly chosen young driver is an A student and is involved in an accident?

16.39 Teenage drivers, continued. Use your work from the previous exercise to find the percent of A students who are involved in accidents. (Start by expressing this as a conditional probability.)

16.40 Freckles and sun exposure. Researchers in Germany examined the link between pigmentation (for example, hair color, eye color, freckles) and reaction to sun exposure. After examining the reaction to midday summer sun exposure for thousands of Caucasian children, they found the following distribution in their sample:[13]

Reaction type to sun	No freckles	Freckles
(I) Always reddens, never tans	79	73
(II) Always reddens, slight tan	581	367
(III) Sometimes reddens, always tans	1025	324
(IV) Never reddens, always tans	1022	135

Rick Barrentine/Corbis

(a) Draw a tree diagram to show the possible outcomes for a Caucasian child in this study. Let's use the results of this study to describe a randomly chosen Caucasian child in Germany.

(b) What is the probability that a child has freckles? What is the probability that a child has type I reaction to sun exposure (always reddens, never tans)?

(c) What is the conditional probability that a child has type I reaction to sun exposure, knowing that this child has freckles? What is the conditional probability that a child has freckles, knowing that this child has type I reaction to sun exposure?

16.41 Family planning. Many parents wish to have at least 1 boy and 1 girl. For simplicity, let's ignore twins and other multiple births and assume that successive births are independent events with a 50-50 chance of that the baby is a boy or a girl.

(a) What is the distribution of the number of boys in a family with 5 children?

(b) What is the probability that a family with 5 children has 5 boys? What is the probability that a family with 5 children has 4 boys and 1 girl?

(c) What is the conditional probability that a couple would have a baby girl next, knowing that they already have 4 boys? Explain how and why this probability is different from the probabilities you calculated in (b).

16.42 **Many tests.** Long ago, a group of psychologists carried out 77 separate significance tests and found that 2 were significant at the 5% level. Suppose that these tests are independent of each other. (In fact, they were not independent, because all involved the same subjects.) If all of the null hypotheses are true, each test has probability 0.05 of being significant at the 5% level. Use the binomial distribution to find the probability that 2 or more of the tests are significant at that level.

16.43 **Brain injuries.** The state of Florida reports that 26% of individuals newly diagnosed with a brain injury are children. A sample survey is designed to interview an SRS of 500 patients newly diagnosed with a brain injury.

(a) What is the actual distribution of the number X in the sample who are children?

(b) What is the probability that 100 or fewer of the patients in the sample are children? 150 or fewer? (Use software or a suitable approximation.)

EESEE CASE STUDIES

The Electronic Encyclopedia of Statistical Examples and Exercises (EESEE) is available on the text CD and Web site. These more elaborate stories, with data, provide settings for longer case studies. Here are some suggestions for EESEE stories that apply the ideas you have learned in Chapters 7 through 15.

16.44 **Anecdotes of Bias.** Answer all of the questions posed about these incidents. (Cautions about sample surveys.)

16.45 **Checkmating and Reading Skills.** Respond to Questions 1 and 3 for this story. (Sampling, data analysis.)

16.46 **Surgery in a Blanket.** Write a response to Questions 1 and 2. (Design of experiments.)

16.47 **Visibility of Highway Signs.** Answer Questions 1, 2, and 3(a) for this study. (Design of experiments, data analysis.)

16.48 **Anecdotes of Significance Testing.** Answer all three questions. (Interpreting *P*-values.)

16.49 **Blinded Knee Doctors.**

(a) Outline the design of this experiment.

(b) You read that "the patients, physicians, and physical therapists were blinded" during the study. What does this mean?

(c) You also read that "the pain scores for Group A were significantly lower than those for Group C but not significantly lower than Group B." What does this mean? What does this finding lead you to conclude about the use of nonsteroidal anti-inflammatory drugs?

Juniors Bildarchiv/Alamy

PART III

INFERENCE ABOUT VARIABLES

INFERENCE ABOUT RELATIONSHIPS

PART III: Inference about Variables

With the principles in hand, we proceed to practice, that is, to inference in fully realistic settings. In the remaining chapters of this book, you will meet many of the most commonly used statistical procedures. We have grouped these procedures into two classes. The first five chapters concern inference about the distribution of a single variable and inference for comparing the distributions of two variables. In Chapters 17 and 18, we analyze data on quantitative variables. Chapters 19 through and 21 concern categorical variables, so that inference begins with counts and proportions of outcomes. The last three chapters deal with inference for relationships among variables. In Chapter 22, both variables are categorical, with data given as a two-way table of counts of outcomes. Chapter 23 considers inference in the setting of regressing a quantitative response variable on a quantitative explanatory variable. In Chapter 24 we meet methods for comparing the mean response in more than two groups. Here, the explanatory variable (group) is categorical and the response variable is quantitative. Chapter 25 reviews this part of the text, along with more comprehensive exercises.

With greater data complexity comes greater reliance on technology. In the last three chapters of Part III you will more often be interpreting the output of statistical software or using software yourself. You can do the calculations needed in Chapter 22 without software or a specialized calculator. In Chapters 23 and 24, the pain is too great and the contribution to learning too small. Fortunately, you can grasp the ideas without step-by-step arithmetic. These chapters introduce elaborate methods on the foundation we have laid without introducing fundamentally new concepts.

The four-step process for approaching a statistical problem can guide much of your work in all Part III chapters. You should review the outlines of the four-step process for a confidence interval (page 361) and for a test of significance (page 372). The statement of an exercise usually does the *State* step for you, leaving the *Formulate, Solve,* and *Conclude* steps for you to complete. We recommend that you first summarize the *State* step in your own words to help organize your thinking. Many examples and exercises in these chapters involve both carrying out inference and thinking about inference in practice. Remember that any inference method is useful only under certain conditions, and that you must judge these conditions before rushing to inference.

430

David A. Northcott/CORBIS

Inference about a Population Mean

This chapter describes confidence intervals and significance tests for the mean μ of a population. We used the z procedures in this setting to introduce the ideas of confidence intervals and tests. Now we discard the unrealistic condition that we know the population standard deviation σ and present procedures for practical use. We also pay more attention to the real-data setting of our work. The details of confidence intervals and tests change only slightly when you don't know σ. More important, you can interpret your results just as before. To emphasize this, Examples 17.2 and 17.3 repeat the most important examples from Chapter 14.

Conditions for inference

Confidence intervals and tests of significance for the mean μ of a Normal population are based on the sample mean $\overline{x}$. Confidence levels and P-values are probabilities calculated from the sampling distribution of $\overline{x}$. Here are the conditions needed for realistic inference about a population mean.

> **CONDITIONS FOR INFERENCE ABOUT A MEAN**
>
> - We can regard our data as a **simple random sample** (SRS) from the population. This condition is very important.
> - Observations from the population have a **Normal distribution** with mean μ and standard deviation σ. Both μ and σ are unknown parameters. In practice, inference procedures can accommodate some deviations from the Normality condition when the sample is large enough.

There is another condition that applies to all of the inference methods in this book: *The population must be much larger than the sample*, say at least 20 times as large.[1] All of our examples and exercises satisfy this condition. Practical settings in which the sample is a large part of the population are rather special, and we will not discuss them.

When the conditions for inference are satisfied, the sample mean $\overline{x}$ has the Normal distribution with mean μ and standard deviation $\sigma/\sqrt{n}$. Because we don't know σ, we estimate it by the sample standard deviation s. We then estimate the standard deviation of $\overline{x}$ by $s/\sqrt{n}$. This quantity is called the *standard error* of the sample mean $\overline{x}$.

> **STANDARD ERROR**
>
> When the standard deviation of a statistic is estimated from data, the result is called the **standard error** of the statistic. The standard error of the sample mean $\overline{x}$ is $s/\sqrt{n}$.

APPLY YOUR KNOWLEDGE

17.1 Neuronal activity. Neurons fire action potentials ("spikes") at different frequencies under different circumstances. A neuron is recorded in an awake-behaving monkey looking at a luminous dot 40 centimeters (cm) away to determine the neuron's baseline activity. Over the course of 72 trials randomly interspaced with other conditions, that neuron is found to have an average baseline activity of $\overline{x} = 27.8$ spikes per second with standard deviation $s = 32.2$ spikes per second. What is the standard error of the mean?

17.2 Rats eating oat bran. In a study of the effect of diet on cholesterol, rats were fed several different diets.[2] One diet had 5% added fiber from oat bran. The study report gives results in the form "mean plus or minus the standard error of the mean." This form is very common in scientific publications. For the 6 rats fed this diet, blood cholesterol levels (in milligrams per deciliter [mg/dl]) were 89.01 ± 5.36. What are $\overline{x}$ and s for these 6 rats?

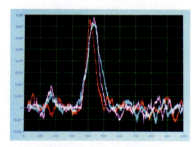

Donald Pye/Alamy

The *t* distributions

If we knew the value of σ, we would base confidence intervals and tests for μ on the one-sample z statistic

$$z = \frac{\overline{x} - \mu}{\sigma/\sqrt{n}}$$

This z statistic has the standard Normal distribution $N(0, 1)$. In practice, we don't know σ, so we substitute the standard error $s/\sqrt{n}$ of $\overline{x}$ for its standard deviation $\sigma/\sqrt{n}$. The statistic that results does not have a Normal distribution. It has a distribution that is new to us, called a *t distribution*.

Better statistics, better beer

The *t* distribution and the *t* inference procedures were invented by William S. Gosset (1876–1937). Gosset worked for the Guinness brewery to make better beer. He used his new *t* procedures to find the best varieties of barley and hops. Gosset's statistical work helped him become head brewer, a more interesting title than professor of statistics. Because of Guinness's strict nondisclosure clause, Gosset published his statistical work under the pen name "Student." You will often see the *t* distribution called "Student's *t*" in his honor.

THE ONE-SAMPLE *t* STATISTIC AND THE *t* DISTRIBUTIONS

Draw an SRS of size n from a large population that has the Normal distribution with mean μ and standard deviation σ. The **one-sample t statistic**

$$t = \frac{\overline{x} - \mu}{s/\sqrt{n}}$$

has the **t distribution** with $n - 1$ degrees of freedom.

The t statistic has the same interpretation as any standardized statistic: It says how far $\overline{x}$ is from its mean (μ) in standard deviation units ($s/\sqrt{n}$).

Sample standard deviations obtained with larger samples are better estimates of the unknown population standard deviations σ. This is reflected in the fact that there is a different t distribution for each sample size. We specify a particular t distribution by giving its **degrees of freedom (df).** When making an inference about a single population mean μ, the t statistic follows a t distribution with $n - 1$ degrees of freedom.[3] (Other inference methods use different degrees of freedom.) We will write the t distribution with $n - 1$ degrees of freedom as $t(n - 1)$ for short.

degrees of freedom

Figure 17.1 compares the density curves of the standard Normal distribution and of the t distributions with 2 and 9 degrees of freedom. The figure illustrates these facts about the t distributions:

- The density curves of the t distributions are similar in shape to the standard Normal curve. They are symmetric about 0, single-peaked, and bell-shaped.

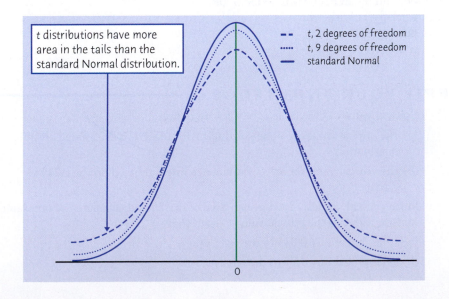

t distributions have more area in the tails than the standard Normal distribution.

- - - t, 2 degrees of freedom
..... t, 9 degrees of freedom
— standard Normal

FIGURE 17.1 Density curves for the t distributions with 2 and 9 degrees of freedom and for the standard Normal distribution. All are symmetric with center 0. The t distributions are somewhat more spread out.

- The spread of the t distributions is a bit greater than that of the standard Normal distribution. The t distributions in Figure 17.1 have more probability in the tails and less in the center than does the standard Normal distribution. This is true because substituting the estimate s for the fixed parameter σ introduces more variation into the statistic. (Therefore, inference will be generally less precise.)

- As the degrees of freedom increase, the t density curve approaches the $N(0, 1)$ curve ever more closely. This happens because s estimates σ more accurately as the sample size increases. So using s in place of σ causes little extra variation when the sample is large.

Table C in the back of the book gives critical values for the t distributions. Each row in the table contains critical values for the t distribution whose degrees of freedom appear at the left of the row. For convenience, we label the table entries both by the confidence level C (in percent) required for confidence intervals and by the one-sided and two-sided P-values for each critical value. You have already used the standard Normal critical values z^* in the bottom row of Table C. By looking down any column, you can check that the t critical values approach the Normal values as the degrees of freedom increase but that the t^* values are always larger than the corresponding z^*. It is the price we pay for replacing the unknown σ with s. As in the case of the Normal table, statistical software makes Table C unnecessary.

EXAMPLE 17.1   *t critical values*

Figure 17.1 shows the density curve for the t distribution with 9 degrees of freedom. What point on this distribution has probability 0.05 to its right? In Table C, look in the df $= 9$ row above one-sided P-value 0.05 and you will find that this critical value is $t^* = 1.833$. To use software, enter the degrees of freedom and the probability you want to the *left*, 0.95 in this case. Here is Minitab's output:

```
Student's t distribution with 9 DF
P(X<=x)        x
  0.95  1.83311
```

APPLY YOUR KNOWLEDGE

17.3   **Critical values.** Use Table C or software to find

   (a) the critical value for a one-sided test with level $\alpha = 0.05$ based on the $t(5)$ distribution.

   (b) the critical value for a 98% confidence interval based on the $t(21)$ distribution.

17.4   **More critical values.** You have an SRS of size 25 and calculate the one-sample t statistic. What is the critical value t^* such that

   (a) t has probability 0.025 to the right of t^*?

   (b) t has probability 0.75 to the left of t^*?

The one-sample *t* confidence interval

To analyze samples from Normal populations with unknown σ, just replace the standard deviation $\sigma/\sqrt{n}$ of $\bar{x}$ by its standard error $s/\sqrt{n}$ in the z procedures of Chapter 14. The confidence interval and test that result are *one-sample t procedures*. Critical values and P-values come from the t distribution with $n - 1$ degrees of freedom. The one-sample t procedures are similar in both reasoning and computational detail to the z procedures.

THE ONE-SAMPLE *t* CONFIDENCE INTERVAL

Draw an SRS of size n from a large population having unknown mean μ. A level C **confidence interval for μ** is

$$\bar{x} \pm t^* \frac{s}{\sqrt{n}}$$

where t^* is the critical value for the $t(n - 1)$ density curve with area C between $-t^*$ and t^*. This interval is exact when the population distribution is Normal and is approximately correct for large n in other cases.

— EXAMPLE 17.2   *Healing of skin wounds* ——————

Let's look again at the biological study we saw in Example 14.3. We follow the four-step process for a confidence interval, outlined on page 361.

STATE: Biologists studying the healing of skin wounds measured the rate at which new cells closed a razor cut made in the skin of an anesthetized newt. Here are data from 18 newts, measured in micrometers (millionths of a meter) per hour (μm/h):[4]

$$\begin{array}{ccccccccc}
29 & 27 & 34 & 40 & 22 & 28 & 14 & 35 & 26 \\
35 & 12 & 30 & 23 & 18 & 11 & 22 & 23 & 33
\end{array}$$

This is one of several sets of measurements made under different conditions. We want to estimate the mean rate for comparison with rates under other conditions.

FORMULATE: We will estimate the mean rate μ for all newts of this species by giving a 95% confidence interval.

SOLVE: We must first check the conditions for inference.

- As in Chapter 14 (page 361), we are willing to regard these newts as an SRS from their species.
- The stemplot in Figure 17.2 does not suggest any strong departures from Normality.

We can proceed to calculation. For these data,

$$\bar{x} = 25.67 \text{ and } s = 8.324$$

The degrees of freedom are $n - 1 = 17$. From Table C we find that for 95% confidence, $t^* = 2.110$. The confidence interval is

$$\begin{aligned}
\bar{x} \pm t^* \frac{s}{\sqrt{n}} &= 25.67 \pm 2.110 \frac{8.324}{\sqrt{18}} \\
&= 25.67 \pm 4.14 \\
&= 21.53 \text{ to } 29.81 \text{ micrometers per hour}
\end{aligned}$$

David A. Northcott/CORBIS

$$\begin{array}{r|l}
1 & 1\ 2\ 4 \\
1 & 8 \\
2 & 2\ 2\ 3\ 3 \\
2 & 6\ 7\ 8\ 9 \\
3 & 0\ 3\ 4 \\
3 & 5\ 5 \\
4 & 0
\end{array}$$

FIGURE 17.2 Stemplot of the healing rates in Example 17.2.

CONCLUDE: We are 95% confident that the mean healing rate for all newts of this species is between 21.53 and 29.81 micrometers per hour.

Our work in Example 17.2 is very similar to what we did in Example 14.3 on page 361. To make the inference realistic, we replaced the assumed $\sigma = 8$ with the $s = 8.324$ calculated from the data and replaced the standard Normal critical value $z^* = 1.960$ with the t critical value $t^* = 2.110$.

The one-sample t confidence interval has the form

$$\text{estimate} \pm t^*\text{SE}_{\text{estimate}}$$

where "SE" stands for "standard error." We will encounter a number of confidence intervals that have this common form. In Example 17.2, the estimate is the sample mean $\overline{x}$, and its standard error is

$$\text{SE}_{\overline{x}} = \frac{s}{\sqrt{n}}$$
$$= \frac{8.324}{\sqrt{18}} = 1.962$$

Software will find $\overline{x}$, s, $\text{SE}_{\overline{x}}$, and the confidence interval from the data. Figure 17.5 on page 440 displays typical software output for Example 17.2.

APPLY YOUR KNOWLEDGE

17.5 Critical values. What critical value t^* from Table C would you use for a confidence interval for the mean of the population in each of the following situations?

(a) A 95% confidence interval based on $n = 10$ observations.

(b) A 99% confidence interval from an SRS of 20 observations.

(c) An 80% confidence interval from a sample of size 7.

17.6 Rats eating oat bran. Exercise 17.2 gave the summary data (mean 89.01, standard error 5.36 mg/dl) for the cholesterol levels of 6 rats fed a diet enriched in fiber from oat bran. Cholesterol levels are usually approximately Normal, and we can regard these 6 rats as an SRS of the population of lab rats fed a diet enriched in oat fiber. Give the 95% confidence interval for the mean cholesterol level μ in this population.

17.7 Hand size. A clever way to determine hand size in three dimensions is to measure the volume (in milliliters [ml]) of water displaced when the hand is dipped in a water container. A study used this method to gather the hand volumes of 12 male university students. Here are the measurements:[5]

| 400 | 360 | 420 | 520 | 460 | 350 | 500 | 420 | 450 | 430 | 395 | 400 |

(a) We can consider this an SRS of all male university students. Make a stemplot. Is there any sign of major deviation from Normality?

(b) Give a 95% confidence interval for the mean hand size.

17.8 Trout habitat. A master's student in wildlife management studied trout habitat in the upper Shavers Fork watershed in West Virginia. The springtime water pH of 29 randomly selected tributary sample sites were found to have the following values:[6]

6.2	6.3	5.0	5.8	4.6	4.7	4.7	5.4	6.2	6.0
5.4	5.9	6.2	6.1	6.0	6.3	6.2	5.8	6.2	6.3
6.3	6.3	6.4	6.5	6.6	6.1	6.3	4.4	6.7	

Use a 90% confidence interval to estimate the mean springtime water pH of the tributary water basin around the Shavers Fork watershed. Follow the four-step process as illustrated in Example 17.2.

Karl Weatherly/Getty Images

The one-sample *t* test

Like the confidence interval, the *t* test is very similar to the *z* test we studied earlier.

THE ONE-SAMPLE *t* TEST

Draw an SRS of size n from a large population having unknown mean μ. To **test the hypothesis $H_0: \mu = \mu_0$**, compute the **one-sample *t* statistic**

$$t = \frac{\overline{x} - \mu_0}{s/\sqrt{n}}$$

In terms of a variable T having the $t(n-1)$ distribution, the *P*-value for a test of H_0 against

$H_a: \mu > \mu_0$ is $P(T \geq t)$

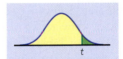

$H_a: \mu < \mu_0$ is $P(T \leq t)$

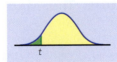

$H_a: \mu \neq \mu_0$ is $2P(T \geq |t|)$

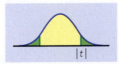

These *P*-values are exact if the population distribution is Normal; they are approximately correct for large n in other cases.

EXAMPLE 17.3 Sweetening colas

Here is a more realistic analysis of the cola-sweetening study from Example 14.5. We follow the four-step process for a significance test, outlined on page 372.

STATE: Cola makers test new recipes for loss of sweetness during storage. Trained tasters rate the sweetness before and after storage. Here are the sweetness losses (sweetness before storage minus sweetness after storage) found by 10 tasters for one new cola recipe:

$$2.0 \quad 0.4 \quad 0.7 \quad 2.0 \quad -0.4 \quad 2.2 \quad -1.3 \quad 1.2 \quad 1.1 \quad 2.3$$

Are these data good evidence that the cola lost sweetness?

FORMULATE: Tasters vary in their perception of sweetness loss. So we ask the question in terms of the mean loss μ for a large population of tasters. The null hypothesis is "no loss," and the alternative hypothesis says "there is a loss."

```
-1 | 3
-0 | 4
 0 | 4 7
 1 | 1 2
 2 | 0 0 2 3
```

FIGURE 17.3 Stemplot of the sweetness losses in Example 17.3.

$$H_0: \mu = 0$$
$$H_a: \mu > 0$$

SOLVE: First check the conditions for inference. As before, we are willing to regard these 10 carefully trained tasters as an SRS from a large population of all trained tasters. Figure 17.3 is a stemplot of the data. We can't judge Normality from just 10 observations; there are no outliers, but the data are somewhat skewed. P-values for the t test may be only approximately accurate.

The basic statistics are

$$\bar{x} = 1.02 \quad \text{and} \quad s = 1.196$$

The one-sample t statistic is

$$t = \frac{\bar{x} - \mu_0}{s/\sqrt{n}} = \frac{1.02 - 0}{1.196/\sqrt{10}}$$
$$= 2.697$$

Since the alternative is one-sided, the P-value for $t = 2.697$ is the area to the right of 2.697 under the t distribution curve with degrees of freedom $n - 1 = 9$. Figure 17.4 shows this area. Software (see Figure 17.6) tells us that $P = 0.0123$.

Without software, we can pin down P between two values by using Table C. Search the df $= 9$ row of Table C for entries that bracket $t = 2.697$.

df = 9		
t^*	2.398	2.821
P	.02	.01

The observed t lies between the critical values for one-sided P-values 0.02 and 0.01.

CONCLUDE: There is quite strong evidence ($0.01 < P < 0.02$) for a loss of sweetness.

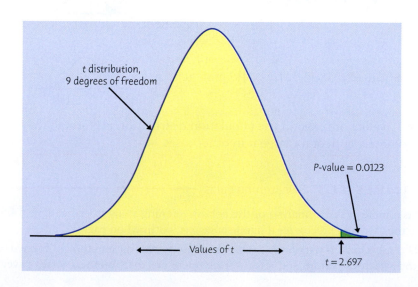

FIGURE 17.4 The P-value (upper tail area) for the one-sided t test in Example 17.3.

APPLY YOUR KNOWLEDGE ———————————

17.9   Is it significant? The one-sample t statistic from a sample of $n = 25$ observations for the two-sided test of

$$H_0: \mu = 64$$
$$H_a: \mu \neq 64$$

has the value $t = 1.12$.

(a) What are the degrees of freedom for t?

(b) Locate the two critical values t^* from Table C that bracket t. What are the two-sided P-values for these two entries?

(c) Is the value $t = 1.12$ statistically significant at the 10% level? At the 5% level?

17.10 Is it significant? The one-sample t statistic for testing

$$H_0: \mu = 0$$
$$H_a: \mu > 0$$

from a sample of $n = 15$ observations has the value $t = 1.82$.

(a) What are the degrees of freedom for this statistic?

(b) Give the two critical values t^* from Table C that bracket t. What are the one-sided P-values for these two entries?

(c) Is the value $t = 1.82$ significant at the 5% level? Is it significant at the 1% level?

17.11 Trout habitat. Do the data of Exercise 17.8 give good reason to think that the springtime water in the tributary water basin around the Shavers Fork watershed is not neutral. (A neutral pH is the pH of pure water, pH 7.) Follow the four-step process as illustrated in Example 17.3.

Using technology

Any technology suitable for statistics will implement the one-sample t procedures. You can read and use almost any output now that you know what to look for. Figure 17.5 displays output for the 95% confidence interval of Example 17.2 from a graphing calculator, two statistical programs, and a spreadsheet program. The TI-83, CrunchIt!, Minitab, and SPSS outputs are straightforward. All four give the estimate $\overline{x}$ and the confidence interval plus a selection of other information. The confidence interval agrees with our hand calculation in Example 17.2. In general, software results are more accurate, because of the rounding in hand calculations. Excel gives several descriptive measures but does not give the confidence interval. The entry labeled "Confidence Level (95.0%)" is the margin of error m, and the first entry, labeled "Mean," is the sample average $\overline{x}$. You can combine these values to compute the confidence interval $\overline{x} \pm m$.

Figure 17.6 displays output for the t test in Example 17.3. The TI-83, CrunchIt!, and Minitab[7] all give the sample mean $\overline{x}$, the t statistic, and its P-value. Accurate P-values are the biggest advantage of software for the t procedures. Note

SPSS

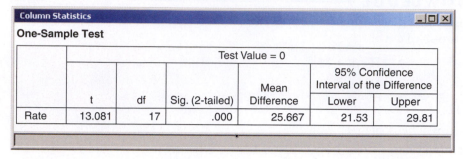

CrunchIt!

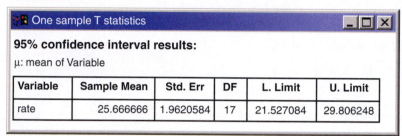

Texas Instruments TI-83 Plus

```
TInterval
 (21.527,29.806)
 x̄=25.6667
 Sx=8.3243
 n=18.0000
```

Minitab

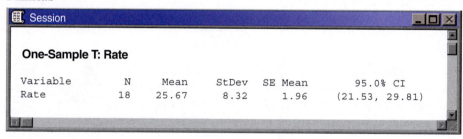

Excel

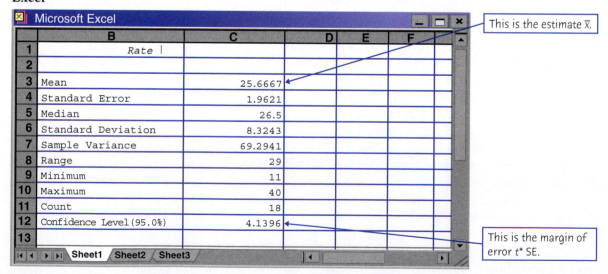

FIGURE 17.5 The *t* confidence interval of Example 17.2: output from a graphing calculator, three statistical programs, and a spreadsheet program.

SPSS

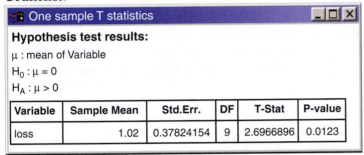

Column Statistics

One-Sample Test

	Test Value = 0					
					95% Confidence Interval of the Difference	
	t	df	Sig. (2-tailed)	Mean Difference	Lower	Upper
Loss	2.697	9	.025	1.0200	.164	1.876

CrunchIt!

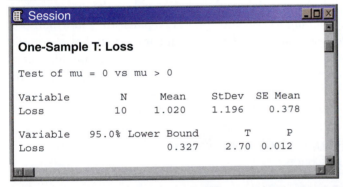

One sample T statistics

Hypothesis test results:

μ : mean of Variable

$H_0 : \mu = 0$

$H_A : \mu > 0$

Variable	Sample Mean	Std.Err.	DF	T-Stat	P-value
loss	1.02	0.37824154	9	2.6966896	0.0123

Minitab

Session

One-Sample T: Loss

Test of mu = 0 vs mu > 0

Variable	N	Mean	StDev	SE Mean
Loss	10	1.020	1.196	0.378

Variable	95.0% Lower Bound	T	P
Loss	0.327	2.70	0.012

Texas Instruments TI-83 Plus

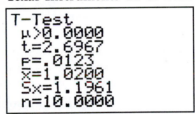

```
T-Test
μ>0.0000
t=2.6967
P=.0123
x̄=1.0200
Sx=1.1961
n=10.0000
```

Excel

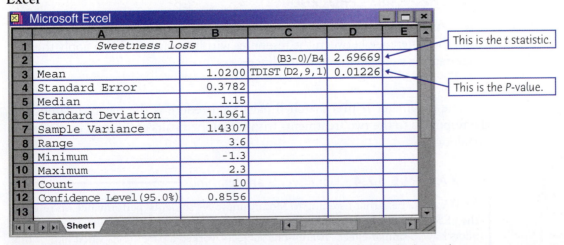

Microsoft Excel

	A	B	C	D	E
1	Sweetness loss				
2			(B3-0)/B4	2.69669	
3	Mean	1.0200	TDIST (D2,9,1)	0.01226	
4	Standard Error	0.3782			
5	Median	1.15			
6	Standard Deviation	1.1961			
7	Sample Variance	1.4307			
8	Range	3.6			
9	Minimum	-1.3			
10	Maximum	2.3			
11	Count	10			
12	Confidence Level (95.0%)	0.8556			
13					

Sheet1

This is the *t* statistic.

This is the *P*-value.

FIGURE 17.6 The *t* test of Example 17.3: output from a graphing calculator, three statistical programs, and a spreadsheet program.

that the statistical program SPSS gives a generic 2-sided P-value called "Sig. (2-tailed)." It is up to you to remember that with a one-sided alternative, the P-value is computed using only one side (or one "tail") of the sampling distribution. Here the true P-value for our test is thus half of the 0.025 displayed. Excel is as usual more awkward than software designed for statistics. It lacks a one-sample t test menu selection but does have a function named TDIST, equivalent to Table C, that gives tail areas under t density curves. The Excel output shows the functions used to obtain the t statistic and its P-value, along with their actual values $t = 2.69669$ and $P = 0.01226$.

Some software, like Minitab and SPSS, round the P-value to three decimal places, which means that you will sometimes see output like this "P-value 0.000." This doesn't mean that the P-value is zero (a P-value cannot be exactly zero or one) but that it is closer to 0.0009 than to 0.001 (in other words, read as "$P < 0.001$").

Matched pairs t procedures

The study of healing in Example 17.2 estimated the mean healing rate for newts under natural conditions, but the researchers then compared results under several conditions. The taste test in Example 17.3 was a matched pairs study in which the same 10 tasters rated before-and-after sweetness. Comparative studies are more convincing than single-sample investigations. For that reason, one-sample inference is less common than comparative inference. However, one common design to compare two treatments makes use of one-sample procedures. In a **matched pairs** *matched pairs design* **design,** subjects are matched in pairs, and each treatment is given to one subject in each pair. Another situation calling for matched pairs is before-and-after observations on the same subjects, as in the taste test of Example 17.3. We first introduced the matched pairs design in Chapter 8.

> **MATCHED PAIRS t PROCEDURES**
>
> To compare the responses to the two treatments in a matched pairs design, find the difference between the responses within each pair. Then apply the one-sample t procedures to these differences.

The parameter μ in a matched pairs t procedure is the mean difference in the responses to the two treatments within matched pairs of subjects in the entire population.

EXAMPLE 17.4 *Floral scents and learning*

STATE: We hear that listening to Mozart improves students' performance on tests. In the EESEE case study "Floral Scents and Learning," investigators asked whether pleasant odors have a similar effect. Twenty-one subjects worked a paper-and-pencil maze while wearing a mask. The mask either was unscented or carried a floral scent. The response

TABLE 17.1   Average time (seconds) to complete a maze

Subject	Unscented	Scented	Difference	Subject	Unscented	Scented	Difference
1	30.60	37.97	−7.37	12	58.93	83.50	−24.57
2	48.43	51.57	−3.14	13	54.47	38.30	16.17
3	60.77	56.67	4.10	14	43.53	51.37	−7.84
4	36.07	40.47	−4.40	15	37.93	29.33	8.60
5	68.47	49.00	19.47	16	43.50	54.27	−10.77
6	32.43	43.23	−10.80	17	87.70	62.73	24.97
7	43.70	44.57	−0.87	18	53.53	58.00	−4.47
8	37.10	28.40	8.70	19	64.30	52.40	11.90
9	31.17	28.23	2.94	20	47.37	53.63	−6.26
10	51.23	68.47	−17.24	21	53.67	47.00	6.67
11	65.40	51.10	14.30				

variable is their average time on three trials. Each subject worked the maze with both masks, in a random order. The randomization is important because subjects tend to improve their times as they work a maze repeatedly. Table 17.1 gives the subjects' average times with both masks. Is there evidence that subjects worked the maze faster wearing the scented mask?

FORMULATE: Take μ to be the mean difference (time unscented minus time scented) in the population of healthy adults. The null hypothesis says that the scents have no effect, and H_a says that unscented times are longer than scented times on the average. So we test the hypotheses

$$H_0: \mu = 0$$
$$H_a: \mu > 0$$

SOLVE: The subjects are not an actual SRS from the population of all healthy adults. But we are willing to regard them as an SRS in their performance on a maze. To analyze the data, subtract the scented time from the unscented time for each subject. The 21 differences form a single sample from the population with unknown mean μ. They appear in the "Difference" column in Table 17.1. Positive differences show that a subject's scented time was shorter than the unscented time. Figure 17.7 is a stemplot of the differences, rounded to the nearest whole second. The distribution is symmetric and reasonably Normal in shape.

The 21 differences have

$$\bar{x} = 0.9567 \text{ and } s = 12.5479$$

The one-sample *t* statistic is therefore

$$t = \frac{\bar{x} - 0}{s/\sqrt{n}} = \frac{0.9567 - 0}{12.5479/\sqrt{21}}$$
$$= 0.349$$

Find the *P*-value from the $t(20)$ distribution. (Remember that the degrees of freedom are 1 less than the sample size.) Table C shows that 0.349 is less than the critical value for one-sided $P = 0.25$. Because the alternative is one-sided, the *P*-value of our test is greater than 0.25. Software gives the value $P = 0.3652$.

```
−2 | 5
−1 | 7 1 1
−0 | 8 7 6 4 4 3 1
 0 | 3 4 7 9 9
 1 | 2 4 6 9
 2 | 5
```

FIGURE 17.7 Stemplot of the time differences in Example 17.4.

df = 20

t^*	0.687	0.860
P	.25	.20

CONCLUDE: The data do not support the claim that floral scents improve performance. The average improvement of the 21 subjects is small, just 0.96 seconds over the 50 seconds that the average subject took when wearing the unscented mask. This small improvement is not statistically significant at even the 10% level.

Example 17.4 illustrates how to turn matched pairs data into single-sample data by calculating a difference within each pair.[8] We are making inferences about a single population, the population of all differences within matched pairs. *It is incorrect to ignore the matching and analyze the data as if we had two samples,* one from subjects who wore unscented masks and a second from subjects who wore scented masks. Inference procedures for comparing two samples assume that the samples are selected independently of each other, which is not true when the same subjects are measured twice. The proper analysis depends on the design used to produce the data. We will describe the procedures used for two independent samples in the next chapter.

CAUTION

APPLY YOUR KNOWLEDGE

Many exercises from this point on ask you to give the P-value of a t test. If you have suitable technology, give the exact P-value. Otherwise, use Table C to give two values between which P lies.

17.12 Magnets for pain relief. A randomized, double-blind experiment studied whether magnetic fields applied over a painful area can reduce pain intensity. The subjects were 50 volunteers with postpolio syndrome who reported muscular or arthritic pain. The pain level when pressing a painful area was graded subjectively on a scale from 0 to 10 (0 is no pain, 10 is maximum pain). Patients were randomly assigned to wear either a magnetic device or a placebo device over the painful area for 45 minutes.[9]

(a) All patients rated their pain before and after application of the device. For the 29 patients assigned to the magnetic group, the mean pain scores before and after treatment were 9.6 and 4.4, respectively. Explain briefly why a matched pairs procedure must be used to assess whether application of the magnetic device significantly lowers pain.

(b) The 21 patients assigned to the placebo device had mean pain scores before and after treatment of 9.5 and 8.4, respectively. Explain briefly why we cannot use a matched pairs procedure to test the hypothesis that the magnetic device is significantly better than the placebo at reducing pain intensity in patients with postpolio syndrome.

STEP

17.13 Does nature heal better? Our bodies have a natural electrical field that is known to help wounds heal. Might stronger or weaker fields speed healing? A series of experiments with newts investigated this question. In one experiment, the two hind limbs of 14 newts were assigned at random to either experimental or control groups. This is a matched pairs design. The electrical field in the experimental limbs was reduced to half its natural value by applying an artificial voltage. The control limbs were not manipulated. Table 17.2 gives the rates at which new cells closed a razor cut in each limb.[10] Is there good evidence that changing the electrical field from its natural level slows healing?

(a) State the hypotheses to be tested. Explain what the parameter μ in your hypotheses stands for.

Magnets for pain relief?

The use of magnetic stones to treat pain is documented as far back as ancient Greece. The National Center for Complementary and Alternative Medicine (NCCAM, a branch of the NIH) estimates that Americans spend about $500 million per year on magnetic devices for pain relief. Yet there is no sound theory of how magnets could influence a person's physiology other than via a placebo effect. NCCAM states that "scientific research so far does not firmly support a conclusion that magnets of any type can relieve pain."

| TABLE 17.2 | Healing rates (micrometers per hour) for newts Reduced voltage condition | | | | |

Newt	Experimental limb	Control limb	Newt	Experimental limb	Control limb
1	24	25	8	33	36
2	23	13	9	28	35
3	47	44	10	28	38
4	42	45	11	21	43
5	26	57	12	27	31
6	46	42	13	25	26
7	38	50	14	45	48

(b) We are willing to regard these 14 newts, which were randomly assigned from a larger group, as an SRS of all newts of this species. Because the *t* test uses the differences within matched pairs, these differences must be at least roughly Normal. Make a stemplot of the 14 differences and comment on its shape.

(c) Complete the *Solve* and *Conclude* steps by carrying out the matched pairs *t* test.

17.14 How much better does nature heal? Give a 90% confidence interval for the difference in healing rates (control minus experimental) in Table 17.2.

Robustness of *t* procedures

The *t* confidence interval and test are exactly correct when the distribution of the population is exactly Normal. No real data are exactly Normal. The usefulness of the *t* procedures in practice therefore depends on how strongly they are affected by lack of Normality.

ROBUST PROCEDURES

A confidence interval or significance test is called **robust** if the confidence level or *P*-value does not change very much when the conditions for use of the procedure are violated.

The condition that the population is Normal rules out outliers, so the presence of outliers shows that this condition is not fulfilled. *The t procedures are not robust against outliers unless the sample is very large, because* x̄ *and* s *are not resistant to outliers.*

Fortunately, the *t* procedures are quite robust against non-Normality of the population except when outliers or strong skewness is present. (Skewness is more serious than other kinds of non-Normality.) As the size of the sample increases, the central limit theorem ensures that the distribution of the sample mean x̄ becomes more nearly Normal and that the *t* distribution becomes more accurate for critical values and *P*-values of the *t* procedures.

Always make a plot to check for skewness and outliers before you use the *t* procedures for small samples. For most purposes, you can safely use the one-sample *t* procedures when $n \geq 15$ unless an outlier or quite strong skewness is present. Here are practical guidelines for inference on a single mean.[11]

USING THE *t* PROCEDURES

- Except in the case of small samples, the condition that the data are an SRS from the population of interest is more important than the condition that the population distribution is Normal.

- *Sample size less than 15:* Use *t* procedures if the data appear close to Normal (roughly symmetric, single peak, no outliers). If the data are skewed or if outliers are present, do not use *t*.

- *Sample size at least 15:* The *t* procedures can be used except in the presence of outliers or strong skewness.

- *Large samples:* The *t* procedures can be used even for clearly skewed distributions when the sample is large, roughly $n \geq 40$.

EXAMPLE 17.5   Can we use *t*?

Figure 17.8 shows plots of several data sets. For which of these can we safely use the *t* procedures?

- Figure 17.8(a) is a histogram of the probability at birth of not surviving to age 40 in developing countries. The data come from the United Nations' Human Development Report 2005 and includes all 121 countries in the developing world. *We have data on the entire population of 121 developing countries, so inference is not needed.* We can calculate the exact mean for the population. There is no uncertainty due to having only a sample from the population and no need for a confidence interval or test.

- Figure 17.8(b) is a stemplot of the percents of nitrogen found in the gas bubbles of 9 specimens of amber from the late Cretaceous era (75 to 95 million years ago).[12] *The data are strongly skewed to the left with possible low outliers, so we cannot trust the t procedures for* n = 9.

- Figure 17.8(c) is a stemplot of the lengths of 23 specimens of the red variety of the tropical flower *Heliconia*.[13] *The histogram is mildly skewed to the right, and there are no outliers. We can use the t distributions for such data.*

- Figure 17.8(d) is a histogram of the heights of the students in a college class. *This distribution is quite symmetric and appears close to Normal. We can use the t procedures for any sample size if the students can be considered an SRS of a larger population.*

APPLY YOUR KNOWLEDGE

17.15 An outlier strikes. Table 17.3 gives data for another experiment from the study of healing rates in newts. The setup is exactly as in Exercise 17.13, except that the electrical field in the experimental limbs was reduced to zero by applying an artificial voltage.

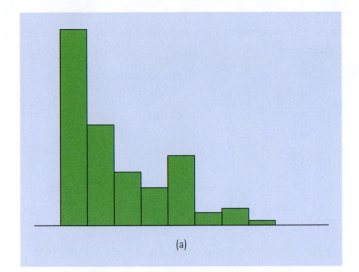

(a)

```
4 | 9
5 | 1
5 |
5 | 4
5 |
5 |
6 | 0
6 | 3 3
6 | 4 4 5
```

(b)

```
37 | 4 8 9
38 | 0 0 1 1 2 2 8 9
39 | 2 6 8
40 | 6 7
41 | 5 7 9 9
42 | 0 2
43 | 1
```

(c)

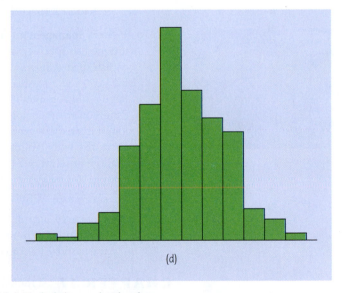

(d)

FIGURE 17.8 Can we use *t* procedures for these data? **(a)** Probability at birth of not surviving to age 40, in developing countries. *No*, this is an entire population (all 121 developing countries), not a sample. **(b)** Percent of nitrogen found in the gas bubbles of 9 specimens of amber from the late Cretaceous era. *No*, there are just 9 observations and strong skewness. **(c)** Lengths of 23 tropical flowers of the same variety. *Yes*, the sample is large enough to overcome the mild skewness. **(d)** Heights of college students. *Yes, for any size sample*, because the distribution is close to Normal.

(a) Make a stemplot of the differences between limbs of the same newt (control limb minus experimental limb). There is a high outlier.

(b) Carry out two *t* tests to see if the mean healing rate is significantly higher in the control limbs, one including all 12 newts and another that omits the outlier. What are the test statistics and their *P*-values? Does the outlier have

TABLE 17.3	Healing rates (micrometers per hour) for newts Zero voltage condition					
Newt	Experimental limb	Control limb	Newt	Experimental limb	Control limb
1	28	36	7	45	39
2	31	41	8	25	56
3	27	39	9	28	33
4	33	42	10	33	20
5	33	44	11	47	49
6	38	39	12	23	30

a strong influence on your conclusion? How to handle outliers is a different challenge in different situations. Refer to the discussion on outliers in Chapter 2 (page 45) for an in-depth discussion.

17.16 Fishery management in Lake Malawi. Lake Malawi is a large African lake with unique fish species essential to the local ecosystem and economy. The endemic catfish species *Bagrus meridionalis*, or kampango, is an important part of the local fishing trade. Small-scale fisheries draw mainly from the shallower water, while large-scale commercial fisheries operate mainly in deeper water. To better manage fishing from an ecological perspective, a study examined the length of kampango found in shallow versus deep water. The study reports the following information about the body lengths of a random sample of male kampango extracted from shallow water: $n = 2131$, $\bar{x} = 39.9$ cm, and standard error 0.25 cm.[14]

(a) We don't have the 2131 individual fish lengths, but use of the t procedures is surely safe. Why?

(b) Give a 99% confidence interval for the mean body length of male kampango swimming in the shallow waters of Lake Malawi. (Be careful: the report gives the standard error of $\bar{x}$, not the standard deviation s.)

Brian Atkinson/Alamy

CHAPTER 17 SUMMARY

Tests and confidence intervals for the mean μ of a Normal population are based on the sample mean $\bar{x}$ of an SRS. Because of the central limit theorem, the resulting procedures are approximately correct for other population distributions when the sample is large.

The standardized sample mean is the **one-sample z statistic**

$$z = \frac{\bar{x} - \mu}{\sigma/\sqrt{n}}$$

If we knew σ, we would use the z statistic and the standard Normal distribution. In practice, we do not know σ. Replace the standard deviation $\sigma/\sqrt{n}$ of $\bar{x}$ by the **standard error** $s/\sqrt{n}$ to get the **one-sample t statistic**

$$t = \frac{\bar{x} - \mu}{s/\sqrt{n}}$$

The t statistic has the **t distribution** with $n - 1$ degrees of freedom.

There is a t distribution for every positive **degrees of freedom.** All are symmetric distributions similar in shape to the standard Normal distribution. The t distribution approaches the $N(0, 1)$ distribution as the degrees of freedom increase.

A level C **confidence interval for the mean μ** of a Normal population is

$$\bar{x} \pm t^* \frac{s}{\sqrt{n}}$$

The **critical value** t^* is chosen so that the t curve with $n - 1$ degrees of freedom has area C between $-t^*$ and t^*.

Significance tests for $H_0: \mu = \mu_0$ are based on the t statistic. Use P-values or fixed significance levels from the $t(n - 1)$ distribution.

Use these one-sample procedures to analyze **matched pairs** data by first taking the difference within each matched pair to produce a single sample.

The t procedures are quite **robust** when the population is non-Normal, especially for larger sample sizes. The t procedures are useful for non-Normal data when $n \geq 15$ unless the data show outliers or strong skewness.

CHECK YOUR SKILLS

17.17 We prefer the t procedures to the z procedures for inference about a population mean because

    (a) z can be used only for large samples.

    (b) z requires that you know the population standard deviation σ.

    (c) z requires that you can regard your data as an SRS from the population.

17.18 You are testing $H_0: \mu = 10$ against $H_a: \mu < 10$ based on an SRS of 20 observations from a Normal population. The data give $\bar{x} = 8$ and $s = 4$. The value of the t statistic is

    (a) -0.5.   (b) -10.   (c) -2.24.

17.19 You are testing $H_0: \mu = 10$ against $H_a: \mu < 10$ based on an SRS of 20 observations from a Normal population. The t statistic is $t = -2.25$. The degrees of freedom for this statistic are

    (a) 19.   (b) 20.   (c) 21.

17.20 The P-value for the statistic in the previous exercise

    (a) falls between 0.01 and 0.02.

    (b) falls between 0.02 and 0.04.

    (c) is greater than 0.25.

17.21 You are testing $H_0: \mu = 500$ against $H_a: \mu \neq 500$ based on an SRS of 12 observations from a Normal population. The t statistic is $t = -1.85$. The P-value for this test

    (a) falls between 0.025 and 0.05.

    (b) falls between 0.05 and 0.10.

    (c) is greater than 0.50.

17.22 You have an SRS of 15 observations from a Normally distributed population. What critical value would you use to obtain a 98% confidence interval for the mean μ of the population?

(a) 2.326.　　　　　(b) 2.602.　　　　　(c) 2.624.

17.23 Data on the blood cholesterol levels of 24 rats (mg/dl) give $\bar{x} = 85$ and $s = 12$. A 95% confidence interval for the mean blood cholesterol of rats under this condition is

(a) 79.9 to 91.1　　　(b) 80.2 to 89.8　　　(c) 82.6 to 87.4

17.24 Which of the following would be most worrisome for the validity of the confidence interval you calculated in the previous exercise?

(a) The presence of a clear outlier in the raw data.

(b) A mild skew in the stemplot of the raw data.

(c) Not knowing the population standard deviation σ.

17.25 Which of these settings does *not* allow use of a matched pairs t procedure?

(a) You interview both the husband and the wife in 64 married couples and ask each about their ideal number of children.

(b) You interview a sample of 64 unmarried male students and another sample of 64 unmarried female students and ask each about their ideal number of children.

(c) You interview 64 female students in their freshman year and again in their senior year and ask each about their ideal number of children.

17.26 Because the t procedures are robust, the most important condition for their safe use is that

(a) the population standard deviation σ is known.

(b) the population distribution is exactly Normal.

(c) the data can be regarded as an SRS from the population.

CHAPTER 17 EXERCISES

17.27 Sharks. Great white sharks are big and hungry. Here are the lengths in feet of 44 great whites:[15]

18.7	12.3	18.6	16.4	15.7	18.3	14.6	15.8	14.9	17.6	12.1
16.4	16.7	17.8	16.2	12.6	17.8	13.8	12.2	15.2	14.7	12.4
13.2	15.8	14.3	16.6	9.4	18.2	13.2	13.6	15.3	16.1	13.5
19.1	16.2	22.8	16.8	13.6	13.2	15.7	19.7	18.7	13.2	16.8

(a) Examine these data for shape, center, spread, and outliers. The distribution is reasonably Normal except for one outlier in each direction. Because these are not extreme and preserve the symmetry of the distribution, use of the t procedures is safe with 44 observations.

(b) Give a 95% confidence interval for the mean length of great white sharks. Based on this interval, is there significant evidence at the 5% level to reject the claim "Great white sharks average 20 feet in length"?

(c) Before accepting the conclusions of (b), you need more information about the data. What would you like to know?

17.28 Alcohol in wine. The alcohol content of wine depends on the grape variety, the way in which the wine is produced from the grapes, the weather, and other influences. Here are data on the percent of alcohol in wine produced from the same grape variety in the same year by 48 winemakers in the same region of Italy:[16]

12.86	12.88	12.81	12.70	12.51	12.60	12.25	12.53	13.49	12.84
12.93	13.36	13.52	13.62	12.25	13.16	13.88	12.87	13.32	13.08
13.50	12.79	13.11	13.23	12.58	13.17	13.84	12.45	14.34	13.48
12.36	13.69	12.85	12.96	13.78	13.73	13.45	12.82	13.58	13.40
12.20	12.77	14.16	13.71	13.40	13.27	13.17	14.13		

(a) Make a histogram of the data, using class width 0.25. The shape of the distribution is a bit irregular, but there are no outliers or strong skewness. There is no reason to avoid use of t procedures for $n = 48$.

(b) Give a 95% confidence interval for the mean alcohol content of wine of this type.

(c) Based on your confidence interval, is the mean alcohol content significantly different at the $\alpha = 0.05$ level from 12%? From 13%?

17.29 Wood lice. Students in a physiology lab collected 50 wood lice ("pill bugs") from the university grounds. Regard these as a random sample of wood lice on the university grounds. Here are the lengths in millimeters:[17]

11.0	11.1	10.1	10.8	9.2	9.0	11.0	9.5	8.0	14.6
9.1	9.7	12.0	12.5	13.1	11.7	8.1	8.7	11.9	10.3
8.1	8.3	8.2	12.1	12.3	10.9	8.9	7.5	9.0	8.2
8.3	9.0	7.9	10.5	10.0	9.8	8.9	11.0	11.0	10.9
11.0	11.5	11.0	13.9	9.3	8.5	10.2	8.3	9.1	7.0

Verify that there are no outliers in the data. What is a 95% confidence interval for the mean body length in this population?

Hornbil Images/Alamy

17.30 A big toe problem. Hallux abducto valgus (call it HAV) is a deformation of the big toe that often requires surgery. Doctors used X-rays to measure the angle (in degrees) of deformity in 38 consecutive patients under the age of 21 who came to a medical center for surgery to correct HAV. The angle is a measure of the seriousness of the deformity. Here are the data:[18]

28	32	25	34	38	26	25	18	30	26	28	13	20
21	17	16	21	23	14	32	25	21	22	20	18	26
16	30	30	20	50	25	26	28	31	38	32	21	

It is reasonable to regard these patients as a random sample of young patients who require HAV surgery. Carry out the *Solve* and *Conclude* steps of a 95% confidence interval for the mean HAV angle in the population of all such patients.

17.31 Conditions for inference. Exercise 17.1 gave the summary data ($\bar{x} = 27.8$, $s = 32.2$ spikes per second) for the baseline activity of a neuron based on 72 separate recordings. Neural activity is expressed in number of action potentials per second and cannot be less than 0. You notice that the standard deviation is larger than the mean. What does it suggest about the shape of the distribution? Is it nonetheless acceptable to calculate a t confidence interval for the population mean? Explain why.

17.32 An outlier's effect. The data in Exercise 17.30 follow a Normal distribution quite closely except for one patient with HAV angle 50 degrees, a high outlier.

(a) Find the 95% confidence interval for the population mean based on the 37 patients who remain after you drop the outlier.

(b) Compare your interval in (a) with your interval from Exercise 17.30. What is the most important effect of removing the outlier? In general, you shouldn't remove an outlier unless you have reason to believe it is an error or unless you are interested only in "typical" cases. For a more in-depth discussion of how to handle outliers in various situations read the discussion on outliers in Chapter 2 (page 45).

17.33 Cockroach metabolism. To study the metabolism of insects, researchers fed cockroaches measured amounts of a sugar solution. After 2, 5, and 10 hours, they dissected some of the cockroaches and measured the amount of sugar in various tissues.[19] Five roaches fed the sugar D-glucose and dissected after 10 hours had the following amounts (in micrograms) of D-glucose in their hindguts:

$$55.95 \quad 68.24 \quad 52.73 \quad 21.50 \quad 23.78$$

The researchers gave a 95% confidence interval for the mean amount of D-glucose in cockroach hindguts under these conditions. The insects are a random sample from a uniform population grown in the laboratory. We therefore expect responses to be Normal. What confidence interval did the researchers give?

17.34 Growing trees faster. The concentration of carbon dioxide (CO_2) in the atmosphere is increasing rapidly due to our use of fossil fuels. Because plants use CO_2 to fuel photosynthesis, more CO_2 may cause trees and other plants to grow faster. An elaborate apparatus allows researchers to pipe extra CO_2 to a 30-meter circle of forest. They selected two nearby circles in each of three parts of a pine forest and randomly chose one of each pair to receive extra CO_2. The response variable is the mean increase in base area for 30 to 40 trees in a circle during a growing season. We measure this in percent increase per year. Here are one year's data:[20]

Pair	Control plot	Treated plot
1	9.752	10.587
2	7.263	9.244
3	5.742	8.675

(a) State the null and alternative hypotheses. Explain clearly why the investigators used a one-sided alternative.

(b) Carry out a test and report your conclusion in simple language.

(c) The investigators used the test you just carried out. Any use of the t procedures with samples this size is risky. Why?

17.35 Fungus in the air. The air in poultry-processing plants often contains fungus spores. Inadequate ventilation can affect the health of the workers. The problem is most serious during the summer. To measure the presence of spores, air samples are pumped to an agar plate and "colony forming units" (CFUs) are counted after an incubation period. Here are data from two locations in a plant that processes

37,000 turkeys per day, taken on 4 days in the summer. The units are CFUs per cubic meter of air.[21]

	Day 1	Day 2	Day 3	Day 4
Kill room	3175	2526	1763	1090
Processing	529	141	362	224

(a) Explain carefully why these are matched pairs data.

(b) The spore count is clearly higher in the kill room. Give sample means and a 90% confidence interval to estimate how much higher. State your conclusion in plain English.

(c) You will often see the t procedures used for data like these. You should regard the results as only rough approximations. Why?

17.36 Blood pressure. Researchers measured the seated systolic blood pressure of 27 healthy white males. The resulting publication reports $\bar{x} = 114.9$ and $s = 9.3$.

(a) Give a 95% confidence interval for the mean blood pressure in the population from which the subjects were recruited.

(b) What conditions for the population and the study design are required by the procedure you used in (a)? Which of these conditions are important for the validity of the procedure in this case?

17.37 The placebo effect. The placebo effect is particularly strong in patients with Parkinson's disease. To understand the workings of the placebo effect, scientists measure activity at a key point in the brain when patients receive a placebo that they think is an active drug and also when no treatment is given.[22] The same 6 patients are measured both with and without the placebo, at different times.

(a) Explain why the proper procedure to compare the mean response to placebo with that to control (no treatment) is a matched pairs t test.

(b) The six differences (treatment minus control) had $\bar{x} = -0.326$ and $s = 0.181$. Is there significant evidence of a difference between treatment and control?

17.38 How much oil? The environmental impact of oil drilling goes beyond the devastating consequences of major oil spills. Animal and plant species are displaced by the extraction activity, and over time, local ecosystems are affected by low but ongoing contamination of soil and water sources. Public debate centers on the economic benefits of drilling against the ecological and health risks. An important piece of information in deciding whether to drill more wells in a given field is how much oil these wells would ultimately produce. Here are the estimated total amounts of oil recovered from 64 wells in the Devonian Richmond Dolomite area of the Michigan basin, in thousands of barrels:[23]

21.71	53.2	46.4	42.7	50.4	97.7	103.1	51.9
43.4	69.5	156.5	34.6	37.9	12.9	2.5	31.4
79.5	26.9	18.5	14.7	32.9	196	24.9	118.2
82.2	35.1	47.6	54.2	63.1	69.8	57.4	65.6
56.4	49.4	44.9	34.6	92.2	37.0	58.8	21.3
36.6	64.9	14.8	17.6	29.1	61.4	38.6	32.5
12.0	28.3	204.9	44.5	10.3	37.7	33.7	81.1
12.1	20.1	30.5	7.1	10.1	18.0	3.0	2.0

Take these wells to be an SRS of wells in this area.

(a) Give a 95% t confidence interval for the mean amount of oil recovered from all wells in this area.

(b) Make a graph of the data. The distribution is very skewed, with several high outliers. A computer-intensive method that gives accurate confidence intervals without assuming any specific shape for the distribution gives a 95% confidence interval of 40.28 to 60.32. How does the t interval compare with this? Should the t procedures be used with these data?

17.39 Weeds among the corn. Velvetleaf is a particularly annoying weed in cornfields. It produces lots of seeds, and the seeds wait in the soil for years until conditions are right. How many seeds do velvetleaf plants produce? Here are counts from 28 plants that came up in a cornfield when no herbicide was used:[24]

2450	2504	2114	1110	2137	8015	1623	1531	2008	1716
721	863	1136	2819	1911	2101	1051	218	1711	164
2228	363	5973	1050	1961	1809	130	880		

We would like to give a confidence interval for the mean number of seeds produced by velvetleaf plants. Alas, the t interval can't be safely used for these data. Why not?

*The following exercises ask you to answer questions from data without having the steps outlined as part of the exercise. Follow the **Formulate, Solve,** and **Conclude** steps of the four-step process illustrated in Examples 17.2, 17.3, and 17.4. It may be helpful to restate in your own words the **State** information given in the exercise.*

17.40 Natural weed control? Fortunately, we aren't really interested in the number of seeds velvetleaf plants produce (see Exercise 17.39). The velvetleaf seed beetle feeds on the seeds and might be a natural weed control. Here are the total seeds, seeds infected by the beetle, and percent of seeds infected for 28 velvetleaf plants:

Seeds	2450	2504	2114	1110	2137	8015	1623	1531	2008	1716
Infected	135	101	76	24	121	189	31	44	73	12
Percent	5.5	4.0	3.6	2.2	5.7	2.4	1.9	2.9	3.6	0.7
Seeds	721	863	1136	2819	1911	2101	1051	218	1711	164
Infected	27	40	41	79	82	85	42	0	64	7
Percent	3.7	4.6	3.6	2.8	4.3	4.0	4.0	0.0	3.7	4.3
Seeds	2228	363	5973	1050	1961	1809	130	880		
Infected	156	31	240	91	137	92	5	23		
Percent	7.0	8.5	4.0	8.7	7.0	5.1	3.8	2.6		

Do a complete analysis of the percent of seeds infected by the beetle. Include a 90% confidence interval for the mean percent infected in the population of all velvetleaf plants. Do you think that the beetle is very helpful in controlling the weed? Follow the four-step process as illustrated in Example 17.2.

17.41 Evolution and relative fitness. Can bacteria evolve a preference for the pH of their environment? An evolutionary biologist examined the relative fitness of *Escherichia coli* bacteria grown for 2000 generations (about 300 days) at stressful acidic pH 5.5 and their parental generation, grown and preserved at pH 7.2. Both types were later grown together in an acidic medium, and their relative fitness was computed. The experiment was replicated with 6 different lines of *E. coli*, giving the following relative fitness values:[25]

1.24   1.22   1.23   1.24   1.18   1.09

A relative fitness of 1 indicates that both bacteria types are equally fit. A relative fitness larger than 1 indicates that the acid-evolved line is more fit than the parental line kept at neutral pH when both are grown in acidic conditions (that is, the acid-evolved bacteria grew the most). Do the data provide evidence that bacteria evolved in acidic pH are better adapted to acidic conditions? Do a complete analysis, following the four-step process as illustrated in Example 17.3.

17.42 Stimulating bone formation. The effect of phosphate supplementation on bone formation was assessed in 6 healthy adult dogs. For each dog, bone formation was measured for a 12-week period of phosphate supplementation as well as for a 12-week control period. Here are the results in percent growth per year:[26]

Dog	1	2	3	4	5	6
Control	1.73	3.37	3.59	2.05	1.86	3.6
Phosphate	8.16	4.58	3.98	5.24	3.04	7.03

(a) Explain clearly why the matched pairs *t* test is the proper choice for this experimental design.

(b) The authors concluded that phosphate supplementation significantly stimulates bone formation, suggesting that typical plasma phosphate levels in adult dogs are suboptimal for new bone formation. Do a complete analysis to verify this statement. Follow the four-step process as illustrated in Example 17.4.

17.43 Task performance. The design of controls and instruments affects how easily people can use them. A student project investigated this effect by asking 25 right-handed students to turn with their right hands a knob that moved an indicator by screw action. There were two identical instruments, one with a clockwise thread and the other with a counterclockwise thread. Table 17.4 (page 456) gives the times in milliseconds each subject took to move the indicator a fixed distance.[27]

(a) Each of the 25 students used both instruments. Discuss briefly how you would use randomization in arranging the experiment.

(b) The project hoped to show that right-handed people find clockwise threads easier to use. Do an analysis that leads to a conclusion about this issue. Can you conclude that the time difference is due to the subjects' right-handedness?

17.44 Comparing two drugs. Makers of generic drugs must show that they do not differ significantly from the "reference" drugs that they imitate. One aspect in which drugs might differ is the extent of their absorption in the blood. Table 17.5 gives data taken from 20 healthy nonsmoking male subjects for one pair of drugs.[28] This is a matched pairs design. Numbers 1 to 20 were assigned at random to the subjects. Subjects 1 to 10 received the generic drug first, and Subjects 11 to 20 received the reference drug first. In all cases, a washout period separated the two drugs so that the first had disappeared from the blood before the subject took the second. Do the drugs differ significantly in absorption?

TABLE 17.4   Performance times (milliseconds)

Subject	Clockwise	Counterclockwise	Subject	Clockwise	Counterclockwise
1	113	137	14	107	87
2	105	105	15	118	166
3	130	133	16	103	146
4	101	108	17	111	123
5	138	115	18	104	135
6	118	170	19	111	112
7	87	103	20	89	93
8	116	145	21	78	76
9	75	78	22	100	116
10	96	107	23	89	78
11	122	84	24	85	101
12	103	148	25	88	123
13	116	147			

TABLE 17.5   Absorption extent for two versions of a drug

Subject	Reference	Generic	Subject	Reference	Generic
15	4108	1755	4	2344	2738
3	2526	1138	16	1864	2302
9	2779	1613	6	1022	1284
13	3852	2254	10	2256	3052
12	1833	1310	5	938	1287
8	2463	2120	7	1339	1930
18	2059	1851	14	1262	1964
20	1709	1878	11	1438	2549
17	1829	1682	1	1735	3340
2	2594	2613	19	1020	3050

17.45 Practical significance? Give a 90% confidence interval for the mean time advantage of clockwise over counterclockwise threads in the setting of Exercise 17.43.

(a) The brain works with reaction times in the order of 100 milliseconds. Is the difference in mean performance time practically significant from a biological point of view?

(b) For individuals sporadically using the type of device used in the study, is the difference in mean performance time practically significant?

(c) Do you think that the time saved on average would be of practical importance if the task were performed many times—for example, by an assembly-line worker? Explain, however, why you cannot use the results of this study to infer anything about the difference in performance time for individuals trained to perform these tasks over and over.

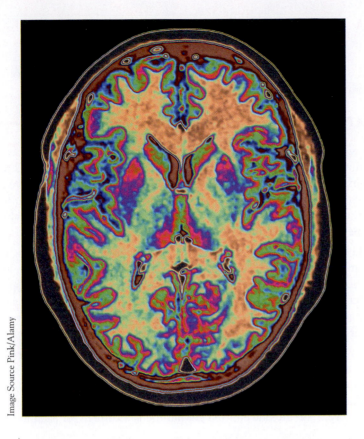

Image Source Pink/Alamy

Two-Sample Problems

Comparing two populations or two treatments is one of the most common situations encountered in statistical practice. We call such situations *two-sample problems*.

TWO-SAMPLE PROBLEMS

- The goal of inference is to compare the responses to two treatments or to compare the characteristics of two populations.
- We have a separate sample from each treatment or each population.

Two-sample problems

A two-sample problem can arise from a randomized comparative experiment that randomly divides subjects into two groups and exposes each group to a different treatment. Comparing random samples selected separately from two populations is also a two-sample problem. Unlike the matched pairs designs studied earlier, there is no matching of the individuals in the two samples, and the two samples can be

Can I trust this study?

An anesthesiologist thought he might ease menopausal hot flashes by injecting a local anesthetic directly into a specific nerve in the neck, hoping to influence the brain's temperature regulation center. He found that almost all of the 22 women who elected to have the procedure done experienced some benefit, which lasted from weeks to months. That sounds promising, and the results were reported on NBC. But there were no controls and only 22 women, who were not selected at random. Medical standards require randomized comparative experiments and statistically significant results. Only then can we be confident that a new procedure really helps.

of different sizes. Inference procedures for two-sample data differ from those for matched pairs. Here are some typical two-sample problems.

EXAMPLE 18.1   Two-sample problems

(a) Does regular physical therapy help with lower back pain? A randomized experiment assigned patients with lower back pain to two groups: 142 received an examination and advice from a physical therapist; another 144 received regular physical therapy for up to five weeks. After a year, the change in their level of disability (0% to 100%) was assessed by a doctor who did not know which treatment the patients had received.

(b) A field biologist observes gender-based behavior in wild chimpanzees. Twelve randomly chosen young chimpanzees are tagged remotely with a dart, and their behavior is monitored. The amount of time each young chimpanzee spends in contact with its mother is recorded. After determining the gender of each young chimpanzee, the biologist compares the amount of time spent in contact with the mother by male and female young chimpanzees.

(c) A physiologist compares the fermentation rates of yeast metabolizing glucose (a simple carbohydrate) or starch (a complex carbohydrate). Live yeast suspensions are placed in 20 fermentation flasks. Ten of the flasks contain a solution of water and glucose, and the other 10 flasks contain a solution of water and starch. The volume of carbon dioxide emitted per minute is measured for each flask.

We may wish to compare either the *centers* or the *spreads* of the two groups in a two-sample setting. This chapter emphasizes the most common inference procedures, those for comparing two population means. We comment briefly on the issue of comparing spreads (standard deviations), where simple inference is much less satisfactory.

APPLY YOUR KNOWLEDGE

Which data design? *Each situation described in Exercises 18.1 to 18.4 requires inference about a mean or means. Identify each as involving (1) a single sample, (2) matched pairs, or (3) two independent samples. The procedures of Chapter 17 apply to designs (1) and (2). We are about to learn procedures for (3).*

18.1   **Acid rain.** You obtain historical records from 24 hydrological stations in California and examine their data on the acidity of rainwater for the years 2005 and 1995. Acidity (or basicity) is measured by pH on a scale of 0 to 14. (The pH of distilled water is 7.0.) You want to estimate the average change in acidity of rainwater in California between the year 1995 and the year 2005.

18.2   **Agricultural pests.** Agricultural pests can be controlled to some extent either by using pesticides to kill them or by introducing a large number of sterilized males to diminish the species' reproductive potential. Ten large cornfields are treated with pesticides, and 10 have sterilized males introduced. Corn yield at harvest time is then compared for the two treatments.

18.3   **Chemical analysis.** To check a new analytical method, a chemist obtains a reference specimen of known concentration from the National Institute of Standards and Technology. She then makes 20 measurements of the concentration of this specimen with the new method and checks for bias by comparing the mean result with the known concentration.

18.4 **Chemical analysis again.** Another chemist is checking the same new method. He has no reference specimen, but a familiar analytic method is available. He wants to know if the new and old methods agree. He takes a specimen of unknown concentration and measures the concentration 10 times with the new method and 10 times with the old method.

Comparing two population means

We can examine two-sample data graphically by comparing boxplots, stemplots (for small samples), or histograms (for larger samples). Now we will learn confidence intervals and tests in this setting. When both population distributions are symmetric, and especially when they are at least approximately Normal, a comparison of the mean responses in the two populations is the most common goal of inference. Here are the conditions for inference.

CONDITIONS FOR INFERENCE COMPARING TWO MEANS

- We have **two SRSs,** from two distinct populations. The samples are **independent.** That is, one sample has no influence on the other (matching violates independence, for example). We measure the same variable for both samples.

- Both populations are **Normally distributed.** The means and standard deviations of the populations are unknown. In practice, it is enough that the distributions have similar shapes and that the data have no strong outliers.

Call the variable we measure x_1 in the first population and x_2 in the second, because the variable may have different distributions in the two populations. Here is the notation we will use to describe the two populations:

Population	Variable	Population Mean	Population Standard deviation
1	x_1	μ_1	σ_1
2	x_2	μ_2	σ_2

There are four unknown parameters, the two means and the two standard deviations. The subscripts remind us which population a parameter describes. We want to compare the two population means, either by giving a confidence interval for their difference $\mu_1 - \mu_2$ or by testing the hypothesis of no difference, $H_0: \mu_1 = \mu_2$, which is the same as $H_0: \mu_1 - \mu_2 = 0$.

We use the sample means and standard deviations to estimate the unknown parameters. Again, subscripts remind us which sample a statistic comes from. Here is the notation that describes the samples:

Population	Sample size	Sample mean	Sample standard deviation
1	n_1	$\overline{x}_1$	s_1
2	n_2	$\overline{x}_2$	s_2

To do inference about the difference $\mu_1 - \mu_2$ between the means of the two populations, we start from the difference $\overline{x}_1 - \overline{x}_2$ between the means of the two samples.

EXAMPLE 18.2 Does polyester decay?

STATE: How quickly do synthetic fabrics such as polyester decay in landfills? A researcher buried polyester strips in the soil for different lengths of time, then dug up the strips and measured the force required to break them. Breaking strength is easy to measure and is a good indicator of decay. Lower strength means the fabric has decayed.

Part of the study buried 10 strips of polyester fabric in well-drained soil in the summer. Five of the strips, chosen at random, were dug up after 2 weeks; the other 5 were dug up after 16 weeks. Here are the breaking strengths in pounds:[1]

Sample 1 (2 weeks)	118	126	126	120	129
Sample 2 (16 weeks)	124	98	110	140	110

We suspect that decay increases over time. Do the data give good evidence that mean breaking strength is less after 16 weeks than after 2 weeks?

FORMULATE: This is a two-sample setting. We want to compare the mean breaking strengths in the entire population of polyester fabric, μ_1 for fabric buried for 2 weeks and μ_2 for fabric buried for 16 weeks. So we will test the hypotheses

$$H_0: \mu_1 = \mu_2 \text{ (that is, } \mu_1 - \mu_2 = 0)$$
$$H_a: \mu_1 > \mu_2 \text{ (that is, } \mu_1 - \mu_2 > 0)$$

SOLVE (FIRST STEPS): Are the conditions for inference met? Because of the randomization, we are willing to regard the two groups of fabric strips as two independent SRSs from large populations of fabric. Although the samples are small, we check for serious non-Normality by examining the data. Figure 18.1 is a back-to-back stemplot of the responses. The 16-week group is much more spread out. As far as we can tell from so few observations, there are no departures from Normality that violate the conditions for comparing two means.

From the data, calculate the summary statistics:

2 weeks		16 weeks
	9	8
	10	
8	11	0 0
9 6 6 0	12	4
	13	
	14	0

FIGURE 18.1 Back-to-back stemplot of the breaking strength data from Example 18.2.

Stephen Wilkes/Getty Images

Group	Treatment	n	$\overline{x}$	s
1	2 weeks	5	123.80	4.60
2	16 weeks	5	116.40	16.09

The fabric that was buried longer has somewhat lower mean strength, along with more variation. The observed difference in mean strengths is

$$\overline{x}_1 - \overline{x}_2 = 123.80 - 116.40 = 7.40 \text{ pounds}$$

To complete the *Solve* step, we must learn the details of inference comparing two means.

Two-sample *t* procedures

To assess the significance of the observed difference between the means of our two samples, we follow a familiar path. Whether an observed difference is surprising depends on the spread of the observations as well as on the two means. Widely different means can arise just by chance if the individual observations vary a great deal. How much the difference $\overline{x}_1 - \overline{x}_2$ can vary from one random sampling to another is given by its sampling distribution.

When two random variables are Normally distributed, the new variable "difference" also follows a Normal distribution, centered on the difference of the two variables' means and with variance equal to the sum of the two variables' variances. We already know from Chapter 14 that the sampling distributions of $\overline{x}_1$ and $\overline{x}_2$ have standard deviations $\sigma_1 / \sqrt{n_1}$ and $\sigma_2 / \sqrt{n_2}$, respectively. Therefore, when we look at the difference $\overline{x}_1 - \overline{x}_2$, the standard deviation of its sampling distribution is

$$\sqrt{\frac{\sigma_1^2}{n_1} + \frac{\sigma_2^2}{n_2}}$$

This standard deviation gets larger as either population gets more variable, that is, as σ_1 or σ_2 increases. It gets smaller as the sample sizes n_1 and n_2 increase.

Because we don't know σ_1 and σ_2, we estimate them by the sample standard deviations s_1 and s_2. The result is the **standard error,** or estimated standard deviation, of the difference in sample means: *standard error*

$$SE = \sqrt{\frac{s_1^2}{n_1} + \frac{s_2^2}{n_2}}$$

When we standardize the estimate, we get

$$t = \frac{(\overline{x}_1 - \overline{x}_2) - (\mu_1 - \mu_2)}{SE}$$

but, because in a typical two-sample test $\mu_1 - \mu_2 = 0$, the result is the **two-sample**   *two-sample t statistic*
t **statistic:**

$$t = \frac{\overline{x}_1 - \overline{x}_2}{SE}$$

The statistic *t* has the same interpretation as any z or *t* statistic: It says how far the difference $\overline{x}_1 - \overline{x}_2$ is from 0 $(\mu_1 - \mu_2)$ in standard deviation units. (Exceptionally, the null hypothesis may define $\mu_1 - \mu_2$ as a value other than zero and the *t* statistic would be standardized using the full standardization formula.)

The two-sample *t* statistic has approximately a *t* distribution. It does not have exactly a *t* distribution even if the populations are both exactly Normal. In practice, however, the approximation is very accurate. There is a catch: The degrees of freedom of the *t* distribution we want to use are calculated from the data by a somewhat messy formula; moreover, the degrees of freedom need not be a whole number. There are two practical options for using the two-sample t procedures:

Option 1. With software, use the statistic t with accurate critical values from the approximating t distribution.

Option 2. Without software, use the statistic t with critical values from the t distribution with degrees of freedom equal to the smaller of $n_1 - 1$ and $n_2 - 1$. These procedures are always conservative for any two Normal populations.

The two options are the same except for the degrees of freedom used for t critical values and P-values. The Using Technology section (page 468) illustrates how software uses Option 1. Some details of Option 1 appear in a later section on page 472. We recommend that you use Option 1 unless you are working without software. Here is a description of the Option 2 procedures that includes a statement of just how they are "conservative." The formulas are the same as in Option 1; only the degrees of freedom differ.

THE TWO-SAMPLE t PROCEDURES (OPTION 2)

Draw an SRS of size n_1 from a large Normal population with unknown mean μ_1, and draw an independent SRS of size n_2 from another large Normal population with unknown mean μ_2. A level C **confidence interval for $\boldsymbol{\mu_1 - \mu_2}$** is given by

$$(\overline{x}_1 - \overline{x}_2) \pm t^* \sqrt{\frac{s_1^2}{n_1} + \frac{s_2^2}{n_2}}$$

Here t^* is the critical value with area C between $-t^*$ and t^* under the t density curve with degrees of freedom equal to the smaller of $n_1 - 1$ and $n_2 - 1$. This critical value gives a conservative margin of error *as large as or larger than* is needed for confidence level C, no matter what the population standard deviations may be.

To **test the hypothesis H_0: $\boldsymbol{\mu_1 = \mu_2}$**, calculate the **two-sample t statistic**

$$t = \frac{\overline{x}_1 - \overline{x}_2}{\sqrt{\dfrac{s_1^2}{n_1} + \dfrac{s_2^2}{n_2}}}$$

Find P-values from the t distribution with degrees of freedom equal to the smaller of $n_1 - 1$ and $n_2 - 1$. This distribution gives a conservative P-value *equal to or greater than* the true P-value, no matter what the population standard deviations may be.

The Option 2 two-sample t procedures always err on the safe side. They report *wider* confidence intervals and *higher* P-values than the more accurate Option 1 method. As the sample sizes increase, confidence levels and P-values from Option 2 become more accurate. The gap between what Option 2 reports and the truth is quite small unless the sample sizes are both small and unequal.[2] If you cannot find a particular degrees of freedom in Table C, pick the closest smaller value in the table. Smaller degrees of freedom result in more conservative *P*-values and confidence intervals and err on the safe side.

EXAMPLE 18.3   Does polyester decay?

We can now complete Example 18.2.

SOLVE (INFERENCE):   The test statistic for the null hypothesis $H_0: \mu_1 = \mu_2$ is

$$t = \frac{\overline{x}_1 - \overline{x}_2}{\sqrt{\dfrac{s_1^2}{n_1} + \dfrac{s_2^2}{n_2}}}$$

$$= \frac{123.8 - 116.4}{\sqrt{\dfrac{4.60^2}{5} + \dfrac{16.09^2}{5}}}$$

$$= \frac{7.4}{7.484} = 0.9889$$

Software (Option 1) gives one-sided *P*-value $P = 0.1857$.

Without software, use the conservative Option 2. Because $n_1 - 1 = 4$ and $n_2 - 1 = 4$, there are 4 degrees of freedom. Because H_a is one-sided on the high side, the *P*-value is the area to the right of $t = 0.9889$ under the $t(4)$ curve. Figure 18.2 illustrates this *P*-value. Table C shows that it lies between 0.15 and 0.20.

CONCLUDE: The experiment did not find convincing evidence that polyester decays more in 16 weeks than in 2 weeks ($P > 0.15$).

df = 4		
t^*	0.941	1.190
P	0.20	0.15

Sample size strongly influences the *P*-value of a test. An effect that fails to be significant at a level α in a small sample may be significant at the same level in a larger sample. Therefore, our result is statistically inconclusive. Beyond statistical significance, effect size also matters. The sample data showed buried polyester decaying slowly. The mean breaking strength dropped only from 123.8 pounds to 116.4 pounds between 2 and 16 weeks, which isn't statistically significant.

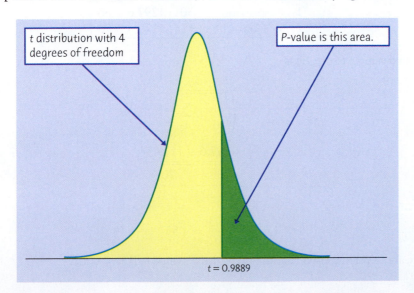

FIGURE 18.2 The *P*-value as the upper-tail area in Example 18.3. The conservative Option 2 leads to the *t* distribution with 4 degrees of freedom.

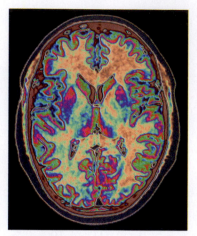

Image Source Pink/Alamy

EXAMPLE 18.4 Brain size and autism

STATE: Is autism marked by different brain growth patterns in early life, even before the diagnosis is made? Studies have linked brain size in infants and toddlers to a number of future ailments, including autism. One study looked at the brain sizes of 30 autistic boys and 12 nonautistic boys (control) who all had received an MRI scan as toddlers. Here are their whole-brain volumes in milliliters:[3]

Autistic									
1311	1250	1292	1419	1401	1297	1202	1336	1308	1353
1515	1461	1365	1364	1362	1303	1278	1247	1333	1340
1319	1286	1223	1241	1229	1209	1171	1154	1128	1230

Control									
1040	1180	1207	1179	1115	1133	1298	1263	1194	1198
1230	1114								

FORMULATE: We had no specific direction for the difference in brain volumes before looking at the data, so the alternative is two-sided. We will test the hypotheses

$$H_0: \mu_1 = \mu_2 \text{ (that is, } \mu_1 - \mu_2 = 0)$$
$$H_a: \mu_1 \neq \mu_2 \text{ (that is, } \mu_1 - \mu_2 \neq 0)$$

SOLVE: The two samples can be regarded as SRSs from two populations, although there are clear limitations due to the lack of availability of young children for biomedical studies. Figure 18.3 shows the two data sets in a back-to-back stemplot and in Normal quantile plots.[4] The stemplot in Figure 18.3(a) shows no deviation from Normality for the 30 autistic boys and a very mild possible outlier for the 12 control boys. The quantile plots in Figures 18.3(b) and 18.3(c) suggest that a Normal distribution is a reasonable model for both groups. Thus, the t procedures should be valid. Here are the calculations leading to the two-sample t-test:

Group	Condition	n	$\bar{x}$	s
1	Autistic	30	1297.6	88.4
2	Control	12	1179.3	70.7

The two-sample t statistic is

$$t = \frac{\bar{x}_1 - \bar{x}_2}{\sqrt{\dfrac{s_1^2}{n_1} + \dfrac{s_2^2}{n_2}}}$$

$$= \frac{1297.6 - 1179.3}{\sqrt{\dfrac{88.4^2}{30} + \dfrac{70.7^2}{12}}}$$

$$= \frac{118.3}{26.02} = 4.55$$

Software (Option 1) says that the two-sided P-value is $P = 0.0001$.

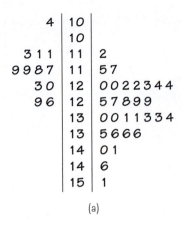

```
         4 │ 10
           │ 10
      3 1 1 │ 11 │ 2
    9 9 8 7 │ 11 │ 5 7
        3 0 │ 12 │ 0 0 2 2 3 4 4
        9 6 │ 12 │ 5 7 8 9 9
           │ 13 │ 0 0 1 1 3 3 4
           │ 13 │ 5 6 6 6
           │ 14 │ 0 1
           │ 14 │ 6
           │ 15 │ 1
```

(a)

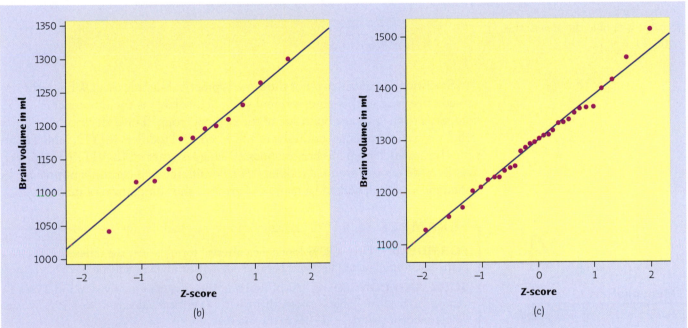

(b)   (c)

FIGURE 18.3 Brain volumes from Example 18.4 displayed in a back-to-back stemplot (a) and in Normal quantile plots for the control group (b) and for the autistic group (c).

Without software, use Option 2 to find a conservative *P*-value. There are 11 degrees of freedom, the smaller of

$$n_1 - 1 = 30 - 1 = 29 \quad \text{and} \quad n_2 - 1 = 12 - 1 = 11$$

Figure 18.4 illustrates the *P*-value. Find it by comparing 4.55 with the two-sided critical values for the *t*(11) distribution.

Notice that our *t* statistic is larger than the largest *t* critical in Table C for degrees of freedom 11. When this happens, simply conclude that your *P*-value is smaller than the smallest *P*-value provided by the table. Here we conclude that the *P*-value for our test is less than 0.001 (a 2-sided *P*).

CONCLUDE: The data give very strong evidence (*P* < 0.001) that autistic boys have larger brains on average than nonautistic boys during the toddler years.

df = 11

t*	4.437
P	0.001

t distribution with 11 degrees of freedom

P-value is 2 times this area.

$t = -4.55$ $t = 4.55$

FIGURE 18.4 The P-value in Example 18.4. Because the alternative is two-sided, the P-value is double the area to the right of $t = 4.55$.

Statistical significance is not the same as practical significance. What is the observed effect size in this observational study? How large is the difference $\bar{x}_1 - \bar{x}_2$ compared to the average brain size of the control group? The difference $\bar{x}_1 - \bar{x}_2 = 118.3$ ml, and the average brain size of the 12 control toddlers is 1179.3 ml. Therefore, the autistic group has brains about 10% larger (that's 118.3/1179.3) on average than the control group of toddlers. A 10% difference is certainly not trivial. A confidence interval will add a margin of error to the comparison of means.

STEP 4

Meta-analysis

Small samples have large margins of error. Large samples are expensive. Often we can find several studies of the same issue; if we could combine their results, we would have a large sample with a small margin of error. That is the idea of "meta-analysis." Of course, we can't just lump the studies together, because of differences in design and quality. Statisticians have more sophisticated ways of combining the results. Meta-analysis has been applied to issues ranging from the effect of secondhand smoke to whether Echinacea can prevent the common cold.

── **EXAMPLE 18.5**   *The autistic brain: how much larger?* ──

FORMULATE: Give a 90% confidence interval for $\mu_1 - \mu_2$, the difference in mean brain size during the toddler years between all autistic and nonautistic boys.

SOLVE AND CONCLUDE: As in Example 18.4, the conservative Option 2 uses 11 degrees of freedom. Table C shows that the $t(11)$ critical value is $t^* = 1.796$. We are 90% confident that $\mu_1 - \mu_2$ lies in the interval

$$(\bar{x}_1 - \bar{x}_2) \pm t^* \sqrt{\frac{s_1^2}{n_1} + \frac{s_2^2}{n_2}} = (1297.6 - 1179.3) \pm 1.796 \sqrt{\frac{88.4^2}{30} + \frac{70.7^2}{12}}$$

$$= 118.3 \pm 46.7$$

$$= 71.6 \text{ to } 165.0$$

Notice that, because 0 lies outside the 90% confidence interval, we can reject H_0: $\mu_1 = \mu_2$ in favor of the two-sided alternative at the $\alpha = 0.10$ level of significance. Remember that a confidence interval can also help you test hypotheses.

The authors concluded that abnormal brain development in autism may occur prior to 2 or 3 years of age and that future research should explore ways to limit or prevent the full expression of these abnormalities. However, because this is an observational study, it is not possible to conclude that autism is indeed the cause of the observed difference in brain volumes. Confounding variables might explain the difference, especially since the two groups were recruited separately. The researcher honestly disclosed this and other limitations of the study. Disclosing the limitations of a study design or of a statistical procedure is an important part of the ethical conduct of research.

APPLY YOUR KNOWLEDGE

18.5 **Whelks on the Pacific coast.** Published reports of statistical analyses are often very terse. A knowledge of basic statistics helps you decode what you read. In a study of the presence of whelks along the Pacific coast, investigators put down a frame that covers 0.25 square meter and counted the whelks on the sea bottom inside the frame. They did this at 7 locations off California and 6 locations off Oregon. The report says that whelk densities "were twice as high in Oregon as in California (mean ± SEM, 26.9 ± 1.56 versus 11.9 ± 2.68 whelks per 0.25 m^2, Oregon versus California, respectively; Student's t test, $P < 0.001$)."[5]

Peter Egerton

(a) SEM stands for the standard error of the mean, $s/\sqrt{n}$. Fill in the values in this summary table:

Group	Location	n	$\bar{x}$	s
1	Oregon	?	?	?
2	California	?	?	?

(b) What degrees of freedom would you use in the conservative two-sample t procedures to compare Oregon and California?

18.6 **Echinacea for the common cold?** Echinacea is widely used as an herbal remedy for the common cold, but does it work? In a double-blind experiment, healthy volunteers agreed to be exposed to common-cold-causing rhinovirus type 39 and have their symptoms monitored. The volunteers were randomly assigned to take either a placebo or an echinacea supplement daily from 7 days before till 5 days after viral exposure. A symptom score was recorded for each subject over the 5 days following exposure, with higher scores indicating more severe symptoms. The published results reported the mean ± SEM for both groups as 13.21 ± 1.91 (echinacea) and 15.05 ± 1.43 (placebo).[6]

(a) The two-sample t statistic for $\bar{x}_1 - \bar{x}_2$ was $t = -0.771$. You can draw a conclusion from this t without using a table and even without knowing the sizes of the samples (remember to specify your null and alternative hypotheses first). What is your conclusion? Why don't you need the sample sizes and a table?

(b) In fact, 52 subjects were assigned to the echinacea treatment and 103 to the placebo. Fill in the values in this summary table:

Group	Treatment	n	$\bar{x}$	s
1	Echinacea	?	?	?
2	Placebo	?	?	?

What degrees of freedom would you use in the conservative two-sample t procedures recommended for use without software? What P-value would you get from Table C?

18.7 **Logging in the rain forest.** "Conservationists have despaired over destruction of tropical rain forest by logging, clearing, and burning." These words begin a report on a statistical study of the effects of logging in Borneo.[7] Here are data on the number of tree species in 12 unlogged forest plots and 9 similar plots logged 8 years earlier:

Unlogged	22	18	22	20	15	21	13	13	19	13	19	15
Logged	17	4	18	14	18	15	15	10	12			

(a) The study report says, "Loggers were unaware that the effects of logging would be assessed." Why is this important? The study report also explains why the plots can be considered to be randomly assigned.

(b) Does logging significantly reduce the mean number of species in a plot after 8 years? Follow the four-step process as illustrated in Examples 18.2 and 18.3.

18.8 **Logging in the rain forest, continued.** Use the data in the previous exercise to give a 90% confidence interval for the difference in mean number of species between unlogged and logged plots.

Using technology

Software should use Option 1 for the degrees of freedom to give accurate confidence intervals and P-values. However, there is some variation in software output due to how precisely the degrees of freedom are computed for Option 1. Figure 18.5 displays output from a graphing calculator, three statistical programs, and a spreadsheet program for the test and 90% confidence interval of Example 18.3. All five claim to use Option 1. The two-sample t statistic is exactly as in Example 18.3, $t = 0.9889$. You can find this in all five outputs (Minitab rounds to 0.99 and SPSS to 0.989). The different technologies use different methods to find the P-value for $t = 0.9889$.

- The TI-83, SPSS, and CrunchIt! are the most accurate, because they use the t distribution with 4.65 degrees of freedom. The P-value is 0.1857 (although SPSS provides the two-sided P-value, "Sig. (2-tailed)," which must be cut in half for this one-sided test).

- Minitab and Excel both round the degrees of freedom to a whole number. Minitab truncates the exact degrees of freedom to the next smaller whole number (df = 4.65 is truncated to df = 4) and therefore always errs a bit on the conservative side, whereas Excel rounds the exact degrees of freedom to the nearest whole number (df = 4.65 becomes df = 5). As a result, Minitab offers $P = 0.1895$ (half of the two-sided value 0.379) and Excel shows $P = 0.1841$.

These slight differences are not crucial for your conclusions, because it is the order of magnitude of the P-value that really matters, not its exact value. Even "between 0.15 and 0.20" from Table C is close enough for practical purposes. Also, the differences seen in various software outputs get even smaller for larger sample

Texas Instruments TI-83 Plus

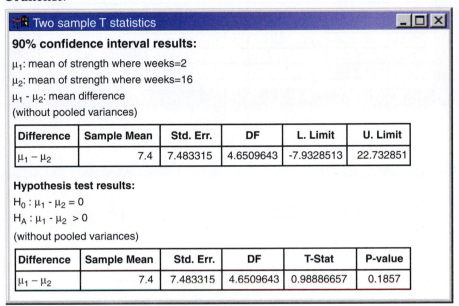

```
2-SampTInt          2-SampTTest
 (-7.933,22.733)     μ1>μ2
 df=4.6510           t=.9889
 x̄1=123.8000         P=.1857
 x̄2=116.4000         df=4.6510
 Sx1=4.6043          x̄1=123.8000
↓Sx2=16.0873        ↓x̄2=116.4000
```

CrunchIt!

Two sample T statistics

90% confidence interval results:

μ_1: mean of strength where weeks=2

μ_2: mean of strength where weeks=16

$\mu_1 - \mu_2$: mean difference

(without pooled variances)

Difference	Sample Mean	Std. Err.	DF	L. Limit	U. Limit
$\mu_1 - \mu_2$	7.4	7.483315	4.6509643	-7.9328513	22.732851

Hypothesis test results:

$H_0 : \mu_1 - \mu_2 = 0$

$H_A : \mu_1 - \mu_2 > 0$

(without pooled variances)

Difference	Sample Mean	Std. Err.	DF	T-Stat	P-value
$\mu_1 - \mu_2$	7.4	7.483315	4.6509643	0.98886657	0.1857

Minitab

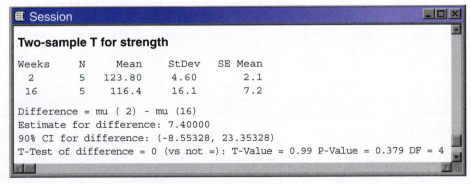

Session

Two-sample T for strength

```
Weeks   N    Mean   StDev   SE Mean
  2     5   123.80   4.60     2.1
 16     5   116.4   16.1      7.2

Difference = mu ( 2) - mu (16)
Estimate for difference: 7.40000
90% CI for difference: (-8.55328, 23.35328)
T-Test of difference = 0 (vs not =): T-Value = 0.99 P-Value = 0.379 DF = 4
```

FIGURE 18.5 The two-sample *t* procedures applied to the polyester decay data of Example 18.3: output from a graphing calculator, three statistical programs, and a spreadsheet program (*continued*).

sizes, because they have larger degrees of freedom, resulting in a proportionally smaller impact of rounding errors.

Excel's label for the test, "Two-Sample Assuming Unequal Variances," is seriously misleading. *The two-sample* t *procedures we have described work whether or not the two populations have the same variance.* The SPSS output is more appropriately

CAUTION

Excel

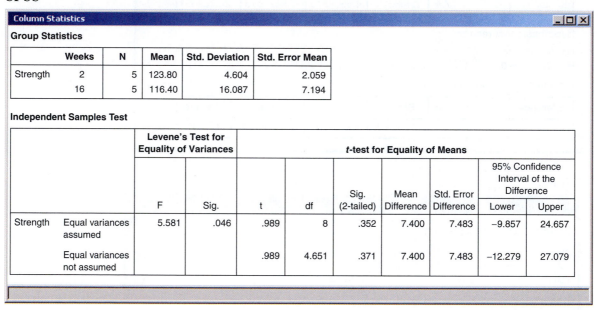

SPSS

Column Statistics

Group Statistics

	Weeks	N	Mean	Std. Deviation	Std. Error Mean
Strength	2	5	123.80	4.604	2.059
	16	5	116.40	16.087	7.194

Independent Samples Test

		Levene's Test for Equality of Variances		t-test for Equality of Means						
									95% Confidence Interval of the Difference	
		F	Sig.	t	df	Sig. (2-tailed)	Mean Difference	Std. Error Difference	Lower	Upper
Strength	Equal variances assumed	5.581	.046	.989	8	.352	7.400	7.483	−9.857	24.657
	Equal variances not assumed			.989	4.651	.371	7.400	7.483	−12.279	27.079

FIGURE 18.5 (*continued*)

labeled: "Equal variances not assumed." There is a variant of the two-sample *t*-test we have described that works only when the two variances are equal. We discuss this method briefly on page 474, but you should always prefer the test that does not assume equal variances, because it will give correct results no matter what the population variances are.

Last, some software outputs show both the one-sided ("one-tail") and two-sided ("two-tail") *P*-values or just the two-sided *P*-value. Remember to specify your null and alternative hypotheses even before examining the sample data and then select or work out the one- or two-sided P-value that matches your alternative.

Robustness again

The two-sample t procedures are more robust than the one-sample t methods, particularly when the distributions are not symmetric. When the sizes of the two samples are equal and the two populations being compared have distributions with similar shapes, probability values from the t table are quite accurate for a broad range of distributions when the sample sizes are as small as $n_1 = n_2 = 5$.[8] When the two population distributions have different shapes, larger samples are needed.

As a guide to practice, adapt the guidelines given on page 446 for the use of one-sample t procedures to two-sample procedures by replacing "sample size" with "sum of the sample sizes," $n_1 + n_2$. These guidelines err on the side of safety, especially when the two samples are of equal size. *In planning a two-sample study, choose equal sample sizes whenever possible. The two-sample* t *procedures are most robust against non-Normality in this case.*

APPLY YOUR KNOWLEDGE

18.9 Bone loss in nursing mothers. Exercise 2.36 (page 62) gives the percent change in the mineral content of the spine for 47 mothers during three months of nursing a baby and for a control group of 22 women of similar age who were neither pregnant nor lactating.

(a) What two populations did the investigators want to compare? We must be willing to regard the women recruited for this observational study as SRSs from these populations.

(b) Do these data give good evidence that the average bone mineral loss is higher in the population of nursing mothers? Complete the *Formulate, Solve,* and *Conclude* steps of the four-step process as illustrated in Examples 18.2 and 18.3.

18.10 Weeds among the corn. Lamb's quarter is a common weed that interferes with the growth of corn. An agriculture researcher planted corn with the same density in identical small plots of ground. The plots were then weeded by hand to allow a fixed density of lamb's quarter plant. No other weed was allowed to grow. Here are the corn yields (in bushels per acre) for experimental plots controlled to have 1 weed per meter of planted corn (corn is always planted in rows) and 3 weeds per meter.[9]

| 1 weed/meter | 166.2 | 157.3 | 166.7 | 161.1 |
| 3 weeds/meter | 158.6 | 176.4 | 153.1 | 156.0 |

Explain carefully why a two-sample t confidence interval for the difference in mean yields may not be accurate.

Details of the *t* approximation*

The exact distribution of the two-sample *t* statistic is not a *t* distribution. Moreover, the distribution changes as the unknown population standard deviations σ_1 and σ_2 change. However, an excellent approximation is available. We call this Option 1 for *t* procedures.

APPROXIMATE DISTRIBUTION OF THE TWO-SAMPLE *t* STATISTIC

The distribution of the two-sample *t* statistic is very close to the *t* distribution with degrees of freedom df given by

$$df = \frac{\left(\dfrac{s_1^2}{n_1} + \dfrac{s_2^2}{n_2}\right)^2}{\dfrac{1}{n_1 - 1}\left(\dfrac{s_1^2}{n_1}\right)^2 + \dfrac{1}{n_2 - 1}\left(\dfrac{s_2^2}{n_2}\right)^2}$$

This approximation is accurate when both sample sizes n_1 and n_2 are 5 or larger.

The *t* procedures remain exactly as before except that we use the *t* distribution with df degrees of freedom to give critical values and *P*-values.

EXAMPLE 18.6   Does polyester decay?

In the experiment of Examples 18.2 and 18.3, the data on buried polyester fabric gave

Group	Treatment	n	$\overline{x}$	s
1	2 weeks	5	123.80	4.60
2	16 weeks	5	116.40	16.09

The two-sample *t* test statistic calculated from these values is $t = 0.9889$.

The one-sided *P*-value is the area to the right of 0.9889 under a *t* density curve, as in Figure 18.2. The conservative Option 2 uses the *t* distribution with 4 degrees of freedom. Option 1 finds a very accurate *P*-value by using the *t* distribution with degrees of freedom df given by

$$df = \frac{\left(\dfrac{4.60^2}{5} + \dfrac{16.09^2}{5}\right)^2}{\dfrac{1}{4}\left(\dfrac{4.60^2}{5}\right)^2 + \dfrac{1}{4}\left(\dfrac{16.09^2}{5}\right)^2}$$

$$= \frac{3137.08}{674.71} = 4.65$$

These degrees of freedom appear in the output from the TI-83, SPSS, and CrunchIt! in Figure 18.5.

---

*This section can be omitted unless you are using software and wish to understand what the software does.

The degrees of freedom df is generally not a whole number. It is always at least as large as the smaller of $n_1 - 1$ and $n_2 - 1$. The larger degrees of freedom that result from Option 1 give slightly shorter confidence intervals and slightly smaller *P*-values than the conservative Option 2 produces. There is a *t* distribution for any positive degrees of freedom, even though Table C contains entries only for whole-number degrees of freedom.

The difference between the *t* procedures using Options 1 and 2 is rarely of practical importance. That is why we recommend the simpler, conservative Option 2 for inference without software. With software, the more accurate Option 1 procedures are painless.

APPLY YOUR KNOWLEDGE

18.11 DDT poisoning. In a randomized comparative experiment, researchers compared 6 white rats poisoned with DDT with a control group of 6 unpoisoned rats. Electrical measurements of nerve activity are the main clue to the nature of DDT poisoning. When a nerve is stimulated, its electrical response shows a sharp spike followed by a much smaller second spike. The experiment found that the second spike is larger in rats fed DDT than in normal rats.[10]

The researchers measured the height of the second spike as a percent of the first spike when a nerve in the rat's leg was stimulated. Here are the results.

Poisoned	12.207	16.869	25.050	22.429	8.456	20.589
Unpoisoned	11.074	9.686	12.064	9.351	8.182	6.642

Figure 18.6 shows the CrunchIt! output for the two-sample *t* test. What are $\bar{x}_i$ and s_i for the two samples? Starting from these values, find the *t* test statistic and its degrees of freedom. Your work should agree with Figure 18.6.

18.12 Children's mental health. The Piers-Harris Children's Self-Concept Scale is one of the most widely used measures of mental health in children and adolescents, assessing the subjects self-perception of their own behavior and attributes. A study of the self-concept of seventh-grade students asked if girls and

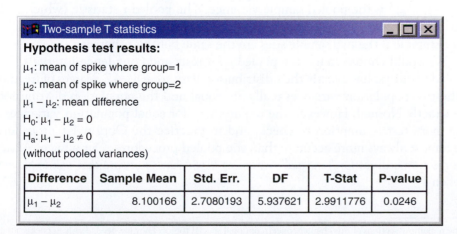

FIGURE 18.6 Two-sample *t* output from CrunchIt! for Exercise 18.11.

boys differ in mean score on this scale. Software gave the following summary results.[11]

Gender	n	Mean	Std dev	Std err	t	df	P
G	31	55.5161	12.6961	2.2803	-0.8276	62.8	0.4110
B	47	57.9149	12.2649	1.7890			

Starting from the sample means and standard deviations, verify each of these entries: the standard errors of the means; the degrees of freedom for two-sample t; the value of t.

18.13 DDT poisoning, continued. Do poisoned rats differ significantly from unpoisoned rats in the study of Exercise 18.11? Write a summary in a sentence or two, including t, df, P, and a conclusion. Use the output in Figure 18.6.

18.14 Children's mental health, continued. Write a sentence or two summarizing the comparison of girls and boys in Exercise 18.12, as if you were preparing a report for publication. Use the output in Exercise 18.12.

Avoid the pooled two-sample t procedures*

Most software packages, including those illustrated in Figure 18.5, offer a choice of two-sample t statistics. One is often labeled for "unequal" variances, the other for "equal" variances. The "unequal" variance procedure is our two-sample t. *This test is valid whether or not the population variances are equal.* The other choice is a special version of the two-sample t statistic that assumes that the two populations have the same variance. It averages (the statistical term is "pools") the two sample variances to estimate the common population variance. The resulting statistic is called the *pooled two-sample* t *statistic*. It has the equation

$$t_{pooled} = \frac{\overline{x}_1 - \overline{x}_2}{s_{pooled}\sqrt{\dfrac{1}{n_1} + \dfrac{1}{n_2}}} = \frac{\overline{x}_1 - \overline{x}_2}{\sqrt{\dfrac{(n_1 - 1)s_1^2 + (n_2 - 1)s_2^2}{n_1 + n_2 - 2}}\sqrt{\dfrac{1}{n_1} + \dfrac{1}{n_2}}}$$

where s_{pooled}^2 is the pooled sample variance. The pooled t statistic (which is directly related to the ANOVA procedure we will discuss in Chapter 24) is equal to our t statistic if the two sample sizes are the same but not otherwise.

We could choose to use the pooled t for tests and confidence intervals. The pooled t statistic has exactly the t distribution with $n_1 + n_2 - 2$ degrees of freedom *if* the two population variances really are equal and the population distributions are exactly Normal. However, the requirement for equal population variances is one more test assumption to check, and in practice the Option 1 t procedures are almost always more accurate than the pooled procedures. Our advice: *Always use the* t *procedures for "unequal" variances if you have software that will implement Option 1.*

CAUTION

---

*The remaining sections of this chapter concern optional special topics. They are needed only as background for Chapter 24 on one-way ANOVA.

Avoid inference about standard deviations*

Two basic features of a distribution are its center and its spread. In a Normal population, we measure the center by the mean and the spread by the standard deviation. We use the t procedures for inference about population means for Normal populations, and we know that t procedures are widely useful for non-Normal populations as well. It is natural to turn next to inference about the standard deviations of Normal populations. Our advice here is short and clear: Don't do it without expert advice.

There are methods for inference about the standard deviations of Normal populations. The most common such method is the F test for comparing the spread of two Normal populations. *Unlike the t procedures for means, the F test is extremely sensitive to non-Normal distributions.* This lack of robustness does not improve in large samples. It is difficult in practice to tell whether a significant F-value is evidence of unequal population spreads or simply a sign that the populations are not Normal.

You can see intuitively why an inference procedure for the spread of a Normal population would work very poorly for non-Normal distributions. We saw back in Chapter 2 that the standard deviation is a natural measure of spread for Normal distributions but not for distributions in general. In fact, because skewed distributions have unequally spread tails, no single numerical measure does a good job of describing the spread of a skewed distribution. In summary, the standard deviation is not always a useful parameter, and even when it is (for symmetric distributions), the results of inference are not always trustworthy. Consequently, *we do not recommend trying to do inference about population standard deviations in basic statistical practice.*[12]

CHAPTER 18 SUMMARY ───────────────

The data in a **two-sample problem** are two independent SRSs, each drawn from a separate population.

Tests and confidence intervals for the difference between the means μ_1 and μ_2 of two Normal populations start from the difference $\overline{x}_1 - \overline{x}_2$ between the two sample means. Because of the central limit theorem, the resulting procedures are approximately correct for other population distributions when the sample sizes are large.

Draw independent SRSs of sizes n_1 and n_2 from two Normal populations with parameters μ_1, σ_1 and μ_2, σ_2. The **two-sample t statistic** is

$$t = \frac{(\overline{x}_1 - \overline{x}_2) - (\mu_1 - \mu_2)}{\sqrt{\dfrac{s_1^2}{n_1} + \dfrac{s_2^2}{n_2}}}$$

The statistic t has approximately a t distribution.

There are two choices for the **degrees of freedom** of the two-sample t statistic.

Option 1: Software produces accurate probability values using degrees of freedom calculated from the data. Option 2: For conservative inference procedures, use degrees of freedom equal to the smaller of $n_1 - 1$ and $n_2 - 1$.

The **confidence interval for $\mu_1 - \mu_2$** is

$$(\bar{x}_1 - \bar{x}_2) \pm t^* \sqrt{\frac{s_1^2}{n_1} + \frac{s_2^2}{n_2}}$$

The critical value t^* from Option 1 gives a confidence level very close to the desired level C. Option 2 produces a margin of error at least as wide as is needed for the desired level C.

Significance tests for H_0: $\mu_1 = \mu_2$ are based on

$$t = \frac{\bar{x}_1 - \bar{x}_2}{\sqrt{\frac{s_1^2}{n_1} + \frac{s_2^2}{n_2}}}$$

P-values calculated from Option 1 are very accurate. Option 2 P-values are always at least as large as the true P.

The guidelines for practical use of two-sample t procedures are similar to those for one-sample t procedures. Equal sample sizes are recommended.

CHECK YOUR SKILLS

18.15 How many calories does a fruit juice popsicle labeled "no sugar added" pack? A random sample of 10 popsicles of a particular brand is selected and the calories per serving measured. To get a confidence interval for the mean calories per serving of all no-sugar-added popsicles from this brand, you would use

(a) the one-sample t interval.

(b) the matched pairs t interval.

(c) the two-sample t interval.

18.16 A study is designed to compare the amount of vitamin C (in milligrams [mg] per serving) in oranges that reach the stores either 1 day after being picked or 3 days after being picked. A random sample of 15 oranges is taken for each group. To test whether there is a difference in the mean vitamin C amount of all oranges reaching stores either 1 day or 3 days after being picked, you would use

(a) the one-sample t test.

(b) the matched pairs t test.

(c) the two-sample t test.

18.17 There are two common methods for measuring the concentration of a pollutant in fish tissue. Do the two methods differ on the average? You apply both methods to one sample of 18 carp and use

(a) the one-sample t test.

(b) the matched pairs t test.

(c) the two-sample t test.

18.18 A study of the effects of exercise used rats bred to have high or low capacity for exercise. There were 8 high-capacity and 8 low-capacity rats. To compare the mean blood pressure of the two types of rats using the conservative Option 2 t procedures, the correct degrees of freedom is

    (a) 7.   (b) 14.   (c) 15.

18.19 The 8 high-capacity rats from the previous question had mean blood pressure 89 with standard deviation 9; the 8 low-capacity rats had mean blood pressure 105 with standard deviation 13. (Blood pressure is measured in millimeters of mercury [mm Hg].) The two-sample t statistic for comparing the population means has value

    (a) 0.5.   (b) 2.86.   (c) 9.65.

Use the following information for Exercises 18.20 to 18.23. A study of road rage asked samples of 596 men and 523 women about their behavior while driving. Based on their answers, each subject was assigned a road rage score on a scale of 0 to 20.

18.20 The subjects were chosen by random digit dialing of telephone numbers. Are the conditions for two-sample t inference satisfied?

    (a) Maybe: The SRS condition is OK but we need to look at the data to check Normality.

    (b) No: Scores in a range between 0 and 20 can't be Normal.

    (c) Yes: The SRS condition is OK and large sample sizes make the Normality condition unnecessary.

18.21 We suspect that men are more prone to road rage than women. To see if this is true, test these hypotheses for the mean road rage scores of all male and female drivers:

    (a) H_0: $\mu_M = \mu_F$ versus H_a: $\mu_M > \mu_F$.

    (b) H_0: $\mu_M = \mu_F$ versus H_a: $\mu_M \neq \mu_F$.

    (c) H_0: $\mu_M = \mu_F$ versus H_a: $\mu_M < \mu_F$.

18.22 The two-sample t statistic for the road rage study (male mean minus female mean) is $t = 3.18$. The P-value for testing the hypotheses from the previous exercise satisfies

    (a) $0.001 < P < 0.005$.   (b) $0.0005 < P < 0.001$.   (c) $0.001 < P < 0.002$.

18.23 To calculate a 95% confidence interval for $\mu_M - \mu_F$, you would use the conservative t critical

    (a) 1.96.   (b) 1.984.   (c) 12.71.

CHAPTER 18 EXERCISES ──────────────

In exercises that call for two-sample t procedures, use Option 1 if you have technology that implements that method. Otherwise, use Option 2 (degrees of freedom = the smaller of $n_1 - 1$ and $n_2 - 1$). Many of these exercises ask you to think about issues of statistical practice as well as to carry out procedures.

18.24 **More nutritious corn.** Ordinary corn doesn't have as much of the amino acid lysine as animals need in their feed. Plant scientists have developed varieties of corn that have increased amounts of lysine. In a test of the quality of high-lysine corn as animal feed, an experimental group of 20 one-day-old male chicks ate a ration containing the new corn. A control group of another 20 chicks received a ration that was identical except that it contained ordinary corn. Here are the weight gains (in grams) after 21 days.[13]

Control				Experimental			
380	321	366	356	361	447	401	375
283	349	402	462	434	403	393	426
356	410	329	399	406	318	467	407
350	384	316	272	427	420	477	392
345	455	360	431	430	339	410	326

Is there good evidence that chicks fed high-lysine corn gain more weight? Follow the four-step process as illustrated in Examples 18.2 and 18.3. That is, state hypotheses, make graphs to examine the data, discuss the conditions for inference, carry out a test, and state your conclusion.

18.25 **IQ scores for boys and girls.** Here are the IQ test scores of 31 seventh-grade girls in a midwest school district.[14]

114	100	104	89	102	91	114	114	103	105
108	130	120	132	111	128	118	119	86	72
111	103	74	112	107	103	98	96	112	112	93

The IQ test scores of 47 seventh-grade boys in the same district are

111	107	100	107	115	111	97	112	104	106	113
109	113	128	128	118	113	124	127	136	106	123
124	126	116	127	119	97	102	110	120	103	115
93	123	79	119	110	110	107	105	105	110	77
90	114	106								

(a) Make stemplots or histograms of both sets of data. Because the distributions are reasonably symmetric with no extreme outliers, the *t* procedures will work well.

(b) Treat these data as SRSs from all seventh-grade students in the district. Is there good evidence that girls and boys differ in their mean IQ scores?

18.26 **More nutritious corn, continued.**

(a) Use the data in Exercise 18.24 to give a 90% confidence interval for the difference in mean weight gain after 21 days for chicks fed high-lysine corn or ordinary corn.

(b) Give a 90% confidence interval for the mean weight gain after 21 days for chicks fed high-lysine corn.

18.27 **IQ scores for boys and girls, continued.** Use the data in Exercise 18.25 to give a 95% confidence interval for the difference between the mean IQ scores of all boys and all girls in the district.

18.28 **Compressing soil.** Farmers know that driving heavy equipment on wet soil compresses the soil and injures future crops. Here are data on the "penetrability" of the same type of soil at two levels of compression.[15] Penetrability is a measure of how much resistance plant roots will meet when they try to grow through the soil.

(a) Make stemplots to investigate the shape of the distributions. The penetrabilities for intermediate soil are skewed to the right and have a high outlier. Returning to the source of the data shows that the outlying sample had unusually low soil density, so that it belongs in a different "loose soil" class. We are justified in removing the outlier.

Compressed soil									
2.86	2.68	2.92	2.82	2.76	2.81	2.78	3.08	2.94	2.86
3.08	2.82	2.78	2.98	3.00	2.78	2.96	2.90	3.18	3.16

Intermediate soil									
3.14	3.38	3.10	3.40	3.38	3.14	3.18	3.26	2.96	3.02
3.54	3.36	3.18	3.12	3.86	2.92	3.46	3.44	3.62	4.26

(b) We suspect that the penetrability of compressed soil is less than that of intermediate soil. Do the data (with the outlier removed) support this suspicion? Carry out the appropriate test and state your conclusion.

(c) Give a 95% confidence interval for the decrease in penetrability of compressed soil relative to intermediate soil. (Omit the outlier.)

18.29 Fungus in the air. The air in poultry-processing plants often contains fungus spores. Inadequate ventilation can affect the health of the workers. The problem is most serious during the summer and least serious during the winter. To measure the presence of spores, air samples are pumped to an agar plate and "colony-forming units (CFUs)" are counted after an incubation period. Here are data from the "kill room" of a plant that processes 37,000 turkeys per day, taken on four separate days in the summer and in the winter. The units are CFUs per cubic meter of air.[16]

Summer	3175	2526	1763	1090
Winter	384	104	251	97

The counts are clearly much higher in the summer. Give a 90% confidence interval to estimate how much higher the mean count is during the summer. Follow the four-step process as illustrated in Example 18.5.

Paula Bronstein/Getty Images

18.30 Do subliminal messages work? A "subliminal" message is below our threshold of awareness but may nonetheless influence us. A study looked at the effect of subliminal messages on math skills. The messages were flashed on a screen too rapidly to be consciously read. Twenty-eight students who had failed the mathematics part of the City University of New York Skills Assessment Test were randomly assigned to receive daily either a positive subliminal message ("Each day I am getting better in math.") or a neutral subliminal message ("People are walking on the street.") All students participated in a summer program designed to raise their math skills, and all took the assessment test again at the end of the program. Table 18.1 gives data on the subjects' scores before and after the program.[17] Is there good evidence that the positive message brought about a greater improvement in math scores than the neutral message? How large is the mean difference in gains between the two groups? (Use 90% confidence.)

Exercises 18.31 to 18.37 are based on summary statistics rather than raw data. This information is typically all that is presented in published reports. Inference procedures can be calculated by hand from the summaries. You must trust that the authors understood the conditions for inference and verified that they apply. This isn't always true.

18.31 Pharmacokinetics of a drug. Avandia (rosiglitazone maleate) is an oral antidiabetic drug produced by the pharmaceutical company GlaxoSmithKline. Before a drug can be prescribed, we must know how the body absorbs and excretes

TABLE 18.1	Mathematics skills scores before and after a subliminal message		

Positive Message		Neutral Message	
Before	After	Before	After
18	24	18	29
18	25	24	29
21	33	20	24
18	29	18	26
18	33	24	38
20	36	22	27
23	34	15	22
23	36	19	31
21	34		
17	27		

it. Patients were given a single dose of either 1 mg or 2 mg of rosiglitazone, and the maximum plasma concentration of the drug (in ng/ml) was assessed.[18]

Treatment	n	$\overline{x}$	s
1 mg	32	76	13
2 mg	32	156	42

Is there significant evidence that maximum plasma concentration is dose dependent?

18.32 More on pharmacokinetics. The study in the previous exercise also looked at how fast the body gets rid of the drug by measuring the elimination half-life (in hours). Here are the results.

Treatment	n	$\overline{x}$	s
1 mg	32	3.16	0.72
2 mg	32	3.15	0.39

Is there significant evidence of a difference in elimination half-life depending on drug dosage? Write your conclusions from this and the previous exercise in the context of the drug's pharmacokinetic properties.

18.33 Extraterrestrial handedness? Many kinds of molecules have "left-handed" and "right-handed" versions. Some classes of molecules found in life on earth are almost entirely left-handed. Did this left-handedness precede the origin of life? To find out, scientists analyzed meteorites from space. To correct for bias in the delicate analysis, they also analyzed standard compounds known to have equal proportions of left-handed and right-handed forms. Here are the results for the percents of left-handed forms of one molecule in two analyses:[19]

Analysis	Meteorite			Standard		
	n	$\overline{x}$	s	n	$\overline{x}$	s
1	5	52.6	0.5	14	48.8	1.9
2	10	51.7	0.4	13	49.0	1.3

The researchers used the *t* test to see if the meteorite had a significantly higher percent of left-handed molecules than the standard. Carry out the tests for both analyses and report the results. The researchers concluded, "The observations suggest that organic matter of extraterrestrial origin could have played an essential role in the origin of terrestrial life."

18.34 Beetles in oats. In a study of cereal leaf beetle damage in oats, researchers measured the number of beetle larvae per stem in small plots of oats after randomly applying one of two treatments: no pesticide or malathion at the rate of 0.25 lb per acre. The data appear roughly Normal. Here are the summary statistics.[20]

Group	Treatment	n	$\overline{x}$	s
1	Control	13	3.47	1.21
2	Malathion	14	1.36	0.52

Holt Studios International/Alamy

Is there significant evidence at the 1% level that malathion reduces the mean number of larvae per stem?

18.35 Acupuncture for treating migraines. Acupuncture is widely used to prevent migraine attacks, but does it work? A study investigated the effectiveness of acupuncture compared with sham acupuncture and with no acupuncture (patients left on the wait list) in individuals with migraines. Patients were randomly assigned to one of the three "treatments" for a period of eight weeks, after which they monitored their migraine attacks over the next three weeks. Here are summary data about the number of days in which each patient used pain medication for migraines or headaches during the three treatment evaluation weeks:[21]

Acupuncture			Sham			Wait list		
n	$\overline{x}$	s	n	$\overline{x}$	s	n	$\overline{x}$	s
132	3.2	3.0	76	3.4	2.9	64	4.4	3.6

(a) Let's first ask if patients who received the acupuncture treatment had significantly fewer pain medication days than patients who stayed on the wait list. What do you conclude?

(b) Give a 95% confidence interval for the mean difference. Describe the effect size. Would you consider recommending acupuncture for the preventive treatment of headaches?

18.36 Acupuncture for treating migraines, continued. The placebo effect has been shown to be particularly strong in the treatment of pain.

(a) Is there significant evidence that patients who received the acupuncture treatment had significantly fewer pain medication days than patients who received the sham acupuncture (needles inserted at random locations)? What do you conclude?

(b) Explain what makes this study a randomized, controlled experiment. Finding statistical significance in Exercise 18.35 does not imply that acupuncture is causing the improvement. Explain briefly why.

18.37 Acupuncture for treating migraines: critique. The initial pool of volunteers for this study was larger. A few individuals dropped out for various reasons after the study started. Here are the retention rates for all three groups:

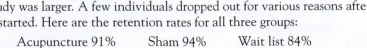

Acupuncture 91%   Sham 94%   Wait list 84%

Comparatively more patients dropped out of the study in the wait list group. This could be just chance; or maybe individuals with fewer migraines lost patience after being left on the wait list for several weeks; or maybe those who least believed in the effectiveness of acupuncture decided to drop out. How could it affect the conclusions of the study?

*The following exercises ask you to answer real questions from real data without having your work outlined in the exercise statement. Follow the **Formulate, Solve,** and **Conclude** steps of the four-step process. It may be helpful to restate in your own words the **State** information given in the exercise.*

18.38 Tropical flowers. Different varieties of the tropical flower *Heliconia* are fertilized by different species of hummingbirds. Over time, the lengths of the flowers and the form of the hummingbirds' beaks have evolved to match each other. Here are data on the lengths in millimeters of two color varieties of the same species of flower on the island of Dominica:[22]

H. caribaea red							
41.90	42.01	41.93	43.09	41.47	41.69	39.78	40.57
39.63	42.18	40.66	37.87	39.16	37.40	38.20	38.07
38.10	37.97	38.79	38.23	38.87	37.78	38.01	

H. caribaea yellow							
36.78	37.02	36.52	36.11	36.03	35.45	38.13	37.1
35.17	36.82	36.66	35.68	36.03	34.57	34.63	

Is there good evidence that the mean lengths of the two varieties differ? Estimate the difference between the population means. (Use 95% confidence.) Then estimate the mean flower length for each species separately.

18.39 Ink toxicity. The National Toxicology Program evaluates the toxicity of chemicals found in manufacturing, in consumer products, or in the environment after disposal. Toxicity is assessed through a battery of tests. Here are some results from a study of the toxicity of black newsprint ink in 7-week-old female rats. The rats' fur was locally clipped twice a week for 13 weeks. One group of rats received a dermal application of ink right after each clipping, and a control group of rats was left untreated. Table 18.2 contains the body weights of female rats at the beginning of the study and at the end of the 13 weeks.[23]

(a) Verify that the two experimental groups are not significantly different at the beginning of the study.

(b) Is there good evidence that ink application impairs growth in female rats between 7 and 20 weeks of age? Estimate the difference between the population mean weight gain. (Use 95% confidence.)

18.40 Magnets for pain relief. A randomized, double-blind experiment studied whether magnetic fields applied over a painful area can reduce pain intensity. The subjects were 50 volunteers with postpolio syndrome who reported muscular or arthritic pain. The pain level when pressing a painful area was graded subjectively on a scale from 0 to 10 (0 is no pain, 10 is maximum pain). Patients were randomly assigned to wear either a magnetic device or a placebo device over the painful area for 45 minutes. Here is a summary of the pain scores for this experiment, expressed as means ± standard deviations:[24]

TABLE 18.2   Body weights (g) at beginning and end of ink toxicity study

Control Group		Treatment Group	
Week 0	Week 13	Week 0	Week 13
111.2	191.6	107.3	187.0
105.4	191.2	116.7	189.5
110.8	210.7	112.2	179.2
105.6	185.2	103.4	172.2
106.1	195.0	113.2	178.7
104.4	188.3	110.6	180.9
114.0	188.4	110.6	188.3
115.1	195.6	100.5	188.9
109.2	204.6	106.3	183.1
111.3	195.7	112.5	184.5

	Magnetic device $(n = 29)$	Placebo device $(n = 21)$
Pretreatment	9.6 ± 0.7	9.5 ± 0.8
Posttreatment	4.4 ± 3.1	8.4 ± 1.8
Change	5.2 ± 3.2	1.1 ± 1.6

(a) Is there good evidence that the magnetic device is better than a placebo equivalent at reducing pain?

(b) How much reduction in pain is achieved with the magnetic device? Give a 95% confidence interval for the mean difference in pain scores before and after treatment among patients given the magnetic device. What procedure did you use and why?

18.41 **Magnets for pain relief, continued.** Conclusions from any study should be more comprehensive than a simple statement of significance or a confidence interval.

(a) The random assignment of subjects to treatments can sometimes lead, by chance, to an unbalanced split of subjects. Is there evidence of a significant difference in mean pain scores between the two groups at the beginning of the experiment?

(b) In all your calculations, you have been using summary data from the experiment. What would you like to know about the raw data to support the legitimacy of your statistical results?

Do birds learn to time their breeding? *Blue titmice eat caterpillars. The birds would like lots of caterpillars around when they have young to feed, but they breed earlier than peak caterpillar season. Do the birds learn from one year's experience when they time breeding the next year? Researchers randomly assigned 7 pairs of birds to have the natural caterpillar supply supplemented while feeding their young and another 6 pairs to serve as a control group relying on natural food supply. The next year, they measured how many days after the caterpillar peak the birds produced their nestlings.[25] Exercises 18.42 to 18.44 are based on this experiment.*

18.42 Did the randomization produce similar groups? First, compare the two groups in the first year. The only difference should be the chance effect of the random assignment. The study report says, "In the experimental year, the degree of synchronization did not differ between food-supplemented and control females." For this comparison, the report gives $t = -1.05$. What type of t statistic (paired or two-sample) is this? Show that this t leads to the quoted conclusion.

18.43 Did the treatment have an effect? The investigators expected the control group to adjust their breeding date the next year, whereas the well-fed supplemented group had no reason to change. The report continues, "But in the following year food-supplemented females were more out of synchrony with the caterpillar peak than the controls." Here are the data (days behind the caterpillar peak):

Control	4.6	2.3	7.7	6.0	4.6	−1.2	
Supplemented	15.5	11.3	5.4	16.5	11.3	11.4	7.7

Carry out a t test and show that it leads to the quoted conclusion.

18.44 Year-to-year comparison. Rather than comparing the two groups in each year, we could compare the behavior of each group in the first and second years. The study report says: "Our main prediction was that females receiving additional food in the nestling period should not change laying date the next year, whereas controls, which (in our area) breed too late in their first year, were expected to advance their laying date in the second year."

Comparing days behind the caterpillar peak in Years 1 and 2 gave $t = 0.63$ for the control group and $t = -2.63$ for the supplemented group. Are these paired or two-sample t statistics? What are the degrees of freedom for each t? Show that these t-values do *not* agree with the prediction.

18.45 The cost of sex. Evolution theory predicts a physiological cost of increased reproduction that translates into shorter lifespan. A study examined the cost of sexual activity in male *Drosophila* by comparing males allowed to be sexually active with those that were not. Table 18.3 gives the longevity in days for a sample of males randomly assigned to one or the other condition.[26] Do a complete analysis that reports on

(a) the longevity of sexually active male *Drosophila*.

(b) the longevity of sexually inactive male *Drosophila*.

(c) a comparison of the longevity of sexually active and sexually inactive male *Drosophila*.

TABLE 18.3   Longevity (days) of male *Drosophila*

Sexually inactive					Sexually active				
35	49	64	70	76	16	30	35	42	54
37	56	65	70	76	19	33	35	44	54
39	56	65	70	81	19	34	40	46	54
46	64	65	76	85	26	34	42	46	56
46	64	70	76		30	34	42	46	61

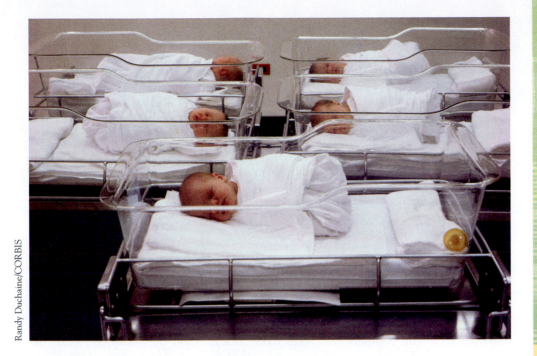

Randy Duchaine/CORBIS

Inference about a Population Proportion

Our discussion of statistical inference up to this point has concerned making inferences about population *means*. Now we turn to questions about the *proportion* of some outcome in a population when studying a categorical variable. Here are some examples that call for inference about population proportions.

EXAMPLE 19.1 Risky behavior in the age of AIDS

How common is behavior that puts people at risk of AIDS? The National AIDS Behavioral Surveys interviewed a random sample of 2673 adult heterosexuals. Of these, 170 had more than one sexual partner in the past year. That's 6.36% of the sample.[1] Based on these data, what can we say about the percent of all adult heterosexuals who have multiple partners in a year? We want to *estimate a single population proportion*. This chapter concerns inference about one proportion.

EXAMPLE 19.2 Treatment for Type I diabetes

Chronic conditions can lead to medical complications even more life impairing than the disease itself. The Diabetes Control and Complications Trial (DCCT) is the largest long-term study of type I diabetes (early-onset, or insulin-dependent, diabetes). Volunteers were randomly assigned to either a standard treatment or a more intensive treatment aimed at maintaining blood glucose as close to normal as possible. One complication of diabetes can be progressive damage of the retina (retinopathy), which can eventually lead to blindness. Among the patients without retinopathy at the beginning of the study, 24% in the standard treatment group developed retinopathy but only 7% in the intensive treatment group did.[2] Is this significant evidence that the proportions of diabetes patients who develop retinopathy differ in the populations of all patients receiving the standard treatment and all patients receiving the intensive treatment? We want to *compare two population proportions*. This is the topic of Chapter 20.

To do inference about a population mean μ, we use the mean $\overline{x}$ of a random sample from the population. The reasoning of inference starts with the sampling distribution of $\overline{x}$. Now we follow the same pattern, replacing means with proportions.

The sample proportion $\hat{p}$

We are interested in the unknown proportion p of a population that has some outcome. For convenience, call the outcome we are looking for a "success." In Example 19.1, the population is adult heterosexuals and the parameter p is the proportion who have had more than one sexual partner in the past year. To estimate p, the National AIDS Behavioral Surveys used random dialing of telephone numbers to contact a sample of 2673 people. Of these, 170 said they had multiple sexual partners. The statistic that estimates the parameter p is the **sample proportion.**

sample proportion

$$\hat{p} = \frac{\text{number of successes in the sample}}{\text{total number of individuals in the sample}}$$

$$= \frac{170}{2673} = 0.0636$$

Read the sample proportion $\hat{p}$ as "p-hat."

How good is the statistic $\hat{p}$ as an estimate of the parameter p? To find out, we ask, "What would happen if we took many samples?" The sampling distribution of $\hat{p}$ answers this question.

We described the properties of the sampling distribution of $\hat{p}$ in Chapter 13 (page 344) and Figure 13.6 (page 346) summarizes them graphically. The behavior of sample proportions $\hat{p}$ is similar to the behavior of sample means $\overline{x}$. When the sample size n is large, the sampling distribution is approximately Normal. The larger the sample, the more nearly Normal the distribution is. *Don't use the Normal approximation of the distribution of* $\hat{p}$ *when the sample size* n *is small.*

The mean of the sampling distribution of $\hat{p}$ is the true value of the population proportion p. That is, $\hat{p}$ is an unbiased estimator of p. The standard deviation of $\hat{p}$ is $\sqrt{p(1-p)/n}$ and gets smaller as the sample size n gets larger, so that estimation is likely to be more accurate when the sample is larger. As is the case for $\overline{x}$, the standard deviation gets smaller only at the rate $\sqrt{n}$. We need four times as many observations to cut the standard deviation in half and 100 times as many observations to cut it 10 times.

The Normal approximation of the sampling distribution of $\hat{p}$ *is least accurate when* p *is close to 0 or 1. If* $p = 0$, successes are impossible; every sample has $\hat{p} = 0$;

and there is no Normal distribution in sight. In the same way, the approximation works poorly when p is close to 1. In practice, this means that we need larger n for values of p near 0 or 1.

Inference about a population proportion p starts with using the sample proportion $\hat{p}$ to estimate p. Confidence levels and P-values are probabilities calculated from the sampling distribution of $\hat{p}$. We will consider only situations that allow us to use the Normal approximation of this sampling distribution. Here is a summary of the conditions we need.

CONDITIONS FOR INFERENCE ABOUT A PROPORTION

- We can regard our data as a **simple random sample** (SRS) from the population. This is, as usual, the most important condition.

- The **sample size n is large enough** to ensure that the distribution of $\hat{p}$ is close to Normal. We will see that different inference procedures require different answers to the question "How large is large enough?"

Remember also that *all of our inference procedures require that the population be much larger than the sample.*[3] This condition is usually satisfied in practice and is satisfied in all of our examples and exercises.

APPLY YOUR KNOWLEDGE

19.1 Antibiotic resistance. A sample of 1714 cultures from individuals in Florida diagnosed with a strep infection were tested for resistance to the antibiotic penicillin; 973 showed partial or complete resistance to the antibiotic.[4]

(a) Describe the population and explain in words what the parameter p is.

(b) Give the numerical value of the statistic $\hat{p}$ that estimates p.

19.2 How common is teen smoking? A 2004 Gallup survey interviewed a nationally representative random sample of 785 teenagers aged 13 to 17; 71 acknowledged having smoked in the past week.[5]

(a) What is the population studied and what parameter are we interested in?

(b) Give the numerical value of the statistic $\hat{p}$ that estimates p.

19.3 Student drinking. The College Alcohol Study interviewed an SRS of 14,941 college students about their drinking habits. Suppose that half of all college students "drink to get drunk" at least once in a while. That is, $p = 0.5$.

(a) What are the mean and standard deviation of the proportion $\hat{p}$ of the sample who drink to get drunk?

(b) Are the conditions for inference satisfied? Explain why or why not.

19.4 No inference. A local television station conducts a call-in poll about a proposed city tax increase to buy natural areas and protect them from development. Of the 2372 calls, 1921 support the proposal. We can't use these data as the basis for inference about the proportion of all citizens who support the tax increase. Why not?

Who is a smoker?

When estimating a proportion p, be sure you know what counts as a "success." The news says that 20% of adolescents smoke. Shocking. It turns out that this is the percent who smoked at least once in the past month. If we say that a smoker is someone who smoked on at least 20 of the past 30 days and smoked at least half a pack on those days, fewer than 4% of adolescents qualify.

Large-sample confidence intervals for a proportion

To estimate a population proportion p, use the sample proportion $\hat{p}$. If our conditions for inference apply, the sampling distribution of $\hat{p}$ is close to Normal with mean p and standard deviation $\sqrt{p(1-p)/n}$. To obtain a level C confidence interval for p, we would like to use

$$\hat{p} \pm z^* \sqrt{\frac{p(1-p)}{n}}$$

with the critical value z^* chosen to cover the central area C under the standard Normal curve. Figure 19.1 shows why.

The major challenge of confidence intervals for a population proportion is that we don't know the value of p. One approach consists of replacing the standard *standard error of $\hat{p}$* deviation with the **standard error of $\hat{p}$**

$$SE = \sqrt{\frac{\hat{p}(1-\hat{p})}{n}}$$

to get the confidence interval

$$\hat{p} \pm z^* \sqrt{\frac{\hat{p}(1-\hat{p})}{n}}$$

This interval has the form

$$\text{estimate} \pm z^* SE_{\text{estimate}}$$

Notice that we *don't* change z^* to t^* when we replace the standard deviation with the standard error. When the sample mean $\bar{x}$ estimates the population

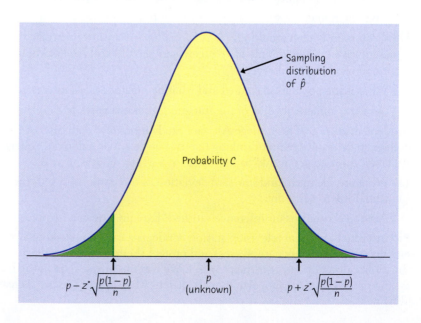

FIGURE 19.1 With probability C, $\hat{p}$ lies within $\pm z^* \sqrt{p(1-p)/n}$ of the unknown population proportion p. That is to say that in these samples p lies within $\pm z^* \sqrt{p(1-p)/n}$ of $\hat{p}$.

Sampling distribution of $\hat{p}$

Probability C

$p - z^* \sqrt{\dfrac{p(1-p)}{n}}$ $\underset{\text{(unknown)}}{p}$ $p + z^* \sqrt{\dfrac{p(1-p)}{n}}$

mean μ, a separate parameter σ describes the spread of the distribution of $\bar{x}$. We separately estimate σ, and this leads to a t distribution. When the sample proportion $\hat{p}$ estimates the population proportion p, the spread depends on p, not on a separate parameter. There is no t distribution—we just make the Normal approximation less accurate when we replace p in the standard deviation with $\hat{p}$.

However, this aproach has been shown experimentally to be less accurate than we would like it to be, especially for smaller sample sizes. *We do not recommend using this method, because the resulting confidence interval can be trusted only for quite large samples.* In addition, "how large" must take into account the fact that n must be larger if the sample proportion $\hat{p}$ suggests that p may be close to 0 or 1. Statisticians have shown that, as a guideline, this "large sample method" should not be used if the numbers of successes ($n\hat{p}$) and failures ($n[1 - \hat{p}]$) in the sample are not both at least 15.[6]

LARGE-SAMPLE CONFIDENCE INTERVAL FOR A POPULATION PROPORTION—*NOT RECOMMENDED*

Draw an SRS of size n from a large population that contains an unknown proportion p of successes. An approximate level C **confidence interval for *p*** is

$$\hat{p} \pm z^* \sqrt{\frac{\hat{p}(1 - \hat{p})}{n}}$$

where z^* is the critical value for the standard Normal density curve with area C between $-z^*$ and z^*.

This interval should not be used unless the numbers of successes and failures in the sample are both at least 15. We recommend not using it at all.

Although we do not recommend the large-sample method for computing a confidence interval for a population proportion, you may still come across publications that used it. Here is an example to help you understand how the method works.

EXAMPLE 19.3   *Estimating risky behavior*

The four-step process for any confidence interval is outlined on page 361.

STATE: The National AIDS Behavioral Surveys found that 170 of a sample of 2673 adult heterosexuals had multiple partners. That is, $\hat{p} = 0.0636$. What can we say about the population of all adult heterosexuals?

FORMULATE: We will give a 99% confidence interval to estimate the proportion p of all adult heterosexuals who have multiple partners.

SOLVE: First verify the conditions for inference:

*   The sampling design was a complex stratified sample, and the survey used inference procedures for that design. The overall effect is close to an SRS, however.

• The sample is large enough: The numbers of successes (170) and failures (2503) in the sample are both much larger than 15.

The sample size condition is easily satisfied. The condition that the sample be an SRS is only approximately met.

A 99% confidence interval for the proportion p of all adult heterosexuals with multiple partners uses the standard Normal critical value $z^* = 2.576$. The confidence interval is

$$\hat{p} \pm z^* \sqrt{\frac{\hat{p}(1-\hat{p})}{n}} = 0.0636 \pm 2.576\sqrt{\frac{(0.0636)(0.9364)}{2673}}$$

$$= 0.0636 \pm 0.0122$$

$$= 0.0514 \text{ to } 0.0758$$

CONCLUDE: We are 99% confident that the percent of adult heterosexuals who have had more than one sexual partner in the past year lies between about 5% and 7.6%.

As usual, the practical problems of a large sample survey weaken our confidence in the AIDS survey's conclusions. Only people in households with telephones could be reached. This is acceptable for surveys of the general population, because about 95% of American households have telephones. However, some groups at high risk for AIDS, like intravenous drug users, often don't live in settled households and therefore are under represented in the sample. About 30% of the people reached refused to cooperate. A nonresponse rate of 30% is not unusual in large sample surveys, but it may cause some bias if those who refuse differ systematically from those who cooperate. The survey used statistical methods that adjust for unequal response rates in different groups. Finally, some respondents may not have told the truth when asked about their sexual behavior. The survey team tried hard to make respondents feel comfortable. For example, Hispanic women were interviewed only by Hispanic women, and Spanish speakers were interviewed by Spanish speakers with the same regional accent (for example, Cuban, Mexican, or Puerto Rican). Nonetheless, the survey report says that some bias is probably present:

> It is more likely that the present figures are underestimates; some respondents may underreport their numbers of sexual partners and intravenous drug use because of embarrassment and fear of reprisal, or they may forget or not know details of their own or of their partner's HIV risk and their antibody testing history.[7]

Reading the report of a large study like the National AIDS Behavioral Surveys reminds us that statistics in practice involves much more than formulas for inference.

APPLY YOUR KNOWLEDGE

19.5 Antibiotic resistance. Exercise 19.1 gave data about the resistance of strep cultures to the antibiotic penicillin. Give a 95% confidence interval for the proportion of strep cultures from Florida patients showing partial or complete resistance to penicillin. Follow the four-step process as illustrated in Example 19.3.

19.6 **How common is teen smoking?** Exercise 19.2 gave the results of a 2004 Gallup survey about teenage smoking. Give a 95% confidence interval for the proportion of teenagers in the United States who would acknowledge having smoked in the past week. Follow the four-step process as illustrated in Example 19.3.

19.7 **No confidence interval.** In the National AIDS Behavioral Surveys sample of 2673 adult heterosexuals, 0.2% (that's 0.002 as a decimal fraction) had both received a blood transfusion and had a sexual partner from a group at high risk of AIDS. Explain why we can't use the large-sample confidence interval to estimate the proportion p in the population who share these two risk factors.

Accurate confidence intervals for a proportion

The confidence interval $\hat{p} \pm z^* \sqrt{\hat{p}(1 - \hat{p})/n}$ for a sample proportion p is easy to calculate. It is also easy to understand, because it rests directly on the approximately Normal distribution of $\hat{p}$. Unfortunately, confidence levels from this interval are often quite inaccurate unless the sample is very large. The actual confidence level is usually *less* than the confidence level you asked for in choosing the critical value z^*. That's bad. What is worse, accuracy does not consistently get better as the sample size n increases. There are "lucky" and "unlucky" combinations of the sample size n and the true population proportion p.

Fortunately, there is a simple modification that has been shown experimentally to successfully improve the accuracy of the confidence interval. We call it the "plus four" method, because all you need to do is *add four imaginary observations, two successes and two failures*. With the added observations, the **plus four estimate** of p is

plus four estimate

$$\tilde{p} = \frac{\text{number of successes in the sample} + 2}{n + 4}$$

The formula for the confidence interval is exactly as before, with the new sample size and number of successes.[8] You do not need software that offers the plus four interval—just enter the new sample size (actual size + 4) and number of successes (actual number + 2) into the large-sample procedure.

PLUS FOUR CONFIDENCE INTERVAL FOR A PROPORTION

Draw an SRS of size n from a large population that contains an unknown proportion p of successes. To get the **plus four confidence interval for p,** add four imaginary observations, two successes and two failures. Then use the large-sample confidence interval with the new sample size ($n + 4$) and count of successes (actual count + 2).

Use this interval when the confidence level is at least 90% and the sample size n is at least 10.

EXAMPLE 19.4   Blinding in medical trials

STATE: Many medical trials randomly assign patients to either an active treatment or a placebo. These trials are always double-blind. Sometimes the patients can tell whether or not they are getting the active treatment. This defeats the purpose of blinding. Reports of medical research usually ignore this problem. Investigators looked at a random sample of 97 articles reporting on placebo-controlled randomized trials in the top five general medical journals. Only 7 of the 97 discussed the success of blinding—and in 5 of these, the blinding was imperfect.[9] What proportion of all such studies discuss the success of blinding?

FORMULATE: Take p to be the proportion of articles that discuss the success of blinding. Give a 95% confidence interval for p.

SOLVE: The conditions for use of the large-sample interval are not met, because there are fewer than 15 successes in the sample. Add two successes and two failures to the original data. The plus four estimate of p is

$$\tilde{p} = \frac{7+2}{97+4} = \frac{9}{101} = 0.0891$$

The plus four confidence interval is the same as the large-sample interval based on 9 successes in 101 observations. Here it is:

$$\tilde{p} \pm z^* \sqrt{\frac{\tilde{p}(1-\tilde{p})}{n+4}} = 0.0891 \pm 1.960 \sqrt{\frac{(0.0891)(0.9109)}{101}}$$

$$= 0.0891 \pm 0.0556$$

$$= 0.0335 \text{ to } 0.1447$$

CONCLUDE: We estimate with 95% confidence that between about 3.4% and 14.5% of all such articles discuss whether the blinding succeeded.

For comparison, the ordinary sample proportion is

$$\hat{p} = \frac{7}{97} = 0.0722$$

The plus four estimate $\tilde{p} = 0.0891$ in Example 19.4 is farther away from zero than $\hat{p} = 0.0722$. The plus four estimate gains its added accuracy by always moving toward 0.5 and away from 0 or 1, whichever is closer. This is particularly helpful when the sample contains only a few successes or a few failures. The numerical difference between a large-sample interval and the corresponding plus four interval is often small. Remember that the confidence level is the probability that the interval will catch the true population proportion *in very many uses*. Small differences every time add up to accurate confidence levels from plus four versus inaccurate levels from large-sample intervals.

How much more accurate is the plus four interval? Computer studies have asked how large n must be to guarantee an accurate confidence interval. If $p = 0.1$, for example, the answer is $n \geq 646$ for the large-sample interval and $n \geq 11$ for the plus four interval when we want a 95% confidence level.[10] The consensus of computational and theoretical studies is that plus four is much better than the large-sample interval for many combinations of n and p. **We recommend that you always use the plus four interval.**

APPLY YOUR KNOWLEDGE

19.8 Antibiotic resistance. In Exercise 19.5 you gave a 95% confidence interval for the proportion of strep cultures from Florida patients showing partial or complete resistance to penicillin. This estimate was based on the finding of 973 cultures with partial or complete resistance out of a random sample of 1714 throat cultures.

    (a) What is the plus four estimate $\tilde{p}$ for this population proportion? Give the plus four 95% confidence interval for p.

    (b) How do the two confidence intervals compare? Explain why the two methods give similar results when the sample size is large. For large samples, both methods give similar and accurate estimates. For smaller samples, the plus four estimate is always more accurate.

19.9 Whelks and mussels. Sample surveys usually contact large samples, so we can use the large-sample confidence interval if the sample design is close to an SRS. Scientific studies often use small samples that require the plus four method. For example, the small round holes you often see in sea shells were drilled by other sea creatures, who ate the former owners of the shells. Whelks often drill into mussels, but this behavior appears to be more or less common in different locations. Investigators collected whelk eggs from the coast of Oregon, raised the whelks in the laboratory, and then put each whelk in a container with some delicious mussels. Only 9 of 98 whelks drilled into mussels.[11]

Andrew J. Martinez/Photo Researchers, Inc.

    (a) Why can't we use the large-sample confidence interval?

    (b) Give the plus four 90% confidence interval for the proportion of Oregon whelks that will spontaneously drill mussels.

19.10 High-risk behavior. In the National AIDS Behavioral Surveys sample of 2673 adult heterosexuals, 5 respondents had both received a blood transfusion and had a sexual partner from a group at high risk of AIDS. Exercise 19.7 asked you to explain why we should not use the large-sample confidence interval for the proportion p in the population who share these two risk factors.

    (a) The plus four method adds four observations, two successes and two failures. What are the sample size and the count of successes after you do this? What is the plus four estimate $\tilde{p}$ of p?

    (b) Give the plus four 95% confidence interval for p.

Choosing the sample size

In planning a study, we may want to choose a sample size that will allow us to estimate the parameter within a given margin of error. We saw earlier (page 397) how to do this for a population mean. The method is similar for estimating a population proportion.

    The margin of error in the large-sample confidence interval for p is

$$m = z^* \sqrt{\frac{\hat{p}(1-\hat{p})}{n}}$$

Here z^* is the standard Normal critical value for the level of confidence we want. Because the margin of error involves the sample proportion of successes $\hat{p}$, we need to guess this value when choosing n. Call our guess p^*. Here are two ways to get p^*:

1. Use a guess p^* based on a pilot study or on past experience with similar studies. You can do several calculations to cover the range of values of $\hat{p}$ you might get.

2. Use $p^* = 0.5$ as the guess. The margin of error m is largest when $\hat{p} = 0.5$, so this guess is conservative in the sense that if we get any other $\hat{p}$ when we do our study, we will get a margin of error smaller than planned.

Once you have a guess p^*, the recipe for the margin of error can be solved to give the sample size n needed. Here is the result for the large-sample confidence interval. For simplicity, use this result even if you plan to use the plus four interval.

SAMPLE SIZE FOR DESIRED MARGIN OF ERROR

The level C confidence interval for a population proportion p will have a margin of error approximately equal to a specified value m when the sample size is

$$n = \left(\frac{z^*}{m}\right)^2 p^*(1 - p^*)$$

where p^* is a guessed value for the sample proportion. The margin of error will be less than or equal to m if you take the guess p^* to be 0.5.

Which method for finding the guess p^* should you use? The n you get doesn't change much when you change p^* as long as p^* is not too far from 0.5. You can use the conservative guess $p^* = 0.5$ if you expect the true $\hat{p}$ to be roughly between 0.3 and 0.7. If the true $\hat{p}$ is close to 0 or 1, using $p^* = 0.5$ as your guess will yield a sample much larger than you need. Try to use a better guess from a pilot study when you suspect that $\hat{p}$ will be less than 0.3 or greater than 0.7.

Digital Vision Photography/Veer

STEP

— EXAMPLE 19.5 Planning a survey —————————

STATE: Flossing helps prevent tooth decay and gum disease. Yet reports suggest that millions of Americans never floss. You are planning a sample survey to determine what percent of students at your university never floss. You obtain permission to use the University roster to contact an SRS of students. You want to estimate the proportion p of students who never floss with 95% confidence and a margin of error no greater than 4% (0.04). How large a sample do you need?

FORMULATE: Find the sample size n needed for margin of error $m = 0.04$ and 95% confidence. Based on your readings on the topic, you would guess that the proportion of individuals who never floss is between 30% and 70% for any given population. You can use the guess $p^* = 0.5$.

SOLVE: The sample size you need is

$$n = \left(\frac{1.96}{0.04}\right)^2 (0.5)(1 - 0.5) = 600.25$$

Round the result up to $n = 601$. (Rounding down would yield a margin of error slightly greater than 0.04.)

CONCLUDE: An SRS of 601 students from your university is adequate for a margin of error $\pm 4\%$ with a confidence level of 95%.

What sample size would you have needed for a margin of error of 3% rather than 4%? The answer is, after rounding up

$$n = \left(\frac{1.96}{0.03}\right)^2 (0.5)(1 - 0.5) = 1068$$

For a 2% margin of error, the sample size you would have needed is

$$n = \left(\frac{1.96}{0.02}\right)^2 (0.5)(1 - 0.5) = 2401$$

As usual, smaller margins of error call for larger samples. Remember, though, that the population of interest must be much larger than the sample in order to conduct inference for proportions legitimately. Are more than 24,000 students enrolled at your university? If not, you cannot legitimately calculate a plus four or large-sample confidence interval with only a 2% margin of error. Inference requires taking everything into consideration.

Notice also how large a sample size is required for a 2% or 3% margin of error. When dealing with proportions, the margin of error is always a value well below 1. Because it goes in the denominator of the sample size equation and gets squared, it drives the sample size up. You will find that, typically, studies of categorical variables (for example, clinical trials looking for improvement vs. no improvement) tend to have rather large sample sizes compared to studies of quantitative variables.

APPLY YOUR KNOWLEDGE

19.11 Canadians and doctor-assisted suicide. A Gallup Poll asked a sample of Canadian adults if they thought the law should allow doctors to end the life of a patient who is in great pain and near death if the patient makes a request in writing. The poll included 270 people in Quebec, 221 of whom agreed that doctor-assisted suicide should be allowed.[12]

   (a) What is the margin of error of the large-sample 95% confidence interval for the proportion of all Quebec adults who would allow doctor-assisted suicide?

   (b) How large a sample is needed to get the common ± 3 percentage point margin of error? Use the previous sample as a pilot study to get p^*.

19.12 Can you taste PTC? PTC is a substance that has a strong bitter taste for some people and is tasteless for others. The ability to taste PTC is inherited. About 75% of Italians can taste PTC, for example. You want to estimate the proportion of Americans with at least one Italian grandparent who can taste PTC. Starting with the 75% estimate for Italians, how large a sample must you collect in order to estimate the proportion of PTC tasters within ±0.04 with 90% confidence?

Significance tests for a proportion

We now turn to tests of significance and will assume that the conditions for inference we described earlier are met. When the null hypothesis H_0: $p = p_0$ is true, the sampling distribution of the sample proportion $\hat{p}$ is approximately Normal, centered on p_0 with standard deviation $\sqrt{p_0(1 - p_0)/n}$, as illustrated in Figure 19.2.

The test statistic for the null hypothesis H_0: $p = p_0$ is the sample proportion $\hat{p}$ standardized using the value p_0 specified by H_0,

$$z = \frac{\hat{p} - p_0}{\sqrt{\dfrac{p_0(1 - p_0)}{n}}}$$

This z statistic has approximately the standard Normal distribution when H_0 is true. P-values therefore come from the standard Normal distribution. Because H_0 gives us a value for p, the inaccuracy that plagues the large-sample confidence interval does not affect tests (and therefore no adjustment is required). Here is the procedure for tests.

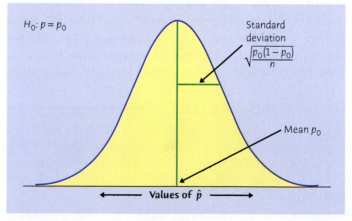

FIGURE 19.2 When H_0: $p = p_0$ is true, the sampling distribution of $\hat{p}$ is approximately Normal with mean p_0 and standard deviation $\sqrt{p_0(1 - p_0)/n}$.

SIGNIFICANCE TESTS FOR A PROPORTION

Draw an SRS of size n from a large population that contains an unknown proportion p of successes. To **test the hypothesis H_0: $p = p_0$**, compute the z statistic

$$z = \frac{\hat{p} - p_0}{\sqrt{\dfrac{p_0(1 - p_0)}{n}}}$$

In terms of a variable Z having the standard Normal distribution, the approximate P-value for a test of H_0 against

H_a: $p > p_0$ is $P(Z \geq z)$

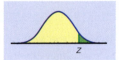

H_a: $p < p_0$ is $P(Z \leq z)$

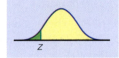

H_a: $p \neq p_0$ is $2P(Z \geq |z|)$

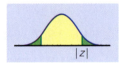

Use this test when the sample size n is so large that both np_0 and $n(1 - p_0)$ are 10 or more.[13]

─── **EXAMPLE 19.6** *Are boys biologically more likely?* ───

The four-step process for any significance test is outlined on page 372.

STATE: We saw in Example 9.3 (page 221) that there are slightly more boys than girls born in the United States every year even though genetic inheritance predicts a 50-50 split of both genders among embryos. This imbalance could reflect a biological mechanism (genetic or physiologic) favoring the birth of boys, or it could arise from a differential attitude of parents toward having further children based on the sex of their existing children. Parental involvement, however, could not explain a gender imbalance among firstborn children. A study looked at the sex of the firstborn child of a random sample of couples with two children. The result was 13,173 boys among 25,468 firstborns.[14] The sample proportion of boys was

$$\hat{p} = \frac{13{,}173}{25{,}468} = 0.5172$$

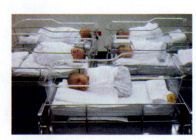

Randy Duchaine/CORBIS

Boys do make up more than half of the sample, but of course we don't expect a perfect 50-50 split in a random sample. Is this sample evidence that boys are more common than girls in the entire population of firstborns?

FORMULATE: Take p to be the proportion of boys among all firstborn children of American mothers. We want to test the hypotheses

$$H_0: p = 0.5$$
$$H_a: p > 0.5$$

SOLVE: The conditions for inference are met, so we can go on to the z test statistic:

$$z = \frac{\hat{p} - p_0}{\sqrt{\dfrac{p_0(1 - p_0)}{n}}}$$
$$= \frac{0.5172 - 0.5}{\sqrt{\dfrac{(0.5)(0.5)}{25,468}}} = 5.49$$

The P-value is the area under the standard Normal curve to the right of $z = 5.49$. We know that this is a very small area. Table C shows that $P < 0.0005$. Software (see Figure 19.3) tells us that in fact $P = 0.00000002$.

CONCLUDE: There is very strong evidence that more than half of firstborns are boys ($P < 0.0005$). This supports the idea that there is a biological basis for the observed gender imbalance at birth.

Kids on bikes

In the most recent year for which data are available, 77% of children killed in bicycle accidents were boys. You might take these data as a sample and start from $\hat{p} = 0.77$ to do inference about bicycle deaths in the near future. What you should not do is conclude from these data that boys on bikes are in greater danger than girls. We don't know how many boys and how many girls ride bikes—it may be that most fatalities are boys because most riders are boys.

—— **EXAMPLE 19.7** *Estimating the chance of a firstborn boy* ——

With 13,173 successes in 25,468 trials, the large-sample and plus four estimates of p are almost identical. Both are 0.5172 to four decimal places. So both methods give the 99% confidence interval

$$\hat{p} \pm z^* \sqrt{\hat{p}(1 - \hat{p})/n} = 0.5172 \pm 2.576 \sqrt{(0.5172)(0.4828)/25,468}$$
$$= 0.5172 \pm 0.0081$$
$$= 0.5091 \text{ to } 0.5253$$

using z^* 2.576 from Table C. We are 99% confident that between about 51% and 52.5% of firstborns are boys.

The confidence interval is more informative than the test in Example 19.6, which tells us only that more than half are boys. The proportion of boys among firstborns is significantly larger than 0.5, but it is not a lot larger. Theories of evolution hypothesize that sex imbalance at birth compensates for higher mortality rates of one gender until reproductive age so that, by the time organisms are sexually mature, the sex ratio is 50-50.

APPLY YOUR KNOWLEDGE ———————

19.13 Sex ratio in geese. A field biologist examined the sex ratio at birth of the lesser snow geese. A random sample of nests containing four eggs was taken. For each egg resulting in a live gosling, the laying order and the gender of the gosling were recorded. Of the 27 successfully hatched first eggs, 17 were male.[15] Is there significant evidence that the proportion of male goslings is not 50%? Follow the four-step process as illustrated in Example 19.6.

CrunchIt!

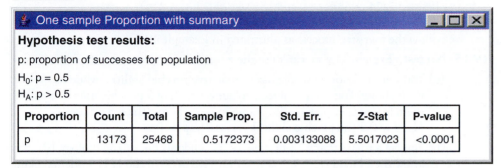

One sample Proportion with summary

Hypothesis test results:

p: proportion of successes for population

H_0: p = 0.5
H_A: p > 0.5

Proportion	Count	Total	Sample Prop.	Std. Err.	Z-Stat	P-value
p	13173	25468	0.5172373	0.003133088	5.5017023	<0.0001

CrunchIt!

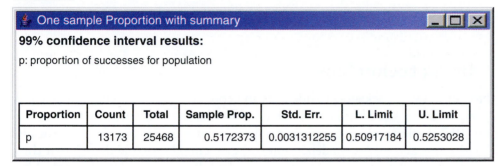

One sample Proportion with summary

99% confidence interval results:

p: proportion of successes for population

Proportion	Count	Total	Sample Prop.	Std. Err.	L. Limit	U. Limit
p	13173	25468	0.5172373	0.0031312255	0.50917184	0.5253028

Minitab

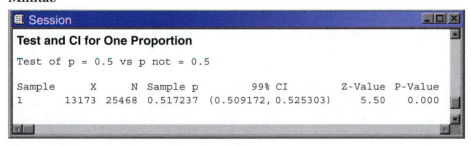

Session

Test and CI for One Proportion

```
Test of p = 0.5 vs p not = 0.5

Sample     X      N  Sample p        99% CI         Z-Value  P-Value
1      13173  25468  0.517237  (0.509172, 0.525303)    5.50    0.000
```

Texas Instruments TI-83 Plus

```
1-PropZTest          1-PropZInt
 prop>.5              (.50917,.5253)
 z=5.501702408        p̂=.5172373174
 p=1.8853523E-8       n=25468
 p̂=.5172373174
 n=25468
```

FIGURE 19.3 Test of hypothesis and 99% confidence interval for the firstborn data: output from a graphing calculator and two statistical programs.

19.14 Formulating the theory of inheritance. Gregor Mendel has been called the "father of genetics," having formulated laws of inheritance even before the function of DNA was discovered. In one famous experiment with peas, he crossed pure breeds of plants producing smooth peas and plants producing wrinkled peas.

The first-generation peas were all smooth. In the second generation (F2), he obtained 5474 smooth peas and 1850 wrinkled peas. Do these data agree with the conclusion of 75% dominant trait (in this case "smooth") occurrence in F2? Follow the four-step process as illustrated in Example 19.6.

19.15 **No test.** Explain why we can't use the z test for a proportion in these situations:

(a) You want to know if people have predictive psychic abilities when answering their phone. You ask 10 subjects to guess which of 5 possible friends is calling them before they pick up the phone in order to test the hypotheses $H_0: p = 0.2$ (no psychic ability) versus $H_a: p > 0.2$.

(b) You bet a friend that 90% or more of the students taking the Biology 101 class know their blood type. You contact an SRS of 80 of the 1211 students taking Biology 101 this term to test the hypotheses $H_0: p = 0.9$ versus $H_a: p > 0.9$.

Using technology

Not all software packages provide tests and confidence intervals for a sample proportion. Excel and SPSS, for instance, do not have an easy way to do that. Figure 19.3 displays output from a graphing calculator and two statistical programs for the test and 99% confidence interval of Examples 19.6 and 19.7. All three use the Normal approximation for the sampling distribution of $\hat{p}$, although this is something you must specify with Minitab.

All three programs give the same numerical values except for rounding differences. Because Minitab rounds the P-value to 3 decimal places, the output shows "P-Value 0.000," which does not imply that the P-value is zero but just close to zero (< 0.001). And because Minitab gives the two-sided P-value, we can conclude that the correct P-value for our one-sided alternative is less than 0.0005. Remember to always specify your null and alternative hypotheses even before examining the sample data and then select or work out the one- or two-sided P-value that matches your alternative.

Last, you do not need additional technology to use the plus four method for a confidence interval for p. In Example 19.7, the counts are so large that the large-sample and the plus four methods would give identical results for all practical purposes. However, if you need a confidence interval for a smaller sample and want to use the plus four method, simply type in the count of successes plus 2 and the sample size plus 4 in the software input window.

CHAPTER 19 SUMMARY

Tests and confidence intervals for a population proportion p when the data are an SRS of size n are based on the **sample proportion $\hat{p}$.**

When n is large, $\hat{p}$ has approximately the Normal distribution with mean p and standard deviation $\sqrt{p(1 - p)/n}$.

The level C **large-sample confidence interval for p** is

$$\hat{p} \pm z^* \sqrt{\frac{\hat{p}(1 - \hat{p})}{n}}$$

where z^* is the critical value for the standard Normal curve with area C between $-z^*$ and z^*.

The true confidence level of the large-sample interval can be substantially less than the planned level C unless the sample is very large. *We recommend using the plus four interval instead.*

To get a more accurate confidence interval, add four imaginary observations, two successes and two failures, to your sample. Then use the same formula for the confidence interval. This is the **plus four confidence interval.** Use this interval in practice for confidence level 90% or higher and sample size n at least 10.

The **sample size** needed to obtain a confidence interval with approximate margin of error m for a population proportion is

$$n = \left(\frac{z^*}{m}\right)^2 p^*(1 - p^*)$$

where p^* is a guessed value for the sample proportion $\hat{p}$ and z^* is the standard Normal critical point for the level of confidence you want. If you use $p^* = 0.5$ in this formula, the margin of error of the interval will be less than or equal to m no matter what the value of $\hat{p}$ is.

Significance tests for H_0: $p = p_0$ are based on the z statistic

$$z = \frac{\hat{p} - p_0}{\sqrt{\dfrac{p_0(1 - p_0)}{n}}}$$

with P-values calculated from the standard Normal distribution. Use this test in practice when $np_0 \geq 10$ and $n(1 - p_0) \geq 10$.

CHECK YOUR SKILLS

19.16 Suppose that a population has 30% of individuals displaying a particular phenotype. In repeated random samples of 400 individuals, the sample proportion $\hat{p}$ would follow a Normal distribution with mean

(a) 20.   (b) 0.3.   (c) 0.023.

19.17 The standard deviation of the distribution of $\hat{p}$ in the previous exercise is about

(a) 0.0011.   (b) 0.3.   (c) 0.023.

Sports Illustrated *asked a random sample of 757 Division I college athletes, "Do you believe performance-enhancing drugs are a problem in college sports?" Of the 757 athletes interviewed, 273 said "Yes." Use this information to answer Exercises 19.18 through 19.20.*

19.18 The sample proportion $\hat{p}$ who said "Yes" is

(a) 36.   (b) 2.77.   (c) 0.36.

19.19 Based on the *Sports Illustrated* sample, the 95% plus four confidence interval for the proportion of all Division I athletes who think performance-enhancing drugs are a problem is

(a) 0.36 ± 0.017.   (b) 0.36 ± 0.034.   (c) 0.36 ± 0.00030.

19.20 How many athletes must be interviewed to estimate that population proportion to within ± 0.02 with 95% confidence? Use 0.5 as the conservative guess for p.

(a) $n = 25$ (b) $n = 1225$ (c) $n = 2401$

A farmer wants to test the effectiveness of a pest control method in allowing strawberry blooms to yield marketable strawberries. Of a random sample of 100 blooms, 77 yield marketable strawberries. Is this evidence that two thirds or more of all blooms grown with this pest control method end up as marketable strawberries? Use this information to answer Exercises 19.21 through 19.23.

19.21 The hypotheses for a test to answer this question are

(a) H_0: $p = 0.667$, H_a: $p > 0.667$.

(b) H_0: $p = 0.667$, H_a: $p \neq 0.667$.

(c) H_0: $p = 0.77$, H_a: $p \neq 0.77$.

19.22 The value of the z statistic for this test is about

(a) $z = 1.03$.   (b) $z = 2.19$.   (c) $z = 2.46$.

19.23 The P-value of this test is about

(a) $P = 0.014$.   (b) $P = 0.029$.   (c) $P = 0.047$.

19.24 A Harris Poll found that 54% of American adults do not think that human beings developed from earlier species. The poll's margin of error was 3%. This means that

(a) the poll used a method that gets an answer within 3% of the truth about the population 95% of the time.

(b) we can be sure that the percent of all adults who feel this way is between 51% and 57%.

(c) if Harris takes another poll using the same method, the results of the second poll will lie between 51% and 57%.

19.25 The Harris Poll in the previous exercise probably had about 50% nonresponse. The nonresponse could cause the survey result to be in error. The error due to nonresponse

(a) is in addition to the margin of error found in Exercise 19.24.

(b) is included in the margin of error found in Exercise 19.24.

(c) can be ignored because it isn't random.

CHAPTER 19 EXERCISES

We recommend using the plus four method for all confidence intervals for a proportion. However, the large-sample method is acceptable when the guidelines for its use are met.

19.26 Detecting genetically modified soybeans. Most soybeans grown in the United States are genetically modified to, for example, resist pests and so reduce use of pesticides. Because some nations do not accept genetically modified (GM) foods, grain-handling facilities routinely test soybean shipments for the presence of GM beans. In a study of the accuracy of these tests, researchers submitted lots of soybeans containing 1% of GM beans to 23 randomly selected facilities. Eighteen detected the GM beans.[16]

(a) Show that the conditions for the large-sample confidence interval are not met. Show that the conditions for the plus four interval are met.

(b) Use the plus four method to give a 90% confidence interval for the percent of all grain-handling facilities that will correctly detect 1% of GM beans in a shipment.

19.27 White-eyed fruit flies. Wild-type fruit flies have red eyes, but a recessive mutation produces white-eyed individuals. A researcher wants to assess the frequency of heterozygous individuals among red-eyed fruit flies. A heterozygous red-eyed fly crossed with a white-eyed mutant will have mixed progeny. Of the 100 red-eyed fruit flies crossed with white-eyed mutants, 11 produced mixed progeny.

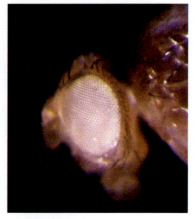

Terry Gleason/Visuals Unlimited

(a) Show that the conditions for the large-sample confidence interval are not met. Show that the conditions for the plus four interval are met.

(b) Give the plus four 95% confidence interval for the proportion of heterozygotes in the population of red-eyed fruit flies studied.

19.28 Looking for health information online. In a 2005 survey, the Pew Research Center found that 58% of the 1931 adult Internet users interviewed said that the Internet has helped them in some way with obtaining health information.[17]

(a) Verify that the conditions for inference are met. Give a 99% confidence interval for the proportion of all adult Internet users who feel this way.

(b) Can you generalize this finding to all American adults? Explain why or why not. Do you think that the proportion who would say that the Internet helped them obtain health information is higher among all adult Internet users or among all adult Americans? Explain your answer.

19.29 White-eyed fruit flies, continued. Go back to the experiment of Exercise 19.27. How many red-eyed fruit flies should be used in this experiment to obtain a 95% confidence interval with a margin of error no more than 3%? Use 20% as a conservative (worst-case) estimate of the percent of heterozygotes in the population.

19.30 Color blindness. Color blindness, or dyschromatopsia, is a form of genetic deficiency in color perception. The condition is much more prevalent among males than females, pointing to a genetic connection with the X chromosome. The frequency of dyschromatopsia in the Caucasian American male population is about 8%. However, it is thought that this proportion might be smaller among males of other ethnicities. We want to estimate the proportion of Asian American males who are color-blind. How large a sample size do we need in order to obtain a 95% confidence interval with a margin of error no greater than 2%, or 0.02? Use 0.1 for p^*.

19.31 Fast food for baby. In 2003, the Gerber company sponsored a large survey of the eating habits of American infants and toddlers. Among the many questions parents were asked was whether their child had eaten fried potatoes on the day before the interview. Among the 679 infants 9 to 11 months old, 9% had eaten fried potatoes that day.[18] Verify that the conditions for inference are met and give the plus four 95% confidence interval for the population proportion of 9 to 11-month-old infants who eat fried potatoes on a given day.

19.32 **Side effects.** An experiment on the side effects of pain relievers assigned arthritis patients to take one of several over-the-counter pain medications. Of the 440 patients who took one brand of pain reliever, 23 suffered some "adverse symptom."

(a) If 10% of all patients suffer adverse symptoms, what would be the sampling distribution of the proportion with adverse symptoms in a sample of 440 patients?

(b) Does the experiment provide strong evidence that fewer than 10% of patients who take this medication have adverse symptoms? Verify that the conditions for inference are met. State the hypotheses, calculate the test statistic, and then obtain and interpret the P-value.

19.33 **Do chemists have more girls?** Some people think that chemists are more likely than other parents to have female children. (Perhaps chemists are exposed to something in their laboratories that affects the sex of their children.) The Washington State Department of Health lists the parents' occupations on birth certificates. Between 1980 and 1990, 555 children were born to fathers who were chemists. Of these births, 273 were girls. During this period, 48.8% of all births in Washington State were girls.[19]

(a) Is there evidence that the proportion of girls born to chemists is higher than the proportion for the whole state? State the hypotheses, calculate the test statistic, and then obtain and interpret the P-value.

(b) What assumption are you making when conducting this test? Do you feel that the assumption is entirely justified? Explain your reasoning.

19.34 **Matched pairs.** One-sample procedures for proportions, like those for means, are used to analyze data from matched pairs designs. Here is an example.

Each of 50 subjects tastes two unmarked cups of coffee and says which he or she prefers. One cup in each pair contains instant coffee, the other fresh-brewed coffee. Thirty-one of the subjects prefer the fresh-brewed coffee. Take p to be the proportion of the population who would prefer fresh-brewed coffee in a blind tasting.

(a) Test the claim that a majority of people prefer the taste of fresh-brewed coffee. State hypotheses and report the z statistic and its P-value. Is your result significant at the 5% level? What is your practical conclusion?

(b) Find a 90% confidence interval for p.

(c) When you do an experiment like this, in what order should you present the two cups of coffee to the subjects?

In responding to Exercises 19.35 through 19.41, follow the **Formulate, Solve,** *and* **Conclude** *steps of the four-step process. It may be helpful to restate in your own words the* **State** *information given in the exercise.*

19.35 **Lead in tap water.** The U.S. Environmental Protection Agency (EPA) defines the Action Level for lead in tap water as 15 parts per billion (ppb; 15 ppb = 0.015 mg/l). This is a level above which further analysis and public education campaigns must be done, but not yet a level posing a real health risk. A random sample of 10,341 water samples from the Washington, D.C., area found that 7496 water samples had lead content up to 15 ppb.[20] What is the proportion of all possible water samples in D.C. that would exceed the Action Level? Use a 95% confidence level.

19.36 **Student drinking.** The College Alcohol Study interviewed a sample of 14,941 college students about their drinking habits. The sample was stratified using 140 colleges as strata, but the overall effect is close to an SRS of students. The response rate was between 60% and 70% at most colleges. This is quite good for a national sample, though nonresponse is as usual the biggest weakness of this survey. Of the students in the sample, 10,010 supported cracking down on underage drinking.[21] Estimate with 99% confidence the proportion of all college students who feel this way. Then describe briefly what uncertainty is and isn't included in this estimate.

19.37 **Condom usage.** The National AIDS Behavioral Surveys (Example 19.1) also interviewed a sample of adults in the cities where AIDS is most common. This sample included 803 heterosexuals who reported having more than one sexual partner in the past year. We can consider this an SRS of size 803 from the population of all heterosexuals in high-risk cities who have multiple partners. These people risk infection with the AIDS virus. Yet 304 of the respondents said they never use condoms. Is this strong evidence that more than one-third of this population never use condoms?

19.38 **Online publishing.** Publishing scientific papers online is fast, and the papers can be long. Publishing in a paper journal means that the paper will live forever in libraries. The *British Medical Journal* combines the two: it prints short and readable versions, with longer versions available online. Is this OK with authors? The journal asked a random sample of 104 of its recent authors several questions.[22] One question was "Should the journal continue using this system?" In the sample, 72 said "Yes." What proportion of all authors would say "Yes" if asked? (Estimate with 95% confidence.) Do the data give good evidence that more than two-thirds (67%) of authors support continuing this system? Answer both questions with appropriate inference methods.

19.39 **More online publishing.** The previous exercise describes a survey of authors of papers in a medical journal. Another question in the survey asked whether authors would accept a stronger move toward online publishing: "As an author, how acceptable would it be for us to publish only the abstract of papers in the paper journal and continue to put the full long version on our website?" Of the 104 authors in the sample, 65 said "Not at all acceptable." What proportion of all authors feel that abstract-only publishing is not acceptable? (Estimate with 95% confidence.) Do the data provide good evidence that more than half of all authors feel that abstract-only publishing is not acceptable?

19.40 **Fast food for baby, continued.** The 2003 Gerber survey cited in Excercise 19.31 also found that among the 316 toddlers 19 to 24 months old who were surveyed, 82 had eaten fried potatoes the day before the interview. Is this evidence that more than one third of toddlers 19 to 24 months old eat fried potatoes on a given day? Estimate with 95% confidence the proportion of toddlers 19 to 24 months old who eat fried potatoes on a given day. Answer both questions with appropriate inference methods.

19.41 **Alternative medicine.** A nationwide random survey of 1500 adults asked about attitudes toward "alternative medicine," such as acupuncture, massage therapy, and herbal therapy. Among the respondents, 660 said they would use alternative medicine if traditional medicine was not producing the results they wanted.[23] Calculate a 95% confidence interval for the proportion of all adults who would use alternative medicine, and write a short paragraph for a news report based on the survey.

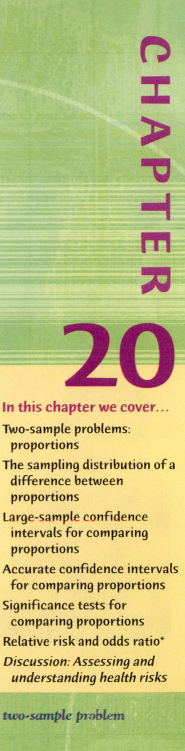

two-sample problem

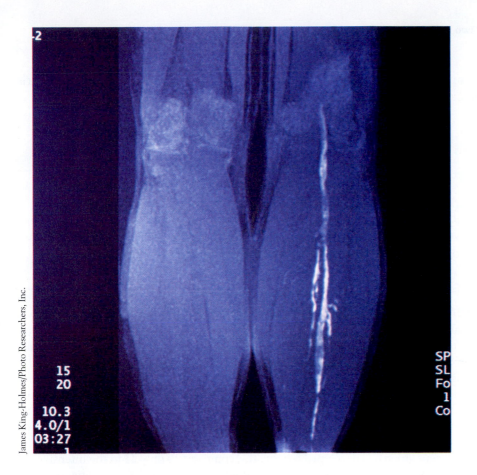

James King-Holmes/Photo Researchers, Inc.

Comparing Two Proportions

In a **two-sample problem,** we want to compare two populations or the responses to two treatments based on two independent samples. When the comparison involves the *means* of two populations, we use the two-sample t methods of Chapter 18. Now we turn to methods to compare the *proportions* of successes in two populations.

Two-sample problems: proportions

We will use notation similar to that used in our study of two-sample t statistics. The groups we want to compare are Population 1 and Population 2. We have a separate SRS from each population or responses from two treatments in a randomized comparative experiment. A subscript shows which group a parameter or statistic describes. Here is our notation:

Population	Population proportion	Sample size	Sample proportion
1	p_1	n_1	$\hat{p}_1$
2	p_2	n_2	$\hat{p}_2$

We compare the populations by doing inference about the difference $p_1 - p_2$ between the population proportions. The statistic that estimates this difference is the difference between the two sample proportions, $\hat{p}_1 - \hat{p}_2$.

EXAMPLE 20.1 How to treat type I diabetes

STATE: Chronic diseases, such as type I diabetes (early-onset or insulin-dependent diabetes), can lead to severe medical complications. For a long time, the medical community was divided on whether or not to aggressively control patients' blood glucose levels. The Diabetes Control and Complications Trial (DCCT) randomly assigned volunteers with type I diabetes without retinopathy (damage to the retina that can lead to blindness) either to a conventional treatment or to a more intensive treatment aimed at maintaining blood glucose level as close to normal as possible. The health of 378 patients in the conventional care group and 348 in the intensive care group was closely monitored for about 6 years. By the end of the study, 91 patients in the conventional treatment group had developed retinopathy, compared to only 23 in the intensive treatment group.[1] Is this good evidence that the proportions of diabetes patients developing retinopathy differ in the two populations? How large is the difference between the proportions developing retinopathy when patients are given one of the two treatment alternatives?

FORMULATE: Take patients receiving the intensive treatment to be Population 1 and patients receiving the conventional treatment to be Population 2. The population proportions that develop retinopathy within 6 years are p_1 for the conventional treatment and p_2 for the intensive treatment. We want to test the hypotheses

$$H_0: p_1 = p_2 \text{ (the same as } H_0: p_1 - p_2 = 0)$$
$$H_a: p_1 \neq p_2 \text{ (the same as } H_a: p_1 - p_2 \neq 0)$$

Given the medical knowledge at the time there was no reason to anticipate better results with one treatment or the other; therefore, the alternative hypothesis should be two-sided. We also want to give a confidence interval for the difference $p_1 - p_2$.

SOLVE: Inference about population proportions is based on the sample proportions

$$\hat{p}_1 = \frac{91}{378} = 0.2407 \quad \text{(conventional)}$$

$$\hat{p}_2 = \frac{23}{348} = 0.0661 \quad \text{(intensive)}$$

We see that about 24% of patients following the conventional treatment but only about 7% of patients following the intensive treatment developed retinopathy during the 6 years of the study. Because the samples are large and the sample proportions are quite different, we expect that a test will be highly significant. So we will concentrate on the

Tony Rusecki/Alamy

confidence interval (which will also answer whether or not the test is significant). To estimate $p_1 - p_2$, start from the difference of sample proportions

$$\hat{p}_1 - \hat{p}_2 = 0.2407 - 0.0661 = 0.1746$$

To complete the *Solve* step, we must know how this difference behaves.

The sampling distribution of a difference between proportions

To use $\hat{p}_1 - \hat{p}_2$ for inference, we must know its sampling distribution. We discussed in Chapter 18 that when two random variables are Normally distributed, the new variable "difference" also follows a Normal distribution, centered on the difference of the two variables' means and with variance equal to the sum of the two variables' variances. In Chapter 19 we learned that the sampling distribution of $\hat{p}$ is approximately Normal when the sample size n is large, with mean and standard deviation p and $\sqrt{p(1-p)/n}$, respectively. So here are the facts on which we rely when conducting inference for the difference between two population proportions:

- When the samples are large, the distribution of $\hat{p}_1 - \hat{p}_2$ is **approximately Normal.**

- The **mean** of the sampling distribution is $p_1 - p_2$. That is, the difference between sample proportions is an unbiased estimator of the difference between population proportions.

- The **standard deviation** of the distribution is

$$\sqrt{\frac{p_1(1-p_1)}{n_1} + \frac{p_2(1-p_2)}{n_2}}$$

Figure 20.1 displays the distribution of $\hat{p}_1 - \hat{p}_2$. The standard deviation of $\hat{p}_1 - \hat{p}_2$ involves the unknown parameters p_1 and p_2. Just as in the previous chapter,

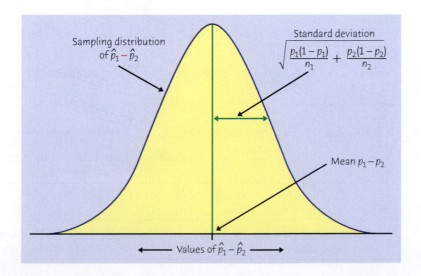

Sounds good—but no comparison

Most women have mammograms to check for breast cancer once they reach middle age. Could a fancier test do a better job of finding cancers early? PET scans are a fancier (and more expensive) test. Doctors used PET scans on 14 women with tumors and got the detailed diagnosis right in 12 cases. That's promising. But there were no controls, and 14 cases are not statistically significant. Medical standards require randomized comparative experiments and statistically significant results. Only then can we be confident that the fancy test really is better.

FIGURE 20.1 Select independent SRSs from two populations having proportions of successes p_1 and p_2. The proportions of successes in the two samples are $\hat{p}_1$ and $\hat{p}_2$. When the samples are large, the sampling distribution of the difference $\hat{p}_1 - \hat{p}_2$ is approximately Normal.

we must replace these by estimates in order to do inference. And just as in the previous chapter, we do this a bit differently for confidence intervals and for tests.

Large-sample confidence intervals for comparing proportions

standard error

To obtain a confidence interval, one option is to replace the population proportions p_1 and p_2 in the standard deviation with the sample proportions. The result is the **standard error** of the statistic $\hat{p}_1 - \hat{p}_2$:

$$\text{SE} = \sqrt{\frac{\hat{p}_1(1 - \hat{p}_1)}{n_1} + \frac{\hat{p}_2(1 - \hat{p}_2)}{n_2}}$$

The confidence interval has the same form we encountered in the previous chapter,

$$\text{estimate} \pm z^*\text{SE}_{\text{estimate}}$$

> **LARGE-SAMPLE CONFIDENCE INTERVAL FOR COMPARING TWO PROPORTIONS**
>
> Draw an SRS of size n_1 from a large population having proportion p_1 of successes, and draw an independent SRS of size n_2 from another large population having proportion p_2 of successes. When n_1 and n_2 are large, an approximate level C **confidence interval for $p_1 - p_2$** is
>
> $$(\hat{p}_1 - \hat{p}_2) \pm z^*\text{SE}$$
>
> In this formula, the standard error SE of $\hat{p}_1 - \hat{p}_2$ is
>
> $$\text{SE} = \sqrt{\frac{\hat{p}_1(1 - \hat{p}_1)}{n_1} + \frac{\hat{p}_2(1 - \hat{p}_2)}{n_2}}$$
>
> and z^* is the critical value for the standard Normal density curve with area C between $-z^*$ and z^*.
>
> Use this interval only when the numbers of successes and failures are each 10 or more in both samples.

EXAMPLE 20.2   How to treat type I diabetes

We can now complete Example 20.1. Here is a summary of the basic information:

Population	Population description	Sample size	Number of successes	Sample proportion
1	conventional	$n_1 = 378$	91	$\hat{p}_1 = 91/378 = 0.2407$
2	intensive	$n_2 = 348$	23	$\hat{p}_2 = 23/348 = 0.0661$

SOLVE: We will give a 95% confidence interval for $p_1 - p_2$, the difference between the proportions developing retinopathy when diabetes patients are given one of the two treatment alternatives. To check that the large-sample confidence interval is safe, look at the counts of successes and failures in the two samples. All of these four counts are much larger than 10, so the large-sample method will be accurate. The standard error is

$$
\begin{aligned}
\text{SE} &= \sqrt{\frac{\hat{p}_1(1 - \hat{p}_1)}{n_1} + \frac{\hat{p}_2(1 - \hat{p}_2)}{n_2}} \\
&= \sqrt{\frac{(0.2407)(0.7593)}{378} + \frac{(0.0661)(0.9339)}{348}} \\
&= \sqrt{0.000661} = 0.02571
\end{aligned}
$$

The 95% confidence interval is

$$
\begin{aligned}
(\hat{p}_1 - \hat{p}_2) \pm z^*\text{SE} &= (0.2407 - 0.0661) \pm (1.960)(0.02571) \\
&= 0.1746 \pm 0.0504 \\
&= 0.1242 \text{ to } 0.2250
\end{aligned}
$$

CONCLUDE: We are 95% confident that the percent of diabetes patients who would develop retinopathy over a 6 years period is between 12.5 and 22.5 percentage points higher when following a conventional treatment than when following an intensive treatment.

Figure 20.2 displays software output for Example 20.2 from a graphing calculator and two statistical software programs (not all software packages provide a simple way to obtain confidence intervals for proportions). As usual, you can understand the output even without knowledge of the program that produced it. Minitab gives the test as well as the confidence interval, confirming that the difference between the two treatment types is highly significant.

Like the large-sample confidence interval for a single proportion p, *the large-sample interval for* $p_1 - p_2$ *generally has a true confidence level less than the level you asked for.* The inaccuracy is not as serious as in the one-sample case, however, at least if our guidelines for use are followed.

APPLY YOUR KNOWLEDGE

20.1   More about the DCCT. The Diabetes Control and Complications Trial, described in Example 20.1, also followed diabetes patients diagnosed with retinopathy *before* joining the study. They too were randomly assigned to one of the two treatments and monitored for 6 years. The study found that 143 of the 352 patients assigned to the conventional treatment showed a sustained progression of their original retinopathy. In contrast, only 77 of the 363 patients assigned to the intensive treatment had sustained retinopathy progression.[2] Give a 95% confidence interval for the difference between the proportions of patients with retinopathy progression when diabetes patients with pre-existing retinopathy receive either conventional or intensive treatment. Follow the four-step process as illustrated in Examples 20.1 and 20.2.

Texas Instruments TI-83 or TI-84

```
2-PropZInt
  (.12426,.22504)
  p̂₁=.2407407407
  p̂₂=.066091954
  n₁=378
  n₂=348
```

CrunchIt!

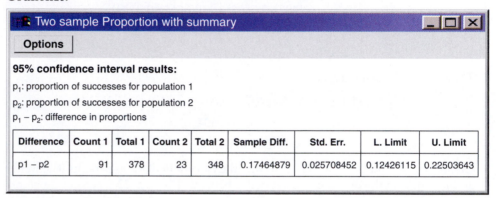

Two sample Proportion with summary

Options

95% confidence interval results:

p_1: proportion of successes for population 1

p_2: proportion of successes for population 2

$p_1 - p_2$: difference in proportions

Difference	Count 1	Total 1	Count 2	Total 2	Sample Diff.	Std. Err.	L. Limit	U. Limit
p1 – p2	91	378	23	348	0.17464879	0.025708452	0.12426115	0.22503643

Minitab

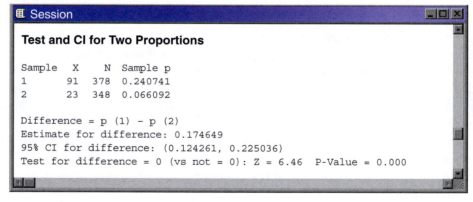

Session

Test and CI for Two Proportions

```
Sample   X    N  Sample p
1       91  378  0.240741
2       23  348  0.066092


Difference = p (1) - p (2)
Estimate for difference: 0.174649
95% CI for difference: (0.124261, 0.225036)
Test for difference = 0 (vs not = 0): Z = 6.46  P-Value = 0.000
```

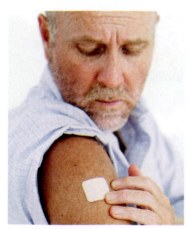

Stockbyte Platinum/Alamy

FIGURE 20.2 Output from the TI-83 graphing calculator, CrunchIt!, and Minitab for the 95% confidence interval of Example 20.2.

20.2   **How to quit smoking.** Nicotine patches are often used to help smokers quit. Does giving medicine to fight depression help? A randomized double-blind experiment assigned 244 smokers who wanted to stop to receive nicotine patches and another 245 to receive both a patch and the antidepression drug bupropion. After a year, 40 subjects in the nicotine patch group and 87 in the patch-plus-drug group had abstained from smoking.[3] Give a 99% confidence interval for the difference (treatment minus control) in the proportion of smokers who quit. Follow the four-step process as illustrated in Examples 20.1 and 20.2.

Accurate confidence intervals for comparing proportions

As in the previous chapter, another option to obtain a confidence interval for $p_1 - p_2$ is to add four imaginary observations to the sample data. This simple adjustment has been shown to greatly improve the accuracy of the confidence interval compared with the large-sample method.[4]

PLUS FOUR CONFIDENCE INTERVAL FOR COMPARING TWO PROPORTIONS

Draw independent SRSs from two populations with population proportions of successes p_1 and p_2. To get the **plus four confidence interval for the difference $p_1 - p_2$,** add four imaginary observations, one success and one failure in each of the two samples. Then use the large-sample confidence interval with the new sample sizes (actual sample sizes + 2) and counts of successes (actual counts + 1).

Use this interval when the sample size is at least 5 in each group, with any counts of successes and failures.

If your software does not offer the plus four method, just enter the new plus four sample sizes and success counts into the large-sample procedure.

EXAMPLE 20.3   Shrubs that withstand fire

STATE: Some shrubs can resprout from their roots after their tops are destroyed. Fire is a serious threat to shrubs in dry climates, as it can injure the roots as well as destroy the tops. One study of resprouting took place in a dry area of Mexico.[5] The investigators randomly assigned shrubs to treatment and control groups. They clipped the tops of all the shrubs. They then applied a propane torch to the stumps of the treatment group to simulate a fire. A shrub is a success if it resprouts. Here are the data for the shrub *Xerospirea hartwegiana:*

Population	Population description	Sample size	Number of successes	Sample proportion
1	control	$n_1 = 12$	12	$\hat{p}_1 = 12/12 = 1.000$
2	treatment	$n_2 = 12$	8	$\hat{p}_2 = 8/12 = 0.667$

How much does burning reduce the proportion of shrubs of this species that resprout?

FORMULATE: Give a 90% confidence interval for the difference of population proportions, $p_1 - p_2$.

SOLVE: The conditions for the large-sample interval are not met. In fact, there are *no* failures in the control group. We will use the plus four method. Add four imaginary

observations. The new data summary is

Population	Population description	Sample size	Number of successes	Plus four sample proportion
1	control	$n_1 + 2 = 14$	$12 + 1 = 13$	$\tilde{p}_1 = 13/14 = 0.9286$
2	treatment	$n_2 + 2 = 14$	$8 + 1 = 9$	$\tilde{p}_2 = 9/14 = 0.6429$

The standard error based on the new facts is

$$\mathrm{SE} = \sqrt{\frac{\tilde{p}_1(1 - \tilde{p}_1)}{n_1 + 2} + \frac{\tilde{p}_2(1 - \tilde{p}_2)}{n_2 + 2}}$$

$$= \sqrt{\frac{(0.9286)(0.0714)}{14} + \frac{(0.6429)(0.3571)}{14}}$$

$$= \sqrt{0.02113} = 0.1454$$

The plus four 90% confidence interval, using $z^*1.645$ from Table C, is

$$(\tilde{p}_1 - \tilde{p}_2) \pm z^*\mathrm{SE} = (0.9286 - 0.6429) \pm (1.645)(0.1454)$$

$$= 0.2857 \pm 0.2392$$

$$= 0.047 \text{ to } 0.525$$

CONCLUDE: We are 90% confident that burning reduces the percent of these shrubs that resprout by between 4.7% and 52.5%.

The plus four interval may be conservative (that is, the true confidence level may be *higher* than you asked for) for very small samples and population p's close to 0 or 1, as in this example. It is generally much more accurate than the large-sample interval when the samples are small. Nevertheless, the plus four interval in Example 20.3 cannot save us from the fact that small samples produce wide confidence intervals.

APPLY YOUR KNOWLEDGE

20.3   Echinacea for the common cold? Echinacea is widely used as an herbal remedy for the common cold, but does it work? In a double-blind experiment, healthy volunteers agreed to be exposed to common-cold-causing rhinovirus type 39 and have their symptoms monitored. The volunteers were randomly assigned to take either a placebo or an echinacea supplement daily for 5 days following viral exposure. Among the 103 subjects taking a placebo, 88 developed a cold, whereas 44 of the 48 subjects taking echinacea developed a cold.[6]

(a) Explain why the large-sample confidence interval is not appropriate for these data.

(b) Give the 95% plus four confidence interval for the difference in proportion of individuals developing a cold after viral exposure between the echinacea treatment and the placebo. Follow the four-step process as illustrated in Example 20.3.

imagebroker/Alamy

20.4   In-line skaters. A study of injuries to in-line skaters used data from the National Electronic Injury Surveillance System, which collects data from a random sample of hospital emergency rooms. The researchers interviewed 161 people who came to emergency rooms with injuries from in-line skating. Wrist injuries (mostly fractures) were the most common.[7]

(a) The interviews found that 53 people were wearing wrist guards and 6 of these had wrist injuries. Of the 108 who did not wear wrist guards, 45 had wrist injuries. Why should we not use the large-sample confidence interval for these data?

(b) Give the plus four 95% confidence interval for the difference between the two population proportions of wrist injuries. State carefully what populations your inference compares. We would like to draw conclusions about all in-line skaters, but we have data only for injured skaters.

Significance tests for comparing proportions

An observed difference between two sample proportions can reflect an actual difference between the populations, or it may just be due to chance variation in random sampling. Significance tests help us decide if the effect we see in the samples is really there in the populations. The null hypothesis says that there is no difference between the two populations:

$$H_0: p_1 = p_2$$

The alternative hypothesis says what kind of difference we expect.

EXAMPLE 20.4   *Sex ratio and hatching order in geese*

STATE: Is sex determination truly a random process? The laws of genetics would suggest so, because of the random distribution of X and Y chromosomes among sperm cells. However, physiological or environmental factors could favor one gender over the other under specific circumstances. A field biologist surveyed a random sample of lesser snow geese nests containing exactly four eggs to study sex ratios by laying order. Of the 52 live goslings hatched from early eggs (eggs laid either first or second), 19 were female. Of the 43 live goslings hatched from late eggs (eggs laid either third or last), 31 were female.[8] Is there reason to think that different proportions of females exist among the early eggs versus the late eggs in geese nests containing four eggs?

FORMULATE: Take the early eggs to be Population 1 and the late eggs to be Population 2. We had no direction for the difference in mind before looking at the data, so we have a two-sided alternative:

$$H_0: p_1 = p_2$$
$$H_a: p_1 \neq p_2$$

SOLVE: The eggs in a single SRS can be treated as if they were separate SRSs of early and late eggs. The sample proportions that are female are

$$\hat{p}_1 = \frac{19}{52} = 0.365 \text{ (early eggs)}$$

$$\hat{p}_2 = \frac{31}{43} = 0.721 \text{ (late eggs)}$$

That is, about 37% of the early eggs but as many as 72% of the late eggs are female. Is this apparent difference statistically significant? To continue the solution, we must learn the proper test.

To do a test, standardize $\hat{p}_1 - \hat{p}_2$ to get a z statistic. If H_0 is true, all the observations in both samples come from a single population of eggs, of which a single unknown proportion p would be females. So instead of estimating p_1 and p_2 separately, we pool the two samples and use the overall sample proportion to estimate the single population parameter p. Call this the **pooled sample proportion.** It is

pooled sample proportion

$$\hat{p} = \frac{\text{number of successes in both samples combined}}{\text{number of individuals in both samples combined}}$$

Use $\hat{p}$ in place of both $\hat{p}_1$ and $\hat{p}_2$ in the expression for the standard error SE of $\hat{p}_1 - \hat{p}_2$ to get a z statistic that has the standard Normal distribution when H_0 is true. Here is the test.

SIGNIFICANCE TEST FOR COMPARING TWO PROPORTIONS

Draw an SRS of size n_1 from a large population having proportion p_1 of successes and draw an independent SRS of size n_2 from another large population having proportion p_2 of successes. To **test the hypothesis H_0: $p_1 = p_2$,** first find the pooled proportion $\hat{p}$ of successes in both samples combined. Then compute the z statistic

$$z = \frac{\hat{p}_1 - \hat{p}_2}{\sqrt{\hat{p}(1 - \hat{p})\left(\dfrac{1}{n_1} + \dfrac{1}{n_2}\right)}}$$

In terms of a variable Z having the standard Normal distribution, the P-value for a test of H_0 against

H_a: $p_1 > p_2$ is $P(Z \geq z)$

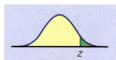

H_a: $p_1 < p_2$ is $P(Z \leq z)$

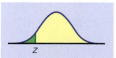

H_a: $p_1 \neq p_2$ is $2P(Z \geq |z|)$

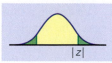

Use this test when the counts of successes and failures are each 5 or more in both samples.[9]

EXAMPLE 20.5 *Sex ratio and hatching order, continued* ———————

SOLVE: The data come from an SRS, and the counts of successes and failures are all larger than 5. The pooled proportion of female goslings is

$$\hat{p} = \frac{\text{number of female goslings among early and late eggs combined}}{\text{number of early and late eggs combined}}$$

$$= \frac{19 + 31}{52 + 43} = \frac{50}{95} = 0.5263$$

The z test statistic is

$$z = \frac{\hat{p}_1 - \hat{p}_2}{\sqrt{\hat{p}(1 - \hat{p})\left(\dfrac{1}{n_1} + \dfrac{1}{n_2}\right)}}$$

$$= \frac{0.365 - 0.721}{\sqrt{(0.5263)(0.4737)\left(\dfrac{1}{52} + \dfrac{1}{43}\right)}}$$

$$= \frac{-0.3555}{0.1029} = -3.455$$

The two-sided P-value is the area under the standard Normal curve more than 3.455 distant from 0 in either direction. Figure 20.3 shows this area. Figure 20.4 shows the output from the TI-83 graphing calculator, indicating that $P = 0.0006$.

Without software, you can use the bottom row of Table C (standard Normal critical values) to approximate P with no calculations: The positive value of z is 3.455, which

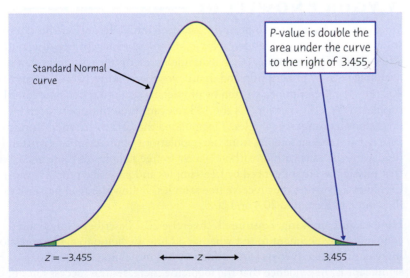

FIGURE 20.3 The P-value for the two-sided test of Example 20.5.

Texas Instruments TI-83 or TI-84

```
2-PropZTest
 p₁≠p₂
 z=-3.45463185
 p=5.5113992E-4
 p̂₁=.3653846154
 p̂₂=.7219302326
 p̂=.5263157895
 n₁=52
 n₁=43
```

FIGURE 20.4 Output from the TI-83 graphing calculator for Example 20.5.

is larger than the critical z value 3.291. Therefore, the two-sided P-value is less than 0.001.

CONCLUDE: There is very strong evidence ($P < 0.001$) of a difference between early and late eggs. Female goslings are less common among eggs laid first or second than among the later sets of eggs in lesser snow geese.

The study in this example selected a single random sample of nests containing four eggs. To get two samples, we divided the single sample by egg-laying order: early (first and second eggs) versus late (third and last eggs). Because only eggs resulting in a live gosling were considered, we did not know the two sample sizes n_1 and n_2 until after the data were in hand. The two-sample z procedures for comparing proportions are valid in such situations. This is an important fact about these methods.

APPLY YOUR KNOWLEDGE

20.5   Do free samples influence prescription decisions? To answer this question, a study randomly assigned resident physicians to either accept free drug samples or refuse them. The researchers then examined prescriptions for drugs that are available both by prescription and in an over-the-counter (OTC) version. Of the 202 relevant prescriptions written by physicians accepting free samples, 51 stipulated the OTC version. Of the 188 relevant prescriptions written by physicians refusing free samples, 73 stipulated the OTC version.[10] Consider these data as two random samples from the populations of prescriptions written by resident physicians with or without access to free samples. Is there good evidence that physicians are influenced by free samples and prescribe proportionally fewer OTC versions when they receive free samples? Follow the four-step process as illustrated in Examples 20.4 and 20.5.

20.6   How to quit smoking, continued. Exercise 20.2 describes a randomized comparative experiment to test whether adding medicine to fight depression increases the effectiveness of nicotine patches in helping smokers to quit. How significant is the evidence that the medicine increases the success rate? Follow the four-step process as illustrated in Examples 20.4 and 20.5.

Relative risk and odds ratio*

Categorical variables are particularly widespread in the health sciences. Many studies describe the presence or absence of a disease in a particular population, or an improvement versus lack of improvement after treatment. In fact, the objective of most scientific reports in this field is to study either risk factors or treatment efficacy.

Studies evaluating the potential impact of a risk factor look for the association between exposure to the risk factor and a negative health outcome in sample individuals. As discussed in Chapter 7, these epidemiological studies can have different designs, but they are typically observational. Exercise 20.4 describes one such study comparing injuries among in-line skaters wearing wrist guards or not.

Studies evaluating the efficacy of a treatment compare the outcome of a sample given the treatment to the outcome of a sample given a placebo or a control. These are typically experimental studies. We covered their design in detail in Chapter 8. The Diabetes Control and Complications Trial described in Example 20.1 is one such experiment.

Both types examine a categorical variable among two populations, but they rarely report the difference in sample proportions as we have studied in this chapter so far. Instead, you would typically read about a "relative risk" or an "odds ratio" computed from the sample data. The following section gives a brief introduction to these topics.

Relative risk and odds ratio. Both observational and experimental studies in the health sciences typically compare a sample from a population of interest with a sample from a control population. The comparison is then represented by either a relative risk or an odds ratio.

Chapter 9 introduced the concepts of risk and odds in probability. When comparing two groups, the relative risk RR is simply the ratio of the two risks with the control group in the denominator. Likewise, the odds ratio OR is the ratio of both odds with the control group in the denominator. So if a study finds that 50% of individuals in the study group had some symptoms compared with only 25% in the control group, the relative risk would be $RR = 50\%/25\% = 2$ and the odds ratio would be $OR = (1/1)/(1/3) = 3$. Relative risk is easy to interpret—individuals in the study group are twice as likely to have symptoms as similar control individuals. Interpreting the odds ratio is more challenging, because it compares unfamiliar odds rather than probabilities. Odds ratios can also give an exaggerated impression of the relative difference between both groups, particularly when the events studied are not rare. However, odds ratios offer mathematical advantages for advanced statistical analysis. Odds ratios are found most often in observational, case-control, and historical control epidemiological studies, while relative risks are typically used in prospective cohort studies and randomized controlled experiments.

---

*The remainder of this chapter presents more advanced material that is not needed to read the rest of the book. Relative risk and odds ratio are, however, important concepts in the health sciences.

> ## RELATIVE RISK AND ODDS RATIO IN THE HEALTH SCIENCES
>
> Draw an SRS of size n_1 from a large population having proportion p_1 of a medical outcome, and draw an independent SRS of size n_2 from another large population having proportion p_2 of that outcome. When the first sample represents the group of interest and the second sample the reference, or control, group:
>
> The relative risk for the medical outcome in the group of interest compared with the reference group is
>
> $$RR = \frac{\hat{p}_1}{\hat{p}_2}$$
>
> The relative risk is sometimes also called the hazard ratio, or HR.
>
> The odds ratio for the medical outcome in the group of interest compared with the reference group is
>
> $$OR = \frac{odds_1}{odds_2} = \frac{\hat{p}_1(1 - \hat{p}_2)}{\hat{p}_2(1 - \hat{p}_1)}$$

James King-Holmes/Photo Researchers, Inc.

┌─ **EXAMPLE 20.6** *Preventing blood clots in immobilized patients* ─────

Patients immobilized for a substantial amount of time can develop deep vein thrombosis (DVT), a blood clot in a leg or pelvis vein. DVT can have serious adverse health effects and can be difficult to diagnose. On its website, drug manufacturer Pfizer reports the outcome of a study looking at the effectiveness of the drug Fragmin (dalteparin) compared with that of a placebo in preventing DVT in immobilized patients.

In a double-blind, multinational study, severely immobilized patients were randomly assigned to receive daily subcutaneous injections of either Fragmin or a placebo for 12 to 14 days and were followed for 90 days. The results, in number of patients experiencing a complication from DVT (including death), are summarized in the table below.

	Treatment outcome		
	Complication	No complication	Sample size
Fragmin	42	1476	1518
Placebo	73	1400	1473

The proportion of subjects experiencing DVT complications in the two samples are:

$$\hat{p}_{Fragmin} = 42/1518 = 0.0277$$
$$\hat{p}_{placebo} = 73/1473 = 0.0496$$

The relative risk of DVT complications is the ratio of the two sample proportions, with the placebo group in the denominator.

$$RR = \frac{\hat{p}_{Fragmin}}{\hat{p}_{placebo}} = \frac{0.0277}{0.0496} = 0.558$$

That is, patients receiving the Fragmin injections are less likely to develop complications from DVT than patients in the control group.

The odds of a subject's experiencing DVT complications in the two samples are

$$odds_{Fragmin} = 42/1476, \text{ or about 1 to 35 against}$$

$$odds_{placebo} = 73/1400, \text{ or about 1 to 19 against}$$

The odds ratio of DVT complications is the ratio of these two odds with the placebo group in the denominator, but it can also be computed from the two sample proportions.

$$OR = \frac{odds_{Fragmin}}{odds_{placebo}} = \frac{42/1476}{73/1400} = 0.546$$

$$= \frac{\hat{p}_{Fragmin}(1 - \hat{p}_{placebo})}{\hat{p}_{placebo}(1 - \hat{p}_{Fragmin})} = \frac{0.0277 * (1 - 0.0496)}{0.0496 * (1 - 0.277)} = 0.546$$

This value is quite similar to the RR value. When samples are large and the outcome studied is rare (here 3 to 5 percent), RR and OR are very similar.

What do the values for an RR or an OR mean? A relative risk of 1 means that there is no difference in risk between the two groups. However, relative risk is computed from sample data and is only an estimate of the true population risks. Therefore, an RR different from 1 does not necessarily imply a difference between the two populations. How large or how small a relative risk—or an odds ratio—needs to be in order to be statistically significant depends on the sampling distributions of these statistics.

Because relative risk and odds ratio are expressed as ratios of random variables, their sampling distributions are not symmetric. Therefore, confidence intervals for the population parameter are not symmetric and cannot be expressed as "estimate ± margin of error." For instance, in the Fragmin study of Example 20.6, special software intended for the analysis of medical data gives the following confidence intervals:

Statistic	Value	Low 95% CI	High 95% CI
Odds ratio	0.546	0.364	0.816
Relative risk	0.558	0.378	0.823

The calculations required to obtain such confidence intervals are beyond the scope of this introductory statistics textbook. The interpretation of a confidence interval or a P-value is the same, however, whether we study an average, a proportion, an odds ratio, or a relative risk. As always, *statistical significance does not imply relevance or importance. When sample sizes are very large, statistical significance can be reached for even very small effects. And observational studies always carry potential confounding factors that make a clear-cut interpretation of cause and effect between a studied risk factor and an observed outcome impossible.*

In the Fragmin experiment, because the 95% confidence interval for RR contains only values less than 1, we can conclude with 95% certainty that severely immobilized patients given Fragmin are less likely (lower risk) to experience DVT complications than similar patients given a placebo.

APPLY YOUR KNOWLEDGE

20.7 **Aspirin and heart attacks.** The Physicians' Health Study randomly assigned 22,071 healthy male physicians at least 40 years old to take either an aspirin every other day or a placebo pill every other day. Of the 11,037 physicians who took aspirin regularly for 5 years, 10 had a fatal heart attack. Of the 11,034 who took the placebo, 26 suffered a fatal heart attack.[11]

(a) What are the proportions of physicians who suffered a fatal heart attack in the two groups?

(b) Calculate the relative risk of a fatal heart attack for physicians taking aspirin compared with those taking a placebo. Explain what this value means concretely.

(c) Calculate the odds ratio for a fatal heart attack, comparing physicians taking aspirin to those taking a placebo. How does this value compare with the RR you calculated in (b)?

20.8 **Aspirin and heart attacks, continued.** The Physicians' Health Study also recorded the number of nonfatal heart attacks in the two groups. Of the 11,037 physicians who took aspirin regularly for 5 years, 129 had a nonfatal heart attack. Of the 11,034 who took the placebo, 213 suffered a nonfatal heart attack.

(a) What are the proportions of physicians who suffered a nonfatal heart attack in the two groups?

(b) Calculate the relative risk of a nonfatal heart attack, for physicians taking aspirin compared with those taking a placebo. Explain what this value means concretely.

(c) Calculate the odds ratio for a nonfatal heart attack, comparing physicians taking aspirin to those taking a placebo. How does this value compare with the RR you calculated in (b)?

Effect size: relative versus absolute measures. The relative risk gives a measure of how effective a treatment is or how influential a risk factor is in a group compared with a control group without the treatment or the risk factor. One topic of debate in the biomedical community centers on how to interpret it. We will compare the relative risk to other measures of treatment effectiveness in comparative, randomized experiments.

RELATIVE RISK REDUCTION, ABSOLUTE RISK REDUCTION, AND NUMBER NEEDED TO TREAT

In a randomized clinical trial comparing the proportion of a negative medical outcome in a group provided a medical treatment and in a control group:

The relative risk reduction RRR for the treatment group compared with the control group is the reduction in risk expressed as a fraction of the control group's risk:

$$RRR = \frac{\hat{p}_{control} - \hat{p}_{treatment}}{\hat{p}_{control}}$$

The absolute risk reduction ARR for the treatment group compared with the control group is simply the difference in risks:

$$ARR = \hat{p}_{control} - \hat{p}_{treatment}$$

The number needed to treat NNT represents the expected number of subjects who must be treated before 1 subject in the treatment group can be spared the negative medical outcome compared with what would have been expected in the control group. It is given by

$$NNT = \frac{1}{ARR} = \frac{1}{\hat{p}_{control} - \hat{p}_{treatment}}$$

when ARR is a strictly positive value. (When studying treatment side effects, ARR is typically negative and the number needed to harm, NNH, is reported instead.)

─── **EXAMPLE 20.7**   Preventing blood clots: relative risk reduction ───

Example 20.6 described the efficacy of Fragmin for preventing complications from DVT in immobilized patients compared with those receiving a placebo. Here are the proportions of adverse events in the two groups.

$$\hat{p}_{Fragmin} = 42/1518 = 0.0277$$
$$\hat{p}_{placebo} = 73/1473 = 0.0496$$

The Fragmin website reports a relative risk reduction of 44%. How was this number computed?

$$RRR = \frac{\hat{p}_{placebo} - \hat{p}_{Fragmin}}{\hat{p}_{placebo}} = \frac{0.0496 - 0.0277}{0.0496} = 0.442$$

That is, immobilized patients treated with Fragmin have a 44% lower probability of suffering complications from DVT than similar patients given a placebo instead.

How does this value compare with the absolute risk reduction for this study?

$$ARR = \hat{p}_{placebo} - \hat{p}_{Fragmin} = 0.0496 - 0.0277 = 0.0219$$

Compared with receiving a placebo, receiving Fragmin injections helps prevent DVT complications in about 2% of all immobilized patients.

> The number needed to treat, *NNT*, is gradually becoming a standard in communicating clinical results. In the Fragmin study it is
>
> $$NNT = \frac{1}{ARR} = \frac{1}{0.0219} = 45.7$$
>
> That is, we expect that on average for every 46 immobilized patients treated with Fragmin instead of a placebo, DVT complications can be prevented in 1 patient.

The 2% *ARR* is radically different from the 44% *RRR*, and both give quite a different impression of treatment success. *RRR* tends to inflate the impression of success, but *ARR* can be difficult to comprehend. There is a growing trend toward always providing the *NNT*, as it gives a clear, practical medical interpretation of the relative benefit of treatment. Of course, the decision to provide a preventive or curative treatment depends not only on the relative benefit of one particular treatment but also on such things as the availability, costs, and benefits of alternative treatments, the number and seriousness of potential adverse events, and the seriousness and consequences of the condition (for example, death versus minor discomfort).

APPLY YOUR KNOWLEDGE

20.9   **Aspirin and heart attacks.** Go back to Exercise 20.7 about the Physicians' Health Study.

(a) Calculate the absolute and relative risk reduction in fatal heart attacks for physicians taking aspirin compared with those taking a placebo. Explain briefly what these two values represent.

(b) Calculate the number needed to treat with regular aspirin compared with a placebo to prevent 1 fatal heart attack. What does this value tell you?

20.10   **Aspirin and heart attacks.** Go back to Exercise 20.8 about the Physicians' Health Study.

(a) Calculate the absolute and relative risk reduction in nonfatal heart attacks for physicians taking aspirin compared with those taking a placebo. Explain briefly what these two values represent.

(b) Calculate the number needed to treat with regular aspirin compared with a placebo to prevent one nonfatal heart attack. What does this value tell you?

DISCUSSION: Assessing and understanding health risks

We have all heard news stories about factors that may increase or decrease the risk of getting a particular disease. At times, press releases even seem conflicting. For instance, we hear that coffee may reduce the risk of heart disease but also that it may increase the risk of a heart attack. So where does this information come from? And what does it really mean? Epidemiological studies and clinical trials are two major tools used to study health risk. Here we discuss how health risks are assessed in these two different contexts and what information they really convey.

Studying risk factors in epidemiological studies

Epidemiology is the science of understanding disease patterns through observations. Epidemiologists examine data to see what behaviors, locations, or exposures are associated with various diseases and causes of death. Because these studies are observational, confounding is a common problem and causality cannot be directly established. As discussed in Chapter 7, there are several types of epidemiological studies. Prospective cohort studies are the least prone to confounding errors, because subject selection does not depend on the factors studied. Despite their weaknesses, epidemiological studies can be very effective when many different studies converge to the same conclusion, as with tobacco smoking and the risk of developing lung and several other cancers.

Because confounding can never be completely ruled out in observational studies, finding a significant difference between some groups does not necessarily imply that a risk factor has been uncovered. This is especially true if the effect found is small. For example, a Harvard study found a two-and-a-half times higher risk of pancreatic cancer among coffee drinkers than among non–coffee drinkers. However, the association was not substantiated by later research from this group and countless others. By comparison, smoking multiplies the risk of cancer 20 to 30 times. On the other hand, even a small increase that shows up consistently in many studies with different designs can be credible and may have a large societal impact if the risk is widespread. Meta-analysis is an advanced statistical technique developed to find significant trends in the results of many different studies of the same variable. It has become a fairly common and important tool of epidemiology.

The information about new health risks that reaches the general public depends on scientific publications and how they are reported in the media. Unfortunately, studies that find no evidence of an association are often unreported or altogether abandoned. Significant results can be published more easily and in more prestigious papers and are much more likely to be funded and reported in the news. This is an important bias that cannot be compensated for by running a meta-analysis. A related and potentially devastating bias is due to financial conflicts of interest. Studies show that results may go unreported or conclusions may be unreasonably speculative as a consequence. For example, a 1998 observational study of 12 patients speculated about a possible link between the measles-mumps-rubella (MMR) vaccine and autism, creating quite a stir. As a result, many parents refused to have their children given the MMR vaccine. It was later revealed that the lead author had received funding from lawyers attempting class action suits against the vaccine manufacturer. Many well-designed studies since have found no evidence of a link between the MMR vaccine and autism, yet public skepticism persists. The public can be scared off easily, but it may take a whole generation to undo the damage.

Health risks and treatment options in clinical trials

Clinical trials are randomized controlled experiments involving human subjects. For obvious ethical reasons, clinical trials are limited to studying ways to reduce risk and not risk factors themselves. Because subjects are randomly assigned to various treatments and controls, differences in response can be directly attributed to differences in treatment. However, study participants are typically volunteers, and they may not always represent the target population adequately. The undercoverage of women and minority groups in clinical trials, for instance, is a serious issue, because trial results are often generalized to the whole target population despite their poor representation.

Patients are increasingly involved in treatment decisions, especially since the appearance of Internet health sites and pharmaceutical advertising. So how do patients make sense of claims about risk reduction from a given treatment? Studies show that our understanding of risk is strongly influenced by the way it is presented and framed. Like physicians and funding institutions, patients are much more likely to favor treatment when the information about risk is presented in a relative way, for instance, by using relative risk reduction *RRR*. Patients also tend to choose treatment more often when risk is described in terms of negative outcomes (death or disease) rather than as a positive outcome (survival or remaining event free).

Let's examine a clinical trial for pravastatin, a cholesterol-lowering drug and one of the highest-selling prescription drugs in the United States (under its brand names). The study can be summarized with the statement that pravastatin leads to a 22% reduction of the risk of coronary death (relative risk reduction *RRR*), or by saying that the reduction represents 0.9 percentage points (absolute risk reduction *ARR*) or that 111 patients need to take pravastatin daily for five years to avoid 1 coronary death (number needed to treat *NNT*). These three statements focus on the negative outcome, coronary death. The information can also be reframed to say that taking the drug daily for five years improves the probability of *avoiding* coronary death from 95.9% to 96.8%. All four summaries reflect the same data, but they give very different impressions. Historically, the relative risk reduction has been predominant, largely pushed by the pharmaceutical industry. However, there is a growing recognition of the need to provide a more comprehensive and balanced description of health risks.

One reason why the way risk is presented can be so influential is that patients have difficulty grasping the difference between average risk and personal outcomes. They may perceive a stated percent risk reduction as a guaranteed benefit. But there is no such guarantee. In fact, only a fraction of all patients treated will avoid developing the disease thanks to the treatment; all other patients would have the same outcome (disease or not) whether or not they took the treatment. There is no way to tell which it will be for any given patient. At the same time, some proportion of all patients treated will experience negative side effects because of the treatment.

Patients should also understand that clinical trials are conducted over a defined time period and that this impacts risk calculations. For instance, a woman's risk of breast cancer over the next 10 years is substantially lower than her risk of breast cancer over a lifetime, yet this distinction is rarely emphasized. Long-term adverse events due to treatment are often monitored more extensively in phase IV clinical trials, after a treatment has been made publicly available. This is the dilemma between bringing the benefits of a new treatment to the target population as soon as possible and remaining cautious about side effects that may be too rare to see in limited trials or may take many years to develop. The withdrawal of the pain reliever VIOXX and warnings about suicide attempts among teenagers taking antidepressants are unfortunate examples of side effects discovered too late.

Health claims and you

The take-home message is that being properly informed is critical to understanding health risks. Clearing up misconceptions about risk and risk reduction should become an important part of the patient-physician dialog, but this will also require educating physicians about the meaning of risk and how to clearly communicate it. If the information matters to you, you should find the original publication and assess the study design and its possible flaws. No single study is entirely conclusive by itself, so you may also want to examine the findings from related studies.

Last, you should know that all over-the-counter herbal treatments are available for sale without ever having to undergo the rigors of a clinical trial to establish their effectiveness or safety. Their claims to reduce certain health risks do not have to be substantiated scientifically. And, being sold without prescription, they also come without medical or pharmaceutical counsel. You can find some objective information on the National Center for Complementary and Alternative Medicine website at `nccam.nih.gov`. A few brands do carry a quality seal, but this typically represents only a voluntary evaluation of manufacturing standards (for example, to attest lack of dangerous contaminants). Buyers beware!

A blanket presumption for dietary supplements

The Food and Drug Administration (FDA) regulates dietary supplements (vitamin, mineral, herb) as foods rather than drugs. Under the stated presumption that these are good for you, the Dietary Supplement Health and Education Act of 1994 exempted dietary supplements from existing premarket safety evaluations (which still apply to other foods) and shifted the burden of proof to the FDA for issues of safety and false or misleading claims. Manufacturers are only required to follow health claims with this warning: "This statement has not been evaluated by the Food and Drug Administration. This product is not intended to diagnose, treat, cure, or prevent any disease."

CHAPTER 20 SUMMARY

The data in a **two-sample problem** are two independent SRSs, each drawn from a separate population.

Tests and confidence intervals to compare the proportions p_1 and p_2 of successes in the two populations are based on the difference $\hat{p}_1 - \hat{p}_2$ between the sample proportions of successes in the two SRSs.

When the sample sizes n_1 and n_2 are large, the sampling distribution of $\hat{p}_1 - \hat{p}_2$ is close to Normal with mean $p_1 - p_2$.

The level C **large-sample confidence interval for $p_1 - p_2$** is

$$(\hat{p}_1 - \hat{p}_2) \pm z^* \text{SE}$$

where the standard error of $\hat{p}_1 - \hat{p}_2$ is

$$SE = \sqrt{\frac{\hat{p}_1(1 - \hat{p}_1)}{n_1} + \frac{\hat{p}_2(1 - \hat{p}_2)}{n_2}}$$

and z^* is a standard Normal critical value.

The true confidence level of the large-sample interval can be substantially less than the planned level C. Use this interval only if the counts of successes and failures in both samples are 10 or greater.

To get a more accurate confidence interval, add four imaginary observations, one success and one failure in each sample. Then use the same formula for the confidence interval. This is the **plus four confidence interval.** You can use it whenever both samples have 5 or more observations.

Significance tests for H_0: $p_1 = p_2$ use the **pooled sample proportion**

$$\hat{p} = \frac{\text{number of successes in both samples combined}}{\text{number of individuals in both samples combined}}$$

and the z statistic

$$z = \frac{\hat{p}_1 - \hat{p}_2}{\sqrt{\hat{p}(1 - \hat{p})\left(\frac{1}{n_1} + \frac{1}{n_2}\right)}}$$

P-values come from the standard Normal distribution. Use this test when there are 5 or more successes and 5 or more failures in each sample.

In the health sciences, the value $\hat{p}_2 - \hat{p}_1$ is called the absolute risk reduction (ARR) for a medical outcome in a target population (1) compared with a reference population (2). Other measures for comparing proportions are also used in this context:

• The **relative risk** is

$$RR = \frac{\hat{p}_1}{\hat{p}_2}$$

• The **odds ratio** is

$$OR = \frac{\hat{p}_1(1 - \hat{p}_2)}{\hat{p}_2(1 - \hat{p}_1)}$$

• The **relative risk reduction** is

$$RRR = \frac{\hat{p}_2 - \hat{p}_1}{\hat{p}_2}$$

• The **number needed to treat** is

$$NNT = \frac{1}{ARR} = \frac{1}{\hat{p}_2 - \hat{p}_1}$$

Odds ratios are found most often in observational, retrospective studies, while *RR*, *RRR*, and *NNT* are typically used in prospective designs, such as randomized controlled experiments. The number needed to treat represents the number of subjects that must be treated before 1 subject in the treatment group can be spared a negative medical outcome compared with what would have been expected in the control group.

CHECK YOUR SKILLS

Exercises 19.31 and 19.40 described a large survey of the eating habits of American infants and toddlers. Among the 679 infants 9 to 11 months old surveyed, 61 had eaten fried potatoes on a given day. Among the 316 toddlers 19 to 24 months old surveyed, 82 had eaten fried potatoes. Exercises 20.11 to 20.15 are based on this survey.

20.11 Take p_I and p_T to be, respectively, the proportions of all infants and toddlers in the cited age ranges who eat fried potatoes on a given day. We conjectured before seeing the data that toddlers would be more likely to eat fried potatoes than infants. The hypotheses to be tested are

(a) H_0: $p_I = p_T$ versus H_a: $p_I \neq p_T$.

(b) H_0: $p_I = p_T$ versus H_a: $p_I > p_T$.

(c) H_0: $p_I = p_T$ versus H_a: $p_I < p_T$.

20.12 The sample proportions of infants and toddlers who ate fried potatoes are about

(a) $\hat{p}_I = 0.09$ and $\hat{p}_T = 0.26$.

(b) $\hat{p}_I = 0.09$ and $\hat{p}_T = 0.91$.

(c) $\hat{p}_I = 0.26$ and $\hat{p}_T = 0.74$.

20.13 The pooled sample proportion of infants and toddlers who ate fried potatoes is about

(a) $\hat{p} = 0.35$. (b) $\hat{p} = 0.14$. (c) $\hat{p} = 0.17$.

20.14 The z statistic for a test comparing the proportions of infants and toddlers who eat fried potatoes on a given day is about (in absolute value)

(a) $z = 7.12$. (b) $z = 6.02$. (c) $z = 3.01$.

20.15 The 95% large-sample confidence interval for the difference $p_T - p_I$ in the proportions of infants and toddlers who eat fried potatoes on a given day is about

(a) 0.17 ± 0.95. (b) 0.17 ± 0.053. (c) 0.17 ± 0.039.

In an experiment to learn if substance M can help restore memory, the brains of 20 rats were treated to damage their memories. The rats were trained to run a maze. After a day, 10 rats were given M and 7 of them succeeded in the maze; only 2 of the 10 control rats were successful. Exercises 20.16 to 20.18 are based on this experiment.

20.16 The z test for "no difference" in this case

(a) may be inaccurate because the populations are too small.

(b) may be inaccurate because some counts of successes and failures are too small.

(c) is reasonably accurate because the conditions for inference are met.

20.17 The plus four 90% confidence interval for the difference between the proportion of rats that succeed when given M and the proportion that succeed without it has for its center the value

(a) 0.455. (b) 0.417. (c) 0.5.

20.18 The plus four 90% confidence interval described above has for its margin of error the value

(a) 0.312. (b) 0.304. (c) 0.185.

Glycoprotein IIb/IIIa inhibitors are platelet aggregation inhibitors used during angioplasty for the treatment of heart attacks. You read that eptifibatide is an effective and less expensive GPIIb/IIIa inhibitor. A randomized, double-blind clinical trial found that 5.4% of subjects in the angioplasty with eptifibatide group and 9.2% in the angioplasty with placebo group had either died or experienced another heart attack within 48 hours following treatment.[12] Exercises 20.19 and 20.20 are based on this study.

20.19 **(Optional)** The relative risk reduction *RRR* is

(a) 0.587. (b) 0.413. (c) 0.704.

20.20 **(Optional)** The number needed to treat *NNT* is approximately

(a) 4. (b) 9. (c) 26.

CHAPTER 20 EXERCISES

We recommend using the plus four method for all confidence intervals for proportions. However, the large-sample method is acceptable when the guidelines for its use are met.

20.21 **Genetically altered mice.** Genetic influences on cancer can be studied by manipulating the genetic makeup of mice. One of the processes that turn genes on or off (so to speak) in particular locations is called "DNA methylation." Do low levels of this process help cause tumors? Compare mice altered to have low levels with normal mice. Of 33 mice with lowered levels of DNA methylation, 23 developed tumors. None of the control group of 18 normal mice developed tumors in the same time period.[13]

(a) Explain why we cannot safely use either the large-sample confidence interval or the test for comparing the proportions of normal and altered mice that develop tumors.

(b) The plus four method adds two observations, a success and a failure, to each sample. What are the sample sizes and the numbers of mice with tumors after you do this? Give a plus four 99% confidence interval for the difference in the proportions of the two populations that develop tumors.

(c) Based on your confidence interval, is the difference between normal and altered mice significant at the 1% level?

20.22 **Drug testing in schools.** In 2002 the Supreme Court ruled that schools could require random drug tests of students participating in competitive after-school activities such as athletics. Does drug testing reduce use of illegal drugs? A study compared two similar high schools in Oregon. Wahtonka High School tested athletes at random, and Warrenton High School did not. In a confidential survey, 7 of 135 athletes at Wahtonka and 27 of 141 athletes at Warrenton said they were using drugs.[14] Regard these athletes as SRSs from the populations of athletes at similar schools with and without drug testing.

(a) You should not use the large-sample confidence interval. Why not?

(b) The plus four method adds two observations, a success and a failure, to each sample. What are the sample sizes and the numbers of drug users after you do this?

(c) Give the plus four 95% confidence interval for the difference between the proportions of athletes using drugs at schools with and without testing.

20.23 Treating AIDS. The drug AZT was the first drug that seemed effective in delaying the onset of AIDS in people infected with HIV. Evidence for AZT's effectiveness came from a large randomized comparative experiment. The subjects were 1300 HIV-positive volunteers who had not yet developed AIDS. The study assigned 435 of the subjects at random to take 500 milligrams of AZT each day and another 435 to take a placebo. (The others were assigned to a higher dose of AZT, but we will compare only the first two groups.) At the end of the study, 38 of the placebo subjects and 17 of the AZT subjects had developed AIDS. We want to test the claim that taking AZT lowers the proportion of infected people who will develop AIDS in a given period of time.

(a) State hypotheses and check that you can safely use the z procedures.

(b) How significant is the evidence that AZT is effective?

(c) The experiment was double-blind. Explain what this means.

Comment: Medical experiments on treatments for AIDS and other fatal diseases raise hard ethical questions. Some people argue that because AIDS is always fatal, infected people should get any drug that has any hope of helping them. The counter-argument is that then we would never find out which drugs really work. The placebo patients in this study were given AZT as soon as results indicated that AZT was clearly more effective.

20.24 Drug testing in schools, continued. Exercise 20.22 describes a study that compared the proportions of athletes who use illegal drugs in two similar high schools, one that tests for drugs and one that does not. Drug testing is intended to reduce use of drugs. Do the data give good reason to think that drug use among athletes is lower in schools that test for drugs? State hypotheses, find the test statistic, and use either software or the bottom row of Table C for the P-value. Be sure to state your conclusion. (Because the study is not an experiment, the conclusion depends on the condition that athletes in these two schools can be considered SRSs from all similar schools.)

Call a statistician. *Does involving a statistician to help with statistical methods improve the chance that a medical research paper will be published? A study of a random sample of papers submitted to two medical journals found that 135 of 190 papers that lacked statistical assistance were rejected without even being reviewed in detail. In contrast, 293 of the 514 papers with statistical help were sent back without review.[15] Exercises 20.25 to 20.27 are based on this study.*

20.25 Does statistical help make a difference? Is there a significant difference in the proportions of papers with and without statistical help that are rejected without review? Use software or the bottom row of Table C to get a P-value. (This observational study does not establish causation: studies that include statistical help may also be better in other ways than those that do not.)

20.26 How often are statisticians involved? Give a 95% confidence interval for the proportion of papers submitted to these journals that include help from a statistician.

20.27 How big a difference? Give a 95% confidence interval for the difference between the proportions of papers rejected without review when a statistician is and is not involved in the research.

20.28 Sport and arthritis. A study in Sweden compared former elite soccer players with people of the same age who had played soccer but not at the elite level. Of the 71 former elite soccer players surveyed, 10 had developed arthritis of the hip or knee by their mid-50s, compared with only 9 of the 215 recreational soccer players.[16]

(a) Does it appear that elite soccer players are more likely to develop arthritis of the hip or knee than comparable recreational soccer players? State hypotheses, find the test statistic, and use either software or the bottom row of Table C for the *P*-value. Be sure to state your conclusion. (Because the study is not an experiment, the conclusion depends on how comparable the two groups are.)

(b) Explain why you should not use the large-sample method to calculate a confidence interval in this case. The plus four method adds two observations, a success and a failure, to each sample. What are the sample sizes and the numbers of drug users after you do this? Give the plus four 95% confidence interval for the difference in the proportion of ex-players with arthritis among the two populations.

20.29 Detecting genetically modified soybeans. Exercise 19.26 (page 502) describes a study in which batches of soybeans containing some genetically modified (GM) beans were submitted to 23 grain-handling facilities. When batches contained 1% of GM beans, 18 of the facilities detected the presence of GM beans. Only 7 of these 23 facilities detected GM beans when they made up 0.1% of the beans in the batches. Explain why we *cannot* use the methods of this chapter to compare the proportions of facilities that will detect the two levels of GM soybeans.

*In responding to Exercises 20.30 to 20.40, follow the **Formulate, Solve,** and **Conclude** steps of the four-step process. It may be helpful to restate in your own words the **State** information given in the exercise.*

20.30 Sickle-cell and malaria. Sickle-cell anemia is a hereditary chronic blood disease that is extremely severe when an individual carries two copies of the defective gene. It is particularly common in countries plagued by malaria, a parasitic infection transmitted by mosquitoes. A study in Africa tested 543 children for the sickle-cell gene and also for malaria. In all, 136 of the children had the sickle-cell gene and 36 of these had severe malaria infections. The other 407 children lacked the sickle-cell gene, and 152 of them had severe malaria infections.[17]

(a) Give a 95% confidence interval for the proportion of all children in the population studied who have the sickle-cell gene.

(b) Is there good evidence that the proportion of severe malaria infections is lower among children with the sickle-cell trait? Does this support the notion that the sickle-cell trait provides some protection against malaria?

20.31　Altruism in prairie dogs. Is altruistic behavior influenced by its potential cost? Prairie dogs are social rodents who warn each other of a predator's presence with barking calls. A researcher examined whether close proximity to a predator results in less frequent alarm calls because the callers bear more personal risk. The "predator," a stuffed badger controlled remotely, was placed multiple times randomly either near to or far from a prairie dog colony. The number of warning calls made by isolated foraging colony members is displayed in the following table:[18]

Tyler Mallory/Alamy

	Distance to predator	
	Near	Far
Alert call	29	55
No call	80	78

Do the data support the researcher's hypothesis?

20.32　Altruism in prairie dogs, continued. The researcher in the previous exercise also hypothesized that if prairie dogs make fewer calls when they feel individually more at risk, it would be reflected in their behavior. That is, individuals remaining above ground would make a higher proportion of alert calls than individuals seeking cover upon appearance of a predator. Here are the study's findings:

	Behavioral response	
	Seek cover	Stay above ground
Alert call	11	95
No call	63	121

Do the data support the researcher's second hypothesis?

20.33　Lyme disease. Lyme disease is spread in the northeastern United States by infected ticks. The ticks are infected mainly by feeding on mice, so more mice result in more infected ticks. The mouse population in turn rises and falls with the abundance of acorns, their favored food. Experimenters studied two similar forest areas in a year when the acorn crop failed. They added hundreds of thousands of acorns to one area to imitate an abundant acorn crop, while leaving the other area untouched. The next spring, 54 of the 72 mice trapped in the first area were in breeding condition, versus 10 of the 17 mice trapped in the second area.[19] Estimate the difference between the proportions of mice ready to breed in good acorn years and bad acorn years. (Use 90% confidence. Be sure to justify your choice of confidence interval.)

20.34　Aflatoxicosis in Kenya. Aflatoxins are toxic compounds secreted by fungus found most often in damaged crops. Kenya experienced an outbreak of aflatoxicosis in 2004 that resulted in several hundred cases of liver failure, including 125 deaths. The Kenyan Ministry of Health suspected that improper maize (corn) storage conditions were at least in part responsible for the outbreak. A random sample of 27 patients with aflatoxicosis and 43 healthy controls were asked whether they stored their maize in the house or in a dedicated granary. Prolonged storage in the house exposes crops to levels of humidity that foster fungal growth. The study found that 22 of the patients and 23 of the controls had stored maize in the house.[20] Choose carefully a valid statistical method to compare the effectiveness of both storage methods against aflatoxicosis. Do your results support the ministry's hypothesis?

20.35 Side effects of medication. A study of "adverse symptoms" in users of over-the-counter pain relief medications assigned subjects at random to one of two common pain relievers: acetaminophen and ibuprofen. (Both of these pain relievers are sold under various brand names, sometimes combined with other ingredients.) In all, 650 subjects took acetaminophen, and 44 experienced some adverse symptom. Of the 347 subjects who took ibuprofen, 49 had an adverse symptom. How strong is the evidence that the two pain relievers differ in the proportion of people who experience an adverse symptom?

20.36 Preventing strokes. Aspirin prevents blood from clotting and so helps prevent strokes. The Second European Stroke Prevention Study asked whether adding another anticlotting drug named dipyridamole would be more effective for patients who had already had a stroke. Here are the data on strokes and deaths during the two years of the study:[21]

	Number of patients	Number of strokes	Number of deaths
Aspirin alone	1649	206	182
Aspirin + dipyridamole	1650	157	185

(a) The study was a randomized comparative experiment. Outline the design of the study.

(b) Is there a significant difference in the proportion of strokes in the two groups?

(c) Is there a significant difference in death rates for the two groups?

20.37 Duct tape for wart removal. A study compared the effectiveness of duct tape compared with cryotherapy with liquid nitrogen in the treatment of common warts in children and young adults. A total of 61 patients ages 3 to 22 were randomly assigned to either treatment (duct tape applied directly on warts continuously for up to 2 months or up to 6 applications of liquid nitrogen every 2 or 3 weeks). Of the 26 patients treated with duct tape, 22 showed complete wart remission. In comparison, 15 of the 25 patients treated with cryotherapy reached complete wart remission.[22]

(a) Explain why the study cannot be double blind.

(b) Choose carefully a valid statistical method to compare the effectiveness of both methods. What do you conclude?

20.38 Treating prostate disease. A large study used records from Canada's national health care system to compare the effectiveness of two ways to treat prostate disease. The two treatments are traditional surgery and a new method that does not require surgery. The records described many patients whose doctors had chosen each method. The study found that patients treated with the new method were significantly more likely to die within 8 years.[23] Further study of the data showed that this conclusion was wrong. The extra deaths among patients who got the new treatment could be explained by lurking variables. What lurking variables might be confounded with a doctor's choice of surgical or nonsurgical treatment?

20.39 **Carcinogenicity of electromagnetic fields.** The U.S. National Toxicology Program studies the toxicity and carcinogenicity of agents potentially causing a risk to human health. Electromagnetic fields around electrical installations are harmless in theory, but they might have a long-term effect on health—for instance, in triggering tumors. Observational studies in human populations have suggested a potential association. One study looked at the occurrence of cancers in otherwise healthy rats after two years of daily exposure to 60-Hz electromagnetic fields. Several groups were studied, each exposed to a different electromagnetic intensity, as well as a control group kept in the same conditions but not exposed to any electromagnetic field. Here are the number of rats with a tumor after the 2-year study period for the control group and for the group exposed to 2 gauss, which is approximately 1000-fold the level considered high exposure for humans.[24]

	Male rats	Female rats
Control: no exposure	16 (N = 99)	19 (N = 100)
2 gauss exposure	30 (N = 100)	22 (N = 100)

In rats not exposed to an electromagnetic field, is there significant evidence of a difference in the rate of tumor between males and females?

20.40 **Carcinogenicity study, continued.** Use the data from the previous exercise to answer the following questions:

(a) Is there significant evidence of a difference in the tumor rates in male rats either exposed to 2 gauss or not exposed?

(b) Is there significant evidence of a difference in the tumor rates in female rats either exposed to 2 gauss or not exposed?

(c) Write a brief summary of your findings from this and the previous exercise.

The following exercises concern the optional material on relative risk and odds ratio in the health sciences.

20.41 **Treating AIDS.** Go back to Exercise 20.23, which describes the first study of AZT efficiency for delaying the onset of AIDS.

(a) Calculate and interpret the odds ratio and relative risk for the AZT group compared with the placebo group. How do they compare?

(b) Calculate and interpret the absolute and the relative risk reduction for onset of AIDS.

(c) Calculate the number needed to treat with AZT compared with a placebo to prevent 1 onset of AIDS.

20.42 **Cardiovascular prevention.** The Heart Outcomes Prevention Evaluation (HOPE) trial studied the effectiveness of angiotensin-converting enzyme (ACE) inhibitors in cardiovascular disease prevention. Part of the study assigned diabetic patients over age 55 at random to receive either the ACE inhibitor ramipril daily for 5 years or a placebo. Of the 1808 subjects taking ramipril daily, 112 died of a cardiovascular accident during the 5-year study period. In contrast, 172 of the subjects receiving a placebo died of a cardiovascular accident.[25]

(a) Calculate the relative risk reduction, and write a short statement using this measure to summarize the study findings.

(b) Calculate the number needed to treat to save 1 life from cardiovascular death, and write a short statement using this measure to summarize the study findings.

(c) Some studies also report the number of pills that must be taken to save 1 life. Calculate this value considering, for simplicity, that each subject in the treatment group took 1 pill every day for 5 years.

20.43 **The DCCT and diabetes treatment.** Go back to Example 20.1, comparing the effectiveness of a conventional treatment with that of an intensive treatment in preventing complications from type I diabetes.

(a) Give the absolute and the relative risk reduction for the development of retinopathy in this study. Also compute the number needed to treat.

(b) Choose one of the values you calculated in (a), and write a brief statement explaining the study findings to a patient.

20.44 **The DCCT and diabetes treatment, continued.** Exercise 20.1 describes the results of the DCCT study for individuals who already suffered from retinopathy at the beginning of the study, comparing the effectiveness of the conventional treatment with that of the intensive treatment in preventing retinopathy progression.

(a) Give the absolute and the relative risk reduction for the progression of retinopathy in this study. Also compute the number needed to treat.

(b) Choose one of the values you calculated in (a), and write a brief statement explaining the study findings to a patient.

Bettmann/CORBIS

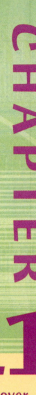

The Chi-Square Test for Goodness of Fit

In Chapter 19 we used the one-sample z procedures to obtain confidence intervals and test hypotheses about the proportion of successes in a population. The definition of "success" was arbitrary and simply referred to the outcome of interest, with any other outcome constituting a "failure." Such dichotomous labeling is sometimes an obvious choice, as in Example 19.6, in which we examined the gender, either male or female, of first-born children. In other situations, a more detailed description of possible outcomes would be helpful. For instance, we might want to study the response to a new treatment by counting the number of subjects receiving it whose health status improved, remained stable, or deteriorated. In this chapter, we introduce a new statistical method, the chi-square test, that allows testing of hypotheses about a categorical variable with two *or more* levels. We will see in the next chapter that this versatile method can also be adapted to study the relationship between two independent categorical variables.

Hypotheses for goodness of fit

We often have a choice in defining a response variable. For instance, if we suspect that births are less common on the weekend, we could take a random sample of births and record whether they occurred on a weekday or on a weekend. However, this distinction might not be fine enough. Saturday and Sunday could have very

Weekend outing?

You work all week. Then it rains on the weekend. Can there really be a statistical truth behind our perception that the weather is against us? At least on the East Coast of the United States, the answer is "Yes." Going back to 1946, it seems that 22% more precipitation falls on Sundays than on Mondays. The likely explanation is that the pollution from all those workday cars and trucks forms the seeds for raindrops—with just enough delay to cause rain on the weekend.

different birth rates, but this would be masked by pooling them into a single outcome. Likewise, Monday and Friday might differ substantially from midweek days. Therefore, a better way to look for any nonrandom pattern in the distribution of births would be to consider all 7 days of the week.

EXAMPLE 21.1   Never on Sunday?

A random sample of 700 births from local records shows this distribution across the days of the week.

Day	Sun.	Mon.	Tue.	Wed.	Thu.	Fri.	Sat.
Births	84	110	124	104	94	112	72

As expected, the two smallest counts of births are on Saturday and Sunday. But do these data give significant evidence that local births are not equally likely on all days of the week?

The null hypothesis says that births *are* evenly distributed. To state the hypotheses carefully, write the discrete probability distribution for days of birth:

Day	Sun.	Mon.	Tue.	Wed.	Thu.	Fri.	Sat.
Probability	p_1	p_2	p_3	p_4	p_5	p_6	p_7

The null hypothesis says that the probabilities are the same on all days. In that case, each day would get one-seventh of all births. That is, all 7 probabilities must be 1/7. So the null hypothesis is

$$H_0: p_1 = p_2 = p_3 = p_4 = p_5 = p_6 = p_7 = \frac{1}{7}$$

Beware of wanting to state the null hypothesis in terms of the sample proportions. This is a mistake often made by students. The null hypothesis must reflect your assumptions, not the data used to test it. The alternative hypothesis says that days are *not* all equally probable:

$$H_a: \text{not all } p_i = \frac{1}{7}$$

The alternative hypothesis is in essence "many-sided" and simply says that H_0 is not true.

In this example, we hypothesized under H_0 that births would be equally distributed across all 7 days of the week. However, we could have specified any distribution of our choice for the null hypothesis, as long as there was a sound biological argument for it. Here is an example in which it would not make sense to set a null hypothesis of equal proportions.

EXAMPLE 21.2   Genetics of seed color

Epistasis is the control of a phenotype by two or more interacting genes. In a dominant epistatic model, one gene can mask the effect of the second gene, leading to the expression of one main phenotype and two rarer phenotype variants.

Geneticists examined the distribution of seed coat color in cultivated amaranth grains, *Amaranthus caudatus*. Crossing black-seeded and pale-seeded *A. caudatus* populations gave the following counts of black, brown, and pale seeds in a second generation (F2):[1]

Seed coat color	black	brown	pale
Seed count	321	77	31

According to genetics laws, dominant epistasis should lead to a 12:3:1 distribution in F2. That is, out of 16 individuals, 12 would be expected to express the dominant phenotype (black), 3 the intermediary phenotype (brown), and only 1 the recessive phenotype (pale). We want to know if seed color could follow a dominant epistatic model. The null hypothesis is therefore:

$$H_0:\ p_{black} = \frac{12}{16} = \frac{3}{4} \quad \text{and} \quad p_{brown} = \frac{3}{16} \quad \text{and} \quad p_{pale} = \frac{1}{16}$$

Again, the alternative hypothesis is simply that H_0 is not true.

In Example 21.1, H_0 postulates equal proportions of births for the seven days (a uniform distribution), whereas in Example 21.2, the three seed colors are assumed to have different proportions under H_0. In both cases we want to test whether a categorical variable (day of birth or seed color) has a particular distribution outlined by H_0. The statistical test we will use for that is the chi-square test for *goodness of fit*. The idea is that the test assesses whether the observed counts "fit" the distribution outlined by H_0.

APPLY YOUR KNOWLEDGE

21.1   Saving birds from windows. Many birds are injured or killed by flying into windows. It appears that birds don't see windows. Can tilting windows down so that they reflect earth rather than sky reduce bird strikes? Researchers placed six windows at the edge of a woods: two vertical, two tilted 20 degrees, and two tilted 40 degrees. During the next four months, there were 53 bird strikes, 31 on the vertical windows, 14 on the 20-degree windows, and 8 on the 40-degree windows.[2] Does the tilt have an effect? State the null and alternative hypotheses.

21.2   Formulating the theory of inheritance. Gregor Mendel has been called the "father of genetics," having formulated laws of inheritance even before the function of DNA was discovered. In one famous experiment with peas, he crossed pure breeds of plants producing smooth peas and plants producing wrinkled peas. The first-generation peas were all smooth. In the second generation (F2), he obtained 5474 smooth-pea plants and 1850 wrinkled-pea plants. Do these data agree with the conclusion that F2 is made up of 75% dominant-trait (in this case "smooth") and 25% recessive-trait ("wrinkled") pea plants? State the null and alternative hypotheses.

Bettmann/CORBIS

The chi-square test for goodness of fit

We have stated the null and alternative hypotheses, and now we need to test whether the observed results differ significantly from expectations under H_0. To test H_0, we compare the observed counts with the *expected counts*, the counts we would expect—except for random variation—if H_0 were true. If the observed counts are far from the expected counts, that is evidence against H_0.

The expected count in n independent trials for a particular outcome with probability p is simply np. (We covered this in great detail in Chapters 9, 10, and 12.) This doesn't mean that we expect to see exactly np counts of that outcome in n trials. Rather, we'd expect np counts on average over the long run if we were to repeat the n trials many, many times.

EXPECTED COUNTS

A categorical variable has k possible outcomes, with probabilities p_1, p_2, p_3, ..., p_k. That is, p_i is the probability of the ith outcome. We have n independent observations from this categorical variable.

To test the null hypothesis that the probabilities have specified values

$$H_0: p_1 = p_{1_0}, \quad p_2 = p_{2_0}, \quad \ldots, \quad p_k = p_{k_0}$$

first compute an expected count for each of the k outcomes as follows.

$$\text{expected count of outcome } i = np_{i_0}$$

EXAMPLE 21.3   Never on Sunday? Expected counts

In Example 21.1, the observations are counts of births for each of the 7 days of the week. That is, the outcomes are days of the week, with $k = 7$.

The null hypothesis says that the probability of a birth on the ith day is $p_{i_0} = 1/7$ for all days. There were a total of 700 births in our sample. If H_0 were true, we would expect to see an even distribution of these 700 births across all 7 days of the week; that is:

$$\text{expected count}_i = np_{i_0} = 700 \times \frac{1}{7} = 100$$

Each of the seven expected counts is equal to 100.

Expected counts do not have to be round numbers, though. In fact, if there had been only 699 observations instead of 700, the expected counts would each have been equal to $699/7 = 99.86$.

The statistical test that tells us whether the observed differences between the 7 weekdays are statistically significant compares the observed and expected counts. The test statistic that makes this comparison is the *chi-square statistic*.

CHI-SQUARE STATISTIC

The **chi-square statistic** is a measure of how far observed counts are from expected counts under the null hypothesis. The formula for the statistic is

$$X^2 = \sum \frac{(\text{observed count} - \text{expected count})^2}{\text{expected count}}$$

$$= \frac{(\text{observed}_1 - \text{expected}_1)^2}{\text{expected}_1} + \cdots + \frac{(\text{observed}_k - \text{expected}_k)^2}{\text{expected}_k}$$

where k is the number of different outcomes the categorical variable can take. Each of the k terms in the sum is called a *chi-square component*.

The chi-square statistic is a sum of terms, one for each possible outcome. In the birth example, 84 babies were born on Sunday. The expected count for each of the seven days is 100. So the term of the chi-square statistic for the Sunday outcome is

$$\frac{(\text{observed count} - \text{expected count})^2}{\text{expected count}} = \frac{(84 - 100)^2}{100}$$

$$= \frac{256}{100} = 2.56$$

To compute the chi-square statistic, we need to calculate each of the seven chi-square components and then sum them. That is,

$$X^2 = \sum \frac{(\text{observed count} - \text{expected count})^2}{\text{expected count}}$$

$$= \frac{(84 - 100)^2}{100} + \frac{(110 - 100)^2}{100} + \frac{(124 - 100)^2}{100} + \frac{(104 - 100)^2}{100}$$

$$+ \frac{(94 - 100)^2}{100} + \frac{(112 - 100)^2}{100} + \frac{(72 - 100)^2}{100}$$

$$= 19.12$$

In this particular example, each outcome has the same expected count because the null hypothesis assumed equal probability for all 7 days. When H_0 does not assume a uniform distribution, the expected counts vary across outcomes and this is reflected in the chi-square calculations. Here is an example.

EXAMPLE 21.4   Genetics of seed color: calculating X^2

Example 21.2 described the seed color of 429 second-generation A. *caudatus* after crossing black-seeded and pale-seeded populations. The null hypothesis, based on a dominant epistatic model of genetic inheritance, assumed that:

$$H_0: \ p_{black_0} = \frac{12}{16} = \frac{3}{4} \ \text{ and } \ p_{brown_0} = \frac{3}{16} \ \text{ and } \ p_{pale_0} = \frac{1}{16}$$

If H_0 were true, we would expect the distribution of seed color to be:

$$\text{black seeds:}\quad np_{black_0} = 429 \times 3/4 = 321.75$$
$$\text{brown seeds:}\quad np_{brown_0} = 429 \times 3/16 = 80.4375$$
$$\text{pale seeds:}\quad np_{pale_0} = 429 \times 1/16 = 26.8125$$

We can now calculate the chi-square statistic to test H_0.

$$X^2 = \sum \frac{(\text{observed count} - \text{expected count})^2}{\text{expected count}}$$

$$= \frac{(321 - 321.75)^2}{321.75} + \frac{(77 - 80.4375)^2}{80.4375} + \frac{(31 - 26.8125)^2}{26.8125}$$

$$= 0.8027$$

Think of the chi-square statistic, X^2, as a measure of the distance of the observed counts from the expected counts. Like any distance, it is always zero or positive, and it is zero only when the observed counts are exactly equal to the expected counts. Small values of X^2 represent small deviations from H_0 that do not provide sufficient evidence to reject H_0. Inversely, large values of X^2 are evidence against H_0, because they say that the observed counts are far from what we would expect if H_0 were true. *Although the alternative hypothesis H_a is many-sided, the chi-square test is nondirectional,* because any violation of H_0 tends to produce a large value of X^2.

APPLY YOUR KNOWLEDGE

21.3 Saving birds from windows, continued. Exercise 21.1 described an experiment designed to figure out whether tilting windows down so that they reflect earth rather than sky can help reduce accidental bird strikes. Calculate the expected counts in each of the three conditions and compute the chi-square statistic.

21.4 Formulating the theory of inheritance. Go back to Exercise 21.2, describing Mendel's crossing of smooth and wrinkled peas. What are the expected counts for each of the two phenotypes? Use these expected counts to compute the chi-square statistic.

Carol Bloomfield/Alamy

Using technology

As usual, after calculating a statistic we want the probability of finding a statistic at least as extreme as the one obtained, if H_0 were true. We first turn to technology for this computation. Figure 21.1 shows output for the chi-square goodness of fit test for the birth data in Example 21.1 from three statistical programs and a spreadsheet program.

Minitab

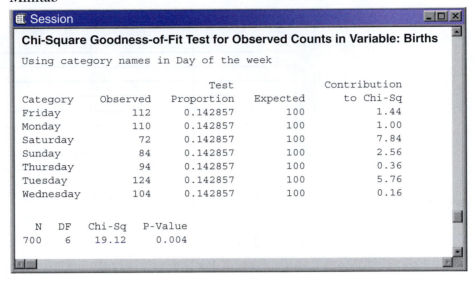

Chi-Square Goodness-of-Fit Test for Observed Counts in Variable: Births

Using category names in Day of the week

Category	Observed	Test Proportion	Expected	Contribution to Chi-Sq
Friday	112	0.142857	100	1.44
Monday	110	0.142857	100	1.00
Saturday	72	0.142857	100	7.84
Sunday	84	0.142857	100	2.56
Thursday	94	0.142857	100	0.36
Tuesday	124	0.142857	100	5.76
Wednesday	104	0.142857	100	0.16

N	DF	Chi-Sq	P-Value
700	6	19.12	0.004

CrunchIt!

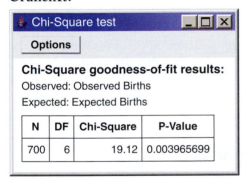

Chi-Square goodness-of-fit results:

Observed: Observed Births
Expected: Expected Births

N	DF	Chi-Square	P-Value
700	6	19.12	0.003965699

SPSS

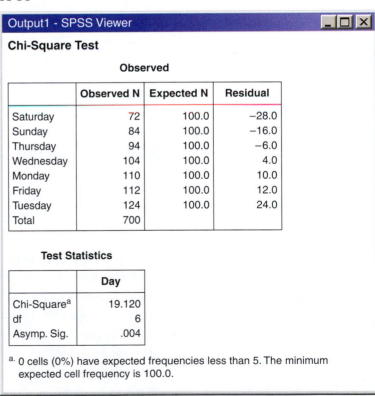

Output1 - SPSS Viewer

Chi-Square Test

Observed

	Observed N	Expected N	Residual
Saturday	72	100.0	−28.0
Sunday	84	100.0	−16.0
Thursday	94	100.0	−6.0
Wednesday	104	100.0	4.0
Monday	110	100.0	10.0
Friday	112	100.0	12.0
Tuesday	124	100.0	24.0
Total	700		

Test Statistics

	Day
Chi-Square[a]	19.120
df	6
Asymp. Sig.	.004

[a.] 0 cells (0%) have expected frequencies less than 5. The minimum expected cell frequency is 100.0.

FIGURE 21.1 Output from CrunchIt!, Minitab, SPSS, and Excel for the large birth data set for Example 21.5 (*continued*).

Excel

	A	B	C
1	Day	Observed Births	Expected Births
2	Sunday	84	100
3	Monday	110	100
4	Tuesday	124	100
5	Wednesday	104	100
6	Thursday	94	100
7	Friday	112	100
8	Saturday	72	100
9			
10			
11	0.003965699 =CHITEST(B2:B8, C2:C8)		

FIGURE 21.1 (*continued*)

EXAMPLE 21.5 Never on Sunday? Chi-square from software ─────

The outputs differ in the type and amount of information they give. All except the Excel spreadsheet tell us that the chi-square statistic is $X^2 = 19.12$. All give a P-value of about 0.004. That is, there is very strong evidence that the distribution of births is *not* evenly ("uniformly") distributed across all 7 days of the week. The data observed are not consistent with a null hypothesis of equal proportions.

Minitab and SPSS provide additional information. SPSS offers some remarks about expected counts that are related to the validity of this chi-square test, while Minitab gives the details of the chi-square calculations by providing the chi-square components. We will discuss their respective relevance in the following sections.

Interpreting chi-square results

The chi-square test is the overall test for detecting departures from a distribution model assumed under H_0. **When the test is significant,** it is important to look at the data to understand the nature of the distribution. Here are three ways to look at the data after significance has been established.

- **Compare appropriate percents:** which outcomes occur in percents quite different from those hypothesized in H_0?

- **Compare observed and expected counts:** which outcomes have more or fewer observations than we would expect if H_0 were true?

- **Look at the chi-square components:** which outcomes contribute the most to the value of X^2?

EXAMPLE 21.6 Never on Sunday? Conclusions ──────────────

We found in Example 21.5 a significant departure from the uniform model assuming an equal probability of births across all 7 days of the week. Because the test was significant, we can look deeper to explain this finding.

Look at the seven X^2 components provided by Minitab in Figure 21.1. About 40% of the value of X^2 (7.84 out of 19.12) comes from just one outcome. This points to the most important difference between the observations and the uniform model: a really low count of births on Saturday, representing the largest difference between any observed and expected counts. Most of the rest of X^2 comes from two other outcomes: a high count of births on Tuesday and a low count of births on Sunday.

Computing the percentage of births occurring on each day, we get

Day	Sun.	Mon.	Tue.	Wed.	Thu.	Fri.	Sat.
Percent	12%	16%	18%	15%	13%	16%	10%

With 18% of the 700 births, Tuesday has nearly twice as many births as Saturday, with only 10% of births, and one and a half times as many births as Sunday, with 12% of births.

Thus, we can conclude that the main difference from a uniform birth model is that weekend births are particularly uncommon, while Tuesday appears to have the highest frequency of births.

Looking at the details of the X^2 test is appropriate only when the test has been shown to be statistically significant. If not, any deviation from the model assumed in H_0 would represent only the kind of random variations we would expect to see when H_0 was true. That is, differences between chi-square components or differences between sample proportions are meaningless when the overall X^2 test is not significant.

We have discussed how to interpret a significant X^2 test. Now we turn to interpreting lack of significance. *Lack of significance in any statistical test means that there is not enough evidence to reject the null hypothesis. It does not, however, imply that the null hypothesis is true.* This is particularly important to remember when the goodness of fit test is used in an attempt to confirm the suspicion that a variable has a particular distribution stated under H_0.

EXAMPLE 21.7 *Genetics of seed color, continued*

Example 21.2 described the seed color of 429 second-generation A. *caudatus* plants after crossing black-seeded and pale-seeded populations. The null hypothesis, based on a dominant epistatic model of genetic inheritances, assumed that

$$H_0: p_{black_0} = \frac{12}{16} = \frac{3}{4} \text{ and } p_{brown_0} = \frac{3}{16} \text{ and } p_{pale_0} = \frac{1}{16}$$

Using software, we find that the P-value for this test is 0.669. This P-value is not significant, and therefore, we cannot reject the null hypothesis of a distribution of seed color based on a dominant epistatic model. The genetic determination of seed color in A. *caudatus* **could** indeed represent a case of dominant epistasis.

Ken Lucas/Visuals Unlimited

The fact that we cannot reject H_0 does not imply that it is true. We cannot claim that seed color does follow a dominant epistatic model of genetic inheritance, only that such a model is possible. That is, the data gathered are *consistent with* dominant epistasis.

Always remember that a large *P*-value can arise under any of the following conditions.

* H_0 is indeed true. Only further scientific investigation can confirm this.
* H_0 is not actually true but is too close to the real population distribution to tell them apart statistically.
* H_0 is definitely not true, but the sample size is too small or the variability too great to reach significance. Example 21.8 illustrates this case.

EXAMPLE 21.8   More on birth days

Based on a sample of 700 births in Example 21.6, we clearly *rejected* the null hypothesis that births are uniformly distributed over the 7 days of the week. Figure 21.2 shows the SPSS output for the same test using only preliminary data based on a smaller sample of 140 births. With a *P*-value of 0.269, the test is not significant, and we cannot reject H_0. It would be a mistake to conclude that births are uniformly distributed over the week. Instead, we should conclude that the preliminary data did not give convincing evidence that births are not equally likely on all days of the week.

SPSS

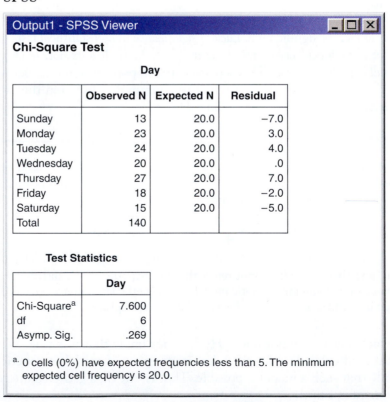

FIGURE 21.2 SPSS output for the smaller birth data set for Example 21.8.

APPLY YOUR KNOWLEDGE

21.5 **Saving birds from windows, continued.** Exercise 21.1 described an experiment designed to figure out whether tilting windows down so that they reflect earth rather than sky could help reduce accidental bird strikes. Software gives a *P*-value less than 0.001. Interpret this result and draw a conclusion about the efficacy of tilting windows. Refer to your earlier calculations in Exercise 21.3.

21.6 **Formulating the theory of inheritance.** Go back to Exercise 21.2, describing Mendel's crossing of wrinkled and smooth peas. Software gives a *P*-value of 0.608. Interpret this result. Can you conclude that the distribution of wrinkled and smooth peas in F2 follows a dominant-recessive model (75%-25%)?

Conditions for the chi-square test

The chi-square test for goodness of fit is a one-sample test for categorical data. As with all the statistical tests we have studied, this test is valid only when certain conditions are met. First of all, the data must satisfy a *multinomial setting*.

THE MULTINOMIAL SETTING

1. There is a fixed number *n* of observations.
2. The *n* observations are all **independent.** That is, knowing the result of one observation does not change the probabilities we assign to other observations.
3. Each observation falls into just one of a finite number *k* of complementary and mutually exclusive outcomes.
4. The probability of a given outcome is the same for each observation.

If you covered optional Chapter 12, you will notice the similarity to the binomial setting. In fact, the binomial setting is a particular case of the multinomial setting. In Example 21.1, each birth falls on one of the 7 days of the week, the 7 possible outcomes for this variable. Births are independent events: Knowing which day one particular birth occurred says nothing about the day of another birth.

The second set of requirements for a valid test is related to issues of sample size. The chi-square test, like the *z* procedures, is an approximate method that becomes more accurate as the counts for each outcome get larger. We must therefore check that the counts are large enough to trust the *P*-value. Fortunately, the chi-square approximation is accurate for quite modest counts. Here is a practical guideline.[3]

COUNTS REQUIRED FOR THE CHI-SQUARE TEST

You can safely use the chi-square test with critical values from the chi-square distribution when no more than 20% of the *expected counts* are less than 5 and all individual *expected counts* are 1 or greater.

Bird fatalities

According to the U.S. Fish and Wildlife Service (FWS), more birds die because of window collisions than because of any other factor associated with human activity. The FWS estimates that anywhere from 100 million to 1 billion birds are killed by window strikes each year, representing up to 5% of the North American fall bird population. Actual deaths are likely underestimated by the general public, because some injured birds fly away before dying from injuries sustained and natural predators feed on injured and dying birds.

Note that the guideline uses *expected* counts. The expected counts for the day of birth study of Example 21.1 appear in the Minitab and SPSS outputs in Figure 21.1. The smallest expected count is 100, so the data easily meet the guideline for safe use of chi-square.

Now that we have covered the requirements for the chi-square goodness of fit test, we have all the information needed to conduct such a test from beginning to end. Here is an example treated using the 4-step approach.

━ **EXAMPLE 21.9** More on the genetics of seed color ━━━━━━━━

STATE: In Example 21.2, we tested the hypothesis of a dominant epistatic model for the seed color of amaranth, expressed as black, brown, and pale phenotypes. The chi-square test was not significant, implying that the data were consistent with such a genetic model—*but not proving* that this was indeed the genetic basis of amaranth seed color inheritance.

A simpler, single-gene mode of inheritance can also produce three phenotypes in F2. When one gene locus has two codominant alleles, crossing pure breeds of the dominant and recessive traits gives rise to the two original traits as well as an intermediate phenotype in F2. The expected ratios for such a model are 1:2:1. That is, we would expect to see 1/4 (25%) of the dominant phenotype, 2/4 (50%) of the intermediate phenotype, and 1/4 (25%) of the recessive phenotype in F2. Example 21.2 showed the results in F2 of crossing black- and pale-seeded amaranth plants.

FORMULATE: Carry out a chi-square test for

$$H_0: p_{black} = 1/4 \text{ and } p_{brown} = 2/4 = 1/2 \text{ and } p_{pale} = 1/4$$

versus H_a: H_0 is not true. If the test is significant, compare observed outcome percents, observed versus expected counts, or chi-square components to describe the nature of the distribution.

SOLVE: First, check the guideline for use of chi-square. The data represent a single sample with 3 complementary and mutually exclusive outcomes. The expected counts appear in the Minitab output in Figure 21.3. All of the expected counts are quite large; therefore, we can safely use chi-square. The output shows that the test is highly significant ($X^2 = 568.357$, $P = 0.000$). Note that Minitab rounds the P-value to three

Minitab

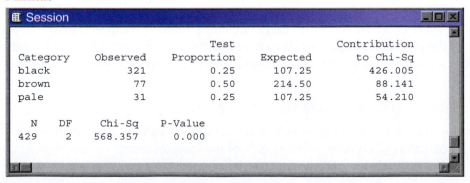

Category	Observed	Test Proportion	Expected	Contribution to Chi-Sq
black	321	0.25	107.25	426.005
brown	77	0.50	214.50	88.141
pale	31	0.25	107.25	54.210

N	DF	Chi-Sq	P-Value
429	2	568.357	0.000

FIGURE 21.3 Minitab output for Example 21.9.

decimal places, which means that $P < 0.001$ (a P-value is never exactly zero). The chi-square components indicate that black-seeded plants, in particular, are much too frequent for such a model.

CONCLUDE: We find that the F2 generation resulting from crossing black- and pale-seeded amaranth plants is composed of about 75% black-, 18% brown-, and 7% pale-seeded plants. A single-gene theory with codominant alleles is clearly not a good model of genetic inheritance of seed color in amaranth (1:2:1 model, $P < 0.001$). Previous analysis of the data with a dominant epistatic model of genetic inheritance provided a good fit for the data (12:3:1 model, $P = 0.37$), suggesting that seed color in amaranth is governed by two separate gene loci.

APPLY YOUR KNOWLEDGE

21.7 **Saving birds from windows, continued.** Go back to Exercise 21.1, describing an experiment with tilted windows to help reduce accidental bird strikes. Explain why a chi-square procedure is appropriate.

21.8 **Formulating the theory of inheritance, continued.** Go back to Exercise 21.2, describing Mendel's crossing of wrinkled and smooth peas. Explain why a chi-square procedure is appropriate.

21.9 **Are the conditions met?** You want to run a chi-square test for goodness of fit on multinomial data from one random sample. Here are the observed and expected counts.

Observed	3	8	9	4
Expected	6	6	6	6

Are the conditions met for this test? Explain your answer.

21.10 **Are the conditions met?** You want to run a chi-square test for goodness of fit on multinomial data from one random sample. Here are the observed and expected counts:

Observed	7	9	7	13
Expected	4	8	8	16

Are the conditions met for this test? Explain your answer.

The chi-square distributions*

Software usually finds P-values for us. The P-value for a chi-square test comes from comparing the value of the chi-square statistic with critical values for a *chi-square distribution*.

---

*The remainder of the material in this chapter is optional.

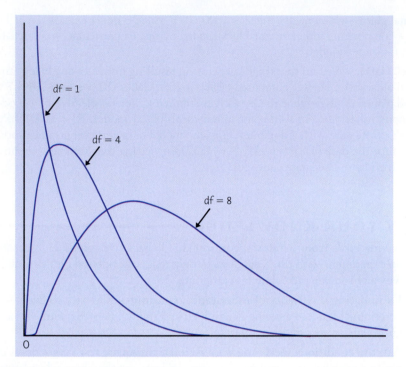

FIGURE 21.4 Density curves for the chi-square distributions with 1, 4, and 8 degrees of freedom. Chi-square distributions take only positive values and are right-skewed.

THE CHI-SQUARE DISTRIBUTIONS

The **chi-square distributions** are a family of distributions that take only positive values and are skewed to the right. A specific chi-square distribution is specified by giving its **degrees of freedom.**

The chi-square goodness of fit test involving k outcomes uses critical values from the chi-square distribution with $k - 1$ degrees of freedom. The P-value is the area to the right of X^2 under the density curve of this chi-square distribution.

Figure 21.4 shows the density curves for three members of the chi-square family of distributions. As the degrees of freedom increase, the density curves become less skewed and larger values become more probable. Table D in the back of the book gives critical values for chi-square distributions. You can use Table D if you do not have software that gives you P-values for a chi-square test.

EXAMPLE 21.10 Never on Sunday? Using the chi-square table

Example 21.1 described a test for the equal distribution of births over the 7 days of the week. That is, $k = 7$. The chi-square statistic therefore has $k - 1 = 6$ degrees of freedom. Three of the outputs in Figure 21.1 provide the degrees of freedom.

The observed value of the chi-square statistic is $X^2 = 19.12$. Look in the df = 6 row of Table D. The value $X^2 = 19.12$ falls between the 0.005 and 0.0025 critical values of the chi-square distribution with 6 degrees of freedom.

Remember that the chi-square test is always nondirectional. So the P-value of $X^2 = 19.12$ is between 0.005 and 0.0025. The outputs in Figure 21.1 show that the P-value is about 0.004.

df = 6

p	.005	.0025
x^*	18.55	20.25

APPLY YOUR KNOWLEDGE

21.11 Saving birds from windows, continued. Go back to Exercise 21.1, describing an experiment with tilted widows to help reduce accidental bird strikes. What are the degrees of freedom for this test? Use Table D to find the critical values closest to the chi-square statistic you calculated in Exercise 21.3.

21.12 Formulating the theory of inheritance, continued. Go back to Exercise 21.2, describing Mendel's crossing of wrinkled and smooth peas. What are the degrees of freedom for this test? Use Table D to find the critical values closest to the chi-square statistic you calculated in Exercise 21.4.

The chi-square test and the one-sample z test*

The chi-square goodness of fit test can be used to evaluate the distribution of a categorical random variable in a one-sample setting. When the random variable can take only one of two values—arbitrarily labeled success or failure—we can run either a chi-square goodness of fit test with one degree of freedom or a one-sample z test as described in Chapter 19. The null hypotheses tested are

$$H_0: p_{success} = p_0$$

for the one-sample z test and

$$H_0: p_{success} = p_0, \quad p_{failure} = 1 - p_0$$

for the chi-square goodness of fit test.

These two tests always agree. In fact, the chi-square statistic X^2 is just the square of the z statistic, and the P-value for X^2 is exactly the same as the two-sided P-value for z. We recommend using the one-sample z test, because it gives you the choice of a one-sided test and it is related to a confidence interval for the population proportion of success p.

APPLY YOUR KNOWLEDGE

21.13 Formulating the theory of inheritance. Exercise 21.2 described Mendel's crossing of wrinkled and smooth peas. To test the hypothesis that the data fit a 75%-25% dominant-recessive genetic model, you computed the chi-square statistic in Exercise 21.4 and were told in Exercise 21.6 that $P = 0.608$. Test the same null hypothesis using a one-sample z test.

(a) How do the X^2 and the z statistics compare for these two tests?

(b) How do the P-value for the chi-square test and the two-sided P-value for the z test compare?

CHAPTER 21 SUMMARY

The **chi-square test for goodness of fit** tests the null hypothesis H_0 that the data come from a population with a given distribution. It specifies a proportion for each of the k possible outcomes in the distribution, and the sum of all these proportions equals 1. The alternative hypothesis H_a simply says that H_0 is not true.

The test compares the observed counts of observations with the counts that would be expected if H_0 were true. With n total observations, the **expected count** for any given outcome i is

$$\text{expected count}_i = n \times p_{i_0}$$

The **chi-square statistic** is a sum of components computed separately for each of the k possible outcomes in the distribution

$$X^2 = \sum \frac{(\text{observed count}_i - \text{expected count}_i)^2}{\text{expected count}_i}$$

The chi-square test compares the value of the statistic X^2 with critical values from the **chi-square distribution** with $k - 1$ **degrees of freedom**. Large values of X^2 are evidence against H_0, so the P-value is the area under the chi-square density curve to the right of X^2.

The chi-square distribution is an approximation to the distribution of the statistic X^2. You can safely use this approximation when all *expected counts* are at least 1 and no more than 20% are less than 5.

When the chi-square test finds a statistically significant departure from the hypothesized distribution, do data analysis to describe the nature of the distribution. You can do this by comparing the percents of each outcome, comparing the observed counts with the expected counts, and looking for the largest **components of the chi-square statistic.**

CHECK YOUR SKILLS

Example 21.2 showed a case of dominant epistasis with a 12:3:1 phenotypic ratio. A cross of white and green summer squash plants gives the following number of squash in the second generation (F2): 131 white squash, 34 yellow squash, and 10 green squash. Are these data consistent with a 12:3:1 dominant epistatic model of genetic inheritance (white being dominant)? Use this information for Exercises 21.14 through 21.21.

21.14 The null hypothesis for the chi-square goodness of fit test is

(a) H_0: the distribution is not uniform.

(b) H_0: the distribution is uniform: $p_{white} = p_{yellow} = p_{green}$.

(c) H_0: $p_{white} = 12/16$ and $p_{yellow} = 3/16$ and $p_{green} = 1/16$.

21.15 The alternative hypothesis is

(a) H_a: the distribution is not uniform.

(b) H_a: $p_{white} \neq p_{yellow} \neq p_{green}$.

(c) H_a: H_0 is not true.

21.16 The expected count of green squash is about

(a) 10.9.   (b) 1.   (c) 0.06.

21.17 The chi-square component for the green squash is about

(a) 10.   (b) 1.   (c) 0.08.

21.18 The degrees of freedom for this chi-square goodness of fit test are

(a) 15.   (b) 3.   (c) 2.

21.19 Software gives a chi-square statistic $X^2 = 0.124$ for this test. From the table of critical values, we can say that the P-value is

(a) less than 0.05.

(b) between 0.05 and 0.1.

(c) greater than 0.1.

21.20 Your conclusion based on this test is that

(a) the test results prove that squash color is determined by dominant epistasis.

(b) the test results are consistent with the idea that squash color is determined by dominant epistasis.

(c) the test results suggest that squash color is not determined by dominant epistasis.

21.21 You can trust the validity of this test because

(a) all expected counts are larger than 5.

(b) all observed counts are larger than 5.

(c) the sample size is larger than 40.

You run a valid chi-square goodness of fit test. Software gives $P < 0.01$ for the hypothesis $H_0: p_1 = p_2 = p_3$. Use this information for Exercises 21.22 and 21.23.

21.22 Your conclusion based on this test is that

(a) the test results are consistent with a uniform distribution.

(b) the test results provide strong evidence that all three proportions are different.

(c) the test results provide strong evidence that the distribution is not uniform.

21.23 Next, you would

(a) replicate the study with a larger sample size.

(b) examine the chi-square components or the observed sample percents.

(c) do nothing further because the test was not significant.

CHAPTER 21 EXERCISES

If you have access to software or a graphing calculator, use it to speed your analysis of the data in these exercises. Exercises 21.24 to 21.28 are suitable for hand calculation if necessary. Exercises 21.29 to 21.33 provide software output for their respective tests.

21.24 Mendel's first law. Exercise 21.2 described one of the crossing experiments that led Mendel to formulate his first law of inheritance, a model with a dominant and a recessive phenotype. The first law says that we can expect a 3:1 ratio, that is, 3/4 with the dominant and 1/4 with the recessive phenotype, in F2. When

Mendel crossed tall and dwarf pea plants, he obtained 787 tall plants and 277 dwarf plants in F2.

(a) Are the conditions for a chi-square goodness of fit test satisfied?

(b) What are the null and alternative hypotheses for this test?

(c) What are the expected counts under the null hypothesis?

(d) Compute the chi-square statistic and use software or Table D to find the P-value. What can you conclude? Do these data support Mendel's first law?

21.25 Mendel's second law. Mendel formulated the law of independent assortment as his second law of inheritance. It involves two genes independently segregated during reproduction, each independently determining one aspect of the phenotype. In one experiment, Mendel crossed pea plants producing yellow, round seeds with pea plants producing green, wrinkled seeds. The first generation resulted in only plants producing yellow, round seeds. Self-crossing of the F1 yielded the following phenotypes in F2:

Phenotype	Yellow, round	Yellow, wrinkled	Green, round	Green, wrinkled
Counts	315	101	108	32

Assuming two independent genes, each with a dominant and a recessive alleles, we would expect to find a 9:3:3:1 phenotypic ratio.

(a) Are the conditions for a chi-square goodness of fit test satisfied?

(b) What are the null and alternative hypotheses for this test?

(c) What are the expected counts under the null hypothesis?

(d) The chi-square statistic for this test is $X^2 = 0.47$. Use software or Table D to find the P value. What can you conclude? Do the data agree with Mendel's second law of genetic inheritance?

21.26 Soybean genetics. A soybean mutant exhibiting a tan-colored seed coat was recently identified. Crossing the mutant with the wild-type, yellow-seeded soybean plant gave all yellow-seeded soybeans in the first generation, F1. Self-pollination of F1 provided in the second generation, F2, 814 yellow-seeded and 251 tan-seeded plants. This pattern suggests a simple dominant-recessive model involving a single genetic mutation (3:1 expected ratio; 3/4 and 1/4 expected proportions).[4] Do the data support this interpretation? Run the appropriate chi-square test and conclude. Follow the four-step process in your answer.

21.27 Frizzled feathers. The Frizzle fowl is a striking variety of chicken with curled feathers. In a 1930 experiment, Launder and Dunn crossed Frizzle fowls with a Leghorn variety exhibiting straight feathers. The first generation (F1) produced all slightly frizzled chicks. When the F1 was interbred, the following characteristics were observed in F2:[5]

Phenotype (feather type)	Observed counts
Frizzled	23
Slightly frizzled	50
Straight	20

Annebicque Bernard/Corbis Sygma

The most likely genetic model for these results is that of a single gene locus with two codominant alleles. Under such a model, we would expect a 1:2:1 ratio in F2. Do the data support a codominance model of inheritance for this feather phenotype? Follow the four-step process in your answer.

21.28 Characteristics of a forest. Forests are complex, evolving ecosystems. For instance, pioneer tree species can be displaced by successional species better adapted to the changing environment. Ecologists mapped a large Canadian forest plot dominated by Douglas fir with an understory of western hemlock and western red cedar. Sapling trees (young trees shorter than 1.3 meter) are indicative of the future of a forest. The 246 sapling trees recorded in this sample forest plot were of the following types:[6]

	Count of sapling trees
Western red cedar (RC)	164
Douglas fir (DF)	0
Western hemlock (WH)	82

Is there an equal proportion of RC, DF, and WH among sapling trees in this forest? What does your analysis suggest about this forest's successional stage? Follow the four-step process in reaching your answer.

21.29 Melanoma. Melanoma is a rare form of skin cancer that accounts for the great majority of skin cancer fatalities. Ultraviolet (UV) exposure is a major risk factor for melanoma. Some body parts are regularly more exposed to the sun than others. Would this be reflected in the distribution of melanoma locations on the body in a population? A random sample of 310 women diagnosed with melanoma was classified according to the known locations of the melanomas on their bodies. Here are the results:[7]

Location	Head/neck	Trunk	Upper limbs	Lower limbs
Count	45	80	34	151

Assuming that these 4 body parts represent roughly equal skin areas, do the data support the hypothesis that melanoma occurs evenly on the body? Figure 21.5 (a) shows the Minitab output for this test. Follow the four-step process in finding your answer.

21.30 More on melanoma. The previous exercise described data about melanoma body locations in a random sample of women. The study also reported data for a random sample of 221 men diagnosed with melanoma on a known body location. Those results are as follows:

Location	Head/neck	Trunk	Upper limbs	Lower limbs
Count	36	139	17	32

Assuming that these 4 body parts represent roughly equal skin areas, do the data support the hypothesis that melanoma occurs evenly on the body? Figure 21.5 (b)

Minitab

```
Session                                                          _□×

                                 Test              Contribution
Category        Observed    Proportion   Expected      to Chi-Sq
Head/neck             45          0.25       77.5        13.6290
Lower limbs          151          0.25       77.5        69.7065
Trunk                 80          0.25       77.5         0.0806
Upper limbs           34          0.25       77.5        24.4161

   N    DF     Chi-Sq     P-Value
 310     3    107.832       0.000
```

(a)

```
Session                                                          _□×

                                 Test              Contribution
Category        Observed    Proportion   Expected      to Chi-Sq
Head/neck             36          0.25         56          7.143
Lower limbs           32          0.25         56         10.286
Trunk                139          0.25         56        123.018
Upper limbs           17          0.25         56         27.161

   N    DF     Chi-Sq     P-Value
 224     3    167.607      0.0001
```

(b)

FIGURE 21.5 Minitab outputs (a) for the melanoma data in women for Exercise 21.29 and (b) for the melanoma data in men for Exercise 21.30.

shows the Minitab output for this test. Follow the four-step process to get your answer.

21.31 Are clinical trials representative of the American population? Individuals of different ethnicities are somewhat more genetically variable than those of the same ethnicity and may respond to treatments differently. The FDA has recently requested that every effort be made to include individuals of all major ethnic backgrounds when enrolling subjects in clinical trials. A randomly chosen article in the *New England Journal of Medicine* describes the efficacy of calcium and vitamin D supplementation in preventing hip and other fractures in healthy postmenopausal women. Here is the breakdown by major ethnic group of the subjects whose ethnicity is known, along with the percent of individuals from each ethnicity in the general American population:[8]

Ethnicity	Count in study	Percent in U.S.
White	30,153	75.6%
Black	3,317	10.8%
Hispanic	1,507	9.1%
Asian or Pacific Islander	722	3.8%
Native American	149	0.7%

Minitab

```
 Session                                              _ □ X

                                     Test              Contribution
  Category          Observed  Proportion  Expected     to Chi-Sq
  Asian-Pacific Isl.     722       0.038    1362.2        300.895
  Black                 3317       0.108    3871.6         79.441
  Hispanic              1507       0.091    3262.2        944.346
  Native American        149       0.007     250.9         41.409
  White                30153       0.756   27101.1        343.682


      N    DF    Chi-Sq   P-Value
  35848     4   1709.77     0.000
```

FIGURE 21.6 Minitab output for Exercise 21.31.

Did this study successfully match the ethnic diversity of the American population? Figure 21.6 shows the Minitab output for this test. What does your analysis suggest? Follow the four-step process in your answer.

21.32 What's your sign? The University of Chicago's General Social Survey (GSS) is the nation's most important social science sample survey. The GSS regularly asks its subjects their astrological sign. Since the 12 zodiac signs evenly divide the calendar year, this information can be used to test whether births are uniformly distributed across the year. Here are the counts of responses in the most recent year this question was asked:[9]

Sign	Aries	Taurus	Gemini	Cancer	Leo	Virgo
Count	225	222	241	240	260	250

Sign	Libra	Scorpio	Sagittarius	Capricorn	Aquarius	Pisces
Count	243	214	200	216	224	244

If births are spread uniformly across the year, we expect all 12 signs to be equally likely. Are they? Software gives $X^2 = 14.392$. Follow the four-step process to find your answer.

21.33 Inbreeding in prairie dogs. A field biologist recorded events of inbreeding in a colony of prairie dogs. Based on a sample of 44 estrous females and 17 sexually mature males, the researcher computed the coefficient of relatedness of all possible male-female pairs in this sample. If individuals avoided inbreeding, we would expect to observe breeding mostly between individuals with little genetic relatedness. On the other hand, if breeding occurred randomly, without regard for the genetic relatedness of the mating couple, we would expect breeding observations among individuals of various relatedness to be proportional to the number of possible pairs with a given coefficient of relatedness. Here is the table of observed and expected breeding pairs in this study:[10]

Coefficient of relatedness	Observed pairs	Expected pairs
$r \geq 1/4$ (siblings)	6	6
$1/4 > r \geq 1/8$	4	5
$1/8 > r \geq 1/16$	18	10
$1/16 > r \geq 1/32$	12	12
$1/32 > r \geq 1/64$	10	11
$1/64 > r \geq 1/128$	3	7
$1/128 > r \geq 0$	8	8
No known kinship	8	10

Software computes $X^2 = 9.377$. Do the data support the hypothesis that breeding among prairie dogs occurs randomly, without regard for genetic relatedness? Follow the four-step process in your answer.

21.34 **Colors of M&M's.** Chocolate has had many properties attributed to it, including that of a mood enhancer. It is highly recommended that you perform this experiment yourself, especially on a long winter evening :)

The M&M's candies Web site `www.mms.com` says that the distribution of colors for M&M's Milk Chocolate Candies is

Color	Yellow	Red	Orange	Brown	Green	Blue
Probability	0.14	0.13	0.20	0.13	0.16	0.24

A package of M&M's contains 58 candies. (The count varies slightly from package to package.) The color counts are

Color	Yellow	Red	Orange	Brown	Green	Blue
Count	8	5	15	7	9	14

(a) How well do the counts from this package fit the claimed distribution? Do a chi-square test for goodness of fit, report the P-value, and conclude.

(b) Now take your own package and record the counts of each color. Do your data fit the distribution of M&M colors listed on the website?

Juniors Bildarchiv/Alamy

The Chi-Square Test for Two-Way Tables

In Chapter 21 we introduced a new statistic, X^2, and used it to test whether a categorical variable follows (or "fits") a particular distribution. This is a situation often found in genetic analysis. However, the most common use of the chi-square statistic is to test the hypothesis that there is *no relationship between two categorical variables*.

In Chapter 20 we used the two-sample z procedures to compare the proportions of successes in two groups, either two populations or two treatment groups in an experiment. In Example 20.4 (page 515) we looked at the sex of lesser snow goose goslings, comparing goslings hatched from nest eggs laid early to those hatched from eggs laid later. That is, we looked at a *relationship* between two categorical variables, sex (male or female) and laying order (early or late). In fact, the complete data set contained information about the exact laying order: first, second,

559

third, and fourth. When there are more than two levels of one factor, or when we want to compare more than two groups, we cannot use simple two-sample z procedures. In this chapter we learn how to organize these more complex problems into two-way tables and how to use the chi-square statistic to ask *whether there is a relationship between two categorical variables*.

Two-way tables

two-way table   We saw in Chapter 5 that we can present data on two categorical variables in a **two-way table** of counts. That's our starting point. Here is an example.

EXAMPLE 22.1 Health care: Canada and the United States

Canada has universal health care. The United States does not, but it often offers more elaborate treatment to patients who have access. For instance, a comparison of random samples of 2600 U.S. and 400 Canadian heart attack patients found that "the Canadian patients typically stayed in the hospital one day longer ($P = 0.009$) than the U.S. patients but had a much lower rate of cardiac catheterization (25 percent vs. 72 percent, $P < 0.001$), coronary angioplasty (11 percent vs. 29 percent, $P < 0.001$), and coronary bypass surgery (3 percent vs. 14 percent, $P < 0.001$)."[1]

The study found significant differences in the care heart attack patients receive in both systems. But does the difference in health care system influence patient outcomes? The study then looked at many outcomes a year after the heart attack. There was no significant difference in the patients' survival rate. Another key outcome was the patients' own assessment of their quality of life relative to what it had been before the heart attack. Here are the data for the patients who survived a year.

Quality of life	Canada	United States
Much better	75	541
Somewhat better	71	498
About the same	96	779
Somewhat worse	50	282
Much worse	19	65
Total	311	2165

The two-way table in Example 22.1 shows the relationship between 2 categorical variables. The explanatory variable is the patient's country, Canada or the United States. The response variable is quality of life a year after a heart attack, with 5 levels. The two-way table gives the counts for all 10 combinations of values of these variables. Each of the 10 counts occupies a **cell** of the table.

cell

It is hard to compare the counts, because the U.S. sample is much larger. Here are the percents of each sample with each outcome.

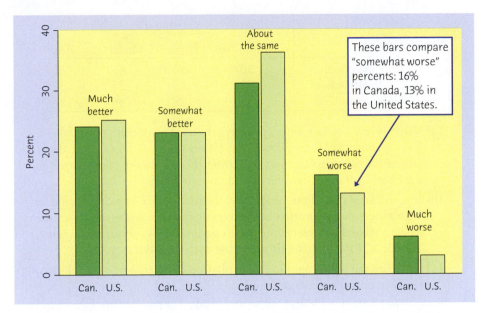

FIGURE 22.1 Bar graph comparing quality of life a year after a heart attack in Canada and the United States, for Example 22.1.

Quality of life	Canada	United States
Much better	24%	25%
Somewhat better	23%	23%
About the same	31%	36%
Somewhat worse	16%	13%
Much worse	6%	3%
Total	100%	100%

In the language of Chapter 5 (page 133), these are the *conditional distributions* of outcomes, given the patients' nationalities. The differences are not large, but slightly higher percents of Canadians thought their quality of life was "somewhat worse" or "much worse." Figure 22.1 compares the two distributions. We want to know if there is a significant difference between the two distributions of outcomes.

APPLY YOUR KNOWLEDGE

22.1   **Smoking among French men.** Smoking remains more common in much of Europe than in the United States. In the United States, there is a strong relationship between education and smoking: Well-educated people are less likely to smoke. Does a similar relationship hold in France? Here is a two-way table of the level of education and smoking status (nonsmoker, former smoker, moderate smoker, heavy smoker) of a sample of 459 French men aged 20 to 60 years.[2] The subjects are a random sample of men who visited a health center for a routine checkup. We are willing to consider them an SRS of men from their region of France.

| | Smoking status | | | |
Education	Nonsmoker	Former	Moderate	Heavy
Primary school	56	54	41	36
Secondary school	37	43	27	32
University	53	28	36	16

(a) What percent of men with a primary school education are nonsmokers? Former smokers? Moderate smokers? Heavy smokers? These percents should add to 100% (aside from roundoff error). They form the conditional distribution of smoking, given a primary education.

(b) In a similar way, find the conditional distributions of smoking among men with a secondary education and among men with a university education. Make a table that presents the three conditional distributions. Be sure to include a "Total" column showing that each row adds to 100%.

(c) Compare the three conditional distributions. Is there any clear relationship between education and smoking?

22.2 **Soccer and arthritis.** A study in Sweden looked at former elite soccer players, people who had played soccer but not at the elite level, and people of the same age who did not play soccer. The table below displays the number of individuals in each group who had arthritis of the hip or knee by their mid-50s.[3]

	Elite	Non-elite	Did not play
Arthritis	10	9	24
No arthritis	61	206	548
Sample size	71	215	572

(a) What percent of former elite soccer players have arthritis? Former non-elite soccer players? Individuals who did not play soccer? These percents should add to 100% (aside from roundoff error). They form the conditional distribution of having arthritis of the hip or knee, given a former level of soccer activity.

(b) Make a bar graph that compares your percents for all three samples. Is there any clear relationship between arthritis of the hip or knee and soccer? Explain your answer.

Royalty-Free/CORBIS

The problem of multiple comparisons

The null hypothesis in Example 22.1 is that there is *no difference* between the distributions of outcomes in Canada and in the United States. Put more generally, the null hypothesis is that there is *no relationship* between two categorical variables,

H_0: there is no relationship between nationality and quality of life

The alternative hypothesis says that there *is* a relationship but does not specify any particular kind of relationship,

H_a: there is some relationship between nationality and quality of life

Any difference between the Canadian and American distributions means that the null hypothesis is false and the alternative hypothesis is true. The alternative hypothesis is not one-sided or two-sided. We might call it "many-sided," because it allows any kind of difference.

Using the z procedures, we might start by comparing the proportions of patients in the two nations who declared they had a "much better" quality of life. We could similarly compare the proportions with each of the other outcomes: five tests in all, with five P-values. This is a bad idea. The P-values belong to each test separately, not to the collection of five tests together. To see why this is an issue, think of the distinction between the probability that a basketball player makes one free throw and the probability that she makes all of five free throws. *When we do many individual tests or confidence intervals, the individual P-values and confidence levels don't tell us how confident we can be in all of the inferences taken together.*

CAUTION

Because of this, it's cheating to pick out the largest of the five differences and then test its significance as if it were the only comparison we had in mind. For example, the "much worse" proportions in Example 22.1 are significantly different ($P = 0.0047$) if we compare just this one outcome. But is it surprising that the *most different* proportions among five outcomes differ by this much? That's a different question.

The problem of how to do many comparisons at once with an overall measure of confidence in all our conclusions is common in statistics. This is the problem of **multiple comparisons.** Statistical methods for dealing with multiple comparisons usually have two steps:

multiple comparisons

1. An *overall test* to see if there is good evidence of *any* differences among the parameters that we want to compare.

2. A detailed *follow-up analysis* to decide which of the parameters differ and to estimate how large the differences are.

The overall test, though more complex than the tests we saw earlier, is often reasonably straightforward. The follow-up analysis can be quite elaborate. In our basic introduction to statistical practice, we will concentrate on the overall test, along with data analysis that points to the nature of the differences.

APPLY YOUR KNOWLEDGE

22.3   Soccer and arthritis. Exercise 22.2 describes a study of arthritis of the hip and knee among individuals with differing soccer backgrounds.

(a) Give three 95% confidence intervals, for the percents of individuals who develop arthritis among former elite soccer players, among non-elite former soccer players, and among people who did not play soccer.

(b) Explain clearly why we are *not* 95% confident that *all three* of these intervals capture their respective population proportions.

22.4   Mating behavior of female prairie dogs. Gunnison's prairie dogs (*Cynomys gunnisoni*) are small social rodents found in the American Midwest. Careful observations in the wild have shown that both males and females can have multiple sex partners during the mating season. Male reproductive success is clearly related to the number of inseminations completed, but is there such a

Juniors Bildarchiv/Alamy

relationship for the females? A field biologist observed the mating behavior and reproductive success of females in one colony of Gunnison's prairie dogs:[4]

| | Number of mother's sexual partners | | | | |
	1	2	3	4	5
Number of copulating females	87	93	61	17	5
Number that gave birth	81	85	61	17	5
Proportion that gave birth	0.93	0.91	1.0	1.0	1.0

(a) Ten different pairwise comparisons of proportions are possible here. Write down the corresponding 10 null hypotheses.

(b) The females that copulated with 2 males or with 3 males represent two extremes in these observations. Give the P-value for a two-sided test comparing these two proportions. Why is it cheating to select the most extreme results and test only these for significance?

(c) If you calculated all 10 P-values, they would not tell you how often the five proportions would be spread this far apart just by chance. Explain why.

Expected counts in two-way tables

Our general null hypothesis H_0 is that there is *no relationship* between the two categorical variables that label the rows and columns of a two-way table. To test H_0, we compare the observed counts in the table with the *expected counts*, the counts we would expect—except for random variation—if H_0 were true. If the observed counts are far from the expected counts, that is evidence against H_0. It is easy to find the expected counts.

EXPECTED COUNTS

The **expected count** in any cell of a two-way table when H_0 is true is

$$\text{expected count} = \frac{\text{row total} \times \text{column total}}{\text{table total}}$$

EXAMPLE 22.2 Observed versus expected counts

Let's find the expected counts for the quality-of-life study. Here is the two-way table with row and column totals:

Quality of life	Canada	United States	Total
Much better	75	541	616
Somewhat better	71	498	569
About the same	96	779	875
Somewhat worse	50	282	332
Much worse	19	65	84
Total	311	2165	2476

The expected count of Canadians with much better quality of life a year after a heart attack is

$$\frac{\text{row 1 total} \times \text{column 1 total}}{\text{table total}} = \frac{(616)(311)}{2476} = 77.37$$

Here is the table of all 10 expected counts:

Quality of life	Canada	United States	Total
Much better	77.37	538.63	616
Somewhat better	71.47	497.53	569
About the same	109.91	765.09	875
Somewhat worse	41.70	290.30	332
Much worse	10.55	73.45	84
Total	311	2165	

As this table shows, *the expected counts have exactly the same row and column totals (aside from roundoff error) as the observed counts.* That's a good way to check your work.

To see how the data diverge from the null hypothesis, compare the observed counts with these expected counts. You see, for example, that 19 Canadians reported much worse quality of life, whereas we would expect only 10.55 to do so, if the null hypothesis were true.

Why the formula works. Where does the formula for an expected cell count come from? As stated in Chapter 21, if we have n independent tries and the probability of a success on each try is p, we expect np successes. (We also covered this in Chapters 9, 10, and 12.) That is, we expect np counts on average over the long run. Think of a basketball player who makes 70% of her free throws in the long run. If she shoots 10 free throws in a game, we expect her to make 70% of them, or 7 of the 10. Of course, she won't make exactly 7 every time she shoots 10 free throws in a game. There is chance variation from game to game. But in the long run, 7 of 10 is what we expect.

Now go back to the count of Canadians with much better quality of life a year after a heart attack. The proportion of all 2476 subjects with much better quality of life is

$$\frac{\text{count of successes}}{\text{table total}} = \frac{\text{row 1 total}}{\text{table total}} = \frac{616}{2476}$$

Think of this as p, the overall proportion of successes. If H_0 is true, we expect (except for random variation) the same proportion of successes in both countries. So the expected count of successes among the 311 Canadians is

$$np = (311)\left(\frac{616}{2476}\right) = 77.37$$

That's the formula in the Expected Counts box.

APPLY YOUR KNOWLEDGE

22.5   Smoking among French men. The two-way table in Exercise 22.1 displays data on the education and smoking behavior of a sample of French men. The null hypothesis says that there is no relationship between these variables. That is, the distribution of smoking is the same for all three levels of education.

(a) Find the expected counts for each smoking status among men with a university education. This is one row of the two-way table of expected counts. Find the row total and verify that it agrees with the row total for the observed counts.

(b) We conjecture that men with a university education smoke less than the null hypothesis calls for. How does comparing the observed and expected counts in this row confirm this conjecture?

22.6   Soccer and arthritis. Exercise 22.2 describes a study comparing the number of individuals in their 50s with arthritis of the hip or knee among individuals who used to play soccer at the elite or the non-elite level and among individuals who never played soccer. The null hypothesis "no relationship" says that in the population of individuals in their 50s, the proportions who have arthritis of the hip or knee are the same for former elite soccer players, former non-elite soccer players, and individuals who did not play soccer.

(a) Find the expected cell counts if this hypothesis is true and display them in a two-way table. Add the row and column totals to your table and check that they agree with the totals for the observed counts.

(b) Are there any large deviations between the observed counts and the expected counts? What kind of relationship between the two variables do these deviations point to?

The chi-square test

The statistical test that tells us whether the observed differences between Canada and the United States are statistically significant compares the observed and expected counts. The test statistic that makes the comparison is the *chi-square statistic*.

CHI-SQUARE STATISTIC

The **chi-square statistic** is a measure of how far the observed counts in a two-way table are from the expected counts. The formula for the statistic is

$$X^2 = \sum \frac{(\text{observed count} - \text{expected count})^2}{\text{expected count}}$$

The sum is over all cells in the table. That is, there are as many terms in the sum as there are cells in the table. Each term in the sum is called a X^2 **component.**

The chi-square statistic is a sum of terms, one for each cell in the table. There are 10 cells in the quality-of-life table on page 565 (2 groups with 5 answers each), and therefore we first need to compute 10 chi-square components separately. Here is how to compute the first one: 75 Canadian patients reported much better quality of life. The expected count for this cell is 77.37. So the chi-square component from this cell is

$$\frac{(\text{observed count} - \text{expected count})^2}{\text{expected count}} = \frac{(75 - 77.37)^2}{77.37}$$

$$= \frac{5.617}{77.37} = 0.073$$

To obtain the chi-square statistic for this test, we must add all 10 components, so X^2 is

$$X^2 = \frac{(75 - 77.4)^2}{77.4} + \frac{(541 - 538.6)^2}{538.6}$$

$$+ \frac{(71 - 71.5)^2}{71.5} + \frac{(498 - 497.5)^2}{497.5}$$

$$+ \frac{(96 - 109.9)^2}{109.9} + \frac{(779 - 765.1)^2}{765.1}$$

$$+ \frac{(50 - 41.7)^2}{41.7} + \frac{(282 - 290.3)^2}{290.3}$$

$$+ \frac{(19 - 10.6)^2}{10.6} + \frac{(65 - 73.4)^2}{73.4}$$

$$= 11.725$$

Think of the chi-square statistic X^2 as a measure of the distance between the observed counts and the expected counts. Like any distance, it is always zero or positive, and it is zero only when the observed counts are exactly equal to the expected counts. Small values of X^2 represent small deviations from H_0 that do not provide sufficient evidence to reject H_0. Inversely, large values of X^2 are evidence against H_0, because they say that the observed counts are far from what we would expect if H_0 were true. *Although the alternative hypothesis H_a is many-sided, the chi-square test is nondirectional*, because any violation of H_0 tends to produce a large value of X^2.

CAUTION

Using technology

Calculating the expected counts and then the chi-square statistic by hand is a bit time consuming. As usual, software saves time and always gets the arithmetic right. Figure 22.2 shows output for the chi-square test for the quality-of-life data from a graphing calculator, three statistical programs, and a spreadsheet program.

CrunchIt!

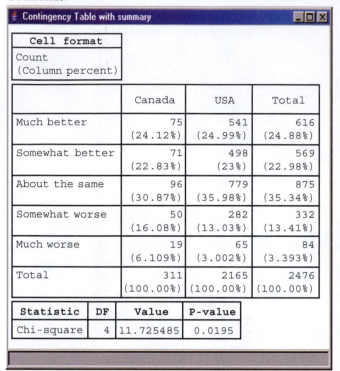

Contingency Table with summary

Cell format			
Count (Column percent)			

	Canada	USA	Total
Much better	75 (24.12%)	541 (24.99%)	616 (24.88%)
Somewhat better	71 (22.83%)	498 (23%)	569 (22.98%)
About the same	96 (30.87%)	779 (35.98%)	875 (35.34%)
Somewhat worse	50 (16.08%)	282 (13.03%)	332 (13.41%)
Much worse	19 (6.109%)	65 (3.002%)	84 (3.393%)
Total	311 (100.00%)	2165 (100.00%)	2476 (100.00%)

Statistic	DF	Value	P-value
Chi-square	4	11.725485	0.0195

TI-83

```
χ²-Test        [B]
χ²=11.7255   [[77.37   538.63…
P=.0195       [71.47   497.53…
df=4.0000     [109.91  765.09…
              [41.70   290.30…
              [10.55   73.45 …
```

Excel

Microsoft Excel - Book1

	A	B	C
1	Observed	Canada	USA
2		75	541
3		71	498
4		96	779
5		50	282
6		19	65
7			
8	Expected	Canada	USA
9		77.37	538.63
10		71.47	497.53
11		109.91	765.09
12		41.7	290.3
13		10.55	73.45
14			
15			
16	CHITEST(B2:C6,B9:C13)		0.019482
17			

Sheet1 / Sheet2 / Sheet3

Minitab

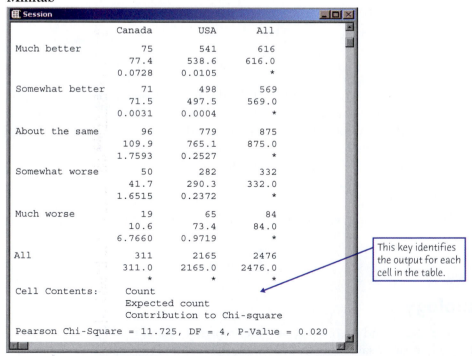

Session

	Canada	USA	All
Much better	75	541	616
	77.4	538.6	616.0
	0.0728	0.0105	*
Somewhat better	71	498	569
	71.5	497.5	569.0
	0.0031	0.0004	*
About the same	96	779	875
	109.9	765.1	875.0
	1.7593	0.2527	*
Somewhat worse	50	282	332
	41.7	290.3	332.0
	1.6515	0.2372	*
Much worse	19	65	84
	10.6	73.4	84.0
	6.7660	0.9719	*
All	311	2165	2476
	311.0	2165.0	2476.0
	*	*	*

Cell Contents: Count
                  Expected count
                  Contribution to Chi-square

Pearson Chi-Square = 11.725, DF = 4, P-Value = 0.020

This key identifies the output for each cell in the table.

FIGURE 22.2 Output from the TI-83 graphing calculator, CrunchIt!, Minitab, SPSS, and Excel for the two-way table in the quality-of-life study (*continued*).

SPSS

Column Statistics   _ □ ×

Quality * Country Crosstabulation

| | | | Country | | |
			Canada	United States	Total
Quality	Much better	Count	75	541	616
		Expected Count	77.4	538.6	616.0
	Somewhat better	Count	71	498	569
		Expected Count	71.5	497.5	569.0
	About the same	Count	96	779	875
		Expected Count	109.9	765.1	875.0
	Somewhat worse	Count	50	282	332
		Expected Count	41.7	290.3	332.0
	Much worse	Count	19	65	84
		Expected Count	10.6	73.4	84.0
Total		Count	311	2165	2476
		Expected Count	311.0	2165.0	2476.0

Chi-Square Tests

	Value	df	Asymp. Sig. (2-sided)
Pearson Chi-Square	11.725[a]	4	.020
Likelihood Ratio	10.435	4	.034
Linear-by-Linear Association	2.790	1	.095
N of Valid Cases	2476		

a. 0 cells (.0%) have expected count less than 5. The minimum expected count is 10.55.

FIGURE 22.2 (*continued*).

EXAMPLE 22.3   Chi-square from software

The outputs differ in the information they give. All except the Excel spreadsheet tell us that the chi-square statistic is $X^2 = 11.725$; all show a P-value of about 0.020. There is quite good evidence that the distributions of outcomes are different in Canada and in the United States.

The three statistical programs repeat the two-way table of observed counts and add the row and column totals. All three offer additional information on request. We asked CrunchIt! to add the column percents that enable us to compare the Canadian and American distributions. The chi-square statistic is a sum of 10 terms, one for each cell in the table. We asked Minitab to give the expected count and the contribution to chi-square for each cell. The top left cell has expected count 77.4 and chi-square term 0.073, just as we calculated. Look at the 10 terms. More than half the value of X^2 (6.766 out of 11.725) comes from just one cell. This points to the most important difference between the two countries: A higher proportion of Canadians report much worse quality of life. Most of the rest of X^2 comes from two other cells: More Canadians report somewhat worse quality of life, and fewer report about the same quality that we would expect if the null hypothesis were true.

Excel is, as usual, more awkward than software designed for statistics. It lacks a menu selection for the chi-square test. You must program the spreadsheet to calculate the expected cell counts and then use the CHITEST worksheet formula. This gives the P-value but not the test statistic itself. You can, of course, program the spreadsheet to find the value of X^2. The Excel output shows the observed and expected cell counts and the P-value.

The chi-square test is the overall test for detecting relationships between two categorical variables. If the test is significant, it is important to look at the data to learn the nature of the relationship. We have three ways to look at the quality-of-life data:

- **Compare appropriate percents:** which outcomes occur in quite different percents of Canadian and American patients? This is the method we learned in Chapter 5.

- **Compare observed and expected cell counts:** which cells have more or fewer observations than we would expect if H_0 were true?

- **Look at the terms of the chi-square statistic:** which cells contribute the most to the value of X^2?

EXAMPLE 22.4   Canada and the United States: conclusions

There is a significant difference between the distributions of quality of life reported by Canadian and by American patients a year after a heart attack. All three ways of comparing the distributions show that the main difference is that a higher proportion of Canadians report that their quality of life is worse than before their heart attack. Other response variables measured in the study agree with this conclusion.

The broader conclusion, however, is controversial. Americans are likely to point to the better outcomes produced by their much more intensive treatment. Canadians reply that the differences are small, that there was no significant difference in survival, and that the American advantage comes at high cost. The resources spent on expensive treatment of heart attack victims could instead be spent on providing basic health care to the many Americans who lack it.

There is an important message here: Although statistical studies shed light on issues of public policy, statistics alone rarely settles complicated questions such as "Which kind of health care system works better?"

APPLY YOUR KNOWLEDGE

22.7 **Smoking among French men.** In Exercises 22.1 and 22.5, you began to analyze data on the smoking status and education of French men. Figure 22.3 displays the Minitab output for the chi-square test applied to these data.

   (a) Starting from the observed and expected counts in the output, calculate the four terms of the chi-square statistic for the bottom row (university

Minitab

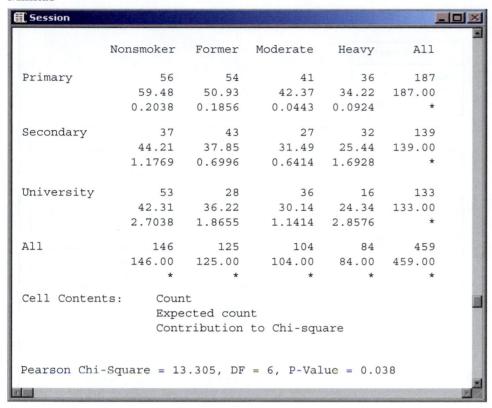

```
Session                                                    _ □ ×

          Nonsmoker   Former   Moderate   Heavy      All

Primary         56       54         41        36      187
             59.48    50.93      42.37     34.22   187.00
            0.2038   0.1856     0.0443    0.0924        *

Secondary       37       43         27        32      139
             44.21    37.85      31.49     25.44   139.00
            1.1769   0.6996     0.6414    1.6928        *

University      53       28         36        16      133
             42.31    36.22      30.14     24.34   133.00
            2.7038   1.8655     1.1414    2.8576        *

All            146      125        104        84      459
            146.00   125.00     104.00     84.00   459.00
                 *        *          *         *        *

Cell Contents:      Count
                    Expected count
                    Contribution to Chi-square


Pearson Chi-Square = 13.305, DF = 6, P-Value = 0.038
```

FIGURE 22.3 Minitab output for the two-way table of education level and smoking status among French men, for Exercise 22.7.

education). Verify that your work agrees with Minitab's "Contribution to Chi-square" aside from roundoff error.

(b) According to Minitab, what are the value of the chi-square statistic X^2 and the P-value of the chi-square test?

(c) Look at the "Contribution to Chi-square" entries in Minitab's display. Which terms contribute the most to X^2? Write a brief summary of the nature and significance of the relationship between education and smoking.

22.8 **Soccer and arthritis.** In Exercises 22.2 and 22.6, you began to analyze data on arthritis for individuals with various sports backgrounds. Figure 22.4 gives CrunchIt! output for these data.

(a) Starting from the observed and expected counts, find the six terms of the chi-square statistic and then the statistic X^2 itself. Check your work against the computer output.

(b) What is the P-value for the test? Explain in simple language what it means to reject H_0 in this setting.

(c) Which cells contribute the most to X^2? What kind of relationship do these terms in combination with the column percents in the table point to?

CrunchIt!

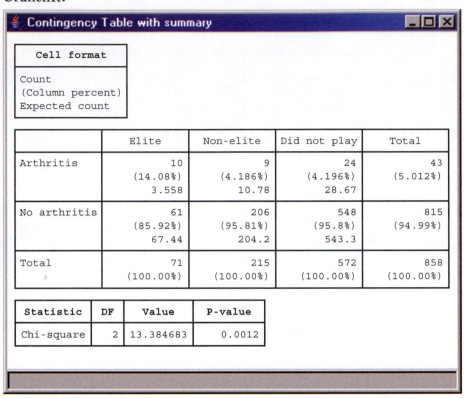

FIGURE 22.4 CrunchIt! output for the two-way table in the study of arthritis among soccer players, for Exercise 22.8.

Conditions for the chi-square test

The chi-square test, like the z procedures for comparing two proportions, is an approximate method that becomes more accurate as the counts in the cells of the table get larger. We must therefore check that the counts are large enough to trust the P-value. Fortunately, the chi-square approximation is accurate for quite modest counts. Here is a practical guideline.[5]

CELL COUNTS REQUIRED FOR THE CHI-SQUARE TEST

You can safely use the chi-square test with critical values from the chi-square distribution when no more than 20% of the *expected counts* are less than 5 and all individual *expected counts* are 1 or greater. In particular, all four expected counts in a 2×2 table should be 5 or greater.

Note that the guideline uses *expected* cell counts. The expected counts for the quality of life study of Example 22.1 appear in the Minitab output in Figure 22.2. The smallest expected count is 10.6, so the data easily meet the guideline for safe use of chi-square.

APPLY YOUR KNOWLEDGE

22.9 Smoking among French men. Figure 22.3 displays Minitab output for data on smoking among French men. Using the information in the output, verify that the data meet the cell count requirement for use of chi-square.

22.10 Soccer and arthritis. Figure 22.4 displays CrunchIt! output for data on arthritis among former soccer players. Using the information in the output, verify that the data meet the cell count requirement for use of chi-square.

22.11 Mating behavior of female prairie dogs. Exercise 22.4 described an observational study of the mating behavior and reproductive success of females in one colony of Gunnison's prairie dogs. Here is a table providing the observed counts (top) and expected counts (bottom) for each cell in the two-way table under a null hypothesis of no relationship between mating behavior and reproductive success.

	\multicolumn{5}{c}{Number of mother's sexual partners}				
	1	2	3	4	5
Gave birth	81	85	61	17	5
	82.37	88.05	57.75	16.1	4.73
Did not give birth	6	8	0	0	0
	4.63	4.95	3.25	0.9	0.27

With a total of 263 observations, this is a rather large data set. Nonetheless, you should *not* perform a chi-square test on these data. Explain why not.

Uses of the chi-square test

Two-way tables can arise in several ways. The study of the quality of life of heart attack patients compared two independent random samples, one in Canada and the other in the United States. The design of the study fixed the sizes of the two samples. The next example illustrates a different setting, in which all the observations come from just one sample.

EXAMPLE 22.5 *Characteristics of a forest*

STATE: Forests are complex, evolving ecosystems. For instance, pioneer tree species can be displaced by successional species better adapted to the changing environment. Ecologists mapped a large Canadian forest plot dominated by Douglas fir with an understory of western hemlock and western red cedar. The two-way table below records all 2050 trees in the plot by species and by life stage. Sapling trees (young trees shorter than 1.3 meter) are indicative of the future of a forest.[6]

	Dead	Live	Sapling
Western red cedar (RC)	48	214	154
Douglas fir (DF)	326	324	2
Western hemlock (WH)	474	420	88

FORMULATE: Carry out a chi-square test for

H_0: there is no relationship between species and life stage

H_a: there is some relationship between these two variables

Compare column percents or observed versus expected cell counts or terms of chi-square to see the nature of the relationship.

SOLVE: First check the guideline for use of chi-square. The expected cell counts appear in the Minitab output in Figure 22.5. The smallest expected count is 49.51, which definitely meets the requirement, so we can safely use chi-square. The output shows that there is a significant relationship ($X^2 = 420.311$, $P < 0.001$). The largest contributor to X^2 comes from the cell for sapling western red cedar (RC) (220 out of 420). There are many more sapling western red cedar trees than would be expected if H_0 were true. The next largest chi-square component is for the RC and dead cell (89 out of 420). Western red cedars appear to have the most unusual pattern, with many more sapling trees and far fewer dead ones.

CONCLUDE: We find that 63% of sapling trees are western red cedar trees, compared with 36% western hemlock and only 1% Douglas fir. The species western red cedar is thus taking over this forest. In addition, western red cedars have the lowest percent of dead trees among all three species, with only about 12% of western red cedar trees in this forest found dead. Together, these data suggest that the western red cedar species is a successional species displacing the established other species in this forest.

Minitab

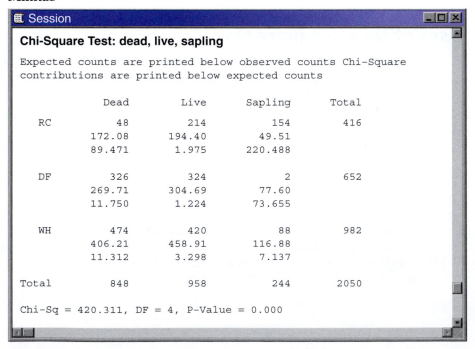

Chi-Square Test: dead, live, sapling

Expected counts are printed below observed counts Chi–Square contributions are printed below expected counts

	Dead	Live	Sapling	Total
RC	48	214	154	416
	172.08	194.40	49.51	
	89.471	1.975	220.488	
DF	326	324	2	652
	269.71	304.69	77.60	
	11.750	1.224	73.655	
WH	474	420	88	982
	406.21	458.91	116.88	
	11.312	3.298	7.137	
Total	848	958	244	2050

Chi–Sq = 420.311, DF = 4, P–Value = 0.000

FIGURE 22.5 Minitab output for the two-way table in the forest study, for Example 22.5.

Pay attention to the nature of the data in Example 22.5:

- We do not have three separate samples of trees, one for each species. We have a single plot of forest with 2050 trees in it, each classified in two ways (species and life stage).

- The data cover *all* of the trees found in this forest plot. We might regard this as a sample of similar forest areas in the same successional stage. But we might also regard these 2050 trees as the entire population of interest rather than a sample from a larger population.

One of the most useful properties of chi-square is that it tests the null hypothesis "the row and column variables are not related to each other" whenever this hypothesis makes sense for a two-way table. It makes sense when we are comparing a categorical response in two or more samples, as when we compared quality of life for patients in Canada and the United States. The hypothesis also makes sense when we have data on two categorical variables for the individuals in a single sample, as when we examined life stages and tree species for a sample of forest area. The hypothesis "no relationship" makes sense even if the single sample is an entire population. Statistical significance has the same meaning in all these settings: "A relationship this strong is not likely to happen just by chance." This makes sense whether the data are a sample or an entire population.

USES OF THE CHI-SQUARE TEST

Use the chi-square test to test the null hypothesis

H_0: there is no relationship between two categorical variables

when you have a two-way table from one of these situations:

- Independent SRSs from each of two or more populations, with each individual classified according to one categorical variable. (The other variable says which sample the individual comes from.)

- A single SRS, with each individual classified according to both of two categorical variables.

APPLY YOUR KNOWLEDGE

22.12 Soccer and arthritis. Exercise 22.2 describes a study of arthritis of the hip and knee among individuals with various sports backgrounds. Did the study use a single SRS or multiple SRSs? Would you say that the study represents its own population or that it hopes to represent a larger population?

22.13 Smoking among French men. Exercise 22.1 describes data on the smoking status and education of French men. Did the study use a single SRS or multiple SRSs? Would you say that the study represents its own population or that it hopes to represent a larger population?

22.14 Hearing impairment in dalmatians. Congenital sensorineural deafness is the most common form of deafness in dogs and is often associated with congenital pigmentation deficiencies. A study of hearing impairment in dogs examined over

SPSS

Column Statistics					

Hearing * EyeColor Crosstabulation

			EyeColor		
			Brown eyed	Blue eyed	Total
Hearing	Deaf	Count	324	102	426
		Expected Count	381.7	44.3	426.0
		% within EyeColor	6.8%	18.4%	8.0%
	Unilaterally impaired	Count	988	179	1167
		Expected Count	1045.8	121.2	1167.0
		% within EyeColor	20.7%	32.3%	21.9%
	Bilateral hearing	Count	3467	273	3740
		Expected Count	3351.5	388.5	3740.0
		% within EyeColor	72.5%	49.3%	70.1%
Total		Count	4779	554	5333
		Expected Count	4779.0	554.0	5333.0
		% within EyeColor	100.0%	100.0%	100.0%

Chi-Square Tests

	Value	df	Asymp. Sig. (2-sided)
Pearson Chi-Square	153.138[a]	2	.000
Likelihood Ratio	133.597	2	.000
Linear-by-Linear Association	153.052	1	.000
N of Valid Cases	5333		

[a.] 0 cells (.0%) have expected count less than 5. The minimum expected count is 44.25.

FIGURE 22.6 SPSS output for the two-way table in the study of hearing impairment among dalmatians, for Exercise 22.14.

5000 dalmatians for both hearing impairment and iris color. Figure 22.6 shows the SPSS output for a chi-square test on these data.[7] Dogs with either one or both of their irises blue (a trait due to low iris pigmentation) were labeled "blue-eyed." This is an example of a single sample classified according to two categorical variables (hearing impairment and iris color).

(a) Describe the differences between the distributions of hearing impairments among brown-eyed and blue-eyed dogs with percents, with a bar graph, and in words.

(b) Verify that the expected cell counts satisfy the requirement for use of chi-square.

(c) Test the null hypothesis that there is no relationship between hearing impairment and iris color. Give a P-value.

(d) How do the observed and expected counts differ? Relate this to your conclusions in (a).

Using tables of critical values*

Software usually finds P-values for us. The P-value for a chi-square test comes from comparing the value of the chi-square statistic with critical values for a *chi-square distribution*.

We saw in Chapter 21 that the chi-square distributions are a family of distributions that take only positive values and are skewed to the right. A specific chi-square distribution is specified by giving its *degrees of freedom*.

> **CHI-SQUARE DISTRIBUTIONS FOR TWO-WAY TABLES**
>
> The chi-square test for a two-way table with r rows and c columns uses critical values from the chi-square distribution with $(r-1)(c-1)$ **degrees of freedom**.
>
> The P-value is the area to the right of X^2 under the density curve of this chi-square distribution.

Figure 21.4 shows the density curves for three members of the chi-square family of distributions. As the degrees of freedom increase, the density curves become less skewed and larger values become more probable. Table D in the back of the book gives critical values for chi-square distributions. You can use Table D if you do not have software that gives you P-values for a chi-square test.

EXAMPLE 22.6   Using the chi-square table

The two-way table of 5 outcomes by 2 countries for the quality-of-life study has 5 rows and 2 columns. That is, $r = 5$ and $c = 2$. The chi-square statistic therefore has

$$(r-1)(c-1) = (5-1)(2-1) = (4)(1) = 4$$

degrees of freedom. Three of the outputs in Figure 22.2 give 4 as the degrees of freedom.

The observed value of the chi-square statistic is $X^2 = 11.725$. Look in the df $= 4$ row of Table D. The value $X^2 = 11.725$ falls between the 0.02 and 0.01 critical values of the chi-square distribution with 4 degrees of freedom. Remember that the chi-square test is always one-sided. So the P-value of $X^2 = 11.725$ is between 0.02 and 0.01. The outputs in Figure 22.2 show that the P-value is 0.0195 or 0.02.

df $= 4$		
p	.02	.01
x^*	11.67	13.28

APPLY YOUR KNOWLEDGE

22.15 Smoking among French men. The Minitab output in Figure 22.3 gives the degrees of freedom for the table of education and smoking status as DF = 6.

(a) Show that this is correct for a table with 3 rows and 4 columns.

---

*The remainder of the material in this chapter is optional.

(b) Minitab gives the chi-square statistic as `Chi-Square = 13.305`. Between which two entries in Table D does this value lie? Verify that Minitab's result `P-Value = 0.038` lies between the tail areas for these values.

22.16 Soccer and arthritis. The CrunchIt! output in Figure 22.4 gives 2 degrees of freedom for the table in Exercise 22.2.

(a) Verify that this is correct.

(b) The computer gives the value of the chi-square statistic as $X^2 = 13.384683$. Between what two entries in Table D does this value lie? What does the table tell you about the P-value?

The chi-square test and the two-sample z test*

One use of the chi-square test is to compare the proportions of successes in any number of groups. When comparing the proportions of successes in only two groups, the data can be summarized in a 2×2 table like this one.

	Success	Failure
Group 1		
Group 2		

We then have two inference procedures available: the two-sample z test from Chapter 20 and the chi-square test with 1 degree of freedom for a 2×2 table. The null hypotheses tested are

$$H_0: p_1 = p_2$$

for the two-sample z test, where p_1 and p_2 are the proportion of successes in each group, and

$$H_0: \text{there is no relationship between the row and column variables}$$

for the chi-square test for two-way tables.

These two tests always agree. In fact, the chi-square statistic X^2 is just the square of the z statistic, and the P-value for X^2 is exactly the same as the two-sided P-value for z. We recommend using the z test to compare two proportions, because it gives you the choice of a one-sided test and is related to a confidence interval for the difference $p_1 - p_2$.

APPLY YOUR KNOWLEDGE ────────────

22.17 Treating ulcers. Gastric freezing was once a recommended treatment for peptic ulcers. Use of gastric freezing stopped after experiments showed it had no effect. One randomized comparative experiment found that 28 of the 82 gastric-freezing patients improved, while 30 of the 78 patients in the placebo group improved.[8] We can test the hypothesis of "no difference" between the two groups in either of two ways: using the two-sample z statistic or using the chi-square statistic.

(a) Check the conditions required for both tests, given in the boxes on pages 516 and 572. The conditions are very similar, as they ought to be.

(b) State the null hypothesis with a two-sided alternative and carry out the z test. What is the P-value, exactly from software or approximately from the bottom row of Table C?

(c) Present the data in a 2 × 2 table. Use the chi-square test to test the hypothesis from (b). Verify that the X^2 statistic is the square of the z statistic. Use software or Table D to verify that the chi-square P-value agrees with the z result (up to the accuracy of the tables if you do not use software).

(d) What do you conclude about the effectiveness of gastric freezing as a treatment for peptic ulcers?

CHAPTER 22 SUMMARY

The **chi-square test** for a two-way table tests the null hypothesis H_0 that there is no relationship between the row variable and the column variable. The alternative hypothesis H_a says that there is some relationship but does not say what kind.

The test compares the observed counts of observations in the cells of the table with the counts that would be expected if H_0 were true. The **expected count** in any cell is

$$\text{expected count} = \frac{\text{row total} \times \text{column total}}{\text{table total}}$$

The **chi-square statistic** is

$$X^2 = \sum \frac{(\text{observed count} - \text{expected count})^2}{\text{expected count}}$$

The chi-square test compares the value of the statistic X^2 with critical values from the **chi-square distribution** with $(r-1)(c-1)$ **degrees of freedom.** Large values of X^2 are evidence against H_0, so the P-value is the area under the chi-square density curve to the right of X^2.

The chi-square distribution is an approximation of the distribution of the statistic X^2. You can safely use this approximation when all expected cell counts are at least 1 and no more than 20% are less than 5.

If the chi-square test finds a statistically significant relationship between the row and column variables in a two-way table, do data analysis to describe the nature of the relationship. You can do this by comparing well-chosen percents, comparing the observed counts with the expected counts, and looking for the largest **terms of the chi-square statistic.**

CHECK YOUR SKILLS

Use the following information for Exercises 22.18 through 22.27. Washington, D.C., has one of the worst rates of HIV/AIDS of all major cities in the United States, with an estimated one in 20 residents infected with the virus. Understanding which groups are most at risk can help target prevention and health programs. It is very difficult to assess the number of individuals who carry the HIV virus. Statistics about individuals living with AIDS (once the disease has progressed to its symptomatic stage) are more reliable.

Dropping out

An experiment found that weight loss is significantly more effective than exercise for reducing high cholesterol and high blood pressure. The 170 subjects were randomly assigned to a weight-loss program, an exercise program, or a control group. Only 111 of the 170 subjects completed their assigned treatment, and the analysis used data from these 111. Did the dropouts create bias? Always ask about details of the data before trusting inference.

The District of Columbia has made it mandatory for health providers to report—anonymously—cases of D.C. residents diagnosed with AIDS. Here is the breakdown by gender and ethnicity as of December 2003.[9]

	Male	Female
White	1106	76
African American	4548	1799
Hispanic	270	35
Asian/Pacific Islander	29	8
Unknown/Other	220	110

22.18 The percent of females living with AIDS who are African American is about

(a) 88.7%. (b) 21.9%. (c) 17.9%.

22.19 The percent of females living with AIDS who are African American is

(a) higher than the percent of males living with AIDS who are African American.

(b) about the same as the percent of males living with AIDS who are African American.

(c) lower than the percent of males living with AIDS who are African American.

22.20 The expected count of females living with AIDS who are African American is about

(a) 7274.3. (b) 1799.0. (c) 1569.5

22.21 The term in the chi-square statistic for the cell of females living with AIDS who are African American is about

(a) 1569.5. (b) 299.3. (c) 33.5.

22.22 The degrees of freedom for the chi-square test for this two-way table are

(a) 4. (b) 8. (c) 10.

22.23 The null hypothesis for the chi-square test for this two-way table is

(a) Equal proportions of females and males living with AIDS are African American.

(b) There is no difference between females and males living with AIDS based on ethnicity.

(c) There are equal numbers of females and males living with AIDS.

22.24 The alternative hypothesis for the chi-square test for this two-way table is

(a) Females and males living with AIDS do not have the same ethnic distribution.

(b) Females living with AIDS are more likely than males living with AIDS to be African American.

(c) Females living with AIDS are less likely than males living with AIDS to be African American.

22.25 Software gives the chi-square statistic $X^2 = 299.307$ for this table. From Table D of critical values, we can say that the P-value is

(a) between 0.0025 and 0.001.

(b) between 0.001 and 0.0005.

(c) less than 0.0005.

22.26 The most important fact that allows us to trust the results of the chi-square test is that

(a) the sample is large, 8201 individuals in all.

(b) all of the cell counts are greater than 5.

(c) all of the expected cell counts are greater than 5.

22.27 This study

(a) is an accurate SRS of the population of DC residents living with AIDS.

(b) represents the entire population of DC residents living with AIDS.

(c) is a reasonable assessment of the entire population of DC residents living with AIDS except, for instance, for some individuals without health insurance.

CHAPTER 22 EXERCISES

If you have access to software or a graphing calculator, use it to speed your analysis of the data in these exercises. Exercises 22.28 to 22.32 are suitable for hand calculation if necessary. About half of the remaining exercises provide software output for their respective tests.

22.28 Did the randomization work? After randomly assigning subjects to treatments in a randomized comparative experiment, we can compare the treatment groups to see how well the randomization worked. We hope to find no significant differences among the groups. A study of how to provide premature infants with a substance essential to their development assigned infants at random to receive one of four types of supplement, called PBM, NLCP, PL-LCP, and TG-LCP.[10]

(a) The subjects were 77 premature infants. Outline the design of the experiment if 20 are assigned to the PBM group and 19 to each of the other treatments.

(b) The random assignment resulted in 9 females in the TG-LCP group and 11 females in each of the other groups. Make a two-way table of groups by gender. What percent of premature infants in each group were females?

(c) Can we safely use the chi-square test? What null and alternative hypotheses does X^2 test? What are the degrees of freedom for this test?

(d) Software gives $X^2 = 0.568$. Is this evidence of a significant difference among the groups? Use Table D to approximate the P-value of the test.

22.29 Do you use cocaine? Sample surveys on sensitive issues can give different results depending on how the question is asked. A University of Wisconsin study divided 2400 respondents into 3 groups at random. All were asked if they had ever used cocaine. One group of 800 was interviewed by phone; 21% said they had used cocaine. Another 800 people were asked the question in a one-on-one personal interview; 25% said "Yes." The remaining 800 were allowed to make an anonymous written response; 28% said "Yes."[11]

(a) Make a two-way table of answers (yes or no) by survey method.

(b) Can we safely use the chi-square test? What null and alternative hypotheses does X^2 test? What are the degrees of freedom for this test?

(c) Compute the chi-square component for each cell and give X^2. Use Table D to approximate the P-value of the test.

(d) What do you conclude from these data?

22.30 Cranberry juice for preventing urinary tract infections. American cranberry (*Vaccinium macrocarpon*) was used medicinally by Native Americans for the treatment of bladder and kidney ailments. A randomized controlled study was designed to test whether regular drinking of cranberry juice can prevent the recurrence of urinary tract infections (UTIs) in women. One hundred and fifty women with a urinary tract infection were treated with an antibiotic and then randomly assigned to one of 3 groups. One group drank cranberry juice concentrate daily for six months; another group took a drink containing *Lactobacillus* (a lactose-fermenting bacterium thought to help inhibit the growth of UTI-causing bacteria) daily for six months; the last group served as the control group and drank neither cranberry juice nor *Lactobacillus* drinks for six months. After six months, the number of women in each group with recurring symptomatic urinary tract infection (defined as one or more new infections) was recorded. Here are the results.[12]

FoodCollection/StockFood

| | Outcome | | |
Treatment	Recurring UTI	No new UTI	Total
Cranberry juice	8	42	50
Lactobacillus drink	19	30	49
Control	18	32	50

(a) Find the percent of women experiencing no new UTI in each of the three treatment groups. Make a graph to compare these percents. Describe the association between treatment and outcome.

(b) Explain in words what the null hypothesis for the chi-square test says about treatment outcome. Verify that the conditions are met for a chi-square test.

(c) Compute the chi-square statistic and find the P-value for this test. Examine the terms of chi-square to confirm the pattern you saw in (a). What is your overall conclusion? Interestingly, recent pharmacological studies showed that some cranberry compounds prevent bacteria from binding to the membranes of host cells.

22.31 Stress and heart attacks. You read a newspaper article that describes a study of whether stress management can help reduce heart attacks. The 107 subjects all had reduced blood flow to the heart and so were at risk of a heart attack. They were assigned at random to one of three groups. The article goes on to say:

One group took a four-month stress management program, another underwent a four-month exercise program and the third received usual heart care from their personal physicians.

In the next three years, only three of the 33 people in the stress management group suffered "cardiac events," defined as a fatal or non-fatal heart attack or a surgical procedure such as a bypass or angioplasty. In the same period, seven of the 34 people in the exercise group and 12 out of the 40 patients in usual care suffered such events.[13]

(a) Use the information in the news article to make a two-way table that describes the study results. What are the success rates of the three treatments in preventing cardiac events?

(b) Find the expected cell counts under the null hypothesis that there is no difference among the treatments. Verify that the expected counts meet our guideline for use of the chi-square test.

(c) Is there a significant difference among the success rates for the three treatments? Compute the chi-square statistic, use Table D to approximate the P-value of the test, and conclude.

*The remaining exercises ask you to follow the **Formulate, Solve,** and **Conclude** steps of the four-part process in your answers. It may be helpful to restate in your own words the **State** information given in the exercise.*

22.32 Sickle-cell and malaria. Sickle-cell anemia is a hereditary condition that is common among blacks and can cause medical problems. Some biologists suggest that the sickle-cell trait protects against malaria. That would explain why it is found in people whose ancestors originally came from Africa, where malaria is common. A study in Africa tested 543 children for the sickle-cell trait and also for malaria. Here are the results.[14]

	Severe malaria infection	
	Yes	No
Sickle-cell trait	36	100
No sickle-cell trait	152	255

Is there good evidence of a relationship between the sickle-cell trait and severe malaria infection?

22.33 Python eggs. How is the hatching of water python eggs influenced by the temperature of the snake's nest? Researchers assigned newly laid eggs to one of three temperatures: hot, neutral, or cold. Hot duplicates the extra warmth provided by the mother python, and cold duplicates the absence of the mother. Here are the data on the number of eggs and the number that hatched:[15]

	Eggs	Hatched
Cold	27	16
Neutral	56	38
Hot	104	75

(a) Make a two-way table of temperature by outcome (hatched or not).

(b) Calculate the percent of eggs in each group that hatched. The researchers anticipated that eggs would not hatch at cold temperatures. Do the data support that anticipation?

(c) Are there significant differences among the proportions of eggs that hatched in the three groups?

22.34 Smoking by students and their parents. How are the smoking habits of students related to their parents' smoking? Figure 22.7 gives the observed and

Minitab

```
🔳 Session                                                    _ ☐ ✕

Chi-Square Test

Expected counts are printed below observed counts

               Smokes        NoSmoke        Total
Both            400           1380          1780
              332.49        1447.51

One             416           1823          2239
              418.22        1820.78

None            188           1168          1356
              253.29        1102.71

Total          1004           4371          5375

Chi-Sq = 13.709 + 3.149 +
          0.012 + 0.003 +
         16.829 + 3.866 = 37.566
DF = 2, P-Value = 0.000
```

FIGURE 22.7 Minitab output for the sample survey responses of Exercise 22.34.

expected counts from a survey of students in eight Arizona high schools. For each student surveyed, the study recorded whether or not the student smoked and whether one, both, or neither parent smoked.[16]

(a) Find the percent of students who smoke in each of the three parent groups. Make a graph to compare these percents. Describe the association between parent smoking and student smoking.

(b) Look at the statistical analysis in Figure 22.7. Write a short description of the study findings.

22.35 **Mediterranean diet and heart attack.** Observational studies have pointed to a potential beneficial effect of a Mediterranean diet in protecting against heart disease. An experiment was performed on 605 survivors of a heart attack, randomly assigning them to follow for 4 years either a diet close to the "prudent diet step 1" of the American Heart Association (AHA) or a Mediterranean-type diet consisting of more bread and grains, more fresh fruit and vegetables, more fish, and less meat and animal fat. Here is the record of subjects who experienced a cardiac event over the course of the study:[17]

	AHA	Mediterranean
Cardiac death	19	6
Nonfatal heart attack	25	8
Neither	259	288

Describe the most important differences in cardiac health between the AHA diet and the Mediterranean diet. Is there a significant overall difference between the two distributions' health outcomes?

22.36 The Mediterranean diet and cancer. Cancer of the colon and rectum is less common in the Mediterranean region than in Western countries in other regions. The Mediterranean diet contains little animal fat and lots of olive oil. Italian researchers compared 1953 patients with colon or rectal cancer with a control group of 4154 patients admitted to the same hospitals for unrelated reasons. They estimated consumption of various foods from a detailed interview, and then classified the patients according to their consumption of olive oil (low, medium, or high).[18] Figure 22.8 gives counts and chi-square output for this study.

(a) Is this study an experiment? Explain your answer.

(b) The investigators report that "less than 4% of cases or controls refused to participate." Why does this fact strengthen our confidence in the results?

(c) The researchers conjectured that high olive oil consumption would be more common among patients without colon cancer or rectal cancer than among patients with these cancers. What do the data say?

22.37 Treating cocaine addiction. Cocaine produces short-term feelings of physical and mental well-being, followed by feelings of tiredness and depression once the drug wears off. This cycle of positive and negative feelings is the root problem of addiction and can be very hard to break. Perhaps giving cocaine addicts a medication that fights depression can help them break the addiction.

Minitab

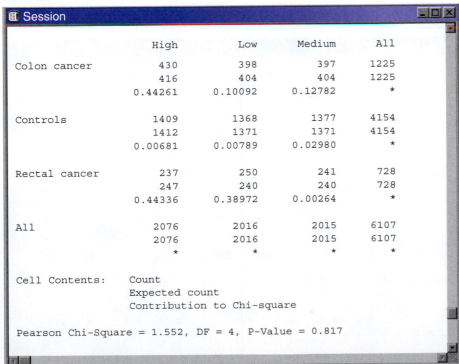

	High	Low	Medium	All
Colon cancer	430	398	397	1225
	416	404	404	1225
	0.44261	0.10092	0.12782	*
Controls	1409	1368	1377	4154
	1412	1371	1371	4154
	0.00681	0.00789	0.02980	*
Rectal cancer	237	250	241	728
	247	240	240	728
	0.44336	0.38972	0.00264	*
All	2076	2016	2015	6107
	2076	2016	2015	6107
	*	*	*	*

```
Cell Contents:      Count
                    Expected count
                    Contribution to Chi-square

Pearson Chi-Square = 1.552, DF = 4, P-Value = 0.817
```

FIGURE 22.8 Minitab output for the epidemiological data of Exercise 22.36.

A three-year study compared an antidepressant called desipramine with lithium (a standard treatment for cocaine addiction) and a placebo. The subjects were 72 chronic users of cocaine who wanted to break their drug habit. Twenty-four of the subjects were randomly assigned to each treatment. Here are the counts and proportions of the subjects who avoided relapse into cocaine use during the study:[19]

Group	Treatment	Subjects	No relapse	Proportion
1	Desipramine	24	14	0.583
2	Lithium	24	6	0.250
3	Placebo	24	4	0.167

Does data analysis suggest that desipramine is more successful than the other two treatments? Are there significant differences among the outcomes for the treatments?

22.38 Preventing strokes. Individuals who have experienced a stroke are at an increased risk of having another stroke. Prevention is particularly important for high-risk populations. The Second European Stroke Prevention Study asked whether daily doses of an anti-blood-clotting agent, such as aspirin or dipyridamole, would help prevent strokes among patients who had just experienced a first stroke. Volunteer patients were randomly assigned to one of 4 treatment options, including a placebo. Researchers recorded how many patients experienced another stroke and how many died of a stroke during the two years of the study. Figure 22.9 gives the observed and expected counts for this

Minitab

```
🗒 Session                                                              _ □ ✗

            Stroke   NoStroke   Total                Death   NoDeath   Total

Placebo       250       1399     1649   Placebo        202      1447    1649
            205.81    1443.19                        189.08   1459.92


Aspirin       206       1443     1649   Aspirin        182      1467    1649
            205.81    1443.19                        189.08   1459.92


Dipyr.        211       1443     1654   Dipyr.         188      1466    1654
            206.44    1447.56                        189.65   1464.35


Both          157       1493     1650   Both           185      1465    1650
            205.94    1444.06                        189.19   1460.81


Total         824       5778     6602   Total          757      5845    6602

ChiSq = 9.487 + 1.353 +                 ChiSq = 0.883 + 0.114 +
        0.000 + 0.000 +                         0.265 + 0.034 +
        0.101 + 0.014 +                         0.014 + 0.002 +
       11.629 + 1.658 = 24.243                  0.093 + 0.012 = 1.418
df = 3, p = 0.000                       df = 3, p = 0.701
```

FIGURE 22.9 Minitab output for the experimental results of Exercise 22.38.

study, along with the chi-square calculations. Is there a relationship between treatment and strokes? Write a careful summary of your overall findings.

22.39 More on preventing strokes. The study described in the previous exercise also recorded the number of patients in each group who died following a stroke during the two years of the study. Is there a relationship between treatment and stroke-related deaths? Write a careful summary of your overall findings.

22.40 Do angry people have more heart disease? People who get angry easily tend to have more heart disease. That's the conclusion of a study that followed a random sample of 12,986 people from three locations for about four years. All subjects were free of heart disease at the beginning of the study. The subjects took the Spielberger Trait Anger Scale test, which measures how prone a person is to sudden anger. Here are data for the 8474 people in the sample who had normal blood pressure.[20] CHD stands for "coronary heart disease." This includes people who had heart attacks and those who needed medical treatment for heart disease.

	Low anger	Moderate anger	High anger	Total
CHD	53	110	27	190
No CHD	3057	4621	606	8284
Total	3110	4731	633	8474

Do these data support the study's conclusion about the relationship between anger and heart disease?

22.41 Echinacea for the common cold? The National Center for Complementary and Alternative Medicine (NCCAM, a branch of NIH) offers on its website comprehensive information about various complementary and alternative healing practices, including a summary of research findings for each practice. Echinacea is a flowering plant related to daisies with a long history of medicinal use, starting with Native Americans. Today it is the top-selling herbal remedy in the United States, mainly used against the common cold. In a double-blind experiment, healthy volunteers agreed to be exposed to common-cold-causing rhinovirus type 39. Three types of echinacea extracts were produced (labeled E1, E2, and E3 in the table below). Subjects received either an echinacea extract or a placebo (P) over two different periods: for 7 consecutive days before viral exposure (infection and symptom prevention) and for 7 consecutive days following exposure (symptom treatment). Among the variables studied were whether or not a subject became infected following viral exposure and whether or not infected subjects developed cold symptoms. Here are the results.[21]

Terry Donnelly/Alamy

Treatment	Infected	Not infected	Cold symptoms	No symptoms
E1 then E1	40	5	25	15
E2 then E2	42	10	24	18
E3 then E3	48	4	24	24
P then E1	43	5	27	16
P then E2	44	4	33	11
P then E3	44	7	28	16
P then P	88	15	58	30

Running a chi-square test for each variable separately gave $X^2 = 4.745$ and $X^2 = 7.120$ for the first and the second variable, respectively. Write a careful summary of the overall findings from this study.

22.42   Melanoma in men and women.   Intense or repetitive sun exposure can lead to melanoma, a rare but severe form of skin cancer. Patterns of sun exposure, however, differ between men and women. A study of cutaneous malignant myeloma in the Italian population recorded the distribution of melanoma by body location for men and for women.[22] Figure 22.10 displays the observed and expected counts for this study and the chi-square test from CrunchIt! Write a careful summary of your overall findings.

CrunchIt!

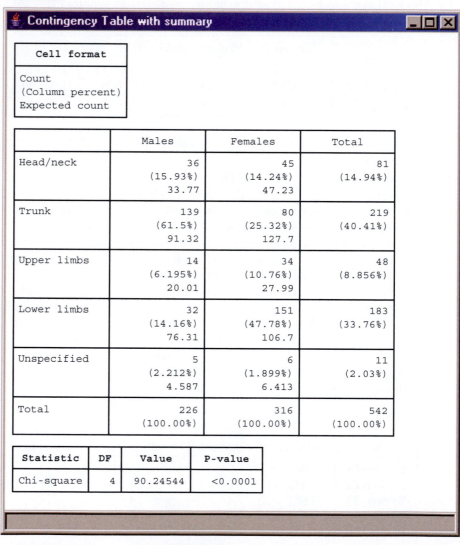

FIGURE 22.10   CrunchIt! output for the epidemiological data of Exercise 22.42.

22.43 Carcinogenicity of electromagnetic fields. The U.S. National Toxicology Program (NTP) studies the toxicity and carcinogenicity of agents potentially causing a risk to human health. Electromagnetic fields around electrical installations are harmless in theory, but they might have a long-term effect on health, for instance, in triggering tumors. Observational studies in human populations have suggested a potential association.

An NTP study looked at the occurrence of cancers in otherwise healthy rats after two years of daily exposure to 60-Hz electromagnetic fields of various intensities. Here are the numbers of male and female rats with a tumor after the two-year study period.[23] Note that 2 Gauss is approximately 1000-fold greater than what is considered high exposure for humans.

	Male rats	Female rats
Control: no exposure	16 (N = 99)	19 (N = 100)
0.02-Gauss exposure	31 (N = 100)	22 (N = 100)
2-Gauss exposure	30 (N = 100)	22 (N = 100)

In rats not exposed to an electromagnetic field (sample size in parentheses), is there significant evidence of a difference in the rates of tumors between males and females?

22.44 Carcinogenicity study, continued. Use the data from the previous exercise to answer the following questions:

(a) Is there significant evidence of a relationship between tumor rate and electromagnetic exposure in male rats?

(b) Is there significant evidence of a relationship between tumor rate and electromagnetic exposure in female rats?

(c) Write a brief summary of your findings from this and the previous exercise.

22.45 Kidney stones. A study compared the success rates of two different procedures for removing kidney stones: open surgery and percutaneous nephrolithotomy (PCNL), a minimally invasive technique. Here are the number of procedures that were successful or not at getting rid of patients' kidney stones for each type of procedure. A separate table is given for patients with small kidney stones and for patients with large stones.[24]

Small Stones	Open surgery	PCNL
Success	81	234
Failure	6	36

Large Stones	Open surgery	PCNL
Success	192	55
Failure	71	25

(a) Give the percent of successful procedures of each type for small kidney stones. Do the same for large kidney stones. What trend emerges?

(b) Is there a significant difference between the two methods for small stones? For large stones? Use either a chi-square or a z test.

PCNL performed worse for *both* small and large kidney stones, yet it did better overall. That sounds impossible. Explain carefully, referring to the data, how this paradox can happen.

22.46 Kidney stones, continued. Going back to the previous exercise, combine the data for small and large stones into a single 2-by-2 table. Now answer the following questions using the pooled data.

(a) Which procedure had the highest overall success rate? Is the difference significant?

Simpson's Paradox

(b) How do your conclusions in this exercise compare to the conclusions in the previous exercise? This is an example of "**Simpson's Paradox,**" when pooling categorical data that are not homogeneous can create a confounding variable and reverse the direction of an association (Chapter 5 describes this paradox in more detail). It is also a reminder that association does not imply causation.

Inference for Regression

When a scatterplot shows a linear relationship between a quantitative explanatory variable x and a quantitative response variable y, we can use the least-squares line fitted to the data to predict y for a given value of x. When the data are a sample from a larger population, we need statistical inference to answer questions like these about the population:

- Is there really a linear relationship between x and y in the population, or might the pattern we see in the scatterplot plausibly arise just by chance?

- How large is the slope (rate of change) that relates y to x in the population, including a margin of error for our estimate of the slope?

- If we use the least-squares line to predict y for a given value of x, how accurate is our prediction (again, with a margin of error)?

This chapter shows you how to answer these questions. Here is an example we will explore.

EXAMPLE 23.1   *Crying and IQ*

STATE: Infants who cry easily may be more easily stimulated than others. This may be a sign of higher IQ. Child development researchers explored the relationship between the crying of infants 4 to 10 days old and their later IQ test scores. A snap of a rubber band on the sole of the foot caused the infants to cry. The researchers recorded the crying and measured its intensity by the number of peaks in the most active 20 seconds. They later measured the children's IQ at 3 years of age using the Stanford-Binet IQ test.

TABLE 23.1 Infants' crying and IQ scores

Crying	IQ	Crying	IQ	Crying	IQ	Crying	IQ
10	87	20	90	17	94	12	94
12	97	16	100	19	103	12	103
9	103	23	103	13	104	14	106
16	106	27	108	18	109	10	109
18	109	15	112	18	112	23	113
15	114	21	114	16	118	9	119
12	119	12	120	19	120	16	124
20	132	15	133	22	135	31	135
16	136	17	141	30	155	22	157
33	159	13	162				

Rune Hellestad/Corbis

Table 23.1 contains data on 38 infants.[1] Do children with higher crying counts tend to have a higher IQ?

FORMULATE: Make a scatterplot. If the relationship appears linear, use correlation and regression to describe it. Finally, ask whether there is a *statistically significant* relationship between crying and IQ.

SOLVE (FIRST STEPS): Chapters 3 and 4 introduced the data analysis that must come before inference. The first steps we take are a review of this data analysis. Figure 23.1 is a *scatterplot* **scatterplot** of the crying data. Plot the explanatory variable (count of crying peaks) horizontally and the response variable (IQ) vertically. Look for the form, direction,

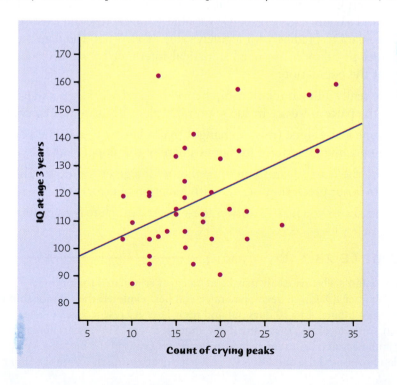

FIGURE 23.1 Scatterplot of the IQ score of infants at 3 years of age against the intensity of their crying soon after birth, with the least-squares regression line.

and strength of the relationship as well as for outliers or other deviations. There is a moderate positive linear relationship, with no extreme outliers or potentially influential observations.

Because the scatterplot shows a roughly linear (straight-line) pattern, the **correlation** describes the direction and strength of the relationship. The correlation between crying and IQ is $r = 0.455$. We are interested in predicting the response from information about the explanatory variable. So we find the **least-squares regression line** for predicting IQ from crying. The equation of the regression line is

$$\hat{y} = a + bx$$
$$= 91.27 + 1.493x$$

correlation

least-squares regression line

CONCLUDE (FIRST STEPS):   Children who cry more vigorously do tend to have higher IQs. Because $r^2 = 0.207$, only about 21% of the variation in IQ scores is explained by crying intensity. Prediction of IQ will not be very accurate. It is nonetheless impressive that behavior soon after birth can even partly predict IQ three years later. Is this observed relationship statistically significant? We must now develop tools for inference in the regression setting.

Conditions for regression inference

We can fit a regression line to *any* data relating two quantitative variables, though the results are useful only if the scatterplot shows a linear pattern. Statistical inference requires more detailed conditions. Because the conclusions of inference always concern some *population*, the conditions describe the population and how the data are produced from it. The slope b and intercept a of the least-squares line are *statistics*. That is, we calculated them from the sample data. These statistics would take somewhat different values if we repeated the study with different infants. To do inference, think of a and b as estimates of unknown *parameters* that describe the population of all infants.

CONDITIONS FOR REGRESSION INFERENCE

We have n observations on an explanatory variable x and a response variable y. Our goal is to study or predict the behavior of y for given values of x.

- For any fixed value of x, the response y varies according to a **Normal distribution.** Repeated responses y are **independent** of each other.

- The mean response μ_y has a **straight-line relationship** with x given by a **population regression line**

$$\mu_y = \alpha + \beta x$$

    The slope β and intercept α are unknown parameters.

- The **standard deviation** of y (call it σ) is the same for all values of x. The value of σ is unknown.

There are thus three population parameters that we must estimate from the data: α, β, and σ.

FIGURE 23.2 The nature of regression data when the conditions for inference are met. The line is the population regression line, which shows how the mean response μ_y changes as the explanatory variable x changes. For any fixed value of x, the observed response y varies according to a Normal distribution having mean μ_y and standard deviation σ.

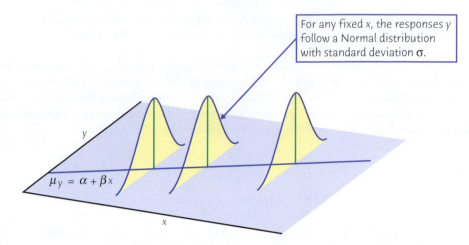

For any fixed x, the responses y follow a Normal distribution with standard deviation σ.

$$\mu_y = \alpha + \beta x$$

First of all, notice the important distinction between x and y. Unlike correlation, regression treats the two quantitative variables very differently. The distinction is particularly important for regression inference, because predictions are valid only for y and the conditions for inference all concern the response variable y.

These conditions say that in the population there is an "on average" straight-line relationship between y and x. The population regression line $\mu_y = \alpha + \beta x$ says that the *mean* response μ_y moves along a straight line as the explanatory variable x changes. We can't observe the population regression line. The values of y that we do observe vary about their means according to a Normal distribution. If we hold x fixed and take many observations on y, the Normal pattern will eventually appear in a stemplot or histogram. In practice, we observe y for many different values of x, so that we see an overall linear pattern formed by points scattered about the population line. The standard deviation σ determines whether the points fall close to the population regression line (small σ) or are widely scattered (large σ).

Figure 23.2 shows the nature of regression data in picture form. The line in the figure is the population regression line. The mean of the response y moves along this line as the explanatory variable x takes different values. The Normal curves show how y will vary when x is held fixed at different values. All of the curves have the same σ, so the variability of y is the same for all values of x. You should check the conditions for inference when you do inference about regression. We will see later how to do that.

Estimating the parameters

The first step in inference is to estimate the unknown parameters α, β, and σ.

ESTIMATING THE POPULATION REGRESSION LINE

When the conditions for regression are met and we calculate the least-squares line $\hat{y} = a + bx$, the slope b of the least-squares line is an unbiased estimator of the population slope β and the intercept a of the least-squares line is an unbiased estimator of the population intercept α.

EXAMPLE 23.2 Crying and IQ: slope and intercept

The data in Figure 23.1 satisfy the condition of scatter about an invisible population regression line reasonably well. The least-squares line is $\hat{y} = 91.27 + 1.493x$. The slope is particularly important. A *slope is a rate of change*. The population slope β says how much higher average IQ is for children with one more peak in their crying measurement. Because $b = 1.493$ estimates the unknown β, we estimate that, on average, IQ is about 1.5 points higher for each added crying peak.

We need the intercept $a = 91.27$ to draw the line, but it has no statistical meaning in this example. No child had fewer than 9 crying peaks, so we have no data near $x = 0$. We suspect that all healthy children would cry when snapped with a rubber band, so that we will never observe $x = 0$.

The remaining parameter is the standard deviation σ, which describes the variability of the response y about the population regression line. The least-squares line estimates the population regression line. So the **residuals** estimate how much *residuals* y varies about the population line. Recall that the residuals are the vertical deviations of the data points from the least-squares line:

$$\text{residual} = \text{observed } y - \text{ predicted } y$$
$$= y - \hat{y}$$

There are n residuals, one for each data point. Because σ is the standard deviation of responses about the population regression line, we estimate it by a sample standard deviation of the residuals. We call this sample standard deviation the *regression standard error* to emphasize that it is estimated from data. The residuals from a least-squares line always have a mean of zero. That simplifies their standard error.

REGRESSION STANDARD ERROR

The **regression standard error** is

$$s = \sqrt{\frac{1}{n-2}\sum \text{residual}^2} = \sqrt{\frac{1}{n-2}\sum (y - \hat{y})^2}$$

Use s to estimate the standard deviation σ of responses about the mean given by the population regression line.

Because we use the regression standard error so often, we just call it s. Notice that s^2 is an average of the squared deviations of the data points from the line, so it qualifies as a variance. We average the squared deviations by dividing by $n - 2$, 2 less than the number of data points. It turns out that if we know s and $n - 2$ of the n residuals, the other two residuals are determined. That is, $n - 2$ are the **degrees of freedom** of s. We first encountered the idea of degrees of freedom in *degrees of freedom* the case of the ordinary sample standard deviation of n observations, which has $n - 1$ degrees of freedom. Now we observe two variables rather than one, and the proper degrees of freedom are $n - 2$ rather than $n - 1$.

Calculating s is unpleasant. You must find the predicted response for each x in your data set, then the residuals, and then s. In practice, you will use software that does this arithmetic instantly. Nonetheless, here is an example to help you understand the standard error s.

EXAMPLE 23.3   Crying and IQ: residuals and standard error ──────

Table 23.1 shows that the first infant studied had 10 crying peaks and a later IQ of 87. The predicted IQ for $x = 10$ is

$$\hat{y} = 91.27 + 1.493x$$
$$= 91.27 + 1.493(10) = 106.2$$

The residual for this observation is

$$\text{residual} = y - \hat{y}$$
$$= 87 - 106.2 = -19.2$$

That is, the observed IQ for this infant lies 19.2 points below the least-squares line on the scatterplot.

Repeat this calculation 37 more times, once for each subject. The 38 residuals are

−19.20	−31.13	−22.65	−15.18	−12.18	−15.15	−16.63	−6.18
−1.70	−22.60	−6.68	−6.17	−9.15	−23.58	−9.14	2.80
−9.14	−1.66	−6.14	−12.60	0.34	−8.62	2.85	14.30
9.82	10.82	0.37	8.85	10.87	19.34	10.89	−2.55
20.85	24.35	18.94	32.89	18.47	51.32		

Check the calculations by verifying that the sum of the residuals is zero. It is 0.04, not quite zero, because of roundoff error. Another reason to use software in regression is that roundoff errors in hand calculation can accumulate to make the results inaccurate.

The variance about the line is

$$s^2 = \frac{1}{n-2} \sum \text{residual}^2$$

$$= \frac{1}{38-2}[(-19.20)^2 + (-31.13)^2 + \cdots + (51.32)^2]$$

$$= \frac{1}{36}(1,023.3) = 306.20$$

Finally, the regression standard error is

$$s = \sqrt{306.20} = 17.50$$

We will study several kinds of inference in the regression setting. The regression standard error s is the key measure of the variability of the responses in regression. It is part of the standard error of all the statistics we will use for inference.

APPLY YOUR KNOWLEDGE ──────────────────

23.1   Coffee and deforestation. Coffee is a leading export from several developing countries. When coffee prices are high, farmers often clear forest to plant more coffee trees. Here are 5 years' data on prices paid to coffee growers in Indonesia and the percent of forest area lost in a national park that lies in a coffee-producing region.[2]

Price (cents per pound)	29	40	54	55	72
Forest lost (percent)	0.49	1.59	1.69	1.82	3.10

(a) Examine the data. Make a scatterplot with coffee price as the explanatory variable. What are the correlation r and the equation of the least-squares regression line? Do you think that coffee price will allow good prediction of forest lost?

(b) Explain in words what the slope β of the population regression line would tell us if we knew it. Based on the data, what are the estimates of β and the intercept α of the population regression line?

(c) Calculate by hand the residuals for the five data points. Check that their sum is 0 (aside from roundoff error). Use the residuals to estimate the standard deviation σ of percents of forest lost about the means given by the population regression line. You have now estimated all three parameters.

Susan E. Degginger/Alamy

Using technology

Basic "two-variable statistics" calculators will find the slope b and intercept a of the least-squares line from keyed-in data. Inference about regression requires, in addition, the regression standard error s. At this point, software or a graphing calculator that includes procedures for regression inference becomes almost essential for practical work.

Figure 23.3 shows regression output for the data of Table 23.1 from a graphing calculator, three statistical programs, and a spreadsheet program. When we entered the data into the programs, we called the explanatory variable "Crycount." The outputs use that label. The TI-83 just uses "x" and "y" to label the explanatory and response variables. You can locate the basic information in all of the outputs. The regression slope is $b = 1.4929$, and the regression intercept is $a = 91.268$. The equation of the least-squares line is therefore (after rounding) just as given in Example 23.1. The regression standard error is $s = 17.4987$, and the squared correlation is $r^2 = 0.207$. Both of these results reflect the rather wide scatter of the points in Figure 23.1 about the least-squares line.

Each output contains other information, some of which we will need shortly and some of which we don't need. In fact, we left out some output to save space. Once you know what to look for, you can find what you want in almost any output and ignore what doesn't interest you.

APPLY YOUR KNOWLEDGE

23.2   Frog mating calls. *Hyla chrysoscelis* is a type of gray tree frog with a distinct mating call. Frogs are cold-blooded animals, and their physiology is affected by variations in temperature. Field biologists wanted to know if the mating song of *H. chrysoscelis* is affected by the temperature of its natural habitat. Here is the mating song frequency as a function of habitat temperature for 20 *H. chrysoscelis* in the wild.[3]

TI-83

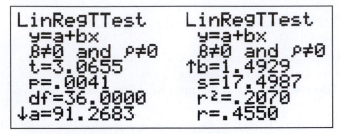

CrunchIt!

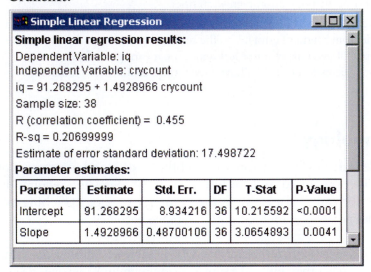

Minitab

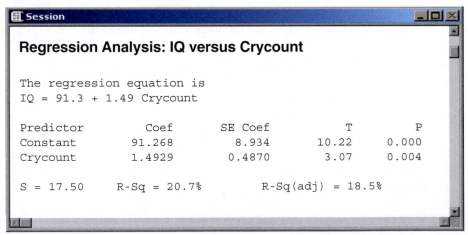

FIGURE 23.3 Regression of IQ on crying peaks: output from a graphing calculator, three statistical programs, and a spreadsheet program (*continued*).

Excel

	A	B	C	D	E	F	G
1	SUMMARY OUTPUT						
2							
3	*Regression statistics*						
4	Multiple R	0.4550					
5	R Square	0.2070					
6	Adjusted R Square	0.1850					
7	Standard Error	17.4987					
8	Observations	38					
9							
10		*Coefficients*	*Standard Error*	*t Stat*	*P-value*	*Lower 95%*	*Upper 95%*
11	Intercept	91.2683	8.9342	10.2156	3.5E-12	73.1489	109.3877
12	Crycount	1.4929	0.4870	3.0655	0.004105	0.5052	2.4806
13							

SPSS

Column Statistics

Coefficients[a]

Model	Unstandardized Coefficients B	Std. Error	Standardized Coefficients Beta	t	Sig.	95% Confidence Interval for B Lower Bound	Upper Bound
1 (Constant)	91.268	8.934		10.216	.000	73.149	109.388
Crycount	1.493	.487	.455	3.065	.004	.505	2.481

a. Dependent Variable: IQscore

FIGURE 23.3 *(continued).*

Temperature (Celsius)	19	21	22	22	23	23	23	23	23	24
Frequency (notes/sec)	38	42	45	45	41	45	48	50	53	51
Temperature (Celsius)	24	24	24	25	25	25	25	26	26	27
Frequency (notes/sec)	48	53	47	53	49	56	53	55	55	54

We want to predict song frequency from habitat temperature. Figure 23.4 shows Minitab regression output for these data.

(a) Make a scatterplot suitable for predicting frequency from temperature. The pattern is linear. What is the squared correlation r^2? Temperature explains a lot of the variations in song frequency.

(b) For regression inference, we must estimate the three parameters α, β, and σ. From the output, what are the estimates of these parameters?

(c) What is the equation of the least-squares regression line of frequency on temperature? Add this line to your plot. We will continue the analysis of these data in later exercises.

23.3 Great Arctic rivers. One effect of global warming is to increase the flow of water into the Arctic Ocean from rivers. Such an increase may have major effects

Morguefile.com

Minitab

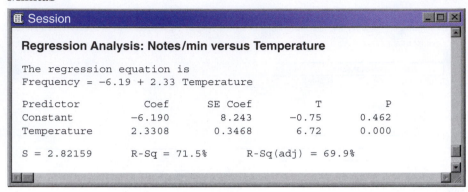

FIGURE 23.4 Minitab output for the frog mating call data, for Exercise 23.2.

on the world's climate. Six rivers (Yenisey, Lena, Ob, Pechora, Kolyma, and Severnaya Dvina) drain two-thirds of the Arctic in Europe and Asia. Several of these are among the largest rivers on earth. Table 23.2 presents the total discharge from these rivers each year from 1936 to 1999.[4] Discharge is measured in cubic kilometers of water. Use software to analyze these data.

(a) Make a scatterplot of river discharge against time. Is there a clear increasing trend? Using technology, find r^2 and briefly interpret its value. There is considerable year-to-year variation, so we wonder if the trend is statistically significant.

(b) As a first step, find the least-squares line and draw it on your plot. Then find the regression standard error s, which measures scatter about this line. We will continue the analysis in later exercises.

TABLE 23.2 Arctic river discharge (cubic kilometers), 1936 to 1999

Year	Discharge	Year	Discharge	Year	Discharge	Year	Discharge
1936	1721	1952	1829	1968	1713	1984	1823
1937	1713	1953	1652	1969	1742	1985	1822
1938	1860	1954	1589	1970	1751	1986	1860
1939	1739	1955	1656	1971	1879	1987	1732
1940	1615	1956	1721	1972	1736	1988	1906
1941	1838	1957	1762	1973	1861	1989	1932
1942	1762	1958	1936	1974	2000	1990	1861
1943	1709	1959	1906	1975	1928	1991	1801
1944	1921	1960	1736	1976	1653	1992	1793
1945	1581	1961	1970	1977	1698	1993	1845
1946	1834	1962	1849	1978	2008	1994	1902
1947	1890	1963	1774	1979	1970	1995	1842
1948	1898	1964	1606	1980	1758	1996	1849
1949	1958	1965	1735	1981	1774	1997	2007
1950	1830	1966	1883	1982	1728	1998	1903
1951	1864	1967	1642	1983	1920	1999	1970

Testing the hypothesis of no linear relationship

Example 23.1 asked, "Do children with higher crying counts tend to have higher IQs?" Data analysis supports this conjecture. But is the positive association statistically significant? That is, is it too strong to have occurred just by chance? To answer this question, test hypotheses about the slope β of the population regression line:

$$H_0: \beta = 0$$
$$H_a: \beta > 0$$

A regression line with slope 0 is horizontal. That is, the mean of y does not change at all when x changes. So H_0 says that there is *no linear relationship* between x and y in the population. Put another way, H_0 says that *linear regression of y on x is of no value for predicting y.*

The test statistic is just the standardized version of the least-squares slope b, using the hypothesized value $\beta = 0$ for the mean of b. It is another t statistic. Here are the details.

SIGNIFICANCE TEST FOR REGRESSION SLOPE

To **test the hypothesis H_0: $\beta = 0$,** compute the t statistic

$$t = \frac{b}{SE_b}$$

In this formula, the standard error of the least-squares slope b is

$$SE_b = \frac{s}{\sqrt{\sum(x - \bar{x})^2}}$$

The sum runs over all observations on the explanatory variable x. In terms of a random variable T having the $t(n - 2)$ distribution, the P-value for a test of H_0 against

$H_a: \beta > 0$ is $P(T \geq t)$

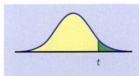

$H_a: \beta < 0$ is $P(T \leq t)$

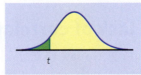

$H_a: \beta \neq 0$ is $2P(T \geq |t|)$

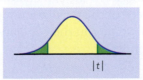

The standard error of b is a multiple of the regression standard error s. The degrees of freedom $n - 2$ are the degrees of freedom of s. Although we give the formula for this standard error, you should not try to calculate it by hand. Regression software gives the standard error SE_b along with b itself.

EXAMPLE 23.4 Crying and IQ: is the relationship significant?

The hypothesis $H_0: \beta = 0$ says that crying has no straight-line relationship with IQ. We conjecture that there is a positive relationship, so we use the one-sided alternative $H_a: \beta > 0$.

Figure 23.1 shows that there is a positive relationship, so it is not surprising that all of the outputs in Figure 23.3 give $t = 3.07$ with two-sided P-value 0.004. The P-value for the one-sided test is half of this, $P = 0.002$. There is very strong evidence that IQ increases as the intensity of crying increases. Remember, however, that strong statistical significance does not imply a strong effect, only that the observed effect is highly unlikely to have arisen just by chance because of the random sampling process.

APPLY YOUR KNOWLEDGE

23.4 **Coffee and deforestation: testing.** Exercise 23.1 presents data on coffee prices and loss of forest in Indonesia. In that exercise, you estimated the parameters using only a two-variable statistics calculator. Software tells us that the least-squares slope is $b = 0.0543$ with standard error $SE_b = 0.0097$.

(a) What is the t statistic for testing $H_0: \beta = 0$?

(b) How many degrees of freedom does t have? Use Table C to approximate the P-value of t against the one-sided alternative $H_a: \beta > 0$. What do you conclude?

23.5 **Great Arctic rivers: testing.** The most important question we ask of the data in Table 23.2 is this: Is the increasing trend visible in your plot (Exercise 23.3) statistically significant? If so, changes in the Arctic may already be affecting the earth's climate. Use software to answer this question. Give a test statistic, its P-value, and the conclusion you draw from the test.

23.6 **Does fast driving waste fuel?** Exercise 3.6 (page 72) gives data on the fuel consumption of a small car at various speeds from 10 to 150 kilometers per hour. Is there significant evidence of straight-line dependence between speed and fuel use? Make a scatterplot and use it to explain the result of your test.

Testing lack of correlation*

Back in Chapter 4, we saw that the least-squares slope b is closely related to the correlation r between the explanatory and response variables x and y. In the same way, the slope β of the population regression line is closely related to the correlation between x and y in the population. In particular, the slope is 0 exactly when the correlation is 0.

---

*This section is optional.

Testing the null hypothesis H_0: $\beta = 0$ is therefore exactly the same as testing that there is *no correlation* between x and y in the population from which we drew our data. You can use the test for zero slope to test the hypothesis of zero correlation between any two quantitative variables. That's a useful trick.

Because correlation also makes sense when there is no explanatory-response distinction, it is handy to be able to test correlation without doing regression. The statistic r follows a particular distribution with $n - 2$ degrees of freedom. We don't cover the specifics of this distribution here, but Table E in the back of the book gives critical values of the sample correlation r under the null hypothesis that the correlation is 0 in the population. Use this table when *both variables* have at least approximately Normal distributions or when the sample size is large.

EXAMPLE 23.5   Hand and body lengths

STATE: While the human body has overall proportions undeniably characteristic of the species, individuals do vary to some extent in size and body shape. Is there a relationship, for instance, between a person's height and hand size? Figure 23.5 displays the body length (in meters [m]) and hand length (in millimeters [mm]) of 21 healthy adult males.[5] The relationship, if any, appears rather weak.

   Neither body length nor hand length can be considered a possible explanatory variable. Rather, genes and environmental factors influence both variables and would be responsible for any relationship that may exist between body and hand lengths. Therefore, we want to test whether body length and hand length are significantly *correlated* in men.

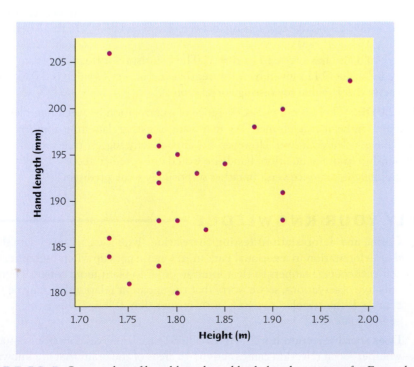

FIGURE 23.5 Scatterplot of hand length and body height in men, for Example 23.5.

FORMULATE: Because we would not realistically anticipate a negative correlation, we test the hypotheses

$$H_0: \text{population correlation} = 0$$
$$H_a: \text{population correlation} > 0$$

SOLVE: The data are displayed in the table below. Stemplots (we don't display them) show that both variables are slightly right-skewed, without outliers, and Normal quantile plots indicate that the deviations from Normality are minor. With 21 observations, a test of hypotheses is safe.

Body length (m)	Hand length (mm)	Body length (m)	Hand length (mm)	Body length (m)	Hand length (mm)
1.75	181	1.80	195	1.82	193
1.91	188	1.78	188	1.73	186
1.73	184	1.80	188	1.85	194
1.80	180	1.83	187	1.77	197
1.78	183	1.91	191	1.98	203
1.88	198	1.78	196	1.91	200
1.78	192	1.78	193	1.73	206

The correlation coefficient for these data is $r = 0.356$. Compare this value to the critical values in the $n = 19$ row of Table E. The closest critical values in that row are 0.3077 and 0.3887, for tail areas 0.10 and 0.05, respectively. H_0 is one-sided, because we clearly wouldn't expect a negative relationship between hand and body lengths. Therefore, the P-value is $0.01 > P > 0.05$.

We could also have tested $H_0: \beta = 0$. The t statistic for this test is $t = 1.66$ with two-sided P-value 0.113 given by software; that is, our test P-value is 0.0565. **This is also exactly the P-value for testing correlation 0.**

CONCLUDE: There is only weak evidence of a correlation between hand length and body length in healthy adult men. We may want to gather data from a larger sample of men to have a clearer answer. However, the data already suggest that even if there was a relationship in the population, hand size would only predict about 13% of variations in overall height, and vice versa (a rather surprisingly weak predictor).

APPLY YOUR KNOWLEDGE

23.7 **Coffee and deforestation: testing correlation.** Exercise 23.1 gives data showing that deforestation in a national park in Indonesia goes up when high prices for coffee encourage farmers to clear forest in order to plant more coffee. There are only five observations, so we worry that the apparent relationship may be just chance. Is the correlation significantly greater than 0? Use Table E to approximate the P-value.

23.8 **Does social rejection hurt?** Exercise 3.38 (page 91) gives data from a study of whether social rejection causes activity in areas of the brain that are known to be activated by physical pain. The explanatory variable is a subject's score on a test of "social distress" after being excluded from an activity. The response variable is activity in an area of the brain that responds to physical pain. Your scatterplot shows a positive linear relationship. The research report gives the correlation r and the P-value for a test of whether r is greater than 0. What are r and the P-value? (You can use Table E, or you can get more accurate P-values for the correlation from regression software.) What do you conclude about the relationship?

Confidence intervals for the regression slope

The slope β of the population regression line is usually the most important parameter in a regression problem. The slope is the rate of change of the mean response as the explanatory variable increases. We often want to estimate β. The slope b of the least-squares line is an unbiased estimator of β. A confidence interval is more useful, because it shows how accurate the estimate b is likely to be. The confidence interval for β has the familiar form

$$\text{estimate} \pm t^*\text{SE}_{\text{estimate}}$$

Because b is our estimate, the confidence interval is $b \pm t^*\text{SE}_b$. Here are the details.

CONFIDENCE INTERVAL FOR REGRESSION SLOPE

A level C **confidence interval for the slope β** of the population regression line is

$$b \pm t^*\text{SE}_b$$

Here, t^* is the critical value for the $t(n-2)$ density curve with area C between $-t^*$ and t^*.

EXAMPLE 23.6 *Crying and IQ: estimating the slope*

The four software outputs in Figure 23.3 give the slope $b = 1.4929$ and also the standard error $\text{SE}_b = 0.4870$. The outputs use a similar arrangement, a table in which each regression coefficient is followed by its standard error. Excel and SPSS also give the lower and upper endpoints of the 95% confidence interval for the population slope β, 0.505 and 2.481.

Once we know b and SE_b, it is easy to find the confidence interval. There are 38 data points, so the degrees of freedom are $n - 2 = 36$. Because Table C does not have a row for df $= 36$, we must use either software or the next smaller degrees of freedom in the table, df $= 30$. Using Table C, we get an approximate value for t^* of 2.042. Using Excel, enter for 95% confidence the corresponding two-tailed area 0.05 and 36 degrees of freedom in the command =TINV(0.05,36) which returns a value of 2.02809 for t^*.

The 95% confidence interval for the population slope β is

$$b \pm t^*SE_b = 1.4929 \pm (2.02809)(0.4870)$$
$$= 1.4929 \pm 0.9877$$
$$= 0.505 \text{ to } 2.481$$

This agrees with Excel and SPSS in Figure 23.3. We are 95% confident that mean IQ increases by between about 0.5 and 2.5 points for each additional peak in crying.

You can find a confidence interval for the intercept α of the population regression line in the same way, using a and SE_a from the "Constant" line of the Minitab output or the "Intercept" line in CrunchIt! or Excel. However, we rarely need to estimate α.

APPLY YOUR KNOWLEDGE

23.9 **Coffee and deforestation: estimating slope.** Exercise 23.1 presents data on coffee prices and loss of forest in Indonesia. Software tells us that the least-squares slope is $b = 0.0543$ with standard error $SE_b = 0.0097$. Because there are only 5 observations, the observed slope b may not be an accurate estimate of the population slope β. Give a 95% confidence interval for β.

23.10 **Frog mating calls: estimating slope.** Exercise 23.2 gives data on the mating song frequency of *H. chrysoscelis* as a function of habitat temperature. We want a 95% confidence interval for the slope of the population regression line. Starting from the information in the Minitab output in Figure 23.4, find this interval. Say in words what the slope of the population regression line tells us about the frequency of this species' mating song in the wild under various temperatures.

23.11 **Great Arctic rivers: estimating slope.** Use the data in Table 23.2 to give a 90% confidence interval for the slope of the population regression of Arctic river discharge on year. Does this interval convince you that discharge is actually increasing over time? Explain your answer.

Inference about prediction

One of the most common reasons to fit a line to data is to predict the response to a particular value of the explanatory variable. This is another setting for regression inference: We want not simply a prediction, but a prediction with a margin of error that describes how accurate the prediction is likely to be.

EXAMPLE 23.7 Beer and blood alcohol

STATE: The EESEE story "Blood Alcohol Content" describes a study in which 16 student volunteers at The Ohio State University drank a randomly assigned number of cans of beer. Thirty minutes later, a police officer measured their blood alcohol content (BAC) in grams of alcohol per deciliter of blood. Here are the data.

Student	1	2	3	4	5	6	7	8
Beers	5	2	9	8	3	7	3	5
BAC	0.10	0.03	0.19	0.12	0.04	0.095	0.07	0.06
Student	9	10	11	12	13	14	15	16
Beers	3	5	4	6	5	7	1	4
BAC	0.02	0.05	0.07	0.10	0.085	0.09	0.01	0.05

Jame Shaffer/The Image Works

The students were equally divided between men and women and differed in weight and usual drinking habits. Many students don't believe that number of drinks predicts BAC well. Steve did not participate in the study, but he thinks he can drive legally 30 minutes after he finishes drinking 5 beers. The legal limit for driving is a BAC of 0.08 in all states. We want to predict Steve's blood alcohol content, using no information except that he drinks 5 beers.

FORMULATE: Regress BAC on number of beers. Use the regression line to predict Steve's BAC. Give a margin of error that allows us to have 95% confidence in our prediction.

SOLVE: The scatterplot in Figure 23.6 and the regression output in Figure 23.7 show that student opinion is wrong: Number of beers predicts blood alcohol content quite well. In fact, $r^2 = 0.80$, so number of beers explains 80% of the observed variation in BAC. To predict Steve's BAC after 5 beers, use the equation of the regression line:

$$\hat{y} = -0.0127 + 0.0180x$$
$$= -0.0127 + 0.0180(5) = 0.077$$

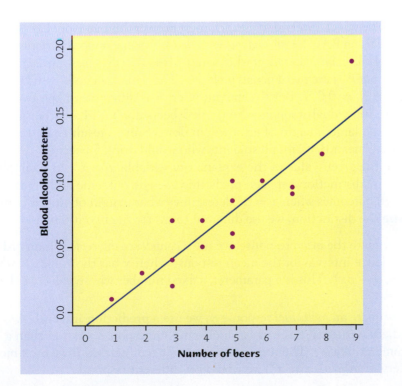

FIGURE 23.6 Scatterplot of students' blood alcohol content against the number of cans of beer consumed, with the least-squares regression line.

CrunchIt!

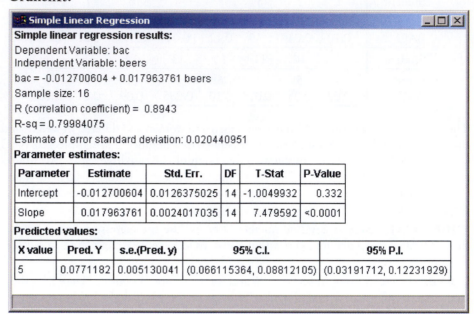

Simple Linear Regression

Simple linear regression results:

Dependent Variable: bac
Independent Variable: beers

bac = -0.012700604 + 0.017963761 beers

Sample size: 16

R (correlation coefficient) = 0.8943

R-sq = 0.79984075

Estimate of error standard deviation: 0.020440951

Parameter estimates:

Parameter	Estimate	Std. Err.	DF	T-Stat	P-Value
Intercept	-0.012700604	0.0126375025	14	-1.0049932	0.332
Slope	0.017963761	0.0024017035	14	7.479592	<0.0001

Predicted values:

X value	Pred. Y	s.e.(Pred. y)	95% C.I.	95% P.I.
5	0.0771182	0.005130041	(0.066115364, 0.08812105)	(0.03191712, 0.12231929)

FIGURE 23.7 CrunchIt! regression output for the blood alcohol content data for Example 23.7.

That's dangerously close to the legal limit 0.08. What about 95% confidence? The "Predicted values" part of the output in Figure 23.7 shows *two* 95% intervals. Which should we use?

To decide which interval to use, you must answer this question: Do you want to predict the *mean* BAC for *all students* who drink 5 beers, or do you want to predict the BAC of *one individual student* who drinks 5 beers? *Both of these predictions may be interesting, but they are two different problems.* The actual, sample-based prediction is the same, $\hat{y} = 0.077$. But the margin of error is different for the two kinds of prediction. Individual students who drink 5 beers don't all have the same BAC. So we need a larger margin of error to pin down Steve's result than we would to estimate the mean BAC for all students who would drink 5 beers.

Write the given value of the explanatory variable x as x^*. In Example 23.7, $x^* = 5$. The distinction between predicting a single outcome and predicting the mean of all outcomes when $x = x^*$ determines what margin of error is correct. To emphasize the distinction, we use different terms for the two intervals.

- To estimate the *mean* response, we use a *confidence interval*. It is an ordinary confidence interval for the mean response when x has the value x^*, which is $\mu_y = \alpha + \beta x^*$. This is a parameter, a fixed number whose value we don't know.

prediction interval
- To estimate an *individual* response y, we use a **prediction interval.** A prediction interval estimates a single random response y rather than a parameter like μ_y. The response y is not a fixed number. If we took more observations with $x = x^*$, we would get different responses.

EXAMPLE 23.8 Beer and blood alcohol: conclusion

Steve is one individual, so we must use the prediction interval. The output in Figure 23.7 helpfully labels the confidence interval as "C.I." and the prediction interval as "P.I." We are 95% confident that Steve's BAC after 5 beers will lie between 0.032 and 0.122. The upper part of that range will get him arrested if he drives. The 95% confidence interval for the mean BAC of all students who drink 5 beers is much narrower, 0.066 to 0.088.

The meaning of a prediction interval is very much like the meaning of a confidence interval. A 95% prediction interval, like a 95% confidence interval, is right 95% of the time in repeated use. "Repeated use" now means that we take an observation on y for each of the n values of x in the original data, and then take one more observation y, with $x = x^*$. Form the prediction interval from the n observations, and then see if it covers the one more y. It will in 95% of all repetitions.

The interpretation of prediction intervals is a minor point. The main point is that it is harder to predict one response than to predict a mean response. Both intervals have the usual form

$$\hat{y} \pm t^* \text{SE}$$

but the prediction interval is wider than the confidence interval, because individuals are more variable than averages. You will rarely need to know the details, because software automates the calculation, but here they are.

CONFIDENCE AND PREDICTION INTERVALS FOR REGRESSION RESPONSE

A level C **confidence interval for the mean response μ_y** when x takes the value x^* is

$$\hat{y} \pm t^* \text{SE}_{\hat{\mu}}$$

The standard error $\text{SE}_{\hat{\mu}}$ is

$$\text{SE}_{\hat{\mu}} = s \sqrt{\frac{1}{n} + \frac{(x^* - \bar{x})^2}{\sum (x - \bar{x})^2}}$$

A level C **prediction interval for a single observation y** when x takes the value x^* is

$$\hat{y} \pm t^* \text{SE}_{\hat{y}}$$

The standard error for prediction $\text{SE}_{\hat{y}}$ is

$$\text{SE}_{\hat{y}} = s \sqrt{1 + \frac{1}{n} + \frac{(x^* - \bar{x})^2}{\sum (x - \bar{x})^2}}$$

In both intervals, t^* is the critical value for the $t(n-2)$ density curve with area C between $-t^*$ and t^*.

Is regression garbage?

No—but garbage can be the setting for regression. The Census Bureau once asked if weighing a neighborhood's garbage would help count its people. So 63 households had their garbage sorted and weighed. It turned out that pounds of plastic in the trash gave the best garbage prediction of the number of people in a neighborhood. The margin of error for a 95% prediction interval in a neighborhood of about 100 households, based on 5 weeks' worth of garbage, was about ±2.5 people. Alas, that is not accurate enough to help the Census Bureau.

CrunchIt!

Simple Linear Regression						
Parameter estimates:						

Parameter	Estimate	Std. Err.	DF	T-Stat	P-Value
Intercept	-0.9763527	0.50608647	3	-1.9292212	0.1493
Slope	0.054287054	0.0097159175	3	5.587435	0.0113

Predicted values:

X value	Pred.Y	s.e.(Pred.y)	95% C.I	95% P.I.
60	2.2808704	0.17194703	(1.7336583, 2.8280828)	(1.1325622, 3.429179)

FIGURE 23.8 Partial CrunchIt! output for the regression of percent of forest lost on coffee price, for Exercise 23.12.

There are two standard errors: $SE_{\hat{\mu}}$ for estimating the mean response μ_y and $SE_{\hat{y}}$ for predicting an individual response y. The only difference between the two standard errors is the extra 1 under the square root sign in the standard error for prediction. The extra 1 makes the prediction interval wider. Both standard errors are multiples of the regression standard error s. The degrees of freedom are again $n - 2$, the degrees of freedom of s.

APPLY YOUR KNOWLEDGE

23.12 Coffee and deforestation: prediction. Exercise 23.1 presents data on coffee prices and loss of forest in Indonesia. If the world coffee price next year is 60 cents per pound, what percent of the national park forest do you predict will be cleared? Figure 23.8 is part of the output from CrunchIt! for prediction when $x^* = 60$.

(a) Which interval in the output is the proper 95% interval for predicting next year's loss of forest?

(b) CrunchIt! gives only one of the two standard errors used in prediction. It is $SE_{\hat{\mu}}$, the standard error for estimating the mean response. Use this fact along with the CrunchIt! output to give a 90% confidence interval for the mean percent of forest lost in years when the coffee price is 60 cents per pound.

23.13 Frog mating calls: prediction. Analysis of the data in Exercise 23.2 shows that mating call frequency in *H. chrysoscelis* varies linearly with habitat temperature. We might want to predict the mean call frequency (in notes per second) at 24 degrees Celsius in the wild. Here is the Minitab output for prediction when $x^* = 24$ degrees Celsius (labeled "New Obs 1" in the output):

Minitab

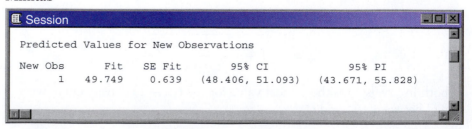

Session				
Predicted Values for New Observations				
New Obs	Fit	SE Fit	95% CI	95% PI
1	49.749	0.639	(48.406, 51.093)	(43.671, 55.828)

(a) Use the regression line from Figure 23.4 to verify that "Fit" is the predicted value for $x^* = 24$. (Start with the results in the "Coef" column of Figure 23.4 to reduce roundoff error.)

(b) What is the 95% interval we want?

Checking the conditions for inference

You can fit a least-squares line to any set of explanatory-response data when both variables are quantitative and show a roughly linear pattern. The regression line can be used to model sample data or even population data (for example, census data). To use regression inference, however, the data must satisfy additional conditions. *Before we can trust the results of inference, we must check the conditions for inference one by one.* There are ways to deal with violations of any of the conditions. If you see a clear violation, get expert advice.

Although the conditions for regression inference are a bit elaborate, it is not hard to check for gross violations. The residuals are a great help. Most regression software will calculate and save the residuals for you. Let's look at each condition in turn.

The observations are independent. In particular, repeated observations on the same individual are not allowed. We should not use ordinary regression to make inferences about the growth of a single child over time, for example.

The relationship is linear in the population. We can't observe the population regression line, so we will almost never see a perfect straight-line relationship in our data. Look at the scatterplot to check that the overall pattern is roughly linear. A plot of the residuals against x magnifies any unusual pattern. Draw a horizontal line at zero on the residual plot to orient your eye. Because the sum of the residuals is always zero, zero is also the mean of the residuals.

The standard deviation of the response about the population line is the same everywhere. Look at the scatterplot again. The scatter of the data points above and below the line should be roughly the same over the entire range of the data. A plot of the residuals against x, with a horizontal line at zero, makes this easier to check. You will sometimes find that as the response y gets larger, so does the scatter of the points about the fitted line. This pattern indicates that rather than remaining fixed, the standard deviation σ about the line is changing with x as the mean response changes with x. When this happens, you cannot trust the results of inference, because there is no fixed σ for s to estimate.

The response varies Normally about the population regression line. We can't observe the population regression line. We can observe the least-squares line and the residuals, which show the variation of the response about the fitted line. The residuals estimate the deviations of the response from the population regression line, so they should follow a Normal distribution. Make a histogram or a stemplot of the residuals and check for clear skewness or other major departures from Normality. You can also assess Normality by making a Normal quantile plot using the residuals. Like other t procedures, inference for regression is (with one exception) not very sensitive to minor lack of Normality, especially when we have many observations. Do beware of influential observations, which move the regression line and can greatly affect the results of inference.

The exception is the prediction interval for a single response y. This interval relies on Normality of individual observations, not just on the approximate Normality of statistics like the slope a and intercept b of the least-squares line. The statistics a and b become more Normal as we take more observations. This contributes to the robustness of regression inference, but it isn't enough for the prediction interval. We will not study methods that carefully check Normality of the residuals, so *you should regard prediction intervals as rough approximations*.

EXAMPLE 23.9 Climate change chases fish north

STATE: As the climate grows warmer, we expect many animal species to move toward the poles in an attempt to maintain their preferred temperature range. Do data on fish in the North Sea confirm this expectation? Here are data for 25 years, 1977 through 2001, on mean winter temperatures at the bottom of the North Sea (degrees Celsius) and the center of the distribution of anglerfish in degrees of north latitude.[6]

Temperature	6.26	6.26	6.27	6.31	6.34	6.32	6.37	6.39	6.42
Latitude	57.20	57.96	57.65	57.59	58.01	59.06	56.85	56.87	57.43
Temperature	6.52	6.68	6.76	6.78	6.89	6.90	6.93	6.98	7.02
Latitude	57.72	57.83	57.87	57.48	58.13	58.52	58.48	57.89	58.71
Temperature	7.09	7.13	7.15	7.29	7.34	7.57	7.65		
Latitude	58.07	58.49	58.28	58.49	58.01	58.57	58.90		

FORMULATE: Regress latitude on temperature. Look for a positive linear relationship and assess its significance. Be sure to check the conditions for regression inference.

SOLVE: The scatterplot in Figure 23.9 shows a clear positive linear relationship. The solid line in the plot is the least-squares regression line of the center of the fish distribution (north latitude) on winter ocean temperature. Software shows that the slope is $b = 0.818$. That is, each degree of ocean warming moves the fish about 0.8 degrees of latitude farther north. The t statistic for testing H_0: $\beta = 0$ is $t = 3.6287$ with one-sided P-value $P = 0.0007$ and $r^2 = 0.364$. There is very strong evidence that the population slope is positive, $\beta > 0$.

CONCLUDE: The data give highly significant evidence that anglerfish have moved north as the ocean has grown warmer. Before relying on this conclusion, we must check the conditions for inference.

The software that did the regression calculations also finds the 25 residuals. In the same order as the observations in Example 23.9, they are

```
-0.3731    0.3869    0.0687   -0.0240    0.3714    1.4378   -0.8131
-0.8095   -0.2740   -0.0658   -0.0867   -0.1121   -0.5185    0.0415
 0.4234    0.3588   -0.2721    0.5152   -0.1821    0.2052   -0.0211
 0.0743   -0.4466   -0.0747    0.1899
```

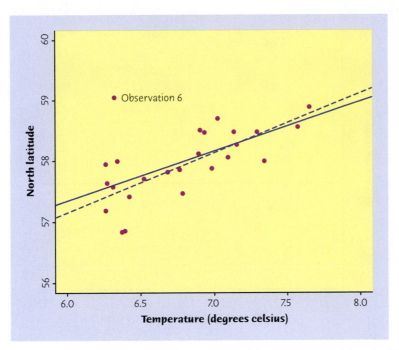

FIGURE 23.9 Plot of the latitude of the center of the distribution of anglerfish in the North Sea against mean winter temperature at the bottom of the sea for Example 23.9. The two regression lines are for the data with (solid) and without (dashed) Observation 6.

Graphs play a central role in checking the conditions for inference. Figure 23.10 plots the residuals against the explanatory variable, sea-bottom temperature. The horizontal line at 0 residual marks the position of the regression line. Both the scatterplot in Figure 23.9 and the residual plot in Figure 23.10 show that Observation 6 is a high outlier.

The observations were taken a year apart, so we are willing to regard them as close to **independent observations.** Except for the outlier, the plots do show **a linear relationship** with roughly **equal variation about the line** for all values of the explanatory variable. A histogram of the residuals (Figure 23.11) shows no strong deviations from a **Normal distribution** except for the high outlier. We conclude that the conditions for regression inference are met except for the presence of the outlier.

How influential is the outlier? The dashed line in Figure 23.9 is the regression line without Observation 6. Because there are several other observations with similar values of temperature, dropping Observation 6 does not move the regression line very much. So the outlier is not influential for regression. It *is* influential for correlation: r^2 increases from 0.364 to 0.584 when we drop Observation 6. *Even though the outlier is not highly influential for the regression line, it strongly influences inference because of its effect on the regression standard error.* The standard error is $s = 0.4734$ with Observation 6 and $s = 0.3622$ without it. When we omit the outlier, the t statistic changes from $t = 3.6287$ to $t = 5.5599$ and the one-sided P-value changes from $P = 0.0007$ to $P < 0.00001$.

CAUTION

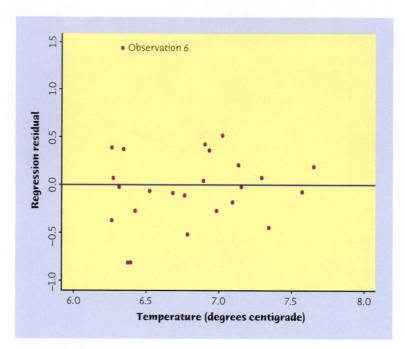

FIGURE 23.10 Residual plot for the regression of latitude on temperature in Example 23.9.

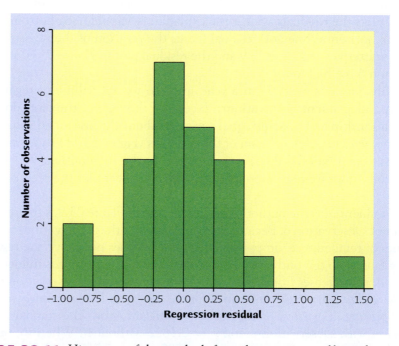

FIGURE 23.11 Histogram of the residuals from the regression of latitude on temperature in Example 23.9.

Fortunately, the outlier does not affect the conclusion we drew from the data. Therefore, we can trust the results of our analysis.

APPLY YOUR KNOWLEDGE

23.14 Crying and IQ: residuals. The residuals for the study of crying and IQ appear in Example 23.3.

(a) Make a stemplot to display the distribution of the residuals. (Round to the nearest whole number first.) Are there outliers or signs of strong departures from Normality?

(b) Make a plot of the residuals against the explanatory variable. Draw a horizontal line at height 0 on your plot. Does the plot show a nonrandom pattern?

23.15 Frog mating calls: residuals. Figure 23.4 gives part of the Minitab output for the data on mating call frequency and temperature in wild *H. chrysoscelis* from Exercise 23.2. Table 23.3 comes from another part of the output. It gives the predicted response $\hat{y}$ and the residual $y - \hat{y}$ for each of the 20 observations. Most statistical software packages provide similar output. Examine the conditions for regression inference one by one.

TABLE 23.3 Frog mating calls: predictions and residuals

Obs.	Temp. x	Freq. y	Prediction $\hat{y}$	Residual $y - \hat{y}$
1	19	38	38.10	−0.10
2	21	42	42.76	−0.76
3	22	45	45.09	−0.09
4	22	45	45.09	−0.09
5	23	41	47.42	−6.42
6	23	45	47.42	−2.42
7	23	48	47.42	0.58
8	23	50	47.42	2.58
9	23	53	47.42	5.58
10	24	51	49.75	1.25
11	24	48	49.75	−1.75
12	24	53	49.75	3.25
13	24	47	49.75	−2.75
14	25	53	52.08	0.92
15	25	49	52.08	−3.08
16	25	56	52.08	3.92
17	25	53	52.08	0.92
18	26	55	54.41	0.59
19	26	55	54.41	0.59
20	27	54	56.74	−2.74

(a) **Independent observations.** The data come from observations of 20 different male *H. chrysoscelis* in the wild. They are therefore independent observations.

(b) **Linear relationship.** Your plot and r^2 from Exercise 23.2 show that the relationship is indeed linear. Residual plots magnify effects. Plot the residuals against temperature. Are there clear deviations from a linear relationship?

(c) **Spread about the line stays the same.** Your plot in (b) shows that it does not. Variation is greater for warmer temperatures. However, the reason for less variation in lower temperature is that there are very few observations. Rather than a problem with constant spread, this suggests that the low temperature observations might be more influential.

(d) **Normal variation about the line.** Make a histogram of the residuals. Do strong skewness or outliers suggest lack of Normality?

CHAPTER 23 SUMMARY

Least-squares regression fits a straight line to data in order to predict a response variable y from an explanatory variable x. Inference about regression requires more conditions.

The **conditions for regression inference** say that there is a **population regression line** $\mu_y = \alpha + \beta x$ that describes how the mean response varies as x changes. The observed response y for any x has a Normal distribution with mean given by the population regression line and with the same standard deviation σ for any value of x. Observations on y are independent.

The **parameters to be estimated** are the intercept α and the slope β of the population regression line, and also the standard deviation σ. The slope a and intercept b of the least-squares line estimate α and β. Use the **regression standard error s** to estimate σ.

The regression standard error s has $n - 2$ **degrees of freedom.** All t procedures in regression inference have $n - 2$ degrees of freedom.

To test **the hypothesis that the slope is zero in the population,** use the t statistic $t = b/SE_b$. This null hypothesis says that straight-line dependence on x has no value for predicting y. In practice, use software to find the slope b of the least-squares line, its standard error SE_b, and the t statistic.

The t test for regression slope is also a test for **the hypothesis that the population correlation between x and y is zero.** To do this test without software, use the sample correlation r and Table E.

Confidence intervals for the slope of the population regression line have the form $b \pm t^* SE_b$.

Confidence intervals for the mean response when x has value x^* have the form $\hat{y} \pm t^* SE_{\hat{\mu}}$. **Prediction intervals** for an individual future response y have a similar form with a larger standard error, $\hat{y} \pm t^* SE_{\hat{y}}$. Software often gives these intervals.

CHECK YOUR SKILLS

Our first example of regression (Example 4.1, page 93) presented data showing that people who increased their nonexercise activity (NEA) when they were deliberately overfed gained less fat than other people. The scatterplot and regression line based on 16 overfed subjects are displayed in Figure 4.1, page 94. Here is part of the Minitab output for regressing fat gain on NEA change in this study, along with prediction for a person adding 400 NEA calories:

```
Predictor            Coef     SE Coef      T      P
Constant           3.5051     0.3036   11.54   0.000
NEA change     -0.0034415  0.0007414   -4.64   0.000

S = 0.739853    R-Sq = 60.6%    R-Sq(adj) = 57.8%


Predicted Values for New Observations
New Obs    Fit   SE Fit      95% CI            95% PI
      1  2.129    0.193  (1.714, 2.543)   (0.488, 3.769)
```

Exercises 23.16 to 23.24 are based on this information.

23.16 The equation of the least-squares regression line for predicting fat gain from NEA change is

(a) fat $= 11.54 - 4.64 \times$ NEA change.

(b) fat $= -0.0034 + 3.5051 \times$ NEA change.

(c) fat $= 3.5051 - 0.0034 \times$ NEA change.

23.17 What is the correlation between fat gain and NEA change?

(a) 0.606   (b) 0.778   (c) −0.778

23.18 Is there significant evidence that fat gain decreases as NEA change increases? To answer this question, we test the hypotheses

(a) H_0: $\beta = 0$ versus H_a: $\beta < 0$.

(b) H_0: $\beta = 0$ versus H_a: $\beta \neq 0$.

(c) H_0: $\alpha = 0$ versus H_a: $\alpha < 0$.

23.19 Minitab shows that the P-value for this test is

(a) 0.7398.   (b) 0.3036.   (c) less than 0.001.

23.20 The regression standard error for these data is

(a) 0.0007.   (b) 0.3036.   (c) 0.7399.

23.21 Confidence intervals and tests for these data use the t distribution with degrees of freedom

(a) 16.   (b) 15.   (c) 14.

23.22 A 95% confidence interval for the population slope β is

(a) -0.0034 ± 0.00145.

(b) -0.0034 ± 0.00159.

(c) -0.0034 ± 0.00344.

23.23 The Minitab output includes a prediction for y when $x^* = 400$. If an overfed adult burned an additional 400 NEA calories, we can be 95% confident that the person's fat gain would be between

(a) 1.71 and 2.54 kilograms (kg).

(b) 1.75 and 2.51 kg.

(c) 0.49 and 3.77 kg.

23.24 If a whole population of overfed adults burned an additional 400 NEA calories, we can be 95% confident that the population mean fat gain would be between

(a) 1.71 and 2.54 kg.   (b) 1.75 and 2.51 kg.   (c) 0.49 and 3.77 kg.

CHAPTER 23 EXERCISES

23.25 **Too much nitrogen?** Intensive agriculture and burning of fossil fuels increase the amount of nitrogen deposited on the land. Too much nitrogen can reduce the variety of plants by favoring rapid growth of some species—think of putting fertilizer on your lawn to help grass choke out weeds. A study of 68 grassland sites in Britain measured nitrogen deposited (kilograms of nitrogen per hectare of land area per year) and also the "richness" of plant species (based on number of species and how abundant each species is). The authors reported a regression analysis as follows.[7]

$$\text{plant species richness} = 23.3 - 0.408 \times \text{nitrogen deposited}$$
$$r^2 = 0.55 \quad P < 0.0001$$

(a) What does the slope $b = -0.408$ say about the effect of increased nitrogen deposits on species richness?

(b) What does $r^2 = 0.55$ add to the information given by the equation of the least-squares line?

(c) What null and alternative hypotheses do you think the P-value refers to? What does this P-value tell you?

23.26 **Beavers and beetles.** Ecologists sometimes find rather strange relationships in our environment. One study seems to show that beavers benefit beetles. The researchers laid out 23 circular plots, each 4 m in diameter, in an area where beavers were cutting down cottonwood trees. In each plot, they measured the number of stumps from trees cut by beavers and the number of clusters of beetle larvae. Here are the data.[8]

Stumps	2	2	1	3	3	4	3	1	2	5	1	3
Beetle larvae	10	30	12	24	36	40	43	11	27	56	18	40

Stumps	2	1	2	2	1	1	4	1	2	1	4
Beetle larvae	25	8	21	14	16	6	54	9	13	14	50

(a) Make a scatterplot that shows how the number of beaver-caused stumps influences the number of beetle larvae clusters. What does your plot show?

(b) Here is part of the Minitab regression output for these data.

Minitab

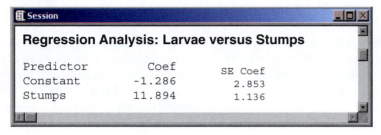

Find the least-squares regression line and draw it on your plot.

(c) Is there strong evidence that beaver stumps help explain beetle larvae counts? State hypotheses, give a test statistic and its P-value, and state your conclusion.

23.27 Prey attract predators. Stable ecosystems should have enough predators to keep prey densities in check. An experiment tested this theory by studying kelp perch and their common predator, the kelp bass, in four large circular pens of identical size. The explanatory variable is the number of perch (the prey) in each enclosure. The response variable is the proportion of perch killed by bass (the predator) in 2 hours when the bass are allowed access to the perch.

Perch	Proportion killed			
10	0.0	0.1	0.3	0.3
20	0.2	0.3	0.3	0.6
40	0.075	0.3	0.6	0.725
60	0.517	0.55	0.7	0.817

(a) Make a scatterplot and describe the relationship.

(b) Figure 23.12 contains Excel's output for the regression. What is the equation of the least-squares line for predicting proportion killed from count of

	A	B	C	D	E	F	G
1	SUMMARY OUTPUT						
2							
3	Regression statistics						
4	Multiple R	0.6821					
5	R Square	0.4652					
6	Adjusted R Square	0.4270					
7	Standard Error	0.1886					
8	Observations	16.0000					
9							
10		Coefficients	Standard Error	t Stat	P-value	Lower 95%	Upper 95%
11	Intercept	0.1205	0.0927	1.2999	0.2146	-0.0783	0.3193
12	Perch	0.0086	0.0025	3.4899	0.0036	0.0033	0.0138
13							

FIGURE 23.12 Partial Excel output for the regression of proportion of perch killed by bass on count of perch in a pen, for Exercises 23.27, 23.29, and 23.31.

perch? What part of this equation shows that more perch do result in a higher proportion being killed by bass? What is the regression standard error s?

23.28 **Heritability of phenotype.** Figure 23.13 shows the relationship between parent nest size and offspring nest size for 100 parent-offspring observations of barn swallows, *Hirundo rustica*.[9] The line is the least-squares regression line. Note that the graph actually plots the logarithm of nest size, because nest sizes, like many measures of volume, are otherwise strongly skewed to the right. Here is a partial Excel regression output for these data.

```
Regression Statistics
Multiple R      0.7229
R Square        0.5226
Standard Error  0.1396
Observations    100
```

	Coefficients	Standard Error	t Stat	P-value
Intercept	0.5764	0.1542		
Parents	0.6762	0.0653		

(a) Describe the relationship from the graph and assess the validity of inference for these data.

(b) Give the value of r^2 and the equation of the least-squares line.

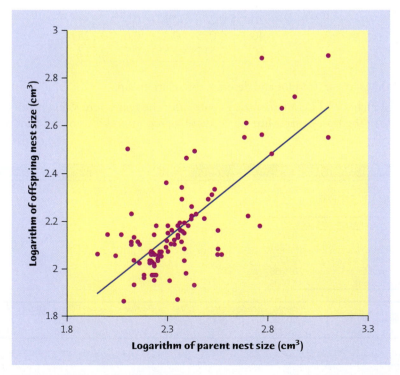

FIGURE 23.13 Plot of the log of offspring nest size against the log of parent nest size, for Exercise 23.28.

(c) Based on the information given, test the hypothesis that there is no straight-line relationship between parent and offspring nest size. Give a test statistic, its approximate P-value, and your conclusion.

23.29 Prey attract predators: estimating the slope.

(a) The Excel output in Figure 23.12 includes a 95% confidence interval for the slope of the population regression line. What is it? Starting from Excel's values of the least-squares slope b and its standard error, verify this confidence interval.

(b) Give a 90% confidence interval for the population slope. As usual, this interval is shorter than the 95% interval.

23.30 Heritability of phenotype: estimating the slope. Regression is used in genetics to assess the heritability of a given phenotype from parents to their offspring. An estimate of heritability can be obtained by plotting the phenotypic value of a sample of offspring against that of their parents and finding the slope of the corresponding regression line. Using the information from Exercise 23.28, give a 95% confidence interval for the true population heritability (slope β) of nest size on a logarithmic scale.

23.31 Prey attract predators: correlation. The Excel output in Figure 23.12 includes the correlation between proportion of perch killed by bass and initial count of perch $r = 0.6821$. Use Table E to say how significant this correlation is for testing zero correlation against positive correlation in the population. Verify that your result is consistent with Excel's two-sided P-value.

23.32 Prey attract predators: residuals. Here are the residuals (rounded to three decimal places) for the regression of proportion of perch killed by bass on initial count of perch.

Perch	10	10	10	10	20	20	20	20
Residual	−0.206	−0.106	0.094	0.094	−0.092	0.008	0.008	0.308

Perch	40	40	40	40	60	60	60	60
Residual	−0.388	−0.163	0.137	0.262	−0.118	−0.085	0.065	0.182

(a) Check the calculation of residuals by finding their sum. What should the sum be? Does the sum have that value (aside from roundoff error)?

(b) Plot the residuals against the initial count of perch (the explanatory variable). Does your plot show a systematically nonlinear relationship? Does it show systematic change in the spread about the regression line?

(c) Make a histogram of the residuals. The pattern, with just 16 observations, is somewhat irregular. A Normal probability plot shows that the distribution of residuals is reasonably Normal, with one small outlier.

23.33 DNA on the ocean floor. We think of DNA as the stuff that stores the genetic code. It turns out that DNA occurs, mainly outside living cells, on the ocean floor. It is important in nourishing seafloor life. Scientists think that this DNA comes from organic matter that settles to the bottom from the top layers of the ocean. "Phytopigments," which come mainly from algae, are a measure of the amount of organic matter that has settled to the bottom. Table 23.4 contains data on concentrations of DNA and phytopigments (both in grams per square

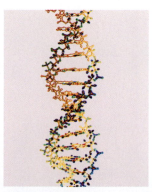

Minoru Toi/Getty Images

TABLE 23.4		DNA and phytopigment concentrations (g/m^2) on the ocean floor					
DNA	Phyto	DNA	Phyto	DNA	Phyto	DNA	Phyto
0.148	0.010	0.276	0.056	0.156	0.032	0.300	0.022
0.108	0.009	0.214	0.023	0.112	0.016	0.116	0.008
0.180	0.008	0.330	0.016	0.280	0.005	0.120	0.004
0.218	0.006	0.240	0.007	0.308	0.005	0.064	0.006
0.152	0.006	0.100	0.006	0.238	0.011	0.228	0.010
0.589	0.050	0.463	0.038	0.461	0.034	0.333	0.020
0.357	0.023	0.382	0.032	0.414	0.034	0.241	0.012
0.458	0.036	0.396	0.033	0.307	0.018	0.236	0.002
0.076	0.001	0.001	0.002	0.009	0	0.099	0
0.187	0.001	0.104	0.004	0.088	0.009	0.072	0.002
0.192	0.005	0.152	0.028	0.152	0.006	0.272	0.004
0.288	0.046	0.232	0.006	0.368	0.003	0.216	0.011
0.248	0.006	0.280	0.002	0.336	0.062	0.320	0.006
0.896	0.055	0.200	0.017	0.408	0.018	0.472	0.017
0.648	0.034	0.384	0.008	0.440	0.042	0.592	0.032
0.392	0.036	0.312	0.002	0.312	0.003	0.208	0.001
0.128	0.001	0.264	0.001	0.264	0.008	0.328	0.010
0.264	0.002	0.376	0.003	0.288	0.001	0.208	0.024
0.224	0.017	0.376	0.010	0.600	0.024	0.168	0.014
0.264	0.018	0.152	0.010	0.184	0.016	0.312	0.017
0.344	0.009	0.184	0.010	0.360	0.010	0.264	0.026
0.464	0.030	0.328	0.028	0.296	0.010	1.056	0.082
0.538	0.055	0.090	0.003	0.130	0.001	0.207	0.001
0.153	0.001	0.206	0.001	0.172	0.001	0.131	0.001
0.095	0	0.307	0.001	0.171	0.001	0.822	0.058
0.901	0.075	0.552	0.040	0.391	0.026	0.172	0.006
0.116	0.003	0.168	0.005	0.074	0	0.100	0.001
0.132	0.005	0.112	0.003	0.121	0.004	0.162	0.001
0.302	0.014	0.179	0.002	0.369	0.023	0.213	0.007

meter [g/m^2]) in 116 ocean locations around the world.[10] Look first at DNA alone. Describe the distribution of DNA concentration and give a confidence interval for the mean concentration. Be sure to explain why your confidence interval is trustworthy in the light of the shape of the distribution. The data show surprisingly high DNA concentration, and this by itself was an important finding.

23.34 **Predicting tree height.** Measuring tree height is not an easy task. How well might trunk diameter predict tree height? A survey of 958 live trees in an old-growth forest in Canada answered this question.[11] Here is part of the Minitab output based on these data for regressing height on diameter, along with prediction for a tree having a diameter of 50 centimeters (cm):

```
Predictor      Coef    SE Coef       T      P
Constant     2.6696     0.1677   15.92  0.000
Diameter   0.550940   0.005058  108.93  0.000

S = 3.67427   R-Sq = 92.5%   R-Sq(adj) = 92.5%

New Obs    Fit  SE Fit       95% CI            95% PI
    1   30.217   0.179  (29.865, 30.569)  (22.997, 37.436)
```

(a) A scatterplot of the data shows a reasonably linear relationship between tree height and diameter. Is this relationship statistically significant? How strong is the relationship?

(b) Give a 95% confidence interval for the height of 1 tree randomly selected from this forest if the tree has a diameter of 50 cm.

(c) Now give a 95% confidence interval for the mean height of all the trees in this forest that have a 50-cm diameter. How does this interval compare with the one you calculated in (b)?

23.35 DNA on the ocean floor, continued. Another conclusion of the study introduced in Exercise 23.33 was that organic matter settling down from the top layers of the ocean is the main source of DNA on the seafloor. An important piece of evidence is the relationship between DNA and phytopigments. Do the data in Table 23.4 give good reason to think that phytopigment concentration helps explain DNA concentration? Describe the data and follow the four-step process in answering this question.

23.36 Sparrowhawk colonies. One of nature's patterns connects the percent of adult birds in a colony that return from the previous year and the number of new adults that join the colony. Here are data for 13 colonies of sparrowhawks.[12]

Percent return x	74	66	81	52	73	62	52	45	62	46	60	46	38
New adults y	5	6	8	11	12	15	16	17	18	18	19	20	20

Earlier (Exercises 3.4, page 69, and 4.4, page 102), you found that there is a moderately strong linear relationship. Figure 23.14 shows part of the Minitab regression output, including a prediction for y when 60% of the previous year's adult birds return.

(a) Write the equation of the least-squares line and use it to check that the "Fit" in the output is the predicted response for $x^* = 60\%$.

(b) Which 95% interval in the output gives us a margin of error for predicting the average number of new birds in colonies to which 60% of the past year's adults return?

23.37 DNA on the ocean floor: residuals. Save the residuals from the regression of DNA concentration on phytopigment concentration (Exercise 23.33). Examine the residuals to see how well the conditions for regression inference are met.

(a) Plot the residuals against phytopigment concentration (the explanatory variable), using vertical limits −1 to 1 to make the pattern clearer. Add a

Minitab

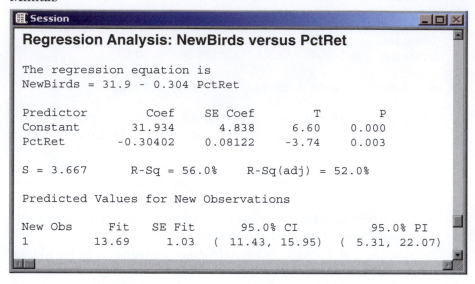

FIGURE 23.14 Partial Minitab output for predicting number of new birds in a sparrowhawk colony from percent of birds returning, for Exercise 23.36.

horizontal line at height 0 to represent the regression line. What do you conclude about the conditions of linear relationship and constant standard deviation?

(b) Make a histogram of the residuals. What do you conclude about Normality?

23.38 **Sparrowhawk colonies: residuals.** The regression of number of new birds that join a sparrowhawk colony on the percent of adult birds in the colony that return from the previous year is an example of data that satisfy the conditions for regression inference well. Here are the residuals for the 13 colonies in Exercise 23.36:

Percent return	74	66	81	52	73	62	52
Residual	−4.44	−5.87	0.69	−5.13	2.26	1.92	−0.13

Percent return	45	62	46	60	46	38
Residual	−1.25	4.92	0.05	5.31	2.05	−0.38

(a) **Independent observations.** Why are the 13 observations independent?

(b) **Linear relationship.** A plot of the residuals against the explanatory variable x magnifies the deviations from the least-squares line. Does the plot show any systematic deviation from a roughly linear pattern?

(c) **Spread about the line stays the same.** Does your plot in (b) show any systematic change in spread as x changes?

(d) **Normal variation about the line.** Make a histogram of the residuals. With only 13 observations, no clear shape emerges. Do strong skewness or outliers suggest lack of Normality?

23.39 Time at the table. Does how long young children remain at the lunch table help predict how much they eat? Here are data on 20 toddlers observed over several months at a nursery school.[13] "Time" is the average number of minutes a child spent at the table when lunch was served. "Calories" is the average number of calories the child consumed during lunch, calculated from careful observation of what the child ate each day.

Time	21.4	30.8	37.7	33.5	32.8	39.5	22.8	34.1	33.9	43.8
Calories	472	498	465	456	423	437	508	431	479	454
Time	42.4	43.1	29.2	31.3	28.6	32.9	30.6	35.1	33.0	43.7
Calories	450	410	504	437	489	436	480	439	444	408

Follow the four-step process in the following analysis.

(a) Describe the relationship in a graph and with a regression line. Be sure to save the regression residuals.

(b) Check the conditions for inference. Parts (a) to (d) of Exercise 23.38 provide a handy outline. Use vertical limits −100 to 100 in your plot of the residuals against time to help you see the pattern.

(c) Give a 95% confidence interval to estimate how rapidly calories consumed changes as time at the table increases.

23.40 Foot problems. Hallux abducto valgus (call it HAV) is a deformation of the big toe that is not common in youth and often requires surgery. Metatarsus adductus (call it MA) is a turning in of the front part of the foot that is common in adolescents and usually corrects itself. Doctors used X-rays to measure the angle (in degrees) of deformity ("HAV angle" and "MA angle") in 38 consecutive patients under the age of 21 who came to a medical center for surgery to correct HAV.[14] The data appear in Table 23.5.

TABLE 23.5	Angle of deformity (degrees) for two types of foot deformity				
HAV angle	MA angle	HAV angle	MA angle	HAV angle	MA angle
28	18	21	15	16	10
32	16	17	16	30	12
25	22	16	10	30	10
34	17	21	7	20	10
38	33	23	11	50	12
26	10	14	15	25	25
25	18	32	12	26	30
18	13	25	16	28	22
30	19	21	16	31	24
26	10	22	18	38	20
28	17	20	10	32	37
13	14	18	15	21	23
20	20	26	16		

Metatarsus adductus may help predict the severity of hallux abducto valgus. The paper that reports this study says, "Linear regression analysis, using the hallux abductus angle as the response variable, demonstrated a significant correlation between the metatarsus adductus and hallux abductus angles." Do a suitable analysis to verify this finding, following the four-step process. (Be sure to check the conditions for inference as part of the *Solve* step. Parts (a) through (d) of Exercise 23.38 provide a handy outline. The authors note that the scatterplot suggests that the variation in y may change as x changes, so they offer a more elaborate analysis as well.)

23.41 Time at the table: prediction. Rachel is a new child at the nursery school of Exercise 23.39. Over several months, Rachel averages 40 minutes at the lunch table. Give a 95% interval to predict Rachel's average calorie consumption at lunch.

23.42 Weeds among the corn. Lamb's-quarter is a common weed that interferes with the growth of corn. An agriculture researcher planted corn at the same rate in 16 small plots of ground, then weeded the plots by hand to allow a fixed number of lamb's-quarter plants to grow in each meter of corn row. No other weeds were allowed to grow. Here are the yields of corn (bushels per acre) in each of the plots.[15]

Weeds per meter	Corn yield	Weeds per meter	Corn yield	Weeds per meter	Corn yield	Weeds per meter	Corn yield
0	166.7	1	166.2	3	158.6	9	162.8
0	172.2	1	157.3	3	176.4	9	142.4
0	165.0	1	166.7	3	153.1	9	162.8
0	176.9	1	161.1	3	156.0	9	162.4

Use software to analyze these data.

(a) Make a scatterplot and find the least-squares line. What percent of the observed variation in corn yield can be explained by a linear relationship between yield and weeds per meter?

(b) Is there good evidence that more weeds reduce corn yield?

(c) Explain from your findings in (a) and (b) why you expect predictions based on this regression to be quite imprecise. Predict the mean corn yield under these experimental conditions when there are 6 weeds per meter of row. If your software allows, give a 95% confidence interval for this mean.

23.43 Revenge! Does revenge feel good? Or do people take revenge just because they are mad about being harmed? Different areas in the brain are active in the two cases, so brain scans can help decide which explanation is correct. Here's a game that supports the first explanation.

Player A is given $10. If he gives it to Player B, it turns into $40. B can keep all of the money or give half to A, who naturally feels that B owes him half. If B keeps all $40, A can take revenge by removing up to $20 of B's ill-gotten gains, at no cost or gain to himself. Scan A's brain at that point, recording activity in the caudate nucleus, a region involved in "making decisions or taking actions that are motivated by anticipated rewards." Only the A players who took $20 from B play again, with different partners B. So all the A players have shown the same level of revenge when cheated by B.

The new B also keeps all the money—but punishing B now costs A $1 for every $2 he takes from B. The researchers predicted that A players who get more kicks from revenge, as measured by caudate activity, will punish B more severely even when it costs them money to do it. Here are data for 11 players:[16]

Caudate activity	−0.057	−0.011	−0.032	−0.025	−0.012	0.028
A takes from B	$0	$0	$5	$5	$10	$10
Caudate activity	−0.002	0.008	0.029	0.037	0.043	
A takes from B	$10	$20	$20	$20	$20	

(a) Make a scatterplot with caudate activity as the explanatory variable. Add the least-squares regression line to your plot to show the overall pattern.

(b) The research report mentions positive correlation and its significance. What is the correlation r? Is it significantly greater than zero?

(c) The nature of the data gives some reason to doubt the accuracy of the significance level. Why?

23.44 Standardized residuals (Optional). Software often calculates **standardized residuals** as well as the actual residuals from regression. Because the standardized residuals have the standard z-score scale, it is easier to judge whether any are extreme. Here are the standardized residuals from Exercise 23.26 (beavers and beetles), rounded to two decimal places:

standardized residuals

−1.99   1.20   0.23 −1.67 0.26 −1.06 1.38   0.06   0.72 −0.40 1.21 0.90
0.40 −0.43 −0.24 −1.36 0.88 −0.75 1.30 −0.26 −1.51   0.55 0.62

(a) Find the mean and standard deviation of the standardized residuals. Why do you expect values close to those you obtain?

(b) Make a stemplot of the standardized residuals. Are there any striking deviations from Normality? The most extreme residual is $z = −1.99$. Would this be surprisingly large if the 23 observations had a Normal distribution? Explain your answer.

(c) Plot the standardized residuals against the explanatory variable. Are there any suspicious patterns?

23.45 Tests for the intercept (Optional). Figure 23.7 gives CrunchIt! output for the regression of blood alcohol content (BAC) on number of beers consumed. The t test for the hypothesis that the population regression line has slope $\beta = 0$ has $P < 0.0001$. The data show a positive linear relationship between BAC and beers. We might expect the intercept α of the population regression line to be 0, because no beers ($x = 0$) should produce no alcohol in the blood ($y = 0$). To test

$$H_0: \alpha = 0$$
$$H_a: \alpha \neq 0$$

we use a t statistic formed by dividing the least-squares intercept a by its standard error SE_a. Locate this statistic in the output of Figure 23.7 and verify that it is in fact a divided by its standard error. What is the P-value? Do the data suggest that the intercept is not 0?

23.46 Confidence intervals for the intercept (Optional). The output in Figure 23.7 allows you to calculate confidence intervals for both the slope β and the intercept α of the population regression line of BAC on beers in the population of all students. Confidence intervals for the intercept α have the familiar form $a \pm t^* SE_a$ with degrees of freedom $n - 2$. What is the 95% confidence interval for the intercept? Does it contain 0, the value we might guess for α?

Larry F. Jernigan/Index Stock

One-Way Analysis of Variance: Comparing Several Means

The two-sample *t* procedures of Chapter 18 compare the means of two populations or the mean responses to two treatments in an experiment. Of course, studies don't always compare just two groups. We need a method for comparing any number of means.

─ **EXAMPLE 24.1** Comparing tropical flowers ───────

STATE: Ethan Temeles of Amherst College, with his colleague W. John Kress, studied the relationship between varieties of the tropical flower *Heliconia* on the island of Dominica and the different species of hummingbirds that fertilize the flowers.[1] Over time, the researchers believe, the lengths of the flowers and the form of the hummingbirds' beaks have evolved to match each other. If that is true, flower varieties fertilized by different hummingbird species should have distinct distributions of length.

Table 24.1 gives length measurements (in millimeters [mm]) for samples of three varieties of *Heliconia*, each fertilized by a different species of hummingbird. Do the three varieties display distinct distributions of length? In particular, are the average lengths of their flowers different?

629

TABLE 24.1	Flower lengths (millimeters) for three *Heliconia* varieties						
H. bihai							
47.12	46.75	46.81	47.12	46.67	47.43	46.44	46.64
48.07	48.34	48.15	50.26	50.12	46.34	46.94	48.36
H. caribaea red							
41.90	42.01	41.93	43.09	41.47	41.69	39.78	40.57
39.63	42.18	40.66	37.87	39.16	37.40	38.20	38.07
38.10	37.97	38.79	38.23	38.87	37.78	38.01	
H. caribaea yellow							
36.78	37.02	36.52	36.11	36.03	35.45	38.13	37.10
35.17	36.82	36.66	35.68	36.03	34.57	34.63	

FORMULATE: Use graphs and numerical descriptions to describe and compare the three distributions of flower length. Finally, ask whether the differences among the mean lengths of the three varieties are *statistically significant*.

SOLVE (FIRST STEPS): Perhaps these data seem familiar. We first saw them in Chapter 2 (page 56), where we compared the distributions. Figure 24.1 repeats a stemplot display from Chapter 2. The lengths have been rounded to the nearest tenth of a millimeter. Here are the summary measures we will use in further analysis:

Sample	Variety	Sample size	Mean length	Standard deviation
1	*bihai*	16	47.60	1.213
2	red	23	39.71	1.799
3	yellow	15	36.18	0.975

```
        bihai              red                yellow
    34 |               34 |               34 | 6 6
    35 |               35 |               35 | 2 5 7
    36 |               36 |               36 | 0 0 1 5 7 8 8
    37 |               37 | 4 8 9         37 | 0 1
    38 |               38 | 0 0 1 1 2 2 8 9   38 | 1
    39 |               39 | 2 6 8         39 |
    40 |               40 | 6 7           40 |
    41 |               41 | 5 7 9 9       41 |
    42 |               42 | 0 2           42 |
    43 |               43 | 1             43 |
    44 |               44 |               44 |
    45 |               45 |               45 |
    46 | 3 4 6 7 8 8 9 46 |               46 |
    47 | 1 1 4         47 |               47 |
    48 | 1 2 3 4       48 |               48 |
    49 |               49 |               49 |
    50 | 1 3           50 |               50 |
```

FIGURE 24.1 Side-by-side stemplots comparing the lengths in millimeters of samples of flowers from three varieties of *Heliconia*, from Table 24.1.

CONCLUDE (FIRST STEPS):   The three varieties differ so much in flower length that there is little overlap among them. In particular, the *bihai* flowers are longer than either the red or the yellow flowers. The mean flower lengths are 47.6 mm for *H. bihai*, 39.7 mm for *H. caribaea* red, and 36.2 mm for *H. caribaea* yellow. Are these observed differences in sample means statistically significant? We must develop a test for comparing more than two population means.

Photo Resource Hawaii/Alamy

Comparing several means

Call the mean lengths for the three populations of flowers μ_1 for *bihai*, μ_2 for red, and μ_3 for yellow. The subscript reminds us which group a parameter or statistic describes. To compare these three population means, we might use the two-sample t test several times:

- Test H_0: $\mu_1 = \mu_2$ to see if the mean length for *bihai* differs from the mean length for red.
- Test H_0: $\mu_1 = \mu_3$ to see if *bihai* differs from yellow.
- Test H_0: $\mu_2 = \mu_3$ to see if red differs from yellow.

The weakness of doing three tests is that we get three P-values, one for each test performed. That doesn't tell us how likely it is that *three* sample means are spread apart as far as these are. It may be that $\bar{x}_1 = 47.60$ and $\bar{x}_3 = 36.18$ are significantly different if we look at just two groups but not significantly different if we know that they are the largest and the smallest means in three groups. (Think of comparing the tallest and shortest person in a large group of people. They may have quite different heights, yet they simply represent two extremes of a continuum of heights.) The problem gets worse as we wish to compare more groups, because we expect the gap between the largest and smallest sample mean to get larger just by chance even if they are all samples from the same population. That is, *we can't safely compare many parameters by doing tests or confidence intervals for two parameters at a time*.

The problem of how to do many comparisons at once with an overall measure of confidence in all our conclusions is common in statistics. This is the problem of **multiple comparisons.** Statistical methods for dealing with multiple comparisons usually have two steps:

multiple comparisons

1. An *overall test* to see if there is good evidence of *any* differences among the parameters that we want to compare.
2. A detailed *follow-up analysis* to decide which of the parameters differ and to estimate how large the differences are.

The overall test, though more complex than the tests we learned about earlier, is often reasonably straightforward. The follow-up analysis can be quite elaborate. This chapter concentrates on the overall test, along with data analysis that points to the nature of the differences. In optional Chapter 26 on CD, we cover the details of some common follow-up analyses used when comparing several means.

The analysis of variance *F* test

We want to test the null hypothesis that there are *no differences* among the mean lengths for the three populations of flowers:

$$H_0: \mu_1 = \mu_2 = \mu_3$$

The alternative hypothesis is that there is *some difference*. That is, not all three population means are equal:

$$H_a: \text{not all of } \mu_1, \ \mu_2, \text{ and } \mu_3 \text{ are equal}$$

H_a simply says that H_0 is not true. The alternative hypothesis is no longer one-sided or two-sided. It is in a sense "many-sided" or nondirectional, because it allows any relationship other than "all three equal." For example, H_a includes the case in which $\mu_2 = \mu_3$ but μ_1 has a different value. The test of H_0 against H_a is called *analysis of variance F test* the **analysis of variance F test.** Analysis of variance is usually abbreviated as ANOVA. The ANOVA *F* test is almost always carried out with software that reports the test statistic and its *P*-value.

EXAMPLE 24.2 *Comparing tropical flowers: ANOVA*

SOLVE (INFERENCE): Software tells us that for the flower length data in Table 24.1, the test statistic is $F = 259.12$ with *P*-value $P < 0.0001$. There is very strong evidence that the three varieties of flowers do not all have the same mean length.

The *F* test does not say *which* of the three means are significantly different. It appears from our preliminary data analysis that *bihai* flowers are distinctly longer than either red or yellow flowers. Red and yellow are closer together, but the red flowers tend to be longer.

CONCLUDE: There is strong evidence ($P < 0.0001$) that the population means are not all equal. The most important difference among the means is that the *bihai* variety has longer flowers than the red and yellow varieties.

Example 24.2 illustrates our approach to comparing means. The ANOVA *F* test (done with software) assesses the evidence for *some* difference among the population means. Even if you strongly expect from looking at the data to find some effect, a formal test is important to guard against being misled by chance variation. We will not do the formal follow-up analysis that is often the most useful part of an ANOVA study. Follow-up analysis would allow us to say which means differ and by how much, with (say) 95% confidence that *all* our conclusions are correct. Optional Chapter 26 on the companion CD and Web sites show you how to do and interpret such analysis. In this chapter, we rely instead on examination of the data to show what differences are present and whether they are large enough to be interesting.

APPLY YOUR KNOWLEDGE

24.1 **Do fruit flies sleep?** Mammals and birds sleep. Insects such as fruit flies rest, but is this rest sleep? Biologists now think that insects do sleep. One experiment gave caffeine to fruit flies to see if it affected their rest. We know that caffeine reduces sleep in mammals, so if it reduces rest in fruit flies, that's another hint that the rest is really sleep. The paper reporting the study contains a graph similar to

Larry F. Jernigan/Index Stock

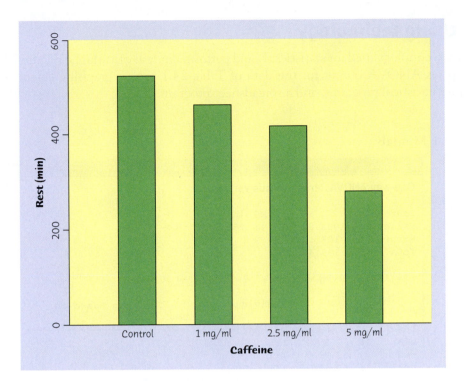

Figure 24.2 and states, "Flies given caffeine obtained less rest during the dark period in a dose-dependent fashion ($n = 36$ per group, $P < 0.0001$)."[2]

(a) The explanatory variable is amount of caffeine, in milligrams per milliliter of blood. The response variable is minutes of rest (measured by an infrared motion sensor) during a 12-hour dark period. Outline the design of this experiment.

(b) The P-value in the report comes from the ANOVA F test. What means does this test compare? State in words the null and alternative hypotheses for the test in this setting. What do the graph and the statistical test together lead you to conclude?

24.2 **Smoking during pregnancy.** Cigarette labels warn pregnant women against smoking. Does nicotine actually reach the fetus, crossing the protective placental barrier? Researchers selected consecutive pregnant women delivering at an Egyptian hospital and categorized them as either active smokers, passive smokers, or nonsmokers. They then analyzed the newborns' meconium for cotinine content, the metabolized form of nicotine. Meconium is a newborn's first stool right after birth, composed of materials ingested by the fetus in utero, and is a good biological marker for fetal exposure to drugs or other chemical agents. Here are the mean meconium cotinine levels (in nanograms per milliliter [ng/ml]) for the three groups.[3]

Active smokers	Passive smokers	Nonsmokers
367.2	263.4	185.0

The research summary states that $F = 10.45$, with $P = 0.01$.

(a) What are the null and alternative hypotheses for the ANOVA F test? Be sure to explain what means the test compares.

(b) Based on the sample means and the F test, what do you conclude?

Using technology

Any technology used for statistics should perform analysis of variance. Figure 24.3 displays ANOVA output for the data of Table 24.1 from a graphing calculator, three statistical programs, and a spreadsheet program.

TI-83

Minitab

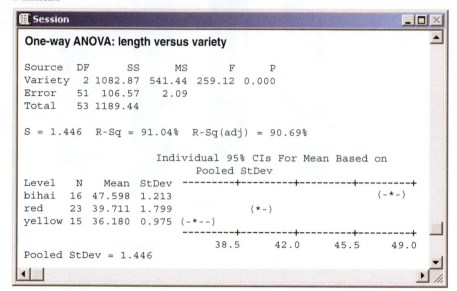

CrunchIt!

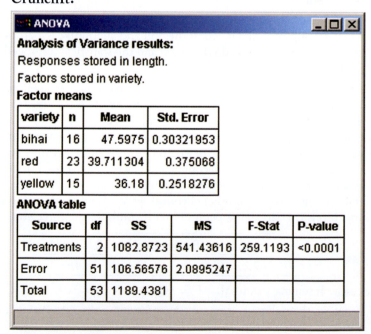

FIGURE 24.3 ANOVA for the flower length data: output from a graphing calculator, three statistical programs, and a spreadsheet program (*continued*).

Excel

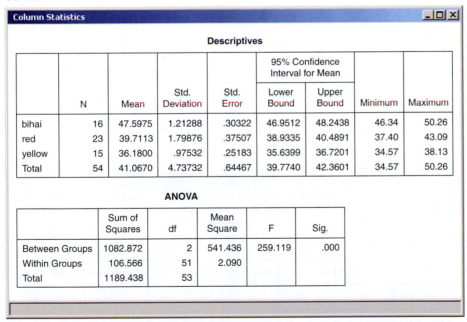

	A	B	C	D	E	F	G
1	Anova: Single Factor						
2							
3	SUMMARY						
4	*Groups*	*Count*	*Sum*	*Average*	*Variance*		
5	bihai	16	761.56	47.5975	1.471073		
6	red	23	913.36	39.7113	3.235548		
7	yellow	15	542.7	36.18	0.951257		
8							
9							
10	ANOVA						
11	*Source of variation*	*SS*	*df*	*MS*	*F*	*P-value*	*F crit*
12	Between Groups	1082.872	2	541.4362	259.1193	1.92E-27	3.178799
13	Within Groups	106.5658	51	2.089525			
14							
15	Total	1189.438	53				

SPSS

Column Statistics

Descriptives

	N	Mean	Std. Deviation	Std. Error	95% Confidence Interval for Mean		Minimum	Maximum
					Lower Bound	Upper Bound		
bihai	16	47.5975	1.21288	.30322	46.9512	48.2438	46.34	50.26
red	23	39.7113	1.79876	.37507	38.9335	40.4891	37.40	43.09
yellow	15	36.1800	.97532	.25183	35.6399	36.7201	34.57	38.13
Total	54	41.0670	4.73732	.64467	39.7740	42.3601	34.57	50.26

ANOVA

	Sum of Squares	df	Mean Square	F	Sig.
Between Groups	1082.872	2	541.436	259.119	.000
Within Groups	106.566	51	2.090		
Total	1189.438	53			

FIGURE 24.3 (*continued*)

The software outputs give the sizes of the three samples and their means. They agree with those in Example 24.1. Minitab and SPSS also give the standard deviations. You should be able to recover the standard deviations from either the variances (Excel) or the standard errors of the means (CrunchIt!). The most important part of all four outputs reports the F test statistic, $F = 259.12$, and its P-value. CrunchIt!, Minitab, and SPSS sensibly report the P-value as 0 to three decimal places, which is all we need to know in practice. Excel and the TI-83 compute a more specific value. The "E−27" in these displays means to move the decimal point 27 digits to the left. There is very strong evidence that the three varieties of flowers do not all have the same mean length.

All outputs also report degrees of freedom (df), sums of squares (SS), and mean squares (MS). We don't need this information now.

Minitab and SPSS also give confidence intervals for all three means that help us see which means differ and by how much. (We will see on page 652 how these are derived.) None of the intervals overlap, and that for *bihai* is well above the other two. These are 95% confidence intervals for each mean separately. We are *not* 95% confident that *all three* intervals cover the three means. This is another example of the perils of multiple comparisons.

APPLY YOUR KNOWLEDGE

24.3 **Logging in the rain forest.** How does logging in a tropical rain forest affect the forest several years later? Researchers compared forest plots in Borneo that had never been logged (Group 1) with similar plots nearby that had been logged 1 year earlier (Group 2) and 8 years earlier (Group 3). Although the study was not an experiment, the authors explain why we can consider the plots to be randomly selected. The data appear in Table 24.2. The variable Trees is the count of trees in a plot; Species is the count of tree species in a plot. The variable Richness is the number of species divided by the number of individual trees, Species/Trees.[4]

(a) Make side-by-side stemplots of Trees for the three groups. Use stems 0, 1, 2, and 3 and split the stems (see Example 24.1). What effects of logging are visible?

(b) Figure 24.4 shows Excel ANOVA output for Trees. What do the group means show about the effects of logging?

	A	B	C	D	E	F	G
1	Anova: Single Factor						
2							
3	SUMMARY						
4	Groups	Count	Sum	Average	Variance		
5	Group 1	12	285	23.75	25.6591		
6	Group 2	12	169	14.0833	24.8106		
7	Group 3	9	142	15.7778	33.1944		
8							
9							
10	ANOVA						
11	Source of variation	SS	df	MS	F	P-value	F crit
12	Between Groups	625.1566	2	312.57828	11.4257	0.000205	3.31583
13	Within Groups	820.7222	30	27.3574			
14							
15	Total	1445.879	32				

Sheet4 / Sheet1 / Sheet2 / Sheet3 /

FIGURE 24.4 Excel output for analysis of variance on the number of trees in forest plots, for Exercise 24.3.

TABLE 24.2	Data from a study of logging in Borneo			
Observation	Group	Trees	Species	Richness
1	1	27	22	0.81481
2	1	22	18	0.81818
3	1	29	22	0.75862
4	1	21	20	0.95238
5	1	19	15	0.78947
6	1	33	21	0.63636
7	1	16	13	0.81250
8	1	20	13	0.65000
9	1	24	19	0.79167
10	1	27	13	0.48148
11	1	28	19	0.67857
12	1	19	15	0.78947
13	2	12	11	0.91667
14	2	12	11	0.91667
15	2	15	14	0.93333
16	2	9	7	0.77778
17	2	20	18	0.90000
18	2	18	15	0.83333
19	2	17	15	0.88235
20	2	14	12	0.85714
21	2	14	13	0.92857
22	2	2	2	1.00000
23	2	17	15	0.88235
24	2	19	8	0.42105
25	3	18	17	0.94444
26	3	4	4	1.00000
27	3	22	18	0.81818
28	3	15	14	0.93333
29	3	18	18	1.00000
30	3	19	15	0.78947
31	3	22	15	0.68182
32	3	12	10	0.83333
33	3	12	12	1.00000

(c) What are the values of the ANOVA F statistic and its P-value? What hypotheses does F test? What conclusions about the effects of logging on number of trees do the data lead to?

24.4 **Dogs, friends, and stress.** If you are a dog lover, perhaps having your dog along reduces the effect of stress. To examine the effect of pets in stressful situations, researchers recruited 45 women who said they were dog lovers. The EESEE story "Stress among Pets and Friends" describes the results. Fifteen of the subjects were randomly assigned to each of three groups to do a stressful task alone (the control

TABLE 24.3	Mean heart rates during stress with a pet (P), with a friend (F), and for the control group (C)				
Group	Rate	Group	Rate	Group	Rate
P	69.169	P	68.862	C	84.738
F	99.692	C	87.231	C	84.877
P	70.169	P	64.169	P	58.692
C	80.369	C	91.754	P	79.662
C	87.446	C	87.785	P	69.231
P	75.985	F	91.354	C	73.277
F	83.400	F	100.877	C	84.523
F	102.154	C	77.800	C	70.877
P	86.446	P	97.538	F	89.815
F	80.277	P	85.000	F	98.200
C	90.015	F	101.062	F	76.908
C	99.046	F	97.046	P	69.538
C	75.477	C	62.646	P	70.077
F	88.015	F	81.600	F	86.985
F	92.492	P	72.262	P	65.446

group), with a good friend present, or with their dog present. The subject's mean heart rate during the task is one measure of the effect of stress. Table 24.3 contains the data.

(a) Make stemplots of the heart rates for the three groups (round to the nearest whole number of beats). Do any of the groups show outliers or extreme skewness?

(b) Figure 24.5 gives the Minitab ANOVA output for these data. Do the mean heart rates for the groups appear to show that the presence of a pet or a friend reduces heart rate during a stressful task?

Minitab

FIGURE 24.5 Minitab output for the data in Table 24.3 on heart rates during stress, for Exercise 24.4. The "Control" group worked alone, the "Friend" group had a friend present, and the "Pet" group had a pet dog present.

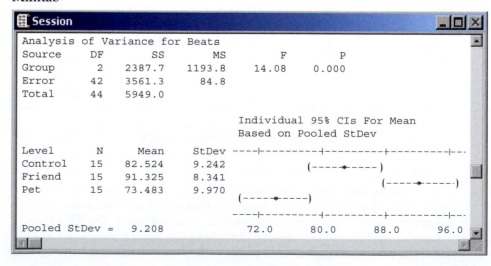

```
Session                                                      _ □ ×
Analysis of Variance for Beats
Source     DF        SS        MS         F         P
Group       2     2387.7    1193.8     14.08     0.000
Error      42     3561.3      84.8
Total      44     5949.0

                                 Individual 95% CIs For Mean
                                 Based on Pooled StDev
Level      N      Mean    StDev  ----+---------+---------+---------+--
Control   15    82.524    9.242                  (-----*-----)
Friend    15    91.325    8.341                        (-----*-----)
Pet       15    73.483    9.970    (-----*-----)
                                 ---+---------+---------+---------+--
Pooled StDev =   9.208            72.0      80.0      88.0      96.0
```

(c) What are the values of the ANOVA F statistic and its P-value? What hypotheses does F test? Briefly describe the conclusions you draw from these data. Did you find anything surprising?

The idea of analysis of variance

The details of ANOVA are a bit heavier computationally than those of other tests we have studied so far. (They appear in an optional section at the end of this chapter.) The main idea of ANOVA is both more accessible and much more important. Here it is: When we ask if a set of sample means gives evidence for differences among the population means, what matters is not how far apart the sample means are but how far apart they are *relative to the variability of individual observations*.

Look at the two sets of boxplots in Figure 24.6. For simplicity, these distributions are all symmetric, so that the mean and median are the same. The centerline in each boxplot is therefore the sample mean. Both sets of boxplots compare three samples with the same three means. Could differences this large easily arise just due to chance, or are they statistically significant?

- The boxplots in Figure 24.6(a) have tall boxes, which show lots of variation among the individuals in each group. With this much variation among individuals, we would not be surprised if another set of samples gave quite different sample means. The observed differences among the sample means could easily happen just by chance.

- The boxplots in Figure 24.6(b) have the same centers as those in Figure 24.6(a), but the boxes are much shorter. That is, there is much less variation among the individuals in each group. It is unlikely that any sample from the first group would have a mean as small as the mean of the second

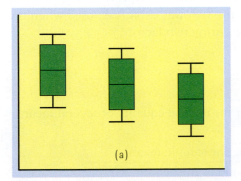
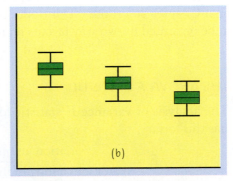

(a)   (b)

FIGURE 24.6 Boxplots for two sets of three samples each. The sample means are the same in (a) and (b) [a boxplot displays the median rather than the mean but the mean and median are the same if the distribution is symmetric, as is the case here]. Analysis of variance will find a more significant difference among the means in (b) because there is less variation among the individuals within those samples.

group. Because means as far apart as those observed would rarely arise just by chance in repeated sampling, they are good evidence of real differences among the means of the three populations we are sampling from.

You can use the *One-Way ANOVA* applet to demonstrate the analysis of variance idea for yourself. The applet allows you to change both the group means and the spread within groups. You can watch the ANOVA *F* statistic and its *P*-value change as you work.

This comparison of the two parts of Figure 24.6 is, of course, rudimentary. It ignores the effect of the sample sizes, an effect that boxplots do not show. *Small differences among sample means can be significant if the samples are large. Large differences among sample means can fail to be significant if the samples are small.* All we can be sure of is that for the same sample size, Figure 24.6(b) will give a much smaller *P*-value than Figure 24.6(a). Despite this caveat, the big idea remains: If sample means are far apart relative to the variation among individuals in the same groups, that's evidence that something other than chance is at work.

THE ANALYSIS OF VARIANCE IDEA

Analysis of variance compares the variation due to specific sources with the variation among individuals who should be similar. In particular, ANOVA tests whether several populations have the same mean by comparing how far apart the sample means are with how much variation there is within the samples.

It is one of the oddities of statistical language that methods for comparing means are named after the variance. The reason is that the test works by comparing two kinds of variation. Analysis of variance is a general method for studying sources of variation in responses. Comparing several means is the simplest form of ANOVA, *one-way ANOVA*   called **one-way ANOVA.** This chapter treats only the one-way ANOVA. Optional Chapter 26 on the CD introduces the two-way ANOVA, used when data can be organized in two-way layouts representing two factors.

THE ANOVA *F* STATISTIC

The **analysis of variance *F* statistic** for testing the equality of several means has this form:

$$F = \frac{\text{variation among the sample means}}{\text{variation among individuals in the same sample}}$$

The numerator is fundamentally a variance calculated using the sample means, while the denominator is fundamentally an average of the sample variances. Both are weighted to take into account different sample sizes. If you want more detail, read the optional section at the end of this chapter.

The F statistic can take only values that are zero or positive. It is zero only when all the sample means are identical and gets larger as they move farther apart. When H_0 is true, the numerator and denominator should have similar values and we would expect F to be small. F is large when the sample means are much more variable than individuals in the same sample. Large values of F are evidence against the null hypothesis H_0 that all population means are the same. Like the alternative hypothesis H_a, the ANOVA F test is nondirectional, because any violation of H_0 tends to produce a large value of F.

APPLY YOUR KNOWLEDGE

24.5 ANOVA compares several means. The *One-Way ANOVA* applet displays the observations in three groups, with the group means highlighted by black dots. When you open or reset the applet, the scale at the bottom of the display shows that for these groups the ANOVA F statistic is $F = 31.74$, with $P < 0.001$. (The P-value is marked by a red dot that moves along the scale.)

    (a) The middle group has larger mean than the other two. Grab its mean point with the mouse. How small can you make F? What did you do to the mean to make F small? Roughly how significant is your small F?

    (b) Starting with the three means aligned from your configuration at the end of (a), drag any one of the group means either up or down. What happens to F? What happens to the P-value? Convince yourself that the same thing happens if you move any one of the means, or if you move one slightly and then another slightly in the opposite direction.

24.6 ANOVA uses within-group variation. Reset the *One-Way ANOVA* applet to its original state. As in Figure 24.6(b), the differences among the three means are highly significant (large F, small P-value), because the observations in each group cluster tightly about the group mean.

    (a) Use the mouse to slide the Pooled Standard Error at the top of the display to the right. You see that the group means do not change but the spread of the observations in each group increases. What happens to F and P as the spread among the observations in each group increases? What are the values of F and P when the slider is all the way to the right? This is similar to Figure 24.6(a): Variation within groups hides the differences among the group means.

    (b) Leave the Pooled Standard Error slider at the extreme right of its scale, so that spread within groups stays fixed. Use the mouse to move the group means apart. What happens to F and P as you do this?

Conditions for ANOVA

Like all inference procedures, ANOVA is valid only in some circumstances. Here are the conditions under which we can use ANOVA to compare population means.

> ### CONDITIONS FOR APPLYING ANOVA
>
> - We have **k independent SRSs,** one from each of k populations.
> - Each of the k populations has a **Normal distribution** with an unknown mean. μ_i is the unknown mean of the ith population. The means may be different in the different populations. The ANOVA F statistic tests the null hypothesis that all of the populations have the same mean:
>
> $$H_0: \mu_1 = \mu_2 = \cdots = \mu_k$$
>
> $$H_a: \text{not all of the } \mu_i \text{ are equal } (H_0 \text{ is not true})$$
>
> - All of the populations have the **same standard deviation** σ, whose value is unknown.
>
> There are $k + 1$ population parameters that we must estimate from the data: the k population means and the standard deviation σ.

The first two requirements are familiar from our study of the two-sample t procedures for comparing two means. As usual, the design of the data production is the most important condition for inference. Biased sampling or confounding can make any inference meaningless. *If we do not actually draw separate SRSs from each population or carry out a randomized comparative experiment, it may be unclear to what population the conclusions of inference apply.* In the life sciences, samples are often not true SRSs but rather individuals randomly chosen from a pool of available individuals. You must judge each use on its merits, a judgment that usually requires some knowledge of the subject of the study in addition to some knowledge of statistics.

Because no real population has an exactly Normal distribution, the usefulness of inference procedures that assume Normality depends on how sensitive they are to departures from Normality. Fortunately, procedures for comparing means are not very sensitive to lack of Normality. The ANOVA F test, like the t procedures, is **robust.** What matters is Normality of the sample means, so ANOVA becomes safer as the sample sizes get larger, because of the central limit theorem effect. Remember to check for outliers that change the value of sample means and for extreme skewness. When there are no outliers and the distributions are roughly symmetric, you can safely use ANOVA for sample sizes as small as 4 or 5.

robustness

The third condition is annoying: ANOVA assumes that the variability of observations, measured by the standard deviation, is the same in all populations. You may recall from Chapter 18 (page 474) that there is a special version of the two-sample t test that assumes equal standard deviations in both populations. The ANOVA F for comparing two means is exactly the square of this special t statistic. We prefer the t test that does not assume equal standard deviations, but for comparing more than two means there is no general alternative to the ANOVA F. It is not easy to check the condition that the populations have equal standard deviations. Statistical tests for equality of standard deviations are very sensitive

Don't touch the plants

We know that confounding can distort inference. We don't always recognize how easy it is to confound data. Consider the innocent scientist who visits plants in the field once a week to measure their size. A study of six plant species found that one touch a week significantly increased leaf damage by insects in two species and significantly decreased damage in another species.

to lack of Normality, so much so that they are of little practical value. You must either seek expert advice or rely on the robustness of ANOVA.

How serious are unequal standard deviations? ANOVA is not too sensitive to violations of the condition, especially when all samples have the same or similar sizes and no sample is very small. When designing a study, try to take samples of about the same size from all the groups you want to compare. The sample standard deviations estimate the population standard deviations, so check before doing ANOVA that the sample standard deviations are similar to each other. We expect some variation among them due to chance. Here is a rule of thumb that is safe in almost all situations.

CHECKING STANDARD DEVIATIONS IN ANOVA

The results of the ANOVA F test are approximately correct when the largest sample standard deviation is no more than twice as large as the smallest sample standard deviation.

EXAMPLE 24.3 *Comparing tropical flowers: conditions for ANOVA*

The study of *Heliconia* blossoms is based on three independent samples that the researchers consider to be random samples from all flowers of these varieties in Dominica. The stemplots in Figure 24.1 show that the *bihai* and red varieties have slightly skewed distributions, but the sample means of samples of sizes 16 and 23 will have distributions that are close to Normal. The sample standard deviations for the three varieties are

$$s_1 = 1.213 \quad s_2 = 1.799 \quad s_3 = 0.975$$

These standard deviations satisfy our rule of thumb:

$$\frac{\text{largest } s}{\text{smallest } s} = \frac{1.799}{0.975} = 1.85$$

We can safely use ANOVA to compare the mean lengths for the three populations.

EXAMPLE 24.4 *Which color attracts beetles best?*

STATE: To detect the presence of harmful insects in farm fields, we can put up boards covered with a sticky material and examine the insects trapped on the boards. Which colors attract insects best? Experimenters placed six boards of each of four colors at random locations in a field of oats and measured the number of cereal leaf beetles trapped. Here are the data.[5]

Board color	Beetles trapped					
Blue	16	11	20	21	14	7
Green	37	32	20	29	37	32
White	21	12	14	17	13	20
Yellow	45	59	48	46	38	47

Holt Studios International/Alamy

FORMULATE: Examine the data to determine the effect of board color on beetles trapped and check that we can safely use ANOVA. If the data allow ANOVA, assess the significance of the observed differences in mean counts of beetles trapped.

SOLVE: Because the samples are small, we plot the data in side-by-side stemplots in Figure 24.7. CrunchIt! output for ANOVA appears in Figure 24.8. The yellow boards attract by far the most beetles ($\bar{x}_4 = 47.2$), with green next ($\bar{x}_2 = 31.2$) and blue and white far behind.

```
    Blue         Green        White        Yellow
  0 | 7        0 |          0 |          0 |
  1 | 146      1 |          1 | 2347      1 |
  2 | 01       2 | 09       2 | 01        2 |
  3 |          3 | 2277     3 |           3 | 8
  4 |          4 |          4 |           4 | 5678
  5 |          5 |          5 |           5 | 9
```

FIGURE 24.7 Side-by-side stemplots comparing the counts of insects attracted by six boards of each of four colors, for Example 24.4.

CrunchIt!

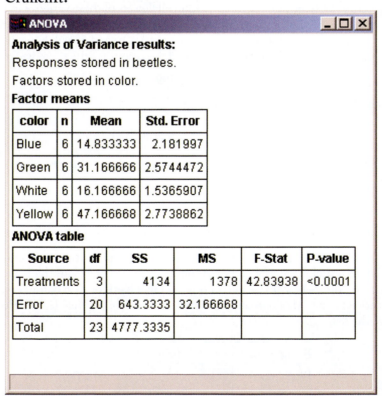

Analysis of Variance results:

Responses stored in beetles.

Factors stored in color.

Factor means

color	n	Mean	Std. Error
Blue	6	14.833333	2.181997
Green	6	31.166666	2.5744472
White	6	16.166666	1.5365907
Yellow	6	47.166668	2.7738862

ANOVA table

Source	df	SS	MS	F-Stat	P-value
Treatments	3	4134	1378	42.83938	<0.0001
Error	20	643.3333	32.166668		
Total	23	4777.3335			

FIGURE 24.8 CrunchIt! ANOVA output for comparing the four board colors in Example 24.4.

Check that we can safely use ANOVA to test equality of the four means. Because the standard error of $\bar{x}$ is $s/\sqrt{n}$, each sample standard deviation is $\sqrt{6}$ times the standard error given by CrunchIt! The largest of the four standard deviations is 6.795, and the smallest is 3.764. The ratio

$$\frac{\text{largest } s}{\text{smallest } s} = \frac{6.795}{3.764} = 1.8$$

is less than 2, so these data satisfy our rule of thumb. The shapes of the four distributions are irregular, as we expect with only 6 observations in each group, but there are no outliers. The ANOVA results will be approximately correct. The F statistic is $F = 42.84$, a large F with $P < 0.0001$.

CONCLUDE: Despite the small samples, the experiment gives very strong evidence of differences among the colors. Yellow boards appear to be best at attracting leaf beetles.

APPLY YOUR KNOWLEDGE

24.7 Checking standard deviations. Verify that the sample standard deviations for these sets of data do allow use of ANOVA to compare the population means.

(a) The counts of trees in Exercise 24.3 and Figure 24.4.

(b) The heart rates of Exercise 24.4 and Figure 24.5.

24.8 Species richness after logging. Table 24.2 gives data on the species richness in rain forest plots, defined as the number of tree species in a plot divided by the number of trees in the plot. ANOVA may not be trustworthy for the richness data. Do data analysis: Make side-by-side stemplots to examine the distributions of the response variable in the three groups, and also compare the standard deviations. What characteristic of the data makes ANOVA risky?

24.9 Smoking during pregnancy. Exercise 24.2 provided the mean meconium cotinine levels (in ng/ml) in newborns of mothers who were active smokers, passive smokers, or nonsmokers. The individuals participating in the study were consecutive pregnant women arriving at one hospital for delivery. Here are the raw data from the study:[6]

Active smokers	490	418	405	328	700	292	295	272	240	232
Passive smokers	254	219	287	257	271	282	148	273	350	293
Nonsmokers	158	163	153	207	211	159	199	187	200	213

(a) Examine each of the three samples. Do the standard deviations satisfy our rule of thumb? What are the overall shapes of the distributions? Are there outliers? Can we safely use ANOVA on these data?

(b) Suppose that the high outlier you found in part (a) was absent. Could we safely use ANOVA on the remaining data? Explain your answer. (It is not hopeless, though. When samples have very different standard deviations, transforming the original scale into a logarithmic scale can help obtain more similar standard deviations. We discussed logarithmic transformations previously, in the context of correlation and regression in Chapters 3 and 4.)

F *distribution*

F **distributions and degrees of freedom**

To find the P-value for the ANOVA F statistic, we must know the sampling distribution of F when the null hypothesis (all population means equal) is true. This sampling distribution is an **F distribution.**

The F distributions are a family of right-skewed distributions with two parameters. The parameters are the degrees of freedom of the numerator and of the denominator of the F statistic. The numerator's degrees of freedom are always mentioned first. *Interchanging the degrees of freedom changes the distribution, so the order is important.* Our brief notation will be $F(\text{df1}, \text{df2})$ for the F distribution with df1 degrees of freedom in the numerator and df2 in the denominator.

EXAMPLE 24.5 *Comparing tropical flowers: the F distribution*

Look again at the software output for the flower length data in Figure 24.3. All four outputs give the degrees of freedom for the F test, labeled "df" or "DF." There are 2 degrees of freedom in the numerator and 51 in the denominator. P-values for the F test therefore come from the F distribution with 2 and 51 degrees of freedom. Figure 24.9 shows the density curve of this distribution. The 5% critical value is 3.179, and the 1% critical value is 5.047. The observed value $F = 259.12$ of the ANOVA F statistic lies far to the right of these values, so the P-value is extremely small.

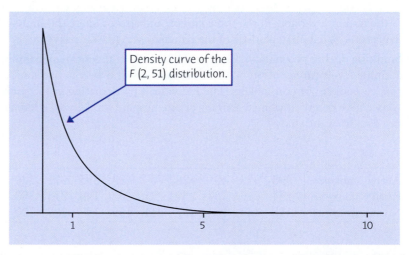

> Density curve of the F (2, 51) distribution.

FIGURE 24.9 The density curve of the F distribution with 2 degrees of freedom in the numerator and 51 degrees of freedom in the denominator, for Example 24.5.

The degrees of freedom of the ANOVA F statistic depend on the number of means we are comparing and the number of observations in each sample. That is, the F test takes into account the number of observations. Here are the details.

DEGREES OF FREEDOM FOR THE *F* TEST

We want to compare the means of k populations. We have an SRS of size n_i from the ith population, so that the total number N of observations in all samples combined is

$$N = n_1 + n_2 + \cdots + n_k$$

If the null hypothesis that all population means are equal is true, the ANOVA F statistic has the F distribution with $k - 1$ degrees of freedom in the numerator and $N - k$ degrees of freedom in the denominator.

EXAMPLE 24.6   Degrees of freedom for F

In Examples 24.1 and 24.2, we compared the mean lengths for three varieties of flowers, so $k = 3$. The three sample sizes are

$$n_1 = 16 \quad n_2 = 23 \quad n_3 = 15$$

The total number of observations is therefore

$$N = 16 + 23 + 15 = 54$$

The ANOVA F test has numerator degrees of freedom

$$k - 1 = 3 - 1 = 2$$

and denominator degrees of freedom

$$N - k = 54 - 3 = 51$$

These are the degrees of freedom given in the outputs in Figure 24.3.

Tables of F critical values are awkward because we need a separate table for every pair of degrees of freedom df1 and df2. Table F in the back of the book contains critical values for F distributions with various pairs of degrees of freedom. The critical values correspond to $p = 0.10, 0.05, 0.025, 0.01$, and 0.001. For example, here are the critical values for the $F(2, 50)$ distribution—a distribution similar to the $F(2, 51)$ distribution shown in Figure 24.9:

p	0.10	0.05	0.025	0.01	0.001
F^*	2.41	3.18	3.97	5.06	7.96

With $F = 259.12$ we can conclude that the test P-value for Example 24.5 is much smaller than 0.001.

You will rarely need Table F, however, because software gives P-values directly.

APPLY YOUR KNOWLEDGE

24.10 Logging in the rain forest, continued. Exercise 24.3 compares the number of tree species in rain forest plots that had never been logged (Group 1) with similar plots nearby that had been logged 1 year earlier (Group 2) and 8 years earlier (Group 3).

(a) What are k, the n_i, and N for these data? Identify these quantities in words and give their numerical values.

(b) Find the degrees of freedom for the ANOVA F statistic. Check your work against the Excel output in Figure 24.4.

(c) For these data, $F = 11.43$. What does Table F tell you about the P-value of this statistic?

24.11 Do fruit flies sleep? Exercise 24.1 describes a study of the effect of caffeine on sleep in fruit flies.

(a) What are the degrees of freedom for this experiment?

(b) Find in Table F the F distribution with the most similar degrees of freedom. Using this distribution, how large would the test F statistic need to be to reach significance at alpha 5%?

(c) The report gives $P < 0.0001$ for this ANOVA test. Based on Table F, we know that the test F statistic must have been larger than what value?

The one-way ANOVA and the pooled two-sample t test*

If the conditions for inference are met, the one-way ANOVA tests the hypothesis that all of k populations have the same mean,

$$H_0: \mu_1 = \mu_2 = \cdots = \mu_k$$
$$H_a: \text{not all of the } \mu_i \text{ are equal}$$

When there are just two populations, the hypotheses become

$$H_0: \mu_1 = \mu_2$$
$$H_a: \mu_1 \neq \mu_2$$

which can also be tested with a two-sample t procedure.

Because ANOVA requires equal population standard deviations, it is most closely related to the pooled two-sample t test. Recall from Chapter 18 (page 474) that the pooled two-sample t statistic is

$$t = \frac{\bar{x}_1 - \bar{x}_2}{s_p \sqrt{\dfrac{1}{n_1} + \dfrac{1}{n_2}}}$$

---

*This more advanced section is optional.

where s_p, the pooled estimate of the population standard deviation σ, is

$$s_p = \sqrt{\frac{(n_1 - 1)s_1^2 + (n_2 - 1)s_2^2}{n_1 + n_2 - 2}}$$

In fact, in the two-sample case the F statistic is exactly the square of the pooled t statistic

$$F = \frac{\text{variation among the sample means}}{\text{variation among individuals within samples}}$$

$$= \frac{(\bar{x}_1 - \bar{x}_2)^2}{s_p^2 \left(\dfrac{1}{n_1} + \dfrac{1}{n_2}\right)} = t^2$$

and both tests give exactly the same two-sided P-value.

We discussed in Chapter 18 the advantages of not having to assume equal population variances. When comparing two means, we recommend avoiding the pooled t procedure, because extensive studies show that t with its accurate degrees of freedom performs essentially as well as pooled t when the standard deviations really are equal and provides notably more accurate P-values when they are not. When we want to compare more than two means, however, there is no simple way to avoid the equal standard deviations condition.

The ANOVA F for more than two samples extends the idea of the two-sample procedure in a straightforward way. In fact, the denominator of F is exactly the weighted average of all the sample variances with degrees of freedom as weights (see for yourself that the sum of all the $n_i - 1$ is $N - k$). The numerator must compare k sample means rather than just 2, so it becomes a bit more complicated.

Details of ANOVA calculations*

Now we will give the general formula for the ANOVA F statistic. We have SRSs from each of k populations. Subscripts from 1 to k tell us which sample a statistic refers to:

Population	Sample size	Sample mean	Sample std. dev.
1	n_1	$\bar{x}_1$	s_1
2	n_2	$\bar{x}_2$	s_2
⋮	⋮	⋮	⋮
k	n_k	$\bar{x}_k$	s_k

---

*This more advanced section is optional if you are using software to find the F statistic.

You can find the F statistic from just the sample sizes n_i, the sample means $\bar{x}_i$, and the sample standard deviations s_i. You don't need to go back to the individual observations.

The ANOVA F statistic has the form

$$F = \frac{\text{variation among the sample means}}{\text{variation among individuals in the same sample}}$$

mean squares   The measures of variation in the numerator and denominator of F are called **mean squares.** A mean square is a more general form of a sample variance. An ordinary sample variance s^2 is an average (or mean) of the squared deviations of observations from their mean, so it qualifies as a "mean square."

The numerator of F is a mean square that measures variation among the k sample means $\bar{x}_1, \bar{x}_2, \ldots, \bar{x}_k$. Call the overall mean response (the mean of all N observations together) $\bar{x}$. You can find $\bar{x}$ from the k sample means by taking

$$\bar{x} = \frac{n_1\bar{x}_1 + n_2\bar{x}_2 + \cdots + n_k\bar{x}_k}{N}$$

The sum of each mean multiplied by the number of observations it represents is the sum of all the individual observations. Dividing this sum by N, the total number of observations, gives the overall mean $\bar{x}$. The numerator mean square in F is an average of the k squared deviations of the means of the samples from $\bar{x}$. We call it

MSG   the **mean square for groups,** abbreviated as MSG.

$$\text{MSG} = \frac{n_1(\bar{x}_1 - \bar{x})^2 + n_2(\bar{x}_2 - \bar{x})^2 + \cdots + n_k(\bar{x}_k - \bar{x})^2}{k - 1}$$

Each squared deviation is weighted by n_i, the number of observations it represents.

The mean square in the denominator of F measures variation among individual observations in the same sample. For any one sample, the sample variance s_i^2 does this job. For all k samples together, we use an average of the individual sample variances. It is again a weighted average in which each s_i^2 is weighted by one fewer than the number of observations it represents, $n_i - 1$. Another way to put this is that each s_i^2 is weighted by its degrees of freedom $n_i - 1$. The resulting mean

MSE   square is called the **mean square for error,** MSE.

$$\text{MSE} = \frac{(n_1 - 1)s_1^2 + (n_2 - 1)s_2^2 + \cdots + (n_k - 1)s_k^2}{N - k}$$

"Error" doesn't mean a mistake has been made. It's a traditional term for chance variation. Here is a summary of the ANOVA test.

THE ANOVA *F* TEST

Draw an independent SRS from each of k populations. The ith population has the $N(\mu_i, \sigma)$ distribution, where σ is the common standard deviation in all the populations. The ith sample has size n_i, sample mean $\bar{x}_i$, and sample standard deviation s_i.

The **ANOVA F statistic** tests the null hypothesis that all k populations have the same mean.

$$H_0: \mu_1 = \mu_2 = \cdots = \mu_k$$
$$H_a: \text{not all of the } \mu_i \text{ are equal } (H_0 \text{ is not true})$$

The statistic is

$$F = \frac{\text{MSG}}{\text{MSE}}$$

The numerator of F is the **mean square for groups**

$$\text{MSG} = \frac{n_1(\bar{x}_1 - \bar{x})^2 + n_2(\bar{x}_2 - \bar{x})^2 + \cdots + n_k(\bar{x}_k - \bar{x})^2}{k - 1}$$

The denominator of F is the **mean square for error**

$$\text{MSE} = \frac{(n_1 - 1)s_1^2 + (n_2 - 1)s_2^2 + \cdots + (n_k - 1)s_k^2}{N - k}$$

When H_0 is true, F has the **F distribution** with $k - 1$ and $N - k$ degrees of freedom.

The denominators in the formulas for MSG and MSE are the two degrees of freedom $k - 1$ and $N - k$ of the F test. The numerators are called **sums of squares,** from their algebraic form. It is usual to present the results of ANOVA in an **ANOVA table.** Output from software usually includes an ANOVA table.

sums of squares

ANOVA table

EXAMPLE 24.7 ANOVA calculations: software

Look again at the five outputs in Figure 24.3. The four software outputs give the ANOVA table. The TI-83, with its small screen, gives the degrees of freedom, sums of squares, and mean squares separately. Each output uses slightly different language to identify the two sources of variation. The basic ANOVA table is

Source of variation	df	SS	MS	F statistic
Variation among samples	2	1082.87	MSG = 541.44	259.12
Variation within samples	51	106.57	MSE = 2.09	

You can check that each mean square MS is the corresponding sum of squares SS divided by its degrees of freedom df. The F statistic is MSG divided by MSE.

Because MSE is an average of the individual sample variances, it is also called the *pooled sample variance*, written as s_p^2. When all k populations have the same population variance σ^2, as ANOVA assumes that they do, s_p^2 estimates the common variance σ^2. The square root of MSE is the **pooled standard deviation** s_p. It estimates the common standard deviation σ of observations in each group. The Minitab and TI-83 outputs in Figure 24.3 give the value $s_p = 1.446$.

pooled standard deviation

The pooled standard deviation s_p is a better estimator of the common σ than any individual sample standard deviation s_i because it combines (pools) the information in all k samples. We can get a confidence interval for any of the means μ_i from the usual form

$$\text{estimate} \pm t^* \text{SE}_{\text{estimate}}$$

using s_p to estimate σ. The confidence interval for μ_i is

$$\bar{x}_i \pm t^* \frac{s_p}{\sqrt{n_i}}$$

Use the critical value t^* from the t distribution with $N - k$ degrees of freedom, because s_p has $N - k$ degrees of freedom. These are the confidence intervals that appear in Minitab ANOVA output.

EXAMPLE 24.8   ANOVA calculations: without software

We can do the ANOVA test comparing the mean lengths of *bihai*, red, and yellow flower varieties using only the sample sizes, sample means, and sample standard deviations. These appear in Example 24.1, but it is easy to find them with a calculator. There are $k = 3$ groups with a total of $N = 54$ flowers.

The overall mean of the 54 lengths in Table 24.1 is

$$\bar{x} = \frac{n_1 \bar{x}_1 + n_2 \bar{x}_2 + n_3 \bar{x}_3}{N}$$

$$= \frac{(16)(47.598) + (23)(39.711) + (15)(36.180)}{54}$$

$$= \frac{2217.621}{54} = 41.067$$

The mean square for groups is

$$\text{MSG} = \frac{n_1(\bar{x}_1 - \bar{x})^2 + n_2(\bar{x}_2 - \bar{x})^2 + n_3(\bar{x}_3 - \bar{x})^2}{k - 1}$$

$$= \frac{1}{3 - 1}[(16)(47.598 - 41.067)^2 + (23)(39.711 - 41.067)^2$$

$$+ (15)(36.180 - 41.067)^2]$$

$$= \frac{1082.996}{2} = 541.50$$

The mean square for error is

$$MSE = \frac{(n_1 - 1)s_1^2 + (n_2 - 1)s_2^2 + (n_3 - 1)s_3^2}{N - k}$$

$$= \frac{(15)(1.213^2) + (22)(1.799^2) + (14)(0.975^2)}{51}$$

$$= \frac{106.580}{51} = 2.09$$

Finally, the ANOVA test statistic is

$$F = \frac{MSG}{MSE} = \frac{541.50}{2.09} = 259.09$$

Our work differs slightly from the output in Figure 24.3 because of roundoff error. We don't recommend doing these calculations, because tedium and roundoff errors cause frequent mistakes.

APPLY YOUR KNOWLEDGE

The calculations of ANOVA use only the sample sizes n_i, the sample means $\bar{x}_i$, and the sample standard deviations s_i. You can therefore re-create the ANOVA calculations when a report gives these summaries but does not give the actual data. These optional exercises ask you to do the ANOVA calculations starting with the summary statistics.

24.12 **An obesity crisis in children?** News reports speak of an emerging crisis of childhood obesity in the United States. The National Health and Nutrition Examination Survey (NHANES) is a government survey run every several years recording a number of vital statistics on a representative sample of Americans. A body mass index (BMI) is computed for each individual in the sample based on the individual's height and weight. Here are sample results for 8-year-old boys over the past 40 years. (SEM is the standard error of the mean.)[7]

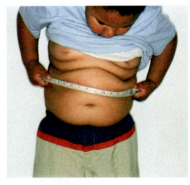

Pat Doyle/CORBIS

Survey	N	$\bar{x}$	SEM
NHES II, 1965	618	16.3	0.1
NHANES II, 1980	145	16.5	0.2
NHANES, 2002	214	18.4	0.4

(a) The distributions of BMIs are somewhat right-skewed. ANOVA is nonetheless safe for these data. Why?

(b) What is the relationship between the SEM and the standard deviation for a sample? Check that the standard deviations satisfy the guideline for ANOVA inference.

(c) Calculate the overall mean response $\bar{x}$, the mean squares MSG and MSE, and the ANOVA F statistic. What are the degrees of freedom for F?

(d) How significant are the differences among the three mean BMIs? The Centers for Disease Control and Prevention (CDC) keeps charts of healthy BMIs for children and teenagers of both sexes. For an 8-year-old boy, a BMI between 13.8 and 17.9 is considered healthy, whereas a BMI between 17.9

and 20 places the child at risk of becoming overweight. Interpret your ANOVA test in this context.

24.13 **Exercise and weight loss.** What conditions help overweight people exercise regularly? A study randomly assigned subjects to three treatments: a single long exercise period 5 days per week; several 10-minute exercise periods 5 days per week; and several 10-minute periods 5 days per week on a home treadmill that was provided to the subjects. The study report contains the following information about weight loss (in kilograms) after six months of treatment:[8]

Treatment	n	$\bar{x}$	s
Long exercise periods	37	10.2	4.2
Short exercise periods	36	9.3	4.5
Short periods with equipment	42	10.2	5.2

(a) Do the standard deviations satisfy the rule of thumb for safe use of ANOVA?

(b) Calculate the overall mean response $\bar{x}$ and the mean square for groups MSG.

(c) Calculate the mean square for error MSE.

(d) Find the ANOVA F statistic and its approximate P-value. Is there evidence that the mean weight losses of people who follow the three exercise programs differ?

24.14 **Did the randomization work?** When assigning patients with existing symptoms to various treatments in a clinical trial, it is important to verify that randomization does not create by chance uneven groups, with different overall levels of symptom severity. A study of low bone mineral density (BMD) among postmenopausal women randomly assigned subjects to one of five treatment groups. BMD at the lumbar spine was measured at the beginning of the study to assess symptom severity. BMD is expressed as a unitless, standardized value; a negative score indicates a BMD less than the normal healthy average of a population of young women. Here are the summarized BMD scores for the 5 groups at the beginning of the study:[9]

Treatment group	N	$\bar{x}$	s
Placebo	46	−2.2	0.7
Denosumab, 6 mg	44	−2.0	0.9
Denosumab, 14 mg	44	−2.0	0.8
Denosumab, 30 mg	41	−2.2	0.7
Alendronate	47	−2.0	0.9

Calculate the ANOVA table and the F statistic. Are the differences in mean BMD scores among the 5 groups significant? What does that say about the randomization process for this study?

CHAPTER 24 SUMMARY

One-way analysis of variance (ANOVA) compares the means of several populations. The **ANOVA F test** tests the overall H_0 that all the populations have the same mean. If the F test shows significant differences, examine the data to see where the differences lie and whether they are large enough to be important.

The **conditions for ANOVA** state that we have an **independent SRS** from each population; that each population has a **Normal distribution;** and that all populations have the **same standard deviation.**

In practice, ANOVA inference is relatively **robust** when the populations are non-Normal, especially when the samples are large. Before doing the F test, check the observations in each sample for outliers or strong skewness. Also verify that the largest sample standard deviation is no more than twice as large as the smallest standard deviation.

When the null hypothesis is true, the **ANOVA F statistic** for comparing k means from a total of N observations in all samples combined has the **F distribution** with $k - 1$ and $N - k$ degrees of freedom.

ANOVA calculations are reported in an **ANOVA table** that gives sums of squares, mean squares, and degrees of freedom for variation among groups and for variation within groups. In practice, we use software to do the calculations.

CHECK YOUR SKILLS

24.15 The purpose of analysis of variance is to compare

(a) the variances of several populations.

(b) the proportions of successes in several populations.

(c) the means of several populations.

24.16 A study of the effects of smoking classifies subjects as nonsmokers, moderate smokers, or heavy smokers. The investigators interview a sample of 200 people in each group. Among the questions is "How many hours do you sleep on a typical night?" The degrees of freedom for the ANOVA F statistic comparing mean hours of sleep are

(a) 2 and 197.   (b) 2 and 597.   (c) 3 and 597.

24.17 The alternative hypothesis for the ANOVA F test in the previous exercise is

(a) the mean hours of sleep in the groups are all the same.

(b) the mean hours of sleep in the groups are all different.

(c) the mean hours of sleep in the groups are not all the same.

The air in poultry processing plants often contains fungus spores. Large spore concentrations can affect the health of the workers. To measure the presence of spores, air samples are pumped to an agar plate and "colony-forming units (CFUs)" are counted after an incubation period. Here are data from the "kill room" of a plant that slaughters 37,000 turkeys per day. Each observation was made on a different day. The units are CFUs per cubic meter of air.[10]

Fall	Winter	Spring	Summer
1231	384	2105	3175
1254	104	701	2526
752	251	2947	1763
1088	97	842	1090

Here is Minitab's output for ANOVA to compare mean CFUs in the four seasons:

```
Source   DF       SS       MS      F      P
Season    3  8236359  2745453   5.38  0.014
Error    12  6124211   510351
Total    15 14360570
```

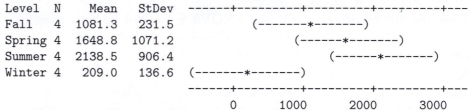

```
                              Individual 95% CIs For Mean Based on
                              Pooled StDev
Level  N    Mean   StDev   ------+---------+---------+---------+---
Fall   4  1081.3   231.5              (-------*-------)
Spring 4  1648.8  1071.2                (------*-------)
Summer 4  2138.5   906.4                   (------*-------)
Winter 4   209.0   136.6   (-------*-------)
                           ------+---------+---------+---------+---
                                 0      1000      2000      3000
```

Exercises 24.18 to 24.23 are based on this study.

24.18 The most striking conclusion from the numerical summaries for the turkey processing plant is that

(a) there appears to be little difference among the seasons.

(b) on average, CFUs are much lower in winter than in other seasons.

(c) the air in the plant is clearly unhealthy.

24.19 We might use the two-sample t procedures to compare summer and winter. The conservative 90% confidence interval for the difference in the two population means is

(a) 1929.5 ± 458.3.   (b) 1929.5 ± 1078.4.   (c) 1929.5 ± 1458.4.

24.20 In all, we would have to give 6 two-sample confidence intervals to compare all pairs of seasons. The weakness of doing this is that

(a) we don't know how confident we can be that all 6 intervals cover the true differences in means.

(b) 90% confidence is OK for one comparison, but it isn't high enough for 6 comparisons done at once.

(c) the conditions for two-sample t inference are not met for all 6 pairs of seasons.

24.21 The conclusion of the ANOVA test is that

(a) there is quite strong evidence ($P = 0.014$) that the mean CFUs are not the same in all four seasons.

(b) there is quite strong evidence ($P = 0.014$) that the mean CFUs are much lower in winter than in any other season.

(c) the data give no evidence ($P = 0.014$) to suggest that mean CFUs differ from season to season.

24.22 Without software, we would compare $F = 5.38$ with critical values from Table F. This comparison shows that

(a) the P-value is greater than 0.1.

(b) the P-value is between 0.01 and 0.025.

(c) the P-value is between 0.001 and 0.01.

24.23 The P-value 0.014 in the output may not be accurate, because the conditions for ANOVA are not satisfied. The most serious violation of the conditions is that

(a) the sample standard deviations are too different.

(b) there is an extreme outlier in the data.

(c) the data can't be regarded as random samples.

CHAPTER 24 EXERCISES

Exercises 24.24 to 24.27 describe situations in which we want to compare the mean responses in several populations. For each setting, identify the populations and the response variable. Then give k, n_i, and N. Finally, give the degrees of freedom of the ANOVA F test.

24.24 Aging and body weight. We typically associate body changes with either youth or old age, when these changes are most obvious. But what happens to the body over the course of active adulthood? The National Health and Nutrition Examination Survey takes representative samples of Americans and records various statistics such as gender, age, and weight. Here are data about the weight (in pounds) of men by age range for the 2002 survey (SEM is the standard error of the mean).[11]

Decade	N	$\bar{x}$	SEM
20s	712	183.4	1.5
30s	704	189.1	2.0
40s	776	196.0	1.6
50s	598	195.4	2.1

24.25 Biological rhythms. Are you a morning person, an evening person, or neither? Does this personality trait affect how well you perform? A sample of 100 students took a psychological test that found 16 morning people, 30 evening people, and 54 who were neither. All the students then took a test of their ability to memorize at 8 A.M. and again at 9 P.M. You analyze the score at 8 A.M. minus the score at 9 P.M.

24.26 An obesity crisis in children? Exercise 24.12 provided the BMIs of 8-year-old boys in three national surveys spanning the last 40 years. According to CDC charts, a healthy BMI for an 8-year-old girl is between 13.5 and 18.3. Here are sample results for 8-year-old girls from the three surveys of Exercise 24.12. (SEM is the standard error of the mean).

Survey	N	$\bar{x}$	SEM
NHES II, 1965	613	16.4	0.1
NHANES II, 1980	125	16.3	0.2
NHANES, 2002	184	18.3	0.5

24.27 A medical study. The Quebec (Canada) Cardiovascular Study recruited men aged 34 to 64 at random from towns in the Quebec City metropolitan area. Of these, 1824 met the criteria (no diabetes, free of heart disease, and so on) for a study of the relationship between being overweight and medical risks. The 719 normal-weight men had mean triglyceride level 1.5 millimoles per liter (mmol/l); the 885 overweight men had mean 1.7 mmol/l; and the 220 obese men had mean 1.9 mmol/l.[12]

24.28 Whose little bird is that? The males of many animal species signal their fitness to females by displays of various kinds. For male barn swallows, the signal is the color of their plumage. Does "better" color really lead to more offspring? A randomized comparative experiment assigned barn swallow pairs who had already produced a clutch of eggs to three groups: 13 males had their color "enhanced" by the investigators; 9 were handled but their color was left alone; and 8 were not touched. DNA testing showed whether the eggs were sired by the male of the pair or by another male. The investigators then destroyed the eggs. Barn swallows usually produce a second brood in such circumstances. Sure enough the males with enhanced color fathered more young in the second brood and males in the other groups fathered fewer. The investigators used ANOVA to compare the three groups.[13]

(a) What are the degrees of freedom for the F statistic that compares the numbers of the male's own young among the eggs in the first brood? The statistic was $F = 0.07$. What can you say about the P-value? What do you conclude?

(b) Of the 30 original pairs, 27 produced a second brood. What are the degrees of freedom for F to compare differences in the number of the male's own young between the first and second broods produced by these 27 pairs? The statistic was $F = 5.45$. How significant is this F?

24.29 Plants defend themselves. When some plants are attacked by leaf-eating insects, they release chemical compounds that attract other insects that prey on the leaf-eaters. An experiment carried out on plants growing naturally in the Utah desert examined the release of these protective compounds.[14] The investigators compared plants attacked by one of 3 types of leaf-eaters and other plants that were undamaged—32 plants of the same species in all, 8 per group. They then measured emissions of several compounds during seven hours. Here are data (mean ± standard error of the mean) for one compound. The emission rate is measured in nanograms per hour (ng/hr).

Group	Emission rate (ng/hr)
Control	9.22 ± 5.93
Hornworm	31.03 ± 8.75
Leaf bug	18.97 ± 6.64
Flea beetle	27.12 ± 8.62

(a) Make a graph that compares the mean emission rates for the four groups. Does it appear that emissions increase when the plant is attacked?

(b) What hypotheses does ANOVA test in this setting?

(c) We do not have the full data. What would you look for in deciding whether you can safely use ANOVA?

(d) What is the relationship between the SEM and the standard deviation for a sample? Do the standard deviations satisfy our rule of thumb for safe use of ANOVA?

24.30 Can you hear these words? To test whether a hearing aid is right for a patient, audiologists play a tape on which words are pronounced at low volume. The patient tries to repeat the words. There are several different lists of words that are supposed to be equally difficult. Are the lists equally difficult when there is background noise? To find out, an experimenter had subjects with normal hearing listen to four lists with a noisy background. The response variable was the percent of the 50 words in a list that the subject repeated correctly. The data set contains 96 responses.[15] Here are two study designs that could produce these data:

Design A. The experimenter assigns 96 subjects to 4 groups at random. Each group of 24 subjects listens to one of the lists. All individuals listen and respond separately.

Design B. The experimenter has 24 subjects. Each subject listens to all four lists in random order. All individuals listen and respond separately.

Does Design A allow use of one-way ANOVA to compare the lists? Does Design B allow use of one-way ANOVA to compare the lists? Briefly explain your answers.

24.31 Nematodes and tomato plants. How do nematodes (microscopic worms) affect plant growth? A botanist prepares 16 identical planting pots and then introduces different numbers of nematodes into the pots. He transplants a tomato seedling into each pot. Here are data on the increase in height of the seedlings (in centimeters) 16 days after planting.[16]

Nematodes	Seedling growth			
0	10.8	9.1	13.5	9.2
1,000	11.1	11.1	8.2	11.3
5,000	5.4	4.6	7.4	5.0
10,000	5.8	5.3	3.2	7.5

Figure 24.10 shows output from SPSS for ANOVA comparing the four groups of pots.

Eye of Science/Photo Researchers, Inc.

SPSS

Column Statistics _ □ ×

Descriptives

growth

	N	Mean	Std. Deviation	Std. Error	95% Confidence Interval for Mean Lower Bound	95% Confidence Interval for Mean Upper Bound	Minimum	Maximum
0	4	10.6500	2.05345	1.02673	7.3825	13.9175	9.10	13.50
1,000	4	10.4250	1.48633	.74316	8.0599	12.7901	8.20	11.30
5,000	4	5.6000	1.24365	.62183	3.6211	7.5789	4.60	7.40
10,000	4	5.4500	1.77106	.88553	2.6318	8.2682	3.20	7.50
Total	16	8.0313	2.98858	.74715	6.4387	9.6238	3.20	13.50

ANOVA

growth

	Sum of Squares	df	Mean Square	F	Sig.
Between Groups	100.647	3	33.549	12.080	.001
Within Groups	33.328	12	2.777		
Total	133.974	15			

FIGURE 24.10 SPSS ANOVA output for comparing the growth of tomato seedlings with different concentrations of nematodes in the soil, for Exercise 24.31.

Minitab

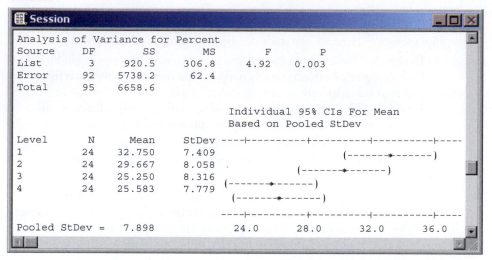

FIGURE 24.11 Minitab ANOVA output for comparing the percents heard correctly in four lists of words, for Exercise 24.32.

(a) Make a table of the means and standard deviations in the groups. Make side-by-side stemplots to compare the treatments. What do the data appear to show about the effect of nematodes on growth? Is use of ANOVA justified?

(b) State H_0 and H_a for the ANOVA test for these data, and explain in words what ANOVA tests in this setting.

(c) Report your overall conclusions about the effect of nematodes on plant growth.

24.32 Can you hear these words? Figure 24.11 displays the Minitab output for one-way ANOVA applied to the hearing data described in Exercise 24.30. The response variable is "Percent," and "List" identifies the four lists of words. Based on this analysis, is there good reason to think that the four lists are not all equally difficult? Write a brief summary of the study findings.

24.33 Evolution in bacteria. Can bacteria evolve a preference for the pH of their environment? An evolutionary biologist took lines of *Escherichia coli* bacteria grown and kept at a near-neutral pH of 7.2 and grew them for 2000 generations (about 300 days) at a stressful acidic pH of 5.5. The original bacteria (kept frozen all that time) and the "acid-evolved" bacteria were then grown together in various test environments, and a relative fitness score was computed. A score of 1 would indicate that both bacteria types were equally fit. A score higher than 1 would indicate that the acid-evolved line was more fit than the ancestral line in that environment (that is, that the acid-evolved bacteria grew the most). Here are the relative fitness scores obtained for test environments of pH 5.5 (acid), 7.2 (neutral), or 8.0 (basic). There are six replicates using different ancestral bacterial lines for each test environment.[17]

Test in acid pH	1.24	1.22	1.23	1.24	1.18	1.09
Test in neutral pH	0.99	0.99	0.98	0.94	0.95	0.95
Test in basic pH	0.56	0.83	0.82	0.72	0.86	0.84

Do the data provide evidence that *E. coli* can evolve adaptations to the pH of its environment over the course of 2000 generations? Follow the four-step process in answering this question. (Although the standard deviations do not quite satisfy our rule of thumb, that rule is conservative and small sample sizes can more easily lead to widely different variances. The analysis might be performed again without the low outlier to confirm the initial interpretation.)

24.34 **Does nature heal best?** Our bodies have a natural electrical field that helps wounds heal. Might higher or lower levels speed healing? An experiment with newts investigated this question. Newts were randomly assigned to five groups. In four of the groups, an electrode applied to one hind limb (chosen at random) changed the natural field, while the other hind limb was not manipulated. Both limbs in the fifth (control) group remained in their natural state.[18]

Table 24.4 gives data from this experiment. The "Group" variable shows the field applied as a multiple of the natural field for each newt. For example, "0.5" is half the natural field, "1" is the natural level (the control group), and "1.5" indicates a field 1.5 times the natural level. "Diff" is the response variable, the difference in the healing rate (in micrometers per hour) of cuts made in the experimental and control limbs of that newt. Negative values mean that the experimental limb healed more slowly. The investigators conjectured that nature heals best, so that changing the field from the natural state (the "1" group) will slow healing.

Do a complete analysis to see whether the groups differ in the effect of the electrical field level on healing. Follow the four-step process in your work.

TABLE 24.4   Effect of electrical field on healing rate in newts

Group	Diff	Group	Diff	Group	Diff	Group	Diff	Group	Diff
0	−10	0.5	−1	1	−7	1.25	1	1.5	−13
0	−12	0.5	10	1	15	1.25	8	1.5	−49
0	−9	0.5	3	1	−4	1.25	−15	1.5	−16
0	−11	0.5	−3	1	−16	1.25	14	1.5	−8
0	−1	0.5	−31	1	−2	1.25	−7	1.5	−2
0	6	0.5	4	1	−13	1.25	−1	1.5	−35
0	−31	0.5	−12	1	5	1.25	11	1.5	−11
0	−5	0.5	−3	1	−4	1.25	8	1.5	−46
0	13	0.5	−7	1	−2	1.25	11	1.5	−22
0	−2	0.5	−10	1	−14	1.25	−4	1.5	2
0	−7	0.5	−22	1	5	1.25	7	1.5	10
0	−8	0.5	−4	1	11	1.25	−14	1.5	−4
		0.5	−1	1	10	1.25	0	1.5	−10
		0.5	−3	1	3	1.25	5	1.5	2
				1	6	1.25	−2	1.5	−5
				1	−1				
				1	13				
				1	−8				

24.35 Does polyester decay? How quickly do synthetic fabrics such as polyester decay in landfills? A researcher buried polyester strips in the soil for different lengths of time, then dug up the strips and measured the force required to break them. Breaking strength is easy to measure and is a good indicator of decay; lower strength means the fabric has decayed.

Part of the study buried 20 polyester strips in well-drained soil in the summer. Five of the strips, chosen at random, were dug up after 2 weeks, five after 4 weeks, and another five after 8 weeks. Here are the breaking strengths in pounds:[19]

2 weeks	118	126	126	120	129
4 weeks	130	120	114	126	128
8 weeks	122	136	128	146	140

The investigator conjectured that buried polyester loses strength over time. Do the data support this conjecture? Follow the four-step process in data analysis and ANOVA. Be sure to check the conditions for ANOVA.

24.36 Neural basis of 3D vision. One way the brain assesses depth in a visual scene is by using the disparity of information received by both retinas. Stereograms, like 3D movies, contain a visual pattern duplicated and shifted a bit horizontally. Each pattern is seen by just one eye through the use of special 3D glasses, introducing retinal disparity. The result is an image that appears to float either in front of or behind the screen, depending on the horizontal disparity introduced.

Some neurons in the primary visual cortex specialize in the perception of depth from binocular disparity. One such neuron was recorded in a monkey shown stereograms of various disparities in random order. Figure 24.12(a) shows the action potentials ("spikes") generated over time (horizontal axis) by that neuron for different disparities. Figure 24.12(b) shows the "tuning curve"

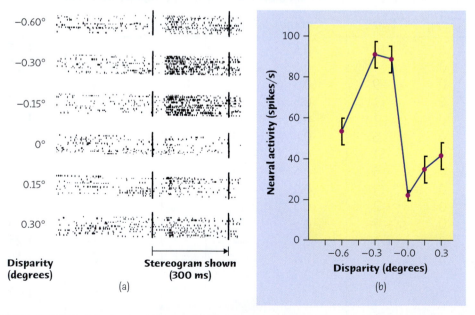

FIGURE 24.12 Action potentials (dots) over time in response to 3D visual stimuli of different disparities (a) and bar graph of the mean response plus or minus standard error of the mean for each disparity value (b), for Exercise 24.36.

representing the mean neural response ($\pm$ SE) for each disparity. Here are the raw data from 14 replications:[20]

Disparity	Neural response to stereogram (spikes/s)						
−0.6	10	46.7	83.3	80	40	36.7	46.7
−0.3	30	110	100	90	70	56.7	90
−0.15	70	93.3	86.7	76.7	103.3	53.3	106.7
0	10	13.3	36.7	16.7	36.7	16.7	20
0.15	13.3	63.3	10	63.3	93.3	16.7	10
0.3	6.7	23.3	50	43.3	76.7	10	46.7
Disparity	Neural response to stereogram (spikes/s)						
−0.6	46.7	70	76.7	20	53.3	66.7	83.3
−0.3	90	130	103.3	60	86.7	126.7	123.3
−0.15	120	100	126.7	30	100	113.3	96.7
0	23.3	23.3	16.7	26.7	23.3	13.3	36.7
0.15	26.7	23.3	70	3.3	23.3	20	40
0.3	83.3	60	26.7	16.7	30	46.7	63.3

A negative disparity corresponds to the impression of a near object. This neuron clearly responds more strongly to objects that appear in the near space. Does your analysis support this interpretation? Use the four-step process to guide your discussion.

24.37 **Compressing soil.** Farmers know that driving heavy equipment on wet soil compresses the soil and injures future crops. Table 2.2 (page 63) gives data on the "penetrability" of the same soil at three levels of compression.[21] Penetrability is a measure of how much resistance plant roots will meet when they try to grow through the soil. Do the data suggest that penetrability decreases as soil is more compressed? Follow the four-step process in data analysis and ANOVA. Be sure to check the conditions for ANOVA, paying special attention to outliers.

STEP

24.38 **Logging in the rain forest: species counts.** Table 24.2 gives data on the number of trees per forest plot, the number of species per plot, and species richness. Exercise 24.3 analyzed the effect of logging on number of trees. Exercise 24.8 concluded that it would be risky to use ANOVA to analyze richness. Use software to analyze the effect of logging on the number of species.

(a) Make a table of the group means and standard deviations. Do the standard deviations satisfy our rule of thumb for safe use of ANOVA? What do the means suggest about the effect of logging on the number of species?

(b) Carry out the ANOVA. Report the F statistic and its P-value and state your conclusion.

24.39 **Plant defenses (optional).** Calculations of ANOVA use only the sample sizes n_i, the sample means $\bar{x}_i$, and the sample standard deviations s_i. You can therefore re-create the ANOVA calculations when a report gives these summaries but does not give the actual data. Use the information in Exercise 24.29 to calculate the ANOVA table (sums of squares, degrees of freedom, mean squares, and F statistic). Note that the report gives the SEM rather than the standard deviation. Are there significant differences among the mean emission rates for the four populations of plants?

24.40 F versus t (optional). Acclimation is a process of adaptation to a persistent change in environment taking place typically over the course of weeks. When goldfish are relocated to a colder aquarium, for instance, their ventilation rate immediately slows down but, after weeks, moves back toward its previous value. This acclimation process can be complete (the rates before and after acclimation are exactly the same) or only partial. Here are ventilation rates (in beats per minute) for goldfish that have been acclimated to either a cold environment (12 degrees Celsius) or a warm environment (22 degrees Celsius):[22]

	Cold-acclimated	Warm-acclimated
n_i	18	18
$\bar{x}_i$	55.8	78.3
s_i	18.4	25.8

A significant difference in ventilation rate between warm- and cold-acclimated goldfish would indicate that acclimation was not complete.

(a) Calculate the two-sample t statistic for testing H_0: $\mu_1 = \mu_2$ against the two-sided alternative. Use the conservative method to find the P-value.

(b) Calculate MSG, MSE, and the ANOVA F statistic for the same hypotheses. What is the P-value of F?

(c) How close are these two P-values?

24.41 ANOVA or chi-square? (optional). Exercise 24.13 describes a randomized, comparative experiment that assigned subjects to three types of exercise programs intended to help them lose weight. Some of the results of this study were analyzed using the chi-square test for two-way tables, and some others were analyzed using one-way ANOVA. For each of the following excerpts from the study report, say which analysis is appropriate and explain how you made your choice.

(a) "Overall, 115 subjects (78% of 148 subjects randomized) completed 18 months of treatment, with no significant difference in attrition rates between the groups ($P = .12$)."

(b) "In analyses using only the 115 subjects who completed 18 months of treatment, there were no significant differences in weight loss at 6 months among the groups."

(c) "The duration of exercise for weeks 1 through 4 was significantly greater in the SB compared with both LB and SBEQ groups ($P < .05$). . . . However, exercise duration was greater in SBEQ compared with both LB and SB groups for months 13 through 18 ($P < .05$)." LB, SB, and SBEQ refer to the three treatment groups described in Exercise 24.13.

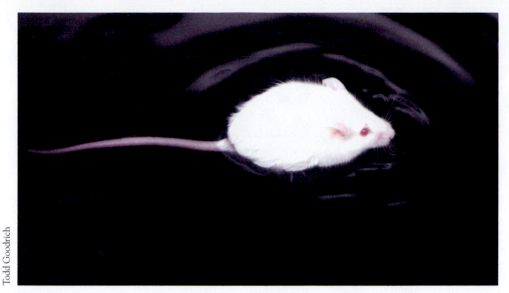

Todd Goodrich

Inference: Part III Review

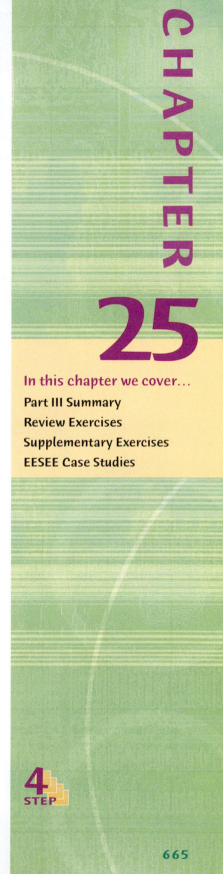
The procedures of Chapters 17 to 24 are among the most common of all statistical inference methods. The Statistics in Summary flowcharts in Figures 25.1 and 25.2 help you decide when to use which one. It is important to do some of the review exercises, because now, for the first time, you must decide which of several inference procedures to use. Learning to recognize problem settings in order to choose the right type of inference is a key step in advancing your mastery of statistics.

The flowchart for means and proportions is based on two key questions:

- **What population parameter** does your problem concern? You should be familiar in detail with inference for *means* (Chapters 17 and 18) and for *proportions* (Chapters 19, 20, and 21). This is the first branch point in the flowchart.

- **What type of design** produced the data? You should be familiar with data from a *single sample*, from *matched pairs*, and from *two independent samples*. This is the second branch point in the flowchart. Your choice here is the same for experiments and for observational studies—inference methods are the same for both. But don't forget that experiments give much better evidence that an effect uncovered by inference can be explained by direct causation.

The flowchart for relationships is based on this simple question: **What type of variables** make up the relationship? You should be familiar in detail with inference for relationships between *two categorical variables* (Chapter 22), between *two quantitative variables* (Chapter 23), and between *a quantitative and a categorical variable* (Chapter 24).

Answering these questions is part of the *Formulate* step in the four-step process. To begin the *Solve* step, follow the flowcharts to the correct procedure. Then ask another question:

- **Are the conditions for this procedure met?** Can you act as if the data come from a random sample or randomized comparative experiment? Do the data

4 STEP

Figure 25.1 Inference about Means and Proportions

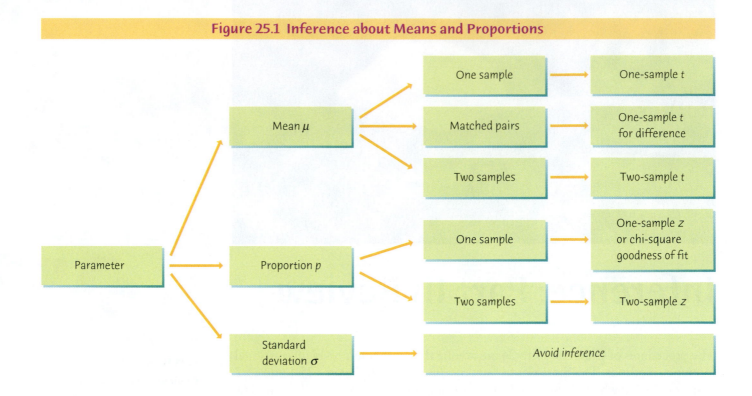

Figure 25.2 Inference for Relationships

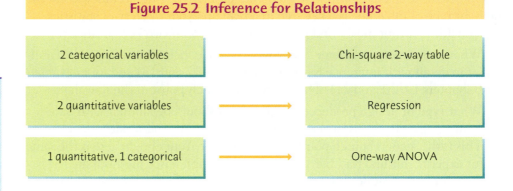

Where are the details?

Papers reporting scientific research must be short. Brevity allows researchers to hide details about their data. Did they choose their subjects in a biased way? Did they report data on only some of their subjects? Did they try several statistical analyses and report the one that looked best? The statistician John Bailar screened more than 4000 papers for the *New England Journal of Medicine*. He says, "When it came to the statistical review, it was often clear that critical information was lacking, and the gaps nearly always had the practical effect of making the authors' conclusions look stronger than they should have."

show extreme outliers or strong skewness that forbid use of inference based on Normality? Do you have enough observations for your intended procedure?

There is, of course, more to doing statistics well than choosing and carrying out the proper procedures. At the end of the book you will find a short outline of "Statistical Thinking Revisited," which will remind you of some of the big ideas that guide statistics in practice, as well as warn of some of the most common pitfalls. Look at that outline now, especially if you are near the end of your study of this book.

PART III SUMMARY

Here are the most important skills you should have acquired from reading Chapters 17 to 24.

A. RECOGNITION

1. Recognize when a problem requires inference about population means (quantitative response variable), population proportions (usually categorical response variable), or relationships.

2. Recognize from the design of a study whether one-sample, matched pairs, or two-sample procedures for one variable are needed, or whether the design assesses the relationship between two variables.

3. Based on recognizing the problem setting, choose among the one- and two-sample t procedures for means and the one- and two-sample z procedures for proportions for one-variable problems. For relationships, choose among the chi-square procedures for two categorical variables, regression procedures for two quantitative variables, and the ANOVA F procedures for a quantitative and a categorical variable.

B. INFERENCE ABOUT ONE MEAN

1. Verify that the t procedures are appropriate in a particular setting. Check the study design and the distribution of the data, and take advantage of robustness against lack of Normality.

2. Recognize when poor study design, outliers, or a small sample from a skewed distribution make the t procedures risky.

3. Use the one-sample t procedure to obtain a confidence interval at a stated level of confidence for the mean μ of a population.

4. Carry out a one-sample t test for the hypothesis that a population mean μ has a specified value against either a one-sided or a two-sided alternative. Use software to find the P-value or Table C to get an approximate value.

5. Recognize matched pairs data and use the t procedures to obtain confidence intervals and to perform tests of significance for such data.

C. COMPARING TWO MEANS

1. Verify that the two-sample t procedures are appropriate in a particular setting. Check the study design and the distribution of the data and take advantage of robustness against lack of Normality.

2. Give a confidence interval for the difference between two means. Use software if you have it. Use the two-sample t statistic with conservative degrees of freedom and Table C if you do not have statistical software.

3. Test the hypothesis that two populations have equal means against either a one-sided or a two-sided alternative. Use software if you have it. Use the two-sample t test with conservative degrees of freedom and Table C if you do not have statistical software.

4. Know that procedures for comparing the standard deviations of two Normal populations are available but that these procedures are risky because they are not at all robust against non-Normal distributions.

D. INFERENCE ABOUT ONE PROPORTION

1. Verify that you can safely use either the large-sample or the plus four z procedures in a particular setting. Check the study design and the guidelines for sample size.

2. Use the large-sample z procedure to give a confidence interval for a population proportion p. Understand that the true confidence level may be substantially less than you ask for unless the sample is very large and the true p is not close to 0 or 1.

3. Use the plus four modification of the z procedure to give a confidence interval for p that is accurate even for small samples and for any value of p.

4. Use the z statistic to carry out a test of significance for the hypothesis $H_0\colon p = p_0$ about a population proportion p against either a one-sided or a two-sided alternative. Use software or Table B to find the P-value, or Table C to get an approximate value.

E. COMPARING TWO PROPORTIONS

1. Verify that you can safely use either the large-sample or the plus four z procedures in a particular setting. Check the study design and the guidelines for sample sizes.

2. Use the large-sample z procedure to give a confidence interval for the difference $p_1 - p_2$ between proportions in two populations based on independent samples from the populations. Understand that the true confidence level may be less than you ask for unless the samples are quite large.

3. Use the plus four modification of the z procedure to give a confidence interval for $p_1 - p_2$ that is accurate even for very small samples and for any values of p_1 and p_2.

4. Use a z statistic to test the hypothesis $H_0\colon p_1 = p_2$ that proportions in two distinct populations are equal. Use software or Table B to find the P-value, or Table C to get an approximate value.

F. CHI-SQUARE TEST FOR GOODNESS OF FIT

1. Explain what null hypothesis the chi-square for goodness of fit tests.

2. Locate the chi-square statistic, its P-value, and other useful facts (percents, expected counts, terms of chi-square) in output from your software or calculator.

3. Use the expected counts to check whether you can safely use the chi-square test.

4. If the test is significant, compare percents, compare observed with expected counts, or look for the largest terms of the chi-square statistic to see what deviations from the null hypothesis are most important.

G. CHI-SQUARE TEST FOR TWO-WAY TABLES

1. Know how to organize data representing two categorical variables as a two-way table of counts of outcomes.

2. Use percents to describe the relationship between any two categorical variables, starting from the counts in a two-way table.

3. Locate the chi-square statistic, its P-value, and other useful facts (row or column percents, expected counts, terms of chi-square) in output from your software or calculator.

4. Use the expected counts to check whether you can safely use the chi-square test.

5. Explain what null hypothesis the chi-square statistic tests in a specific two-way table.

6. If the test is significant, compare percents, compare observed with expected cell counts, or look for the largest terms of the chi-square statistic to see what deviations from the null hypothesis are most important.

H. CHI-SQUARE COMPUTATIONS BY HAND

1. When doing a chi-square test by hand, first calculate all expected counts. Check whether you can safely use the chi-square test.

2. Know that expected counts for a goodness of fit test are based on the null hypothesis and sample sizes, whereas those for a two-way test are based on the observed counts displayed in the two-way table.

3. Calculate the terms of the chi-square statistic, as well as those of the overall statistic.

4. Give the degrees of freedom of the chi-square statistic. Use the chi-square critical values in Table D to approximate the P-value of the chi-square test.

I. PRELIMINARIES OF INFERENCE FOR REGRESSION

1. Make a scatterplot to show the relationship between an explanatory and a response variable.

2. Use a calculator or software to find the correlation and the equation of the least-squares regression line.

3. Recognize the regression setting: a straight-line relationship between an explanatory variable x and a response variable y.

4. Recognize which type of inference you need in a particular regression setting.

5. Inspect the data to recognize situations in which inference isn't safe: a nonlinear relationship, influential observations, strongly skewed residuals

in a small sample, or nonconstant variation of the data points about the regression line.

J. INFERENCE FOR REGRESSION USING SOFTWARE OUTPUT

1. Explain in any specific regression setting the meaning of the slope β of the population regression line.

2. Understand software output for regression. Find in the output the slope and intercept of the least-squares line, their standard errors, and the regression standard error.

3. Use that information to carry out tests of H_0: $\beta = 0$ and calculate confidence intervals for β.

4. Explain the distinction between a confidence interval for the mean response and a prediction interval for an individual response.

5. If your software gives output for prediction, use that output to give either confidence or prediction intervals.

K. PRELIMINARIES OF ANOVA

1. Recognize when testing the equality of several means is helpful in understanding data.

2. Recognize when you can safely use ANOVA to compare means. Check the data production, the presence of outliers, and the sample standard deviations for the groups you want to compare.

3. Know that the statistical significance of differences among sample means depends on the sizes of the samples and on how much variation there is within the samples.

L. INTERPRETING ANOVA SOFTWARE OUTPUT

1. Explain what null hypothesis F tests in a specific setting.

2. Locate the F statistic and its P-value on the output of analysis of variance software.

3. Find the degrees of freedom for the F statistic from the number and sizes of the samples. Use Table F of the F distributions to approximate the P-value when software does not give it.

4. If the test is significant, use graphs and descriptive statistics to see what differences among the means are most important. (Optional Chapter 26 on the CD covers the details of some common follow-up tests used when the F test is significant.)

REVIEW EXERCISES

Review exercises are short and straightforward exercises that help you solidify the basic ideas and skills from Chapters 17 to 24.

*For exercises that call for inference, your answers should include the **Formulate, Solve,** and **Conclude** steps of the four-step process. It is helpful to also summarize in your own words the **State** information given in the exercise. For tests of significance, use a table of*

critical values to approximate P-values unless you use software that reports the
P-value. For confidence intervals for proportions, use the plus four procedures.

25.1 **Comparing universities.** You have data on a random sample of students from each of two large universities. Which procedure from the Statistics in Summary flowchart would you use to compare

(a) the percents of students enrolled in a university-sponsored physical activity?

(b) the average amount of time spent in a university-sponsored physical activity in a typical week?

25.2 **Acid rain?** You have data on rainwater collected at 16 locations in the Adirondack Mountains of New York State. One measurement is the acidity of the water, measured by pH on a scale of 0 to 14. (The pH of distilled water is 7.0.) Which procedure from the Statistics in Summary flowchart would you use to estimate the average acidity of rainwater in the Adirondacks?

25.3 **Looking back on love.** How do young adults look back on adolescent romance? Investigators interviewed 40 couples in their mid-twenties. The female and male partners were interviewed separately. Each was asked about their current relationship and also about a romantic relationship that lasted at least two months when they were aged 15 or 16. One response variable was a measure on a numerical scale of how much the attractiveness of the adolescent partner mattered. Which of the tests in the Statistics in Summary flowchart should you use to compare the men and women on this measure?

(a) *t* for means or *z* for proportions?

(b) one-sample, matched pairs, or two-sample?

25.4 **Preventing AIDS through education.** The Multisite HIV Prevention Trial was a randomized comparative experiment to compare the effects of twice-weekly small-group AIDS discussion sessions (the treatment) with a single one-hour session (the control). Which of the procedures in the Statistics in Summary flowcharts should you use to compare the effects of treatment and control on each of the following response variables?

(a) A subject does or does not use condoms six months after the education sessions.

(b) The number of acts of unprotected intercourse by a subject between four and eight months after the sessions.

(c) A subject is or is not infected with a sexually transmitted disease six months after the sessions.

25.5 **Potato chips.** Here is the description of the design of a study that compared subjects' responses to regular and fat-free potato chips:

> *During a given 2-wk period, the participants received the same type of chip (regular or fat-free) each day. After the first 2-wk period, there was a 1-wk washout period in which no testing was performed. There was then another 2-wk period in which the participants received the alternate type of potato chip (regular or fat-free) each day under the same protocol. The order in which the chips were presented was determined by random assignment.*[1]

One response variable was the weight in grams of potato chips that a subject ate. We want to compare the amounts of regular and fat-free chips eaten.

(a) The subjects were 44 women. Explain how the design should use random assignment. Use the *Simple Random Sample* applet, other software, or Table A to do the randomization. If you use Table A, start at line 101 and assign only the first 5 women.

(b) Which test from the Statistics in Summary flowchart should you use to test the null hypothesis of no difference between regular and fat-free chips?

25.6   **Mouse endurance.** A study of the inheritance of speed and endurance in mice found a trade-off between these two characteristics, both of which help mice survive. To test endurance, mice were made to swim in a bucket with a weight attached to their tails. (The mice were rescued when exhausted.) Here are data on endurance in minutes for female and male mice.[2]

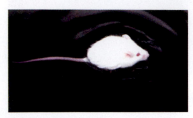

Todd Goodrich

Group	n	Mean	Std. dev.
Female	162	11.4	26.09
Male	135	6.7	6.69

(a) Both sets of endurance data are skewed to the right. Why are *t* procedures nonetheless reasonably accurate for these data?

(b) Give a 95% confidence interval for the mean endurance of female mice swimming.

(c) Give a 95% confidence interval for the mean difference (female minus male) in endurance times.

25.7   **Mouse endurance, continued.** Do the data in the previous exercise show that female mice have significantly higher endurance on average than male mice?

25.8   **An extinct beast.** *Archaeopteryx* is an extinct beast having feathers like a bird but teeth and a long bony tail like a reptile. Here are the lengths in centimeters (cm) of the femur (a leg bone) and the humerus (a bone in the upper arm) for the five fossil specimens that preserve both bones.[3]

Femur (x)	38	56	59	64	74
Humerus (y)	41	63	70	72	84

The strong linear relationship between the lengths of the two bones helped persuade scientists that all five specimens belonged to the same species. Examine the data and describe the relationship both qualitatively and numerically. Make a scatterplot with femur length as the explanatory variable. Do you think that femur length would allow good prediction of humerus length?

25.9   **An extinct beast, continued.** Using software on the data from the previous exercise, we find that the least-squares slope is $b = 1.1969$ with standard error $SE_b = 0.0751$.

(a) Explain in words what the slope β of the true regression line says about *Archaeopteryx*. What is the *t* statistic for testing $H_0: \beta = 0$?

(b) How many degrees of freedom does *t* have? Use Table C to approximate the *P*-value of *t* against the one-sided alternative $H_a: \beta > 0$. What do you conclude?

25.10 Genetically modified foods. Europeans have been more skeptical than Americans about the use of genetic engineering to improve foods. A sample survey gathered responses from random samples of 863 Americans and 12,178 Europeans.[4] (The European sample was larger because Europe is divided into many nations.) Subjects were asked to consider the following issue.

> *Using modern biotechnology in the production of foods, for example to make them higher in protein, keep longer, or change in taste.*

They were asked if they considered this "risky for society." In all, 52% of Americans and 64% of Europeans thought the practice was risky.

(a) It is clear without a formal test that the proportion of the population who consider this use of technology risky is significantly higher in Europe than in the United States. Why is this?

(b) Give a 99% confidence interval for the percent difference between Europe and the United States.

25.11 Genetically modified foods, continued. Give a 95% confidence interval for the proportion of all European adults who consider the use of biotechnology in food production risky.

25.12 Genetically modified foods, continued. Is there convincing evidence that more than half of all adult Americans consider applying biotechnology to the production of foods risky? If you do not use software, use the appropriate table to find the P-value.

25.13 Butterflies mating. Here's how butterflies mate: A male passes to a female a packet of sperm called a spermatophore. Females may mate several times. Will they remate sooner if the first spermatophore they receive is small? Among 20 females who received a large spermatophore (more than 25 milligrams [mg]), the mean time to the next mating was 5.15 days, with standard deviation 0.18 day. For 21 females who received a small spermatophore (about 7 mg), the mean was 4.33 days and the standard deviation was 0.31 day.[5] Is the observed difference in means statistically significant?

Tim Zurowski/CORBIS

25.14 Very-low-birth-weight babies. Starting in the 1970s, medical technology allowed babies with very low birth weight (VLBW, less than 1500 grams (g), or about 3.3 pounds) to survive without major handicaps. It was noticed that these children nonetheless had difficulties in school and as adults. A long-term study has followed 242 VLBW babies to age 20 years, along with a control group of 233 babies from the same population who had normal birth weight.[6]

(a) Is this an experiment or an observational study? Why?

(b) At age 20, 179 of the VLBW group and 193 of the control group had graduated from high school. Is the graduation rate among the VLBW group significantly lower than that for the normal-birth-weight controls?

25.15 Very-low-birth-weight babies, continued. IQ scores were available for 113 men in the VLBW group. The mean IQ was 87.6, and the standard deviation was 15.1. The 106 men in the control group had mean IQ 94.7, with standard deviation 14.9. Is there good evidence that mean IQ is lower among VLBW men than among controls from similar backgrounds?

25.16 Very-low-birth-weight babies, continued. Of the 126 women in the VLBW group, 37 said they had used illegal drugs; 52 of the 124 control group women had done so. The IQ scores for these VLBW women had mean 86.2 (standard deviation 13.4), and the normal-birth-weight controls had mean IQ 89.8 (standard deviation 14.0). Is either of these differences between the two groups statistically significant?

25.17 Do fruit flies sleep? Mammals and birds sleep. Fruit flies show a daily cycle of rest and activity, but does the rest qualify as sleep? Researchers looking at brain activity and behavior finally concluded that fruit flies do sleep. A small part of the study used an infrared motion sensor to see if flies moved in response to vibrations. Here are results for low levels of vibration.[7]

| | Response to vibration? | |
	No	Yes
Fly was walking	10	54
Fly was resting	28	4

Analyze these results. Is there good reason to think that resting flies respond differently from flies that are walking? (That's a sign that the resting flies may actually be sleeping.)

25.18 California brushfires. We often see televised reports of brushfires threatening homes in California. Some people argue that the modern practice of quickly putting out small fires allows fuel to accumulate and so increases the damage done by large fires. A detailed study of historical data suggests that this is wrong—the damage has risen simply because there are more houses in risky areas.[8] As usual, the study report gives statistical information tersely. Here is the summary of a regression of number of fires on decades (9 data points, for the 1910s to the 1990s): "Collectively, since 1910, there has been a highly significant increase ($r^2 = 0.61$, $P < 0.01$) in the number of fires per decade." How would you explain this statement to someone who knows no statistics? Include an explanation of both the description given by r^2 and its statistical significance.

25.19 Cholesterol in dogs. High levels of cholesterol in the blood are not healthy in either humans or dogs. Because a diet rich in saturated fats raises the cholesterol level, it is plausible that dogs owned as pets have higher cholesterol levels than dogs owned by a veterinary research clinic. "Normal" levels of cholesterol based on the clinic's dogs would thus be misleading. A clinic compared healthy dogs it owned with healthy pets brought to the clinic to be neutered. The summary statistics for blood cholesterol levels (milligrams per deciliter of blood [mg/dl]) appear below.[9]

Group	n	$\bar{x}$	s
Pets	26	193	68
Clinic	23	174	44

Is there strong evidence that pets have a higher mean cholesterol level than clinic dogs?

25.20 Pets versus clinic dogs. Using the information in the previous exercise, give a 95% confidence interval for the difference in mean cholesterol levels between pets and clinic dogs.

25.21 **Cholesterol in pet dogs.** Continue your work with the information in Exercise 25.19. Give a 95% confidence interval for the mean cholesterol level in pet dogs.

25.22 **Conditions for inference.** What conditions must be satisfied to justify the procedures you used in Exercise 25.19? In Exercise 25.20? In Exercise 25.21? Assuming that the cholesterol measurements have no outliers and are not strongly skewed, what is the chief threat to the validity of the results of this study?

25.23 **Fish sizes.** Table 25.1 contains data on the size of perch caught in a lake in Finland.[10] Statistical software will help you analyze these data.

(a) How well we can predict the width of a perch from its length?

TABLE 25.1	Measurements on 56 perch						
Observation number	Weight (gs)	Length (cm)	Width (cm)	Observation number	Weight (gs)	Length (cm)	Width (cm)
104	5.9	8.8	1.4	132	197.0	27.0	4.2
105	32.0	14.7	2.0	133	218.0	28.0	4.1
106	40.0	16.0	2.4	134	300.0	28.7	5.1
107	51.5	17.0	2.6	135	260.0	28.9	4.3
108	70.0	18.5	2.9	136	265.0	28.9	4.3
109	100.0	19.2	3.3	137	250.0	28.9	4.6
110	78.0	19.4	3.1	138	250.0	29.4	4.2
111	80.0	20.2	3.1	139	300.0	30.1	4.6
112	85.0	20.8	3.0	140	320.0	31.6	4.8
113	85.0	21.0	2.8	141	514.0	34.0	6.0
114	110.0	22.5	3.6	142	556.0	36.5	6.4
115	115.0	22.5	3.3	143	840.0	37.3	7.8
116	125.0	22.5	3.7	144	685.0	39.0	6.9
117	130.0	22.8	3.5	145	700.0	38.3	6.7
118	120.0	23.5	3.4	146	700.0	39.4	6.3
119	120.0	23.5	3.5	147	690.0	39.3	6.4
120	130.0	23.5	3.5	148	900.0	41.4	7.5
121	135.0	23.5	3.5	149	650.0	41.4	6.0
122	110.0	23.5	4.0	150	820.0	41.3	7.4
123	130.0	24.0	3.6	151	850.0	42.3	7.1
124	150.0	24.0	3.6	152	900.0	42.5	7.2
125	145.0	24.2	3.6	153	1015.0	42.4	7.5
126	150.0	24.5	3.6	154	820.0	42.5	6.6
127	170.0	25.0	3.7	155	1100.0	44.6	6.9
128	225.0	25.5	3.7	156	1000.0	45.2	7.3
129	145.0	25.5	3.8	157	1100.0	45.5	7.4
130	188.0	26.2	4.2	158	1000.0	46.0	8.1
131	180.0	26.5	3.7	159	1000.0	46.6	7.6

(b) The length of a typical perch is about $x^* = 27$ cm. Predict the mean width of such fish and give a 95% confidence interval.

(c) Perch number 143 had six newly eaten fish in its stomach. Examine the residuals. Does fish number 143 have an unusually large residual? What impact might this have on inference?

25.24 **Transforming a variable (Optional).** We can also use the data in Table 25.1 to study the prediction of the weight of a perch from its length.

(a) Describe the pattern of the relationship between weight and length, with length as the explanatory variable.

(b) It is reasonable to expect the one-third power (cube root) of the weight to have a straight-line relationship with length, since animals are three-dimensional. Use your software to create a new variable that is the one-third power of weight. Describe the pattern of this new response variable against length. Is the straight-line pattern now stronger or weaker than that in (a)?

(c) Find the least-squares regression line to predict the new weight variable from length. Predict the mean of the new variable for perch 27 cm long, and give a 95% CI.

SUPPLEMENTARY EXERCISES

*Supplementary exercises apply the skills you have learned in ways that require more thought or more use of technology. Many of these exercises, for example, start from the raw data rather than from data summaries. Remember that the **Solve** step for a statistical problem includes checking the conditions for the inference you plan.*

25.25 **Starting to talk.** At what age do children of English-speaking parents speak their first word? Here are data on 20 infants (ages in months).[11]

15	26	10	9	15	20	18	11	8	20
7	9	10	11	11	10	12	17	11	10

(In fact, the sample contained one more child, who began to speak at 42 months. Child development experts consider this abnormally late, so we dropped the outlier to get a sample of "normal" children. The investigators are willing to treat these data as an SRS.) Is there good evidence that the mean age at first word among all normal children is greater than one year?

25.26 **Starting to talk, continued.** Use the data in the previous exercise to give a 90% confidence interval for the mean age at which children speak their first word.

25.27 **Do parents matter?** A professor asked her sophomore students, "Does either of your parents allow you to drink alcohol around him or her?" and "How many drinks do you typically have per session? (A drink is defined as one 12-oz beer, one 4-oz glass of wine, or one 1-oz shot of liquor.)" Table 25.2 contains the responses of the female students who are not abstainers.[12] The sample is all students in one large sophomore-level class. The class is popular, so we are tentatively willing to regard its members as an SRS of sophomore students at this college. Does the behavior of parents make a significant difference in drinks students have on average?

TABLE 25.2	Drinks per session by female students											
Parent allows student to drink												
2.5	1	2.5	3	1	3	3	3	2.5	2.5	3.5	5	2
7	7	6.5	4	8	6	6	3	6	3	4	7	5
3.5	2	1	5	3	3	6	4	2	7	5	8	1
6	5	2.5	3	4.5	9	5	4	4	3	4	6	4
5	1	5	3	10	7	4	4	4	4	2	2.5	2.5
Parent does not allow student to drink												
9	3.5	3	5	1	1	3	4	4	3	6	5	3
8	4	4	5	7	7	3.5	3	10	4	9	2	7
4	3	1										

25.28 Parents' behavior. We wonder what proportion of female students have at least one parent who allows them to drink around him or her. Table 25.2 contains information about a sample of 94 students. Use this sample to give a 95% confidence interval for this proportion.

25.29 Diabetic mice. The body's natural electrical field helps wounds heal. If diabetes changes this field, that might explain why people with diabetes heal slowly. A study compared normal mice and mice bred to spontaneously develop diabetes. The investigators attached sensors to the right hip and front paw of the mice and measured the difference in electrical potential between these locations. Here are the data, in millivolts.[13]

Diabetic mice						Normal mice				
14.70	13.60	7.40	1.05	10.55	16.40	13.80	9.10	4.95	7.70	9.40
10.00	22.60	15.20	19.60	17.25	18.40	7.20	10.00	14.55	13.30	6.65
9.80	11.70	14.85	14.45	18.25	10.15	9.50	10.40	7.75	8.70	8.85
10.85	10.30	10.45	8.55	8.85	19.20	8.40	8.55	12.60		

(a) Make a stemplot of each sample of potentials. There is a low outlier in the diabetic group. Does it appear that potentials in the two groups differ in a systematic way?

(b) Is there significant evidence of a difference in mean potentials between the two groups?

(c) Repeat your inference without the outlier. Does the outlier affect your conclusion?

25.30 Falling through the ice. Table 25.3 gives the dates on which a wooden tripod fell through the ice of the Tanana River in Alaska for the years 1917 to 2006. The dates determine who wins the Nenana Ice classic contest by guessing the correct date. Give a 95% confidence interval for the mean date on which the tripod falls through the ice. After calculating the interval in the scale used in the table (days from April 20, which is Day 1), translate your result into calendar dates.

TABLE 25.3		Days from April 20 for the Tanana River tripod to fall									
Year	Day	Year	Day	Year	Day	Year	Day	Year	Day	Year	Day
1917	11	1932	12	1947	14	1962	23	1977	17	1992	25
1918	22	1933	19	1948	24	1963	16	1978	11	1993	4
1919	14	1934	11	1949	25	1964	31	1979	11	1994	10
1920	22	1935	26	1950	17	1965	18	1980	10	1995	7
1921	22	1936	11	1951	11	1966	19	1981	11	1996	16
1922	23	1937	23	1952	23	1967	15	1982	21	1997	11
1923	20	1938	17	1953	10	1968	19	1983	10	1998	1
1924	22	1939	10	1954	17	1969	9	1984	20	1999	10
1925	16	1940	1	1955	20	1970	15	1985	23	2000	12
1926	7	1941	14	1956	12	1971	19	1986	19	2001	19
1927	23	1942	11	1957	16	1972	21	1987	16	2002	18
1928	17	1943	9	1958	10	1973	15	1988	8	2003	10
1929	16	1944	15	1959	19	1974	17	1989	12	2004	5
1930	19	1945	27	1960	13	1975	21	1990	5	2005	9
1931	21	1946	16	1961	16	1976	13	1991	12	2006	13

25.31 Keeping crackers from breaking. We don't like to find broken crackers when we open the package. How can manufacturers reduce breaking? One idea is to microwave the crackers for 30 seconds right after baking them. Analyze the following results from two experiments intended to examine this idea.[14] Does microwaving significantly improve indicators of future breaking? How large is the improvement? What do you conclude about the idea of microwaving crackers?

(a) The experimenter randomly assigned 65 newly baked crackers to be microwaved and another 65 to a control group that is not microwaved. Fourteen days after baking, 3 of the 65 microwaved crackers and 57 of the 65 crackers in the control group showed visible cracking which is the starting point for breaks.

(b) The experimenter randomly assigned 20 crackers to be microwaved and another 20 to a control group. After 14 days, he broke the crackers. Here are summaries of the pressure needed to break them, in pounds per square inch.

	Microwave	Control
Mean	139.6	77.0
Standard deviation	33.6	22.6

25.32 Mouse genes. A study of genetic influences on diabetes compared normal mice with similar mice genetically altered to remove a gene called *aP2*. Mice of both types were allowed to become obese by eating a high-fat diet. The researchers then measured the levels of insulin (in nanograms per milliliter [ng/ml]) and glucose (in mg/dl) in their blood plasma. Here are some excerpts from their findings.[15] The normal mice are called "wild-type," and the altered mice are called "*aP2*$^{-/-}$."

Each value is the mean ± SEM of measurements on at least 10 mice. Mean values of each plasma component are compared between aP2$^{-/-}$ mice and wild-type controls by Student's t test (*P < 0.05 and **P < 0.005).

Parameter	Wild type	aP2$^{-/-}$
Insulin (ng/ml)	5.9 ± 0.9	0.75 ± 0.2**
Glucose (mg/dl)	230 ± 25	150 ± 17*

Despite much greater circulating amounts of insulin, the wild-type mice had higher blood glucose than the aP2$^{-/-}$ animals. These results indicate that the absence of aP2 interferes with the development of dietary obesity-induced insulin resistance.

Other biologists are supposed to understand the statistics reported so tersely.

(a) What does "SEM" mean? What is the expression for SEM based on n, $\bar{x}$, and s from a sample?

(b) Which of the tests we have studied did the researchers apply?

(c) Explain to a biologist who knows no statistics what $P < 0.05$ and $P < 0.005$ mean. Which is stronger evidence of a difference between the two types of mice?

25.33 Mouse genes, continued. The report quoted in the previous exercise says only that the sample sizes were "at least 10." Suppose that the results are based on exactly 10 mice of each type. Use the values in the table to find $\bar{x}$ and s for the insulin concentrations in the two types of mice. Carry out a test to assess the significance of the difference in mean insulin concentration. Does your P-value confirm the claim in the report that $P < 0.005$?

25.34 Is wine good for your heart? There is some evidence that drinking moderate amounts of wine helps prevent heart attacks. Table 25.4 gives data on yearly wine consumption (liters of alcohol from drinking wine, per person) and yearly deaths from heart disease (deaths per 100,000 people) in 19 developed nations.[16] Is

TABLE 25.4 Wine consumption and heart attacks

Country	Alcohol from wine	Heart disease deaths	Country	Alcohol from wine	Heart disease deaths
Australia	2.5	211	Netherlands	1.8	167
Austria	3.9	167	New Zealand	1.9	266
Belgium	2.9	131	Norway	0.8	227
Canada	2.4	191	Spain	6.5	86
Denmark	2.9	220	Sweden	1.6	207
Finland	0.8	297	Switzerland	5.8	115
France	9.1	71	United Kingdom	1.3	285
Iceland	0.8	211	United States	1.2	199
Ireland	0.7	300	West Germany	2.7	172
Italy	7.9	107			

there statistically significant evidence that the correlation between wine consumption and heart disease deaths is negative? What reservations might you have about this inference?

25.35 Python eggs. How is the hatching of water python eggs influenced by the temperature of the snake's nest? Researchers assigned newly laid eggs to one of three temperatures: hot, neutral, or cold. Hot duplicates the extra warmth provided by the mother python, and cold duplicates the absence of the mother. Here are the data on the total number of eggs and the number that hatched.[17]

	Eggs	Hatched
Cold	27	16
Neutral	56	38
Hot	104	75

The researchers anticipated that eggs would not hatch at cold temperatures. Do the data support that anticipation? Are there significant differences among the proportions of eggs that hatched in the three groups?

25.36 Weights of newly hatched pythons. The previous study also examined the little pythons that had hatched. The report summarized the data in the common form "mean ± standard error" as follows.

David A. Northcott/CORBIS

Temperature	n	Weight (g) at hatching	Propensity to strike
Cold	16	28.89 ± 8.08	6.40 ± 5.67
Neutral	38	32.93 ± 5.61	5.82 ± 4.24
Hot	75	32.27 ± 4.10	4.30 ± 2.70

Is there evidence that nest temperature affects the mean weight of newly hatched pythons?

25.37 Python strikes. The data in the previous exercise also describe the "propensity to strike" of the hatched pythons at 30 days of age. This is the number of taps on the head with a small brush it takes until the python strikes. (Don't try this with adult pythons.) The data are again summarized in the form "sample mean ± standard error of the mean." Does nest temperature appear to influence propensity to strike?

25.38 Vaccine protection against cervical cancer. Studies worldwide show that almost all cases of cervical cancer are linked to certain oncogenic types of the human papilloma virus (HPV). The pharmaceutical company Merck developed a vaccine ("Gardasil") against the most common oncogenic types of HPV. Data about the vaccine's efficacy are available on its Web site.[18] Worldwide clinical trials followed young women 16 to 26 years of age for 2 to 4 years after vaccination or administration of a placebo for signs of cervical cancer. The results reported are caused by cervical intraepithelial neoplasia (CIN) and adenocarcinoma in situ (AIS) HPV-16 and HPV-18.

	n	Cases of CIN2	Cases of CIN3/AIS
Gardasil	8487	0	0
Placebo	8460	21	32

Do these results find the vaccine effective at preventing cervical cancer? Write a summary describing the study findings and explaining some of its limitations.

EESEE CASE STUDIES

The Electronic Encyclopedia of Statistical Examples and Exercises (EESEE) is available on the text CD and Web site. These more elaborate stories, with data, provide settings for longer case studies. Here are some suggestions for EESEE stories that apply the ideas you have learned in Part III.

25.39 Is Caffeine Dependence Real? Answer Questions 2, 3, 4, and 6 for this case study. (Matched pairs study.)

25.40 Seasonal Weevil Migration. Respond to Question 1. (Proportions.)

25.41 Radar Detectors and Speeding. Read this case study and answer Questions 1, 3, and 5. (Study design, proportions.)

25.42 Leave Survey after the Beep. Carefully answer Question 3; in part (a), use the preferred two-sample t procedure rather than the pooled t. (Two-sample problems, choice of procedure.)

25.43 Passive Smoking and Respiratory Health. Write careful answers to Questions 1, 3, and 4. (Conditions for inference, choice of procedure.)

25.44 Emissions from an Oil Refinery. Answer both questions. (Conditions for inference.)

25.45 Surgery in a Blanket. Read this case study and answer all questions. (Study design, choice of procedure, interpretation of results.)

25.46 Blood Alcohol Content. Read this case study and answer all questions. (Regression.)

Statistical Thinking Revisited ——————

The Thinking Person's Guide to Basic Statistics

We began our study of statistics with a look at "Statistical Thinking." We end with a review in outline form of the most important ideas of basic statistics, combining statistical thinking with your new knowledge of statistical practice. The outline contains some important warnings: Look for the Caution icon.

1. Data Production

- Data basics:

    Individuals (subjects).

    Variables: categorical versus quantitative, units of measurement, explanatory versus response.

    Purpose of study.

- Data production basics:

    Observation versus experiment.

    Simple random samples.

    Completely randomized experiments.

- *Beware: Really bad data production (voluntary response, confounding) can make interpretation impossible.*

- *Beware: Weaknesses in data production (for example, sampling students at only one campus) can make generalizing conclusions difficult.*

2. Data Analysis

- Always plot your data. Look for overall pattern and striking deviations.

- Add numerical descriptions based on what you see.

- *Beware: Averages and other simple descriptions can miss the real story.*

- One quantitative variable:

    Graphs: stemplot, dotplot, time plot, histogram, boxplot.

    Pattern: distribution shape, center, spread; outliers?

    Numerical descriptions: five-number summary or $\overline{x}$ and s.

- Relationships between two quantitative variables:

    Graph: scatterplot.

    Pattern: relationship form, direction, strength; outliers? influential observations?

    Numerical descriptions for linear relationships: correlation, regression line.

    Beware the lurking variable: Correlation does not imply causation.

- *Beware the effects of outliers and influential observations.*

3. The Reasoning of Inference

- Inference uses data to infer conclusions about a wider population.

- When you do inference, you are acting as if your data came from random samples or randomized comparative experiments. *Beware: If they don't, you may have "garbage in, garbage out."*

- Always examine your data before doing inference. Inference often requires a regular pattern, such as roughly Normal with no strong outliers.

- Key idea: "What would happen if we did this many times?"

- Confidence intervals: estimate a population parameter.

   95% confidence: I used a method that captures the true parameter 95% of the time in repeated use.

   Beware: The margin of error of a confidence interval does not include the effects of practical errors such as undercoverage and nonresponse.

STATISTICS IN SUMMARY

Overview of basic inference methods

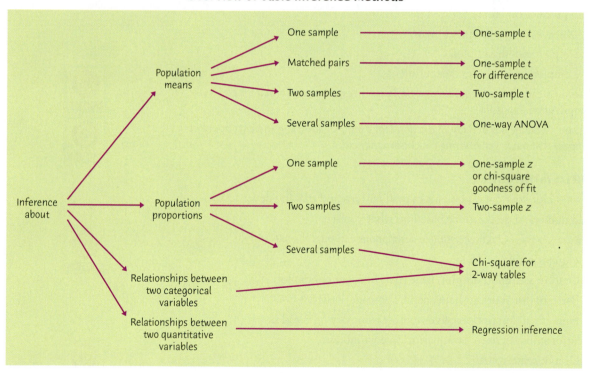

- Significance tests: assess evidence against H_0 in favor of H_a.

   P-value: If H_0 were true, how often would I get an outcome favoring the alternative this strongly? Smaller P is stronger evidence against H_0.

   Statistical significance at the 5% level, $P < 0.05$, means that an outcome this extreme would occur less than 5% of the time if H_0 were true.

Beware: P < 0.05 is not sacred.

Beware: Statistical significance is not the same as practical significance. Large samples can make small effects significant. Small samples can fail to declare large effects significant.

Always try to estimate the size of an effect (for example, with a confidence interval), not just its significance.

- Choose inference procedures by asking "What parameter?" and "What study design?" See the Statistics in Summary overview.

Notes and Data Sources

"About This Book"

1. This summary of the committee's report was unanimously endorsed by the Board of Directors of the American Statistical Association. The full report is G. Cobb, "Teaching statistics," in L. A. Steen (ed.), *Heeding the Call for Change: Suggestions for Curricular Action*, Mathematical Association of America, 1990, pp. 3–43.

2. A summary of the GAISE "Introductory College Course Guidelines," which have also been endorsed by the ASA Board of Directors, appears in *Amstat News*, June 2006, p. 31. See `www.amstat.org/education/gaise` for details.

3. G. Gigerenzer and A. Edwards, "Simple tools for understanding risks: from innumeracy to insight," *BMJ*, 327 (2003), pp. 741–744.

4. L. D. Brown, T. Cai, and A. DasGupta, "Interval estimation for a binomial proportion," *Statistical Science*, 16 (2001), pp. 101–133.

"Statistical Thinking"

1. E. W. Campion, "Editorial: power lines, cancer, and fear," *New England Journal of Medicine*, 337, No. 1 (1997). The study report is M. S. Linet et al., "Residential exposure to magnetic fields and acute lymphoblastic leukemia in children," in the same issue. See also G. Taubes, "Magnetic field-cancer link: will it rest in peace?" *Science*, 277 (1997), p. 29.

2. The data are part of a larger data set in the *Journal of Statistics Education* archive, accessible via the Internet. The original source is P. Brofeldt, "Bidrag till kaennedom on fiskbestondet i vaara sjoear. Laengelmaevesi," in T. H. Jaervi, *Finlands Fiskeriet*, Band 4, *Meddelanden utgivna av fiskerifoereningen i Finland*, Helsinki, 1917. The data were contributed to the archive (with information in English) by Juha Puranen of the University of Helsinki.

3. Contributed by Marigene Arnold of Kalamazoo College.

4. K. Hartmann et al., "Outcomes of routine episiotomy: a systematic review," *Journal of the American Medical Association*, 293 (2005), pp. 2141–2148.

5. We thank Rudi Berkelhamer of the University of California at Irvine for the data. The data are part of a larger set collected for an undergraduate lab exercise in scientific methods.

6. A. C. Nielsen, Jr., "Statistics in marketing," in *Making Statistics More Effective in Schools of Business*, Graduate School of Business, University of Chicago, 1986.

7. H. C. Sox, "Editorial: benefit and harm associated with screening for breast cancer," *New England Journal of Medicine*, 338, No. 16 (1998).

Chapter 1

1. National Health Interview Survey, 2006 data release.

2. The Electronic Encyclopedia of Statistical Examples and Exercises (EESEE) is a free resource for statistical education. Many of the examples in the EESEE database are available on the text CD and Web site.

3. National Center for Health Statistics, `http://www.cdc.gov/nchs`.

4. 2004 Statistical Abstract of the United States, Table 186. Current users are those who used drugs at least once within the month prior to the study.

5. Health, United States, 2005, Table 60, `http://www.cdc.gov/nchs/data/hus/hus05.pdf`.

6. National Center for Health Statistics, *Births: Final Data for 2003*, National Vital Statistics Reports, 54 (2005), No. 2.

7. Data provided by Chris Olsen, who found the information in scuba-diving magazines.

8. Our eyes do respond to area, but not quite linearly. It appears that we perceive the ratio of two bars to be about the 0.7 power of the ratio of their actual areas. See W. S. Cleveland, *The Elements of Graphing Data*, Wadsworth, 1985, pp. 278–284.

9. Data provided by Drina Iglesia, Purdue University. The data are part of a larger study reported in D. D. S. Iglesia, E. J. Cragoe, Jr., and J. W. Vanable, "Electric field strength and epithelialization in the newt (*Notophthalmus viridescens*)," *Journal of Experimental Zoology*, 274 (1996), pp. 56–62.

10. T. Bjerkedal, "Acquisition of resistance in guinea pigs infected with different doses of virulent tubercle bacilli," *American Journal of Hygiene*, 72 (1960), pp. 130–148.

11. The data are part of a larger data set in the *Journal of Statistics Education* archive, accessible via the Internet. The original source is P. Brofeldt, "Bidrag till kaennedom on fiskbestondet i vaara sjoear. Laengelmaevesi," in T. H. Jaervi, *Finlands Fiskeriet*, Band 4, *Meddelanden utgivna av fiskerifoereningen i Finland*, Helsinki, 1917. The data were contributed to the archive (with information in English) by Juha Puranen of the University of Helsinki.

12. Statistical Society of Canada, 2004 case study II. Data contributed by Caroline Davis, Elizabeth Blackmore, Deborah Katzman, and John Fox.

13. D. L. Arsenau, "Comparison of diet management instruction for patients with non-insulin dependent diabetes mellitus: learning activity package vs. group instruction," MS thesis, Purdue University, 1993.

14. R. W. Schaffranek and A. L. Riscassi, *Flow Velocity, Water Temperature, and Conductivity at Selected Locations in Shark River Slough, Everglades National Park, Florida, July 1999–July 2003*, Data Series 110, U.S. Geological Survey, 2004, `water.usgs.gov`.

15. Read from a graph in P. A. Raymond and J. J. Cole, "Increase in the export of alkalinity from North America's largest river," *Science*, 301 (2003), pp. 88–91.

16. Environmental Protection Agency, *Municipal Solid Waste in the United States: 2000 Facts and Figures*, document EPA530-R-02-001, 2002.

17. Read from a graph in K. Emanuel, "Increasing destructiveness of tropical cyclones over the past 30 years," *Nature*, 436 (2005), pp. 686–688.

18. Surveillance, Epidemiology, and End Results (SEER) Program (`www.seer.cancer.gov`), SEER*Stat Databases.

19. See Note 2.

20. National Center for Health Statistics, *Deaths: Preliminary Data for 2003*, National Vital Statistics Reports, 53 (2005), No. 15.

21. See Note 16.

22. Tom Lloyd et al., "Fruit consumption, fitness, and cardiovascular health in female adolescents: the Penn State Young Women's Health Study," *American Journal of Clinical Nutrition*, 67 (1998), pp. 624–630.

23. James T. Fleming, "The measurement of children's perception of difficulty in reading materials," *Research in the Teaching of English*, 1 (1967), pp. 136–156.

24. This data set was provided by Nicolas Fisher.

25. *Statistical Abstract of the United States*, 2004–2005, Table 150.

26. Found online at earthtrends.wri.org.

27. National Oceanic and Atmospheric Administration, www.noaa.gov.

28. We thank Ethan J. Temeles of Amherst College for providing the data. His work is described in E. J. Temeles and W. J. Kress, "Adaptation in a plant-hummingbird association," *Science*, 300 (2003), pp. 630–633.

29. D. M. Fergusson and L. John Horwood, "Cannabis use and traffic accidents in a birth cohort of young adults," *Accident Analysis and Prevention*, 33 (2001), pp. 703–711.

30. 2004 Statistical Abstract of the United States, *Health and Nutrition*, Table 201.

31. Florida Fish and Wildlife Conservation Commission, *Alligator Attacks Fact Sheet*, 2005, www.wildflorida.org.

32. From the Web site of the Bureau of Labor Statistics, www.bls.gov/cpi.

Chapter 2

1. Data collected by Brigitte Baldi.

2. Modified from data provided by Dana Quade, University of North Carolina.

3. See Note 1.

4. This isn't a mathematical theorem. The mean can be less than the median in right-skewed distributions that take only a few values, many of which lie exactly at the median. The rule almost never fails for distributions taking many values, and counter-examples don't appear clearly skewed in graphs even though they may be slightly skewed according to technical measures of skewness. See P. T. von Hippel, "Mean, median, and skew: correcting a textbook rule," *Journal of Statistics Education*, 13, No. 2 (2005), online journal, www.amstat.org/publications/jse.

5. A. Dell'Anno and R. Danovaro, "Extracellular DNA plays a key role in deep-sea ecosystem functioning," *Science*, 309 (2005), p. 2179.

6. See Note 13 from Chapter 1.

7. EESEE story "Acorn Size and Oak Tree Range."

8. See Note 28 from Chapter 1.

9. We thank Charles Cannon of Duke University for providing the data. The study report is C. H. Cannon, D. R. Peart, and M. Leighton, "Tree species diversity in commercially logged Bornean rainforest," *Science*, 281 (1998), pp. 1366–1367.

10. EESEE story "Mercury in Florida's Bass."

11. National Vital Statistics Reports, *Births: Final Data for 2002*, Vol. 52, No. 10, 2003.

12. Modified from M. C. Wilson and R. E. Shade, "Relative attractiveness of various luminescent colors to the cereal leaf beetle and the meadow spittlebug," *Journal of Economic Entomology*, 60 (1967), pp. 578–580.

13. See Note 10 from Chapter 1.

14. Data for 1986 from David Brillinger, University of California, Berkeley. See D. R. Brillinger, "Mapping aggregate birth data," in A. C. Singh and P. Whitridge (eds.), *Analysis of Data in Time*, Statistics Canada, 1990, pp. 77–83. A boxplot similar to Figure 2.5 appears in D. R. Brillinger, "Some examples of random process environmental data analysis," in P. K. Sen and C. R. Rao (eds.), *Handbook of Statistics*, Vol. 18, North Holland, 2000.

15. See Note 11 from Chapter 1.

16. R. A. Berner and G. P. Landis, "Gas bubbles in fossil amber as possible indicators of the major gas composition of ancient air," *Science*, 239 (1988), pp. 1406–1409. The 95% t confidence interval is 54.78 to 64.40. A bootstrap BCa interval is 55.03 to 62.63. So t is reasonably accurate despite the skew and the small sample.

17. M. A. Laskey et al., "Bone changes after 3 mo of lactation: influence of calcium intake, breast-milk output, and vitamin D–receptor genotype," *American Journal of Clinical Nutrition*, 67 (1998), pp. 685–692.

18. P. S. Gupta, "Reaction of plants to the density of soil," *Journal of Ecology*, 21 (1933), pp. 452–474.

Chapter 3

1. Florida Fish and Wildlife Conservation Commission, *Boating accident statistics* and *Manatee mortality*, `www.myfwc.com`.

2. From a graph in B.-E. Saether, S. Engen, and E. Mattysen, "Demographic characteristics and population dynamical patterns of solitary birds," *Science*, 295 (2002), pp. 2070–2073.

3. C. Carbone and J. L. Gittleman, "A common rule for the scaling of carnivore density," *Science*, 295 (2002), pp. 2273–2276.

4. Estimates vary around that half billion mark, with about quarter of that in the United States alone (Worldwatch Institute, AAMA, DRI-WEFA, Global Insight).

5. Based on T. N. Lam, "Estimating fuel consumption from engine size," *Journal of Transportation Engineering*, 111 (1985), pp. 339–357. The data for 10 to 50 km/h are measured; those for 60 km/h and higher are calculated from a model given in the paper and are therefore smoothed.

6. From a graph in L. Partridge and M. Farquhar, "Sexual activity reduces lifespan of male fruitflies," *Nature*, 294 (1981), pp. 580–582.

7. A careful study of this phenomenon is W. S. Cleveland, P. Diaconis, and R. McGill, "Variables on scatterplots look more highly correlated when the scales are increased," *Science*, 216 (1982), pp. 1138–1141.

8. From a graph in T. G. O'Brien and M. F. Kinnaird, "Caffeine and conservation," *Science*, 300 (2003), p. 587.

9. D. M. Etheridge, L. P. Steele, R. J. Francey, and R. L. Langenfelds, "Atmospheric methane between 1000 A.D. and present: Evidence of anthropogenic emissions and climatic variability," *Journal of Geophysical Research*, 103 (1998), p. 15979.

10. These data were originally collected by L. M. Linde of UCLA but were first published by M. R. Mickey, O. J. Dunn, and V. Clark, "Note on the use of stepwise regression in detecting outliers," *Computers and Biomedical Research*, 1 (1967), pp. 105–111. The data have been used by several authors. These are taken from N. R. Draper and J. A. John, "Influential observations and outliers in regression," *Technometrics*, 23 (1981), pp. 21–26.

11. M. A. Houck et al. "Allometric scaling in the earliest fossil bird, *Archaeopteryx lithographica*," *Science*, 247 (1990), pp. 195–198. The authors conclude from a variety of evidence that all specimens represent the same species.

12. M. H. Criqui, University of California, San Diego, reported in the *New York Times*, December 28, 1994.

13. Simplified from W. L. Colville and D. P. McGill, "Effect of rate and method of planting on several plant characters and yield of irrigated corn," *Agronomy Journal*, 54 (1962), pp. 235–238.

14. W. Leutenegger, "Evolution of litter size in primates," *The American Naturalist*, 114 (1979), pp. 525–531.

15. Data from many studies compiled in D. F. Greene and E. A. Johnson, "Estimating the mean annual seed production of trees," *Ecology*, 75 (1994), pp. 642–647.

16. United Nations, "Human development report 2005," Table 10: Survival, and Table 14: Economic performance (World Bank data).

17. From a graph in C. G. Wiklund, "Food as a mechanism of density-dependent regulation of breeding numbers in the merlin *Falco columbarius*," *Ecology*, 82 (2001), pp. 860–867.

18. From a graph in M. Wild, et al., "From dimming to brightening: decadal changes in solar radiation at Earth's surface," *Science*, 308 (2005), pp. 847–850.

19. From a graph in N. I. Eisenberger, M. D. Lieberman, and K. D. Williams, "Does rejection hurt? An fMRI study of social exclusion," *Science*, 302 (2003), pp. 290–292.

Chapter 4

1. From a graph in J. A. Levine, N. L. Eberhardt, and M. D. Jensen, "Role of nonexercise activity thermogenesis in resistance to fat gain in humans," *Science*, 283 (1999), pp. 212–214.

2. Data from G. A. Sacher and E. F. Staffelt, "Relation of gestation time to brain weight for placental mammals: implications for the theory of vertebrate growth," *American Naturalist*, 108 (1974), pp. 593–613. These data were found in F. L. Ramsey and D. W. Schafer, *The Statistical Sleuth: A Course in Methods of Data Analysis*, Duxbury, 1997.

3. University of Oklahoma Police Department, "The Police Notebook," based on material from the National Highway Traffic Safety Administration.

4. Department of Motor Vehicles, "California Driver Handbook." Exact definition varies depending on alcohol content of the particular brand consumed.

5. See Note 2 for Chapter 3.

6. See Note 6 for Chapter 3.

7. From a graph in T. Singer et al., "Empathy for pain involves the affective but not sensory components of pain," *Science*, 303 (2004), pp. 1157–1162. Data for other brain regions showed a stronger correlation and no outliers.

8. Contributed by Marigene Arnold, Kalamazoo College.

9. P. Goldblatt (ed.), *Longitudinal Study: Mortality and Social Organization*, Her Majesty's Stationery Office, 1990. At least, so claims Richard Conniff, in *The Natural History of the Rich*, Norton, 2002, p. 45. The Goldblatt report is not available to us.

10. L. L. Calderon et al., "Risk factors for obesity in Mexican-American girls: dietary factors, anthropometric factors, physical activity, and hours of television viewing," *Journal of the American Dietetic Association*, 96 (1996), pp. 1177–1179.

11. *The Health Consequences of Smoking: 1983*, Public Health Service, Washington, D.C., 1983.

12. World Bank, "World Development Indicators."

13. P. K. Mills, W. L. Beeson, R. L. Phillips, and G. E. Fraser, "Dietary habits and breast cancer incidence among Seventh-Day Adventists," *Cancer*, 64 (1989), pp. 582–590.

14. See previous note.

15. J. D. McLister, "The energetics of male reproductive behavior in the treefrog, *Hyla versicolor*," MS thesis, University of California, Irvine, 2000.

16. G. L. Kooyman et al., "Diving behavior and energetics during foraging cycles in king penguins," *Ecological Monographs*, 62 (1992), pp. 143–163.

17. T. Constable and E. McBean, "BOD/TOC correlations and their application to water quality evaluation," *Water, Air, and Soil Pollution*, 11 (1979), pp. 363–375.

18. See Note 5 from Chapter 1.

19. From a presentation by Charles Knauf, Monroe County (New York) Environmental Health Laboratory.

20. F. Sultan and V. Braitenberg, "Shapes and sizes of different mammalian cerebella. A study in quantitative comparative neuroanatomy." *Journal für Hirnforschung*, 34 (1993), pp. 79–92.

21. Statistical Society of Canada, 2001 case study II. Original publication F. He and R. Duncan, "Density-dependent effects on tree survival in an old-growth Douglas fir forest," *Journal of Ecology*, 88 (2000), pp. 676–688.

22. F. J. Anscombe, "Graphs in statistical analysis," *The American Statistician*, 27 (1973), pp. 17–21.

23. National Toxicology Program, CAS No. 22839-47-0 (2005).

24. From a graph in F. S. Hu et al., "Cyclic variation and solar forcing of Holocene climate in the Alaskan subarctic," *Science*, 301 (2003), pp. 1890–1893.

25. See Note 13 from Chapter 1.

26. From a graph in A. L. Perry et al., "Climate change and distribution shifts in marine fishes," *Science*, 308 (2005), pp. 1912–1915. The explanatory variable is the five-year running mean of winter (December to March) sea-bottom temperature.

27. Occupational Mortality: The Registrar General's Decennial Supplement for England and Wales, 1970–1972, Her Majesty's Stationery Office, London, 1978.

28. F. H. Simmons, "Physiology of the trade-off between fecundity and survival in *Drosophila melanogaster,* as revealed through dietary manipulation," MS thesis, University of California, Irvine, 1996.

29. From a graph in G. D. Martinsen, E. M. Driebe, and T. G. Whitham, "Indirect interactions mediated by changing plant chemistry: beaver browsing benefits beetles," *Ecology,* 79 (1998), pp. 192–200.

Chapter 5

1. T. Kontiokari et al., "Randomised trial of cranberry-lingonberry juice and *Lactobacillus* GG drink for the prevention of urinary tract infections in women," *BMJ,* 322 (2001), pp. 1–5.

2. H. Lindberg, H. Roos, and P. Gardsell, "Prevalence of coxarthritis in former soccer players," *Acta Orthopedica Scandinavica,* 64 (1993), pp. 165–167.

3. J. L. Hoogland, "Why do Gunnison's prairie dogs give anti-predator calls?" *Animal Behavior,* 51 (1996), pp. 871–880.

4. S. Oppe and F. De Charro, "The effect of medical care by a helicopter trauma team on the probability of survival and the quality of life of hospitalized victims," *Accident Analysis & Prevention,* 33 (2001), pp. 129–138. The authors give the data in Example 6.5 as a "theoretical example" to illustrate the need for their more elaborate analysis of actual data using severity scores for each victim.

5. C. R. Charig, D. R. Webb, S. R. Payne, and O. E. Wickham, "Comparison of treatment of renal calculi by operative surgery, percutaneous nephrolithotomy, and extracorporeal shock wave lithotripsy," *BMJ,* 292 (1986), pp. 879–882.

6. Condensed from D. R. Appleton, J. M. French, and M. P. J. Vanderpump, "Ignoring a covariate: an example of Simpson's paradox," *The American Statistician,* 50 (1996), pp. 340–341.

7. R. B. Turner, R. Bauer, K. Woelkart, T. C. Hulsey, and J. D. Gangemi, "An evaluation of *Echinacea angustifolia* in experimental rhinovirus infections," *New England Journal of Medicine,* 353 (2005), pp. 341–348.

8. C. D. Ankney, "Sex-ratio varies with egg sequence in lesser snow geese," *The Auk,* 99 (1982), pp. 662–666.

9. D. A. Marcus, L. Scharff, D. Turk, and L. M. Gourley, "A double-blind provocative study of chocolate as a trigger of headache," *Cephalalgia,* 17 (1997), pp. 855–862.

10. D. M. Barnes, "Breaking the cycle of addiction," *Science,* 241 (1988), pp. 1029–1030.

11. Feeding Infants and Toddlers Study, a Gerber initiative (2002).

12. National Center for Health Statistics, *Deaths: Preliminary Data for 2003,* National Vital Statistics Report, 53 (2005), No. 15.

13. W. Gatling, M. A. Mullee, and R. D. Hill, "The general characteristics of a community-based population," *Practical Diabetes,* 5 (1989), pp. 104–107.

14. R. Shine, T. R. L. Madsen, M. J. Elphick, and P. S. Harlow, "The influence of nest temperatures and maternal brooding on hatchling phenotypes in water pythons," *Ecology,* 78 (1997), pp. 1713–1721.

15. J. E. Williams et al., "Anger proneness predicts coronary heart disease risk," *Circulation,* 101 (2000), pp. 2034–2039.

16. See Note 21 for Chapter 4.

Chapter 6

1. N. Maeno et al., "Growth rates of icicles," *Journal of Glaciology*, 40 (1994), pp. 319–326.

2. J. T. Dwyer et al., "Memory of food intake in the distant past," *American Journal of Epidemiology*, 130 (1989), pp. 1033–1046.

3. Louie H. Yang, "Periodical cicadas as resource pulses in North American forests," *Science*, 306 (2004), pp. 1565–1567. The data are simulated Normal values that match the means and standard deviations reported in this article.

4. *Statistical Abstract of the United States, 2004–2005.*

5. A. S. Banks et al., "Juvenile hallux abducto valgus association with metatarsus adductus," *Journal of the American Podiatric Medical Association*, 84 (1994), pp. 219–224.

6. From a graph in C. Packer et al., "Ecological change, group territoriality, and population dynamics in Serengeti lions," *Science*, 307 (2005), pp. 390–393.

7. T. W. Anderson, "Predator responses, prey refuges, and density-dependent mortality of a marine fish," *Ecology*, 81 (2001), pp. 245–257.

8. From the Nenana Ice Classic Web page, `www.nenanaiceclassic.com`. See R. Sagarin and F. Micheli, "Climate change in nontraditional data sets," *Science*, 294 (2001), p. 811, for a careful discussion.

9. J. W. Grier, "Ban of DDT and subsequent recovery of reproduction in bald eagles," *Science*, 218 (1982), pp. 1232–1234.

10. From a plot in J. J. Ramsey et al., "Energy expenditure, body composition, and glucose metabolism in lean and obese rhesus monkeys treated with ephedrine and caffeine," *American Journal of Clinical Nutrition*, 68 (1998), pp. 42–51.

11. Data provided by Samuel Phillips, Purdue University.

12. Data provided by Robin Bush, University of California, Irvine.

Chapter 7

1. See, for example, M. Enserink, "The vanishing promises of hormone replacement," *Science*, 297 (2002), pp. 325–326; B. Vastag, "Hormone replacement therapy falls out of favor with expert committee," *Journal of the American Medical Association*, 287 (2002), pp. 1923–1924, and J. E. Manson et al., "Estrogen Therapy and Coronary-Artery Calcification," *New England Journal of Medicine*, 356 (2007), pp. 2591–2602.

2. J. C. Barefoot et al., "Alcoholic beverage preference, diet, and health habits in the UNC Alumni Heart Study," *American Journal of Clinical Nutrition*, 76 (2002), pp. 466–472.

3. J. E. Muscat et al., "Handheld cellular telephone use and risk of brain cancer," *Journal of the American Medical Association*, 284 (2000), pp. 3001–3007.

4. Committee on the Science of Climate Change, National Research Council, *Climate Change Science: An Analysis of Some Key Questions*, National Academy Press, 2001, page 17.

5. See Note 28 for Chapter 4.

6. "Birth control methods," found on `WomensHealth.gov`.

7. California Department of Health Services, "Birth, Death, and Fetal Death Records," Table 1.1.

8. Modified Holland (1986) plant community definition.

9. From the Web site of the Gallup Organization, `www.gallup.com`. Press releases remain on this site for only a limited time.

10. R. C. Parker and P. A. Glass, "Preliminary results of double-sample forest inventory of pine and mixed stands with high- and low-density LiDAR," in K. F. Connoe (ed.), *Proceedings of the 12th Biennial Southern Silvicultural Research Conference*, U.S. Department of Agriculture, Forest Service, Southern Research Station, 2004. The researchers actually sampled every 10th plot (a systematic sample).

11. The nonresponse rate for the CPS comes from "Technical notes to household survey data published in *Employment and Earnings*," found on the Bureau of Labor Statistics Web site: `stats.bls.gov/cpshome.htm`. The General Social Survey reports its response rate on its Web site: `www.norc.org/projects/gensoc.asp`. The Pew study is described in G. Flemming and K. Parker, "Race and reluctant respondents: possible consequences of non-response for pre-election surveys," Pew Research Center for the People and the Press, 1997, found at `www.people-press.org`.

12. For more detail on the limits of memory in surveys, see N. M. Bradburn, L. J. Rips, and S. K. Shevell, "Answering autobiographical questions: the impact of memory and inference on surveys," *Science*, 236 (1987), pp. 157–161.

13. The responses on welfare are from a *New York Times*/CBS News Poll reported in the *New York Times*, July 5, 1992. Those for Scotland are from "All set for independence?" *Economist*, September 12, 1998. Many other examples appear in T. W. Smith, "That which we call welfare by any other name would smell sweeter," *Public Opinion Quarterly*, 51 (1987), pp. 75–83.

14. Modified from R. A. Schieber et al., "Risk factors for injuries from in-line skating and the effectiveness of safety gear," *New England Journal of Medicine*, 335 (1996), Internet summary at `www.content.nejm.org`.

15. E. Azziz-Baumgartner et al., "Case-control study of an acute aflatoxicosis outbreak, Kenya, 2004," *Environmental Health Perspectives*, 113 (2005), pp. 1779–1783.

16. K. J. Mukamal et al., "Prior alcohol consumption and mortality following acute myocardial infarction," *Journal of the American Medical Association*, 285 (2001), pp. 1965–1970.

17. C. G. Bacon et al., "A Prospective Study of Risk Factors for Erectile Dysfunction," *Journal of Urology*, 176 (2006), pp. 217–221.

18. L. E. Moses and F. Mosteller, "Safety of anesthetics," in J. M. Tanur et al. (eds.), *Statistics: A Guide to the Unknown*, 3rd ed., Wadsworth, 1989, pp. 15–24.

19. Reported by `www.medicalmarijuanaprocon.org`. Press releases remain on the `www.mason-dixon.com/` site for only a limited time.

20. The Jane Goodall Institute, `www.janegoodall.org/`

21. M. A. Parada et al., "The validity of self-reported seatbelt use: Hispanic and non-Hispanic drivers in El Paso," *Accident Analysis and Prevention*, 33 (2001), pp. 139–143.

22. See Note 17 for Chapter 2.

23. C. Braga et al., "Olive oil, other seasoning fats, and the risk of colorectal carcinoma," *Cancer*, 82 (1998), pp. 448–453.

24. E. Courchesne et al., "Unusual brain growth patterns in early life in patients with autistic disorder: an MRI study," *Neurology,* 57 (2001), pp. 245–254.

25. S. Sutcliffe et al., "A prospective cohort study of red wine consumption and risk of prostate cancer," *International Journal of Cancer,* 120 (2007), pp. 1529–1535.

Chapter 8

1. M. de Lorgeril et al., "Mediterranean dietary pattern in a randomized trial—Prolonged survival and possible reduced cancer rate," *Archives of Internal Medicine,* 158 (1998), pp. 1181–1187.

2. G. L. Cromwell et al., "A comparison of the nutritive value of *opaque-2, floury-2* and normal corn for the chick," *Poultry Science,* 57 (1968), pp. 840–847.

3. L. L. Miao, "Gastric freezing: an example of the evaluation of medical therapy by randomized clinical trials," in J. P. Bunker, B. A. Barnes, and F. Mosteller (eds.), *Costs, Risks, and Benefits of Surgery,* Oxford University Press, 1977, pp. 198–211.

4. See Note 9 for Chapter 5.

5. G. Thiagarajan et al., "Antioxidant properties of green and black tea, and their potential ability to retard the progression of eye lens cataract," *Experimental Eye Research,* 73 (2001), pp. 393–401.

6. D. T. Kirkendall, "Creatinine, carbs, and fluids: how important in soccer nutrition?" *Sports Science Exchange,* 94 (2004), pp. 1–6.

7. See Note 9 for Chapter 1.

8. See Note 19 for Chapter 7.

9. J. H. Kagel, R. C. Battalio, and C. G. Miles, "Marihuana and work performance: results from an experiment," *Journal of Human Resources,* 15 (1980), pp. 373–395. A general discussion of failures of blinding is D. Ferguson et al., "Turning a blind eye: the success of blinding reported in a random sample of randomised, placebo controlled trials," *BMJ,* 328 (2004), p. 432.

10. M. Enserink, "Fickle mice highlight test problems," *Science,* 284 (1999), pp. 1599–1600. There is a full report of the study in the same issue.

11. U.S. Department of Health and Human Services, Public Health Service, National Toxicology Program, Report on Carcinogens, 11th ed., Table B: Agents, Substances, Mixtures, or Exposure Circumstances Delisted from the Report on Carcinogens `http://ntp.niehs.nih.gov/ntp/roc/toc1`.

12. C. A. Warfield, "Controlled-release morphine tablets in patients with chronic cancer pain," *Cancer,* 82 (1998), pp. 2299–2306.

13. Centers for Disease Control and Prevention, "U.S. Public Health Service Syphilis Study at Tuskegee," `http://www.cdc.gov/nchstp/od/tuskegee/`.

14. NPR News Service article," 'My Lobotomy': Howard Dully's journey," from *All Things Considered,* November 16, 2005, `www.npr.org`.

15. K. Hartmann et al., "Outcomes of routine episiotomy: a systematic review," *Journal of the American Medical Association,* 293 (2005), pp. 2141–2148.

16. John Pickrell, "Special reports on key topics in science and technology: instant expert: GM organisms," December 13, 2004, `NewScientist.com`.

17. "President discusses stem cell research," August 9, 2001, `www.whitehouse.gov`.

18. Proposition 71, "Stem Cell Research. Funding. Bonds. Initiative Constitutional Amendment and Statute." Proposition 71 was passed by California voters in November 2004 and resulted in the creation of the California Institute for Regenerative Medicine.

19. J. H. Kagel, R. C. Battalio, and C. G. Miles, "Marijuana and work performance: results from an experiment," *Journal of Human Resources*, 15 (1980), pp. 373–395.

20. Based on E. H. DeLucia et al., "Net primary production of a forest ecosystem with experimental CO_2 enhancement," *Science*, 284 (1999), pp. 1177–1179. The investigators used the block design.

21. E. M. Peters et al., "Vitamin C supplementation reduces the incidence of postrace symptoms of upper-respiratory tract infection in ultramarathon runners," *American Journal of Clinical Nutrition*, 57 (1993), pp. 170–174.

22. Based on P. J. Meunier et al., "The effects of strontium ranelate on the risk of vertebral fracture in women with postmenopausal osteoporosis," *New England Journal of Medicine*, 350 (2004), pp. 459–468.

23. P. R. Solomon et al., "Ginkgo for memory enhancement: a randomized controlled trial," *Journal of the American Medical Association*, 288 (2002), pp. 835–840.

24. R. F. Redburg, "Vitamin E and cardiovascular health," *Journal of the American Medical Association*, 294 (2005), pp. 107–109.

25. The study is described in G. Kolata, "New study finds vitamins are not cancer preventers," *New York Times*, July 21, 1994. Look in the *Journal of the American Medical Association* of the same date for the details.

26. R. C. Shelton et al., "Effectiveness of St. John's wort in major depression," *Journal of the American Medical Association*, 285 (2001), pp. 1978–1986.

27. Steering Committee of the Physicians' Health Study Research Group, "Final report on the aspirin component of the ongoing Physicians' Health Study," *New England Journal of Medicine*, 321 (1989), pp. 129–135.

Chapter 9

1. 2004 Statistical Abstract of the United States, Table No 72.

2. From the American Red Cross website, `http://www.givelife2.org/aboutblood/bloodtypes.asp`. There are some other blood types besides the 8 main types, but they are extremely rare.

3. 2004 Statistical Abstract of the United States, Table No 194.

4. From the Florida Department of Health website (`www.doh.state.fl.us`), Bureau of Epidemiology, "Animal Rabies Confirmed Cases by County and Animal Type, January–December 2004."

5. Discrete sample spaces are most often, but not necessarily, a finite list of integers or individual outcomes. We will see in Chapter 12 the example of the Poisson distribution, a discrete distribution for $X = 0, 1, 2, \ldots$.

6. G. M. Strain, "Deafness prevalence and pigmentation and gender associations in dog breeds at risk," *The Veterinary Journal*, 167 (2004), pp. 23–32.

7. R. Kubey, "Television dependence, diagnosis, and prevention," in T. M. Williams (ed.), *Tuning In to Young Viewers: Social Science Perspectives on Television*, Sage, 1996. Heavy TV viewers exhibit six dependency symptoms: using TV as a sedative;

indiscriminate viewing; feeling loss of control while viewing; feeling angry with themselves for watching so much; being unable to stop watching; and suffering withdrawal when forced to stop watching TV.

8. Based on interviews in 2000 and 2001 by the National Longitudinal Study of Adolescent Health. Found at the Web site of the Carolina Population Center, `www.cpc.unc.edu`.

9. 2004 Statistical Abstract of the United States, Table 193.

10. National Center for Health Statistics, "Health, United States, 2005," p. 38.

11. "Kids and media," A Kaiser Family Foundation Report, 1999.

12. A. Leizorovicz et al., "Randomized, placebo-controlled trial of dalteparin for the prevention of venous thromboembolism in acutely ill medical patients," *Circulation*, 110 (2004), pp. 874–879.

13. From the National Institutes of Health Web site, Disease and Conditions Index (`health.nih.gov`).

14. From the National Cancer Institute Web site, "Fact Sheet: Probability of Breast Cancer in American Women" (`www.cancer.gov`).

15. From `www.bloodbook.com`, "Racial and ethnic distribution of ABO Blood types."

16. "Phenylketonuria," from the Free Health Encyclopedia at `faqs.org/health/`.

17. E. S. Siris et al., "Identification and fracture outcomes of undiagnosed low bone mineral density in postmenopausal women—Results from the National Osteoporosis Risk Assessment," *Journal of the American Medical Association*, 286 (2001), pp. 2815–2822.

18. Health Resources and Services Administration (HRSA), "Providing HIV/AIDS care in a changing environment: hepatitis C and HIV co-infection," September 2003.

19. W. D. Mosher et al., "Use of contraception and use of family planning services in the United States: 1982–2002," *Advance Data from Vital Health Statistics* No. 350 (CDC, 2004).

20. A. N. Dey and B. Bloom, "Summary health statistics for U.S. children: National Health Interview Survey, 2003," National Center for Health Statistics, *Vital Health Statistics*, 10 (2005), Table 17.

21. R. N. Rosenfield et al., "Comparative relationships among eye color, age, and sex in three North American populations of Cooper's hawks," *The Wilson Bulletin*, 115 (2003), pp. 225–230.

Chapter 10

1. See Note 2 for Chapter 9.

2. This is one of several tests discussed in B. M. Branson, "Rapid HIV testing: 2005 update," a presentation by the Centers for Disease Control and Prevention, at `www.cdc.gov`. The Malawi clinic result is reported by B. M. Branson, "Point-of-care rapid tests for HIV antibody," *Journal of Laboratory Medicine*, 27 (2003), pp. 288–295.

3. N. Ranjit et al., "Contraceptive failure in the first two years of use: differences across socioeconomic subgroups," *Family Planning Perspectives*, 33 (2001), pp. 19–27.

4. See Note 3.

5. From the Florida Department of Health Web site (`www.doh.state.fl.us`), Brain and Spinal Cord Injury Program.

6. See Note 21 for Chapter 4.

7. H. Y. Chang et al., "Robustness, scalability, and integration of a wound-response gene expression signature in predicting breast cancer survival," *Proceedings of the National Academy of Sciences*, 102 (2005), 3738–3743.

8. Epidemiologic Catchment Area Survey, 1990, diagnosing mental disorders using a random sample of 18,571 Americans, conducted by the National Institute of Mental Health.

9. See Note 6 for Chapter 9.

10. Royal Statistical Society news release, "Royal Statistical Society concerned by issues raised in Sally Clark case," October 23, 2001, at `www.rss.org.uk`. For background, see an editorial and article in the *Economist*, January 22, 2004. The editorial is titled "The probability of injustice."

11. A. C. Allison and D. F. Clyde, "Malaria in African children with deficient erythrocyte dehydrogenase," *British Medical Journal*, 1 (1961), pp. 1346–1349.

12. L. Naldi et al., "Pigmentary traits, modalities of sun reaction, history of sunburns, and melanocytic nevi as risk factors for cutaneous Malignant Melanoma in the Italian population: results of a collaborative case-control study," *Cancer*, 88 (2000), pp. 2703–2710.

13. W. Uter et al., "Inter-relation between variables determining constitutional UV sensitivity in Caucasian children," *Photodermatology, Photoimmunology and Photomedicine*, 20 (2004), pp. 9–13.

14. Probabilities from trials with 2897 people known to be free of HIV antibodies and 673 people known to be infected, reported in J. R. George, "Alternative specimen sources: methods for confirming positives," 1998 Conference on the Laboratory Science of HIV, found online at the Centers for Disease Control and Prevention, `www.cdc.gov`.

15. G. Gigerenzer and A. Edwards, "Simple tools for understanding risks: from innumeracy to insight," *BMJ*, 327 (2003), pp. 741–744.

16. See Note 14 for Chapter 9.

17. Different countries use different cutoff points for deciding whether to call a result positive or negative, and some countries have each mammogram reviewed by 2 different experts. Thus sensitivity and specificity varies from country to country. See U.S. Preventive Services Task Force, "Screening for breast cancer: recommendations and rationale," February 2002, Agency for Healthcare Research and Quality.

18. See Note 15.

19. A. A. Wright and I. T. Katz, "Home testing for HIV," *The New England Journal of Medicine*, Perspective, 354 (2006), pp. 437–440.

20. NPR News Service, "Silicosis: from public menace to litigation target," March 6, 2006, `www.npr.org`. The Web site also links to the judge's filed opinion, "Silica products liability litigation," MDL Docket No. 1553.

21. Young People's Health Barometer Survey 1997–1998, as reported by the "Collective Expert Evaluation Reports," INSERM, 2001.

22. R. P. Dellavalle et al., "Going, going, gone: lost Internet references," *Science*, 302 (2003), pp. 787–788.

23. S. W. Guo and D. R. Reed, "The genetics of phenylthiocarbamide perception," *Annals of Human Biology,* 28 (2001), pp. 111–142.

24. Polydactyly is not reported systematically, so estimated rates vary. National Center for Health Statistics, Series 20, No. 31 (1996); e-medicine.com, "Supernumerary digit," "polydactyly of the Foot."

25. See Note 2 for Chapter 5.

26. See Note 13.

27. See Note 17 for Chapter 9.

28. See Note 18 for Chapter 9.

29. S. D. Grosse et al., "Newborn screening for cystic fibrosis," Centers for Disease Control and Prevention, *Morbidity and Mortality Weekly Report,* October 15, 2004.

30. S. V. Zagona (ed.), *Studies and Issues in Smoking Behavior,* University of Arizona Press, 1967, pp. 157–180.

31. The probabilities given are realistic, according to the fundraising firm SCM Associates, `scmassoc.com`.

32. 2004 Statistical Abstract of the United States, Table 73.

33. National Tay-Sachs and Allied Diseases Association, `www.ntsad.org`.

Chapter 11

1. Michael Jakob, *Normal Values pocket,* Borm Bruckmeier Publishing, LLC, 2002.

2. Detailed data appear in P. S. Levy et al., *Total Serum Cholesterol Values for Youths 12–17 Years,* Vital and Health Statistics, Series 11, No. 155, National Center for Health Statistics, 1976.

3. R. C. Nelson, C. M. Brooks, and N. L. Pike, "Biomechanical comparison of male and female distance runners," in P. Milvy (ed.), *The Marathon: Physiological, Medical, Epidemiological, and Psychological Studies,* New York Academy of Sciences, 1977, pp. 793–807.

4. See Note 1.

5. Data provided by Darlene Gordon, Purdue University.

6. See Note 1.

7. See Note 17 for Chapter 9.

8. Results for 1988 to 1991 from a large sample survey, reported in National Center for Health Statistics, *Health, United States, 1995,* 1996.

9. Ulric Neisser, "Rising scores on intelligence tests," *American Scientist,* September-October 1997, online edition.

10. Data from the EESEE story "Acorn Size and Oak Tree Range."

11. See Note 6 for Chapter 3.

Chapter 12

1. See Note 18 for Chapter 7.

2. From the Florida Department of Health Web site, Bureau of Epidemiology: "Antibiotic resistance," 2000, Table 4.

3. See Note 2.

4. See Note 3 for Chapter 10.

5. Unlike binomial counts, Poisson counts are open-ended.

6. *Morbidity and Mortality Weekly Report*, MMWR, March 30, 2006, available from the CDC website at `www.cdc.gov`.

7. See Note 6.

8. From the CDC website, `www.cdc.gov`, Division of Bacterial and Mycotic Diseases, "Typhoid Fever."

9. Office of Research and Statistics, South Carolina State Budget and Control Board, "Top 50 Reasons for Emergency Room Visits for Residents of South Carolina."

10. Team USA Olympic Survey, posted on `NBCOlympics.com`, February 2006.

11. United Nations, "Human Development Report 2005," Table 9: Leading global health crises and risks.

12. See Note 17 for Chapter 9.

13. 2004 Statistical Abstract of the United States, Table 194.

Chapter 13

1. Strictly speaking, the formula $\sigma/\sqrt{n}$ for the standard deviation of $\bar{x}$ assumes that we draw an SRS of size n from an *infinite* population. If the population has finite size N, this standard deviation is multiplied by $\sqrt{1 - (n-1)/(N-1)}$. This "finite population correction" approaches 1 as N increases. When the population is at least 20 times as large as the sample, the correction factor is between about 0.97 and 1. It is reasonable to use the simpler form $\sigma/\sqrt{n}$ in these settings.

2. J. H. Catania et al., "Prevalence of AIDS-related risk factors and condom use in the United States," *Science*, 258 (1992), pp. 1101–1106.

3. Strictly speaking, the formula $\sqrt{p(1-p)/n}$ for the standard deviation of $\hat{p}$ assumes that we draw an SRS of size n from an *infinite* population. If the population has finite size N, this standard deviation is multiplied by $\sqrt{1 - (n-1)/(N-1)}$. This "finite population correction" approaches 1 as N increases. When the population is at least 20 times as large as the sample, the correction factor is between about 0.97 and 1. It is reasonable to use the simpler form $\sqrt{p(1-p)/n}$ in these settings. See also Note 2 for Chapter 11.

4. See Note 3 for Chapter 10.

Chapter 14

1. This value of σ is based on the growth curves developed by the National Center for Health Statistics in collaboration with the National Center for Chronic Disease Prevention and Health Promotion (2000).

2. Information from F. L. Rivera-Batiz, "Quantitative literacy and the likelihood of employment among young adults," *Journal of Human Resources*, 27 (1992), pp. 313–328.

3. See Note 9 for Chapter 1.

4. See Note 5 for Chapter 11.

5. See Note 1 for Chapter 11.

6. Data simulated from a Normal distribution with the mean and standard deviation reported by S. Morrison and J. Noyes, "A comparison of two computer fonts: serif versus ornate sans serif," *Usability News*, issue 5.2, 2003, `psychology.wichita.edu/surl/usability_news.html`.

7. *A field guide to the mammals: field marks of all North American species found north of Mexico*, W. H. Burt and R. P. Grossenheider, third edition, Houghton Mifflin company Boston, 1976.

8. Information found on the USDA Web site at `www.ams.usda.gov/howtobuy/eggs.htm`.

9. See Note 3 for Chapter 6.

10. S. L. Webb and S. E. Scanga, "Windstorm disturbance without patch dynamics: twelve years of change in a Minnesota forest," *Ecology*, 82 (2001), pp. 893–897.

11. M. A. Parada et al., "The validity of self-reported seatbelt use: Hispanic and non-Hispanic drivers in El Paso," *Accident Analysis and Prevention*, 33 (2001), pp. 139–143.

Chapter 15

1. You can find detailed information on the CDC Web site about the sampling design and analytic guidelines. Here is an excerpt: "Because NHANES is a complex probability sample, analytic approaches based on data from simple random sample are usually not appropriate. Ignoring the complex design can lead to biased estimates and overstated significance levels. Sample weights and the stratification and clustering of the design must be incorporated into an analysis to get proper estimates and standard errors of estimates."

2. Found online at `www.pbs.org/wgbh/nova/sciencenow/3209/04.html`.

3. E. M. Barsamian, "The rise and fall of internal mammary artery ligation," in J. P. Bunker, B. A. Barnes, and F. Mosteller (eds.), *Costs, Risks, and Benefits of Surgery*, Oxford University Press, 1977, pp. 212–220.

4. S. J. Jachuck, P. Price, and C. L. Bound, "Evaluation of quality of contents of general practice records," *British Medical Journal*, 289 (1984), pp. 26–28.

5. Harvard School of Public Health College Alcohol Study (CAS), April 12, 2001, press release posted on the study website at `www.hsph.harvard.edu/cas`.

6. W. E. Leary, "Cell phones: questions but no answers," *New York Times*, October 26, 1999.

7. R. A. Laven and C. T. Livesey, "The effect of housing and methionine intake on hoof horn hemorrhages in primiparous lactating Holstein cows," *Journal of Dairy Science*, 87 (2004), pp. 1015–1023.

8. P. Weiner et al. "The cumulative effect of long-acting bronchodilators, exercise, and inspiratory muscle training on the perception of dyspnea in patients with advanced COPD," *Chest*, 118 (2000), pp. 672–678.

9. B. J. Marshall and J. R. Warren, "Unidentified curved bacilli in the stomach of patients with gastritis and peptic ulceration," *Lancet*, 1 (1984), pp. 1311–1315; B. J. Marshall et al., "A prospective double-blind trial of duodenal ulcer relapse after eradication of *Campylobacter pylori*," *Lancet*, 2 (1988), pp. 1437–1442; B. J. Marshall et al., "Urea protects *Helicobacter (Campylobacter) pylori* from the bactericidal effect of acid," *Gastroenterology*, 99 (1990), pp. 697–702.

10. See Note 1 for Chapter 11.

11. See Note 11 for Chapter 8.

12. G. Gregoratos et al., "ACC/AHA guidelines for implantation of cardiac pacemakers and antiarrhythmia devices: executive summary," *Circulation*, 97 (1998), pp. 1325–1335.

13. T. M. Blackburn et al., "Avian extinction and mammalian introductions on oceanic islands," *Science*, 305 (2004), pp. 1955–1958.

14. Information about the study can be found on the NIH Web site at `http://diabetes.niddk.nih.gov/dm/pubs/control/`.

15. See Note 7 for Chapter 14.

16. C. Kopp et al., "Modulation of rhythmic brain activity by diazepam: GABAA receptor subtype and state specificity," *Proceedings of the National Academy of Sciences*, 101 (2004), pp. 3674–3679.

17. J. B. Moseley et al., "A controlled trial of arthroscopic surgery for osteoarthritis of the knee," *New England Journal of Medicine*, 347 (2002), pp. 81–88.

Chapter 16

1. D. M. Fergusson and L. J. Horwood, "Cannabis use and traffic accidents in a birth cohort of young adults," *Accident Analysis and Prevention*, 33 (2001), pp. 703–711.

2. Based on a news item "Bee off with you," *Economist*, November 2, 2002, p. 78.

3. P. J. Goodwin et al., "The effect of group psychological support on survival in metastatic breast cancer," *New England Journal of Medicine*, 345 (2001), pp. 1719–1726.

4. M. Park et al., "Recycling endosomes supply AMPA receptors for LTP," *Science*, 305 (2004), pp. 1972–1975.

5. This and similar results of Gallup polls are from the Gallup Organization Web site, `www.gallup.com`.

6. C. S. Fuchs et al., "Alcohol consumption and mortality among women," *New England Journal of Medicine*, 332 (1995), pp. 1245–1250.

7. K. E. Hobbs et al., "Levels and patterns of persistent organochlorines in Minke whale (*Balaenoptera acutorostrata*) stocks from the North Atlantic and European Arctic," *Environmental Pollution*, 121 (2003), pp. 239–252.

8. M. Hack et al., "Outcomes in young adulthood for very-low-birth-weight infants," *New England Journal of Medicine*, 346 (2002), pp. 149–157.

9. Data simulated from a Normal distribution with $\mu = 98.2$ and $\sigma = 0.7$. These values are based on P. A. Mackowiak et al., "A critical appraisal of 98.6 degrees F, the upper limit of the normal body temperature, and other legacies of Carl Reinhold August Wunderlich," *Journal of the American Medical Association*, 268, (1992), pp. 1578–1580.

10. F. J. Sulloway, "Birth order and intelligence," *Science*, 316 (2007), pp. 1711–1712. The Norwegian study is P. Kristensen and T. Bjerkedal, "Explaining the relation between birth order and intelligence," *Science*, 316 (2007), p. 1717.

11. J. F. Swain et al., "Comparison of the effects of oat bran and low-fiber wheat on serum lipoprotein levels and blood pressure," *New England Journal of Medicine*, 322 (1990), pp. 147–152.

12. Population base from the 2005 General Household Survey, at `www.statistics.gov.uk`. Smoking data from Action on Smoking and Health, *Smoking and Health Inequality*, at `www.ash.org.uk`.

13. See Note 13 for Chapter 10.

Chapter 17

1. Note 1 for Chapter 13 explains the reason for this condition in the case of inference about a population mean.

2. I. S. Colon, "Effects of cellulose, oat bran, rice bran, and psyllium on serum and liver cholesterol and fecal steroid excretion in rats," MS thesis, Purdue University, 1992.

3. The degrees of freedom for the one-sample t statistic come from the sample standard deviation s in the denominator of t. We saw in Chapter 2 (page 48) that s has $n-1$ degrees of freedom.

4. See Note 9 for Chapter 1.

5. T. W. McDowell, "An evaluation of vibration and other effects on the accuracy of grip and push force recall," PhD thesis, West Virginia University, 2006.

6. Z. W. Liller, "Spatial variation in brook trout (*salvelinus fontinalis*) population dynamics and juvenile recruitment potential in an Appalachian watershed," MS thesis, West Virginia University, 2006.

7. The Minitab output in Figure 17.6 displays "95%C Lower Bound 0.327." This is not the lower bound of a 95% confidence interval but the lower bound of the upper 95% of the sampling distribution (because the test is one-sided). Don't hesitate to use the software's help menu if you are not entirely sure you understand the output you get.

8. What matters is reducing the original data to a one-sample problem. Taking the difference between the two conditions is the most common approach. One could also, for instance, calculate the percentage change for each pair of data points.

9. C. Vallbona, C. F. Hazlewood, G. Jurida, "Response of pain to static magnetic fields in postpolio patients: a double-blind pilot study," *Archives of Physical Medicine and Rehabilitation*, 78 (1997), pp. 1200–1203.

10. See Note 9 for Chapter 1.

11. For a qualitative discussion explaining why skewness is the most serious violation of the Normal shape condition, see D. D. Boos and J. M. Hughes-Oliver, "How large does n have to be for the Z and t intervals?" *American Statistician*, 54 (2000), pp. 121–128. Our recommendations are based on extensive computer work. See, for example, H. O. Posten, "The robustness of the one-sample t-test over the Pearson system," *Journal of Statistical Computation and Simulation*, 9 (1979), pp. 133–149; and E. S. Pearson and N. W. Please, "Relation between the shape of population distribution and the robustness of four simple test statistics," *Biometrika*, 62 (1975), pp. 223–241.

12. R. A. Berner and G. P. Landis, "Gas bubbles in fossil amber as possible indicators of the major gas composition of ancient air," *Science*, 239 (1988), pp. 1406–1409.

13. See Note 28 for Chapter 1.

14. M. Banda, "Population biology of the catfish *Bagrus meridionalis* from the southern part of Lake Malawi," *Lake Malawi Fisheries Management Symposium Proceedings*, 2001, pp. 200–214.

15. See Note 7 for Chapter 1.

16. Data from the "wine" database in the archive of machine learning data bases at the University of California, Irvine, `ftp.ics.uci.edu/pub/machine-learning-databases`.

17. We thank Rudi Berkelhamer of the University of California at Irvine for the data. The data are part of a larger set collected for an undergraduate lab exercise in scientific methods.

18. See Note 5 for Chapter 6.

19. This example is based on information in D. L. Shankland et al., "The effect of 5-thio-D-glucose on insect development and its absorption by insects," *Journal of Insect Physiology*, 14 (1968), pp. 63–72.

20. Data provided by Jason Hamilton, University of Illinois. The study is reported in E. H. DeLucia et al., "Net primary production of a forest ecosystem with experimental CO_2 enhancement," *Science*, 284 (1999), pp. 1177–1179. Note that resampling methods cannot remove the variation due to random sampling of the original data. No method for inference can be trusted with $n = 3$. In this study, each observation is very costly, so the small n is inevitable.

21. M. W. Peugh, "Field investigation of ventilation and air quality in duck and turkey slaughter plants," MS thesis, Purdue University, 1996.

22. Based on R. de la Fuente-Fernandez et al., "Expectation and dopamine release: mechanism of the placebo effect in Parkinson's disease," *Science*, 293 (2001), pp. 1164–1166.

23. J. M. Jobe and H. Jobe, "A statistical approach for additional infill development," *Energy Exploration and Exploitation*, 18 (2000), pp. 89–103, and Lake Michigan Federation, 2001 publication, "The case against new Great Lakes oil and gas drilling: Michigan fails to clean up oil and gas pollution," found at `www.lakemichigan.org`.

24. H. B. Meyers, "Investigations of the life history of the velvetleaf seed beetle, *Althaeus folkertsi* Kingsolver," MS thesis, Purdue University, 1996. The 95% t interval is 1227.9 to 2507.6. A 95% bootstrap BCa interval is 1444 to 2718, confirming that t inference is inaccurate for these data.

25. Data courtesy of Brad Hughes, Department of Ecology and Evolutionary Biology, University of California, Irvine.

26. W. H. Harris et al., "Stimulation of bone formation in vivo by phosphate supplementation," *Calcified Tissue Research*, 22 (1976), pp. 85–98.

27. Data provided by Timothy Sturm.

28. L. Yuh, "A biopharmaceutical example for undergraduate students," manuscript.

Chapter 18

1. S. Aneja, "Biodeterioration of textile fibers in soil," MS thesis, Purdue University, 1994.

2. Detailed information about the conservative t procedures can be found in P. Leaverton and J. J. Birch, "Small sample power curves for the two-sample location problem," *Technometrics*, 11 (1969), pp. 299–307; in H. Scheffé, "Practical solutions of the Behrens-Fisher problem," *Journal of the American Statistical Association*, 65 (1970), pp. 1501–1508; and in D. J. Best and J. C. W. Rayner, "Welch's approximate solution for the Behrens-Fisher problem," *Technometrics*, 29 (1987), pp. 205–210.

3. E. Courchesne, C. M. Karns, et al., "Unusual brain growth patterns in early life in patients with autistic disorder: an MRI study," *Neurology*, 57 (2001), pp. 245–254.

4. We saw in Chapter 11 that Normal quantile plots are a good way to verify Normality by, in essence, plotting the observed standardized data against expected Normal z scores.

5. E. Sanford et al., "Local selection and latitudinal variation in a marine predator-prey interaction," *Science*, 300 (2003), pp. 1135–1137.

6. See Note 7 for Chapter 5.

7. See Note 9 for Chapter 2.

8. See the extensive simulation studies in H. O. Posten, "The robustness of the two-sample t-test over the Pearson system," *Journal of Statistical Computation and Simulation*, 6 (1978), pp. 295–311, and in H. O. Posten, H. Yeh, and D. B. Owen, "Robustness of the two-sample t-test under violations of the homogeneity assumption," *Communications in Statistics*, 11 (1982), pp. 109–126.

9. Data provided by Samuel Phillips, Purdue University.

10. D. L. Shankland, "Involvement of spinal cord and peripheral nerves in DDT-poisoning syndrome in albino rats," *Toxicology and Applied Pharmacology*, 6 (1964), pp. 197–213.

11. Data provided by Darlene Gordon, Purdue University.

12. The problem of comparing spreads is difficult even with advanced methods. Common distribution-free procedures do not offer a satisfactory alternative to the F test, because they are sensitive to unequal shapes when comparing two distributions. A recent survey of possible approaches is D. D. Boos and C. Brownie, "Comparing variances and other measures of dispersion," *Statistical Science*, 19 (2005), pp. 571–578. See also L. H. Shoemaker, "Fixing the F test for equal variances," *American Statistician*, 57 (2003), pp. 105–114, for adjustments to F that improve its robustness. The adjustments involve data-dependent degrees of freedom, similar in spirit to the Option 1 two-sample t procedures described in this chapter.

13. Based on G. L. Cromwell et al., "A comparison of the nutritive value of *opaque-2*, *floury-2* and normal corn for the chick," *Poultry Science*, 47 (1968), pp. 840–847.

14. Data provided by Darlene Gordon, Purdue University.

15. See Note 18 for Chapter 2.

16. See Note 21 for Chapter 17.

17. Data provided by Warren Page, New York City Technical College, from a study done by John Hudesman.

18. From the Avandia website, `www.avandia.com`, Prescribing information.

19. J. R. Cronin and S. Pizzarello, "Enantiometric excesses in meteoritic amino acids," *Science*, 275 (1997), pp. 951–955.

20. Based on M. C. Wilson et al., "Impact of cereal leaf beetle larvae on yields of oats," *Journal of Economic Entomology*, 62 (1969), pp. 699–702.

21. K. Linde et al., "Acupuncture for patients with migraine: a randomized controlled trial," *JAMA*, 293(2005), pp. 2118–2125.

22. See Note 28 for Chapter 1.

23. TOX-17: Black Newsprint Inks, `ntp.niehs.nih.gov`.

24. See Note 9 for Chapter 17.

25. From a graph in F. Grieco, A. J. van Noordwijk, and M. E. Visser, "Evidence for the effect of learning on timing of reproduction in blue tits," *Science*, 296 (2002), pp. 136–138.

26. See Note 6 for Chapter 3.

Chapter 19

1. J. H. Catania et al., "Prevalence of AIDS-related risk factors and condom use in the United States," *Science*, 258 (1992), pp. 1101–1106.

2. The Diabetes Control and Complications Trial Research Group, "The effect of intensive treatment of diabetes on the development and progression of long-term complications in insulin-dependent diabetes mellitus," *New England Journal of Medicine*, 329(1993), pp. 977–986.

3. Strictly speaking, the formula $\sqrt{p(1-p)/n}$ for the standard deviation of $\hat{p}$ assumes that we draw an SRS of size n from an *infinite* population. If the population has finite size N, this standard deviation is multiplied by $\sqrt{1-(n-1)/(N-1)}$. This "finite population correction" approaches 1 as N increases. When the population is at least 20 times as large as the sample, the correction factor is between about 0.97 and 1. It is reasonable to use the simpler form $\sqrt{p(1-p)/n}$ in these settings. See also Note 1 for Chapter 13.

4. Obtained from the Florida Department of Health's website, `www.doh.state.fl.us`, Epidemiology, Antibiotic resistance.

5. From the Web site of the Gallup Organization, `www.gallup.com`, May 18, 2004, "Number of Teen Smokers Holding Steady."

6. This rule of thumb is based on study of computational results in the papers cited in Note 6 and on discussion with Alan Agresti. We strongly recommend using the plus four interval.

7. The quotation is from page 1104 of the article cited in Note 1.

8. This interval is proposed by A. Agresti and B. A. Coull, "Approximate is better than 'exact' for interval estimation of binomial proportions," *The American Statistician*, 52 (1998), pp. 119–126. There are several yet more accurate but considerably more complex intervals for p that might be used in professional practice. See L. D. Brown, T. Cai, and A. DasGupta, "Interval estimation for a binomial proportion," *Statistical Science*, 16 (2001), pp. 101–133. A detailed theoretical study that uncovers the reason the large-sample interval is inaccurate is L. D. Brown, T. Cai, and A. DasGupta, "Confidence intervals for a binomial proportion and asymptotic expansions," *Annals of Statistics*, 30 (2002), pp. 160–201.

9. D. Fergusson et al., "Turning a blind eye: the success of blinding reported in a random sample of randomised, placebo-controlled trials," *BMJ*, 328 (2004), pp. 432–436.

10. From A. Agresti and B. Caffo, "Simple and effective confidence intervals for proportions and differences of proportions result from adding two successes and two failures," *The American Statistician*, 45 (2000), pp. 280–288. When can the plus four interval be safely used? The answer depends on just how much accuracy you insist on. Brown and coauthors (see Note 6) recommend $n \geq 40$. Agresti and Coull demonstrate that performance is almost always satisfactory in their eyes when $n \geq 5$. Our rule of thumb $n \geq 10$ allows for confidence levels C other than 95% and fits our philosophy of not insisting on more exact results than practice requires. The big point is that plus four is very much more accurate than the standard interval for most values of p and all but very large n.

11. See Note 5 for Chapter 18.

12. G. Edwards and J. Mazzuca, "Three quarters of Canadians support doctor-assisted suicide," Gallup Poll press release, March 24, 1999, at `www.gallup.com`.

13. In fact, P-values for two-sided tests are more accurate than those for one-sided tests. Our rule of thumb is a compromise to avoid the confusion of too many rules.

14. M. A. Carlton and W. D. Stansfield, "Making babies by the flip of a coin?" *American Statistician*, 59 (2005), pp. 180–182.

15. C. Davison, "Sex-ratio varies with egg sequence in Lesser Snow Geese," *The Auk*, 99 (1982), pp. 662–666.

16. J. Fagan et al., "Performance assessment under field conditions of a rapid immunological test for transgenic soybeans," *International Journal of Food Science and Technology*, 36 (2001), pp. 357–367.

17. Pew Internet Project, "Finding answers online in sickness and in health," May 2, 2006, `www.pewinternet.org`.

18. See Note 11 for Chapter 5.

19. E. Ossiander, letter to the editor, *Science*, 257 (1992), p. 1461.

20. "Lead in DC drinking water," available from the EPA website at `www.epa.gov/dclead`.

21. H. Wechsler et al., *Binge Drinking on America's College Campuses*, Harvard School of Public Health, 2001.

22. S. Schroter, H. Barratt, and J. Smith, "Author's perceptions of electronic publishing: two cross-sectional surveys," *BMJ*, 328 (2004), pp. 1350–1353.

23. J. E. Brody, "Alternative medicine makes inroads," *New York Times*, April 28, 1998.

Chapter 20

1. See Note 2 for Chapter 19.

2. See Note 2 for Chapter 19.

3. D. E. Jorenby et al., "A controlled trial of sustained-release bupropion, a nicotine patch, or both for smoking cessation," *New England Journal of Medicine*, 340 (1999), pp. 685–691.

4. The plus four method is by A. Agresti and B. Caffo. See Note 10 for Chapter 19.

5. F. Lloret et al., "Fire and resprouting in Mediterranean ecosystems: insights from an external biogeographical region, the Mexican shrubland," *American Journal of Botany*, 88 (1999), pp. 1655–1661.

6. NCCAM Web site, `nccam.nih.gov/health/echinacea`, and R. B. Turner et al., "An evaluation of Echinacea angustifolia in experimental rhinovirus infections," *New England Journal of Medicine*, 353 (2005), pp. 341–348.

7. Modified from R. A. Schieber et al., "Risk factors for injuries from in-line skating and the effectiveness of safety gear," *New England Journal of Medicine*, 335 (1996), Internet summary at `content.nejm.org`.

8. See Note 15 for Chapter 19.

9. The proper count requirements for the two-sample proportions test of hypothesis should be about expected counts, as for the one sample test (Chapter 19) and for the chi-square tests (Chapters 21 and 22). However, the expected counts requirements become somewhat cumbersome for the two-sample procedure and we provide a simpler alternative focused on actual counts. The reason for this decision is to avoid drowning students in non-critical computational details. Here are the requirements based on expected counts.

   If we have 2 samples of sizes n_1 and n_2 such that $N = n_1 + n_2$ and if we call S the total number of successes and F the total number of failures in both samples combined, then each of the following four expected counts should be greater than 5: $n_1 S/N$, $n_2 S/N$, $n_1 F/N$, and $n_2 F/N$. This implies that the total number of successes S and the total number of failures F should each be equal to or greater than $[5(n_1 + n_2)/(\text{largest of } n_1 \text{ or } n_2)]$. The sample counts requirements described in the text tend to be more conservative than the requirements using expected counts, especially when only one of the two samples is very small. Instructors may opt for a simple and less conservative approach, requiring that the total number of successes S and the total number of failures F should each be 10 or greater.

10. R. F. Adair and L. R. Holmgren, "Do drug samples influence resident prescribing behavior? A randomized trial," *American Journal of Medicine*, 118 (2005), pp. 881–884.

11. Steering Committee of the Physicians' Health Study Research Group, "Final report on the aspirin component of the ongoing Physicians' Health Study," *New England Journal of Medicine*, 321 (1989), pp. 129–135.

12. J. E. Tcheng, "High-dose eptifibatide (Integrilin) in elective coronary stenting: results of the ESPRIT trial," presented at the American College of Cardiology 49th Annual Scientific Session, March 12–15, 2000, Anaheim, Calif.

13. F. Gaudet et al., "Induction of tumors in mice by genomic hypomethylation," *Science*, 300 (2003), pp. 489–492.

14. From an Associated Press dispatch appearing on December 30, 2002. The study report appeared in the *Journal of Adolescent Health*.

15. D. G. Altman, S. N. Goodman, and S. Schroter, "How statistical expertise is used in medical research," *Journal of the American Medical Association*, 287 (2002), pp. 2817–2820.

16. See Note 2 for Chapter 5.

17. A. C. Allison and D. F. Clyde, "Malaria in African children with deficient erythrocyte dehydrogenase," *British Medical Journal*, 1 (1961), pp. 1346–1349.

18. See Note 3 for Chapter 5.

19. C. G. Jones, et al., "Chain reactions linking acorns to gypsy moth outbreaks and Lyme disease risk," *Science*, 279 (1998), pp. 1023–1026.

20. See Note 15 for Chapter 7.

21. M. Enserink, "Fraud and ethics charges hit stroke drug trial," *Science*, 274 (1996), pp. 2004–2005.

22. D. R. Focht III, C. Spicer, and M. P. Fairchok, "The efficacy of duct tape vs. cryotherapy in the treatment of verruca vulgaris (the common wart)," *Archives of Pediatrics and Adolescent Medicine*, 156 (2002), pp. 971–974.

23. Based on C. Anderson, "Measuring what works in health care," *Science*, 263 (1994), pp. 1080–1082.

24. Toxicology and Carcinogenesis Studies of 60-HZ Magnetic Fields in F344/N Rats and B6C3F1 Mice (Whole-body Exposure Studies), TR-488, `ntp.niehs.nih.gov`.

25. "ACE inhibitors as a new frontier in cardiovascular prevention," report from the Heart Outcomes Prevention Evaluation trial, Population Health Research Institute, Hamilton Health Sciences/McMaster University, `www.ccc.mcmaster.ca`.

Chapter 21

1. P. A. Kulakow, H. Hauptli, and S. K. Jain, "Genetics of grain amaranths," *Journal of Heredity*, 1985 pp. 27–30.

2. Based on a news item in *Science*, 305 (2004), p. 1560. The study, by Daniel Klem, appeared in the *Wilson Journal*.

3. There are many computer studies of the accuracy of chi-square critical values for X^2. Our guideline goes back to Cochran (1954). Later work has shown that it is often conservative, in the sense that if the expected cell counts are all similar and the degrees of freedom exceed 1, the chi-square approximation works well for an average expected count as small as 1 or 2. Our guideline protects against dissimilar expected counts. It has the added advantage that it is safe in the 2×2 case, where the chi-square approximation is least good. So our condition is helpful for beginners—there is no single condition that is not conservative and applies to 2×2 and larger tables with similar and dissimilar expected cell counts. There are exact procedures that (with software) should be used for tables that do not satisfy our condition. For a survey, see A. Agresti, "A survey of exact inference for contingency tables," *Statistical Science*, 7 (1992), pp. 131–177.

4. M. Xu and R. G. Palmer, "genetic analysis of 4 new mutants at the unstable k2 Mdh1-n y20 chromosomal region in soybean," *Journal of Heredity*, 97 (2006), pp. 423–427.

5. W. Landauer and L. C. Dunnthe, "Frizzle characters of fowls: its expression and inheritance," *Journal of Heredity*, 21 (1930), pp. 291–305.

6. See Note 21 for Chapter 4.

7. See Note 12 for Chapter 10.

8. R. D. Jackson et al., "Calcium plus vitamin D supplementation and the risk of fractures," *New England Journal of Medicine*, 354 (2006), pp. 669–683.

9. From the GSS data base at the University of Michigan, `webapp.icpsr.umich.edu/GSS`.

10. J. L. Hoogland, "Multiple mating by female prairie dogs," *Animal Behavior*, 55 (1998), pp. 351–359.

Chapter 22

1. D. B. Mark et al., "Use of medical resources and quality of life after acute myocardial infarction in Canada and the United States," *New England Journal of Medicine*, 331 (1994), pp. 1130–1135. See also the discussion in the same journal, 332 (1995), pp. 469–472.

2. K. Marangon et al., "Diet, antioxident status, and smoking habits in French men," *American Journal of Clinical Nutrition*, 67 (1998), pp. 231–239.

3. See Note 2 for Chapter 5.

4. See Note 10 for Chapter 21.

5. See Note 3 for Chapter 21.

6. See Note 21 for Chapter 4.

7. See Note 6 for Chapter 9.

8. See Note 3 for Chapter 8.

9. From the District of Columbia Department of Health, "The HIV/AIDS Epidemiologic Profile for the District of Columbia," December 2003.

10. V. P. Carnielli et al., "Intestinal absorption of long-chain polyunsaturated fatty acids in preterm infants fed breast milk or formula," *American Journal of Clinical Nutrition*, 67 (1998), pp. 97–103.

11. Modified from F. Barringer, "Measuring sexuality through polls can be shaky," *New York Times*, April 25, 1993.

12. See Note 1 for Chapter 5.

13. B. C. Coleman, "Study: heart attack risk cut 74% by stress management," Associated Press dispatch appearing in the Lafayette, Indiana, *Journal and Courier*, October 20, 1997.

14. See Note 11 for Chapter 10.

15. See Note 14 for Chapter 5.

16. See Note 30 for Chapter 10.

17. M. De Longerill et al., "Mediterranean dietary pattern in a randomized trial," *Archives of Internal Medicine*, 158 (1998), pp. 1181–1187.

18. C. Braga et al., "Olive oil, other seasoning fats, and the risk of colorectal carcinoma," *Cancer*, 82 (1998), pp. 448–453.

19. See Note 10 for Chapter 5.

20. See Note 15 for Chapter 5.

21. See Note 7 for Chapter 5, and `nccam.nih.gov/health/echinacea/`.

22. See Note 12 for Chapter 10.

23. National Institute of Environmental Health Sciences, "Toxicology and carcinogenesis studies of 60-Hz magnetic fields in F344/N rats and B6C3F1 mice (whole-body exposure studies)," TR-488, `ntp.niehs.nih.gov`.

24. See Note 5 for Chapter 5.

Chapter 23

1. S. Karelitz et al., "Relation of crying activity in early infancy to speech and intellectual development at age three years," *Child Development*, 35 (1964), pp. 769–777.

2. See Note 8 for Chapter 3.

3. See Note 15 for Chapter 4.

4. From a graph in B. J. Peterson et al., "Increasing river discharge to the Arctic Ocean," *Science*, 298 (2002), pp. 2171–2173.

5. See Note 5 for Chapter 17.

6. From a graph in A. L. Perry, et al., "Climate change and distribution shifts in marine fishes," *Science*, 308 (2005), pp. 1912–1915. The explanatory variable is the five-year running mean of winter (December to March) sea-bottom temperature.

7. C. J. Stevens et al., "Impact of nitrogen deposition on the species richness of grasslands," *Science*, 303 (2004), pp. 1876–1879.

8. See Note 20 for Chapter 3.

9. A. P. Moller, "Rapid change in nest size of a bird related to change in a secondary sexual character," *Behavioral Ecology*, 17 (2006), pp. 108–116.

10. From Table S2 in the online supplement to A. Dell'Anno and R. Danovaro, "Extracellular DNA plays a key role in deep-sea ecosystem functioning," *Science*, 309 (2005), p. 2179.

11. See Note 21 for Chapter 4.

12. See Note 2 for Chapter 3.

13. Based on M. E. Dunshee, "A study of factors affecting the amount and kind of food eaten by nursery school children," *Child Development*, 2 (1931), pp. 163–183. This article gives the means, standard deviations, and correlation for 37 children, from which these data are simulated.

14. See Note 5 for Chapter 6.

15. See Note 11 for Chapter 6.

16. From a plot in D. de Quervain et al., "The neural basis of altruistic punishment," *Science*, 305 (2004), pp. 1254–1258. The description of the study is simplified shamelessly, though it still sounds a bit complicated.

Chapter 24

1. See Note 28 for Chapter 1.

2. Based on the online supplement to P. J. Shaw et al., "Correlates of sleep and waking in *Drosophila melanogaster*," *Science*, 287 (2000), pp. 1834–1837.

3. N. A. Sherif et al., "Detection of cotinine in neonate meconium as a marker for nicotine exposure in utero," *Eastern Mediterranean Health Journal*, 10 (2004), pp. 96–105.

4. See Note 9 for Chapter 2.

5. See Note 12 for Chapter 2.

6. Adapted from a graph in article cited in Note 3.

7. C. L. Ogden et al., "Mean body weight, height, and body mass index, United States 1960–2002," *Advance Data from Vital and Health Statistics, no. 347* (CDC, 2004).

8. J. M. Jakicic et al., "Effects of intermittent exercise and use of home exercise equipment on adherence, weight loss, and fitness in overweight women," *Journal of the American Medical Association*, 282 (1999), pp. 1554–1560.

9. M. R. McClung et al., "Denosumab in postmenopausal women with low bone mineral density," *New England Journal of Medicine*, 354 (2006), pp. 821–831.

10. See Note 21 for Chapter 17.

11. See Note 7.

12. A. C. St.-Pierre et al., "Insulance resistance syndrome, body mass index and the risk of ischemic heart disease," *Canadian Medical Association Journal*, 172 (2005), pp. 1301–1305.

13. R. J. Safran et al., "Dynamic paternity allocation as a function of male plumage color in barn swallows," *Science*, 209 (2005), pp. 2210–2212.

14. Data from the online supplement to A. Kessler and I. T. Baldwin, "Defensive function of herbivore-induced plant volatile emissions in nature," *Science*, 291 (2001), pp. 2141–2144.

15. The data and the full story can be found in the Data and Story Library at `lib.stat.cmu.edu`. The original study is by F. Loven, "A study of interlist equivalency of the CID W-22 word list presented in quiet and in noise," MS thesis, University of Iowa, 1981.

16. Data provided by Matthew Moore.

17. See Note 25 for Chapter 17.

18. See Note 9 for Chapter 1.

19. See Note 1 for Chapter 18.

20. Data provided by Brigitte Stricanne.

21. See Note 18 for Chapter 2.

22. We thank Rudi Berkelhamer of the University of California, Irvine, for the data. The data are part of a larger set collected for an undergraduate lab exercise in scientific methods.

Chapter 25

1. D. L. Miller et al., "Effect of fat-free potato chips with and without nutrition labels on fat and energy intakes," *American Journal of Clinical Nutrition*, 68 (1998), pp. 282–290.

2. M. R. Dohm, J. P. Hayes, and T. Garland, Jr., "Quantitative genetics of sprint running speed and swimming endurance in laboratory house mice (*Mus domesticus*)," *Evolution*, 50 (1996), pp. 1688–1701.

3. See Note 11 for Chapter 3.

4. From the online supplement to G. Gaskell et al., "Worlds apart? The reception of genetically modified foods in Europe and the U.S.," *Science*, 285 (1999), pp. 383–387.

5. K. S. Oberhauser, "Fecundity, lifespan and egg mass in butterflies: effects of male-derived nutrients and female size," *Functional Ecology*, 11 (1997), pp. 166–175.

6. See Note 8 for Chapter 16. The exercises are simplified, in that the measures reported in this paper have been statistically adjusted for "sociodemographic status."

7. See Note 2 for Chapter 24.

8. J. E. Keeley, C. J. Fotheringham, and M. Morais, "Reexamining fire suppression impacts on brushland fire regimes," *Science*, 284 (1999), pp. 1829–1831.

9. From V. D. Bass, W. E. Hoffmann, and J. L. Dorner, "Normal canine lipid profiles and effects of experimentally induced pancreatitis and hepatic necrosis on lipids," *American Journal of Veterinary Research*, 37 (1976), pp. 1355–1357.

10. See Note 11 for Chapter 1.

11. These data were originally collected by L. M. Linde of UCLA but were first published by M. R. Mickey, O. J. Dunn, and V. Clark, "Note on the use of stepwise regression in detecting outliers," *Computers and Biomedical Research*, 1 (1967), pp. 105–111. The data have been used by several authors. I found them in N. R. Draper and J. A. John, "Influential observations and outliers in regression," *Technometrics*, 23 (1981), pp. 21–26.

12. Data provided by Marigene Arnold, Kalamazoo College.

13. Data provided by Corinne Lim, Purdue University, from a student project supervised by Professor Joseph Vanable.

14. S. S. Ahmed, "Effects of microwave drying on checking and mechanical strength of low-moisture baked products," MS thesis, Purdue University, 1994.

15. G. S. Hotamisligil, R. S. Johnson, R. J. Distel, R. Ellis, V. E. Papaioannou, and B. M. Spiegelman, "Uncoupling of obesity from insulin resistance through a targeted mutation in $aP2$, the adipocyte fatty acid binding protein," *Science*, 274 (1996), pp. 1377–1379.

16. M. H. Criqui, University of California, San Diego, reported in the New York Times, December 28, 1994.

17. See Note 14 for Chapter 5.

18. Information obtained on the Web site `www.gardasil.com`, Information for healthcare professionals, Efficacy.

Tables

TABLE A   Random digits

Line								
101	19223	95034	05756	28713	96409	12531	42544	82853
102	73676	47150	99400	01927	27754	42648	82425	36290
103	45467	71709	77558	00095	32863	29485	82226	90056
104	52711	38889	93074	60227	40011	85848	48767	52573
105	95592	94007	69971	91481	60779	53791	17297	59335
106	68417	35013	15529	72765	85089	57067	50211	47487
107	82739	57890	20807	47511	81676	55300	94383	14893
108	60940	72024	17868	24943	61790	90656	87964	18883
109	36009	19365	15412	39638	85453	46816	83485	41979
110	38448	48789	18338	24697	39364	42006	76688	08708
111	81486	69487	60513	09297	00412	71238	27649	39950
112	59636	88804	04634	71197	19352	73089	84898	45785
113	62568	70206	40325	03699	71080	22553	11486	11776
114	45149	32992	75730	66280	03819	56202	02938	70915
115	61041	77684	94322	24709	73698	14526	31893	32592
116	14459	26056	31424	80371	65103	62253	50490	61181
117	38167	98532	62183	70632	23417	26185	41448	75532
118	73190	32533	04470	29669	84407	90785	65956	86382
119	95857	07118	87664	92099	58806	66979	98624	84826
120	35476	55972	39421	65850	04266	35435	43742	11937
121	71487	09984	29077	14863	61683	47052	62224	51025
122	13873	81598	95052	90908	73592	75186	87136	95761
123	54580	81507	27102	56027	55892	33063	41842	81868
124	71035	09001	43367	49497	72719	96758	27611	91596
125	96746	12149	37823	71868	18442	35119	62103	39244
126	96927	19931	36809	74192	77567	88741	48409	41903
127	43909	99477	25330	64359	40085	16925	85117	36071
128	15689	14227	06565	14374	13352	49367	81982	87209
129	36759	58984	68288	22913	18638	54303	00795	08727
130	69051	64817	87174	09517	84534	06489	87201	97245
131	05007	16632	81194	14873	04197	85576	45195	96565
132	68732	55259	84292	08796	43165	93739	31685	97150
133	45740	41807	65561	33302	07051	93623	18132	09547
134	27816	78416	18329	21337	35213	37741	04312	68508
135	66925	55658	39100	78458	11206	19876	87151	31260
136	08421	44753	77377	28744	75592	08563	79140	92454
137	53645	66812	61421	47836	12609	15373	98481	14592
138	66831	68908	40772	21558	47781	33586	79177	06928
139	55588	99404	70708	41098	43563	56934	48394	51719
140	12975	13258	13048	45144	72321	81940	00360	02428
141	96767	35964	23822	96012	94591	65194	50842	53372
142	72829	50232	97892	63408	77919	44575	24870	04178
143	88565	42628	17797	49376	61762	16953	88604	12724
144	62964	88145	83083	69453	46109	59505	69680	00900
145	19687	12633	57857	95806	09931	02150	43163	58636
146	37609	59057	66967	83401	60705	02384	90597	93600
147	54973	86278	88737	74351	47500	84552	19909	67181
148	00694	05977	19664	65441	20903	62371	22725	53340
149	71546	05233	53946	68743	72460	27601	45403	88692
150	07511	88915	41267	16853	84569	79367	32337	03316

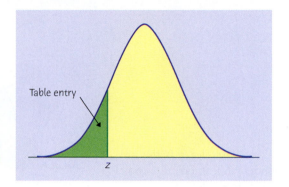

Table entry for z is the area under the standard Normal curve to the left of z.

TABLE B		Standard Normal probabilities								
z	.00	.01	.02	.03	.04	.05	.06	.07	.08	.09
−3.4	.0003	.0003	.0003	.0003	.0003	.0003	.0003	.0003	.0003	.0002
−3.3	.0005	.0005	.0005	.0004	.0004	.0004	.0004	.0004	.0004	.0003
−3.2	.0007	.0007	.0006	.0006	.0006	.0006	.0006	.0005	.0005	.0005
−3.1	.0010	.0009	.0009	.0009	.0008	.0008	.0008	.0008	.0007	.0007
−3.0	.0013	.0013	.0013	.0012	.0012	.0011	.0011	.0011	.0010	.0010
−2.9	.0019	.0018	.0018	.0017	.0016	.0016	.0015	.0015	.0014	.0014
−2.8	.0026	.0025	.0024	.0023	.0023	.0022	.0021	.0021	.0020	.0019
−2.7	.0035	.0034	.0033	.0032	.0031	.0030	.0029	.0028	.0027	.0026
−2.6	.0047	.0045	.0044	.0043	.0041	.0040	.0039	.0038	.0037	.0036
−2.5	.0062	.0060	.0059	.0057	.0055	.0054	.0052	.0051	.0049	.0048
−2.4	.0082	.0080	.0078	.0075	.0073	.0071	.0069	.0068	.0066	.0064
−2.3	.0107	.0104	.0102	.0099	.0096	.0094	.0091	.0089	.0087	.0084
−2.2	.0139	.0136	.0132	.0129	.0125	.0122	.0119	.0116	.0113	.0110
−2.1	.0179	.0174	.0170	.0166	.0162	.0158	.0154	.0150	.0146	.0143
−2.0	.0228	.0222	.0217	.0212	.0207	.0202	.0197	.0192	.0188	.0183
−1.9	.0287	.0281	.0274	.0268	.0262	.0256	.0250	.0244	.0239	.0233
−1.8	.0359	.0351	.0344	.0336	.0329	.0322	.0314	.0307	.0301	.0294
−1.7	.0446	.0436	.0427	.0418	.0409	.0401	.0392	.0384	.0375	.0367
−1.6	.0548	.0537	.0526	.0516	.0505	.0495	.0485	.0475	.0465	.0455
−1.5	.0668	.0655	.0643	.0630	.0618	.0606	.0594	.0582	.0571	.0559
−1.4	.0808	.0793	.0778	.0764	.0749	.0735	.0721	.0708	.0694	.0681
−1.3	.0968	.0951	.0934	.0918	.0901	.0885	.0869	.0853	.0838	.0823
−1.2	.1151	.1131	.1112	.1093	.1075	.1056	.1038	.1020	.1003	.0985
−1.1	.1357	.1335	.1314	.1292	.1271	.1251	.1230	.1210	.1190	.1170
−1.0	.1587	.1562	.1539	.1515	.1492	.1469	.1446	.1423	.1401	.1379
−0.9	.1841	.1814	.1788	.1762	.1736	.1711	.1685	.1660	.1635	.1611
−0.8	.2119	.2090	.2061	.2033	.2005	.1977	.1949	.1922	.1894	.1867
−0.7	.2420	.2389	.2358	.2327	.2296	.2266	.2236	.2206	.2177	.2148
−0.6	.2743	.2709	.2676	.2643	.2611	.2578	.2546	.2514	.2483	.2451
−0.5	.3085	.3050	.3015	.2981	.2946	.2912	.2877	.2843	.2810	.2776
−0.4	.3446	.3409	.3372	.3336	.3300	.3264	.3228	.3192	.3156	.3121
−0.3	.3821	.3783	.3745	.3707	.3669	.3632	.3594	.3557	.3520	.3483
−0.2	.4207	.4168	.4129	.4090	.4052	.4013	.3974	.3936	.3897	.3859
−0.1	.4602	.4562	.4522	.4483	.4443	.4404	.4364	.4325	.4286	.4247
−0.0	.5000	.4960	.4920	.4880	.4840	.4801	.4761	.4721	.4681	.4641

(continued)

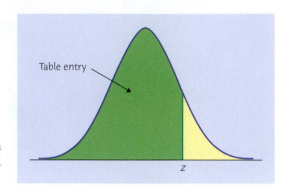

Table entry for z is the area under the standard Normal curve to the left of z.

TABLE B	Standard Normal probabilities (*continued*)								
z	.00	.01	.02	.03	.04	.05	.06	.07	.08	.09
0.0	.5000	.5040	.5080	.5120	.5160	.5199	.5239	.5279	.5319	.5359
0.1	.5398	.5438	.5478	.5517	.5557	.5596	.5636	.5675	.5714	.5753
0.2	.5793	.5832	.5871	.5910	.5948	.5987	.6026	.6064	.6103	.6141
0.3	.6179	.6217	.6255	.6293	.6331	.6368	.6406	.6443	.6480	.6517
0.4	.6554	.6591	.6628	.6664	.6700	.6736	.6772	.6808	.6844	.6879
0.5	.6915	.6950	.6985	.7019	.7054	.7088	.7123	.7157	.7190	.7224
0.6	.7257	.7291	.7324	.7357	.7389	.7422	.7454	.7486	.7517	.7549
0.7	.7580	.7611	.7642	.7673	.7704	.7734	.7764	.7794	.7823	.7852
0.8	.7881	.7910	.7939	.7967	.7995	.8023	.8051	.8078	.8106	.8133
0.9	.8159	.8186	.8212	.8238	.8264	.8289	.8315	.8340	.8365	.8389
1.0	.8413	.8438	.8461	.8485	.8508	.8531	.8554	.8577	.8599	.8621
1.1	.8643	.8665	.8686	.8708	.8729	.8749	.8770	.8790	.8810	.8830
1.2	.8849	.8869	.8888	.8907	.8925	.8944	.8962	.8980	.8997	.9015
1.3	.9032	.9049	.9066	.9082	.9099	.9115	.9131	.9147	.9162	.9177
1.4	.9192	.9207	.9222	.9236	.9251	.9265	.9279	.9292	.9306	.9319
1.5	.9332	.9345	.9357	.9370	.9382	.9394	.9406	.9418	.9429	.9441
1.6	.9452	.9463	.9474	.9484	.9495	.9505	.9515	.9525	.9535	.9545
1.7	.9554	.9564	.9573	.9582	.9591	.9599	.9608	.9616	.9625	.9633
1.8	.9641	.9649	.9656	.9664	.9671	.9678	.9686	.9693	.9699	.9706
1.9	.9713	.9719	.9726	.9732	.9738	.9744	.9750	.9756	.9761	.9767
2.0	.9772	.9778	.9783	.9788	.9793	.9798	.9803	.9808	.9812	.9817
2.1	.9821	.9826	.9830	.9834	.9838	.9842	.9846	.9850	.9854	.9857
2.2	.9861	.9864	.9868	.9871	.9875	.9878	.9881	.9884	.9887	.9890
2.3	.9893	.9896	.9898	.9901	.9904	.9906	.9909	.9911	.9913	.9916
2.4	.9918	.9920	.9922	.9925	.9927	.9929	.9931	.9932	.9934	.9936
2.5	.9938	.9940	.9941	.9943	.9945	.9946	.9948	.9949	.9951	.9952
2.6	.9953	.9955	.9956	.9957	.9959	.9960	.9961	.9962	.9963	.9964
2.7	.9965	.9966	.9967	.9968	.9969	.9970	.9971	.9972	.9973	.9974
2.8	.9974	.9975	.9976	.9977	.9977	.9978	.9979	.9979	.9980	.9981
2.9	.9981	.9982	.9982	.9983	.9984	.9984	.9985	.9985	.9986	.9986
3.0	.9987	.9987	.9987	.9988	.9988	.9989	.9989	.9989	.9990	.9990
3.1	.9990	.9991	.9991	.9991	.9992	.9992	.9992	.9992	.9993	.9993
3.2	.9993	.9993	.9994	.9994	.9994	.9994	.9994	.9995	.9995	.9995
3.3	.9995	.9995	.9995	.9996	.9996	.9996	.9996	.9996	.9996	.9997
3.4	.9997	.9997	.9997	.9997	.9997	.9997	.9997	.9997	.9997	.9998

Table entry for C is the critical value t^* required for confidence level C. To approximate one- and two-sided P-values, compare the value of the t statistic with the critical values of t^* that match the P-values given at the bottom of the table.

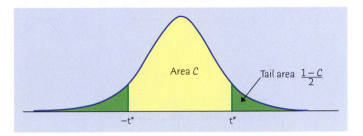

Area C

Tail area $\frac{1-C}{2}$

$-t^*$ t^*

TABLE C	t distribution critical values										

	Confidence level C										
Degrees of freedom	50%	60%	70%	80%	90%	95%	96%	98%	99%	99.5%	99.8%	99.9%
---	---	---	---	---	---	---	---	---	---	---	---	---
1	1.000	1.376	1.963	3.078	6.314	12.710	15.890	31.820	63.660	127.300	318.300	636.600
2	0.816	1.061	1.386	1.886	2.920	4.303	4.849	6.965	9.925	14.090	22.330	31.600
3	0.765	0.978	1.250	1.638	2.353	3.182	3.482	4.541	5.841	7.453	10.210	12.920
4	0.741	0.941	1.190	1.533	2.132	2.776	2.999	3.747	4.604	5.598	7.173	8.610
5	0.727	0.920	1.156	1.476	2.015	2.571	2.757	3.365	4.032	4.773	5.893	6.869
6	0.718	0.906	1.134	1.440	1.943	2.447	2.612	3.143	3.707	4.317	5.208	5.959
7	0.711	0.896	1.119	1.415	1.895	2.365	2.517	2.998	3.499	4.029	4.785	5.408
8	0.706	0.889	1.108	1.397	1.860	2.306	2.449	2.896	3.355	3.833	4.501	5.041
9	0.703	0.883	1.100	1.383	1.833	2.262	2.398	2.821	3.250	3.690	4.297	4.781
10	0.700	0.879	1.093	1.372	1.812	2.228	2.359	2.764	3.169	3.581	4.144	4.587
11	0.697	0.876	1.088	1.363	1.796	2.201	2.328	2.718	3.106	3.497	4.025	4.437
12	0.695	0.873	1.083	1.356	1.782	2.179	2.303	2.681	3.055	3.428	3.930	4.318
13	0.694	0.870	1.079	1.350	1.771	2.160	2.282	2.650	3.012	3.372	3.852	4.221
14	0.692	0.868	1.076	1.345	1.761	2.145	2.264	2.624	2.977	3.326	3.787	4.140
15	0.691	0.866	1.074	1.341	1.753	2.131	2.249	2.602	2.947	3.286	3.733	4.073
16	0.690	0.865	1.071	1.337	1.746	2.120	2.235	2.583	2.921	3.252	3.686	4.015
17	0.689	0.863	1.069	1.333	1.740	2.110	2.224	2.567	2.898	3.222	3.646	3.965
18	0.688	0.862	1.067	1.330	1.734	2.101	2.214	2.552	2.878	3.197	3.611	3.922
19	0.688	0.861	1.066	1.328	1.729	2.093	2.205	2.539	2.861	3.174	3.579	3.883
20	0.687	0.860	1.064	1.325	1.725	2.086	2.197	2.528	2.845	3.153	3.552	3.850
21	0.686	0.859	1.063	1.323	1.721	2.080	2.189	2.518	2.831	3.135	3.527	3.819
22	0.686	0.858	1.061	1.321	1.717	2.074	2.183	2.508	2.819	3.119	3.505	3.792
23	0.685	0.858	1.060	1.319	1.714	2.069	2.177	2.500	2.807	3.104	3.485	3.768
24	0.685	0.857	1.059	1.318	1.711	2.064	2.172	2.492	2.797	3.091	3.467	3.745
25	0.684	0.856	1.058	1.316	1.708	2.060	2.167	2.485	2.787	3.078	3.450	3.725
26	0.684	0.856	1.058	1.315	1.706	2.056	2.162	2.479	2.779	3.067	3.435	3.707
27	0.684	0.855	1.057	1.314	1.703	2.052	2.158	2.473	2.771	3.057	3.421	3.690
28	0.683	0.855	1.056	1.313	1.701	2.048	2.154	2.467	2.763	3.047	3.408	3.674
29	0.683	0.854	1.055	1.311	1.699	2.045	2.150	2.462	2.756	3.038	3.396	3.659
30	0.683	0.854	1.055	1.310	1.697	2.042	2.147	2.457	2.750	3.030	3.385	3.646
40	0.681	0.851	1.050	1.303	1.684	2.021	2.123	2.423	2.704	2.971	3.307	3.551
50	0.679	0.849	1.047	1.299	1.676	2.009	2.109	2.403	2.678	2.937	3.261	3.496
60	0.679	0.848	1.045	1.296	1.671	2.000	2.099	2.390	2.660	2.915	3.232	3.460
80	0.678	0.846	1.043	1.292	1.664	1.990	2.088	2.374	2.639	2.887	3.195	3.416
100	0.677	0.845	1.042	1.290	1.660	1.984	2.081	2.364	2.626	2.871	3.174	3.390
1000	0.675	0.842	1.037	1.282	1.646	1.962	2.056	2.330	2.581	2.813	3.098	3.300
z^*	0.674	0.841	1.036	1.282	1.645	1.960	2.054	2.326	2.576	2.807	3.091	3.291
One-sided P	.25	.20	.15	.10	.05	.025	.02	.01	.005	.0025	.001	.0005
Two-sided P	.50	.40	.30	.20	.10	.05	.04	.02	.01	.005	.002	.001

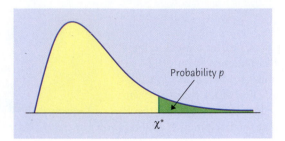

Table entry for p is the critical value χ^* with probability p lying to its right.

Probability p

χ^*

TABLE D	Chi-square distribution critical values

						p						
df	.25	.20	.15	.10	.05	.025	.02	.01	.005	.0025	.001	.0005
1	1.32	1.64	2.07	2.71	3.84	5.02	5.41	6.63	7.88	9.14	10.83	12.12
2	2.77	3.22	3.79	4.61	5.99	7.38	7.82	9.21	10.60	11.98	13.82	15.20
3	4.11	4.64	5.32	6.25	7.81	9.35	9.84	11.34	12.84	14.32	16.27	17.73
4	5.39	5.99	6.74	7.78	9.49	11.14	11.67	13.28	14.86	16.42	18.47	20.00
5	6.63	7.29	8.12	9.24	11.07	12.83	13.39	15.09	16.75	18.39	20.51	22.11
6	7.84	8.56	9.45	10.64	12.59	14.45	15.03	16.81	18.55	20.25	22.46	24.10
7	9.04	9.80	10.75	12.02	14.07	16.01	16.62	18.48	20.28	22.04	24.32	26.02
8	10.22	11.03	12.03	13.36	15.51	17.53	18.17	20.09	21.95	23.77	26.12	27.87
9	11.39	12.24	13.29	14.68	16.92	19.02	19.68	21.67	23.59	25.46	27.88	29.67
10	12.55	13.44	14.53	15.99	18.31	20.48	21.16	23.21	25.19	27.11	29.59	31.42
11	13.70	14.63	15.77	17.28	19.68	21.92	22.62	24.72	26.76	28.73	31.26	33.14
12	14.85	15.81	16.99	18.55	21.03	23.34	24.05	26.22	28.30	30.32	32.91	34.82
13	15.98	16.98	18.20	19.81	22.36	24.74	25.47	27.69	29.82	31.88	34.53	36.48
14	17.12	18.15	19.41	21.06	23.68	26.12	26.87	29.14	31.32	33.43	36.12	38.11
15	18.25	19.31	20.60	22.31	25.00	27.49	28.26	30.58	32.80	34.95	37.70	39.72
16	19.37	20.47	21.79	23.54	26.30	28.85	29.63	32.00	34.27	36.46	39.25	41.31
17	20.49	21.61	22.98	24.77	27.59	30.19	31.00	33.41	35.72	37.95	40.79	42.88
18	21.60	22.76	24.16	25.99	28.87	31.53	32.35	34.81	37.16	39.42	42.31	44.43
19	22.72	23.90	25.33	27.20	30.14	32.85	33.69	36.19	38.58	40.88	43.82	45.97
20	23.83	25.04	26.50	28.41	31.41	34.17	35.02	37.57	40.00	42.34	45.31	47.50
21	24.93	26.17	27.66	29.62	32.67	35.48	36.34	38.93	41.40	43.78	46.80	49.01
22	26.04	27.30	28.82	30.81	33.92	36.78	37.66	40.29	42.80	45.20	48.27	50.51
23	27.14	28.43	29.98	32.01	35.17	38.08	38.97	41.64	44.18	46.62	49.73	52.00
24	28.24	29.55	31.13	33.20	36.42	39.36	40.27	42.98	45.56	48.03	51.18	53.48
25	29.34	30.68	32.28	34.38	37.65	40.65	41.57	44.31	46.93	49.44	52.62	54.95
26	30.43	31.79	33.43	35.56	38.89	41.92	42.86	45.64	48.29	50.83	54.05	56.41
27	31.53	32.91	34.57	36.74	40.11	43.19	44.14	46.96	49.64	52.22	55.48	57.86
28	32.62	34.03	35.71	37.92	41.34	44.46	45.42	48.28	50.99	53.59	56.89	59.30
29	33.71	35.14	36.85	39.09	42.56	45.72	46.69	49.59	52.34	54.97	58.30	60.73
30	34.80	36.25	37.99	40.26	43.77	46.98	47.96	50.89	53.67	56.33	59.70	62.16
40	45.62	47.27	49.24	51.81	55.76	59.34	60.44	63.69	66.77	69.70	73.40	76.09
50	56.33	58.16	60.35	63.17	67.50	71.42	72.61	76.15	79.49	82.66	86.66	89.56
60	66.98	68.97	71.34	74.40	79.08	83.30	84.58	88.38	91.95	95.34	99.61	102.70
80	88.13	90.41	93.11	96.58	101.90	106.60	108.10	112.30	116.30	120.10	124.80	128.30
100	109.10	111.70	114.70	118.50	124.30	129.60	131.10	135.80	140.20	144.30	149.40	153.20

Table entry for p is the critical value r^* of the correlation coefficient r with probability p lying to its right.

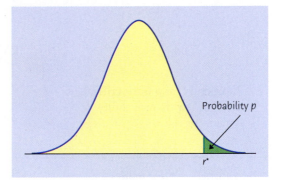

Probability p

r^*

TABLE E	Critical values of the correlation r								
	Upper tail probability p								
n	.20	.10	.05	.025	.02	.01	.005	.0025	.001	.0005
3	0.8090	0.9511	0.9877	0.9969	0.9980	0.9995	0.9999	1.0000	1.0000	1.0000
4	0.6000	0.8000	0.9000	0.9500	0.9600	0.9800	0.9900	0.9950	0.9980	0.9990
5	0.4919	0.6870	0.8054	0.8783	0.8953	0.9343	0.9587	0.9740	0.9859	0.9911
6	0.4257	0.6084	0.7293	0.8114	0.8319	0.8822	0.9172	0.9417	0.9633	0.9741
7	0.3803	0.5509	0.6694	0.7545	0.7766	0.8329	0.8745	0.9056	0.9350	0.9509
8	0.3468	0.5067	0.6215	0.7067	0.7295	0.7887	0.8343	0.8697	0.9049	0.9249
9	0.3208	0.4716	0.5822	0.6664	0.6892	0.7498	0.7977	0.8359	0.8751	0.8983
10	0.2998	0.4428	0.5494	0.6319	0.6546	0.7155	0.7646	0.8046	0.8467	0.8721
11	0.2825	0.4187	0.5214	0.6021	0.6244	0.6851	0.7348	0.7759	0.8199	0.8470
12	0.2678	0.3981	0.4973	0.5760	0.5980	0.6581	0.7079	0.7496	0.7950	0.8233
13	0.2552	0.3802	0.4762	0.5529	0.5745	0.6339	0.6835	0.7255	0.7717	0.8010
14	0.2443	0.3646	0.4575	0.5324	0.5536	0.6120	0.6614	0.7034	0.7501	0.7800
15	0.2346	0.3507	0.4409	0.5140	0.5347	0.5923	0.6411	0.6831	0.7301	0.7604
16	0.2260	0.3383	0.4259	0.4973	0.5177	0.5742	0.6226	0.6643	0.7114	0.7419
17	0.2183	0.3271	0.4124	0.4821	0.5021	0.5577	0.6055	0.6470	0.6940	0.7247
18	0.2113	0.3170	0.4000	0.4683	0.4878	0.5425	0.5897	0.6308	0.6777	0.7084
19	0.2049	0.3077	0.3887	0.4555	0.4747	0.5285	0.5751	0.6158	0.6624	0.6932
20	0.1991	0.2992	0.3783	0.4438	0.4626	0.5155	0.5614	0.6018	0.6481	0.6788
21	0.1938	0.2914	0.3687	0.4329	0.4513	0.5034	0.5487	0.5886	0.6346	0.6652
22	0.1888	0.2841	0.3598	0.4227	0.4409	0.4921	0.5368	0.5763	0.6219	0.6524
23	0.1843	0.2774	0.3515	0.4132	0.4311	0.4815	0.5256	0.5647	0.6099	0.6402
24	0.1800	0.2711	0.3438	0.4044	0.4219	0.4716	0.5151	0.5537	0.5986	0.6287
25	0.1760	0.2653	0.3365	0.3961	0.4133	0.4622	0.5052	0.5434	0.5879	0.6178
26	0.1723	0.2598	0.3297	0.3882	0.4052	0.4534	0.4958	0.5336	0.5776	0.6074
27	0.1688	0.2546	0.3233	0.3809	0.3976	0.4451	0.4869	0.5243	0.5679	0.5974
28	0.1655	0.2497	0.3172	0.3739	0.3904	0.4372	0.4785	0.5154	0.5587	0.5880
29	0.1624	0.2451	0.3115	0.3673	0.3835	0.4297	0.4705	0.5070	0.5499	0.5790
30	0.1594	0.2407	0.3061	0.3610	0.3770	0.4226	0.4629	0.4990	0.5415	0.5703
40	0.1368	0.2070	0.2638	0.3120	0.3261	0.3665	0.4026	0.4353	0.4741	0.5007
50	0.1217	0.1843	0.2353	0.2787	0.2915	0.3281	0.3610	0.3909	0.4267	0.4514
60	0.1106	0.1678	0.2144	0.2542	0.2659	0.2997	0.3301	0.3578	0.3912	0.4143
80	0.0954	0.1448	0.1852	0.2199	0.2301	0.2597	0.2864	0.3109	0.3405	0.3611
100	0.0851	0.1292	0.1654	0.1966	0.2058	0.2324	0.2565	0.2786	0.3054	0.3242
1000	0.0266	0.0406	0.0520	0.0620	0.0650	0.0736	0.0814	0.0887	0.0976	0.1039

Table entry for p is the critical value F^* with probability p lying to its right.

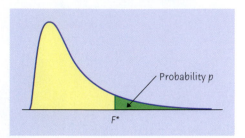

Probability p

F^*

TABLE F	**F distribution critical values**

		Degrees of freedom in the numerator							
	p	1	2	3	4	5	6	7	8
1	.100	39.86	49.50	53.59	55.83	57.24	58.20	58.91	59.44
	.050	161.45	199.50	215.71	224.58	230.16	233.99	236.77	238.88
	.025	647.79	799.50	864.16	899.58	921.85	937.11	948.22	956.66
	.010	4052.20	4999.50	5403.40	5624.60	5763.60	5859	5928.40	5981.10
	.001	405284.00	500000.00	540379.00	562500.00	576405.00	585937.00	592873.00	598144.00
2	.100	8.53	9.00	9.16	9.24	9.29	9.33	.35	9.37
	.050	18.51	19.00	19.16	19.25	19.30	19.33	19.35	19.37
	.025	38.51	39.00	39.17	39.25	39.30	39.33	39.36	39.37
	.010	98.50	99.00	99.17	99.25	99.30	99.33	99.36	99.37
	.001	998.50	999.00	999.17	999.25	999.30	999.33	999.36	999.37
3	.100	5.54	5.46	5.39	5.34	5.31	5.28	5.27	5.25
	.050	10.13	9.55	9.28	9.12	9.01	8.94	8.89	8.85
	.025	17.44	16.04	15.44	15.10	14.88	14.73	14.62	14.54
	.010	34.12	30.82	29.46	28.71	28.24	27.91	27.67	27.49
	.001	167.03	148.50	141.11	137.10	134.58	132.85	131.58	130.62
4	.100	4.54	4.32	4.19	4.11	4.05	4.01	3.98	3.95
	.050	7.71	6.94	6.59	6.39	6.26	6.16	6.09	6.04
	.025	12.22	10.65	9.98	9.60	9.36	9.20	9.07	8.98
	.010	21.20	18.00	16.69	15.98	15.52	15.21	14.98	14.80
	.001	74.14	61.25	56.18	53.44	51.71	50.53	49.66	49.00
5	.100	4.06	3.78	3.62	3.52	3.45	3.40	3.37	3.34
	.050	6.61	5.79	5.41	5.19	5.05	4.95	4.88	4.82
	.025	10.01	8.43	7.76	7.39	7.15	6.98	6.85	6.76
	.010	16.26	13.27	12.06	11.39	10.97	10.67	10.46	10.29
	.001	47.18	37.12	33.20	31.09	29.75	28.83	28.16	27.65
6	.100	3.78	3.46	3.29	3.18	3.11	3.05	3.01	2.98
	.050	5.99	5.14	4.76	4.53	4.39	4.28	4.21	4.15
	.025	8.81	7.26	6.60	6.23	5.99	5.82	5.70	5.60
	.010	13.75	10.92	9.78	9.15	8.75	8.47	8.26	8.10
	.001	35.51	27.00	23.70	21.92	20.80	20.03	19.46	19.03
7	.100	3.59	3.26	3.07	2.96	2.88	2.83	2.78	2.75
	.050	5.59	4.74	4.35	4.12	3.97	3.87	3.79	3.73
	.025	8.07	6.54	5.89	5.52	5.29	5.12	4.99	4.90
	.010	12.25	9.55	8.45	7.85	7.46	7.19	6.99	6.84
	.001	29.25	21.69	18.77	17.20	16.21	15.52	15.02	14.63
8	.100	3.46	3.11	2.92	2.81	2.73	2.67	2.62	2.59
	.050	5.32	4.46	4.07	3.84	3.69	3.58	3.50	3.44
	.025	7.57	6.06	5.42	5.05	4.82	4.65	4.53	4.43
	.010	11.26	8.65	7.59	7.01	6.63	6.37	6.18	6.03
	.001	25.41	18.49	15.83	14.39	13.48	12.86	12.40	12.05

Degrees of freedom in the denominator

Table entry for p is the critical value F^* with probability p lying to its right.

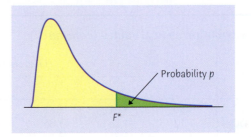

TABLE F		F distribution critical values (*continued*)							
		Degrees of freedom in the numerator							
	p	9	10	15	20	30	60	120	1000
1	.100	59.86	60.19	61.22	61.74	62.26	62.79	63.06	63.30
	.050	240.54	241.88	245.95	248.01	250.10	252.20	253.25	254.19
	.025	963.28	968.63	984.87	993.10	1001.40	1009.80	10140	1017.70
	.010	6022.50	6055.80	6157.30	6208.70	6260.60	63130	6339.40	6362.70
	.001	602284.00	605621.00	615764.00	620908.00	626099.00	631337.00	633972.00	636301.00
2	.100	9.38	9.39	9.42	9.44	9.46	9.47	9.48	9.49
	.050	19.38	19.40	19.43	19.45	19.46	19.48	19.49	19.49
	.025	39.39	39.40	39.43	39.45	39.46	39.48	39.49	39.50
	.010	99.39	99.40	99.43	99.45	99.47	99.48	99.49	99.50
	.001	999.39	999.40	999.43	999.45	999.47	999.48	999.49	999.50
3	.100	5.24	5.23	5.20	5.18	5.17	5.15	5.14	5.13
	.050	8.81	8.79	8.70	8.66	8.62	8.57	8.55	8.53
	.025	14.47	14.42	14.25	14.17	14.08	13.99	13.95	13.91
	.010	27.35	27.23	26.87	26.69	26.50	26.32	26.22	26.14
	.001	129.86	129.25	127.37	126.42	125.45	124.47	123.97	123.53
4	.100	3.94	3.92	3.87	3.84	3.82	3.79	3.78	3.76
	.050	6.00	5.96	5.86	5.80	5.75	5.69	5.66	5.63
	.025	8.90	8.84	8.66	8.56	8.46	8.36	8.31	8.26
	.010	14.66	14.55	14.20	14.02	13.84	13.65	13.56	13.47
	.001	48.47	48.05	46.76	46.10	45.43	44.75	44.40	44.09
5	.100	3.32	3.30	3.24	3.21	3.17	3.14	3.12	3.11
	.050	4.77	4.74	4.62	4.56	4.50	4.43	4.40	4.37
	.025	6.68	6.62	6.43	6.33	6.23	6.12	6.07	6.02
	.010	10.16	10.05	9.72	9.55	9.38	9.20	9.11	9.03
	.001	27.24	26.92	25.91	25.39	24.87	24.33	24.06	23.82
6	.100	2.96	2.94	2.87	2.84	2.80	2.76	2.74	2.72
	.050	4.10	4.06	3.94	3.87	3.81	3.74	3.70	3.67
	.025	5.52	5.46	5.27	5.17	5.07	4.96	4.90	4.86
	.010	7.98	7.87	7.56	7.40	7.23	7.06	6.97	6.89
	.001	18.69	18.41	17.56	17.12	16.67	16.21	15.98	15.77
7	.100	2.72	2.70	2.63	2.59	2.56	2.51	2.49	2.47
	.050	3.68	3.64	3.51	3.44	3.38	3.30	3.27	3.23
	.025	4.82	4.76	4.57	4.47	4.36	4.25	4.20	4.15
	.010	6.72	6.62	6.31	6.16	5.99	5.82	5.74	5.66
	.001	14.33	14.08	13.32	12.93	12.53	12.12	11.91	11.72
8	.100	2.56	2.54	2.46	2.42	2.38	2.34	2.32	2.30
	.050	3.39	3.35	3.22	3.15	3.08	3.01	2.97	2.93
	.025	4.36	4.30	4.10	4.00	3.89	3.78	3.73	3.68
	.010	5.91	5.81	5.52	5.36	5.20	5.03	4.95	4.87
	.001	11.77	11.54	10.84	10.48	10.11	9.73	9.53	9.36

Degrees of freedom in the denominator

(*continued*)

TABLE F		F distribution critical values (*continued*)							
			Degrees of freedom in the numerator						
	p	1	2	3	4	5	6	7	8
9	.100	3.36	3.01	2.81	2.69	2.61	2.55	2.51	2.47
	.050	5.12	4.26	3.86	3.63	3.48	3.37	3.29	3.23
	.025	7.21	5.71	5.08	4.72	4.48	4.32	4.20	4.10
	.010	10.56	8.02	6.99	6.42	6.06	5.80	5.61	5.47
	.001	22.86	16.39	13.90	12.56	11.71	11.13	10.70	10.37
10	.100	3.29	2.92	2.73	2.61	2.52	2.46	2.41	2.38
	.050	4.96	4.10	3.71	3.48	3.33	3.22	3.14	3.07
	.025	6.94	5.46	4.83	4.47	4.24	4.07	3.95	3.85
	.010	10.04	7.56	6.55	5.99	5.64	5.39	5.20	5.06
	.001	21.04	14.91	12.55	11.28	10.48	9.93	9.52	9.20
12	.100	3.18	2.81	2.61	2.48	2.39	2.33	2.28	2.24
	.050	4.75	3.89	3.49	3.26	3.11	3.00	2.91	2.85
	.025	6.55	5.10	4.47	4.12	3.89	3.73	3.61	3.51
	.010	9.33	6.93	5.95	5.41	5.06	4.82	4.64	4.50
	.001	18.64	12.97	10.80	9.63	8.89	8.38	8.00	7.71
15	.100	3.07	2.70	2.49	2.36	2.27	2.21	2.16	2.12
	.050	4.54	3.68	3.29	3.06	2.90	2.79	2.71	2.64
	.025	6.20	4.77	4.15	3.80	3.58	3.41	3.29	3.20
	.010	8.68	6.36	5.42	4.89	4.56	4.32	4.14	4.00
	.001	16.59	11.34	9.34	8.25	7.57	7.09	6.74	6.47
20	.100	2.97	2.59	2.38	2.25	2.16	2.09	2.04	2.00
	.050	4.35	3.49	3.10	2.87	2.71	2.60	2.51	2.45
	.025	5.87	4.46	3.86	3.51	3.29	3.13	3.01	2.91
	.010	8.10	5.85	4.94	4.43	4.10	3.87	3.70	3.56
	.001	14.82	9.95	8.10	7.10	6.46	6.02	5.69	5.44
25	.100	2.92	2.53	2.32	2.18	2.09	2.02	1.97	1.93
	.050	4.24	3.39	2.99	2.76	2.60	2.49	2.40	2.34
	.025	5.69	4.29	3.69	3.35	3.13	2.97	2.85	2.75
	.010	7.77	5.57	4.68	4.18	3.85	3.63	3.46	3.32
	.001	13.88	9.22	7.45	6.49	5.89	5.46	5.15	4.91
50	.100	2.81	2.41	2.20	2.06	1.97	1.90	1.84	1.80
	.050	4.03	3.18	2.79	2.56	2.40	2.29	2.20	2.13
	.025	5.34	3.97	3.39	3.05	2.83	2.67	2.55	2.46
	.010	7.17	5.06	4.20	3.72	3.41	3.19	3.02	2.89
	.001	12.22	7.96	6.34	5.46	4.90	4.51	4.22	4.00
100	.100	2.76	2.36	2.14	2.00	1.91	1.83	1.78	1.73
	.050	3.94	3.09	2.70	2.46	2.31	2.19	2.10	2.03
	.025	5.18	3.83	3.25	2.92	2.70	2.54	2.42	2.32
	.010	6.90	4.82	3.98	3.51	3.21	2.99	2.82	2.69
	.001	11.50	7.41	5.86	5.02	4.48	4.11	3.83	3.61
200	.100	2.73	2.33	2.11	1.97	1.88	1.80	1.75	1.70
	.050	3.89	3.04	2.65	2.42	2.26	2.14	2.06	1.98
	.025	5.10	3.76	3.18	2.85	2.63	2.47	2.35	2.26
	.010	6.76	4.71	3.88	3.41	3.11	2.89	2.73	2.60
	.001	11.15	7.15	5.63	4.81	4.29	3.92	3.65	3.43
1000	.100	2.71	2.31	2.09	1.95	1.85	1.78	1.72	1.68
	.050	3.85	3.00	2.61	2.38	2.22	2.11	2.02	1.95
	.025	5.04	3.70	3.13	2.80	2.58	2.42	2.30	2.20
	.010	6.66	4.63	3.80	3.34	3.04	2.82	2.66	2.53
	.001	10.89	6.96	5.46	4.65	4.14	3.78	3.51	3.30

Degrees of freedom in the denominator

TABLE F   *F distribution critical values (continued)*

		Degrees of freedom in the numerator							
	p	9	10	15	20	30	60	120	1000
9	.100	2.44	2.42	2.34	2.30	2.25	2.21	2.18	2.16
	.050	3.18	3.14	3.01	2.94	2.86	2.79	2.75	2.71
	.025	4.03	3.96	3.77	3.67	3.56	3.45	3.39	3.34
	.010	5.35	5.26	4.96	4.81	4.65	4.48	4.40	4.32
	.001	10.11	9.89	9.24	8.90	8.55	8.19	8.00	7.84
10	.100	2.35	2.32	2.24	2.20	2.16	2.11	2.08	2.06
	.050	3.02	2.98	2.85	2.77	2.70	2.62	2.58	2.54
	.025	3.78	3.72	3.52	3.42	3.31	3.20	3.14	3.09
	.010	4.94	4.85	4.56	4.41	4.25	4.08	4.00	3.92
	.001	8.96	8.75	8.13	7.80	7.47	7.12	6.94	6.78
12	.100	2.21	2.19	2.10	2.06	2.01	1.96	1.93	1.91
	.050	2.80	2.75	2.62	2.54	2.47	2.38	2.34	2.30
	.025	3.44	3.37	3.18	3.07	2.96	2.85	2.79	2.73
	.010	4.39	4.30	4.01	3.86	3.70	3.54	3.45	3.37
	.001	7.48	7.29	6.71	6.40	6.09	5.76	5.59	5.44
15	.100	2.09	2.06	1.97	1.92	1.87	1.82	1.79	1.76
	.050	2.59	2.54	2.40	2.33	2.25	2.16	2.11	2.07
	.025	3.12	3.06	2.86	2.76	2.64	2.52	2.46	2.40
	.010	3.89	3.80	3.52	3.37	3.21	3.05	2.96	2.88
	.001	6.26	6.08	5.54	5.25	4.95	4.64	4.47	4.33
20	.100	1.96	1.94	1.84	1.79	1.74	1.68	1.64	1.61
	.050	2.39	2.35	2.20	2.12	2.04	1.95	1.90	1.85
	.025	2.84	2.77	2.57	2.46	2.35	2.22	2.16	2.09
	.010	3.46	3.37	3.09	2.94	2.78	2.61	2.52	2.43
	.001	5.24	5.08	4.56	4.29	4.00	3.70	3.54	3.40
25	.100	1.89	1.87	1.77	1.72	1.66	1.59	1.56	1.52
	.050	2.28	2.24	2.09	2.01	1.92	1.82	1.77	1.72
	.025	2.68	2.61	2.41	2.30	2.18	2.05	1.98	1.91
	.010	3.22	3.13	2.85	2.70	2.54	2.36	2.27	2.18
	.001	4.71	4.56	4.06	3.79	3.52	3.22	3.06	2.91
50	.100	1.76	1.73	1.63	1.57	1.50	1.42	1.38	1.33
	.050	2.07	2.03	1.87	1.78	1.69	1.58	1.51	1.45
	.025	2.38	2.32	2.11	1.99	1.87	1.72	1.64	1.56
	.010	2.78	2.70	2.42	2.27	2.10	1.91	1.80	1.70
	.001	3.82	3.67	3.20	2.95	2.68	2.38	2.21	2.05
100	.100	1.69	1.66	1.56	1.49	1.42	1.34	1.28	1.22
	.050	1.97	1.93	1.77	1.68	1.57	1.45	1.38	1.30
	.025	2.24	2.18	1.97	1.85	1.71	1.56	1.46	1.36
	.010	2.59	2.50	2.22	2.07	1.89	1.69	1.57	1.45
	.001	3.44	3.30	2.84	2.59	2.32	2.01	1.83	1.64
200	.100	1.66	1.63	1.52	1.46	1.38	1.29	1.23	1.16
	.050	1.93	1.88	1.72	1.62	1.52	1.39	1.30	1.21
	.025	2.18	2.11	1.90	1.78	1.64	1.47	1.37	1.25
	.010	2.50	2.41	2.13	1.97	1.79	1.58	1.45	1.30
	.001	3.26	3.12	2.67	2.42	2.15	1.83	1.64	1.43
1000	.100	1.64	1.61	1.49	1.43	1.35	1.25	1.18	1.08
	.050	1.89	1.84	1.68	1.58	1.47	1.33	1.24	1.11
	.025	2.13	2.06	1.85	1.72	1.58	1.41	1.29	1.13
	.010	2.43	2.34	2.06	1.90	1.72	1.50	1.35	1.16
	.001	3.13	2.99	2.54	2.30	2.02	1.69	1.49	1.22

Degrees of freedom in the denominator

Answers to Selected Exercises ———————

Chapter 1

1.1 Categorical: gender, race, smoker. Quantitative: age, blood pressure, level of calcium.

1.3 **(b)** No, they do not represent pieces of one whole.

1.7 Somewhat symmetric and unimodal.

1.9 **(a)** Yes, there is an outlier. The group did not do so well: Only 4/18 met the goal. **(b)** Midpoint: 146. Spread: 78 to 359, although only one is greater than 271. **(c)** The dotplot and the stemplot are very similar, but the dotplot shows the exact location of each data point and is therefore a little bit more detailed.

1.11 The number of landfills available is decreasing over time. We need to increase recycling before we run out of landfill space.

1.13 (c) **1.14** (c) **1.15** (a) **1.16** (a) **1.17** (b)

1.18 (b) **1.19** (b) **1.20** (b) **1.21** (a) **1.22** (c)

1.23 **(a)** Quantitative. **(b)** Quantitative. **(c)** Categorical. **(d)** Quantitative. **(e)** Categorical.

1.25 **(a)** Lakes **(b)** 5 variables: 4 quantitative and 1 categorical (Age of data).

1.27 **(a)** 7.6 **(b)** 37.4% **(d)** Yes. Together, these are all the categories of garbage type.

1.29 **(b)** No. Each percent represents a different age group separately.

1.31 **(a)** Shape: symmetric without outliers. Center: 105 to 110. Spread: 75 to 155. **(b)** 25% have IQs below 105; 81.7% have IQs between 95 and 135.

1.33 **(a)** The two separate peaks could reflect a bimodal distribution. **(b)** The second histogram appears unimodal, but with a wide peak.

1.35 **(a)** If we looked at emissions per country, larger countries would have higher emissions, even though they may produce less per person. **(b)** The distribution is strongly skewed, centered on approximately 3 metric tons per person and ranging from about 0 to 19.9 metric tons per person. There are three outliers: the United States, Australia, and Canada.

1.37 **(b)** The midpoint for red is about 39, while that for yellow is about 36. Red tends to be longer.

1.39 **(a)** We cannot compare counts from groups of unequal sizes. **(b)** Marijuana usage and accident rates rise (and fall) together.

1.43 **(a)** Prices tend to be highest in late summer/early fall. Prices are lowest in the winter. **(b)** Yes: Prices appear to rise over time.

Chapter 2

2.1 **(a)** The distribution is irregular with 2 high outliers. The spread is from 13 to 86. **(b)** $\bar{x} = 35.8$; 10 observations have value less than the mean.

2.3 Median = 33. The median is less than the mean, which suggests a right skew or high outliers.

2.5   Min = 13, Q_1 = 20, M = 33, Q_3 = 46, max = 86. This reflects what is seen in the stem plot. Note that the maximum is much farther away from the median than the minimum.

2.7   86 is the only outlier.

2.9   (a) 5.4. (b) 0.642.

2.11 (a) Standard deviation and mean should be sufficient, because the histogram is symmetric with a single peak. (b) Sufficient: histogram is roughly symmetric. (c) Not sufficient: histogram is strongly skewed.

2.13 (a)   **2.14** (b)   **2.15** (c)   **2.16** (c)   **2.17** (b)

2.18 (c)   **2.19** (b)   **2.20** (a)   **2.21** (b)   **2.22** (a)

2.23 The data could skewed to the right, have high outliers, or perhaps be bimodal.

2.25 Details are lost, but the overall conclusion is essentially the same.

2.27 M = 2, Q_1 = 1, and Q_3 = 4.

2.29 (a) Symmetric distributions with no outliers. (b) On average, women have lower metabolic rates than men.

2.31 (b) Yellow is by far the most attractive color, followed by green. Blue and white appear to be equally unattractive colors.

2.33 Saturday and Sunday are quite similar and considerably lower than other days.

2.35 (a) The distribution seems left skewed. (b) A five-number summary seems most appropriate. Min = 49.1, Q_1 = 52.9, M = 63.3, Q_2 = 64.45, max = 65.0.

2.37 Compressed: $\bar{x}$ = 2.908, min = 2.68, Q_1 = 2.795, M = 2.88, Q_3 = 2.99, max = 3.18.
Intermediate: $\bar{x}$ = 3.336, min = 2.92, Q_1 = 3.13, M = 3.31, Q_3 = 3.45, max = 4.26.
Loose: $\bar{x}$ = 4.232, min = 3.94, Q_1 = 4.015, M = 4.175, Q_3 = 4.32, max = 4.91.
Soil penetration is greatest for loose soil and least for compressed soil.

2.39 Answers will vary; a range of 9 to 15 digits is typical.

2.41 A simple example could be 0, 0, 0, 1 (Q_3 = 0, $\bar{x}$ = 0.25).

2.43 (a) Min = 5.7, Q_1 = 11.7, M = 12.75, Q_3 = 13.5, max = 17.6. (b) IQR = 1.8, $1.5 \times IQR$ = 2.7, observations less than 9 or greater than 16.2 are potential outliers.

Chapter 3

3.1   (a) Explanatory: Number of calories. Response: Percent body fat. (b) Explore the relationship. (c) Explanatory: Inches of rain. Response: yield of corn. (d) Explore the relationship.

3.3   Weight and different physiology/genetic makeup are two other possible factors.

3.5   Linear, negative, and fairly weak; sparrowhawks are long-lived and territorial.

3.7   (a) Mass is explanatory. Subject number is an index variable and thus contains no data. (b) Moderately strong positive linear relationship. (c) The same relationship seems to hold. Men tend to have higher values and are overall more variable than women.

3.9   No; units do not affect correlation.

3.11 The relationship is clearly not linear; therefore, r is meaningless.

3.13   (a)   **3.14**   (a)   **3.15**   (c)   **3.16**   (b)   **3.17**   (c)

3.18   (a)   **3.19**   (b)   **3.20**   (a)   **3.21**   (b)   **3.22**   (c)

3.23   (a) It is possible, because the data indicate a linear relationship. (b) $r = 0.9941$. This is very strong. (c) No; neither variable is explanatory.

3.25   Correlation is not affected by unit of measure. r would stay the same.

3.27   (a) r will be closer to 1. (b) r will decrease and could even become negative if the point on the far end is dragged down far enough.

3.29   The mean yields are 131.0, 143.2, 146.2, 143.1, and 134.8 bushels/acre. The greatest yield occurs at or around 20,000 plants per acre.

3.31   (a) The order was changed in the groups. (b) It would look like the mirror image. There is no logical order to colors. (c) Numerical values wrongly suggest that the variables are quantitative when they are not.

3.33   (a) They are strongly correlated. The correlation would likely be weaker for individual weights. (b) Animals with multiple births tend to have lighter weights.

3.35   (a) There is a negative linear relationship between GDP and child mortality. $r = -0.5918$. (b) The relationship is similar (slightly less strong), but a single regional point hides a lot of variability from country to country.

3.37   The scatterplot suggests that sunlight has brightened overall, but the increase has not been steady—and may be not entirely linear.

3.39   (a) $r = 1$ for a line. (c) Leave some space above your vertical stack. (d) The curve must be higher at the right than at the left.

Chapter 4

4.1   (a) The slope indicates that for each drink consumed, BAC increases by 0.023. The intercept is zero because if someone doesn't have any drinks, their BAC should be 0. (b) BAC $= 0.023 \times 4 = 0.092$.

4.3   (b) $\bar{x} = 324.75, s_x = 257.66, \bar{y} = 2.3875, s_y = 1.1389, r = -0.7786$.

4.5   (a) $\hat{y} = a + bx$, and $a = \bar{y} - b\bar{x}$. Thus, if we plug in $x = \bar{x}$, we get $\hat{y} = \bar{y} - b\bar{x} + b\bar{x} = \bar{y}$. (b) $b = rs_y/s_x$ If we switch x and y, r remains unchanged, so we would get $1/b$.

4.7   Because the absolute value of the correlation is closer to 1, the data lie closer to the regression line and the regression line is a better predictor.

4.9   (a) Point A is a horizontal outlier, whereas point B is a vertical outlier. (b) Point A has more influence on the regression line. The farther an outlier is from the center, the more leverage it has on the slope of the line.

4.11   More populated cities have a higher demand for pharmacists but also have a higher number of deaths. Everything scales with population size.

4.13   (a) No. There is no biological reason why education would cause breast cancer. (b) Women with higher education tend to have babies later in life.

4.14   (b)   **4.15**   (c)   **4.16**   (c)   **4.17**   (a)   **4.18**   (c)

4.19   (a)   **4.20**   (a)   **4.21**   (a)   **4.22**   (a)   **4.23**   (b)

4.25   (a) On the average, BOD rises by 1.507 mg/l for every 1 mg/l increase in TOC. (b) -55.43 mg/l; extrapolation.

4.27 (a) $y = -0.282x + 37.272$. $r^2 = 0.858$. (b) As soda consumption increases, milk consumption tends to decrease. The high r^2 value tells us that this is a very reliable trend. (c) 26.84 gallons of milk; prediction outside the range of our data.

4.29 89.

4.31 (a) Slope = 0.547. Intercept = 2.791. (b) Prediction = 30.2. A high r^2 (here 0.9216) implies a reliable prediction.

4.33 (a) $y = 0.06078. x - 0.1261$. Prediction = -0.00454. (b) 77.1%.

4.35 (a) They all give similar results; $y = 3.0 + 0.5x$; $r = 0.82$; very close slopes, intercepts, and correlations; $\hat{y} = 8$ for $x = 10$. (c) Only the first relationship is linear and appropriate for regression. The second relationship is strongly curved. The third has a strong outlier. The last has only two values for x.

4.37 (a) Ignoring the outlier, the plot shows a moderately strong negative linear relationship. (b) For all 12 points: $r = -0.3387$. Without the outlier: $r = -0.7866$. The outlier weakens the linear pattern.

4.39 For all 12 points: $\hat{y} = -492.6 - 33.79x$. Without the outlier: $\hat{y} = -1371.6 - 75.52x$. The line is pulled toward the outlier to reduce the very large vertical deviation of this point from the line.

4.41 (a) There is a reasonably strong, positive linear relation between temperature and latitude. (b) $y = 0.818x + 52.45$, $r = 0.6033$. (c) $y = 0.9843x + 51.26$, $r = 0.7343$. (d) The outlier weakens the correlation and makes the slope less steep.

4.43 (a) Residuals: 8.1, 12.1, -0.87, -9.9, 1.1, -3.9, -15.9, 6.1, 3.1. (b) $r = 0$.

4.45 People who are overweight are more likely to be on diets.

4.47 No. It is more probable that those in critical condition are sent to large hospitals.

4.49 Strong, positive linear relationship. $\hat{y} = 10.02x + 11.846$, $r = 0.949$. The data support the hypothesis.

4.51 It makes sense to use a straight line only for prediction on the scatterplot that shows a clear linear relationship without extreme outliers.

Chapter 5

5.1 (a) 858 people. (b) 43 had arthritis. (c) 71 (8.3%) played elite soccer, 215 (25.1%) played non-elite soccer, and 572 (66.7%) did not play.

5.3 There are an infinite amount of possibilities, but $a = 50$, $b = 0$, $c = 10$, $d = 40$, and $a = 40$, $b = 10$, $c = 20$, and $d = 30$ are two solutions.

5.5 (a) Conditional distribution of alarm call: given predator nearby 26.6%, given predator far = 41.3%. (b) When a predator is close, prairie dogs raise the alarm 26.6% percent of the time. When a predator is far away (making it safer for the individual to give a warning call), the prairie dogs raise the alarm 41.3% of the time.

5.7 (a) Overall success rate for open surgery 78.0%. Overall success rate for PCNL 82.6%. (b) Overall success rate for small stones 88.2%. Overall success rate for large stones 72.0%. (c) Open surgery success for small stones 93.1%. PCNL success for small stones 86.7%. Open surgery success for large stones 73.0%. PCNL success for large stones 68.8%. This is Simpson's paradox: PCNL was used more often to treat small stones, which appear to be easier to treat (better chance of successful outcome).

5.9 (a) **5.10** (b) **5.11** (c) **5.12** (a) **5.13** (c)

5.14 (b)   **5.15** (a)   **5.16** (c)   **5.17** (b)   **5.18** (b)

5.19   Marginal for gender: males = 47.4%, females = 52.6%. Marginal for order: 1st = 28.4%, 2nd = 26.3%, 3rd = 25.3%, 4th = 20%.

5.21   Among males, 1st = 37.8%, 2nd = 35.6%, 3rd = 15.6%, 4th = 11.1%. Yes, they should total 100% aside from rounding error.

5.23   **(a)** Overall percent with a headache 28.9%, without a headache 71.1%. **(b)** Conditional distributions: percent of women eating chocolate who got a headache: 17.2%; percent of women eating carob who got a headache: 40.6%. Chocolate triggered headaches less often than carob.

5.25   **(a)** Percent in 9- to 11-month age range who ate fried potatoes: 9.0%. Percent in the 15- to 18-month range: 20.1%. Percent in the 19- to 24-month age range: 25.9%. **(b)** Association does not imply causation.

5.27   **(a)** Survival rates among the younger patients = 99.2% for type 1 and 100% for type 2. Survival rates among the older patients = 54.4% for type 1 and 58.8% for type 2. Older patients are at higher risk of death. **(b)** Overall survival percent for type 1: 70.7%; for type 2: 59.9%. Patients with type 1 diabetes did better overall. **(c)** The lurking variable here is age. Type 2 diabetes rarely occurs in younger people, and diabetes (1 or 2) rarely kills those who are young.

5.29   **(b)** Percents hatching in each cell group: cold, 59.3%; neutral, 67.9%; hot, 72.1%. The cold temperature made hatching less likely.

5.31   Conditional distribution for dead trees: 55.9% western hemlock, 38.4% Douglas fir, and only 5.7% western red cedar. Conditional distribution for live trees: 43.8% western hemlock, 33.8% Douglas fir, and 22.3% western red cedar. Conditional distribution for sapling trees: 63.1% western red cedar, 36.1% western hemlock, and only 0.8% Douglas fir. The western red cedar is in the process of displacing the Douglas fir.

Chapter 6

6.1   For example: Categorical = percent students with undergraduate research experience, Quantitative = average number of months spent doing undergraduate research and number of professors with Nobel Prizes.

6.3   **(a)** Notice that for every 10 minute increment, length is always increasing. **(b)** $y = -2.39 + 0.158x$. **(c)** This makes no sense, but it can be calculated as 225 cm. This value cannot be trusted, because it is extrapolating far beyond the range of values collected.

6.5 $r = 0.217$ is a weak correlation. If people had good memory of what they ate, we would expect a strong correlation.

6.7   **(a)** The stemplot is roughly symmetric. **(b)** $\bar{x} = 0.2426$ and $s = 0.0476$ for all plants; without the extremes, $\bar{x} = 0.2433$ and $s = 0.0396$.

6.9   5.13 billion gallons are for uses not mentioned.

6.11   Full data set: median = 25, $\bar{x} = 25.42$, $s = 7.47$. Without the outlier (HAV = 50): median = 25, $\bar{x} = 24.76$, $s = 6.34$. Outliers have little effect (in this case, none) on medians, a stronger effect on means, and an even stronger effect on standard deviations.

6.13   **(a)** $y = 0.339x + 19.723$. **(b)** 28.198. **(c)** Not very well. Looking at the graph, the points do not seem to lie too close to the line. In numerical terms, $r^2 = 0.0913$.

6.15 The plot seems to be roughly linear. The correlation should be close to -1; there is a very strong correlation, and as days go up, weight goes down.

6.17 -57 g. Soap cannot have negative weight. The problem is that we are applying regression to something outside the time frame of our data, when patterns will change (i.e., we will have no more soap).

6.19 Yes, there is a strong positive linear relationship between number of perch in a pen and proportion killed ($r^2 = 0.465$).

6.21 Min $= 1, Q_1 = 11$, median $= 16, Q_3 = 20$, max $= 31, s = 5.96, \bar{x} = 15.46$. The median date is May 5th.

6.23 Five-number summaries:
Min $= 7, Q_1 = 12, M = 19, Q_3 = 22$, max $= 26$.
Min $= 1, Q_1 = 11, M = 15.5, Q_3 = 19.75$, max $= 27$.
Min $= 9, Q_1 = 13.5, M = 16.5, Q_3 = 19$, max $= 31$.
Min $= 1, Q_1 = 9, M = 11, Q_3 = 18$, max $= 25$.
The median is declining over time, but the overall data are quite variable.

6.25 **(a)** Mean lean body mass: 8.7 kg for lean monkeys and 10.5 kg for obese monkeys. **(b)** For lean monkeys: $\hat{y} = 0.541 + 0.0826x$. For obese monkeys: $\hat{y} = 0.371 + 0.0852x$. Since the slope is about the same but the intercept appears to be different, it would imply that energy increases at about the same rate for both groups of monkeys but that obese monkeys expend less energy per unit of body mass.

6.27 **(a)** $y = -1.099x + 166.483$. This indicates that the more lamb's quarter in a cornfield, the less corn it will produce. **(b)** 159.889.

6.29 As one strain goes up, the others go down. It appears that only one strain is clearly present every year.

Chapter 7

7.1 Observational. The study did not attempt to change the participants in any way.

7.3 **(a)** Observational. **(b)** Explanatory: human activity. Lurking: natural variability in the environment (for example, previous ice ages).

7.5 Actual California population.

7.7 **(a)** Label each one from 0001 to 1410. **(b)** 769, 1315, 94.

7.9 A multistage design divides the population into groups, takes a random sample of these groups, and then takes a random sample of individuals within the groups sampled. An SRS just samples from the entire population, without dividing it into subgroups. Both techniques use randomness for selecting individuals.

7.11 **(a)** The population is all in-line skaters admitted to an ER for skating injuries. The sample size is 161. Conclusions apply only to in-line skaters with injuries, not to all in-line skaters. **(b)** Skaters who wear protective gear are less likely to be included in this study.

7.13 This is a cohort study. The population is individuals who have suffered a severe heart attack. The explanatory variable is the categorical variable alcohol consumption level. The response variable is death within the next four years.

7.14 (b)   **7.15** (b)   **7.16** (c)   **7.17** (a)   **7.18** (a)
7.19 (b)   **7.20** (c)   **7.21** (a)   **7.22** (a)   **7.23** (c)

7.25   (a) Registered voters. (b) An SRS probability sampling design. 732.

7.27   (a) To avoid response bias due to the disturbance of a new presence. (b) If animals are scared or offended by a new, unusual presence, then we cannot observe their natural behavior.

7.29   After labeling them 1 through 40 in alphabetical order, we would select Washburn, Garcia, Rodriguez, Helling, Wallace, Cabrera, Morgan, Husain, Batista, and Nguyen.

7.31   In a stratified random sample, the entire population is broken up into subgroups; then an SRS is taken from each of the individual subgroups, and the SRSs are finally recombined to create the sample. This ensures that all four forest types are represented in the sample in proportion to their representation in the population.

7.33   (a) Assign numerical values 0001 through 5024. Numbers 1388, 0746, 0227, 4001, and 1858 were selected. (b) It is likely that more people really ran the red lights than reported doing so. People usually don't like to admit their flaws.

7.35   (a) The question is clear, but the wording seems to suggest that cell phone usage caused the brain cancer. (b) The question is both unclear and slanted toward an affirmative answer. (c) This question is so unclear and charged with negative words that it is likely to prevent many "Yes" answers.

7.37   Answers will vary.

7.39   (a) This is a case-control study. Two types of hospital patients are compared. (b) Nonresponse can introduce bias. A low nonresponse rate is good. (c) For example, healthy diets are often associated with better living conditions.

7.41   (a) Cohort study. A homogenous group was followed for many years in order to track the effects of alcohol on the risk of prostate cancer. (b) Prostate cancer is rare enough that there might have been too few cases in the study for statistical evaluation.

Chapter 8

8.1   Individuals: pine seedlings. Treatments: full light, 25% light, or 5% light. Response variable: dry weight at the end of the study.

8.3   (a) Observational. (b) They needed a control group to assess the effect of oral contraceptive use.

8.5   (a) The sham acupuncture acts as a placebo. The wait group acts as a control. (b) From the entire group, take an SRS of 101 subjects and assign them the acupuncture treatment. Next, take an SRS of 101 of the remaining subjects and assign them to sham acupuncture. Finally, assign the rest to the wait list.

8.7   (a) Block design. (b) First, divide the crickets into two groups: 7–12 days and 15+ days. Then, within these groups, randomly sample 10 to receive parasites, and compare results.

8.9   (a) Take an SRS of 15 from the entire group of 30 subjects and give them coffee. Compare the two groups. (b) On two separate days, all 30 subjects will drink to BAC 0.08. An SRS of 15 will be given coffee on the first day, whereas the remaining subjects will get coffee on the second day. (c) Randomly assign 6 of the women and 9 of the men to have coffee, and compare the four groups.

8.11   This was not blind, because both the subject and the experimenter knew who was receiving treatment and who was not. Subjects learning the meditation method could experience a placebo effect. More importantly, the experimenter could be biased in his assessment of reported anxiety.

8.13 Answers will vary.

8.14 (a) **8.15** (b) **8.16** (b) **8.17** (c) **8.18** (a)

8.19 (b) **8.20** (a) **8.21** (b) **8.22** (a) **8.23** (a)

8.25 (a) Explanatory: type of operation. Response: survival time. (b) Not an experiment: existing records are simply examined. (c) The type of operation is most likely decided based on the size of the tumor. Larger tumors would be much more dangerous to remove, regardless of type of operation.

8.27 (a) Completely randomized design. Half of the subjects would be randomly sampled and put in the potent group. The other half would be put in the weak group. (b) (Using Table A) Potent group: Dubois, Travers, Cheng, Ullmann, Quinones, Thompson, Fluharty, Lucero, Afifi, and Gerson. Weak group: Abate, Brown, Engel, Gutierrez, Huang, Iselin, Kaplan, McNeill, Morse, and Rosen.

8.29 The first study is an observational, case-control study; the second is an experiment: the exercise plan is assigned randomly.

8.31 (a) Randomly assign the subjects to four groups: group 1, antidepressants and stress management; group 2, antidepressants only; group 3, stress management only; and group 4, control. (b) Using Table A, group 1 = Chai, Hammond, Herrera, Xiang, Irwin, Hurwitz, Reed, Broden, Lucero, and Nho.

8.33 (a) Randomly assign 3 circles to get extra carbon monoxide; the others are a control group. (b) Place pairs of circles close together, and randomly choose one of each pair for treatment.

8.35 (a) A block design.

8.37 (a) Randomly sample half of the subjects and give them the treatment; everybody else gets the placebo pill. Then compare the results. (b) Using Table A: 170, 005, 227, 118, 007.

8.39 The effects of factors other than the nonphysical treatment have been eliminated or accounted for, so that the differences in improvement observed between the subjects can be attributed to the differences in treatments.

8.41 (a) In an observational study, we observe subjects who have chosen to take supplements and compare them with others who do not take supplements. In an experiment, we assign some subjects to take supplements and others to take placebo. (b) Treatments are assigned at random, and a control group is used as a basis for comparison to observe the effects of the treatment. (c) Subjects who choose to take supplements are more likely to make healthy lifestyle choices. When random assignment is used, some of those subjects will take the supplement and some will take the placebo.

8.43 (b) Yes, the results appear trustworthy: The study was a comparative experiment with a placebo control group, subjects were randomly assigned to treatments, double blinding was used to prevent bias, and the variable was clearly defined.

8.45 (a) Randomized: Who got the placebo and who got the treatment was chosen at random. Double blind: Neither the subjects nor the experimenters knew who got the placebo and who got the active treatment. Placebo controlled: The effect of treatment was compared against the effect of getting a placebo.

8.47 (a) Half the subjects are selected using an SRS; they are given the real treatment. The other group is given the placebo. (b) If everyone received the drug, we would not know whether the drug was actually reducing pain or whether the patients were merely experiencing the placebo effect. (c) Once we have enough evidence that the drug performs better than the placebo, there is no longer any reason to give a placebo. Patients' needs must come first.

8.49 The Tanzanian government and the researchers both presented serious arguments. However, the study's purpose was entirely academic. Without patients' informed consent or any benefit to them, the study was not ethically justified and should indeed have been canceled.

8.51 Detailed outcomes will vary.

Chapter 9

9.1 This means that if you repeatedly sampled 100,000 males, on average you would have 13 diagnosed with hemophilia. For one random sample of 100,000 males, you could find more or you could find less than 13 individuals with hemophilia.

9.3 (a) 0. (b) 1. (c) 0.01. (d) 0.6.

9.5 (a) S = {BBB, BBG, BGB, BGG, GBB, GBG, GGB, GGG}. (b) S = {0, 1, 2, 3}. Associated probabilities = $\frac{1}{8}$, $\frac{3}{8}$, $\frac{3}{8}$, and $\frac{1}{8}$, respectively.

9.7 (a) 0.05. (b) 0.3. (c) 0.25.

9.9 (a) The sum of probabilities is 1. (b) $X < 7$ when a subject watches TV less than 7 days a week. $P(X < 7) = 0.43$. (c) $X \geq 1$. $P(X \geq 1) = 0.96$.

9.11 (a) 0.4. (b) 0.4. (c) 0.2. (d) 0.8.

9.13 (a) Discrete. Answers can only be whole numbers. (b) Continuous. Answers can be any value between 0 and 24. (c) Discrete. Answers can only be whole numbers.

9.15 (a) Answers will vary. (b) You would know whether you are generally a cautious or a reckless driver, getting a more accurate view of yourself than just using the population mean. (c) People generally tend to think they are above average; they may also underestimate the dangers of driving.

9.17 (a) Risk = 0.0043, odds = 0.00432. (b) Risk = 0.0365, odds = 0.0379. (c) Probability = risk = 0.127, odds = 0.145. (d) By decade, because risk varies over the lifespan.

9.18 (a)   **9.19** (b)   **9.20** (b)   **9.21** (b)   **9.22** (c)

9.23 (a)   **9.24** (b)   **9.25** (c)   **9.26** (b)   **9.27** (c)

9.29 (a) 0.4592. (b) 0.5408.

9.31 (a) S = {fails, germinates}. (b) S = {all positive numbers}. (c) S = {all numbers}. (d) S = {all positive numbers}.

9.33 (a) S = {cardiovascular disease, cancer, other}. (b) $P(X =$ cancer or cardio) = 0.67, $P(X =$ other) = 0.33.

9.35 (a) 0.04. (b) 0.75.

9.37 (a) All probabilities total to 1. (b) 0.62. (c) P(Taster 1 ranks higher than a 3) = 0.39, P(Taster 2 ranks higher than a 3) = 0.39.

9.39 (a) 0.14. (b) 0.43. (c) 0.75.

9.41 (a) $X < 2$ or $X > 4$. $P(X < 2$ or $X > 4) = 0.17$. (b) "Eye color greater than 2 and less than or equal to 4." $P(2 < X \leq 4) = 0.72$.

9.43 (a) Y is continuous, because it can take any value between 0 and 2. (b) Height $= 0.5$. (c) $P(Y \leq 1) = 0.5$.

9.45 (a) $P(13 \leq X < 16) = 0.77$. (b) $P(X > 13) = 0.88$.

9.47 Actual results will vary.

Chapter 10

10.1 Theoretical probability $= 0.0199$. Actual probability $= 0.185$. A small difference becomes more substantial when compounded.

10.3 (a) P(next two patients are males) $= 0.5625$. (b) P(at least one is female) $= 0.7627$. P(at least one is male) $= 0.9990$.

10.5 P(dead | RC) $= 0.1$, P(dead | DF) $= 0.5$, P(dead | WH) $= 0.479$. The western red cedar is relatively new.

10.7 P(test + and metastases) $= 0.27$. P(test + | metastases) $= 0.9$.

10.9 P(belong and twice a week) $= P$(belong)P(twice | belong) $= (0.1)(0.4) = 0.04$.

10.11 (a) P(malaria | sickle-cell) $= 0.264$, P(malaria | no sickle-cell) $= 0.373$. (b) No, because knowing the sickle-cell status changes the probability of malaria. It tells us that there is a biological link between the two conditions: Malaria is more common among individuals who do not have the sickle-cell trait.

10.13 (a) P(blue eyes | red hair) $= 0.473$. P(blue eyes and red hair) $= 0.0118$. (b) P(freckles | red hair and blue eyes) $= 0.857$. P(red hair and blue eyes and freckles) $= 0.0101$.

10.15 (a) P(red hair) $= .025$, P(blue eyes) $= 0.388$, P(freckles) $= 0.255$. (b) P(freckles and red hair) $= 0.0203$. P(freckles | red hair) $= 0.811$.

10.17 (a)   **10.18** (c)   **10.19** (c)   **10.20** (a)   **10.21** (b)

10.22 (c)   **10.23** (c)   **10.24** (c)   **10.25** (c)   **10.26** (b)

10.27 P(first is O and second is O) $= 0.158$. P(both have same blood type) $= 0.299$.

10.29 0.526.

10.31 $P(X = 0) = 0.0282$. $P(X \geq 1) = 0.9718$.

10.33 (a) 0.810. We are assuming that different players' getting arthritis are independent events. (b) 0.362.

10.35 Because having brown hair and having freckles are not necessarily independent events. (The earlier tree diagram shows that they are definitely not independent.)

10.37 84%.

10.39 1.

10.41 P(NJ job | federal job) $= 0.25$, P(federal job | NJ job) $= 0.125$.

10.43 (a) 0.0651. (b) 0.0605. (c) Fourth $= 0.0563$, fifth $= 0.0524$, kth $= (0.07)(0.93)^{k-1}$.

10.45 1.

10.47 (a) 22.4%. (b) 71.43%.

10.49 (a) A, AB, B. (b) $P(A) = P(B) = 0.25$. $P(AB) = 0.5$.

10.51 (a) 0.25. (b) 0.375.

10.53 (a) P(both carriers | both of European Jewish descent) $= 0.00137$, P(both carriers | mother of European Jewish descent but father not) $= 0.000148$, P(both carriers | neither of European Jewish descent) $= 0.000016$. (b) P(child TS | both parent of European Jewish descent) $= 0.00034$. (c) P(child TS | neither parent of European Jewish descent) $= 0.000004$.

Chapter 11

11.3 (a) 175 ml – 625 ml. (b) 250 ml and less.

11.5 At age 5, z-score $= 0.556$. At age 7, z-score $= 0.476$. John has been growing slightly more slowly than average, because his z score has gone down a little bit. A third measurement one or two years later would help our conclusion.

11.7 (a) 69.1%. (b) 38.3%.

11.9 (a) $z = -0.67$. (b) $z = 0.25$.

11.11 It is close to Normal, because the dots appear to make a straight line.

11.13 (a)   **11.14** (a)   **11.15** (b)   **11.16** (a)   **11.17** (b)

11.18 (c)   **11.19** (b)   **11.20** (c)   **11.21** (a)   **11.22** (b)

11.23 (a) The stemplot looks Normal, although there are two low values. (b) $\bar{x} = 105.84$ and $s = 14.27$. The proportions are about 74.2% and 93.5%; a Normal distribution would have 68% and 95%.

11.25 2.5%.

11.27 (a) 1.22%. (b) 98.78%. (c) 3.84%. (d) 94.94%.

11.29 (a) About 0.845. (b) About 0.385.

11.31 Approximately $z > 1$ for A, $0 < z < 1$ for B, $-1 < z < 0$ for C, $z < -1$ for D and F grades.

11.33 (a) 0.62%. (b) 30.85%.

11.35 (a) 7.93%. (b) About 27.3%.

11.37 (a) 59.87%. (b) 54.7%. (c) > 279 days.

11.39 (a) 2.5%. (b) 24.14%.

11.41 0.675, both above and below.

11.43 Atlantic acorn sizes are clearly not Normal (right skew).

11.45 (a) z-scores for ranks: $-1.46, -1.1, -0.73, -0.37, 0, 0.37, 0.73, 1.1$, and 1.46. (b) The data appear roughly Normal.

11.47 Yellow distribution is closest to Normal.

Chapter 12

12.1 Yes. Reach a live person or not. $n = 15$, $p = 0.2$.

12.3 Yes. Think yes or no. $n = 500$, $p = 0.15$.

12.5 Binomial. X is resistance; $n = 10$, $p = 0.07$. (b) $P(X = 1) = 0.364$, $P(X = 2) = 0.123$. (c) $P(X \geq 1) = 51.6\%$.

Answers to Selected Exercises

12.7 (a) $\mu = 0.7$. (b) $\sigma = 0.807$. (c) This decreases the standard deviation $\sigma_{2000} = 0.223$. As p gets closer to zero, the standard deviation gets closer to zero.

12.9 (a) $P(X \leq 82) = 0.0129$. (b) $P(X \leq 82) = 0.0145$. This is fairly close. (c) It is likely that many of the cases of osteopenia have been missed.

12.11 (a) $\{0, 1, 2, 3, \ldots\}$. (b) Poisson. (c) $\mu = 0.536$, $\sigma = 0.732$.

12.13 (a) $P(X = 0) = 0.585$. (b) $P(X = 1) = 0.314$. (c) $P(X > 1) = 0.101$.

12.15 (b)　　**12.16** (b)　　**12.17** (c)　　**12.18** (a)　　**12.19** (c)

12.20 (b)　　**12.21** (b)　　**12.22** (c)　　**12.23** (c)　　**12.24** (b)

12.25 (a) Binomial is reasonable, $n = 8$. (b) Binomial is reasonable, $n = 100$. (c) Binomial is reasonable, $n = 24$.

12.27 (a) Binomial, $p = 0.708$, $n = 12$. (b) $P(X = 8) = 0.227$ $P(X \geq 8) = 0.744$.

12.29 (a) $n = 20$, $p = 0.25$. (b) 5. (c) $P(X = 5) = 0.202$.

12.31 $n = 6$, $p = 0.75$. (b) $S = \{0, 1, 2, 3, 4, 5, 6\}$ (c) $P(X = 0) = 0.00024$, $P(X = 1) = 0.0044$, $P(X = 2) = 0.03296$, $P(X = 3) = 0.1318$, $P(X = 4) = 0.2966$, $P(X = 5) = 0.3560$, $P(X = 6) = 0.1780$ (d) $\mu = 4.5$, $\sigma = 1.06$.

12.33 (a) $P(X = 0) = 0.0128$, $P(X = 1) = 0.0742$, $P(X \geq 2) = 0.913$. (b) Mean = 338.52, standard deviation = 14.01, $P(X \geq 320) = 0.913$. (c) Situation b.

12.35 (a) 0.014 using software for binomial calculation (Normal approximation is not appropriate here). (b) 3.88×10^{-9}.

12.37 (a) 3.75. (b) 0.0008. (c) 0.0339.

12.39 (a) $S = \{0, 1, 2, 3, 4, 5\}$. (b) $P(X = 0) = 0.2373$, $P(X = 1) = 0.3955$, $P(X = 2) = 0.2637$, $P(X = 3) = 0.0879$, $P(X = 4) = 0.0146$, $P(X = 5) = 0.0010$.

12.41 $P(X \geq 2, 058, 000) \approx 0$. It seems unreasonable to conclude that the probability of having a boy is the same as that of having a girl.

12.43 (a) 0.0651. (b) 0.0605. (c) Fourth = 0.0563, fifth = 0.0524, kth = $(0.07)(0.93)^{k-1}$.

12.45 (a) 0.1175. (b) 0.1175, 0.1175. (c) 0.3525.

12.47 (a) 0.0111. (b) 0.0500. (c) 0.9389.

12.49 $P(X \geq 5) = 0.219$, $P(X \geq 6) = 0.105$, $P(X \geq 7) = 0.045$. Reserving 6 beds for ER admissions would be enough 95.5% of the time.

12.51 (a) Poisson. Mean = 15.8, standard deviation = 3.947. (b) $P(X = 0) = 1.71 \times 10^{-7}$ $P(X \leq 5) = 0.00186$, $P(X \leq 15) = 0.509$, $P(X \leq 25) = 0.990$ $P(X > 25) = 0.01$. (c) $P(X \geq 48) = 1.11 \times 10^{-11}$. This is too unlikely to be due to random, isolated cases. It points to an epidemic. The contaminated food must have been distributed at least in South Dakota and Wisconsin.

Chapter 13

13.1 Both are statistics.

13.3 If 1 of the 12 workers had a serious injury, it would cost more than the premiums collected. For thousands of policies, the average claim should be close to $439.

13.5 (a) 130.2. (b) Sampled 1, 4, 5, 9. Values = 99, 125, 170, 147; $\bar{x} = 135.25$. (c) $\bar{x}$ repeated 9 times: 119.5, 125.75, 141.25, 138, 118, 119, 137.75, 123, 132.5. The center of all these is about 129. **(d)** There are way too many numbers to do by hand.

13.7 **(a)** $\bar{x}$ is not systematically higher or lower than μ. **(b)** With large samples, $\bar{x}$ is more likely to be close to μ.

13.9 **(a)** 0.069. **(b)** 0.0013.

13.11 The mean number of moths per trap in samples of 50 traps is 0.5, and the standard deviation is 0.099. The distribution of this estimator is fairly Normal.

13.13 **(a)** 6% is a parameter, 4% is a statistic. **(b)** Mean = 0.06, standard deviation = 0.0194.

13.15 0.1512.

13.17 (c)   **13.18** (b)   **13.19** (a)   **13.20** (a)   **13.21** (c)

13.22 (a)   **13.23** (c)   **13.24** (b)   **13.25** (c)   **13.26** (b)

13.27 65 inches is a statistic, 64 inches is a parameter.

13.29 **(a)** 0.2812. **(b)** Standard deviation = $2.8/\sqrt{n}$. Required sample size = 32. **(c)** 0.9566.

13.31 **(a)** 1/6 **(b)** 283.

13.33 **(a)** About 0.0228. **(b)** Nearly 0.

13.35 **(a)** About 0.2233 g/mi.

13.37 0.0053.

13.39 For all practical purposes, it is a parameter. The census tracks everyone (except homeless individuals).

13.41 **(a)** The mean proportion of blacks is 0.114. The standard deviation is 0.0082. **(b)** $P(\hat{p} \leq 0.11) = 0.3128$.

13.43 **(a)** Mean = 0.05, standard deviation = $\sqrt{0.05 \times 0.95/n}$. **(b)** $P(\hat{p} \geq 0.07) = 0.0972$. **(c)** $P(\hat{p} \geq 0.08) = 0.0258$, $P(\hat{p} \geq 0.10) = 0.0006$.

Chapter 14

14.1 **(a)** 0.4. **(b)** 0.8 cm. **(d)** 95%.

14.3 **(a)** Normal (unknown μ, $60/\sqrt{840}$). **(b)** (267.9, 276.1). **(c)** When we execute this procedure, we have a 95% chance of capturing the true value μ.

14.5 We are 95% confident that the concentration of active ingredient in this batch is between 0.8327 and 0.8481 g/l.

14.7 **(a)** N(12 g/dl, 0.2263 g/dl). **(b)** 11.3 lies far from the middle of the curve and is therefore unlikely if in fact $\mu = 12$. 11.8 is close to the middle, so it would not be too surprising.

14.9 $H_0 : \mu = 12$, $H_a : \mu < 12$.

14.11 $H_0 : \mu = 64.5$ inches, $H_a : \mu \neq 64.5$ inches.

14.13 $P = 0.1714$.

14.15 **(a)** $P = 0.0010$. This is significant at the α 0.05 and the α 0.01 level. **(b)** $P = 0.1894$. This is not significant at either the α 0.05 or the α 0.01 level. **(c)** A low P-value means the outcome would be very unlikely if the null hypothesis were true.

14.17 $H_0 : \mu = 5$, $H_a : \mu < 5$, $z = 2.7708$, $P = 0.0028$. This is strong evidence that the mean oxygen content is less than 5 mg/l.

14.19 (a) Yes, 10 is within the 95% confidence interval, because the P-value is not low enough to reject at the α 0.05 level. (b) No, 10 is not in the 90% confidence interval, because the P-value is low enough to reject at the $\alpha = 0.10$ level.

14.21 (a) No. (b) Yes.

14.22 (c)   **14.23** (c)   **14.24** (a)   **14.25** (b)   **14.26** (c)

14.27 (a)   **14.28** (b)   **14.29** (b)   **14.30** (b)   **14.31** (a)

14.33 The margin of error says that 95% of surveys should contain the true population proportion, not necessarily the proportion obtained by this survey.

14.35 (a) 2.138. (b) 1.96. (c) 86.9 to 95.3.

14.37 (a) The data are skewed to the left. (b) 29,737.5 to 31,945.5 pounds.

14.39 μ is the population mean difference in blood folic acid (fortified minus unfortified in a matched pair situation). $H_0 : \mu = 0$, $H_a : \mu > 0$.

14.41 (a) Since she had not decided to make it a one-sided test before she collected her data, she cannot change her hypothesis just to fit the data. (b) $P = 0.0358$.

14.43 It is essentially correct.

14.45 The differences in richness and total stem densities between the two areas were so small that they could easily occur by chance if population means were identical.

14.47 Something that occurs "less than once in 100 repetitions" also occurs "less than 5 times in 100 repetitions," but not vice versa.

14.49 μ is the population mean body length of deer mice in forests. $H_0 : \mu = 86$, $H_a : \mu \neq 86$, $P = 0.0171$.

14.51 It is possible to determine statistical significance using both P-values and confidence intervals. However, P-values only compare the parameter to a hypothesis, while confidence intervals actually give a range of estimates for the parameter.

14.53 (a) Yes (P-value very close to zero). (b) 10.0021 to 10.0025. The scale is slightly biased on the heavy side. (c) P-values only tell us that the scale is off, while confidence intervals tell us how much it is off (not too much, in fact).

Chapter 15

15.1 (a) No. The poll is a voluntary survey. (b) No. Inference is still inappropriate, because the poll is a voluntary survey.

15.3 There might be a pattern to these consecutive records, and they come from only one general practice—which may not be representative of other general practices.

15.5 Yes, because the sample is quite large ($n = 72$), although an extreme outlier would still be problematic.

15.7 (a) 4.9. (b) 1.96. (c) As sample size increases, the margin of error decreases.

15.9 (a) $n = 5$, $P = 0.1867$, $n = 15$, $P = 0.0606$, $n = 40$, $P = 0.0057$. (b) As sample size increases, the sampling distribution gets narrower.

15.11 (a) 4.8 ± 0.438. (b) 4.8 ± 0.253. (c) 4.8 ± 0.155.

15.13 (a) No. Since we are running the test so many times, we would expect about 5 individuals to get $P \leq 0.01$. (b) Retest these individuals. Now that there is a smaller group, low P-values are less likely to occur just by chance.

15.15 385.

15.17 (a) 178. (b) 307. A higher confidence level requires a larger sample size.

15.19 There is probability 0.57 that the test will reject the null hypothesis at significance level α 0.05 if the alternative hypothesis is true.

15.21 Power 0.378, 0.749, and 0.981. As sample size increases, power increases. For a test that almost always rejects H_0: when $\mu = 4.7$, we should use $n = 40$.

15.23 (a) H_0: patient is healthy, H_a: patient is ill. Type I error: sending a healthy patient to the doctor, Type II error: clearing a patient who is ill.

15.24 (a)   **15.25** (c)   **15.26** (b)   **15.27** (c)   **15.28** (b)

15.29 (a)   **15.30** (a)   **15.31** (a)   **15.32** (b)   **15.33** (c)

15.35 Were the samples chosen at random? Was nonresponse an issue? What was the sample size? How large was the difference?

15.37 People are likely to lie about their sex lives. The margin of error does not take this into account.

15.39 A significance test answers only question (b).

15.41 The data represent the whole population; therefore, no inference is needed.

15.43 The outlier will have a greater effect on a small sample.

15.45 (a) The margin of error decreases. (b) The P-value gets smaller. (c) The power increases.

15.47 (a) 166. (b) 97.

15.49 (a) There is not enough data to determine whether the outcome was due to chance under the null hypothesis. (b) Small sample size or large variability.

15.51 (a) We cannot conclude because the studies were uncontrolled. (b) This experiment suggests that arthroscopic surgery is no better than a placebo. (c) The lack of significance cannot be explained by low power.

15.53 (a) 0.808. (b) $n = 5$, power $= 0.556$; $n = 15$, power $= 0.925$; $n = 40$, power $= 1.0$. (c) $\mu = 0.5$, power $= 0.468$; $\mu = 1$, power $= 0.933$; $\mu = 1.5$, power $= 0.999$. As the effect size increases, the power increases.

15.55 (a) Power $= 0.922$. (b) Power $= 0.922$. Yes. (c) The power would be higher.

15.57 A Type I error would be to say that there was a correlation when there was none. A Type II error would be to say that there was no correlation when there was one.

Chapter 16

16.1 It is likely to weaken the association, because we expect that marijuana usage will be underreported to a greater degree than accidents caused.

16.3 (a) The control group should have 24 trees with no beehives. (b) Table B gives 53, 64, 56, and 68. (c) The response variable is elephant damage.

16.5 The control is not chosen at random. Bias is likely, since they are all chosen because their parents did not return the form.

16.7 (a) $P(Y > 1) = 0.73$. (b) $P(2 < Y < 4) = 0.3$. (c) $P(Y \neq 2) = 0.67$.

16.9 (a) 0.5899. (b) 384.2 mg.

16.11 (a) Nearly all should be within the range 92 to 452. (b) $\bar{x}$ will nearly always fall within the range 254 to 290.

16.13 (a) A response at least as large at that observed would be seen almost half the time (just by chance) even if there were no effect in the entire population of rats. (b) The differences observed were so small that they would often happen just by chance.

16.15 (a) $H_0 : \mu = 188$, $H_a : \mu < 188$, $z = -1.46$, $P = 0.0721$; significant only at the α 0.10 level. (b) The value 188 doesn't belong to the confidence interval; therefore, the results are not significant at the α 0.05 level for the one-sided alternative.

16.17 $Z = -2.92$ and $P = 0.0018$; small differences can be significant with large sample sizes.

16.19 (a) The survey was a random sample, but only of Americans with a telephone. (b) Nonresponse could be a practical difficulty. The phrasing uses strong language—"should" and "shouldn't"—which might bias the responses.

16.21 Placebos can provide genuine pain relief. (Believing that one will experience relief can lead to actual relief.)

16.23 (a) Factors: storage method (fresh, room temperature for one month, or refrigerated for one month) and preparation method (cooked immediately or after one hour). This makes six treatments (storage-preparation combinations). Response variables: tasters' color and flavor ratings. (b) Randomly allocate n potatoes to each of the six treatment combinations, and then compare ratings. (c) For each taster, randomly choose the order in which the fries are tasted.

16.25 $H_0 : \mu = 100$ ng/g, $H_a : \mu > 100$ ng/g, $z = 14.54$, P is close to 0.

16.27 84.8 to 90.4.

16.29 (a) The treatment was not assigned. (b) Socioeconomic and hereditary factors may also affect IQ.

16.31 (a) There aren't enough data to show that the outcome was very unlikely by chance alone. (b) It is a relatively small effect size. This would explain why previous results failed to reject the null hypothesis. (c) The older siblings in that age range most likely have a higher education and therefore a slightly higher IQ score.

16.33 A Type I error would be to conclude that mean IQ is less than 100 when it really is 100 (or more). A Type II error would be to conclude that the mean IQ is 100 (or more) when it truly is less than 100.

16.35 (a) P (Taster 1 rates higher than taster 2) $= 0.19 = P(2, 1) + P(3, 1) + P(4, 1) + P(5, 1) + P(2, 3) + P(2, 4) + P(2, 5) + P(3, 4) + P(3, 5) + P(4, 5)$; P (Taster 2 rates higher than taster 1) $= 0.19$. (b) $0.1622 = 0.06/0.37$.

16.37 0.036.

16.39 $P(C|A) = 0.10 = 10\%$.

16.41 (a) Binomial $n = 5$, $p = 0.5$. (b) P (5 boys) $= 0.03125$, P (4 boys, 1 girl) $= 0.15625$. (c) P (1 girl | 4 boys) $= 0.5$. This is a conditional probability.

16.43 (a) Binomial $n = 500$, $p = 0.26$. (b) P (100 or fewer) $= 0.0011$, P (150 or fewer) $= 0.9793$.

Chapter 17

17.1 $s = 3.79$.

17.3   (a) 2.015. (b) 2.518.

17.5   (a) 2.262. (b) 2.861. (c) 1.440.

17.7   (a) Only if the students were randomly chosen from the population of all university male students. No major departure from Normality. (b) 425.4 ± 32.4.

17.9   (a) 24. (b) t is between 1.059 and 1.318, so $0.20 < P < 0.30$. (c) $t = 1.12$ is not significant at either level.

17.11   Data are strongly skewed to the left, but the sample size is relatively large ($n = 29$), so the test should be relatively accurate; $P = 4.2 \times 10^{-10}$, extremely significant.

17.13   (a) μ is the mean difference in healing rates, experimental minus control. $H_0 : \mu = 0$, $H_a : \mu < 0$. (b) The stemplot is roughly Normal. (c) $t = -2.02$, df $= 13$, $P = 0.032$.

17.15   (a) The outlier is the only significant departure from Normality. (b) $t = -2.076$, df $= 11$, $P = 0.031$ (all newts); $t = -1.788$, df $= 10$, $P = 0.052$ (without the outlier).

17.17 (b)   **17.18** (c)   **17.19** (a)   **17.20** (a)   **17.21** (b)

17.22 (c)   **17.23** (a)   **17.24** (a)   **17.25** (b)   **17.26** (c)

17.27   (a) $\bar{x} = 15.59$ feet, $s = 2.550$ feet. (b) 14.81 to 16.37 feet. We reject the claim that $\mu = 20$ feet. (c) What population are we examining? Full-grown sharks? Male sharks? Can this be considered an SRS?

17.29 10.04 ± 0.49.

17.31   The data are slightly skewed to the right, but the sample is large enough for the 95% t-confidence interval.

17.33   18.69 to 70.19 μg.

17.35   (a) Weather conditions that change from day to day can affect spore counts. (b) Take the differences, kill room count minus processing count. The interval is 843.2 to 2805.8 CFUs/m^3. (c) The data are counts, which are at best only approximately Normal, and we have only a small sample.

17.37   (a) A subject's responses to the two treatments would not be independent. (b) Yes ($t = -4.41$, $P = 0.0069$).

17.39   The data contain two extreme high outliers, 5973 and 8015. These may distort the t-statistic.

17.41 $H_0 : \mu = 1$, $H_a : \mu > 1$, $t = 8.402$, $P = 0.0002$. However, the data may be skewed to the left, in which case the sample size would not be large enough for inference.

17.43   (a) For each subject, randomly select which knob should be used first. (b) μ is the mean of (right-hand-thread time minus left-hand-thread time); $H_0 : \mu = 0$ sec, $H_a : \mu < 0$ sec, $t = -2.90$, df $= 24$, $P = 0.0039$; right-hand-thread times are less.

17.45 -21.2 to -5.5 ms. (a) This is not biologically significant. (b) This is not relevant for sporadic use. (c) While it might have some practical application for extensive users, the difference would likely be even less pronounced because of the extensive practice.

Chapter 18

18.1 Matched pairs.

18.3 Single sample.

18.5 (a) Oregon: $n = 6, \overline{x} = 26.9, s = 3.82$. California: $n = 7, \overline{x} = 11.9, s = 7.09$.
(b) Use conservative df $= 5$.

18.7 (a) If they had known that the area was to be assessed, they might have logged differently. (b) $H_0 : \mu_u = \mu_1$, $H_a : \mu_u > \mu_1$, t statistic $= 2.11$, $P(\text{df} = 14.793) = 0.0260$, $P(\text{df} = 8) = 0.0337$.

18.9 (a) Breast-feeding woman and other woman. (b) $H_0 : \mu_C = \mu_1$; $H_a : \mu_C < \mu_1$; t statistic $= 8.50$, $P < 0.001$.

18.11 $\overline{x}_1 = 17.6\%$, $\overline{x}_2 = 9.49983\%$, $s_1 = 6.3401\%$, and $s_2 = 1.9501\%$.

18.13 $t = 2.99$, $P = 0.0246$; significant at the 5% level.

18.15 (a) **18.16** (c) **18.17** (b) **18.18** (a) **18.19** (b)

18.20 (c) **18.21** (a) **18.22** (b) **18.23** (b)

18.25 (b) $H_0 : \mu_G = \mu_B$, $H_a : \mu_G \neq \mu_B$, t statistic $= 1.64$, $P = 0.1057$.

18.27 -1.24 to 11.48 (df $= 30$) or -1.12 to 11.35(df $= 56.9$).

18.29 851.0 to 3008.0 CFUs (df $= 3$) or 869.7 to 2989.3 CFUs (df $= 3.136$).

18.31 $H_0 : \mu_1 = \mu_2$, $H_a : \mu_1 < \mu_2$, $t = -10.29$, $P(\text{df} = 36.89) = 1.07 \times 10^{-12}$, $P(\text{df} = 31) < 0.0005$.

18.33 We test $H_0 : \mu_1 = \mu_2$; $H_a : \mu_1 > \mu_2$ for both analyses. For Analysis I, $t = 6.849$, df $= 4$ or 16.51. For Analysis II, $t = 7.066$, df $= 9$ or 14.84. In both cases, P is very small.

18.35 (a) $H_0 : \mu_a = \mu_w$, $H_a : \mu_a < \mu_w$; $t = 2.31$, $P(\text{df} = 106.75) = 0.0115$, $P(\text{df} = 63) = 0.0122$. We can conclude that, on average, people receiving acupuncture had fewer migraines than those on the wait list. (b) 1.2 ± 1.0314. On the low end the reduction is almost nothing, but on the high end it is about 50%.

18.37 If those who had fewer migraines dropped out the most, then those remaining would have a higher migraine rate. Since the wait list had the highest dropout rates, this would imply that the wait list group would have a disproportional increase in migraines. If those who believed least in the acupuncture dropped out, the remaining subjects might feel that only acupuncture can help them.

18.39 (a) Two-sample t-test, $P = 0.992$. (b) Compute the weight difference for each group, then find a two-sample confidence interval: 11.4 ± 6.4 with df 16, 11.4 ± 6.9 with df $= 9$. The control group gained significantly more weight than the group receiving ink applications. The ink appears to be toxic.

18.41 (a) $H_0 : \mu_m = \mu_p$, $H_a : \mu_m \neq \mu_p$, $t = 0.4594$, $P(\text{df} = 37.30) = 0.6484$, $P(\text{df} = 20) = 0.6509$. (b) Were the data skewed or bimodal? Were there any outliers?

18.43 $t = -3.74$; $0.01 < P < 0.02$.

18.45 (a) 95% confidence interval: 38.88 ± 4.95. (b) 62.54 ± 5.97. (c) 95% confidence interval for inactive $-$ active: 23.66 ± 7.56 (df $= 45.11$), 23.66 ± 7.74 (df $= 23$).

Chapter 19

19.1   **(a)** The population was cultures from individuals diagnosed with strep. $p =$ proportion resistant to antibiotic in the population. **(b)** $\hat{p} = 0.568$.

19.3   **(a)** $\mu = 0.5$, $\sigma = 0.0041$ **(b)** Yes. The sample is an SRS, and it is large enough to ensure that $\hat{p}$ is approximately Normally distributed.

19.5 0.568 ± 0.023.

19.7   There were only 5 or 6 "successes" in the sample.

19.9   **(a)** The number of "successes" is only 9. **(b)** 0.1078 ± 0.0505, or 0.0573 to 0.1584.

19.11   **(a)** $\hat{p} = 0.8185$; the margin of error is 0.0460. **(b)** With $p^* = \hat{p}$, we need $n = 635$.

19.13 $H_0 : p = 0.5$, $H_a : p \neq 0.5$, $z = 1.347$, $P = 0.178$.

19.15   **(a)** $np_0 = 2 < 10$. **(b)** $n(1 - p_0) = 8 < 10$.

19.16 (b)   **19.17** (c)   **19.18** (c)   **19.19** (b)   **19.20** (c)

19.21 (a)   **19.22** (b)   **19.23** (a)   **19.24** (a)   **19.25** (a)

19.27   **(a)** The number of successes is less than 15. However, $n > 10$, so the plus four confidence interval is appropriate. **(b)** 0.125 ± 0.0636.

19.29   683.

19.31   Sample and confidence interval sizes are large enough. 0.0922 ± 0.0220.

19.33   **(a)** $H_0 : p_{\text{chem}} = 0.488$, $H_a : p_{\text{chem}} > 0.488$, $z = 0.1834$, $P = 0.4272$. This implies the results are very likely if chemists had the same probability of having a child as the general population. **(b)** We are assuming that Washington chemists represent chemists in general.

19.35 0.7249 ± 0.0086.

19.37   To test $H_0 : p = 1/3$, $H_a : p > 1/3$, we have $\hat{p} = 0.3786$, $z = 2.72$, $P = 0.0033$.

19.39   The interval is 0.5320 to 0.7180 (large sample) or 0.5288 to 0.7119 (plus four). To test $H_0 : p = 0.5$, $H_a : p > 0.5$, $z = 2.55$, $P = 0.0054$.

19.41 0.44 ± 0.0251.

Chapter 20

20.1 0.1941 ± 0.0663.

20.3   **(a)** Not enough of the subjects taking echinacea failed to develop a cold. **(b)** 0.0524 ± 0.1079.

20.5 $H_0 : p_{\text{accept}} = p_{\text{decline}}$; $H_a : p_{\text{accept}} < p_{\text{decline}}$, $z = -2.878$, $P = 0.0020$. Yes, there is significant evidence.

20.7   **(a)** $\hat{p}_{asp} = 0.000906$, $\hat{p}_{plac} = 0.002356$. **(b)** $RR = 0.3845$. Individuals in the aspirin group were 38% less likely to suffer a fatal heart attack compared with individuals in the placebo group. **(c)** $OR = 0.3840$. In this case, OR and RR are very close.

20.9   **(a)** $ARR = -0.00145$, $RRR = -0.6154$. ARR is the difference in risk between treatment and control, RRR is the reduction in risk in the treatment group relative to the risk of the control group. **(b)** 690. This means 690 physicians would need to take aspirin daily for 5 years to save 1 of them from a fatal heart attack compared with taking a placebo daily for 5 years.

20.11 (c)   **20.12** (a)   **20.13** (b)   **20.14** (a)   **20.15** (b)

20.16 (b)   **20.17** (b)   **20.18** (b)   **20.19** (b)   **20.20** (c)

20.21 **(a)** There were no "successes" in the control group. **(b)** Treatment group: $n = 35$ with 24 mice developing tumors. Control group: $n = 20$ with 1 mouse developing a tumor. 0.636 ± 0.238. **(c)** The difference is significant at the 1% level because the 99% confidence interval does not include zero.

20.23 **(a)** $H_0 : p_{AZT} = p_{plac}$, $H_a : p_{AZT} < p_{plac}$. There are enough patients with and without AIDS to use a z procedure for this test. **(b)** $P = 0.0017$. **(c)** This means neither the doctors nor the patients knew who had received the drug and who had received the placebo.

20.25 $H_0 : p_1 = p_2$, $H_a : p_1 \neq p_2$, $z = 3.39$, $P = 0.0006$.

20.27 0.0631 to 0.2179 (plus four: 0.0614 to 0.2158).

20.29 The two samples are not independent, because the same 23 facilities were used for both.

20.31 $H_0 : p_n = p_f$, $H_a : p_n < p_f$, $z = 2.398$, $P = 0.0082$.

20.33 Use the plus four interval, because one count is only 7: -0.0399 to 0.3685.

20.35 $H_0 : p_{acet} = p_{ibu}$, $H_a : p_{acet} \neq p_{ibu}$, $z = 3.802$, $P = 0.00014$.

20.37 **(a)** Both parties will know whether liquid nitrogen or duct tape was applied. **(b)** A plus four 95% confidence interval is the only valid test: 0.2288 ± 0.233. Because the interval does not contain the value zero, we can conclude that the two treatments yield significantly different results ($P < 0.05$), with duct tape being more efficient.

20.39 $H_0 : p_m = p_f$, $H_a : p_m \neq p_f$, $z = 0.5258$, $P = 0.599$, no significant difference.

20.41 **(a)** $OR = 0.4249$, $RR = 0.4474$. **(b)** $ARR = 0.0483$, $RRR = 0.5526$. **(c)** $NNT = 21$.

20.43 $ARR = 0.1746$, $RRR = 0.7255$, $NNT = 6$.

Chapter 21

21.1 $H_0 : p_0 = p_{20} = p_{40}$, H_a: H_0 is not true.

21.3 Expected $= 53/3 = 17.67$, $X^2 = 16.11$.

21.5 Not all window angles are equally likely to produce accidental bird strikes.

21.7 Data fall into a multinomial setting, expected counts are all larger than 5.

21.9 Yes. Data fall into a multinomial setting, expected counts are all 6.

21.11 df $= 2$, closest critical value is 15.20, $P < 0.0005$.

21.13 **(a)** $z = -0.514$, $X^2 = 0.263 = z^2$. **(b)** $P = 0.608$ in both cases (2-sided z test).

21.14 (c)   **21.15** (c)   **21.16** (a)   **21.17** (c)   **21.18** (c)

21.19 (c)   **21.20** (b)   **21.21** (a)   **21.22** (c)   **21.23** (b)

21.25 **(a)** Yes, counts are high enough. **(b)** $H_0 : p_{YR} = 9/16$, $p_{YW} = 3/16$, $p_{GR} = 3/16$, $p_{GW} = 1/16$, H_a: H_0 is not true. **(c)** 312.75, 104.25, 104.25, and 34.75 **(d)** $P = 0.9254$. The data are consistent with Mendel's second law.

21.27 $H_0 : p_F = 1/4$, $p_{SF} = 1/2$, $p_S = 1/4$, H_a: H_0 is not true, $X^2 = 0.7204$, df $= 2$, $P = 0.6975$. The data are consistent with a codominance model of inheritance.

21.29 $H_0: p_H = p_T = p_U = p_L = 1/4$, H_a: H_0 is not true. $X^2 = 107.83$, $P < 0.0005$. Highly significant evidence that melanoma sites are not all equally likely in women.

21.31 H_0: probabilities match U.S. percents, H_a: H_0 is not true. $X^2 = 1709.8$, df $= 4$, $P < 0.0005$. Highly significant evidence that the ethic breakdown in the population targeted by this study does not match the U.S. ethnic breakdown.

21.33 H_0: values of p_i are given by the model of random breeding; H_a: H_0 is not true. $X^2 = 9.377$, df $= 7$, $P = 0.2267$. The data are consistent with the hypothesis that prairie dogs breed randomly.

Chapter 22

22.1 (a) 29.9%, 28.9%, 21.9%, and 19.3%. (b) 26.6%, 30.9%, 19.4%, 23.0%, 39.8%, 21.1%, 27.1%, and 12.0%. (c) University-educated men seem to be more likely to be nonsmokers or moderate smokers.

22.3 (a) Elite $= 0.1408 \pm 0.0809$, non-elite $= 0.0419 \pm 0.0268$, did not play $= 0.0420 \pm 0.0164$. (b) Individual confidence intervals ensure only that one parameter falls within the confidence interval with the chosen confidence level.

22.5 (a) 42.31, 36.22, 30.14, and 24.34. The sum might differ slightly from 133 due to rounding. (b) The heavy-smoker count is less than expected, but the moderate-smoker count is high.

22.7 (b) $X^2 = 13.305$, $P = 0.038$. (c) University-educated men contribute three of the four largest terms.

22.9 All expected cell counts are well over 5.

22.11 A chi-square test is not appropriate, because there are cells with expected counts less than 1.

22.13 The study used a single SRS, which was hoped to represent a larger population.

22.15 (a) $(3 - 1)(4 - 1) = 6$. (b) $12.59 < X^2 < 14.45$, so $0.025 < P < 0.05$.

22.17 (a) For chi-square, all cells have expected counts above 5. Likewise, the number of failures and successes is more than 5 in all groups for the z test. (b) $H_0: p_{GF} = p_{Plac}$; $H_a: p_{GF} \neq p_{Plac}$, $z = 0.568$, $P = 0.5703$. (c) $X^2 = 0.322$, $P = 0.5703$. (d) The data are consistent with the notion that gastric freezing is no better than a placebo as a treatment for gastric ulcers.

22.18 (a)   **22.19** (a)   **22.20** (c)   **22.21** (c)   **22.22** (a)

22.23 (b)   **22.24** (a)   **22.25** (c)   **22.26** (c)   **22.27** (c)

22.29 (b) Yes. H_0: no relationship between answer type and interview type, df $= 2$. (c) $X^2 = 10.619$, $P = 0.0049$. (d) There is significant evidence that interviewing method influences respondent answers.

22.31 (a) Stress $= 0.909$, exercise $= 0.794$, regular treatment $= 0.700$. (b) Expected heart attacks: stress $= 6.78$, exercise $= 6.99$, regular treatment $= 8.22$. All cells have expected counts above 5. (c) $X^2 = 4.84$ and $P = 0.0889$. This is not significant at the 0.05 level.

22.33 (b) Cold $= 59.3\%$, neutral $= 67.9\%$, hot $= 72.1\%$. The data show that some eggs do hatch at cold temperatures. (c) $H_0: p_{cold} = p_{neutral} = p_{hot}$, H_a: H_0 is not true, $X^2 = 1.703$, df $= 2$, $P = 0.4267$. Temperature did not play a significant role in hatching success.

22.35 We can see that those on the Mediterranean diet suffered fewer heart attacks, both fatal and nonfatal. H_0: probability of heart attack, fatal and nonfatal, is unrelated to diet, $X^2 = 17.053$, df $= 2$, $P = 0.0002$. The evidence is indeed highly significant.

22.37 The data suggest that desipramine is more successful. H_0: no relationship between relapse and treatment type, $X^2 = 10.5$, df $= 2$, $P = 0.0052$. The evidence is strongly significant.

22.39 The chi-square test found no significant relationship between treatment type and death following a stroke during the two-year treatment period ($P = 0.701$). It appears that none of the active treatments would do better than a placebo in preventing stroke-related death.

22.41 Neither test was statistically significant ($P = 0.577$ and $P = 0.3099$, respectively). The echinacea extracts used as treatment or preventively do not appear to do better than a placebo at preventing the common cold or reducing the number of cold symptoms.

22.43 H_0: no association between gender and tumor rate, $X^2 = 0.2765$, df $= 1$, $P = 0.5990$. Not significant.

22.45 (a) Small stones: open surgery $= 93.1\%$, PCNL $= 86.7\%$. Large stones: open surgery $= 73.0\%$, PCNL $= 68.8\%$. Open surgery appears to perform better for both small and large stones. (b) Small stones: $X^2 = 2.626$, $z = 1.621$, $P = 0.1051$. Large stones: $X^2 = 0.5507$, $z = 0.7421$, $P = 0.4580$. The differences in outcomes are not significant for the two procedures, whether the stones are small or large. This is Simpson's paradox: PCNL was used more often for small stones, which have a higher rate of cure regardless of treatment.

Chapter 23

23.1 (a) The plot shows a positive linear association. $r = 0.955$, $\hat{y} = -0.9764 + 0.05429x$. (b) The slope β is the rate at which deforestation changes with increases in the price of coffee. The estimates of β and α are the two coefficients in the regression equation: slope $b = 0.05429$, and intercept $a = -0.9764$. (c) Residuals: $-0.1080, 0.3949, -0.2651, -0.1894$, and 0.1677. The squares of the residuals add up to 0.3019, so $s = \sqrt{0.3019/3} = 0.3172$.

23.3 (a) $r^2 = 11.2\%$. (b) $\hat{y} = -2057 + 1.97x$, $s = 104.0$ km^3 of water.

23.5 $b = 1.9662$ and SE$_b = 0.7037$. Therefore, $t = b/\text{SE}_b = 2.79$, df $= n - 2 = 62$, $P = 0.007$, significant.

23.7 $r = 0.9552$; Table E with $n = 5$ gives $0.01 > P > 0.005$.

23.9 With df $= n - 2 = 3$, $t^* = 3.182$, and the 95% confidence interval is 0.0234 to 0.0852.

23.11 0.79 to 3.14 km^3/year; this interval does not contain 0.

23.13 (b) $(48.406, 51.093)$.

23.15 (b) Residuals are more or less randomly spread, but there are very few observations for the lower temperatures. (d) The histogram shows Normality except for one low outlier.

23.16 (c)   **23.17** (c)   **23.18** (a)   **23.19** (c)   **23.20** (c)

23.21 (c)   **23.22** (b)   **23.23** (c)   **23.24** (a)

23.25 (a) $b = -0.408$. For every additional kilogram of nitrogen per hectare, the species richness measure decreases by 0.408 on average. (b) $r^2 = 0.55$. The linear relationship accounts for 55% of the observed variation in species richness. (c) $H_0 : \beta = 0$ versus $H_a : \beta \neq 0$ (or $\beta < 0$). The small P says that there is very strong evidence that the population slope β is negative, that is, that species richness does decrease as the amount of nitrogen deposited increases.

23.27 (a) Positive, linear relationship of moderate strength. (b) $\hat{y} = 0.1205 + 0.0086x$. The slope is positive. $s = 0.1886$.

23.29 (a) Excel's 95% confidence interval for β is 0.0033 to 0.0138; this agrees (except for roundoff error) with $b \pm t^* \, SE_b = 0.0086 \pm (2.145)(0.0025)$. (b) With df $= 14$ and $t^* = 1.761$, the 90% confidence interval is 0.0042 to 0.0130.

23.31 $0.6643 < r < 0.7114$, so the one-sided P-value is $0.001 < P < 0.0025$. Excel's P-value (0.0036) is for a two-sided test.

23.33 The distribution is skewed to the right, but the sample is large, so t procedures should be safe. $\bar{x} \approx 0.2781$ g/m^2, and $s = 0.1803$ g/m^2. The 95% confidence interval for μ is 0.2449 to 0.3113 g/m^2.

23.35 The scatterplot shows a positive association. The regression equation is $\hat{y} = 0.1523 + 8.1676x$, and $r^2 = 0.606$. The slope is significantly different from 0 ($t = 13.25$, df $= 114$, $P < 0.001$).

23.37 (a) There is considerably greater scatter for larger values of the explanatory variable. (b) The distribution of residuals looks reasonably Normal.

23.39 (a) $\hat{y} = 560.65 - 3.0771x$. (b) The scatterplot appears to be roughly linear and does not suggest that the standard deviation changes. The distribution of residuals has a slightly irregular appearance but is not markedly non-Normal. (c) -4.8625 to -1.2917 calories per minute.

23.41 386 to 489 calories.

23.43 (b) $r = 0.8382$, $P = 0.0006$, significant. (c) Amount taken is the response variable, but it only takes a few values. For inference purposes, the response variable should vary continuously.

23.45 $P = 0.332$. No significant evidence that the intercept is not zero.

Chapter 24

24.1 (a) Randomly allocate 36 flies to each of four groups, which receive varying dosages of caffeine; observe length of rest periods. (b) H_0: all groups have the same mean rest period, H_a: at least one group has a different mean rest period. The significant P-value leads us to conclude that caffeine reduces the length of the rest period.

24.3 (a) The stemplots show no extreme outliers or strong skewness (given the small sample sizes). (b) The means suggest that logging reduces the number of trees per plot and that recovery is slow. (c) $F = 11.43$, $P = 0.000205$, $H_0 : \mu_1 = \mu_2 = \mu_3$, H_a: not all the means are the same.

24.5 (a) When the three means are similar, F is very small and P is near 1. (b) Moving any means increases F and decreases P.

24.7 (a) The ratio of standard deviation is about 1.16. (b) The ratio of standard deviations is about 1.20.

24.9 (a) These standard deviations do not follow our rule of thumb: The smallest standard deviation is 24.23, the largest is 143.67. The distributions appear skewed to the right. The value 700 is an extreme outlier. Thus, ANOVA results would not be reliable. (b) Even after removing 700, the standard deviation is still more than twice the standard deviation of the nonsmokers.

24.11 (a) 3 for treatment, 140 for error. (b) 2.70. (c) 5.86.

24.13 (a) Yes: the ratio of standard deviations is 1.24. (b) $\bar{x} = 9.92$, MSG $= 10.0$. (c) MSE $= 21.9$. (d) $F = 0.46$. With df $= 2$ and 112 (use 2 and 100 in the table), we find $P > 0.100$.

24.15 (c)   **24.16** (b)   **24.17** (c)   **24.18** (b)   **24.19** (b)

24.20 (a)   **24.21** (a)   **24.22** (b)   **24.23** (a)

24.25 Populations: morning people, evening people, neither. Response variable: difference in memorization scores. $k = 3, n_1 = 16, n_2 = 30, n_3 = 54, N = 100$; dfs $= 2$ and 97.

24.27 Populations: normal-weight men, overweight men, obese men. Response variable: triglyceride level. $k = 3, n_1 = 719, n_2 = 885, n_3 = 220, N = 1824$; dfs $= 2$ and 1821.

24.29 (a) Yes; the mean control emission rate is half of the smallest of the others. (b) H_0: all groups have the same mean emission rate, H_a: at least one group has a different mean emission rate. (c) Are the data Normally distributed? Were these random samples? (d) $s = \text{SEM} \times \sqrt{8}$. The rule-of-thumb ratio is 1.4755.

24.31 (a) Means: 10.65, 10.425, 5.60, and 5.45 cm; standard deviations: 2.053, 1.486, 1.244, and 1.771 cm. The means and the stemplots suggest that the presence of too many nematodes reduces growth. ANOVA seems safe. (b) $H_0 : \mu_1 = \cdots = \mu_4$; H_a: not all the means are the same. We test whether nematodes affect mean plant growth. (c) $F = 12.08$, dfs $= 3$ and 12, $P = 0.001$. The first two levels are similar, as are the last two. Somewhere between 1000 and 5000 nematodes, the tomato plants are hurt by the worms.

24.33 H_0: relative fitness is not influenced by previous pH exposure. Full data set: standard deviations 0.058, 0.023, and 0.115, $F = 48.571$, dfs 2 and 15, $P = 2.8 \times 10^{-7}$. Without the low outlier (0.56): standard deviations 0.058, 0.023, and 0.055, $F = 93.329$, dfs 2 and 14, $P = 8.1 \times 10^{-9}$. Highly significant.

24.35 The means are 123.8, 123.6, and 134.4 lb; they do not consistently decrease. The standard deviations are 4.6043, 6.5422, and 9.5289 lb. $9.5289/4.6043 = 2.07$; ANOVA is risky. If we do ANOVA anyway, $F = 3.70$ (dfs $= 2$ and 12) and $P = 0.056$; the differences are not significant.

24.37 Means $= 2.908, 3.336$, and 4.232, standard deviations $= 0.1390, 0.3193$, and 0.2713. ANOVA: dfs 2 and 57, $F = 140.53$, $P < 0.00001$. If we remove the outliers: means $= 2.908, 3.287$, and 4.157, standard deviations $= 0.1390$, 0.2401, and 0.1545. ANOVA: dfs 2 and 54, $F = 229.24$, $P < 0.00001$.

24.39 $\bar{x} = 21.585$, MSG $= 745.5$, $MSE = 460.2$, dfs $= 3$ and 28, $F = 1.62$, not significant.

24.41 (a) A chi-square test. (b) ANOVA. (c) ANOVA.

Chapter 25

25.1 (a) Two-sample z for proportions. (b) Two-sample t for means.

25.3 (a) t for means. (b) Matched pairs.

25.5 (a) Label the subjects from 01 to 44 and choose 22 subjects to eat regular chips first. From line 101 of Table A: 19, 22, 39, 34, and 05. (b) matched pairs t test for the means (one-sample t test on the differences).

25.7 $H_0 : \mu_f = \mu_m$, $H_a : \mu_f > \mu_m$, $t = 2.21$, $0.01 < P < 0.02$ (df $= 100$) or $P = 0.0143$(df $= 186.02$).

25.9 β says that the humerus in the *Archaeopteryx* species is on average 1.1969 times the length of the femur. $t = 15.94$. (b) df $= 3$, $P < 0.001$, significant.

25.11 0.6315 to 0.6485. (If we assume that $0.64 = 7794/12.178$, we can find a plus four interval: 0.6314 to 0.6485.)

25.13 $H_0 : \mu_1 = \mu_2$, $H_a : \mu_1 > \mu_2$; $t = 10.4$, $P < 0.0001$ (df $= 19$ or 32.39).

25.15 $H_0 : \mu_1 = \mu_2$, $H_a : \mu_1 < \mu_2$; $t = -3.50$, $P < 0.0005$ (df $= 100$ or 216.4).

25.17 $H_0 : p_1 = p_2$, $H_a : p_1 \neq p_2$, $z = 6.79$, P is very small.

25.19 $H_0 : \mu_{\text{pets}} = \mu_{\text{clinic}}$, $H_a : \mu_{\text{pets}} > \mu_{\text{clinic}}$, $t = 1.17$, $0.10 < P < 0.15$ (df $= 22$) or $P = 0.1234$ (df $= 43.3$), not significant.

25.21 df $= 25$, 193 ± 27.47.

25.23 (a) The relationship is clearly and linear and very strong, r^2 is 0.95. (b) 4.270 ± 0.118. (c) Fish 143 is very wide for its length and has a larger residual than other points; it can make inference less accurate.

25.25 $H_0 : \mu = 12$, $H_a : \mu > 12$, $t = 0.907$, df $= 19$, $P = 0.1879$, not significant.

25.27 $H_0 : \mu_1 = \mu_2$; $H_a : \mu_1 \neq \mu_2$; $t = -0.727$, $0.4 < P < 0.5$ (df $= 28$) or $P = 0.47$ (df $= 46.19$).

25.29 (a) The diabetic potentials appear to be large. (b) $H_0 : \mu_D = \mu_N$, $H_a : \mu_D \neq \mu_N$, $t = 3.077$, $0.005 < P < 0.01$ (df $= 17$) or $P = 0.0039$ (df $= 36.60$). (c) $t = 3.841$; $0.001 < P < 0.002$ (df $= 16$) or $P = 0.0005$ (df $= 37.15$).

25.31 (a) The plus four confidence interval for 2 proportions is the only acceptable procedure for sample size. Difference in proportions: -0.8060 ± 0.0994. (b) Two-sample t confidence interval for means: difference (44.18, 81.02) with *df* 33.3.

25.33 $H_0 : \mu_G = \mu_I$, $H_a : \mu_G \neq \mu_I$; $t = 5.59$, and either df $= 9.89$, $P = 0.00012$ or df $= 9$, $P = 0.00017$, highly significant.

25.35 H_0: no relationship between temperature and hatching outcome, $X^2 = 1.703$, *df* $= 2$, $P = 0.437$, not significant.

25.37 $H_0 : \mu_C = \mu_N = \mu_H$, $F = 0.080$, df $= 2, 126$, $P = 0.923$, not significant.

Data Table Index ——————————————

Index

SYMBOLS

μ	population mean
$\overline{x}$	sample mean
σ^2	population variance
s^2	sample variance
σ	population standard deviation
s	sample standard deviation
Q_1, Q_3	first and third quartiles in a sample
p	population proportion
$\hat{p}$	sample proportion
$\mu_y = \alpha + \beta x$	α slope and β y-intercept of a population regression line; μ_y population mean response for a given value of x
$y = a + bx$	a slope and b y-intercept of a sample regression line
r	sample correlation coefficient for a linear relationship
r^2	sample coefficient of determination for a linear relationship
$P(A)$	probability of event A
$P(A|B)$	probability of event A, given that event B has occurred or is true
$N(\mu, \sigma)$	Normal distribution with mean μ and standard deviation σ
$N(0, 1)$	standard Normal distribution
z	standardized score under a Normal distribution
t	standardized score under a t distribution
SEM	standard error of the mean
SE	standard error of an estimate
df	degrees of freedom
m	margin of error in a confidence interval
H_0	null hypothesis
H_a	alternative hypothesis
α	chosen significance level and probability of a Type I error
β	probability of a Type II error
X^2	chi-square statistic
F	F statistic

Table entry for C is the critical value t^* required for confidence level C. To approximate one- and two-sided P-values, compare the value of the t statistic with the critical values of t^* that match the P-values given at the bottom of the table.

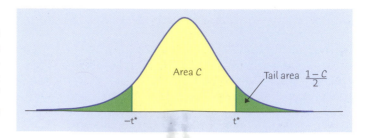

TABLE C	**t distribution critical values**											
	Confidence level C											
Degrees of freedom	50%	60%	70%	80%	90%	95%	96%	98%	99%	99.5%	99.8%	99.9%
1	1.000	1.376	1.963	3.078	6.314	12.710	15.890	31.820	63.660	127.30	318.30	636.60
2	0.816	1.061	1.386	1.886	2.920	4.303	4.849	6.965	9.925	14.09	22.33	31.60
3	0.765	0.978	1.250	1.638	2.353	3.182	3.482	4.541	5.841	7.453	10.21	12.92
4	0.741	0.941	1.190	1.533	2.132	2.776	2.999	3.747	4.604	5.598	7.173	8.610
5	0.727	0.920	1.156	1.476	2.015	2.571	2.757	3.365	4.032	4.773	5.893	6.869
6	0.718	0.906	1.134	1.440	1.943	2.447	2.612	3.143	3.707	4.317	5.208	5.959
7	0.711	0.896	1.119	1.415	1.895	2.365	2.517	2.998	3.499	4.029	4.785	5.408
8	0.706	0.889	1.108	1.397	1.860	2.306	2.449	2.896	3.355	3.833	4.501	5.041
9	0.703	0.883	1.100	1.383	1.833	2.262	2.398	2.821	3.250	3.690	4.297	4.781
10	0.700	0.879	1.093	1.372	1.812	2.228	2.359	2.764	3.169	3.581	4.144	4.587
11	0.697	0.876	1.088	1.363	1.796	2.201	2.328	2.718	3.106	3.497	4.025	4.437
12	0.695	0.873	1.083	1.356	1.782	2.179	2.303	2.681	3.055	3.428	3.930	4.318
13	0.694	0.870	1.079	1.350	1.771	2.160	2.282	2.650	3.012	3.372	3.852	4.221
14	0.692	0.868	1.076	1.345	1.761	2.145	2.264	2.624	2.977	3.326	3.787	4.140
15	0.691	0.866	1.074	1.341	1.753	2.131	2.249	2.602	2.947	3.286	3.733	4.073
16	0.690	0.865	1.071	1.337	1.746	2.120	2.235	2.583	2.921	3.252	3.686	4.015
17	0.689	0.863	1.069	1.333	1.740	2.110	2.224	2.567	2.898	3.222	3.646	3.965
18	0.688	0.862	1.067	1.330	1.734	2.101	2.214	2.552	2.878	3.197	3.611	3.922
19	0.688	0.861	1.066	1.328	1.729	2.093	2.205	2.539	2.861	3.174	3.579	3.883
20	0.687	0.860	1.064	1.325	1.725	2.086	2.197	2.528	2.845	3.153	3.552	3.850
21	0.686	0.859	1.063	1.323	1.721	2.080	2.189	2.518	2.831	3.135	3.527	3.819
22	0.686	0.858	1.061	1.321	1.717	2.074	2.183	2.508	2.819	3.119	3.505	3.792
23	0.685	0.858	1.060	1.319	1.714	2.069	2.177	2.500	2.807	3.104	3.485	3.768
24	0.685	0.857	1.059	1.318	1.711	2.064	2.172	2.492	2.797	3.091	3.467	3.745
25	0.684	0.856	1.058	1.316	1.708	2.060	2.167	2.485	2.787	3.078	3.450	3.725
26	0.684	0.856	1.058	1.315	1.706	2.056	2.162	2.479	2.779	3.067	3.435	3.707
27	0.684	0.855	1.057	1.314	1.703	2.052	2.158	2.473	2.771	3.057	3.421	3.690
28	0.683	0.855	1.056	1.313	1.701	2.048	2.154	2.467	2.763	3.047	3.408	3.674
29	0.683	0.854	1.055	1.311	1.699	2.045	2.150	2.462	2.756	3.038	3.396	3.659
30	0.683	0.854	1.055	1.310	1.697	2.042	2.147	2.457	2.750	3.030	3.385	3.646
40	0.681	0.851	1.050	1.303	1.684	2.021	2.123	2.423	2.704	2.971	3.307	3.551
50	0.679	0.849	1.047	1.299	1.676	2.009	2.109	2.403	2.678	2.937	3.261	3.496
60	0.679	0.848	1.045	1.296	1.671	2.000	2.099	2.390	2.660	2.915	3.232	3.460
80	0.678	0.846	1.043	1.292	1.664	1.990	2.088	2.374	2.639	2.887	3.195	3.416
100	0.677	0.845	1.042	1.290	1.660	1.984	2.081	2.364	2.626	2.871	3.174	3.390
1000	0.675	0.842	1.037	1.282	1.646	1.962	2.056	2.330	2.581	2.813	3.098	3.300
z^*	0.674	0.841	1.036	1.282	1.645	1.960	2.054	2.326	2.576	2.807	3.091	3.291
One-sided P	.25	.20	.15	.10	.05	.025	.02	.01	.005	.0025	.001	.0005
Two-sided P	.50	.40	.30	.20	.10	.05	.04	.02	.01	.005	.002	.001